SYMMETRY AND ENERGY BANDS IN CRYSTALS

John C. Slater

Institute Professor Emeritus, Massachusetts Institute of Technology
and Graduate Research Professor of Physics and Chemistry
University of Florida

DOVER PUBLICATIONS, INC.
NEW YORK

Copyright © 1972 by Dover Publications, Inc.
Copyright © 1965 by John C. Slater.
All rights reserved under Pan American and International Copyright Conventions.

Published in Canada by General Publishing Company, Ltd., 30 Lesmill Road, Don Mills, Toronto, Ontario.
Published in the United Kingdom by Constable and Company, Ltd., 10 Orange Street, London WC 2.

This Dover edition, first published in 1972, is an unabridged and corrected republication of the work originally published by McGraw-Hill Book Company in 1965 as Volume 2 of the three-volume work entitled *Quantum Theory of Molecules and Solids*. This edition contains a new Preface by the author.

International Standard Book Number: 0-486-60573-6
Library of Congress Catalog Card Number: 72-81282

Manufactured in the United States of America
Dover Publications, Inc.
180 Varick Street
New York, N. Y. 10014

Preface to the Dover Edition

This volume was originally published in 1965 as Volume 2 of the McGraw-Hill Book Company's series *Quantum Theory of Molecules and Solids*. The present edition differs from the first in that the errors and misprints of which I am aware have been corrected herein. The most serious of those was in Sec. A3-7, in which through an oversight the discussion of the space group D_{4h}^{14} actually referred to the space group D_{4h}^{9}. I am much indebted to Drs. J. G. Gay, W. A. Albers, Jr., and F. J. Arlinghaus, of the Research Laboratory of the General Motors Corporation, for having pointed this out, and for having supplied the corrections.

In the preface to the first edition of this volume, which follows, it is explained that the series *Quantum Theory of Molecules and Solids* was expected eventually to run to four volumes. The third volume, "Insulators, Semiconductors, and Metals," was published in 1967, and deals with lattice vibrations of crystals, the interaction of lattice vibrations with electrons, the resulting theory of conductivity, including conductivity in a magnetic field, and the many experimental methods of investigating crystals based on these methods. It includes further the interaction of electromagnetic waves with crystals, including optical and dielectric behavior and the interrelations of these topics with energy-band theory. Thus the more experimental aspects of energy-band theory are treated in that volume, rather than in the present one.

The fourth volume, "Self Consistent Field for Molecules and Solids," which is now in press by McGraw-Hill, treats not only the present more sophisticated view of this topic, which is touched on in the earlier volumes, but also the cohesive energy of crystals, including the many-electron aspect of solids; excitons, and optical excitation; and the magnetic properties, paramagnetism, ferromagnetism, antiferromagnetism, and related types of behavior. These four volumes, together with my "Quantum Theory of Atomic Structure," published by McGraw-Hill in 1960, furnish a coordinated treatment of the quantum theory of atoms, molecules, and solids, but each of the volumes, like the present one, is self-contained and not dependent on the others.

JOHN C. SLATER

Gainesville, Florida
May, 1972

Preface to the First Edition

In the preface to Volume 1 of this work,*"Electronic Structure of Molecules," I outlined the general idea behind the series of which the present volume is the second, and the probable contents of the later volumes of the series, still to appear. I explained that many of the fundamental ideas encountered in the study of molecules and solids could be best treated in their simplest applications, to the molecular problem. Such physical ideas as the nature of the chemical bond, and such mathematical techniques as the group theory, and the use of linear combinations of atomic orbitals to approximate molecular orbitals, can best be introduced in connection with their application to molecules. I explained in that earlier preface that the later volumes of the series would treat solids, and that Volume 2 would include the theory of energy bands, along with a discussion of experimental methods of investigating the electronic energy levels of solids.

In the course of writing the present volume, it became very clear that it would grow too long if both the calculation of energy bands and the experimental implications of this calculation were included. I completed, in fact, a manuscript treating this whole range of material, but it was much too long for a single volume. Consequently I decided to restrict this Volume 2 to the topics indicated in the title which I have given it, "Symmetry and Energy Bands in Crystals," treating the calculation of energy bands, but postponing the experimental methods of investigating them. At the moment, with Volume 2 completed, and Volume 3 just begun, it is my intention to include in Volume 3 the lattice vibrations of crystals, the interaction of lattice vibrations with electrons, the resulting theory of conductivity, including conductivity in a magnetic field, and the many experimental methods of investigating crystals based on these methods. Further, I hope to include the interaction of electromagnetic waves with crystals, including optical and dielectric behavior, and the interrelations of these topics with energy-band theory. In other words, the experimental aspects of energy-band theory, which I had hoped to include in Volume 2, will be postponed until

---

* As explained in the Preface to the Dover Edition, "Symmetry and Energy Bands in Crystals" was originally published as Volume 2 of a multi-volume series entitled *Quantum Theory of Molecules and Solids*. None of the other volumes in this series is presently available as a Dover reprint edition.

Volume 3. The main topics of Volume 4 are now intended to be the cohesive energy of crystals, including the many-electron aspect of solids; and the magnetic properties, paramagnetism, ferromagnetism, diamagnetism, and superconductivity. It is clear that both of these volumes will be long, and before the program is completed, it will not be surprising if one or both of these projected volumes may grow into more than one. I mention these topics so that the reader of the present volume will understand why many subjects of great current interest in solid-state theory are omitted; they will come in later volumes.

A number of writers of recent books dealing with aspects of the quantum theory of solids have been struggling with the same problem which I have been facing, namely, the enormously rapid growth of the field. They have felt the impossibility of handling the whole of the theory of solids adequately in a single volume, and have either limited themselves to very restricted parts of the subject, or have explicitly omitted important features of a complete treatment, such for instance as an adequate bibliography. My way out of this difficulty, as I emphasized in the preface to Volume 1, is to keep on writing volumes until as much of the field is covered as I wish to treat. I shall not attempt to duplicate all the existing detailed treatises handling specialized fields. On the other hand, at the time of this writing, there is no text which goes nearly as thoroughly into the theory of energy bands as I have tried to do in the present volume.

It would hardly have been possible to write such a text on energy bands until very recently. Energy-band calculations of a hitherto unattainable accuracy have been coming out in the last year or two, and I am indebted to several of my colleagues and students for permission to present calculations which are still unpublished at the time of writing this preface. In particular, the series of calculations on the elements from argon through gallium, carried out by Wood, Burdick, Mattheiss, and Hanus in my group, and the calculations on the alkali metals by Kenney, show that the augmented-plane-wave method which we are using is capable of great and detailed accuracy. Work, partly reported in this volume, by Johnson, Conklin, and Pratt on lead telluride, treated relativistically, is showing the possibility of remarkably good agreement between the most detailed calculations of energy bands and many types of experimental results dealing with semiconductors. It appears likely that we must look forward to an extremely fruitful development of the calculation of energy bands in sufficient detail to correlate with a great mass of recent experimental results. As I mentioned above, it would hardly have been possible until the present to go far enough in presenting these detailed calculations to indicate the possibilities in the method.

There are some specific points connected with the present volume which deserve further comment. In the first place, I have given an unusually complete description of many types of crystal structure, and many individual compounds. These carry the subject far beyond those materials whose energy bands have been computed, or are likely to be computed in the near future. I have done this partly to suggest to the workers in the field the great variety of crystalline solids which are of importance, and to which the theory should extend as it develops further. There is too much tendency to feel that elements or compounds with three or four of the simple and familiar structures provide almost all there is. Another reason why I have given this treatment of crystals and their structure is that many more complicated materials are of importance at present for their magnetic or optical properties, and their structures, symmetry, and so on, must be understood for study of those properties. We shall be taking up such properties in later volumes of this series, and the first few chapters of the present volume are intended as an introduction to those, as well as to the theory of energy bands.

Another point related to this is that I have included a more thorough treatment of the space groups and their implications regarding the symmetry of wave functions than is found in other texts. The method which I have outlined for handling the symmetry of wave functions demands complete information regarding the matrix elements of the irreducible representations of the space groups. I have included this material for enough space groups to handle a great many structures and compounds, and in enough detail so that the reader will find no difficulty in extending the procedure to other cases. At the time when I was writing these sections, I was not aware of the book by O. V. Kovalev, entitled "Irreducible Representations of the Space Groups," published by the Ukrainian Academy of Science in 1961. This text unfortunately is not yet available in English, and its notation and procedures are sufficiently different from those which I have used so that it is somewhat complicated to carry through the parallelism between the two, though it can be done. If this text should later be made available in English, it would be helpful in treating further space groups beyond those which I have treated.

The group-theoretical methods which I have used, in Volume 1 and the present volume, differ in a number of respects from the more conventional treatment of space groups which we owe to Seitz and other writers. My colleagues have looked into the relation between the two methods, and have established that they are essentially equivalent. However, my notation, as well as procedures, seem to me much more adaptable to actual use than the more conventional methods. Group theory has tended too much to be the exclusive province of a very few workers who had learned it thoroughly. This seems to me to be unnecessary and unfortunate, and I feel that a considerable part of the reason is the rather cumbersome notation and rather difficult descriptions of the method which are found in the current literature. I hope that study of the methods in the present text will make the results familiar to a wider audience than has been able to follow them in the past.

In the course of my career, I have trained a considerable number of students who have gone on to work in the theory of energy bands, and it has seemed to me that without a thorough grounding in the underlying foundation of the field, such as I have endeavored to give in this volume, it is almost impossible to do useful work. I have tried not merely to talk about the theory, but to go into its methods and techniques far enough so that a worker coming into the field from outside could go far enough by himself to be able to do useful work. The one real element of training which he must have, in addition to what I have included here, is thorough familiarity with the methods of digital computation. This is fortunately becoming sufficiently widespread so that plenty of students have it.

As in earlier volumes, I have included an extensive bibliography, not only of papers referred to, but of those bearing on the subject. Too many bibliographies in review articles and other places lose most of their usefulness by not including the titles of references, as well as author and journal references. As a result of this, there are really very few useful bibliographies in existence in the field of the theory of molecules and solids. The bibliographies in the further volumes of this series, dealing with partially experimental as well as purely theoretical papers, will be even more extensive than those which have already come out. As has been remarked, the bibliography under these circumstances grows nearly as long as the rest of the book. There is no point in apologizing for this. The literature is very extensive, and without some sort of guide to it, it is practically useless. So far, I have not thought of a more useful type of guide than a series of such bibliographies as will be found in the various volumes of this series.

As in the preceding volumes on atoms and molecules which I have written, I am greatly indebted to many members, present and past, of the Solid-State and Molecular Theory Group at M.I.T., for a great deal of help in connection with the work reported on in this book. Earlier in this preface I have named some of those whose work has contributed greatly. Particularly my colleagues Profs. G. F. Koster and J. H. Wood have been of great assistance in all phases of the work. And again I should thank the National Science Foundation, the Office of Naval Research, and the Lincoln Laboratory for present and past help to the Solid-State and Molecular Theory Group, without which the results described here would have been impossible to achieve..

JOHN C. SLATER

Contents

Chapter 1. Crystals and Their Symmetry Properties — 1

- 1-1. The Crystalline Nature of Solids — 1
- 1-2. The Symmetry Operations of Crystals — 3
- 1-3. Space Groups — 6
- 1-4. The Crystal Systems and Bravais Lattices — 10
- 1-5. Primitive Unit Cells for the Bravais Lattices — 14

Chapter 2. Space Groups for Structures of the Elements — 24

- 2-1. Survey of Crystal Structures of the Elements — 24
- 2-2. The Body- and Face-centered Cubic Structures — 27
- 2-3. The Hexagonal Close-packed Structure — 32
- 2-4. The Diamond Structure — 40
- 2-5. The Graphite Structure — 44
- 2-6. The Arsenic Structure — 45
- 2-7. The Selenium Structure — 49
- 2-8. The Iodine Structure — 50

Chapter 3. Space Groups for Structures of Compounds — 54

- 3-1. General Survey of Crystal Structures — 54
- 3-2. Structures of Simple Binary Compounds, RX — 56
- 3-3. Structures of Some Compounds of Composition RX_2 — 62
- 3-4. Structures of Compounds in the Cubic System — 68
- 3-5. The Calcite, Corundum, and Ilmenite Structures — 77
- 3-6. Additional Trigonal, Hexagonal, and Orthorhombic Structures — 85

Chapter 4. Atomic Radii and the Chemical Bond — 95

- 4-1. Ionic and Atomic Radii — 95

4-2.	The Physical Significance of Atomic and Ionic Radii	101
4-3.	Interaction of Negative Ions in Crystals	108
4-4.	The Charge Distribution and Potential in a Crystal	111
4-5.	Potentials for the Self-consistent Field in Crystals	115

Chapter 5. The Symmetry of Electronic Wave Functions in Crystals   118

5-1.	The Reciprocal Lattice	118
5-2.	Reciprocal Lattices and Brillouin Zones for Eight Bravais Lattices	121
5-3.	The Effect of Rotational Operations on a Plane Wave	126
5-4.	Stars of Wave Vectors	130
5-5.	Symmetrized Plane Waves and Irreducible Representations	133
5-6.	General Discussion of Symmetry of Wave Functions	140

Chapter 6. Plane-wave Expansions of Wave Functions in Crystals; The One-dimensional Case   144

6-1.	Fourier Expansions of Periodic Functions	144
6-2.	Periodic Boundary Conditions	146
6-3.	Schrödinger's Equation for the Periodic-potential Problem	148
6-4.	Momentum Eigenfunctions	151
6-5.	The Wannier Functions	154
6-6.	Energy Bands in the One-dimensional Case	158

Chapter 7. Plane-wave Expansion in the Two- and Three-dimensional Cases   167

7-1.	The Two-dimensional Square Lattice	167
7-2.	Symmetry Properties of the Wave Functions	174
7-3.	Brillouin Zones	177
7-4.	The Momentum Eigenfunctions for the Two-dimensional Case	182
7-5.	Modified Momentum Eigenfunctions and Wannier Functions	187
7-6.	The Relation of Brillouin Zones to Bragg's Law	195
7-7.	The Three-dimensional Case, and the Plane-wave Expansion for Real Crystals	199

Chapter 8. The Tight-binding and Orthogonalized Plane-wave Methods   203

8-1.	The General Idea of the LCAO and OPW Methods	203
8-2.	Energy Bands of Sodium	206

CONTENTS

8-3.	Wave Functions and Momentum Eigenfunctions for Sodium	209
8-4.	Matrix Elements for the LCAO and OPW Methods	218
8-5.	Crystal Symmetry and the LCAO and OPW Methods	224

Chapter 9. The Cellular and Augmented-plane-wave Methods   228

9-1.	The Wigner-Seitz and Cellular Methods	228
9-2.	The Augmented-plane-wave Method	233
9-3.	The Scattered-wave Method	236
9-4.	The Free-electron Approximation and the Scattered-wave Method	241
9-5.	Other Related Methods	246

Chapter 10. Calculations of Energy Bands in Crystals   250

10-1.	The Free-electron Approximation	250
10-2.	Free-electron Energy Bands in Crystals	253
10-3.	Energy Bands of Typical Elements: Monovalent, Divalent, and Trivalent	257
10-4.	Energy Bands of Tetravalent Elements and 3-5 Compounds	272
10-5.	Energy Bands of Transition Elements	280
10-6.	Energy Bands of Elements with Special Crystal Structures	287
10-7.	Energy Bands for Alkali Halide Crystals and Other Compounds	291
10-8.	Bibliography of Energy-band Calculations	300

Appendix 1. Interatomic Distances and Crystal Structures   307

Appendix 2. Crystal Parameters   334

Appendix 3. Symmetry Properties and Projection Operators for 20 Space Groups   347

A3-1.	The Space Groups D_{6h}^4 ($P6_3/mmc$) and C_{6v}^4 ($P6_3mc$)	347
A3-2.	The Space Groups O_h^5 ($Fm3m$), O_h^7 ($Fd3m$), and T_d^2 ($F\bar{4}3m$)	368
A3-3.	The Space Groups O_h^9 ($Im3m$), O_h^{10} ($Ia3d$), and T_h^7 ($Ia3$)	384
A3-4.	The Space Groups O_h^1 ($Pm3m$) and T_h^6 ($Pa3$)	408
A3-5.	The Space Groups C_{3i}^2 ($R\bar{3}$), D_3^7 ($R32$), D_{3d}^5 ($R\bar{3}m$), and D_{3d}^6 ($R\bar{3}c$)	417
A3-6.	The Space Groups D_{3d}^3 ($P\bar{3}m1$), and D_3^4 ($P3_121$)	426
A3-7.	The Space Group D_{4h}^{14} ($P4_2/mnm$)	429
A3-8.	The Space Group D_{2h}^{16} ($Pnma$)	439
A3-9.	The Space Group D_{2h}^{18} ($Cmca$)	446

CONTENTS

Appendix 4.	The Momentum Eigenfunction and Fourier Expansion of the Potential	452
Appendix 5.	Power-series Expansions of Energy Bands near Symmetry Points	455
Appendix 6.	Matrix Elements for the Augmented-plane-wave Method	461
Appendix 7.	The Free-electron Approximation	466
Appendix 8.	Binding Energy of Diamond, by Method of Schmid	470
Appendix 9.	Spin-Orbit and Relativistic Effects in Energy Bands	472
A9-1.	Spin-Orbit Effects in Atoms	472
A9-2.	Spin-Orbit Interaction in a Crystal Using the APW Method	478
A9-3.	Splitting of P-like State by Spin-Orbit Interaction	480
A9-4.	Double Groups and Their Applications	485
A9-5.	The Effect of a Rotation on the Spin Functions	490
A9-6.	Examples of Rotations of Coordinates	494
A9-7.	The Double Groups C_{Nv}	500
A9-8.	Cubic Double Groups	506
A9-9.	Basis Functions and Irreducible Representations for the Double Groups T_d and O_h	512
A9-10.	Double Point Groups and Double Space Groups	514

BIBLIOGRAPHY 521

INDEX 559

1

Crystals and Their Symmetry Properties

1-1. The Crystalline Nature of Solids. Our task in the present volume is to study the electronic wave functions and energy levels in solids. We shall assume, as in our study of molecules in Volume 1, that we are dealing with the Born-Oppenheimer approximation; that is, we keep the nuclei fixed, and solve for the motion of the electrons in the field of the stationary nuclei. The energy of the electronic system, as a function of nuclear position, can be used as a potential energy function for studying the nuclear motion, as is always found with the Born-Oppenheimer approximation. For certain arrangements of the nuclei, the energy will have a minimum, so that the energy will be a quadratic function of the nuclear displacements from this equilibrium configuration. Harmonic oscillations will be possible around this equilibrium state; these are the oscillations concerned in the temperature agitation of the solid, specific heat, and related problems. All these problems relating to atomic vibrations will be postponed to a later volume of this work.

The equilibrium configuration, then, will come for an electronic state of minimum energy, and this is the state to which the crystal will come at the absolute zero of temperature (aside from the small zero-point vibrations of the atoms, required by the quantum-mechanical treatment of the linear oscillator). In almost all our discussion we shall limit ourselves to this equilibrium state; only occasionally shall we consider the energy of the solid as a function of nuclear positions. The analogous thing with diatomic molecules would be to limit ourselves to discussing the electrons at the equilibrium interatomic distances. As we see from Volume 1,* this is as far as most molecular theory has to be carried. It is only in a few cases, as in the hydrogen molecule and hydrogen molecular ion, that we have a complete treatment as a function of internuclear distance. It would be a good thing, in the theory of solids, if we had a complete treat-

---

* When we refer to Volume 1, this will mean Volume 1 of the present work, "Quantum Theory of Molecules and Solids," McGraw-Hill Book Company, New York, 1963, dealing with Electronic Structure of Molecules.

ment of the energy as a function of nuclear positions, for then we could look for the atomic positions leading to minimum energy, and hence find how the atoms would actually arrange themselves in real solids. Furthermore, by studying the quadratic change of energy with atomic displacements from the equilibrium positions, we could find the force constants for the elastic vibrations of the crystal, and hence find the frequencies of the normal modes of vibration. Unfortunately, this represents a program which has been only slightly worked out so far, and hence lies largely in the future.

If we are not to find the equilibrium positions of the atoms from a priori calculations, we must take the experimental facts as giving the observed positions. We shall therefore start our discussion with a study of the arrangement of atoms in solids. This is a subject which has been pursued vigorously since 1913, the date at which the diffraction of x rays by solids was discovered by von Laue and his associates,[1] a discovery which was at once applied to the study of crystal structure by W. H. and W. L. Bragg,[2] and later by a great number of other workers. By now the structures—that is, the atomic positions—of a great many crystalline solids have been determined by x-ray diffraction, and we shall give in the next chapters a short account of a few of the enormously many facts which have been uncovered by the x-ray crystallographers in their work during the approximately fifty years which have elapsed since the fundamental work of von Laue and the Braggs.

The first experimental fact about the nature of solids is that by far the largest number are crystalline. The characteristic feature of a crystal is not so much the regularity of its outer shape, when it is grown or cleaved, as it is the regularity of the atomic arrangement. This regularity was suspected before the start of x-ray diffraction. It was an obviously attractive hypothesis to suppose that the definite angles observed between the various cleavage planes of a crystal were tied up with the angles between various planes which can be set up if, for example, we pile spheres together in close packing. It is a familiar fact that we can terminate such a pile of spheres with various planes, making quite definite angles. Even quantitative agreement between the observed angles of simple crystals and the angles of these planes was found for some simple crystals before x-ray diffraction became available.

The reality of the atomic regularity in crystals was first convincingly

[1] M. von Laue, *Ann. Physik*, **41**:989 (1913). W. Friedrich, P. Knipping, and M. von Laue, *Ann. Physik*, **41**:971 (1913).

[2] W. H. Bragg, *Physik. Z.*, **14**:472 (1913); *Phil. Mag.*, **27**:881 (1914). W. L. Bragg, *Proc. Cambridge Phil. Soc.*, **17**:43 (1913). W. H. Bragg and W. L. Bragg, "X-rays and Crystal Structure," 5th ed., G. Bell & Sons, Ltd., London, 1925. W. L. Bragg, "The Crystalline State," The Macmillan Company, New York, 1934.

demonstrated, however, by the work of the x-ray crystallographers. The fundamental fact of crystalline structure on an atomic scale is the existence of unit cells in the crystal. These are small volumes, all of identical shape, size, orientation, and constitution, which taken together fill all the crystal. Each unit cell contains an integral number of units of the molecular composition of the crystal. In the simplest cases of crystal structures of the elements, each unit cell may contain only one atom. A few of the elements, however, have complicated structures with many atoms in the unit cell. Similarly in compounds there are some cases where only one molecular constituent of the crystal is found in a unit cell, others where several or many are found. In some cases, such as for example solid O_2 or Br_2 or I_2, the atoms in the solid are definitely grouped into diatomic molecules, which show up clearly from the x-ray work. In other cases, such as for example NaCl, we cannot in any sense group together one Na and one Cl in the crystal and say that they form a molecule. Each atom of one type proves to be surrounded equally closely in NaCl by six of the other type. Whether we have a molecular crystal or not, however, the unit cell must still contain an integral number of molecular constituents. Thus, the crystal of I_2 contains two I_2 molecules, or four I atoms, per unit cell, while the crystal of NaCl contains one Na and one Cl atom per unit cell.

The problem of describing the various structures found for different types of crystals is very extensive, and it requires several volumes to present the known facts regarding them.[1] Consequently we shall have to be very incomplete in our treatment of crystal structure, but we shall try to give the reader a general acquaintance with the field, enough so that he can proceed further on his own initiative.

1-2. The Symmetry Operations of Crystals. It is hardly possible to describe any crystals except the very simplest ones without some knowledge of the various symmetry properties which the crystals can have. This is best discussed in terms of the group theory; the reader will be assumed to have an acquaintance with group theory such as is provided in Volume 1 of this work. The first aspect of the symmetry properties of a crystal arises from the translational operations. If we start at a given point of one unit cell, and make a translation to a corresponding point of another unit cell, we may call the resulting translation vector R_n. Then

---

[1] The standard reference for detailed information about crystal structures is R. W. G. Wyckoff, "Crystal Structures," Interscience Publishers, Inc., New York, 1948 and later. This book is published in a loose-leaf form, and supplementary pages have been issued periodically, to bring it up to date. It is in several volumes. For a more descriptive discussion of inorganic crystals, together with the chemical interpretation of their structures, a recommended book is A. F. Wells, "Structural Inorganic Chemistry," 3d ed., Oxford University Press, Fair Lawn, N.J., 1962.

we see that translation through any of the vectors $\mathbf{R}_n$ is a symmetry operation of the crystal, in that, when an infinite crystal is translated through that vector, the charge density and all other properties are unchanged. There are other symmetry operations as well, such as rotations and reflections with respect to specific axes and planes, and the study of all the possible types of symmetry which we can have by combination of the various possible rotations and reflections with the translations is the foundation of the geometrical aspect of crystallography, which we shall first consider.

The translation vectors $\mathbf{R}_n$ to equivalent points of all unit cells can be written as linear combinations of three primitive vectors $\mathbf{t}_1$, $\mathbf{t}_2$, $\mathbf{t}_3$, the coefficients of the linear combinations being integers; that is,

$$\mathbf{R}_n = n_1 \mathbf{t}_1 + n_2 \mathbf{t}_2 + n_3 \mathbf{t}_3 \tag{1-1}$$

where n_1, n_2, n_3 are integers. It is clear that these vectors $\mathbf{t}_1$, $\mathbf{t}_2$, $\mathbf{t}_3$ cannot be unique, for we could use three linear combinations of them, with integral coefficients, and any linear combinations of these new $\mathbf{t}$'s, with integral coefficients, would represent some of the same vectors $\mathbf{R}_n$ given in Eq. (1-1). It is also clear that the size of the unit cell is not unique; if we took a unit cell twice as large as an original one, for instance, it would still have all the properties we have described as defining a unit cell, but it would ordinarily not be as convenient a cell to use, since it would contain twice as many atoms as the smaller cell, and in general, the smaller the cell, the more convenient it is. We generally use the smallest possible unit cell, sometimes called the primitive unit cell, though sometimes for reasons of symmetry a larger unit cell is more convenient. This will become clear later from examples.

We shall denote the operation of translation through the vector $\mathbf{R}_n$, without rotation or reflection, by a symbol $\{R_1|\mathbf{R}_n\}$. Here the symbol R_1 stands for the identical rotational or reflection operation (the operation denoted by X_0 for the group C_{Nv} in Volume 1, or by R_1 for the group T_d or O_h, for instance); for the operations involving rotations or reflections we shall later replace the symbol R_1 by the appropriate symbol for the rotational or reflection operation. The significance of the translational operation is then that

$$\{R_1|\mathbf{R}_n\}\psi(\mathbf{r}) = \psi(\mathbf{r} + \mathbf{R}_n) \tag{1-2}$$

where $\psi(\mathbf{r})$ is a function of the vector position $\mathbf{r}$, whose rectangular coordinates are x, y, z. We see that the translational operators all commute with each other:

$$\begin{aligned}\{R_1|\mathbf{R}_n\}\{R_1|\mathbf{R}_m\}\psi(\mathbf{r}) &= \psi(\mathbf{r} + \mathbf{R}_n + \mathbf{R}_m) \\ &= \{R_1|\mathbf{R}_m\}\{R_1|\mathbf{R}_n\}\psi(\mathbf{r})\end{aligned} \tag{1-3}$$

Hence the group of translations is Abelian, and all irreducible representations are one-dimensional, with diagonal matrix elements for the translational operators. These operators, as we shall see later when we study them more carefully, have a close analogy to the operators of the group C_N, which we studied in Volume 1.

The symmetry operations of translation through the vectors $\mathbf{R}_n$ carry immediate implications regarding the molecular-orbital solutions of a crystal problem. We approach the problem of finding the motion of the electrons in a crystal by the self-consistent-field method. We shall regard that method for the present in a simple, elementary way, not considering the sophistication of the Hartree-Fock equations. We merely assume that we are dealing with some sort of averaged potential in which the electrons move, arising from the nucleus and the other electrons, and our fundamental problem is to solve Schrödinger's equation for the motion of an electron in such a potential. This potential will show the symmetry of the crystal. Hence the symmetry operations of translation will commute with this potential, and with the Hamiltonian. Since they also commute with each other, we can choose wave functions which simultaneously diagonalize all the symmetry operations, as well as the Hamiltonian, and this is almost always done.

In other words, translation through the vector $\mathbf{t}_1$ must multiply the wave function by one constant, through $\mathbf{t}_2$ by a second, through $\mathbf{t}_3$ by a third, and through the vector $n_1\mathbf{t}_1 + n_2\mathbf{t}_2 + n_3\mathbf{t}_3$ by the first constant raised to the n_1 power times the second raised to the n_2 times the third to the n_3. If both the initial and final functions are normalized, the absolute value of these constants must be unity, so that the three constants may be written in the form of e raised to a pure imaginary power. We may then find a vector $\mathbf{k}$ such that these constants may be written in the form $e^{i\mathbf{k}\cdot\mathbf{t}_1}$, $e^{i\mathbf{k}\cdot\mathbf{t}_2}$, $e^{i\mathbf{k}\cdot\mathbf{t}_3}$. When we multiply the first constant raised to the n_1 power by the second to the n_2 and the third to the n_3 power, the resulting quantity is $e^{i\mathbf{k}\cdot\mathbf{R}_n}$, where $\mathbf{R}_n = n_1\mathbf{t}_1 + n_2\mathbf{t}_2 + n_3\mathbf{t}_3$ as before. Hence we have proved the important theorem: we may write the solutions of the periodic potential problem in such a form that the value of the function, at a given point of the unit cell which is displaced from a given unit cell by the vector $\mathbf{R}_n$, equals the value of that function at the corresponding point of the undisplaced unit cell, multiplied by the factor $e^{i\mathbf{k}\cdot\mathbf{R}_n}$, where $\mathbf{k}$ is a real constant vector, called the wave vector, or propagation constant. We may remark parenthetically that our proof that $\mathbf{k}$ is real is based on the assumption that the wave function can be normalized. We sometimes, in the study of surface states, have occasion to examine wave functions which cannot be normalized, and for which this proof breaks down. Even in such a case the same results hold formally, except that $\mathbf{k}$ in general can be complex.

If $u(\mathbf{r})$ is the wave function as a function of a vector position $\mathbf{r}$, our result is then

$$u(\mathbf{r} + \mathbf{R}_n) = e^{i\mathbf{k}\cdot\mathbf{R}_n} u(\mathbf{r}) \qquad (1\text{-}4)$$

This important theorem, arising merely from the periodicity of the lattice, was first emphasized by Bloch,[1] and is often known by his name. Such results are well known in mathematical physics, and arise whenever we have similar problems of wave propagation in periodic media. In its purely mathematical form, the corresponding theorem is Floquet's theorem, which we have already met in Volume 1, Sec. 8-5. As shown in that section, we can state the theorem in an alternative form which is often useful. We can write $u(\mathbf{r})$ in the form

$$u(\mathbf{r}) = e^{i\mathbf{k}\cdot\mathbf{r}} w(\mathbf{r}) \qquad (1\text{-}5)$$

Then the requirement of Eq. (1-4) demands that $w(\mathbf{r} + \mathbf{R}_n) = w(\mathbf{r})$, or w is a periodic function of $\mathbf{r}$, having the same value in each unit cell of the crystal.

In addition to the translational operations which we have been discussing in this section, we have also rotations and reflections. The combination of these two types of symmetry operations results in the existence of what are called space groups, involving much more complication than we find in the discussion of the symmetry of molecules in Volume 1. We shall go on in the next section to the nature of space groups.

1-3. Space Groups. The groups of rotations and reflections, such as we met in Volume 1, are called point groups. They have the characteristic that they are operations which leave one point, say, the origin, unchanged. If we denote such an operation by R_i, which could stand for instance for one of the X_i's or one of the Y_i's in the group C_{Nv}, then it performs a linear operation on the rectangular coordinates x, y, z, which we may more conveniently denote for the present purpose by x_1, x_2, x_3. If we use the symbol R_i to represent the rotation and reflection, without a translation, then we have by definition

$$R_i \psi(x_1, x_2, x_3) = \psi(x_1', x_2', x_3') \qquad (1\text{-}6)$$

where the quantities x_p' are linear functions of the x_q's, given by

$$x_p' = \Sigma(q) \alpha_{pq}^i x_q \qquad (1\text{-}7)$$

The matrix of quantities α_{pq}^i describes the operation R_i of the point group. These are real quantities, satisfying the orthogonality conditions

$$\begin{aligned} \Sigma(p) \alpha_{pq}^i \alpha_{pr}^i &= \delta_{qr} \\ \Sigma(q) \alpha_{pq}^i \alpha_{rq}^i &= \delta_{pr} \end{aligned} \qquad (1\text{-}8)$$

If the operation of the point group corresponds to a rotation, the deter-

---
[1] F. Bloch, *Z. Physik*, **52**:555 (1928).

minant of the α's equals 1. If it corresponds to an improper rotation, namely, a rotation plus inversion, or a reflection, the determinant of the α's equals -1. As a simplifying notation, we may denote the vector form of Eq. (1-7) by

$$\mathbf{r}' = \alpha^i \mathbf{r} \qquad (1\text{-}9)$$

where $\mathbf{r}$, $\mathbf{r}'$ are vectors, and α^i is the matrix or tensor involved in Eq. (1-7).

We may now set up a space group by combining the operations of the translations discussed in Sec. 1-2, and the point group which we have just described. We shall denote the operators of such a space group by symbols $\{R_i|\mathbf{R}_n\}$. The corresponding operator is defined by the equation

$$\{R_i|\mathbf{R}_n\}\psi(r) = \psi(\alpha^i \mathbf{r} + \boldsymbol{\tau}^i + \mathbf{R}_n) \qquad (1\text{-}10)$$

In this equation, the quantity $\alpha^i \mathbf{r}$ is defined by Eqs. (1-7) and (1-9), $\mathbf{R}_n$ is the translation given by Eq. (1-1), and $\boldsymbol{\tau}^i$ is a so-called nonprimitive translation, which can be different for each operator of the point group, but which does not depend on n. For the special case $i = 1$, where R_i represents the identity, $\boldsymbol{\tau}^i$ has to be zero, but in some space groups, though not all, $\boldsymbol{\tau}^i$ is different from zero for some of the operators of the point group. Space groups which have $\boldsymbol{\tau}^i = 0$ for all values of i are called symmorphic space groups; the others are called nonsymmorphic. We shall see the significance of these nonprimitive translations from later examples.

Let us now consider the requirements which must be fulfilled if these operators defined by Eq. (1-10) are to form a group. First, as we remember, there must be a multiplication table: the successive application of two operations of the group must give an operation of the group. To investigate this, we let an operator $\{R_j|\mathbf{R}_m\}$ operate on an arbitrary function ψ, and then let $\{R_i|\mathbf{R}_n\}$ operate on the resulting function. We first have

$$\{R_j|\mathbf{R}_m\}\psi(x_1,x_2,x_3) = \psi[\Sigma(q)\alpha^j_{1q}x_q + \tau^j_1 + R_{m1},$$
$$\Sigma(r)\alpha^j_{2r}x_r + \tau^j_2 + R_{m2}, \Sigma(s)\alpha^j_{3s}x_s + \tau^j_3 + R_{m3}] \qquad (1\text{-}11)$$

in which we have used Eqs. (1-6), (1-7), (1-9), and (1-10). We now wish to apply the operator $\{R_i|\mathbf{R}_n\}$ to this function, which means that wherever x_p appears we are to replace it by $\Sigma(u)\alpha^i_{pu}x_u + \tau^i_p + R_{np}$. Hence we find

$$\{R_i|\mathbf{R}_n\}\{R_j|\mathbf{R}_m\}\psi(x_1,x_2,x_3)$$
$$= \psi[\Sigma(qu)\alpha^j_{1q}\alpha^i_{qu}x_u + \Sigma(q)\alpha^j_{1q}(\tau^i_q + R_{nq}) + \tau^j_1 + R_{m1},$$
$$\Sigma(rv)\alpha^j_{2r}\alpha^i_{rv}x_v + \Sigma(r)\alpha^j_{2r}(\tau^i_r + R_{nr}) + \tau^j_2 + R_{m2},$$
$$\Sigma(sw)\alpha^j_{3s}\alpha^i_{sw}x_w + \Sigma(s)\alpha^j_{3s}(\tau^i_s + R_{ns}) + \tau^j_3 + R_{m3}]$$
$$= \psi\{\Sigma(u)\alpha^k_{1u}x_u + [\Sigma(q)\alpha^j_{1q}\tau^i_q + \tau^j_1] + [\Sigma(q)\alpha^j_{1q}R_{nq} + R_{m1}],$$
$$\Sigma(v)\alpha^k_{2v}x_v + [\Sigma(r)\alpha^j_{2r}\tau^i_r + \tau^j_2] + [\Sigma(r)\alpha^j_{2r}R_{nr} + R_{m2}],$$
$$\Sigma(w)\alpha^k_{3w}x_w + [\Sigma(s)\alpha^j_{3s}\tau^i_s + \tau^j_3] + [\Sigma(s)\alpha^j_{3s}R_{ns} + R_{m3}]\} \qquad (1\text{-}12)$$

where
$$\alpha^k_{pu} = \Sigma(q)\alpha^j_{pq}\alpha^i_{qu} \qquad (1\text{-}13)$$

Now let us consider Eq. (1-12). If our operators are to have a multiplication table, the argument of the function ψ in this equation must have the form given by Eq. (1-11), for some operator $\{R_k|R_l\}$ of the group. This requires first that the quantities $\Sigma(u)\alpha_{1u}^k x_u$, etc., occurring in Eq. (1-12), have the same form as the quantities $\Sigma(q)\alpha_{1q}^j x_q$, etc., found in Eq. (1-11), for some operation of the point group. To see that this is the case, let us consider the α's which determine the point group. The expression $\Sigma(q)\alpha_{pq}^j \alpha_{qu}^i$, of Eq. (1-13), is the ordinary matrix product of the matrices α_{pq}^j and α_{qu}^i. The matrices representing the linear transformations of coordinates, of the form of Eq. (1-7), form a representation of the point group. We recall that it is a fundamental property of group theory that matrices forming a representation of a group multiply by matrix multiplication the way the operators multiply. Hence the matrix product of the matrices of any two operators of a point group must be the matrix of another operator of the point group. In other words, our matrices α_{pu}^k of Eq. (1-13) must form the matrices for some operators of the point group, so that we have completed the first part of our proof that the combined operator given in Eq. (1-12) is an operator of the space group.

The second part of our proof must involve the statement that the translations found in Eq. (1-12), of which the first component is

$$[\Sigma(q)\alpha_{1q}^j \tau_q^i + \tau_1^j] + [\Sigma(q)\alpha_{1q}^j R_{nq} + R_{m1}] \tag{1-14}$$

must be of the form

$$\tau_1^k + R_{l1} \tag{1-15}$$

where τ_1^k is the first component of the nonprimitive translation associated with the kth operation of the point group, in our definition of the space group, as given in Eq. (1-10), and where R_{l1} is the first component of one of the lattice translations. These conditions are not automatically fulfilled for any arbitrary sets of primitive and nonprimitive translations. Rather, they put great restrictions on the point groups, on the nature of the primitive translation vectors t_1, t_2, t_3, and the nonprimitive translations τ^i, which can occur in space groups. In the first place, the quantity $\Sigma(q)\alpha_{1q}^j R_{nq} + R_{m1}$ must form the first component of a lattice translation. This means that the operation R_j of the point group, operating according to Eq. (1-7) on one of the vectors R_n, must transform this into another vector of the set given in Eq. (1-1), so that addition of R_m to it will again give one of the vectors of the set. This is one of the fundamental requirements of a space group: the point group must transform the vectors R_n into the same set of vectors. Similarly, we have a requirement which must be fulfilled by the nonprimitive translations τ^i: the quantity $\Sigma(q)\alpha_{1q}^j \tau_q^i + \tau_1^j$ must either be equal to τ_1^k, as given by Eq. (1-15), or at most must differ from it by one of the vectors of the form R_{l1}.

These restrictions on the primitive and nonprimitive translations which can be associated with a given point group limit greatly the number of possible space groups which can be set up. For instance, if the point group is one of the cubic groups, it seems rather obvious that the fundamental translation vectors t_1, t_2, t_3 must either be vectors of equal magnitude along the x, y, and z axes or at least must be closely related to them. Careful study shows that there are just 230 possible space groups permitted by these restrictions. These are made up by combining 14 different sets of vectors t_1, t_2, t_3, leading to lattices of points called the 14 Bravais lattices, with 32 possible point groups. These point groups are the following, in the notation of Volume 1, Appendix 12: one containing only the identity, called C_1; one containing only the identity and the inversion, called C_i; and C_2, C_{1h} (alternatively called C_s), C_{2h}, D_2, C_{2v}, D_{2h}, C_4, S_4, C_{4h}, D_4, C_{4v}, D_{2d}, D_{4h}, C_3, S_6 (alternatively called C_{3i}), D_3, C_{3v}, D_{3d}, C_6, C_{3h}, C_{6h}, D_6, C_{6v}, D_{3h}, D_{6h}, T, T_h, O, T_d, and O_h. Each of these 32 point groups is said to form a crystal class. In the next section we shall discuss the 14 Bravais lattices, and shall give preliminary information about the nature of the 230 space groups. In Chaps. 2 and 3 we shall discuss some 20 of the important space groups in detail, with enough explanation so that the reader will be able to study the remaining space groups by himself, from appropriate references.

We have now discussed the first requirement that the operations of the space group should in fact form a group: they lead to a multiplication table. We shall not go into the detailed proofs regarding the possible t's and τ's, but in the later discussion which we have just mentioned we shall give numerous examples of actual space groups, in which the necessary requirements for formation of a group are met. The other requirements for a group—the inclusion of the identity, of the inverse operation to each operation, and the associative law—can also be shown to hold for the 230 space groups which we have just mentioned.

Before we leave the definitions of the space groups, we should state that the definition of Eq. (1-10) is not that adopted by all writers on space groups.[1] Thus, some writers use the alternative definition

$$\{R_i|R_n\}\psi(\alpha^i r + \tau^i + R_n) = \psi(r) \qquad (1\text{-}16)$$

which defines each of the operators as the inverse operator to that given by Eq. (1-10), as we can see by letting the inverse of $\{R_i|R_n\}$ operate on both sides of Eq. (1-10) or (1-16). We shall give later, in Sec. 5-3, our

---

[1] See for example F. Seitz, *Z. Krist.*, **88**:433 (1934), **90**:289 (1935), **91**:336 (1935), **94**:100 (1936), in which the fundamentals of the group-theoretical treatment of crystals are presented, including the discussion of why there are just 230 space groups. See also G. F. Koster, Space Groups and Their Representations, *Solid State Phys.*, **5**:173 (1957). Reprinted 1964.

reasons for preferring the definition of Eq. (1-10). It is obvious that no fundamental differences can arise from our different definition, for it only names the operations differently.

1-4. The Crystal Systems and Bravais Lattices. The 14 Bravais lattices are classified into 7 systems, called the triclinic, monoclinic, orthorhombic, tetragonal, trigonal, hexagonal, and cubic, which we shall first describe, later explaining how they are subdivided into the 14 Bravais lattices. In the triclinic system, the vectors t_1, t_2, t_3 are all of different lengths, and no two of them are at right angles to each other. It is the least symmetrical of any of the systems. In the monoclinic system, the three vectors are still all of different lengths, but one of the vectors, usually taken to be t_3, is at right angles to the plane of the other two. In the orthorhombic system, the three vectors are again of different lengths, but all are at right angles to each other. In the tetragonal system, two of the vectors, generally taken to be t_1 and t_2, have equal magnitude, t_3 has a different magnitude, but all three vectors are at right angles to each other. In the trigonal system all three vectors are of the same magnitude, but the angles between t_1 and t_2, t_2 and t_3, and t_3 and t_1, though equal to each other, are different from right angles. In the hexagonal system, two vectors t_1 and t_2 are of equal magnitude and at angles of 120° to each other, while the third vector t_3 is of different magnitude, and is at right angles to the plane of t_1 and t_2. Finally, in the cubic system, the three vectors t_1, t_2, t_3 are of equal magnitude and at right angles to each other. These descriptions must in some cases be revised, as we shall show in the next section, for the unit cells defined by these vectors are not always primitive, and in the case of some of the Bravais lattices, we must choose the vectors differently from the way just described, to get primitive unit cells.

Each of the 32 point groups enumerated at the end of the preceding section can be found associated with only one of the crystal systems. Specifically, C_1 and C_i are associated with the triclinic system; C_2, C_s, and C_{2h} with the monoclinic system; D_2, C_{2v}, D_{2h} with the orthorhombic system; C_4, S_4, C_{4h}, D_4, C_{4v}, D_{2d}, D_{4h} with the tetragonal system; C_3, C_{3i}, D_3, C_{3v}, D_{3d} with the trigonal system; C_6, C_{3h}, C_{6h}, D_6, C_{6v}, D_{3h}, D_{6h} with the hexagonal system; and T, T_h, O, T_d, and O_h with the cubic system. There is a standard notation for the 230 space groups, discussed together with details of the groups in the reference work International Tables for X-Ray Crystallography, published for the International Union of Crystallography by the Kynoch Press, Birmingham, England, in 1952. This notation is given in Table 1-1.

It is obvious that, since there are only 32 crystal classes or point groups, and 230 space groups, some of the crystal classes must have many space groups associated with them. One notation for the space groups gives

Table 1-1
Symbols for the 230 space groups. Notation is described in the text

Triclinic
1. C_1^1 ($P1$)
2. C_i^1 ($P\bar{1}$)

Monoclinic
3. C_2^1 ($P2$)
4. C_2^2 ($P2_1$)
5. C_2^3 ($B2$)
6. C_s^1 (Pm)
7. C_s^2 (Pb)
8. C_s^3 (Bm)
9. C_s^4 (Bb)
10. C_{2h}^1 ($P2/m$)
11. C_{2h}^2 ($P2_1/m$)
12. C_{2h}^3 ($B2/m$)
13. C_{2h}^4 ($P2/b$)
14. C_{2h}^5 ($P2_1/b$)
15. C_{2h}^6 ($B2/b$)

Orthorhombic
16. D_2^1 ($P222$)
17. D_2^2 ($P222_1$)
18. D_2^3 ($P2_12_12$)
19. D_2^4 ($P2_12_12_1$)
20. D_2^5 ($C222_1$)
21. D_2^6 ($C222$)
22. D_2^7 ($F222$)
23. D_2^8 ($I222$)
24. D_2^9 ($I2_12_12_1$)
25. C_{2v}^1 ($Pmm2$)
26. C_{2v}^2 ($Pmc2_1$)
27. C_{2v}^3 ($Pcc2$)
28. C_{2v}^4 ($Pma2$)
29. C_{2v}^5 ($Pca2_1$)
30. C_{2v}^6 ($Pnc2$)
31. C_{2v}^7 ($Pmn2_1$)
32. C_{2v}^8 ($Pba2$)
33. C_{2v}^9 ($Pna2_1$)
34. C_{2v}^{10} ($Pnn2$)
35. C_{2v}^{11} ($Cmm2$)
36. C_{2v}^{12} ($Cmc2_1$)
37. C_{2v}^{13} ($Ccc2$)
38. C_{2v}^{14} ($Amm2$)
39. C_{2v}^{15} ($Abm2$)
40. C_{2v}^{16} ($Ama2$)
41. C_{2v}^{17} ($Aba2$)
42. C_{2v}^{18} ($Fmm2$)
43. C_{2v}^{19} ($Fdd2$)
44. C_{2v}^{20} ($Imm2$)
45. C_{2v}^{21} ($Iba2$)
46. C_{2v}^{22} ($Ima2$)
47. D_{2h}^1 ($Pmmm$)
48. D_{2h}^2 ($Pnnn$)
49. D_{2h}^3 ($Pccm$)
50. D_{2h}^4 ($Pban$)
51. D_{2h}^5 ($Pmma$)
52. D_{2h}^6 ($Pnna$)
53. D_{2h}^7 ($Pmna$)
54. D_{2h}^8 ($Pcca$)
55. D_{2h}^9 ($Pbam$)
56. D_{2h}^{10} ($Pccn$)
57. D_{2h}^{11} ($Pbcm$)
58. D_{2h}^{12} ($Pnnm$)
59. D_{2h}^{13} ($Pmmn$)
60. D_{2h}^{14} ($Pbcn$)
61. D_{2h}^{15} ($Pbca$)
62. D_{2h}^{16} ($Pnma$)
63. D_{2h}^{17} ($Cmcm$)
64. D_{2h}^{18} ($Cmca$)
65. D_{2h}^{19} ($Cmmm$)
66. D_{2h}^{20} ($Cccm$)
67. D_{2h}^{21} ($Cmma$)
68. D_{2h}^{22} ($Ccca$)
69. D_{2h}^{23} ($Fmmm$)
70. D_{2h}^{24} ($Fddd$)
71. D_{2h}^{25} ($Immm$)
72. D_{2h}^{26} ($Ibam$)
73. D_{2h}^{27} ($Ibca$)
74. D_{2h}^{28} ($Imma$)

Tetragonal
75. C_4^1 ($P4$)
76. C_4^2 ($P4_1$)
77. C_4^3 ($P4_2$)
78. C_4^4 ($P4_3$)
79. C_4^5 ($I4$)
80. C_4^6 ($I4_1$)
81. S_4^1 ($P\bar{4}$)
82. S_4^2 ($I\bar{4}$)
83. C_{4h}^1 ($P4/m$)
84. C_{4h}^2 ($P4_2/m$)
85. C_{4h}^3 ($P4/n$)
86. C_{4h}^4 ($P4_2/n$)
87. C_{4h}^5 ($I4/m$)
88. C_{4h}^6 ($I4_1/a$)
89. D_4^1 ($P422$)
90. D_4^2 ($P42_12$)
91. D_4^3 ($P4_122$)
92. D_4^4 ($P4_12_12$)
93. D_4^5 ($P4_222$)
94. D_4^6 ($P4_22_12$)
95. D_4^7 ($P4_322$)
96. D_4^8 ($P4_32_12$)
97. D_4^9 ($I422$)
98. D_4^{10} ($I4_122$)
99. C_{4v}^1 ($P4mm$)
100. C_{4v}^2 ($P4bm$)
101. C_{4v}^3 ($P4_2cm$)
102. C_{4v}^4 ($P4_2nm$)
103. C_{4v}^5 ($P4cc$)
104. C_{4v}^6 ($P4nc$)
105. C_{4v}^7 ($P4_2mc$)
106. C_{4v}^8 ($P4_2bc$)
107. C_{4v}^9 ($I4mm$)
108. C_{4v}^{10} ($I4cm$)
109. C_{4v}^{11} ($I4_1md$)
110. C_{4v}^{12} ($I4_1cd$)
111. D_{2d}^1 ($P\bar{4}2m$)
112. D_{2d}^2 ($P\bar{4}2c$)
113. D_{2d}^3 ($P\bar{4}2_1m$)

Table 1-1 (Continued)

114. D_{2d}^{4} ($P\bar{4}2_{1}c$)
115. D_{2d}^{5} ($P\bar{4}m2$)
116. D_{2d}^{6} ($P\bar{4}c2$)
117. D_{2d}^{7} ($P\bar{4}b2$)
118. D_{2d}^{8} ($P\bar{4}n2$)
119. D_{2d}^{9} ($I\bar{4}m2$)
120. D_{2d}^{10} ($I\bar{4}c2$)
121. D_{2d}^{11} ($I\bar{4}2m$)
122. D_{2d}^{12} ($I\bar{4}2d$)
123. D_{4h}^{1} ($P4/mmm$)
124. D_{4h}^{2} ($P4/mcc$)
125. D_{4h}^{3} ($P4/nbm$)
126. D_{4h}^{4} ($P4/nnc$)
127. D_{4h}^{5} ($P4/mbm$)
128. D_{4h}^{6} ($P4/mnc$)
129. D_{4h}^{7} ($P4/nmm$)
130. D_{4h}^{8} ($P4/ncc$)
131. D_{4h}^{9} ($P4_{2}/mmc$)
132. D_{4h}^{10} ($P4_{2}/mcm$)
133. D_{4h}^{11} ($P4_{2}/nbc$)
134. D_{4h}^{12} ($P4_{2}/nnm$)
135. D_{4h}^{13} ($P4_{2}/mbc$)
136. D_{4h}^{14} ($P4_{2}/mnm$)
137. D_{4h}^{15} ($P4_{2}/nmc$)
138. D_{4h}^{16} ($P4_{2}/ncm$)
139. D_{4h}^{17} ($I4/mmm$)
140. D_{4h}^{18} ($I4/mcm$)
141. D_{4h}^{19} ($I4_{1}/amd$)
142. D_{4h}^{20} ($I4_{1}/acd$)

Trigonal

143. C_{3}^{1} ($P3$)
144. C_{3}^{2} ($P3_{1}$)
145. C_{3}^{3} ($P3_{2}$)
146. C_{3}^{4} ($R3$)
147. C_{3i}^{1} ($P\bar{3}$)
148. C_{3i}^{2} ($R\bar{3}$)
149. D_{3}^{1} ($P312$)
150. D_{3}^{2} ($P321$)
151. D_{3}^{3} ($P3_{1}12$)
152. D_{3}^{4} ($P3_{1}21$)
153. D_{3}^{5} ($P3_{2}12$)
154. D_{3}^{6} ($P3_{2}21$)
155. D_{3}^{7} ($R32$)
156. C_{3v}^{1} ($P3m1$)
157. C_{3v}^{2} ($P31m$)
158. C_{3v}^{3} ($P3c1$)
159. C_{3v}^{4} ($P31c$)
160. C_{3v}^{5} ($R3m$)
161. C_{3v}^{6} ($R3c$)
162. D_{3d}^{1} ($P\bar{3}1m$)
163. D_{3d}^{2} ($P\bar{3}1c$)
164. D_{3d}^{3} ($P\bar{3}m1$)
165. D_{3d}^{4} ($P\bar{3}c1$)
166. D_{3d}^{5} ($R\bar{3}m$)
167. D_{3d}^{6} ($R\bar{3}c$)

Hexagonal

168. C_{6}^{1} ($P6$)
169. C_{6}^{2} ($P6_{1}$)
170. C_{6}^{3} ($P6_{5}$)
171. C_{6}^{4} ($P6_{2}$)
172. C_{6}^{5} ($P6_{4}$)
173. C_{6}^{6} ($P6_{3}$)
174. C_{3h}^{1} ($P\bar{6}$)
175. C_{6h}^{1} ($P6/m$)
176. C_{6h}^{2} ($P6_{3}/m$)
177. D_{6}^{1} ($P622$)
178. D_{6}^{2} ($P6_{1}22$)
179. D_{6}^{3} ($P6_{5}22$)
180. D_{6}^{4} ($P6_{2}22$)
181. D_{6}^{5} ($P6_{4}22$)
182. D_{6}^{6} ($P6_{3}22$)
183. C_{6v}^{1} ($P6mm$)
184. C_{6v}^{2} ($P6cc$)
185. C_{6v}^{3} ($P6_{3}cm$)
186. C_{6v}^{4} ($P6_{3}mc$)
187. D_{3h}^{1} ($P\bar{6}m2$)
188. D_{3h}^{2} ($P\bar{6}c2$)
189. D_{3h}^{3} ($P\bar{6}2m$)
190. D_{3h}^{4} ($P\bar{6}2c$)
191. D_{6h}^{1} ($P6/mmm$)
192. D_{6h}^{2} ($P6/mcc$)
193. D_{6h}^{3} ($P6_{3}/mcm$)
194. D_{6h}^{4} ($P6_{3}/mmc$)

Cubic

195. T^{1} ($P23$)
196. T^{2} ($F23$)
197. T^{3} ($I23$)
198. T^{4} ($P2_{1}3$)
199. T^{5} ($I2_{1}3$)
200. T_{h}^{1} ($Pm3$)
201. T_{h}^{2} ($Pn3$)
202. T_{h}^{3} ($Fm3$)
203. T_{h}^{4} ($Fd3$)
204. T_{h}^{5} ($Im3$)
205. T_{h}^{6} ($Pa3$)
206. T_{p}^{7} ($Ia3$)
207. O^{1} ($P432$)
208. O^{2} ($P4_{2}32$)
209. O^{3} ($F432$)
210. O^{4} ($F4_{1}32$)
211. O^{5} ($I432$)
212. O^{6} ($P4_{3}32$)
213. O^{7} ($P4_{1}32$)
214. O^{8} ($I4_{1}32$)
215. T_{d}^{1} ($P\bar{4}3m$)
216. T_{d}^{2} ($F\bar{4}3m$)
217. T_{d}^{3} ($I\bar{4}3m$)
218. T_{d}^{4} ($P\bar{4}3n$)
219. T_{d}^{5} ($F\bar{4}3c$)
220. T_{d}^{6} ($I\bar{4}3d$)
221. O_{h}^{1} ($Pm3m$)
222. O_{h}^{2} ($Pn3n$)
223. O_{h}^{3} ($Pm3n$)
224. O_{h}^{4} ($Pn3m$)
225. O_{h}^{5} ($Fm3m$)
226. O_{h}^{6} ($Fm3c$)
227. O_{h}^{7} ($Fd3m$)
228. O_{h}^{8} ($Fd3c$)
229. O_{h}^{9} ($Im3m$)
230. O_{h}^{10} ($Ia3d$)

the Schoenflies symbol for the point group, as given earlier in this section, and then numbers the various space groups having this point group by superscripts. For instance, there are 10 space groups associated with the point group O_h; they are denoted as $O_h^1 \cdots O_h^{10}$. The various space groups associated with a given point group differ in two possible ways. First, they can be associated with the various Bravais lattices connected with the crystal system in question. Second, they can have different nonprimitive translations associated with some of their operations. The second notation given for a space group in Table 1-1, such as $(Pm3m)$ for the space group O_h^1, describes these additional features, but in a rather complicated way which we shall not go into.

We shall now consider the various Bravais lattices associated with each crystal system. We shall use the opposite order to that of Table 1-1 for our discussion, starting with the cubic system, which is the most familiar. There are three Bravais lattices in the cubic system: the simple cubic, the face-centered cubic, and the body-centered cubic lattices. The simple cubic lattice has no features aside from the three primitive vectors t_1, t_2, t_3 of equal magnitude and at right angles to each other. The vectors $R_n = n_1 t_1 + n_2 t_2 + n_3 t_3$ terminate on a simple cubic lattice of points, of the most elementary variety. In the face-centered cubic lattice, however, we have not only the points at the corners of the cube, but also points in the center of each face, which are equivalent crystallographically. In the body-centered cubic lattice, both the points at the corners of the cubes and the points at their centers are equivalent. These three Bravais lattices are indicated by the capital letter in the second symbol of Table 1-1: this symbol is P for the simple cubic lattice, F for face-centered cubic, I for body-centered cubic. Thus we see from Table 1-1 that O_h^1 $(Pm3m)$, O_h^2 $(Pn3n)$, O_h^3 $(Pm3n)$, and O_h^4 $(Pn3m)$ are associated with the simple cubic lattice; O_h^5 $(Fm3m)$, O_h^6 $(Fm3c)$, O_h^7 $(Fd3m)$, and O_h^8 $(Fd3c)$ with the face-centered cubic lattice; and O_h^9 $(Im3m)$ and O_h^{10} $(Ia3d)$ with the body-centered cubic lattice. In a similar way we can find which Bravais lattice occurs for each of the space groups in the cubic system.

In the hexagonal system, the only lattice which occurs is the simple hexagonal, determined by the vectors t_1 and t_2 of equal magnitude and making an angle of 120° with each other, with t_3 of different magnitude at right angles to t_1 and t_2. This is indicated by the symbol P in the symbols for the space groups of the hexagonal system. In the trigonal system, there are two possible lattices. One is the hexagonal lattice just described, again denoted by P. The other is the trigonal or rhombohedral lattice, denoted by R, determined by vectors t_1, t_2, t_3, all of the same magnitude, but making equal angles with each other, not equal to 90°.

There are two lattices for the tetragonal system. First is the simple tetragonal lattice, denoted by P. Second is the body-centered tetragonal lattice, denoted by I, in which the point at the center of the tetragonal unit cell is crystallographically equivalent to the corners of the cell.

The orthorhombic system has four lattices associated with it. First there is the simple orthorhombic lattice, denoted by P, in which we have three vectors $\mathbf{t}_1$, $\mathbf{t}_2$, $\mathbf{t}_3$, all of different magnitudes, but at right angles to each other. There are the face-centered orthorhombic, denoted by F, and the body-centered orthorhombic, denoted by I, in which, as we should expect, the points at the centers of the faces or the point at the body center of the unit cell, respectively, are equivalent to the corners of the cell. The fourth lattice is the one-face-centered orthorhombic lattice, in which the centers of one set of faces are equivalent to the corners of the cell, but not the centers of the other two sets of faces. This last lattice is denoted by A, B, or C, depending on which set of faces is face-centered.

The monoclinic system contains two lattices. The first, the simple monoclinic lattice, has the vectors $\mathbf{t}_1$ and $\mathbf{t}_2$ of different magnitude with an arbitrary angle between them, but the third vector $\mathbf{t}_3$, of still different length, at right angles to $\mathbf{t}_1$ and $\mathbf{t}_2$. It is denoted by P. The other is the one-face-centered monoclinic lattice, denoted by B, in which the point in the center of one face, ordinarily chosen to be that bounded by the vectors $\mathbf{t}_1$ and $\mathbf{t}_3$, is equivalent to the corners of that face.

The triclinic system has only a primitive lattice, denoted by P, with the vectors $\mathbf{t}_1$, $\mathbf{t}_2$, $\mathbf{t}_3$, all different in magnitude and at arbitrary angles to each other.

We thus have enumerated the 14 Bravais lattices. To make them more comprehensible to the reader, we proceed in the next section to find which of the unit cells we have been describing are primitive, to set up the vectors $\mathbf{t}_1$, $\mathbf{t}_2$, $\mathbf{t}_3$ for the primitive unit cells for most of them in terms of their rectangular components, and to discuss the unit cells more in detail.

1-5. Primitive Unit Cells for the Bravais Lattices. A lattice is a collection of points, one per unit cell, at equivalent positions in each unit cell; for instance, the points at the extremities of the vectors $\mathbf{R}_n$ given by Eq. (1-1). We can choose the unit cells in any way provided they contain one lattice point each, and satisfy the other conditions required for a unit cell: that they be identical, and fill all space. In some cases it is obvious how to choose the unit cells, but not in others. We shall now consider this point.

If we start with the simple cubic lattice, we have vectors $\mathbf{t}_1$, $\mathbf{t}_2$, $\mathbf{t}_3$, given by

$$\mathbf{t}_1 = \mathbf{i}a \qquad \mathbf{t}_2 = \mathbf{j}a \qquad \mathbf{t}_3 = \mathbf{k}a \tag{1-17}$$

where $\mathbf{i}$, $\mathbf{j}$, $\mathbf{k}$ are unit vectors along the x, y, and z axes, respectively, and

a is the side of the cube. Lattice points are found for $x = n_1 a$, $y = n_2 a$, $z = n_3 a$. We may choose as the unit cell the cube of side a with the origin at a corner. For some purposes it is more convenient to take the cell with the origin at its center; this has the advantage that the unit cell is invariant under any of the operations of the point group, T, T_h, O, T_d, or O_h as the case may be, consisting of rotations about the origin, at the center of the cube, combined with reflections in planes passing through the origin. In other words, the unit cell chosen in this way shows the full symmetry of the point group.

The other two cubic Bravais lattices, the face-centered and body-centered cubic lattices, are more complicated. Let us start with the

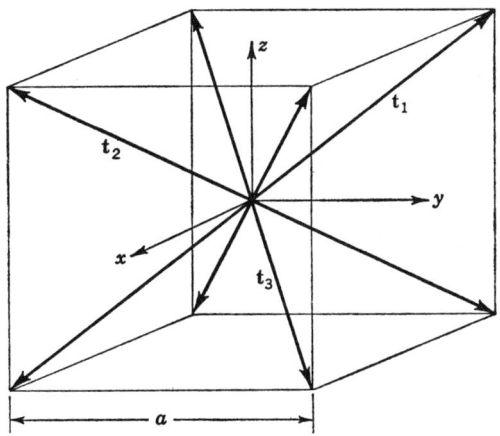

FIG. 1-1. Cube of side a, with atom in center, and atoms at corners, for body-centered cubic structure. The vectors point from the central atom to the eight nearest neighbors. Vectors t_1, t_2, t_3 are indicated.

body-centered case. If we take a cube of side a, with the origin at the corner of the cube, we have lattice points for $x = n_1 a, y = n_2 a, z = n_3 a$, where the n's are integers, as for the simple cubic case; but also for $x = (n_1 + \frac{1}{2})a$, $y = (n_2 + \frac{1}{2})a$, $z = (n_3 + \frac{1}{2})a$, where again the n's are integers. For a given set of n's, we then find two lattice points, not one. Hence the cubic cells are not primitive unit cells: each one holds two lattice points, which are crystallographically equivalent to each other. It is then possible to choose a primitive unit cell half as large as the cube with side a, and to find vectors t_1, t_2, t_3 appropriate for this smaller unit cell.

The simplest way to choose these vectors is illustrated in Fig. 1-1. Here we show a cube of side a, with the origin at the center, and equivalent lattice points at the origin and the corners of the cube. The vectors

t_1, t_2, t_3 must point from the origin to three other lattice points, chosen so that the volume of the unit cell is as small as possible. The choice made in Fig. 1-1 is

$$t_1 = \frac{a}{2}(-i + j + k) \qquad t_2 = \frac{a}{2}(i - j + k) \qquad t_3 = \frac{a}{2}(i + j - k) \qquad (1\text{-}18)$$

which lead from the origin to three of the eight corners of the cubic cell. We may, if we choose, take as the unit cell the parallelepiped whose sides are parallel to t_1, t_2, and t_3 of Eq. (1-18). This unit cell has a volume given by $t_1 \cdot (t_2 \times t_3)$, which we find to be $a^3/2$, or half the volume of the cube, as it should be.

This parallelepiped, however, does not show the symmetry of the point group, and it would be more satisfactory if we could find a unit cell which did show this symmetry. The way to do this was pointed out by Wigner and Seitz.[1] These writers were discussing the energy bands of metallic sodium, which crystallizes in the body-centered cubic structure; that is, a structure in which an atom is placed at each lattice point of a body-centered cubic lattice. They wished to surround each atom (that is, each lattice point) by a polyhedron, such that these polyhedra filled all space, thereby fulfilling the requirements for forming unit cells, and at the same time showed the symmetry of the cubic point groups. They proceeded in the following way. They drew the lines joining the origin to each of the other lattice points of the body-centered cubic lattice, and set up the planes which were the perpendicular bisectors of each of these lines. The volume enclosed by the planes close to the origin has the form shown in Fig. 1-2. This is known as a Wigner-Seitz cell.

It is easy to interpret the form of the cell of Fig. 1-2. Each atom or lattice point has eight nearest neighbors, along the directions of the vectors $\pm i \pm j \pm k$, the body diagonals of the cube, called the 111 directions. The distance to such a neighbor is $(\sqrt{3}/2)a$. We set up planes perpendicular to these directions, halfway out to the nearest neighbors, or distant from the origin by the amount $(\sqrt{3}/4)a$. These planes form the boundaries of the Wigner-Seitz cell in the eight hexagonal faces shown in Fig. 1-2. The second nearest neighbors are the six distant by the amount a, along $\pm i$, $\pm j$, $\pm k$, called the 100 directions. The perpendicularly bisecting planes are the faces of the cube, at distances $a/2$ from the origin. These planes form the boundaries of the Wigner-Seitz cell in the six square faces, rotated 45° from the orientations of the cube faces, shown in Fig. 1-2. The planes forming the perpendicular bisectors of lines from the origin to all more distant lattice points, or atoms, lie outside the cell shown in Fig. 1-2.

[1] E. Wigner and F. Seitz, *Phys. Rev.*, **43**:804 (1933).

A simple method of constructing the Wigner-Seitz cell would be to place the centers of rubber balloons at the lattice points, and inflate the balloons until they touched each other over their whole surface. They would take up the form of the Wigner-Seitz cell. It is an interesting historical fact that a similar method of constructing these cells (for the face-centered cubic structure) was used as early as the eighteenth century: dried peas were packed in a container with the suitable arrangement, water was let in until the peas swelled so as to fill the whole space, and they were then removed, showing the shape of the Wigner-Seitz cell for the face-centered cubic structure.[1]

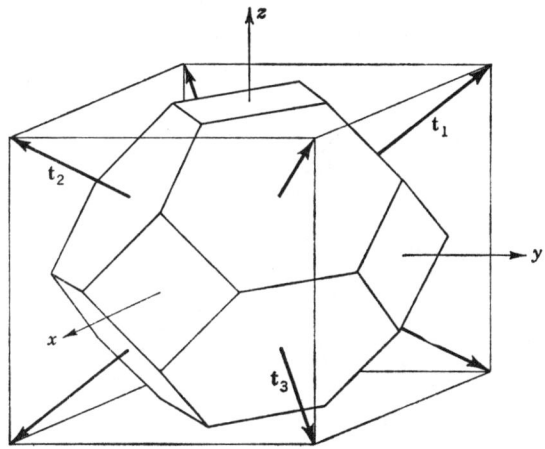

Fig. 1-2. Wigner-Seitz cell for the body-centered cubic structure.

Wigner and Seitz made the construction we have described in Fig. 1-2 in order to surround each atom of a body-centered cubic structure by a polyhedron as nearly spherical in shape as possible. What they secured was a unit cell which can just as well be used for any crystal having the body-centered cubic Bravais lattice, showing the complete symmetry of the point group, in that the cell transforms into itself under any of the operations of the cubic groups T, T_h, O, T_d, or O_h, as the case may be.

We can follow the same method for the other Bravais lattices. Let us next consider the face-centered cubic lattice. We again start with a cube of edges a. We have lattice points not only at $(n_1\mathbf{i} + n_2\mathbf{j} + n_3\mathbf{k})a$, but also at $[(n_1 + \frac{1}{2})\mathbf{i} + (n_2 + \frac{1}{2})\mathbf{j} + n_3\mathbf{k}]a$, $[(n_1 + \frac{1}{2})\mathbf{i} + n_2\mathbf{j} + (n_3 + \frac{1}{2})\mathbf{k}]a$, $[n_1\mathbf{i} + (n_2 + \frac{1}{2})\mathbf{j} + (n_3 + \frac{1}{2})\mathbf{k}]a$. In other words, the

---

[1] See H. S. M. Coxeter, The Problem of Packing a Number of Equal Nonoverlapping Circles on a Sphere, *Trans. N.Y. Acad. Sci.*, ser. II, **24**:320 (1962), for a reference to this experiment. I am indebted to Prof. G. F. Koster for pointing out this interesting reference to me.

cubic cells contain four lattice points, and hence are not primitive, since all lattice points are crystallographically equivalent. In Fig. 1-3 we show the vectors from the origin to the 12 nearest neighbors of a given lattice point, those located at $(\pm i \pm j)a/2$, $(\pm i \pm k)a/2$, and $(\pm j \pm k)a/2$. We may conveniently choose the three vectors

$$t_1 = \frac{a}{2}(j+k) \qquad t_2 = \frac{a}{2}(i+k) \qquad t_3 = \frac{a}{2}(i+j) \qquad (1\text{-}19)$$

as the primitive vectors for the face-centered cubic lattice. We verify that the volume of the parallelopiped bounded by these vectors, $t_1 \cdot (t_2 \times t_3)$, is $a^3/4$, so that it has a fourth the volume of the cube, as it

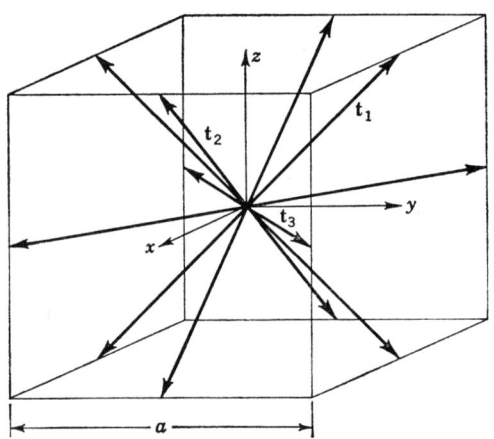

Fig. 1-3. Cube of side a, with atom in center, and atoms at centers of cube edges, representing the 12 nearest neighbors of the atom at the origin in the face-centered cubic structure. Vectors t_1, t_2, t_3 are indicated.

should be for the unit cell. For the Wigner-Seitz construction we form the planes which are the perpendicular bisectors of the vectors from the origin to the 12 nearest neighboring lattice points, as shown in Fig. 1-3, and we find that these planes alone form the boundaries of the Wigner-Seitz cell, shown in Fig. 1-4. All other planes forming perpendicular bisectors of vectors to more distant lattice points lie outside this cell. This Wigner-Seitz cell, like that for the body-centered cubic structure, remains unchanged under any of the operations of the cubic point groups.

The situation of the hexagonal lattice, which we take up next, is somewhat different from those met in the cubic system. The vectors t_1 and t_2, shown in Fig. 1-5, taken together with t_3, at right angles to the plane of the paper, enclose a parallelopiped which is a primitive unit cell, in that only one lattice point is associated with each unit cell. However, it does not show the symmetry of the hexagonal or trigonal point groups. Let

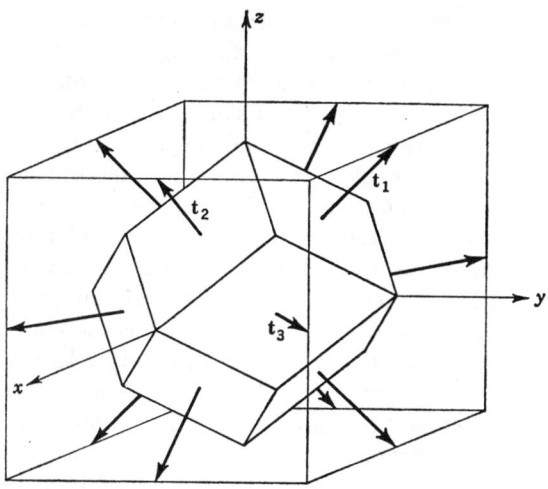

Fig. 1-4. Wigner-Seitz cell for the face-centered cubic structure.

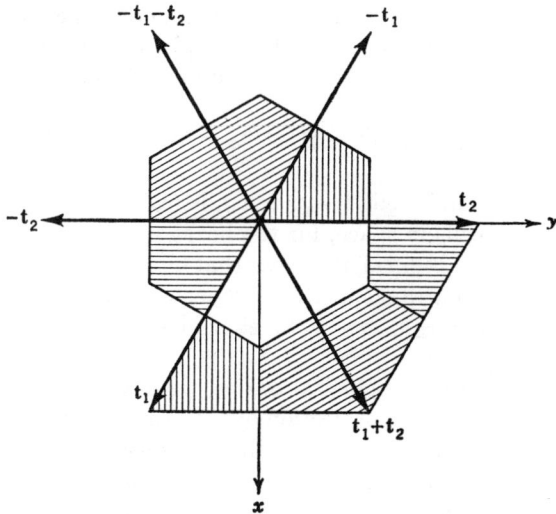

Fig. 1-5. Wigner-Seitz cell for the hexagonal lattice, projected on the xy plane. Shading indicates how segments of the parallelogram outside the hexagon can be fitted into the hexagon, to demonstrate that the parallelogram and the hexagon have equal areas.

us carry out the Wigner-Seitz construction to get a symmetrical unit cell. We set up the six vectors t_1, $t_1 + t_2$, t_2, $-t_1$, $-t_1 - t_2$, $-t_2$, shown in Fig. 1-5. These point to the six nearest lattice points to the origin, in the plane of the paper. The planes forming the perpendicular bisectors of these six vectors intersect the plane of the paper in the regular hexagon shown in the figure. The additional planes parallel to the paper, and at distances up and down from the paper by half the magnitude of the vector t_3, together with the planes indicated in Fig. 1-5, bound a hexagonal prism which is the Wigner-Seitz unit cell for this case. It obviously shows the complete hexagonal symmetry. It is particularly simple in this case to see that the volume of the Wigner-Seitz unit cell is the same as that of the parallelopiped enclosed by the vectors t_1, t_2, and t_3. In Fig. 1-5 we have indicated by shading how segments of the parallelogram can be fitted into the hexagon, so as to give a simple geometrical proof of the equality of volumes. Similar constructions can be carried out for the Wigner-Seitz cells of the body- and face-centered cubic lattices, but they are harder to indicate graphically.

It is convenient to have expressions for the vectors t_1, t_2, and t_3 in rectangular coordinates. In the hexagonal system it is conventional to denote the length of the vectors t_1 and t_2 by a, and of t_3 by c. We shall take the x axis vertically downward in the plane of the paper, the y axis to the right, and the z axis up normal to the plane of the paper, so that the three form a right-handed coordinate system. Then we shall assume

$$t_1 = \frac{a}{2}(\sqrt{3}\,\mathbf{i} - \mathbf{j}) \qquad t_2 = a\mathbf{j} \qquad t_3 = c\mathbf{k} \tag{1-20}$$

The volume of the unit cell, $t_1 \cdot (t_2 \times t_3)$, is $(\sqrt{3}/2)a^2 c$. The packing of the Wigner-Seitz cells in space, for the hexagonal lattice, is very simple. In the xy plane, as in Fig. 1-5, it is clear that hexagons pack together in the familiar honeycomb arrangement. A layer of hexagonal prisms is then found at $z = 0$, $z = \pm c$, $z = \pm 2c$, etc., identical layers being stacked above each other.

In the trigonal or rhombohedral lattice, we have three vectors t_1, t_2, t_3, of equal magnitude. The projection of t_1 on the xy plane is usually taken to be in the direction of the x axis, with the projections of t_2 and t_3 at 120° and 240° to x, but these vectors have equal z components as well. Thus we may write these vectors as

$$t_1 = s\mathbf{i} + r\mathbf{k} \qquad t_2 = \frac{s}{2}(-\mathbf{i} + \sqrt{3}\,\mathbf{j}) + r\mathbf{k} \qquad t_3 = \frac{s}{2}(-\mathbf{i} - \sqrt{3}\,\mathbf{j}) + r\mathbf{k} \tag{1-21}$$

where s and r are constants. It is customary in the trigonal system to denote the length of the vectors t_1, t_2, t_3 by a, and the angle between any two of them by α. We then find

$$a^2 = s^2 + r^2 \qquad \cos\alpha = \frac{-s^2/2 + r^2}{s^2 + r^2}$$
$$s = \frac{2a}{\sqrt{3}} \sin\frac{\alpha}{2} \qquad r = \sqrt{a^2 - s^2} \qquad (1\text{-}22)$$

By use of Eq. (1-22) we can find s and r in case a and α are given.

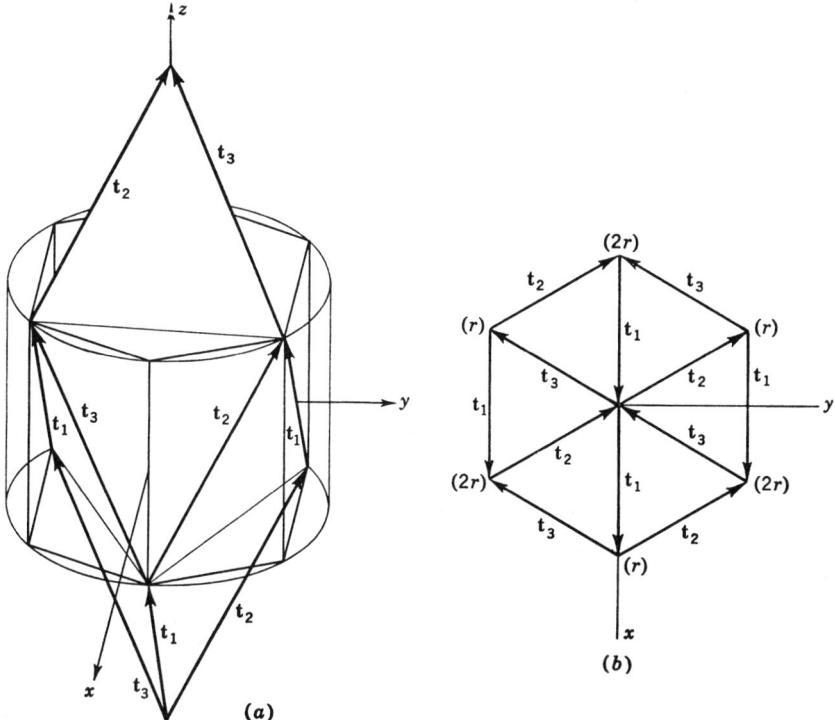

FIG. 1-6. (a) Unit cell for the rhombohedral or trigonal Bravais lattice, including a cylinder and a hexagonal prism for help in visualizing the geometry. (b) Projection of Fig. 1-6a looking downward along $-z$ axis. Heights of corners above bottom corner of parallelopiped indicated by (r), $(2r)$. Center corresponds to heights 0, $3r$.

The vectors t_1, t_2, t_3, and the parallelopiped formed from them, are shown in Fig. 1-6. We show also a hexagonal prism, and cylinder, in this figure, to help in visualizing the geometrical situation. If we choose the origin at the center of the parallelopiped, as shown in the figure, then this parallelopiped as it stands is invariant under the point-group operations C_3, C_{3i}, C_{3v}, D_3, or D_{3d} of the trigonal system, as a hexagonal unit cell also is. Consequently we do not need to carry out a Wigner-Seitz construction to get a unit cell of proper symmetry; we can use the parallelopiped of Fig. 1-6.

Koster[1] (*loc. cit.*) shows the results of the Wigner-Seitz construction, yielding more complicated polyhedrons than that of Fig. 1-6. Koster also points out that this case reduces to other lattices which we have already discussed, for special values of the angle α, or the ratio r/a. Thus, if α equals 90°, we obviously reduce to the simple cubic lattice; for this reason, $\alpha = 90°$ is specifically excluded in defining the trigonal or rhombohedral lattice. But we notice that the vectors t_1, t_2, t_3 for the body-centered and face-centered cubic lattices, which we have discussed earlier, also form special cases of the rhombohedral lattice. The body-centered case comes when $r/s = 1/(2\sqrt{2})$, the face-centered case when $r/s = \sqrt{2}$, and the simple cubic case when $r/s = 1/\sqrt{2}$.

Next we come to the tetragonal system, for which we recall that there are two Bravais lattices, the simple tetragonal and the body-centered tetragonal. For the simple tetragonal lattice, we may take the vectors to be

$$t_1 = a\mathbf{i} \qquad t_2 = a\mathbf{j} \qquad t_3 = c\mathbf{k} \qquad (1\text{-}23)$$

where a and c are different. If we take a unit cell of the same shape, but with the origin at the center, we have achieved the required symmetry for this case. For the body-centered tetragonal, we proceed much as for the body-centered cubic lattice. We may take

$$t_1 = \frac{a}{2}(-\mathbf{i} + \mathbf{j}) + \frac{c}{2}\mathbf{k} \qquad t_2 = \frac{a}{2}(\mathbf{i} - \mathbf{j}) + \frac{c}{2}\mathbf{k}$$

$$t_3 = \frac{a}{2}(\mathbf{i} + \mathbf{j}) - \frac{c}{2}\mathbf{k} \qquad (1\text{-}24)$$

which reduces to the values of Eq. (1-18) if c becomes equal to a. If we carry out a Wigner-Seitz construction, starting with the eight vectors similar to those of Eq. (1-23), pointing to similar lattice points, and setting up their perpendicular bisecting planes, as well as the bisecting planes of the vectors to the second nearest neighbors, we get Wigner-Seitz cells as shown in Fig. 1-7, given by Koster (*loc. cit.*). We have different cases, depending on whether $c > a\sqrt{2}$ (the upper case) or $c < a\sqrt{2}$ (the lower case). The body-centered cubic case, coming when $c = a$, is a special case of the lower figure in Fig. 1-7.

For the orthorhombic, monoclinic, and triclinic systems, the Wigner-Seitz cells become more complicated, and as in the case of the body-centered tetragonal lattice, illustrated in Fig. 1-7, the Wigner-Seitz cells have different shapes depending on the relative lengths of the various

---

[1] The reader comparing with Koster's treatment will note that he interchanges our vectors t_1 and t_2, and in some cases uses left-hand axes.

vectors. These cases are taken up by Koster (*loc. cit.*), and we refer the reader to his paper for further details. In our later work we shall discuss some particular examples of several of these systems and Bravais lattices.

We have now obtained a general idea of the nature of the unit cells and Bravais lattices, and have worked out the detailed behavior for 7 out of the

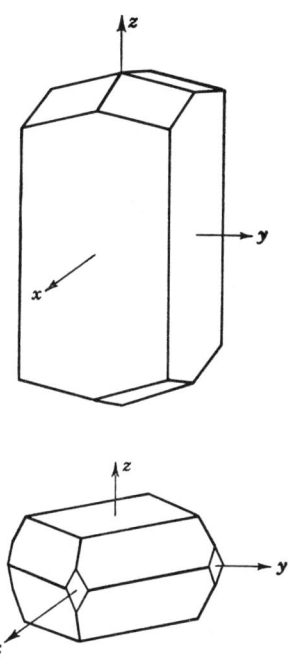

FIG. 1-7. Wigner-Seitz cells for body-centered tetragonal Bravais lattice, following Koster. Upper figure, $c > a \sqrt{2}$. Lower figure, $c < a \sqrt{2}$.

14 Bravais lattices. We must now study the relations of the Bravais lattices and the translation operations to the point groups; that is, we must study the specific behavior of the space groups. We go on to this problem in the next two chapters, illustrating a number of the space groups by use of specific examples.

2

Space Groups for Structures of the Elements

2-1. Survey of Crystal Structures of the Elements. Out of the 230 space groups mentioned in the preceding chapter, a relatively small number are found in the structures of a great majority of the elements and compounds. We shall choose some 20 space groups for detailed discussion, illustrating the general principles by means of these specific examples, and naturally we shall pick space groups which occur widely among actual structures. In the present chapter we shall take up the crystal structures of the elements, and a detailed treatment of several of the important space groups which we meet in these structures; the next chapter proceeds in a similar way with the compounds.

A list of the elements, with information about their crystal structure, is given in Table 2-1. Out of all the elements, the great majority are metals. Those which are not are hydrogen; the inert gases helium, neon, argon, krypton, xenon, and radon; and the elements carbon, nitrogen, oxygen, fluorine, silicon, phosphorus, sulfur, chlorine, germanium, selenium, bromine, tin (one form), tellurium, and iodine.

The inert gases form crystals held together by Van der Waals forces, the type of weak attractive force, felt at large distances, which appears in Van der Waals' equation for an imperfect gas. We have not considered the explanation of Van der Waals forces in Volume 1; this discussion will be postponed to a later volume, where we shall show that they arise from a mutual polarization of each of the atoms by its neighboring atom. The elements hydrogen, nitrogen, oxygen, fluorine, and chlorine form diatomic molecules at room temperature, and their crystals are molecular crystals, composed of the molecules formed much as they are in the free condition, and held together by Van der Waals forces. Bromine and iodine also form crystals composed of diatomic molecules, similar to fluorine and chlorine, though these crystals have higher melting points than fluorine and chlorine. Gallium, which melts around room temperature, though a metal, has a structure similar to iodine, but the distance between the two atoms in a diatomic molecule is only slightly less than that from an atom

Table 2-1

Crystal structures and interatomic distances in elements (Angstroms). Abbreviations: bc, body-centered cubic; fc, face-centered cubic; hx, hexagonal; di, diamond; gr, graphite; As, arsenic; Se, selenium; I, iodine; *, other structure. In the hexagonal crystals, the first distance is to the closest atom in the basal plane; the second, to the closest atom out of the plane.

H*
0.74

He hx	Ne fc	A fc	Kr fc	Xe fc	
3.51, 3.57	3.20	3.84	4.03	4.42	
Li bc, fc, hx	Na bc, hx	K bc	Rb bc	Cs bc	
3.04, 3.11	3.72	4.51	4.88	5.24	
3.11, 3.11	3.77, 3.77				
Be hx	Mg hx	Ca fc, hx	Sr fc, hx	Ba bc	
2.29, 2.23	3.21, 3.20	3.92	4.30	4.35	
		3.98, 3.99	4.32, 4.32		
B*	Al fc	Sc fc, hx	Y hx	La fc, hx	
1.75–1.80	2.86	3.20	3.64, 3.56	3.74	
		3.31, 3.24		3.75, 3.73	
				rare earths	
		Ti hx, bc	Zr hx, bc	Hf hx	
		2.95, 2.90	3.23, 3.18	3.20, 3.13	
		2.86	3.13		
		V bc	Nb bc	Ta bc	
		2.63	2.86	2.86	
		Cr bc, hx	Mo bc	W bc	
		2.50	2.73	2.74	
		2.71, 2.72			
		Mn*	Tc hx	Re hx	
		2.24, 3.00	2.74, 2.70	2.76, 2.74	
		Fe bc, fc	Ru hx	Os hx	
		2.48, 2.54	2.70, 2.65	2.74, 2.66	
		Co fc, hx	Rh fc	Ir fc	
		2.51	2.69	2.71	
		2.50, 2.49			
		Ni fc, hx	Pd fc	Pt fc	
		2.49	2.74	2.77	
		2.65, 2.65			
		Cu fc	Ag fc	Au fc	
		2.56	2.89	2.88	
		Zn hx	Cd hx	Hg*	
		2.66, 2.91	2.98, 3.29	3.00	
		Ga I	In*	Tl bc, hx	
		2.44, 2.76	3.24, 3.36	3.36	
				3.47, 3.41	

Table 2-1 (Continued)

C di, gr	Si di	Ge di	Sn di,*	Pb fc
1.54, 1.42	2.35	2.45	2.80, 3.02	3.50
N*	P*	As As	Sb As	Bi As
1.10	2.18	2.50	2.90	3.10
O*	S*	Se Se	Te Se	
1.21	2.10	2.32	2.86	
F*	Cl*	Br I	I I	
1.42	2.02	2.42	2.68	

Rare earths	Radioactive elements
Ce, fc, hx: 3.64, 3.65, 3.65	Th, fc: 3.57
Pr, fc, hx: 3.65, 3.67, 3.64	U, bc: 3.01. Also other structures
Nd, hx: 3.66, 3.63	Np, bc: 3.05. Also other structures
. . . .	Pu, fc, bc: 3.28, 3.15
Gd, hx: 3.63, 3.58	
Tb, hx: 3.59, 3.55	
Dy, hx: 3.58, 3.51	
Ho, hx: 3.56, 3.49	
Er, hx: 3.56, 3.47	
Tu, hx: 3.53, 3.45	
Yb, fc: 3.86	

in one molecule to that in a neighboring molecule, so that the diatomic nature of the crystal is not very marked.

The elements sulfur, selenium, and tellurium form chain structures, held together by covalent bonds. These elements, like oxygen, are divalent, but instead of forming diatomic molecules with double bonds, like O_2, the two bonds attach each atom to two neighbors, forming chains. The bonds, in other words, are similar to the two oxygen-hydrogen bonds in H_2O, and as in that molecule, they make an angle of something greater than a right angle with each other. In selenium and tellurium these chains, coiled to form helices, extend through the crystal, which consists of a stack of parallel helices. In sulfur, there are several forms, one of which consists of rings S_8, with covalent bonds holding the atoms together. In each of these cases the chains or rings are held together by Van der Waals forces.

The elements phosphorus, arsenic, antimony, and bismuth act in their solid forms somewhat as if they had three covalent bonds. The atoms form sheets, each atom being held more closely to three nearest neighbors than to their other neighbors. The bonds form a pyramid, something like the three bonds formed in NH_3, nitrogen being chemically similar to P, As, Sb, and Bi. Of these elements, phosphorus is a nonmetal, while the other three barely show metallic properties. The remaining nonmetallic elements, carbon, silicon, germanium, together with one form of tin, form crystals having the diamond structure, in which each atom is held to four neighbors by covalent bonds, similar to the tetrahedral bonds of carbon in methane, and in many organic compounds. These substances are insulators or semiconductors, of great technical importance. Carbon also forms the graphite structure, a layer structure, with the atoms arranged in a hexagonal array in each layer. The bonds here are similar to those in the benzene molecule. Graphite is a metal, but just on the border of having nonmetallic properties; it occupies a special place, not exactly identical with any other material, though having some resemblance to arsenic, antimony, and bismuth, in that it is almost on the border between metallic and nonmetallic substances.

The rest of the elements are well-defined metals, and by far the larger number of them crystallize in one of three structures: the body-centered and face-centered cubic structures, and the hexagonal structure. There is no well-understood rule stating which structure a given metal will have, and a number of them exist in different polymorphic modifications, having different structures at different temperatures and pressures. A few elements, notably boron (perhaps a semiconductor rather than a metal), manganese, uranium, and a few others, have peculiar and complicated structures.

Out of all these structures found among the elements, we shall choose the following ones for detailed discussion: the body-centered cubic, with space group O_h^9 ($Im3m$); the face-centered cubic, with space group O_h^5 ($Fm3m$); the hexagonal, with space group D_{6h}^4 ($P6_3/mmc$); diamond, with space group O_h^7 ($Fd3m$); graphite, with space group C_{6v}^4 ($P6_3mc$); arsenic, with space group D_{3d}^5 ($R\bar{3}m$); selenium, with space group D_3^4 ($P3_121$) or D_3^6 ($P3_221$); and iodine, with space group D_{2h}^{18} ($Cmca$). We shall not only go into the properties of these space groups, but shall also discuss various features of the structures, which are of interest in connection with the packing of the atoms in the crystals, and other properties.

2-2. The Body- and Face-centered Cubic Structures. In Chap. 1 we have already studied the body-centered and face-centered cubic Bravais lattices, and have stated that the body-centered and face-centered cubic structures of the elements come when an atom is placed at the origin of each unit cell in one or the other of the lattices. This really is all we need to say to define the structures, but we shall describe more in detail the operations of the space groups, as well as some features of the structures.

We recall that a space group is made up by combining the rotation and reflection operations of a point group, with translations. In the case of the symmorphic space groups, the translations are merely the quantities $\mathbf{R}_n = n_1\mathbf{t}_1 + n_2\mathbf{t}_2 + n_3\mathbf{t}_3$, and both the space groups concerned here, namely, O_h^9 ($Im3m$) and O_h^5 ($Fm3m$), are symmorphic. To write down the operations of the space group in detail, we must have the operations of the point group. For the point group O_h, with which we are concerned here, these operations are given in Volume 1, Table A12-9, which we reproduce in Table 2-2. This table gives the 24 operations of the group T_d; the operations met in O_h, and not in T_d, are numbered $R'_1 \cdots R'_{24}$, and each primed operation is identical with the corresponding unprimed operation, except for an additional inversion. Thus, for instance, $R'_1\psi(x,y,z) = \psi(-x,-y,-z)$; $R'_2\psi(x,y,z) = \psi(-x,y,z)$; etc.

Table 2-2
Operations of the group O_h. The operations tabulated are those of the group T_d. The additional operations met in O_h, and not in T_d, are indicated by primed symbols, as $R'_1 \cdots R'_{24}$, found as indicated in the text.

$$R_1\psi(x,y,z) = \psi(x,y,z) \qquad R_{13}\psi(x,y,z) = \psi(-x,z,-y)$$
$$R_2\psi(x,y,z) = \psi(x,-y,-z) \qquad R_{14}\psi(x,y,z) = \psi(-x,-z,y)$$
$$R_3\psi(x,y,z) = \psi(-x,y,-z) \qquad R_{15}\psi(x,y,z) = \psi(-z,-y,x)$$
$$R_4\psi(x,y,z) = \psi(-x,-y,z) \qquad R_{16}\psi(x,y,z) = \psi(z,-y,-x)$$
$$R_5\psi(x,y,z) = \psi(y,z,x) \qquad R_{17}\psi(x,y,z) = \psi(y,-x,-z)$$
$$R_6\psi(x,y,z) = \psi(-y,z,-x) \qquad R_{18}\psi(x,y,z) = \psi(-y,x,-z)$$
$$R_7\psi(x,y,z) = \psi(-y,-z,x) \qquad R_{19}\psi(x,y,z) = \psi(x,z,y)$$
$$R_8\psi(x,y,z) = \psi(y,-z,-x) \qquad R_{20}\psi(x,y,z) = \psi(x,-z,-y)$$
$$R_9\psi(x,y,z) = \psi(z,x,y) \qquad R_{21}\psi(x,y,z) = \psi(z,y,x)$$
$$R_{10}\psi(x,y,z) = \psi(-z,-x,y) \qquad R_{22}\psi(x,y,z) = \psi(-z,y,-x)$$
$$R_{11}\psi(x,y,z) = \psi(z,-x,-y) \qquad R_{23}\psi(x,y,z) = \psi(y,x,z)$$
$$R_{12}\psi(x,y,z) = \psi(-z,x,-y) \qquad R_{24}\psi(x,y,z) = \psi(-y,-x,z)$$

In the space group, we combine one of these 48 operations with a translation. These translations are expressed differently, depending on whether we are dealing with the body-centered or the face-centered cubic case. For the body-centered case, we have seen the value of the vectors $\mathbf{t}_1$, $\mathbf{t}_2$, $\mathbf{t}_3$ in Eq. (1-18). We then find that the vector $\mathbf{R}_n$ equals

$$\mathbf{R}_n = n_1\mathbf{t}_1 + n_2\mathbf{t}_2 + n_3\mathbf{t}_3$$
$$= n_1\frac{a}{2}(-\mathbf{i}+\mathbf{j}+\mathbf{k}) + n_2\frac{a}{2}(\mathbf{i}-\mathbf{j}+\mathbf{k}) + n_3\frac{a}{2}(\mathbf{i}+\mathbf{j}-\mathbf{k})$$
$$= \frac{a}{2}[(-n_1+n_2+n_3)\mathbf{i} + (n_1-n_2+n_3)\mathbf{j} + (n_1+n_2-n_3)\mathbf{k}] \quad (2\text{-}1)$$

Hence in the notation of Sec. 1-3 we may write the first two operations of the space group for O_h^9 ($Im3m$) as follows:

$$\{R_1|\mathbf{R}_n\}\psi(x,y,z) = \psi\left[x + \frac{a}{2}(-n_1+n_2+n_3),\right.$$
$$\left. y + \frac{a}{2}(n_1-n_2+n_3), z + \frac{a}{2}(n_1+n_2-n_3)\right]$$
$$\{R_2|\mathbf{R}_n\}\psi(x,y,z) = \psi\left[x + \frac{a}{2}(-n_1+n_2+n_3),\right. \quad (2\text{-}2)$$
$$\left. -y + \frac{a}{2}(n_1-n_2+n_3), -z + \frac{a}{2}(n_1+n_2-n_3)\right]$$

Similarly for the face-centered cubic case found in the space group O_h^5 ($Fm3m$), we use Eq. (1-19), and find

$$\{R_1|\mathbf{R}_n\}\psi(x,y,z)$$
$$= \psi\left[x + \frac{a}{2}(n_2 + n_3),\, y + \frac{a}{2}(n_1 + n_3),\, z + \frac{a}{2}(n_1 + n_2)\right]$$
$$\{R_2|\mathbf{R}_n\}\psi(x,y,z) \qquad (2\text{-}3)$$
$$= \psi\left[x + \frac{a}{2}(n_2 + n_3),\, -y + \frac{a}{2}(n_1 + n_3),\, -z + \frac{a}{2}(n_1 + n_2)\right]$$

These operations are of the form indicated in Eqs. (1-6), (1-7), and (1-10), in which the nonprimitive translation τ^i is zero, since we are dealing with a symmorphic group.

If an atom were placed at a position x, y, z, in which the coordinates x, y, z had no special properties (such as being zero, or having $x = y$, or being at the boundary of the unit cell, etc.), then the application of the 48 operations of the point group would produce 48 equivalent points within the unit cell, at each of which an atom would have to be located, according to symmetry. Inclusion of the operations of the space group, with $\mathbf{R}_n \neq 0$, would place 48 equivalent points in each unit cell throughout space. Such a position is called a general position in the cell. For a space group with as many operations in the point group as the cubic groups, it is rather rare to find any atoms located at general positions, for as we have just seen, this would imply at least 48 atoms per unit cell. In some of the space groups of lower symmetry, with far fewer operations in the point group, it is common to have atoms at general positions, and we shall find such cases later in our study.

In contrast to these general positions, there are special positions which allow far fewer atoms per unit cell. In particular, for the two space groups we are considering, the origin forms a special position which allows just one atom per unit cell, and this is the case with the structures we are considering. To verify this, we note that if we apply any one of the 48 operations of the point group to the position with $x = 0$, $y = 0$, $z = 0$, we come out with the same value, so that the origin, or an atom at the origin, is carried into itself by any operation of the point group, or into an atom at the origin of another unit cell by any operation of the space group.

We shall later meet examples of other structures having the same space groups, in which atoms are located at special positions other than the origin. As an example, suppose we had an atom at an arbitrary distance from the origin along the x axis: $x = u$, $y = z = 0$. The 48 operations of the point group will carry this point into one of the six positions $x = \pm u$, $y = z = 0$; $x = 0$, $y = \pm u$, $z = 0$; $x = y = 0$, $z = \pm u$. Or if we had an atom at a position on the body diagonal, $x = y = z = u$, there would be eight equivalent positions, at $x = \pm u$, $y = \pm u$, $z = \pm u$. These are two examples of special positions intermediate in complexity between the origin and a general position. In the International Tables for X-Ray Crystallography, quoted in Sec. 1-4, the various special positions are enumerated for each of the 230 space groups. It is very convenient to have this information, in case one is analyzing a crystal. If, for instance, we knew that in the body- or face-centered cubic space groups we had eight atoms of a given type in a unit cell, they would have to be found at special positions permitting 8, rather than 48, atoms per unit cell.

One of the interesting features concerning the body-centered and face-centered cubic structures is the closeness of packing of the atoms. We shall find later that to a rather good approximation atoms behave like spheres, which are in contact with each other in the crystal. In a body-centered cubic crystal, each atom has 8 neighbors, and in a face-centered cubic case each has 12 neighbors. It seems reasonable from this that the atoms will be more closely packed, with more per unit volume, in the face-centered than in the body-centered cubic structure. This, as a matter of fact, is the case. We have seen in Sec. 1-5 that the volume of the primitive unit cell in the body-centered cubic case is $a^3/2$, and in the face-centered cubic case it is $a^3/4$. The distance between neighboring atoms in the body-centered case is $a\sqrt{3}/2$, and in the face-centered case it is $a/\sqrt{2}$. If we were to set up radii equal to half these interatomic distances, and find the volume of the spheres representing the atoms, these volumes would be

$$\frac{4}{3}\pi \left(\frac{a\sqrt{3}}{4}\right)^3 = \frac{\sqrt{3}}{16}\pi a^3$$

for the body-centered case, and

$$\frac{4}{3}\pi \left(\frac{a}{2\sqrt{2}}\right)^3 = \frac{1}{12\sqrt{2}}\pi a^3$$

for the face-centered case. The fraction of the unit cell occupied by the sphere is then

$$\frac{\sqrt{3}}{16}\frac{\pi a^3}{a^3/2} = \frac{\sqrt{3}\,\pi}{8} = 0.680$$

for the body-centered case, and

$$\frac{1}{12\sqrt{2}}\frac{\pi a^3}{a^3/4} = \frac{\pi}{3\sqrt{2}} = 0.742$$

for the face-centered case. This shows, as we expected, that the face-centered cubic structure is more closely packed than the body-centered cubic case.

The face-centered cubic structure has the spheres packed as closely as they can be: a rigid sphere can be surrounded at most by 12 equally spaced neighbors, since in this case each sphere touches all its neighbors.

We can illustrate the packing of these spheres as in Fig. 2-1. Here we show in (a) a cube of the face-centered cubic structure, drawn as is usually

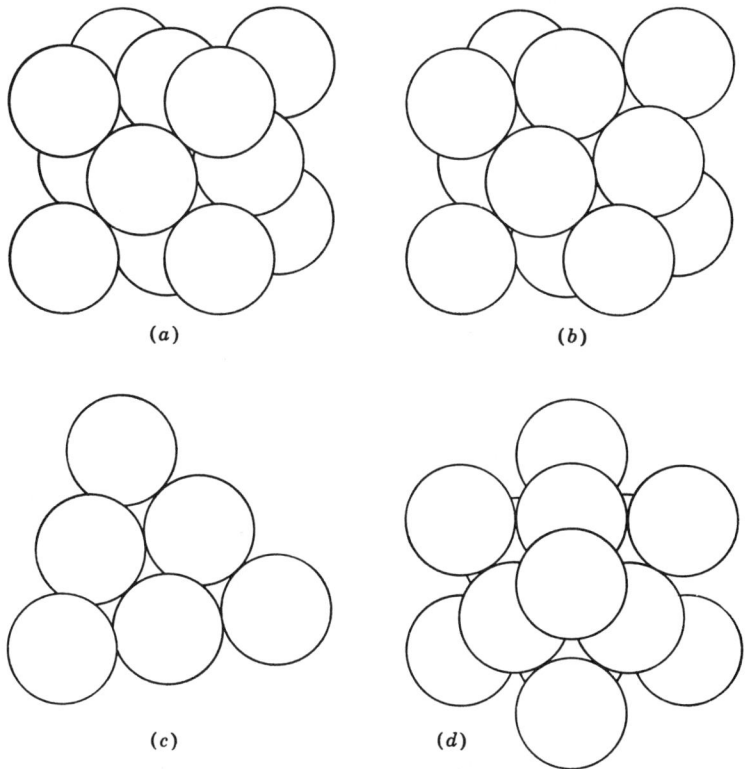

Fig. 2-1. (a) Cube of atoms, face-centered cubic structure, in perspective. (b) Cube with one atom removed, to show next layer of atoms. (c) Next layer of atoms removed from structure of Fig. 2-1b. (d) Face-centered cubic structure looking down along trigonal axis, showing three layers of atoms.

done, with atoms in the corners of the cubes, and at the centers of the faces. In (b) we remove one atom, exposing a layer consisting of six atoms in the form of an equilateral triangle, perpendicular to the so-called trigonal, or 111, axis, that is, the axis whose x, y, and z components are equal, or the body diagonal of the cube. In (c) we remove this exposed layer, showing another similar but displaced layer below it. This shows how the atoms lie in planes with a structure of equilateral triangles, or regular hexagons: each atom in one of these planes, if it is extended far enough, has six equally spaced neighbors in the same plane, at vertices of a regular hexagon. In (d) we show the appearance of the structure looking down along the 111 axis. We see that there are three types of planes which we encounter as we go down through the crystal, differing from each other by displacements in the plane. The structure is made up of a sequence of these three types of planes, arranged in order 123123123. On account of the close packing of this structure, it is often referred to as cubic close packing.

We notice from Fig. 2-1d that there are two ways in which one plane of close-packed spheres can lie on top of another similar plane. Thus, the one atom in the top plane shown could be located as it is, or in one of the three adjacent depressions, directly over one of the atoms in the bottom layer shown. If it were in the latter position, we should have only two types of planes alternating, in the sequence 121212. This gives the hexagonal close-packed structure, again with 12 nearest neighbors for each atom, and with the volume per atom the same as in the face-centered cubic structure. The arrangement of the atoms is as shown in Fig. 2-2. In this figure, we are looking down along a trigonal axis, as in Fig. 2-1d. Successive layers are removed as we go from top to bottom in the figure.

2-3. The Hexagonal Close-packed Structure. The symmetry situation of the hexagonal close-packed structure is entirely different from that of cubic close packing. In the latter case, there are four separate threefold axes, all equivalent to each other, corresponding to the cubic group O_h. In the hexagonal case, however, there is only one such axis. The space group is D_{6h}^4 ($P6_3/mmc$), with the point group D_{6h}, and the hexagonal Bravais lattice described in Sec. 1-5. Let us describe the properties of this space group, and then state the positions of the atoms.

We have stated in Eq. (1-20) that the vectors t_1, t_2, t_3 for the hexagonal Bravais lattice are given by the expressions $t_1 = (a/2)(\sqrt{3}\ \mathbf{i} - \mathbf{j})$, $t_2 = a\mathbf{j}$, $t_3 = c\mathbf{k}$. In setting up the expressions for the space group, we must next have the formulas for the operations of the point group. In Volume 1, Eq. (8-3), we have stated that the operations of the group C_{Nv} are given by $X_q\psi(\phi) = \psi(\phi + 2\pi q/N)$, $Y_q\psi(\phi) = \psi(-\phi + 2\pi q/N)$. In Eq. (A12-63) of Volume 1, we consider the operations of the group D_{Nh}, and show that the operations X_q and Y_q are to be supplemented by other operations X_q' and Y_q', identical with X_q and Y_q as far as their effect on ϕ is concerned, but in which z is changed into $-z$.

For the present purpose, it is not convenient to use cylindrical coordinates, as we did in Volume 1. The reason is that we wish to describe the translational operations by means of rectangular coordinates, or by use of the primitive translational vectors t_1, t_2, t_3, and this is not compatible with the use of cylindrical coordinates. We need to express our operations in a form equivalent to that of Eq. (1-7), $x'_p = \Sigma(q)\alpha^i_{pq}x_q$, using rectangular coordinates. Let us then see how to transform the expressions for the operations of the point group D_{6h} into rectangular coordinates.

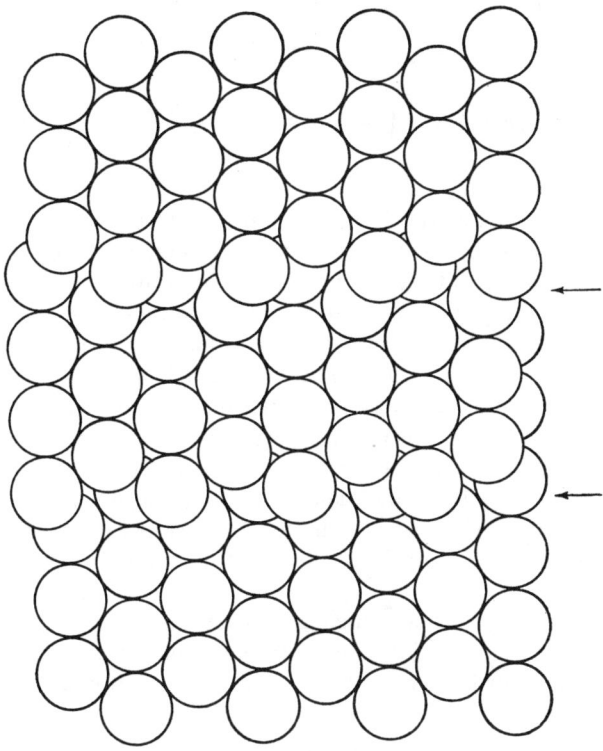

Fig. 2-2. The hexagonal close-packed structure, looking down along the trigonal axis. Arrows show places where layers are removed.

We first consider the unprimed operations, which do not involve the coordinate z. We have $X_q\psi(\mathbf{r}) = \psi(\mathbf{r}')$, where $\mathbf{r}$, $\mathbf{r}'$ are vector positions, related to each other in polar coordinates by the fact that $\mathbf{r}'$ corresponds to the same value of the radius as $\mathbf{r}$, but its angle ϕ is increased by the amount $\pi q/3$. If we write this in terms of rectangular coordinates, in which $\mathbf{r} = \mathbf{i}x + \mathbf{j}y + \mathbf{k}z$, we have $x = r\cos\phi$, $y = r\sin\phi$. We then have $\mathbf{r}' = \mathbf{i}x' + \mathbf{j}y' + \mathbf{k}z$, where

$$x' = r\cos(\phi + \pi q/3) \qquad y' = r\sin(\phi + \pi q/3)$$

If we expand the sine and cosine, we find

$$x' = x \cos \frac{\pi q}{3} - y \sin \frac{\pi q}{3} \qquad y' = x \sin \frac{\pi q}{3} + y \cos \frac{\pi q}{3} \qquad (2\text{-}4)$$

We then find for the X operations of the point group D_{6h}

$$\begin{aligned}
X_0 \psi(x,y,z) &= \psi(x,y,z) \\
X_{\pm 1} \psi(x,y,z) &= \psi\left(\frac{x}{2} \mp \frac{y\sqrt{3}}{2},\ \pm \frac{x\sqrt{3}}{2} + \frac{y}{2},\ z\right) \\
X_{\pm 2} \psi(x,y,z) &= \psi\left(-\frac{x}{2} \mp \frac{y\sqrt{3}}{2},\ \pm \frac{x\sqrt{3}}{2} - \frac{y}{2},\ z\right) \\
X_3 \psi(x,y,z) &= \psi(-x,-y,z)
\end{aligned} \qquad (2\text{-}5)$$

These expressions are of the form given in Eqs. (1-6) and (1-7), where the X's take the place of some of the R_i's, and the coefficients α^i_{pq} are no longer either zero or unity, as they are for the cubic groups.

For the y operations, we must carry out a reflection in which ϕ is changed to $-\phi$, as well as a rotation. We shall find it convenient here in the hexagonal group, as in the point groups of Volume 1, to regard ϕ as being measured from the x axis, so that the reflection changes y into $-y$, leaving x unchanged. As we should expect from this fact, the operations Y in rectangular coordinates have the same effect as the corresponding operations X, except that the sign of y is to be changed wherever it appears on the right side of Eq. (2-5). Finally, the primed operations have the same expression as the unprimed ones, except that the sign of z is to be changed.[1]

There is another method which is usually more useful for expressing the results of these operations: to write the vectors $\mathbf{r}$ and $\mathbf{r}'$, not in terms of rectangular coordinates, but in terms of the fundamental vectors $\mathbf{t}_1$, $\mathbf{t}_2$, $\mathbf{t}_3$. We may write

$$\mathbf{r} = \xi \mathbf{t}_1 + \eta \mathbf{t}_2 + \zeta \mathbf{t}_3 \qquad (2\text{-}6)$$

where ξ, η, ζ are dimensionless quantities. [In the International Tables of Crystallography, the symbols x, y, z are used in place of ξ, η, and ζ in Eq. (2-6). We prefer to save x, y, z for describing rectangular coordinates.] Similarly, we can write $\mathbf{r}'$ in the form of Eq. (2-6). We can then

---

[1] The standard treatment of this case was given by C. Herring, *J. Franklin Inst.*, **233**:525 (1942). He uses different symbols for the operations of the point group from those we are using. To assist the reader who wishes to compare our results with those of Herring, we give in the table below the connection between our symbols for the operations of the point group D_{6h} and those of Herring.

$$\begin{array}{llllll}
X_0\ \epsilon & X_{\pm 1}\delta_6,\delta_6^{-1} & X_{\pm 2}\delta_3,\delta_3^{-1} & X_3\delta_2 & Y_{0,\pm 2}\rho'_i & Y_{\pm 1,3}\rho''_i \\
X'_0\ \rho & X'_{\pm 1}\sigma_6,\sigma_6^{-1} & X'_{\pm 2}\sigma_3,\sigma_3^{-1} & X'_3 i & Y'_{0,\pm 2}\delta'_{2i} & Y'_{\pm 1,3}\delta''_{2i}
\end{array}$$

easily establish the equations

$$X_0\psi(r) = \psi(\xi t_1 + \eta t_2 + \zeta t_3)$$
$$X_1\psi(r) = \psi[(\xi - \eta)t_1 + \xi t_2 + \zeta t_3]$$
$$X_{-1}\psi(r) = \psi[\eta t_1 + (-\xi + \eta)t_2 + \zeta t_3]$$
$$X_2\psi(r) = \psi[-\eta t_1 + (\xi - \eta)t_2 + \zeta t_3]$$
$$X_{-2}\psi(r) = \psi[(-\xi + \eta)t_1 - \xi t_2 + \zeta t_3]$$
$$X_3\psi(r) = \psi(-\xi t_1 - \eta t_2 + \zeta t_3)$$
$$Y_0\psi(r) = \psi[\xi t_1 + (\xi - \eta)t_2 + \zeta t_3] \quad (2\text{-}7)$$
$$Y_1\psi(r) = \psi(\eta t_1 + \xi t_2 + \zeta t_3)$$
$$Y_{-1}\psi(r) = \psi[(\xi - \eta)t_1 - \eta t_2 + \zeta t_3]$$
$$Y_2\psi(r) = \psi[(-\xi + \eta)t_1 + \eta t_2 + \zeta t_3]$$
$$Y_{-2}\psi(r) = \psi(-\eta t_1 - \xi t_2 + \zeta t_3)$$
$$Y_3\psi(r) = \psi[-\xi t_1 + (-\xi + \eta)t_2 + \zeta t_3]$$

We can derive Eqs. (2-7) straightforwardly, by expanding both sides in terms of rectangular coordinates. It is simpler, however, to use the properties of the hexagon, as illustrated in Fig. 1-5. In Fig. 2-3 we show

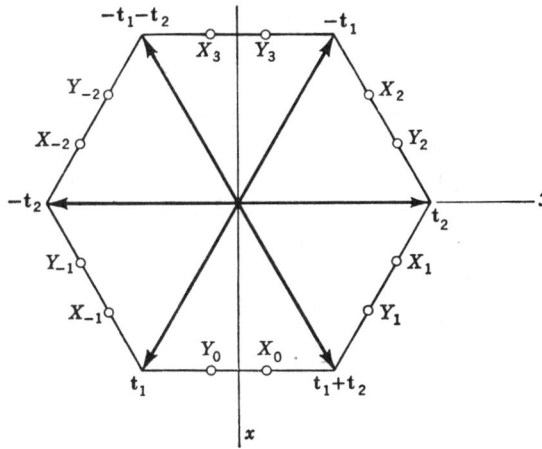

FIG. 2-3. Diagram illustrating points to which the point marked X_0 is transformed by the rotation and reflection operations of the group D_{6h}.

the vectors t_1 and t_2, as in Fig. 1-5, and the other vectors pointing to the vertices of a regular hexagon. We also show a point labeled X_0, and the points to which this is transformed by the various operations of the group D_{6h} or C_{6v} (considering only the operations in which z is unchanged). This diagram is essentially equivalent to Fig. 8-2 of Volume 1. Now let us use this diagram to find, for instance, the effect of the operator X_1. This rotates the point through 60° in a positive direction. The coordinates of the point labeled X_1 could then be set up by substituting the vector $t_1 + t_2$ for t_1, and $-t_1$ for t_2. In other words, for this case we have

$$\mathbf{r}' = \xi(t_1 + t_2) + \eta(-t_1) = (\xi - \eta)t_1 + \xi t_2 \qquad (2\text{-}8)$$

in agreement with Eq. (2-7) for this case. Similarly, the operation Y_0 reflects in the x axis; that is, it changes t_1 into $t_1 + t_2$, and t_2 into $-t_2$. Hence, for this case we have

$$\mathbf{r}' = \xi(t_1 + t_2) + \eta(-t_2) = \xi t_1 + (\xi - \eta)t_2 \qquad (2\text{-}9)$$

again in agreement with Eq. (2-7). The other cases of Eq. (2-7) can be derived equally easily.

We have now investigated the point group, and can next consider the space group D_{6h}^4. We may use this as an example of the method by which space groups are described in the International Tables of Crystallography. Under each space group, a list is given in the Tables of the points to which a given point is transformed, by the operations of the space group not involving any translation $R_n = n_1 t_1 + n_2 t_2 + n_3 t_3$. These transformed points are given in terms of the vector t_1, t_2, t_3; that is, each one is given by listing the transformed values of ξ, η, ζ (denoted in the International Tables by x, y, z). Thus, corresponding to $\{X_1|0\}$, the Tables would list merely $\xi - \eta$, ξ, ζ (or rather, $x - y$, x, z). Except for one thing, the resulting sets of values would be those appearing as the coefficients of t_1, t_2, t_3 on the right side of Eq. (2-7), supplemented by the other 12 quantities in which the sign of ζ is changed. The way in which the values differ is in the existence of a nonprimitive translation associated with some of the operations of the point group. Let us see what the International Tables tell us about this case.

The transformed points, as listed for the group D_{6h}^4 in the International Tables, are as follows:

$$\begin{aligned}
&xyz;\ \bar{y},\ x-y,\ z;\ -x+y,\ \bar{x},\ z;\ \bar{y},\ \bar{x},\ z;\ x,\ x-y,\ z;\ -x+y,\ y,\ z \\
&\bar{x},\ \bar{y},\ \bar{z};\ y,\ -x+y,\ \bar{z};\ x-y,\ x,\ \bar{z};\ y,\ x,\ \bar{z};\ \bar{x},\ -x+y,\ \bar{z};\ x-y,\ \bar{y},\ \bar{z} \\
&\bar{x},\ \bar{y},\ \tfrac{1}{2}+z;\ y,\ -x+y,\ \tfrac{1}{2}+z;\ x-y,\ x,\ \tfrac{1}{2}+z \\
&x,\ y,\ \tfrac{1}{2}-z;\ \bar{y},\ x-y,\ \tfrac{1}{2}-z;\ -x+y,\ \bar{x},\ \tfrac{1}{2}-z \qquad\qquad (2\text{-}10) \\
&y,\ x,\ \tfrac{1}{2}+z;\ \bar{x},\ -x+y,\ \tfrac{1}{2}+z;\ x-y,\ \bar{y},\ \tfrac{1}{2}+z \\
&\bar{y},\ \bar{x},\ \tfrac{1}{2}-z;\ x,\ x-y,\ \tfrac{1}{2}-z;\ -x+y,\ y,\ \tfrac{1}{2}-z
\end{aligned}$$

In these expressions, a bar over a symbol, as in $\bar{y}$, is used to indicate the negative, $-y$. When we remember that x, y, z of the International Tables are equivalent to ξ, η, ζ of Eq. (2-7), we see that aside from the quantities $\tfrac{1}{2}$, which account for the nonprimitive translation, the operations not involving the term $\tfrac{1}{2}$ are in succession

$$X_0,\ X_{\pm 2},\ Y_{-2},\ Y_0,\ Y_2,\ X'_3,\ X'_{\mp 1},\ Y'_1,\ Y'_3,\ Y'_{-1} \qquad (2\text{-}11)$$

Sec. 2-3] **SPACE GROUPS FOR STRUCTURES OF THE ELEMENTS** 37

and those involving the term $\frac{1}{2}$, or the nonprimitive translation, are

$$X_3,\ X_{\mp 1},\ X'_0,\ X'_{\pm 2},\ Y_1,\ Y_3,\ Y_{-1},\ Y'_{-2},\ Y'_0,\ Y'_2 \qquad (2\text{-}12)$$

We are now in position to combine this information and write down explicitly the formulas for the operations of the space group. In terms of the vectors t_1, t_2, t_3, we have

$$\begin{aligned}
\{X_0|R_n\}\psi(\mathbf{r}) &= \psi[(\xi + n_1)t_1 + (\eta + n_2)t_2 + (\zeta + n_3)t_3] \\
\{X_1|R_n\}\psi(\mathbf{r}) &= \psi[(\xi - \eta + n_1)t_1 + (\xi + n_2)t_2 + (\zeta + n_3 + \tfrac{1}{2})t_3] \\
\{X_{-1}|R_n\}\psi(\mathbf{r}) &= \psi[(\eta + n_1)t_1 + (-\xi + \eta + n_2)t_2 + (\zeta + n_3 + \tfrac{1}{2})t_3] \\
\{X_2|R_n\}\psi(\mathbf{r}) &= \psi[(-\eta + n_1)t_1 + (\xi - \eta + n_2)t_2 + (\zeta + n_3)t_3] \\
\{X_{-2}|R_n\}\psi(\mathbf{r}) &= \psi[(-\xi + \eta + n_1)t_1 + (-\xi + n_2)t_2 + (\zeta + n_3)t_3] \\
\{X_3|R_n\}\psi(\mathbf{r}) &= \psi[(-\xi + n_1)t_1 + (-\eta + n_2)t_2 + (\zeta + n_3 + \tfrac{1}{2})t_3] \\
\{Y_0|R_n\}\psi(\mathbf{r}) &= \psi[(\xi + n_1)t_1 + (\xi - \eta + n_2)t_2 + (\zeta + n_3)t_3] \\
\{Y_1|R_n\}\psi(\mathbf{r}) &= \psi[(\eta + n_1)t_1 + (\xi + n_2)t_2 + (\zeta + n_3 + \tfrac{1}{2})t_3] \\
\{Y_{-1}|R_n\}\psi(\mathbf{r}) &= \psi[(\xi - \eta + n_1)t_1 + (-\eta + n_2)t_2 + (\zeta + n_3 + \tfrac{1}{2})t_3] \\
\{Y_2|R_n\}\psi(\mathbf{r}) &= \psi[(-\xi + \eta + n_1)t_1 + (\eta + n_2)t_2 + (\zeta + n_3)t_3] \\
\{Y_{-2}|R_n\}\psi(\mathbf{r}) &= \psi[(-\eta + n_1)t_1 + (-\xi + n_2)t_2 + (\zeta + n_3)t_3] \\
\{Y_3|R_n\}\psi(\mathbf{r}) &= \psi[(-\xi + n_1)t_1 + (-\xi + \eta + n_2)t_2 + (\zeta + n_3 + \tfrac{1}{2})t_3]
\end{aligned}$$
$$(2\text{-}13)$$

The expressions for the primed operators are like those for the unprimed ones, given in Eq. (2-13), except that $(\zeta + n_3)$ is to be replaced by $(-\zeta + n_3 + \frac{1}{2})$, and $(\zeta + n_3 + \frac{1}{2})$ is to be replaced by $(-\zeta + n_3)$.

We can rewrite t_1, t_2, t_3 in Eq. (2-13) in terms of their rectangular components, using Eq. (1-20). When we do this, we obtain expressions following from those of Eq. (2-5), with the addition of the quantities

$$n_1 \frac{\sqrt{3}}{2} a,\ \left(-\frac{n_1}{2} + n_2\right) a,\ n_3 c \qquad (2\text{-}14)$$

respectively, to x, y, and z, for the operations not involving nonprimitive translations, and an additional $c/2$ to z for those involving primitive translations. These expressions are not generally as useful as those of Eq. (2-13).

Now that we have found the operations of the space group D_{6h}^4, we can study its application to the hexagonal close-packed structure. In this structure there are two atoms per unit cell, in the special positions $\xi = \frac{1}{3}$, $\eta = \frac{2}{3}$, $\zeta = \frac{1}{4}$, and $\xi = \frac{2}{3}$, $\eta = \frac{1}{3}$, $\zeta = \frac{3}{4}$. If we apply the operations of the space group enumerated in Eq. (2-13) to either of these positions, we find that the position is transformed into itself, or the corresponding point in another unit cell, by each of the operations not involving a nonprimitive translation, and into the other position, or the corresponding point in another unit cell, by operations involving a nonprimitive translation. For instance, if we start with the position $\tfrac{1}{3}$, $\tfrac{2}{3}$, $\tfrac{1}{4}$, and apply the operation $\{X_1|R_n\}$ to it, we find

$$\psi[(-\tfrac{1}{3}+n_1)t_1 + (\tfrac{1}{3}+n_2)t_2 + (\tfrac{3}{4}+n_3)t_3]$$

which can be written in the form

$$\psi[(\tfrac{2}{3}+n_1-1)t_1 + (\tfrac{1}{3}+n_2)t_2 + (\tfrac{3}{4}+n_3)t_3]$$

which is the second position $\tfrac{2}{3}$, $\tfrac{1}{3}$, $\tfrac{3}{4}$, translated by the vector $(n_1-1)t_1 + n_2t_2 + n_3t_3$. In other words, if we have an atom at one of

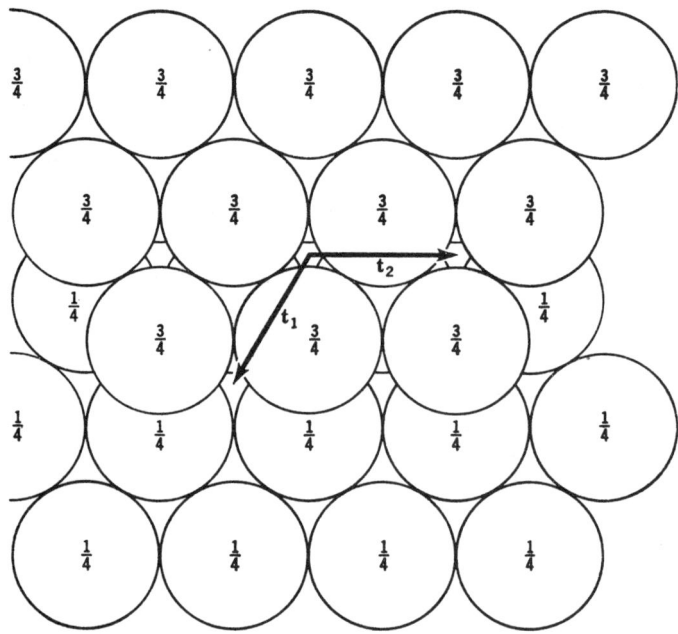

Fig. 2-4. Vectors t_1 and t_2 for the hexagonal close-packed structure, and atomic positions in the planes at heights $c/4$ and $3c/4$, respectively.

these positions in the unit cell, we must also have an atom at the other, but no further atoms are required.

The atomic positions in the two planes whose z coordinates are $c/4$, $3c/4$, are given in Fig. 2-4. We see that these have the symmetry already discussed in Fig. 2-2. In other words, we verify that the hexagonal close-packed structure is to be described group-theoretically as we have done it in the present section. It is interesting to consider the symmetry of the environment of the atoms in the crystal, and also the symmetry found at the origin of coordinates, which is not located at an atom. Let us first consider the origin.

If we consider the surroundings of the origin, in Fig. 2-4, we see that we have the symmetry D_{3d}. We recall from Volume 1, Appendix 12, that with this type of symmetry, we may have atoms at angles 0, $\pm 120°$, for a positive value of z, and similarly atoms at $\pm 60°$, 180°, for an equal negative value of z. This is what we have: at the first three angles with respect to the x axis, pointing vertically downward in Fig. 2-4, we have atoms corresponding to $\zeta = \frac{3}{4}$, or $-\frac{1}{4}$, and at the second three angles we have atoms for $\zeta = \frac{1}{4}$. It must be, then, that any one of the operators of the group D_{3d} will transform the environment of the origin into itself. These operators are given in Volume 1, Eq. (A12-64), in the form

$$X_q\psi(r,\phi,z) = \psi\left[r,\, \phi + \frac{\pi q}{3},\, (-1)^q z\right]$$
$$Y_q\psi(r,\phi,z) = \psi\left[r,\, -\phi + \frac{\pi q}{3},\, (-1)^q z\right]$$
(2-15)

Here q is to go over six values, which we may take to be 0, ± 1, ± 2, 3. Thus these operators of the group D_{3d} are

$$X_0\psi(r,\phi,z) = \psi(r,\phi,z)$$
$$X_{\pm 1}\psi(r,\phi,z) = \psi\left(r,\, \phi \pm \frac{\pi}{3},\, -z\right)$$
$$X_{\pm 2}\psi(r,\phi,z) = \psi\left(r,\, \phi \pm \frac{2\pi}{3},\, z\right)$$
$$X_3\psi(r,\phi,z) = \psi(r,\, \phi + \pi,\, -z)$$
(2-16)

with similar operators for the Y's, obtained by changing the sign of ϕ on the right-hand side. These, as we see by comparison with our case of D_{6h}, are the operators which for that group would be denoted as X_0, $X'_{\pm 1}$, $X_{\pm 2}$, X'_3, Y_0, $Y'_{\pm 1}$, $Y_{\pm 2}$, Y'_3. In other words, they are the operations enumerated in Eq. (2-11), those which do not involve a nonprimitive translation in the space group D_{6h}^4.

The operators of the space group associated with a nonprimitive translation of amount $c/2$ along the z axis, then, are those which appear in the point group D_{6h}, but are not present in D_{3d}. The easiest way to describe these operations is that they consist of the operators of the group D_{3d}, but supplemented by a rotation of 60°, plus a nonprimitive translation. In other words, we may infer that the environment of the point $\xi = 0$, $\eta = 0$, $\zeta = \frac{1}{2}$, directly above the origin and halfway up the cell, is like that at the origin, but rotated through an angle of 60°. Examination of Fig. 2-4 shows that this is indeed the case.

Study of Fig. 2-4 will show that the symmetry of the environment of one of the atoms, unlike that around the origin, is of type D_{3h}. It is possible to set up a description of the group D_{6h}^4 in which the origin is located at one of the atoms, rather than at the point used in Fig. 2-4, which is not at any one of the atoms. If this is done, the operations of the space group which are not associated with nonprimitive translations are those occurring in the group D_{3h}. The remaining operations of the space group are those of the point group D_{6h} not occurring in D_{3h}, supplemented by a nonprimitive translation just sufficient to carry us from one atom, the one at the origin, to the other atom in the unit cell. Information regarding the symmetry at all special positions of the space group is given in the International Tables; thus in particular in the present case, it is stated (though in different notation) that the symmetry at the points 0, 0, 0 and 0, 0, ½ is of type D_{3d}, and at the points ⅓, ⅔, ¼ and ⅔, ⅓, ¾ it is of type D_{3h}. The International Tables also enumerate the symmetry found at all other special positions within the unit cell, for all 230 space groups.

There is one additional point to take up concerning the hexagonal close-packed structure. If we really have close packing, the ratio c/a must have a particular value. The distance from an atom to the nearest neighbor in the same plane (same value of z) is a. The distance to the nearest neighbor out of the plane is $(a^2/3 + c^2/4)^{1/2}$. If we have the close-packed structure, this distance equals a, from which

$$a = \left(\frac{a^2}{3} + \frac{c^2}{4}\right)^{1/2} \qquad \frac{c}{a} = \sqrt{8/3} = 1.6330 \qquad (2\text{-}17)$$

The volume of the unit cell is $a^2c\sqrt{3}/2$, so that the volume per atom is $a^2c\sqrt{3}/4$. In many metals, the c/a ratio is very close to the value 1.633 characteristic of close packing, but in a few, notably zinc and cadmium, it departs widely from this value: the planes are spaced too widely along the z axis.

2-4. The Diamond Structure. In Fig. 2-5 we show one way of looking at the diamond structure. We start with a cube like the fundamental cube of a face-centered cubic lattice. Inside this we insert atoms of another similar lattice, shifted along the 111 direction from the original lattice by the amounts $a/4$ along each of the three coordinate axes, where a is the side of the cube. We see from the figure that we can set up cubes half as large as the fundamental cube along each of the edges, and half of these cubes have an atom at the center, and at alternate corners. The other cubes are empty, having no atom at the center. From this way of looking at the structure, we see that it has the same Bravais lattice and unit cell as the face-centered cubic structure, but has two atoms per unit cell, rather than one.

Another way to describe the environment of an atom in the diamond structure is to fix our attention on one of the small cubes just described.

Sec. 2-4] SPACE GROUPS FOR STRUCTURES OF THE ELEMENTS 41

In half the cubes, we have an atom at the center, and atoms at four out of the eight corners. The cube is like that of the body-centered cubic structure, but with only every other site filled, so that the packing is only half as dense as in the body-centered cubic structure. The fraction of the volume filled by spherical atoms is only $\pi \sqrt{3}/16 = 0.340$. It is obvious that spheres attempting close packing would never arrange themselves in the diamond structure. It is only the tetrahedral bonds met in this structure, coming off in definite directions, and showing the property of saturation, such that only four bonds can be formed from a given atom,

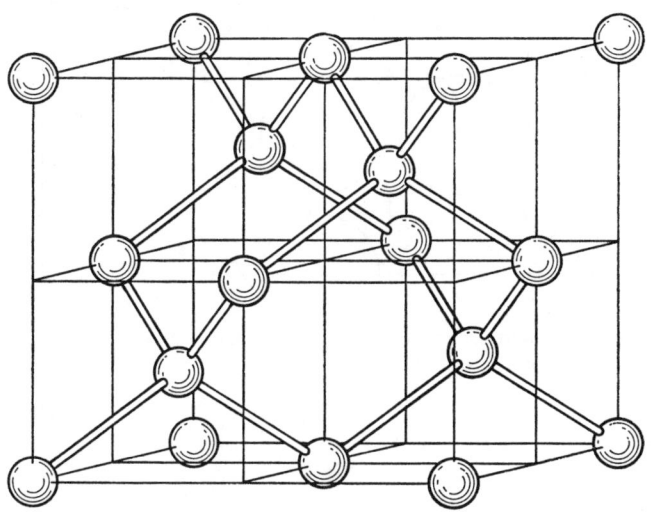

FIG. 2-5. Atomic positions in the diamond structure.

which will lead to this structure. We shall see later the features of the wave functions which lead to this tetrahedral bond formation; they are essentially the same features which lead to tetrahedral bonds in methane.

The structure of this crystal, as of many, can be well understood by a diagram looking down along one of the cubic axes, representing the atoms as spheres in contact with each other, and labeling each of the atoms with its height in multiples of a. This diagram is shown in Fig. 2-6. It is the type of diagram which we have used to illustrate the hexagonal close-packed structure in Fig. 2-4.

Still another way of illustrating the structure shows the carbon tetrahedra more directly. This is to take the crystal, and orient it so that the 111 direction is vertical. Then we have a set of tetrahedra, erected on top of each other, as shown in Fig. 2-7. In this figure we see very plainly the fact that there are two atoms per unit cell: the atoms in one plane in the figure have bonds extending vertically upward to the next higher

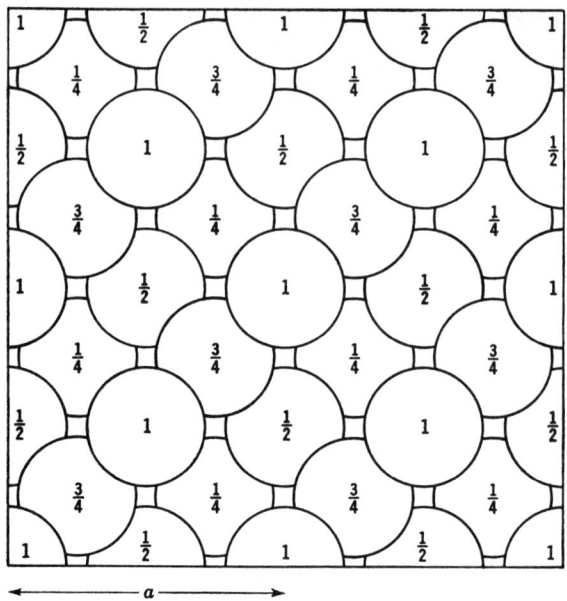

Fig. 2-6. The diamond structure, looking along one of the cubic axes. Heights of atoms, in multiples of a, are shown.

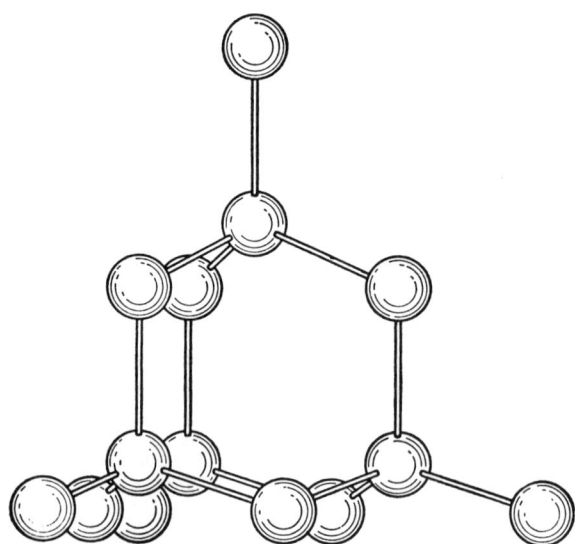

Fig. 2-7. Perspective view of atomic arrangement in diamond, with 111 axis vertical.

plane, while the atoms in the other plane have bonds extending vertically downward. The atoms of one type cannot be found from those of the other type by a simple translation, which would be required if they were in equivalent positions in the crystal. Rather, one can get an atom of one type from that of the other type by an inversion about the atom at the origin of Fig. 2-5, followed by a translation carrying this atom to the atom at $a/4$, $a/4$, $a/4$. Alternatively, one can get an atom of one type from that of the other type by an inversion about the midpoint between the two atoms.

Now that we have considered the qualitative nature of the diamond structure, we can describe it in terms of group theory. As we have mentioned, the space group is O_h^7 ($Fd3m$). The Bravais lattice is that of the face-centered cubic type, and the point group is O_h. The only difference between this case and the group O_h^5 ($Fm3m$) met in the face-centered cubic structure comes from the nonprimitive translations met in the group O_h^7 ($Fd3m$) encountered in the diamond structure. In the International Tables there are two alternative methods of describing the space group, one of which places the origin at an atom, while the other places it midway between the two atoms forming a bond. The former is the more convenient for the present purpose. Then we can describe the operations of the space group very simply, in terms of what we already know about the face-centered cubic structure. Out of the 48 operations of the point group, those which are associated with the operation T_d, namely, those enumerated explicitly in Table 2-2, are associated with no nonprimitive translation; the remaining operations, numbered $R_1' \cdots R_{24}'$ in the notation we have used for the operation O_h, are associated with the nonprimitive translation $(a/4)(\mathbf{i} + \mathbf{j} + \mathbf{k})$, which carries us from the atom at the origin to the other atom in the primitive unit cell. We see, then, that the environment of an atom has the symmetry T_d, as is clear from its tetrahedral structure; the remaining operations are associated with the nonprimitive translation, and with an inversion, verifying our statement in the preceding paragraph that we can get from the environment of one atom to that of the other by combining an inversion and a nonprimitive translation.

If the origin were chosen according to the other method used in the International Tables, namely, at the midpoint of a bond, we should find the D_{3d} type of symmetry about this point; only the operations of the group O_h which were included in the group D_{3d} for rotation about the 111 direction, plus its other associated operations, would have no nonprimitive translations. This would be the convenient origin to use if we were particularly interested in the equivalent orbital located on a bond, as in the method of Hurley, Lennard-Jones, and Pople, which we shall take up for this case in Appendix 8.

2-5. The Graphite Structure. The graphite crystal consists of carbon atoms arranged at the corners of an array of regular hexagons. The positions of the atoms in a sheet are then as given in Fig. 2-8. The next sheet, a long distance from the first, is displaced downward as shown in the figure; the sheet below that is like the first one shown. There is a

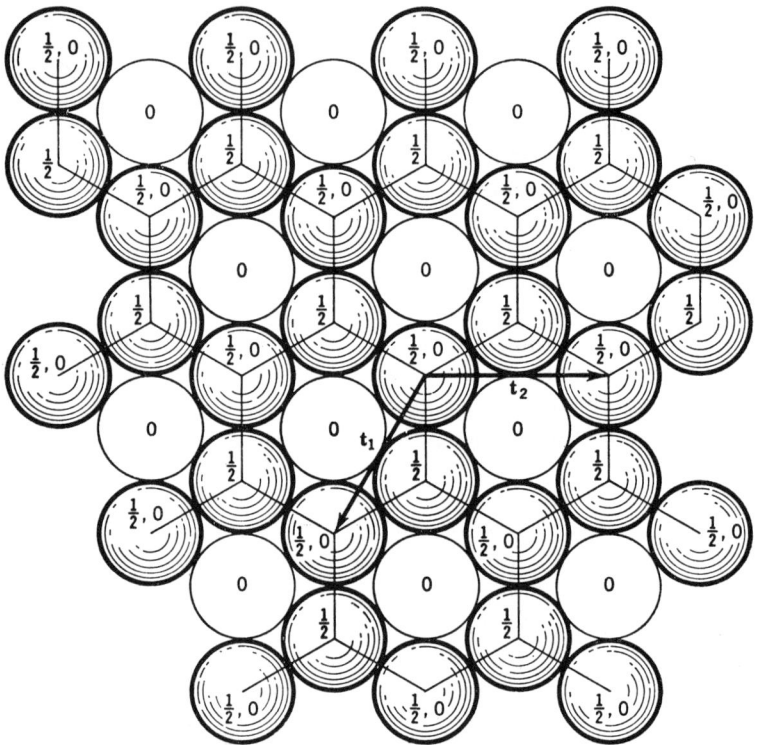

Fig. 2-8. The graphite structure, looking along the hexagonal axis. Heights of atoms are shown, as multiples of c. The atoms in the upper plane, at height $c/2$, are shown shaded, and are connected by a hexagonal framework, to make them easier to see.

close relationship between this structure and the hexagonal close-packed structure, but they are not identical. The successive sheets are so far apart that they are held together only by Van der Waals forces. This affects the physical properties of graphite, which is known to be a good lubricant: one sheet slides easily over another, the internal forces within a sheet being strong, but the sheets being very weakly held together. The atoms within a sheet, on the other hand, are held together much like the atoms in a benzene molecule.

The space group, as we have mentioned, is C_{6v}^4 ($P6_3mc$), closely related to the space group D_{6h}^4 ($P6_3/mmc$) found for the hexagonal close-packed structure. The Bravais lattice is identical; we show the vectors t_1 and t_2 in Fig. 2-8, which resembles Fig. 2-4. The only difference between the two space groups is in the point group, which is C_{6v} instead of D_{6h}, so that the primed operations are missing. The unprimed operations, however, are as in Eq. (2-13), the nonprimitive translations being the same in both groups.

There are four atoms in the unit cell, located for ξ, η, ζ equal, respectively, to 0, 0, 0; 0, 0, $\frac{1}{2}$; $\frac{1}{3}$, $\frac{2}{3}$, 0; and $\frac{2}{3}$, $\frac{1}{3}$, $\frac{1}{2}$. The first two are the atoms which are arranged above each other in Fig. 2-8; the others are arranged as in the hexagonal close-packed structure, except that their height in the unit cell comes at 0 and $\frac{1}{2}$ of the height of the cell, respectively, rather than $\frac{1}{4}$ and $\frac{3}{4}$ as in the hexagonal close-packed structure. We can verify that these latter positions are consistent with the group C_{6v}^4, but not with D_{6h}^4. For instance, the operator $\{X_0'|\mathbf{R}_n\}$ for the case $n_1 = n_2 = n_3 = 0$ leaves ξ and η unchanged, but transforms ζ into $-\zeta + \frac{1}{2}$. Thus it would transform the position $\frac{1}{3}$, $\frac{2}{3}$, 0 into $\frac{1}{3}$, $\frac{2}{3}$, $\frac{1}{2}$, which is not one of the positions occupied by atoms in graphite. This is in contrast to the position $\frac{1}{3}$, $\frac{2}{3}$, $\frac{1}{4}$ found in the hexagonal close-packed structure, which is transformed into itself by the operator $\{X_0'|\mathbf{R}_n\}$.

It is interesting to see how the atoms are arranged in the Wigner-Seitz type of unit cell, like that shown in Fig. 1-5, in the graphite structure, and for comparison in the hexagonal close-packed structure. In the hexagonal close-packed structure, atoms are found along the edges pointing in the directions 0, $\pm 120°$ with respect to the x axis, at heights $\frac{3}{4}$, and along the three remaining edges at heights $\frac{1}{4}$. Each of these six atoms is shared between three adjacent unit cells, so that this leaves two atoms per unit cell, as we should have. In the graphite structure, the atoms along the edges in the directions 0, $\pm 120°$ are at height $\frac{1}{2}$, those along the three remaining edges at height 0. But in addition to these, we have atoms along the axis of the hexagonal prism, at heights 0 and $\frac{1}{2}$. Thus we have the required four atoms per unit cell.

2-6. The Arsenic Structure. The space group for the arsenic structure, as we have mentioned, is D_{3d}^5 ($R\bar{3}m$), with a rhombohedral Bravais lattice, the unit cell being as shown in Fig. 1-6. This space group is symmorphic: there are no nonprimitive translations. It is a sufficiently simple case so that it will not be necessary to write out the operations of the group analytically, as we have done in previous cases. There are two atoms per unit cell, located at the special positions $\pm u(\mathbf{t}_1 + \mathbf{t}_2 + \mathbf{t}_3)$. Since according to Eq. (1-21) we have $\mathbf{t}_1 + \mathbf{t}_2 + \mathbf{t}_3 = 3r\mathbf{k}$, we see that the two atoms are located along the axis of rotation, spaced equally above and below the vertical apexes of the rhombohedral unit cell as shown in Fig. 1-6.

It is easier to visualize the structure from the projection, as shown in Fig. 1-6b. There we show three sets of heights for points, namely, r, $2r$, and zero or $3r$. Each of these of course extends to form a network of points throughout space, as is shown in Fig. 2-9, where we label the three sets of points. If we project the structure down on the xy plane, atoms will be located at the various points of this figure.

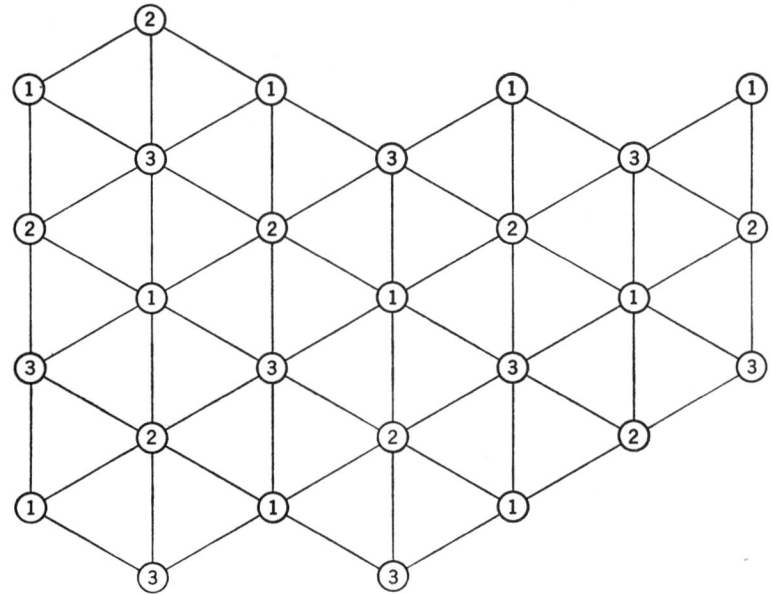

FIG. 2-9. Network of points for the rhombohedral lattice, as in arsenic. Arsenic atoms are located at points marked 1, 2, 3, respectively, at the heights given in Table 2-3.

It is now helpful to find the heights of the atoms at each of the three types of points. In Table 2-3 we give these heights, both in terms of the

Table 2-3

Heights of atoms above the origin, in the three types of points in the arsenic structure shown in Fig. 2-9. Points 1 are at the origin or above it, points 2 displaced from it by the projection of the vector t_1 in the xy plane, and points 3 displaced by the projection of the vector $t_1 + t_2$. In arsenic, $r = 3.50$ A, $u = 0.226$.

1	2	3
$3ru = 2.38$ A	$r - 3ru = 1.12$ A	$2r - 3ru = 4.62$ A
$3r - 3ru = 8.12$ A	$r + 3ru = 5.88$ A	$2r + 3ru = 9.38$ A
$3r + 3ru = 12.88$ A	$4r - 3ru = 11.62$ A	$5r - 3ru = 15.12$ A
. . . .	. . . .	. . . .

parameters, and in Angstroms for the particular case of arsenic, for which

Sec. 2-6]   SPACE GROUPS FOR STRUCTURES OF THE ELEMENTS   47

$a = 4.13$ A, $\alpha = 54°10'$, $r = 3.50$ A, $s = 2.17$ A, $u = 0.226$. When we examine Table 2-3, we see that the sheets of atoms at point 1, height $3ru = 2.38$ A, and at point 2, height $r - 3ru = 1.12$ A, are only 1.26 A apart. Each of these planes is distant by an amount 2.24 A from its next nearest plane. The same sequence persists as we go up through the lattice: the sheets at height $2r - 3ru = 4.62$ A and $r + 3ru = 5.88$ A of points 3 and 2, respectively, are 1.26 A apart; the sheets at height $3r - 3ru = 8.12$ A and $2r + 2ru = 9.38$ A at the same distance, and so on, each pair of sheets being distant by 2.24 A from the next pair.

In Fig. 2-10 we show spherical atoms in a pair of closely-spaced sheets of this type. We see that each atom in the upper sheet of such a pair

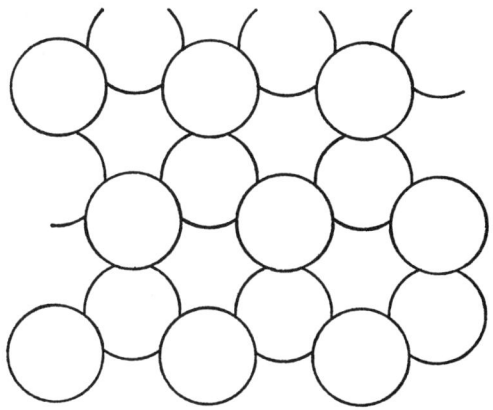

FIG. 2-10. Two adjacent sheets of atoms in the arsenic structure.

has three neighbors in the lower pair, and vice versa; thus we understand the way in which each arsenic atom is bonded to three neighbors. The distance from one atom to its three neighbors in the adjacent sheet is 2.51 A, but the distance to the three closest neighbors in the more distant plane is 3.14 A, considerably larger. The angle between the bonds from one atom to the three neighbors in the adjacent sheet is about 97°. If the two sheets were moved apart, the angle would decrease to 90°, and at the same time the bond lengths to the neighbors in the more distant plane would decrease to equal those to the neighbors in the adjacent plane, and the structure would become simple cubic. On the other hand, if the two sheets were moved together, the angle between bonds would increase to 120°, and the composite sheet, as we can readily see from Fig. 2-10, would become like a single sheet of atoms in graphite.

It is interesting to compare the arsenic structure with the graphite structure. We have three types of double sheets of atoms in arsenic, forming a sequence 123123123, reminding us of the sheets in the face-centered cubic structure, whereas in graphite we have only two types of sheets, in the sequence 121212, as in the hexagonal close-packed structure. One might reasonably ask, cannot graphite also exist in a form with the sequence 123123 of planes, as well as with its ordinary sequence 121212? There is some evidence that it can. Lines have been observed in the x-ray diffraction patterns of some samples of graphite which have been interpreted in terms of such a structure. It is described as an example of the arsenic structure, but with a value of u such that the two closely spaced sheets in the arsenic structure coincide. As we can see from Table 2-3,

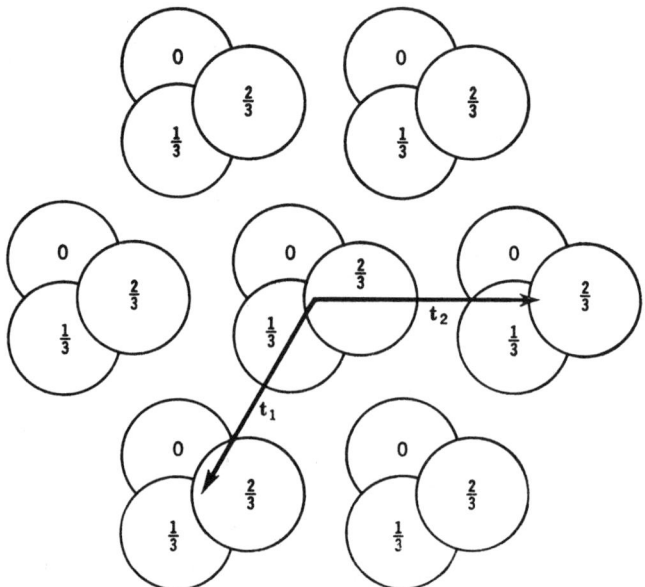

Fig. 2-11. The selenium structure, looking along the hexagonal axis.

this demands that $3ru = r - 3ru$, or $u = \frac{1}{6}$, which is the value found for this type of graphite, called β-graphite.

While this form of graphite may exist, it seems somewhat more likely that rather than having a regular sequence 123123123, one has a crystal which is preponderately 121212, but occasionally with a sequence 123. Such a situation is called a stacking fault. Either arrangement would work equally well if we were geometrically piling layers on top of each other, and we should naturally expect that the energy difference between the two structures would be small enough so that it would not be unlikely to find either sequence in an actual crystal. This is an example of a widespread phenomenon, namely, that when there are two equally reasonable

Sec. 2-7] SPACE GROUPS FOR STRUCTURES OF THE ELEMENTS 49

ways to stack adjacent layers, the crystal often makes errors in growth, resulting in the wrong sequence of layers.

2-7. The Selenium Structure. Selenium and tellurium crystallize in a structure with the space group D_3^4 ($P3_121$) or D_3^6 ($P3_221$); these two space groups differ only in that one of them describes a parallel arrangement of helices with the symmetry of right-hand screws, and the other with left-hand screws. Two such related space groups are called enantiomorphs. One cannot distinguish easily by x rays between the two cases. We shall describe the space group D_3^4 ($P3_121$), which corresponds to a right-hand screw. In Fig. 2-11 we show the arrangement of the atoms, looking down along the $-z$ axis. The right-hand-screw nature of the helices is clear from the figure. In Fig. 2-12 we show a view looking at the side of one of the helices. The spheres used in Figs. 2-11 and 2-12 are of a proper size to touch, and it is clear from Fig. 2-11 how far apart neighboring helices are; as we have mentioned earlier, they are held together only by Van der Waals forces.

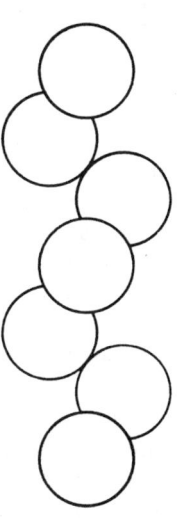

Fig. 2-12. Side view of single helix of atoms in selenium.

The Bravais lattice for this space group is the hexagonal lattice (indicated by the symbol P in the description of the space group). In Fig. 2-11 we show the vectors t_1 and t_2; t_3 is perpendicular to the plane of the paper. If the position is given by $\xi t_1 + \eta t_2 + \zeta t_3$, as we have done before with the hexagonal unit cell, the three atoms in the unit cell are located at special positions for which ξ, η, ζ are equal, respectively, to u, 0, $\frac{1}{3}$; 0, u, $\frac{2}{3}$; and $-u$, $-u$, 0. For selenium, the value of the length a of vectors t_1 and t_2 is found to be 4.36 A; the length c of the vector t_3 is 4.95 A; u is 0.217. For tellurium the corresponding values are 4.47 A, 5.91 A, and 0.269. Figures 2-11 and 2-12 correspond to selenium.

We recall that the operations of the point group D_3, in cylindrical coordinates, are given by

$$
\begin{aligned}
X_0\psi(r,\phi,z) &= \psi(r,\phi,z) \\
X_{\pm 1}\psi(r,\phi,z) &= \psi\left(r,\; \phi \pm \frac{2\pi}{3},\; z\right) \\
Y_0'\psi(r,\phi,z) &= \psi(r,-\phi,-z) \\
Y_{\pm 1}'\psi(r,\phi,z) &= \psi\left(r,\; -\phi \pm \frac{2\pi}{3},\; -z\right)
\end{aligned}
\tag{2-18}
$$

To get agreement with the notation of the International Tables, we take the axis $t_1 + t_2$ as the reflection axis, rather than using the x axis as in Sec. 2-3 where we discussed the hexagonal structure. We may set up the operations of the point group by the method used in deriving Eq. (2-7). We have nonprimitive translations in this space group, equal to $t_3/3$ for the operations X_1 and Y'_1, $-t_3/3$ for X_{-1} and Y'_{-1}. Hence for the operations of the space group we have

$$\{X_0|R_n\}\psi(\mathbf{r}) = \psi[(\xi + n_1)\mathbf{t}_1 + (\eta + n_2)\mathbf{t}_2 + (\zeta + n_3)\mathbf{t}_3]$$
$$\{X_1|R_n\}\psi(\mathbf{r}) = \psi[(-\eta + n_1)\mathbf{t}_1 + (\xi - \eta + n_2)\mathbf{t}_2 + (\zeta + n_3 + \tfrac{1}{3})\mathbf{t}_3]$$
$$\{X_{-1}|R_n\}\psi(\mathbf{r}) = \psi[(-\xi + \eta + n_1)\mathbf{t}_1 + (-\xi + n_2)\mathbf{t}_2 + (\zeta + n_3 - \tfrac{1}{3})\mathbf{t}_3]$$
$$\{Y'_0|R_n\}\psi(\mathbf{r}) = \psi[(\eta + n_1)\mathbf{t}_1 + (\xi + n_2)\mathbf{t}_2 + (-\zeta + n_3)\mathbf{t}_3]$$
$$\{Y'_1|R_n\}\psi(\mathbf{r}) = \psi[(-\xi + n_1)\mathbf{t}_1 + (-\xi + \eta + n_2)\mathbf{t}_2 + (-\zeta + n_3 + \tfrac{1}{3})\mathbf{t}_3]$$
$$\{Y'_{-1}|R_n\}\psi(\mathbf{r}) = \psi[(\xi - \eta + n_1)\mathbf{t}_1 + (-\eta + n_2)\mathbf{t}_2 + (-\zeta + n_3 - \tfrac{1}{3})\mathbf{t}_3]$$
(2-19)

We can now take the special positions of the atoms, substitute them into these operations, and verify that each one of the three positions transforms into one of these positions, either in the same or another unit cell, under the action of the operations. Hence we verify the validity of the assignment of atoms to special positions, in the selenium structure.

2-8. The Iodine Structure. The elements bromine, iodine, gallium, and one form of phosphorus, the so-called black phosphorus, all crystallize with a structure having the space group D_{2h}^{18} (*Cmca*), belonging to the orthorhombic system. As we have mentioned earlier, we have diatomic molecules, well defined in the cases of bromine and iodine, but with the interatomic distance within the molecule hardly greater than that between atoms in other molecules in the cases of gallium and phosphorus. We shall use gallium as our example. In Fig. 2-13 we show the structure looking down along the $-x$ axis onto the yz plane. We have atoms at two heights, namely, 0 and $a/2$, where a is the length of the vector $\mathbf{t}_1$ along the x axis. In Fig. 2-13 we show both planes. It is clear that the molecule centered at $x = a/2$, $y = b/2$, $z = 0$, where b is the length of the vector $\mathbf{t}_2$, is in a position equivalent to that centered at the origin. In other words, we have here an example of the one-face-centered orthorhombic Bravais lattice. We may choose vectors $\mathbf{t}_1$, $\mathbf{t}_2$, $\mathbf{t}_3$ for the orthorhombic cell, at right angles to each other, and along x, y, z, respectively; they have different magnitudes, a, b, c, respectively. But the corners, and center, of the faces in the xy plane are equivalent to each other, as we have just seen. This is shown by the symbol C in the description of the space group; if the symbol had been A or B, it would have been the yz or zx plane, respectively, which was face-centered. Hence our rectangular cell of sides a, b, c is not primitive.

There are eight atoms in this cell of sides a, b, c, four in each of the two planes shown in Fig. 2-13. However, on account of the face-centered arrangement, we can choose a smaller unit cell, with only four atoms in it. This can be illustrated better in Fig. 2-14, in which we look down on the crystal along the $-z$ axis, looking at the xy plane. Here we clearly see the face-centered arrangement, and we show primitive vectors $\mathbf{t}_1$ and $\mathbf{t}_2$ which can be used in place of the vectors $\mathbf{i}a$, and $\mathbf{j}b$, of the orthorhombic cell. These vectors have the values

$$\mathbf{t}_1 = \mathbf{i}\frac{a}{2} + \mathbf{j}\frac{b}{2} \quad \mathbf{t}_2 = -\mathbf{i}\frac{a}{2} + \mathbf{j}\frac{b}{2} \quad \mathbf{t}_3 = \mathbf{k}c \quad (2\text{-}20)$$

We verify that the volume of the parallelopiped defined by these unit vectors, $\mathbf{t}_1 \cdot (\mathbf{t}_2 \times \mathbf{t}_3)$, equals $abc/2$, half the volume of the orthorhombic

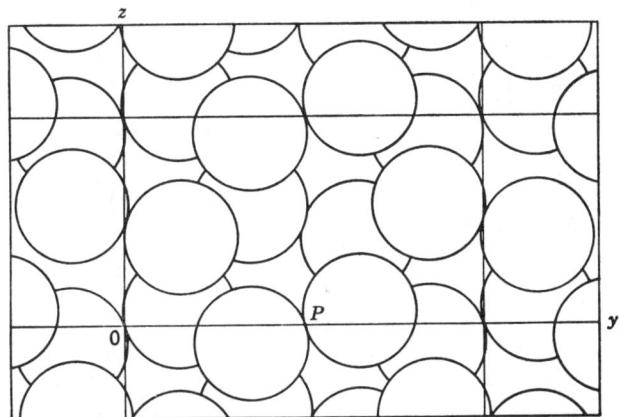

FIG. 2-13. The structure of gallium, looking down along the $-x$ direction. The upper layer of atoms shown is at height $a/2$, the lower layer at height zero. Abscissa, y; ordinate, z. The rectangle shown inside the boundary of the figure is the yz section of the orthorhombic unit cell. The origin is at its lower left corner, labeled 0. The center of the molecule at $x = a/2$, $y = b/2$, $c = 0$ is labeled P.

cell. By setting up planes which are the perpendicular bisectors of the vectors $\mathbf{t}_1$, $\mathbf{t}_2$, and the other vectors to nearest neighboring lattice points in the xy plane, we can get a Wigner-Seitz cell, shown in Fig. 2-14. This is a hexagonal prism, but not a regular hexagon. However, with the dimensions actually found for gallium, it is very close to a regular hexagon. This cell contains four atoms, as we have mentioned earlier: two clearly located within it, and four located on the surfaces of the hexagonal prism, each shared between two prisms (they come very nearly, but not quite, at the corners).

The four atoms in the unit cell may be defined by the positions

$$\begin{aligned} u b\mathbf{j} + vc\mathbf{k} &= u(\mathbf{t}_1 + \mathbf{t}_2) + v\mathbf{t}_3 \\ -ub\mathbf{j} - vc\mathbf{k} &= u(-\mathbf{t}_1 - \mathbf{t}_2) - v\mathbf{t}_3 \\ (\tfrac{1}{2} + u)b\mathbf{j} + (\tfrac{1}{2} - v)c\mathbf{k} &= (\tfrac{1}{2} + u)(\mathbf{t}_1 + \mathbf{t}_2) + (\tfrac{1}{2} - v)\mathbf{t}_3 \\ (\tfrac{1}{2} - u)b\mathbf{j} + (\tfrac{1}{2} + v)c\mathbf{k} &= (\tfrac{1}{2} - u)(\mathbf{t}_1 + \mathbf{t}_2) + (\tfrac{1}{2} + v)\mathbf{t}_3 \end{aligned} \quad (2\text{-}21)$$

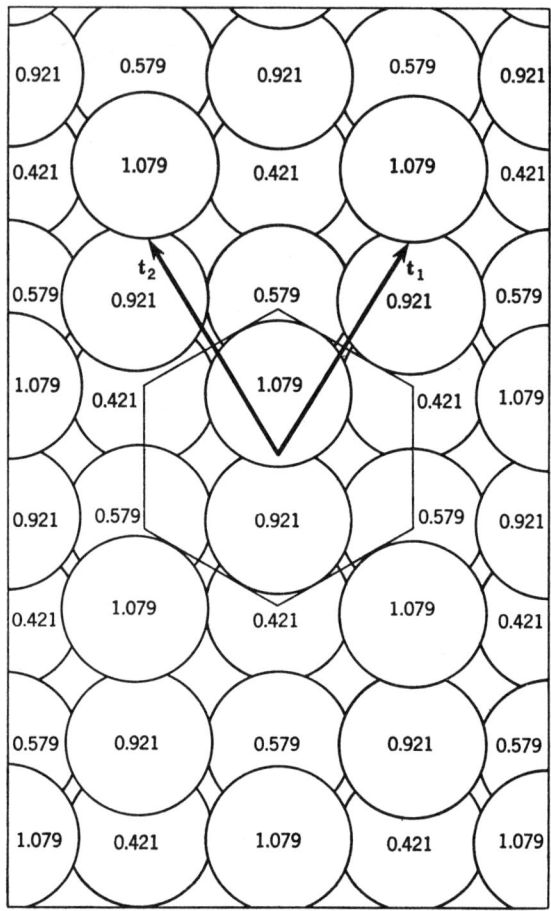

Fig. 2-14. The structure of gallium, looking along the $-z$ direction. Abscissa, x; ordinate, y. The vectors t_1 and t_2, and the Wigner-Seitz unit cell, are shown. Heights of atoms are given as multiples of c.

The first two form the molecule centered on the origin, the last two the molecule in the center of the rectangle shown in Fig. 2-13, for $x = 0$. For the case of gallium, one has the parameters $a = 4.49$ A, $b = 7.63$ A, $c = 4.51$ A, $u = 0.1525$, $v = 0.0785$.[1]

We must now consider the operations of the point group and space

---

[1] For references to the original work, see J. C. Slater, G. F. Koster, and J. H. Wood, *Phys. Rev.*, **126**:1307 (1962). The notation of this paper is different from the present discussion, which follows the International Tables. The coordinates which we are denoting as x, y, z in the present discussion are called y, z, x, respectively, in the paper cited, and are called x, z, y, respectively, by Wyckoff.

Sec. 2-8] SPACE GROUPS FOR STRUCTURES OF THE ELEMENTS 53

group, and verify that the positions of the atoms given in Eq. (2-21) are legitimate special positions for this space group. The point group D_{2h} is a specially simple one if we consider it in rectangular rather than polar coordinates. The eight operations of the group are those in which x, or y, or z can change sign, with any one of the eight possible combinations of change of sign. In the space group, the four operations in which either y and z, but not both, changes sign, are combined with a nonprimitive translation $b\mathbf{j}/2 + c\mathbf{k}/2$. We may then write the effect of the eight operations of the space group on a function of x, y, z, in the following way:

$$\{R_1|\mathbf{R}_n\}\psi(x,y,z) = \psi[x + (n_1 - n_2)a/2, \; y + (n_1 + n_2)b/2, \; z + n_3c]$$
$$\{R_2|\mathbf{R}_n\}\psi(x,y,z) = \psi[x + (n_1 - n_2)a/2, \; -y + (n_1 + n_2)b/2, \; -z + n_3c]$$
$$\{R_3|\mathbf{R}_n\}\psi(x,y,z) = \psi[-x + (n_1 - n_2)a/2,$$
$$y + (n_1 + n_2 + 1)b/2, \; -z + (n_3 + \tfrac{1}{2})c]$$
$$\{R_4|\mathbf{R}_n\}\psi(x,y,z) = \psi[-x + (n_1 - n_2)a/2,$$
$$-y + (n_1 + n_2 + 1)b/2, \; z + (n_3 + \tfrac{1}{2})c]$$
$$\{R_5|\mathbf{R}_n\}\psi(x,y,z) = \psi[-x + (n_1 - n_2)a/2, \; -y + (n_1 + n_2)b/2, \; -z + n_3c]$$
$$\{R_6|\mathbf{R}_n\}\psi(x,y,z) = \psi[-x + (n_1 - n_2)a/2, \; y + (n_1 + n_2)b/2, \; z + n_3c]$$
$$\{R_7|\mathbf{R}_n\}\psi(x,y,z) = \psi[x + (n_1 - n_2)a/2,$$
$$-y + (n_1 + n_2 + 1)b/2, \; z + (n_3 + \tfrac{1}{2})c]$$
$$\{R_8|\mathbf{R}_n\}\psi(x,y,z) = \psi[x + (n_1 - n_2)a/2,$$
$$y + (n_1 + n_2 + 1)b/2, \; -z + (n_3 + \tfrac{1}{2})c]$$
$$(2\text{-}22)$$

We can now verify at once that if we take $x = 0$, $y = ub$, $z = vc$, the operations of Eq. (2-22) transform this point into one of the positions given by Eq. (2-21), either in the original unit cell or a displaced unit cell. Hence the special positions of Eq. (2-21) are suitable positions for atoms in this group.

3
Space Groups for Structures of Compounds

3-1. General Survey of Crystal Structures. Very large numbers of crystal structures, both of elements and compounds, have been determined during the fifty years of x-ray crystallographic investigations. In Appendix 1 we give a certain amount of information about approximately a thousand selected substances. This Appendix is particularly set up to illustrate a fact, pointed out in 1920 by W. L. Bragg,[1] which we shall discuss in detail in Chap. 4. This fact is that the observed interatomic distances in a crystal between two atoms which are forming a bond can be approximately written as the sum of atomic radii for the two atoms, these radii being determined empirically by study of the experimental measurements. Another way of stating the same thing is that we may build up crystal models out of spheres representing the various atoms, and the spheres representing bonding atoms will approximately touch. This simple idea of Bragg's has been greatly complicated in later years, as we shall describe in Chap. 4, by introducing different sorts of radii, depending on whether we have ionic or covalent bonds, on the number of bonds formed by a given atom, and various other features. We shall point out in Chap. 4 the advantage of returning to Bragg's simple scheme of having a unique radius for each type of atom, to be used in every sort of compound. In Table 3-1 we give the radii which we shall use, close to Bragg's values, but somewhat refined as a result of using a much larger amount of empirical material than was available at the time of Bragg's work.

Since Appendix 1 is set up with the object of investigating bonds, it is arranged by first listing the bond concerned, as for instance Li—O; the more electropositive atom appears first, and atoms are arranged in order of atomic number, so that it is easy to locate any given bond in the table. Next we list the compound in which the bond is observed. Since many compounds have bonds of more than one type, they will often appear more than once in the table. We next give, in many cases, abbreviations

---

[1] W. L. Bragg, *Phil. Mag.*, **40**:169 (1920).

Table 3-1
Atomic radii in Angstroms (the inert gases do not appear, since they do not form bonds)

					Cs 2.60	Fr
					Ba 2.15	Ra 2.15
					La 1.95	Ac 1.95
					Ce 1.85	Th 1.80
					Pr 1.85	Pa 1.80
					Nd 1.85	U 1.75
					Pm 1.85	Np 1.75
			K 2.20	Rb 2.35	Sm 1.85	Pu 1.75
			Ca 1.80	Sr 2.00	Eu 1.85	Am 1.75
			Sc 1.60	Y 1.80	Gd 1.80	
			Ti 1.40	Zr 1.55	Tb 1.75	
			V 1.35	Nb 1.45	Dy 1.75	
	Li 1.45	Na 1.80	Cr 1.40	Mo 1.45	Ho 1.75	
	Be 1.05	Mg 1.50	Mn 1.40	Tc 1.35	Er 1.75	
	B 0.85	Al 1.25	Fe 1.40	Ru 1.30	Tu 1.75	
H 0.25	C 0.70	Si 1.10	Co 1.35	Rh 1.35	Yb 1.75	
	N 0.65	P 1.00	Ni 1.35	Pd 1.40	Lu 1.75	
	O 0.60	S 1.00	Cu 1.35	Ag 1.60	Hf 1.55	
	F 0.50	Cl 1.00	Zn 1.35	Cd 1.55	Ta 1.45	
			Ga 1.30	In 1.55	W 1.35	
			Ge 1.25	Sn 1.45	Re 1.35	
			As 1.15	Sb 1.45	Os 1.30	
			Se 1.15	Te 1.40	Ir 1.35	
			Br 1.15	I 1.40	Pt 1.35	
					Au 1.35	
					Hg 1.50	
					Tl 1.90	
					Pb 1.80	
					Bi 1.60	
					Po 1.90	
					At	

indicating the type of crystal structure; for instance, bcc for the body-centered cubic structure of elements, or fcc for the face-centered cubic structure, as in Table 2-1, which includes information repeated in Table 3-1. Then follows the space group, using the notation of Table 1-1. Next is a reference to Wyckoff's work, mentioned in Chap. 1, giving a place where additional information about the structure in question can be found, including references to the original literature. We then have the observed bond length, in Angstroms (with a range of values in case all bonds of this type in the crystal do not have the same length), and finally the sum of the atomic radii of Table 3-1, for comparison with the observed bond lengths. The average magnitude of the deviation between observed bond lengths and the sum of the radii is approximately 0.12 A, which indicates the limit of accuracy of the empirical additivity of atomic radii in crystals. In many cases the deviations are less than this, but these are balanced by occasional much greater deviations. It is not possible to choose radii which reduce the average deviation much below this value. Bragg's radii gave an average deviation of only about 0.06 A in comparison with the observed distances with which he compared them; but he omitted many of the cases where large deviations are inevitable, while we have included them. The radii of Table 3-1 are approximately as good as Bragg's for the type of case for which he made a comparison.

We shall postpone discussion of the significance of the additivity of the radii until Chap. 4, but in the meantime we can use Appendix 1 in two ways. First, it serves as an index to a great many compounds. Second, it suggests that in making models of crystal structures, we may well use spheres whose radii are the atomic radii of Table 3-1 to represent the atoms. We shall use these radii in the diagrams of crystal structures given in this chapter.

When we examine the space groups encountered in Appendix 1, we find that some are represented much oftener than others, and consequently we shall choose those which appear most often to illustrate the general principles of space groups. Of course, the frequency of occurrence of a space group depends on the selection of crystals which we have made, and our selection is not entirely representative of those occurring in nature. It is found that complicated crystals, with many atoms in the unit cell, more often than not have space groups of low symmetry, such as those of the triclinic or monoclinic systems, whereas those with relatively few atoms in the unit cell more often have space groups of high symmetry, such as the cubic, trigonal, or hexagonal. In our list we have tended to favor the simpler crystals; therefore one gets an exaggerated impression, from reading the list, of the importance of the more symmetrical space groups.

In spite of this bias in our choice of crystals, we shall consider crystals in five of the seven systems (all except the triclinic and monoclinic); and in 8 of the 14 Bravais lattices. Out of all the 230 space groups, we shall discuss 20; but these between them account for something like 80 percent of the structures listed in Appendix 1. In Table 3-2 we list these 20 space groups, together with the structures which we shall discuss under each of them, and the section in which the discussion is given. In Appendix 2 we list many crystals having each of these structures, together with the various constants determining their structure. We shall now proceed with the discussion of the structures of compounds listed in Table 3-2, starting with the simple binary compounds.

3-2. Structures of Simple Binary Compounds, RX. The best-known compounds of the type RX, where R stands for an electropositive element, X an electronegative, are the alkali halides, the fluorides, chlorides,

Table 3-2
Space groups discussed in detail in Chaps. 2 and 3 and Appendix 2

Orthorhombic
    62. D_{2h}^{16} (*Pnma*), aragonite, Sec. 3-6
    64. D_{2h}^{18} (*Cmca*), iodine, gallium, Sec. 2-8

Tetragonal
    136. D_{4h}^{14} ($P4_2/mnm$), rutile, Sec. 3-3

Trigonal
    148. C_{3i}^2 ($R\bar{3}$), ilmenite, Sec. 3-5
    152. D_3^4 ($P3_121$), or 154. D_3^6 ($P3_221$), selenium, Sec. 2-7
    155. D_3^7 ($R32$), AlF_3, Sec. 3-6
    164. D_{3d}^3 ($P\bar{3}m1$), CdI_2, Sec. 3-3. La_2O_3, Sec. 3-6. AlB_2, Sec. 3-6
    166. D_{3d}^5 ($R\bar{3}m$), arsenic, Sec. 2-6. $CdCl_2$, Sec. 3-3.   LaOF, Sec. 3-6
    167. D_{3d}^6 ($R\bar{3}c$), calcite, corundum, Sec. 3-5

Hexagonal
    186. C_{6v}^4 ($P6_3mc$), graphite, Sec. 2-5.   Wurtzite, Sec. 3-2
    194. D_{6h}^4 ($P6_3/mmc$), hexagonal close-packed, Sec. 2-3.   NiAs, Sec. 3-2. Na_3As, Sec. 3-6

Cubic
    205. T_h^6 ($Pa3$), pyrite, Sec. 3-3
    206. T_h^7 ($Ia3$), Tl_2O_3, Sec. 3-4
    216. T_d^2 ($F\bar{4}3m$), zinc-blende, Sec. 3-2
    221. O_h^1 ($Pm3m$), CsCl, Sec. 3-2.   Perovskite, Sec. 3-4
    225. O_h^5 ($Fm3m$), face-centered cubic, Sec. 2-2.   NaCl, Sec. 3-2.   Fluorite, Sec. 3-3
    227. O_h^7 ($Fd3m$), diamond, Sec. 2-4.   Spinel, Sec. 3-4
    229. O_h^9 ($Im3m$), body-centered cubic, Sec. 2-2
    230. O_h^{10} ($Ia3d$), garnet, Sec. 3-4

bromides, and iodides of lithium, sodium, potassium, rubidium, and cesium. These all have one of two structures, named for NaCl and CsCl, examples of the two. The sodium chloride structure is one of the most familiar in the whole range of crystal structures.

It is commonly described by setting up a simple cubic lattice of points, and saying that alternate points going along the x, y, or z direction are occupied by Na and Cl. We realize, however, that points occupied by two different sorts of atoms cannot be equivalent in the lattice. If we consider the sodium atoms or ions only in NaCl, they lie on a face-centered cubic lattice. If one of them lies at the origin, and if the distance between adjacent sodiums along the x or y or z axis is a, the sodiums by themselves are arranged just as in the face-centered cubic structure. But the chlorines also have face-centered cubic symmetry, but shifted along the x or y or z axis by the distance $a/2$. We can thus take one atom of each type per primitive unit cell of the face-centered cubic structure. The symmetry is that of the space group O_h^5 ($Fm3m$), the only difference between this and the structure of the elements being the presence of the atom of one type at the origin, and of the other type at the position, say, $ai/2$.

The other structure found for the alkali halides, the cesium chloride structure, is even simpler. This resembles a body-centered cubic structure, but with atoms of one type, say cesium, at the corners of the cubes, and those of the other type, say chlorine, at the centers. Since these points are not equivalent, the symmetry is not that of the body-centered structure, but rather is simple cubic, with the space group O_h^1 ($Pm3m$). This is so simple that we do not have to tabulate the operations of the space group: we have the 48 operations of the point group O_h, and the simple cubic Bravais lattice, with no nonprimitive translations. We have two atoms per unit cell, in special positions: say, a cesium at the origin, and a chlorine at the position $(a/2)(\mathbf{i} + \mathbf{j} + \mathbf{k})$, which we can abbreviate ½, ½, ½.

Many simple binary compounds of the form RX crystallize in one or the other of the two structures just described. Thus, aside from the alkali halides, many of the oxides, sulfides, selenides, and tellurides of the divalent elements Mg, Ca, Sr, and Ba have the sodium chloride structure; a good many nitrides, phosphides, arsenides, and similar compounds of antimony and bismuth, formed from trivalent positive ions from the transition elements and the rare earth group, also have the sodium chloride structure. There are numerous other compounds as well with this structure, as we see from Appendix 2. With the cesium chloride structure, we have not only a number of the alkali halides formed from the heavier alkalies, but also numerous intermetallic compounds, such as LiAg, LiHg, LiTl, BeCo, BeCu, CuZn, and many others. In some of these cases, such as CuZn, there is often a change at a high temperature to a disordered phase, in which the atoms are still found on sites of the body-centered cubic lattice, but with atoms of either type found randomly at either of the sites. Such order-disorder transformations have been studied a great deal in thermodynamics and statistical mechanics.

There are several other structures which are very common among the binary compounds of the form RX. The commonest are the zinc-blende and wurtzite structures, named for the two forms of the substance ZnS, and the nickel arsenide structure, named for NiAs. We shall describe these structures in detail. The zinc-blende structure is closely related to the diamond structure: if one replaces one of the two atoms in the diamond unit cell by an atom R, the other by an atom X, one has the zinc-blende structure. One finds it in such compounds as GaAs, called a 3-5 compound, in which Ga has three outer electrons, one less than Ge, and As has one electron more, or five, so that the crystal as a whole has the same number of electrons as Ge, which has the diamond structure. In the whole series of compounds GaAs, ZnSe, and CuBr, all of which have the same number of electrons, but of the types 3-5, 2-6, and 1-7, respectively, one finds the zinc-blende structure. Some compounds of the same type have the alternative wurtzite structure, which is closely allied to the zinc-blende structure. Here, as in the zinc-blende structure, each atom is surrounded tetrahedrally by four atoms of the other type, but instead of forming a cubic structure, as in zinc-blende, it forms a hexagonal structure. The relation between the two is much like that between the cubic and hexagonal close-packed structures of the elements. The nature of the wurtzite structure is illustrated in Fig. 3-1.

Let us now describe the space groups to which these two structures, the zinc-blende and wurtzite, belong. The zinc-blende structure belongs to the space group T_d^2 ($F\bar{4}3m$), which we have not so far discussed. As we

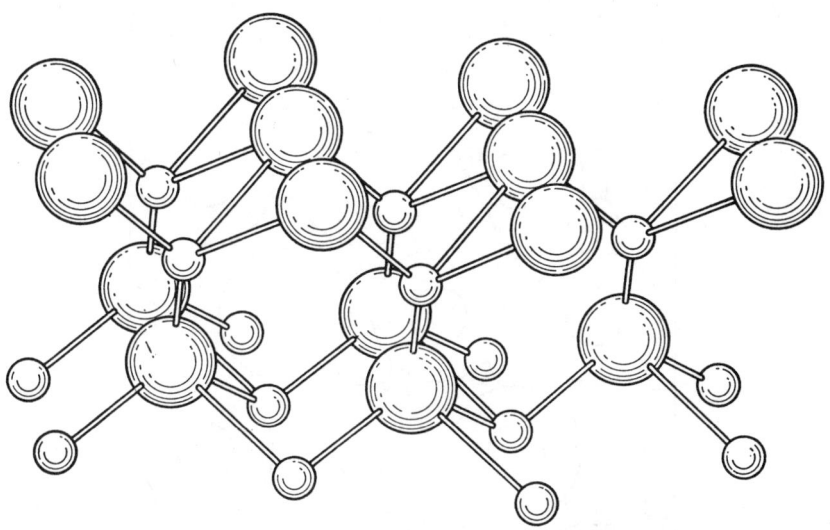

FIG. 3-1. Perspective of part of the wurtzite crystal. The diameters of the spheres representing the two types of atoms have no significance in terms of atomic radii.

see from Table 1-1, this falls in the cubic system, with the face-centered cubic Bravais lattice, but its point group is only T_d, in accordance with the tetrahedral environment of each atom, instead of O_h as in the cubic crystals we have so far considered. The group is symmorphic. The operations of the space group are like those for the face-centered cubic, indicated in Eq. (2-3), except that the primed operations of the point group, which occur for O_h but not T_d, are to be omitted. Alternatively, it is like O_h^7 ($Fd3m$), the space group found with the diamond structure, but again the primed operations are to be omitted. In the zinc-blende structure, atoms of type R, say zinc, are found at the origin, and those of the other type X, say sulfur, at the position $(a/4)(\mathbf{i} + \mathbf{j} + \mathbf{k})$, or ¼, ¼, ¼. Unlike the diamond structure, these two types of positions of course are not equivalent.

Wurtzite has the space group C_{6v}^4 ($P6_3mc$), which we have already discussed in connection with graphite, in Sec. 2-5. There are two atoms per unit cell of each type. We may take atoms of type R at positions for which ξ, η, ζ are equal to $\frac{1}{3}$, $\frac{2}{3}$, 0 and $\frac{2}{3}$, $\frac{1}{3}$, $\frac{1}{2}$, and atoms of type X at positions $\frac{1}{3}$, $\frac{2}{3}$, u, and $\frac{2}{3}$, $\frac{1}{3}$, $u + \frac{1}{2}$, which are suitable special positions

FIG. 3-2. Structure of ZnO (wurtzite structure), looking down along the threefold axis. The large circles represent zinc atoms, at height 0. The small circles between them represent oxygens at height $-\frac{1}{8}a$, and the small circles above the zinc atoms are oxygens at height $\frac{3}{8}a$. These represent the three lowest layers of atoms shown in Fig. 3-1. Vectors t_1 and t_2 are shown. The radii of the spheres are the atomic radii of Table 3-1.

for this group. In all crystals of the wurtzite structure, the c/a ratio is very close to the value $\sqrt{8/3} = 1.633$ found in the hexagonal close-packed structure, and the parameter u is very close to $0.375 = \frac{3}{8}$. If the parameters have exactly these values, the tetrahedra surrounding a given atom are regular. In Fig. 3-2 we show three layers of atoms, illustrated by the close packing of spheres; as we see from Fig. 3-1, the next-higher layer consists of zinc atoms at height $a/2$, located over the oxygen atoms shown in the figure at height $-a/8$.

The nickel arsenide structure, according to some descriptions, has the space group C_{6v}^4 ($P6_3mc$), but according to others it is D_{6h}^4 ($P6_3/mmc$). It has two NiAs groups per unit cell. The nickel atoms are at the special positions for which ξ, η, ζ are equal to 0, 0, 0, and 0, 0, $\tfrac{1}{2}$, respectively. The arsenic atoms are at $\tfrac{1}{3}$, $\tfrac{2}{3}$, u, and $\tfrac{2}{3}$, $\tfrac{1}{3}$, $u + \tfrac{1}{2}$, where u is very closely equal to $\tfrac{1}{4}$. If u were not exactly equal to $\tfrac{1}{4}$, we should have to

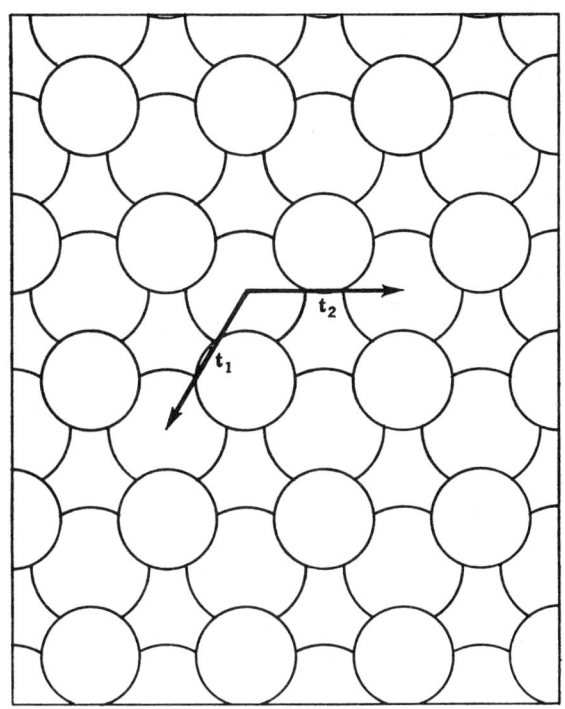

Fig. 3-3. The NiAs structure. The atoms shown are the nickels at 0, 0, $\tfrac{1}{2}$, represented by the larger spheres, and the arsenics at $\tfrac{2}{3}$, $\tfrac{1}{3}$, $\tfrac{3}{4}$. The radii of the spheres are the atomic radii. Between the atoms shown by spheres, one looks down at arsenics at $\tfrac{1}{3}$, $\tfrac{2}{3}$, $\tfrac{1}{4}$, whose bounding spheres cannot be seen in the figure.

describe it according to the space group C_{6v}^4 ($P6_3mc$). However, if it is exactly $\tfrac{1}{4}$, as it seems to be, the arsenic atoms are at suitable special positions for D_{6h}^4 ($P6_3/mmc$), which explains why either space group can be used in this case. It is often true that if atoms are in suitably chosen positions in a space group of lower symmetry, the space group reduces to one of higher symmetry, and this is an example of this situation.

In Fig. 3-3 we show the structure of NiAs itself, looking down along the hexagonal axis. The radii of the spheres are the atomic radii of Table 3-1, the nickel being larger than the arsenic. We see that each nickel atom is surrounded by six arsenics, its environment having the symmetry D_{3d}, and each arsenic is surrounded by six nickels, the environment having the symmetry D_{3h}, consistent with the discussion of the special positions of the group D_{6h}^4 ($P6_3/mmc$) in Sec. 2-3. In addition, each nickel atom has two neighboring nickels, one above and one below it, at a distance of $c/2$. These nickels are close enough to bond to each other, so that in Appendix 1 we list not only the Ni—As bond, but also the Ni—Ni bond, in this compound. The compounds crystallizing in this structure are mostly those formed from an element from the iron or other transition group in the periodic system, and an atom from the set S, Se, Te, As, Sb, Bi, and Sn. These compounds have rather metallic properties, and interesting antiferromagnetic behavior.

3-3. Structures of Some Compounds of Composition RX_2. Many compounds have the composition RX_2, the simplest ones being formed from a divalent electropositive element, and a monovalent electronegative one. An example is CaF_2, fluorite. There are a number of simple structures for such compounds, of which we shall take up five most common ones. The first we consider is that found for fluorite itself, and known by its name.

The Bravais lattice for the fluorite structure is face-centered cubic, with the same space group O_h^5 ($Fm3m$) which we have discussed in detail in Sec. 2-2. We have atoms of Ca, or more generally of R, at the points of a face-centered cubic Bravais lattice. The atoms of F, or more generally of X, are found at the centers of cubes half as large as the fundamental cube along each of the edges, as in diamond, and as indicated in Fig. 2-5. However, instead of having only half of the cubes filled, as in diamond or zinc-blende, they are all filled, so that there are twice as many atoms X as there are atoms R. On account of having all the cubes occupied by X atoms, the environment of an atom R is cubic rather than tetrahedral, with the full symmetry O_h, which is the reason why the crystal has the same space group O_h^5 ($Fm3m$) as the face-centered cubic structure of the elements. Putting things in mathematical form, the atom R is in the special position 0, 0, 0, and the two atoms X are in the special positions given by $(a/4)(\mathbf{i} + \mathbf{j} + \mathbf{k})$ and $(3a/4)(\mathbf{i} + \mathbf{j} + \mathbf{k})$. While the atoms R are in a cubic environment, with eight neighbors, the atoms X are in tetrahedral environment, as in zinc-blende or diamond.

In addition to the compounds like CaF_2, there are a good many others which have the fluorite structure. An important group of these have a modification called the antifluorite structure, in which the electropositive element is monovalent, the electronegative one divalent, as in Li_2O. The structure is the same as the fluorite, except with the roles of the two types of atoms interchanged. Among compounds with this structure are the oxides, sulfides, selenides, and tellurides of the alkali metals. We also find the antifluorite structure for compounds of a divalent electropositive

Sec. 3-3] SPACE GROUPS FOR STRUCTURES OF COMPOUNDS

element with a tetravalent electronegative element. Examples are Mg_2Si, Mg_2Ge, and the similar compounds of magnesium with tin and lead. A number of intermetallic compounds as well form the fluorite structure, such as, for instance, $PtAl_2$, $PtGa_2$, and others.

There is no trace of molecular formation in CaF_2; as we have seen, each Ca atom is surrounded equally by eight fluorines, each fluorine by four calciums. Quite a different situation is found in the structure of pyrite, FeS_2, and of a considerable number of compounds of transition metals

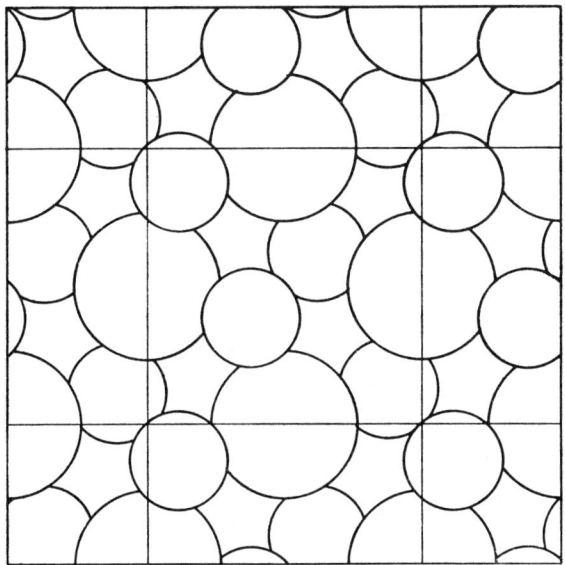

Fig. 3-4. Layers of the pyrite structure, FeS_2. The larger spheres represent the irons, and are at height $\frac{1}{2}a$. The upper sulfurs, represented by the small spheres, are at height $0.614a$, and the lower sulfurs are at height $0.386a$. The square inside the figure represents the cubic unit cell.

with sulfur, selenium, tellurium, arsenic, antimony, and bismuth, which crystallize with the same structure. These are essentially structures formed from the electropositive atoms, and from diatomic molecules like S_2. We show the structure in Fig. 3-4, which shows that if we replaced the S_2 molecules by single atoms located at the midpoint of their bonds, they and the Fe atoms would have a NaCl structure. The symmetry is more complicated than that of NaCl, however, for the axes of the various S_2 molecules point along different ones of the four directions similar to 111.

The space group for the pyrite structure is T_h^6 ($Pa3$), which we have not so far met. It belongs to the cubic system, with the simple cubic Bravais lattice, but the set of nonprimitive translations going with the operations of the point group T_h is quite complicated. We refer to Table 2-2 for the operations of this point group. The first 12 operations of T_h are identical with the first 12 of T_d; the remaining 12 are composed of the first 12 plus an inversion. Then we have the nonprimitive translation $(a/2)(\mathbf{i} + \mathbf{j})$ going with the operations $R_2, R_8, R_{11}, R'_2, R'_8, R'_{11}$; $(a/2)(\mathbf{j} + \mathbf{k})$ going with $R_3, R_6, R_{12}, R'_3, R'_6, R'_{12}$; $(a/2)(\mathbf{i} + \mathbf{k})$ going with $R_4, R_7, R_{10}, R'_4, R'_7, R'_{10}$; and no nonprimitive translations going with the remaining operations $R_1, R_5, R_9, R'_1, R'_5, R'_9$. The atoms of type R are located at the

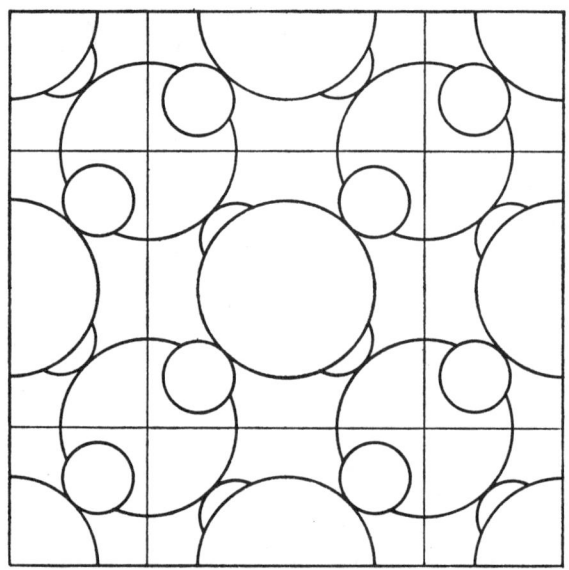

FIG. 3-5. The rutile structure, TiO$_2$, looking down along the z axis. The upper titanium atoms (large spheres) and oxygens are at height $c/2$; the lower ones are at height 0. The square inside the figure indicates the square cross section of the tetragonal unit cell in the xy plane.

special positions $a(\xi\mathbf{i} + \eta\mathbf{j} + \zeta\mathbf{k})$ for which ξ, η, ζ are given by 0, 0, 0; 0, ½, ½; ½, 0, ½; and ½, ½, 0. These are the positions which would be occupied in a face-centered cubic structure. The atoms X are at the special positions $\pm(u, u, u; u + ½, ½ - u, -u; -u, u + ½, ½ - u; ½ - u, -u, u + ½)$, where for the particular case of FeS$_2$ the parameter u equals 0.386.

Next we shall consider the rutile structure, named for one of the modifications of TiO$_2$. This has the space group D_{4h}^{14} ($P4_2/mnm$), belonging to the tetragonal system, with a simple tetragonal Bravais lattice. The structure is shown in Fig. 3-5. We see that each titanium atom has six oxygen neighbors, two in the plane above, two in the same plane, and two in the plane below. Each oxygen has three titanium neighbors. Each titanium also is close enough to its titanium neighbors above and below it to form bonds.

Let us consider the operations of the space group D_{4h}^{14} ($P4_2/mnm$), which we have not previously encountered. We recall that in the simple tetragonal lattice, the vectors t_1, t_2, t_3 are, respectively, $a\mathbf{i}$, $a\mathbf{j}$, $c\mathbf{k}$. The 16 operations of the point group D_{4h} can be conveniently written in terms of the rectangular coordinates. There are 8 unprimed operations, in which z does not change, as given below:

$$
\begin{aligned}
X_0\psi(x,y,z) &= \psi(x,y,z) & Y_0\psi(x,y,z) &= \psi(x,-y,z) \\
X_1\psi(x,y,z) &= \overline{\psi(-y,x,z)} & Y_1\psi(x,y,z) &= \psi(y,x,z) \\
X_{-1}\psi(x,y,z) &= \overline{\psi(y,-x,z)} & Y_{-1}\psi(x,y,z) &= \psi(-y,-x,z) \\
X_2\psi(x,y,z) &= \psi(-x,-y,z) & Y_2\psi(x,y,z) &= \psi(-x,y,z)
\end{aligned}
\qquad (3\text{-}1)
$$

The corresponding primed operations are those in which z is replaced by $-z$ on the right-hand side of the equations. We can now describe the operations of the space group by saying that the operations of the point group with even subscripts are not accompanied by any nonprimitive translation, while those with odd subscripts have the nonprimitive translation $(a/2)(\mathbf{i} + \mathbf{j}) + (c/2)\mathbf{k}$. The operations which have no nonprimitive translation are those of the point group D_{2h}.

The atoms of the type R, or Ti, are then found at the special positions $a(\xi\mathbf{i} + \eta\mathbf{j}) + c\zeta\mathbf{k}$, for which ξ, η, ζ, are 0, 0, 0 and $\frac{1}{2}, \frac{1}{2}, \frac{1}{2}$; the atoms X, or O, are in the positions $\pm(u, u, 0; u + \frac{1}{2}, \frac{1}{2} - u, \frac{1}{2})$. For TiO_2, the parameter u is about 0.31, and the sides of the unit cell are $a = 4.49$ A, $c = 2.89$ A. The symmetry of the environment of a Ti atom is D_{2h}.

The other two structures which we shall describe for the crystals of composition RX_2 are found in CdI_2 and $CdCl_2$, respectively, which we shall denote by the names of these compounds. They both belong to the trigonal system, and have the space groups D_{3d}^3 ($P\bar{3}m1$) and D_{3d}^5 ($R\bar{3}m$), respectively. We show the structure of CdI_2 in Figs. 3-6 and 3-7. In Fig. 3-6 we are looking down along the threefold axis. We see a sheet of Cd atoms, each surrounded by six iodines, the environment of the Cd having the symmetry D_{3d}, with three iodines above the sheet, three below, touching each Cd. This leaves a sheet of iodines above and below each sheet of cadmiums. These structures are then stacked one above another, so that sheets of iodines are adjacent to each other. They are too far apart, however, to form bonds, as is shown in Fig. 3-7, in which we are looking horizontally at the structure, along the vector t_1 of the hexagonal Bravais lattice, which we have in this space group. From Fig. 3-7 it is clear that this is a layer structure, the layers being held together presumably by Van der Waals forces.

The space group D_{3d}^3 ($P\bar{3}m1$), which we have with this structure, is one which we have not met previously. As just stated, we have the hexagonal Bravais lattice, and of course the point group D_{3d}. There are no nonprimitive translations with this space group, which then is simple enough so that we do not have to write down its operations. The Cd, or R, atom is found at the origin, explaining the D_{3d} symmetry which we find for its

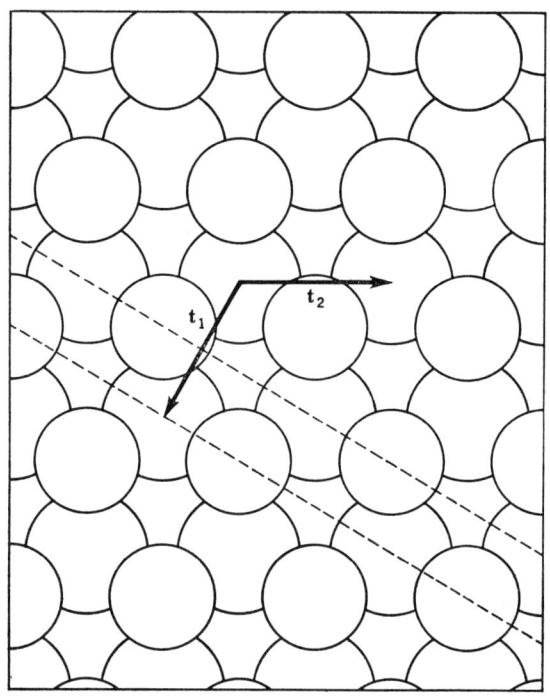

FIG. 3-6. The CdI_2 structure, looking down along the trigonal axis. The slightly smaller spheres in the upper layer are the iodines, at height $c/4$, positions $\frac{1}{3}$, $\frac{2}{3}$, $\frac{1}{4}$. The atoms in the lower layer are the cadmiums, at the origin. Between these spheres one can look down to other iodines at $\frac{2}{3}$, $\frac{1}{3}$, $-\frac{1}{4}$, for which the spheres are hidden by the atoms drawn. The dashed lines pass through the centers of the atoms shown in Fig. 3-7.

environment, and the I, or X, atoms are at the special positions $\frac{1}{3}$, $\frac{2}{3}$, u, and $\frac{2}{3}$, $\frac{1}{3}$, $-u$, with u equal approximately to 0.25. For CdI_2, the parameters are $a = 4.24$ A, $c = 6.84$ A. A good many compounds crystallize in this structure, both bromides and iodides of divalent elements, and sulfides, selenides, and tellurides of tetravalent elements, principally from the transition groups.

The $CdCl_2$ structure, like the CdI_2 structure, is composed of layers, essentially identical with those of CdI_2. As in that case, they are stacked

Sec. 3-3] SPACE GROUPS FOR STRUCTURES OF COMPOUNDS 67

far enough apart so that they are held together only by Van der Waals forces. However, they are stacked differently. Instead of having all the layers directly above each other, as in CdI_2, each layer is laterally displaced with respect to the one below it, so that if we look at Fig. 3-6, the centers of the Cd atoms would be above the three positions occupied, respectively, by a Cd and the two I's, in Fig. 3-6. We then have three repeating layers, much as in the face-centered cubic structure, and in fact

FIG. 3-7. The CdI_2 structure, looking along the $-t_1$ direction. The upper and lower sheets of atoms shown are those indicated by the dashed lines in Fig. 3-6.

the Cl's approximate this structure quite closely. The space group is the same one, D_{3d}^5 ($R\bar{3}m$), which we have already discussed for the arsenic structure in Sec. 2-6. The Cd atoms are at the origin of the rhombohedral Bravais lattice, and the Cl atoms at $\pm u(t_1 + t_2 + t_3)$, where u is approximately $\frac{1}{4}$. That is, the Cl's are at the same special positions as the arsenic atoms in the arsenic structure, but the dimensions of the unit cell are such that the angles between lines connecting Cl atoms are very nearly a right angle, a possibility which we discussed in Sec. 2-6. The magnitude of the vectors t_1, t_2, t_3 of the rhombohedral unit cell is 6.23 A, the angle α between them is 36°2', and the quantities r and s of Sec. 1-5 are $r = 5.82$ A, $s = 2.22$ A.

3-4. Structures of Compounds in the Cubic System.

We have now discussed the structures of a number of compounds of the compositions RX and RX_2. Once we go beyond these simple cases, we have a bewildering number of structures to consider, of which we shall choose a small number of important ones, as we have mentioned earlier. There is no ideal way of classifying these structures. For want of a better method, we shall classify according to crystal systems, starting with the cubic crystals in the present section, and working down to structures of lower symmetry. We shall take up four types of cubic crystals: those with the perovskite structure, named for perovskite, the compound $CaTiO_3$; those with the spinel structure, named for spinel, Al_2MgO_4; those with the structure of Tl_2O_3; and those with the garnet structure, an example being $Ca_3Al_2(SiO_4)_3$. These have the space groups O_h^1 ($Pm3m$), O_h^7 ($Fd3m$) T_h^7 ($Ia3$), and O_h^{10} ($Ia3d$), respectively. Since we have already encountered the first two of these four space groups, we shall have an easy time describing the first two structures.

The perovskite structure has all atoms at special positions of a simple cubic unit cell. The Ca atoms are at the corners, the Ti at the body center, and the O's at the face centers. This means that all the distances are determined in terms of one parameter, the side a of the unit cell. Three interatomic distances in the structure all are approximately given by the sums of the atomic radii: in perovskite itself, these are the Ca—O, Ti—O, and Ca—Ti distances. The first equals $a/\sqrt{2}$, the second is $a/2$, and the third is $a\sqrt{3}/2$. With the value of a found in $CaTiO_3$, namely, 3.84 A, we have 2.72 A, 1.92 A, and 3.34 A, respectively, for these three values, to be compared with the sums of the radii, which are 2.40, 2.00, and 3.20 A, respectively. The agreement is of about the same quality for the considerable number of other crystals showing the perovskite structure, as we see from Appendix 1, but for the perovskite structures as a whole the agreement is considerably poorer than for most types of crystals.

With the constraint which the structure puts on the relation between the interatomic distances, some of these bonds prove to be too short, others too long, for the ideal, and we should expect that there would be a tendency for the crystal to distort. This may be the explanation of the fact that some of these crystals are ferroelectric. They are not exactly cubic, but show slight elongations along one or another axis, thus changing to a lower type of symmetry. This brings with it a displacement of the atoms, or ions, from the positions which they would have in the cubic structure, and results in a permanent electric dipole moment for the crystal. Some of these crystals, such as, for instance, $BaTiO_3$, have several different structures of this type in different temperature ranges below a critical temperature, above which they assume the cubic form. The low-temperature ferroelectric structures are, however, only a slight departure from the cubic form. The constants given in Appendix 1 for some of these compounds are not for the distorted form, but for the cubic modification.

Among the many crystals having this structure, most of them have two electropositive elements and oxygen, as perovskite does. Some of them, however, have the oxygen replaced by fluorine, chlorine, or occasionally bromine.

The next cubic structure which we shall consider is that of spinel, Al_2MgO_4, an example of a large class of important compounds, including the ferrites, which show interesting magnetic properties. This is a quite complicated structure; we show it in Fig. 3-8, where we look down into the crystal. We recall that the space group is O_h^7 ($Fd3m$), the same found for the diamond structure, and this gives a clue to the behavior. The magnesium atoms are located at the positions which the carbons would have in diamond, namely, at the lattice points of a face-centered cubic Bravais lattice, including the origin, and the special position $(a/4)(\mathbf{i} + \mathbf{j} + \mathbf{k})$, or as it is ordinarily abbreviated, the position $\frac{1}{4}, \frac{1}{4}, \frac{1}{4}$. Thus there are two magnesium atoms per primitive unit cell, or eight per cube of side a. In Fig. 3-8, we can see clearly two of the magnesium atoms in the interior of the cube; the two others lie lower down, and do not show, and those in the face-center positions on the face of the cube also do not show very clearly, though those at height $a/2$ are indicated in the figure.

Magnesiums then occupy the centers of four out of the eight small cubes of side $a/2$ into which the larger cube can be subdivided. Each of the other four small cubes, which would be unoccupied in diamond, contains four aluminum atoms, two of these cubes and their contents being clearly visible in the figure. We then have 16 aluminum atoms in the cube of side a, consistent with the chemical structure. The primitive unit cell, one-fourth as large as the cubic cell of side a, contains four aluminum atoms. The aluminum atoms are at the special positions $\frac{5}{8}$, $\frac{5}{8}, \frac{5}{8}; \frac{5}{8}, \frac{7}{8}, \frac{7}{8}; \frac{7}{8}, \frac{5}{8}, \frac{7}{8};$ and $\frac{7}{8}, \frac{7}{8}, \frac{5}{8}$.

In the cubic unit cell there are also 32 oxygen atoms to be accounted for. These are at the special positions $a(\xi\mathbf{i} + \eta\mathbf{j} + \zeta\mathbf{k})$, where ξ, η, ζ are given by $u,u,u; u, -u, -u; -u, u, -u; -u, -u, u; \frac{1}{4} - u, \frac{1}{4} - u, \frac{1}{4} - u; \frac{1}{4} - u, \frac{1}{4} + u, \frac{1}{4} + u; \frac{1}{4} + u, \frac{1}{4} - u, \frac{1}{4} + u; \frac{1}{4} + u, \frac{1}{4} - u$; these eight positions in the primitive unit cell going into 32 in the cubic cell. The parameter u in this case is equal to 0.390. When we locate these atoms in the unit cell, as shown in the figure, we see that each magnesium is surrounded tetrahedrally by four oxygens. Each aluminum is surrounded by six oxygens, forming an approximately regular octahedral array about it; that is, the six are at equal distances, and approximately along the $\pm x, \pm y, \pm z$ directions of a coordinate axis located at the aluminum. To see the departures from the regularity of this octahedral arrangement, we may consider the aluminum atom in the lower left-hand corner of the cube in Fig. 3-8. Its rectangular coordinates, in units of a, are $\frac{1}{8}$, $\frac{1}{8}$, $\frac{5}{8}$, or 0.125, 0.125, 0.625. The oxygens

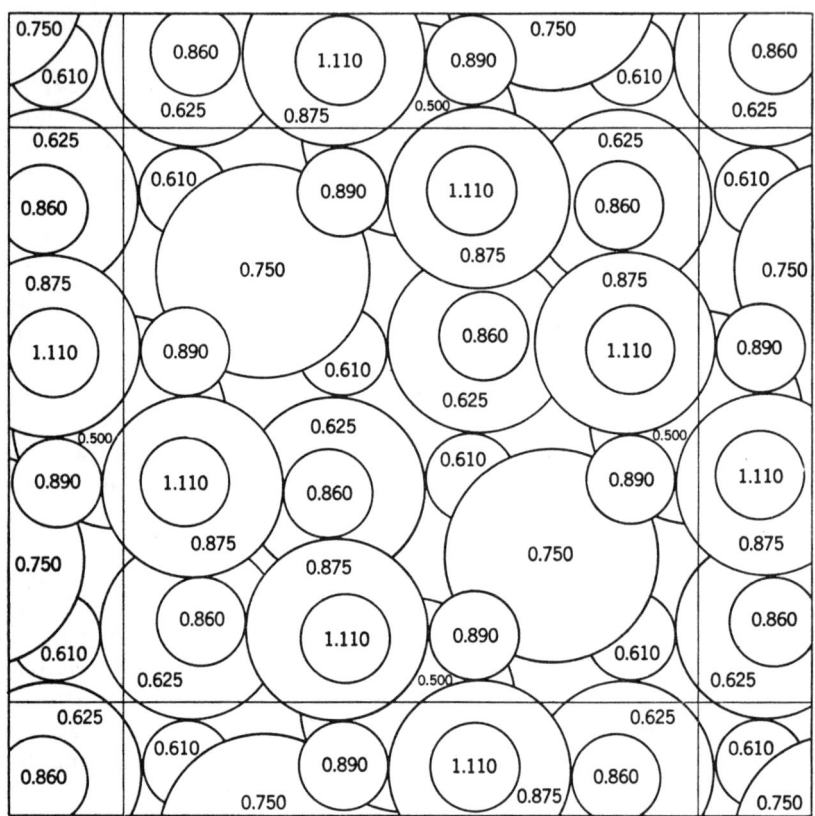

Fig. 3-8. Diagram of spinel structure Al_2MgO_4, indicating atoms by spheres with atomic radii. Largest spheres are Mg, next are Al, smallest are O. Heights are indicated, in fractions of the side of the cubic unit cell. Note how closely the spheres touch, indicating the accuracy of the atomic radii in describing the octahedral arrangement of the oxygens around the aluminums, and the tetrahedral arrangement around the magnesiums. Cubic unit cell is indicated by the square inside the borders of the figure.

which are approximately in line with it along the $\pm x$ directions have coordinates $0.890 = -0.110$, 0.110, 0.610, and 0.360, 0.140, 0.640, out of line by 0.015 with the aluminums in both y and z, but equally spaced along x by the amount ± 0.235. Similarly, those approximately in line along $\pm y$ are at 0.110, 0.890, 0.610, and 0.140, 0.360, and 0.640, and those approximately in line along $\pm z$ at 0.110, 0.110, 0.390, and 0.140, 0.140, 0.860, with the same amount of lack of alinement in each case. We note that the oxygens which form part of the tetrahedral environment of a magnesium also form part of the octahedral environment of an aluminum, so that each oxygen has bonds to both aluminums and magnesiums; examination of the figure shows that each oxygen is bonded to three aluminums and one magnesium.

There are many different substances having the spinel structure. We note that in Al_2MgO_4, the trivalent atoms Al occupy the octahedral sites (that is, with octahedral oxygen arrangement surrounding them), while the divalent magnesiums occupy the tetrahedral sites. Some spinels have different assignment of the atoms with different valencies to the different sites: half the trivalent atoms are located on the tetrahedral sites, and the other half of the trivalent atoms, and the divalent ones, are distributed at random on the octahedral sites. These are called the inverse spinels. An example is Fe_2MgO_4, in which half of the trivalent iron atoms are on the tetrahedral sites, half on the octahedral sites. Another example of the inverse spinel structure is Fe_3O_4, in which divalent iron takes the place of the Mg in Fe_2MgO_4. The trivalent ions are half on the tetrahedral, half on the octahedral sites. These facts regarding the distribution of the ions are established by magnetic properties, and particularly by neutron diffraction, which can give information about the arrangement of the individual magnetic ions in the crystal. These substances, called ferrites, have been very important in the development of our ideas regarding permanent magnetism; the type of magnetism found with them is called ferrimagnetism. We shall go into this subject in a later volume.

An interesting feature of the spinel structure is the arrangement of the oxygens. We have described it so far only in relation to the positions of the metallic atoms. However, if we consider only the oxygens, we find that if the parameter u were only 0.375, instead of the value 0.390 which it has in Al_2MgO_4, the arrangement of the oxygens would be that of a face-centered cubic structure. In all the spinel type of crystals where u has been measured, its value is in the range 0.38–0.39, very slightly larger than the value for the cubic packing of the oxygens. It is this slight discrepancy that keeps the surroundings of the Al atoms from being precisely octahedral. In the next chapter we shall take up the implications of the approximate face-centered cubic structure of the oxygens, and shall point out that though they have this structure, the interatomic distance is much greater than the sum of the oxygen atomic radii, as is obvious from Fig. 3-8. It is clear, in other words, that the oxygens are not bonding to each other in a face-centered structure.

72 QUANTUM THEORY OF MOLECULES AND SOLIDS [Chap. 3

The third cubic structure which we consider is that of Tl_2O_3. This structure is shown in Fig. 3-9; the dimensions chosen for the spheres are not those for Tl_2O_3 itself, for which the parameters have not been very accurately measured, but for a mixed oxide, $(Fe,Mn)_2O_3$, in which we have Fe and Mn atoms arranged at random on the metallic sites. It happens

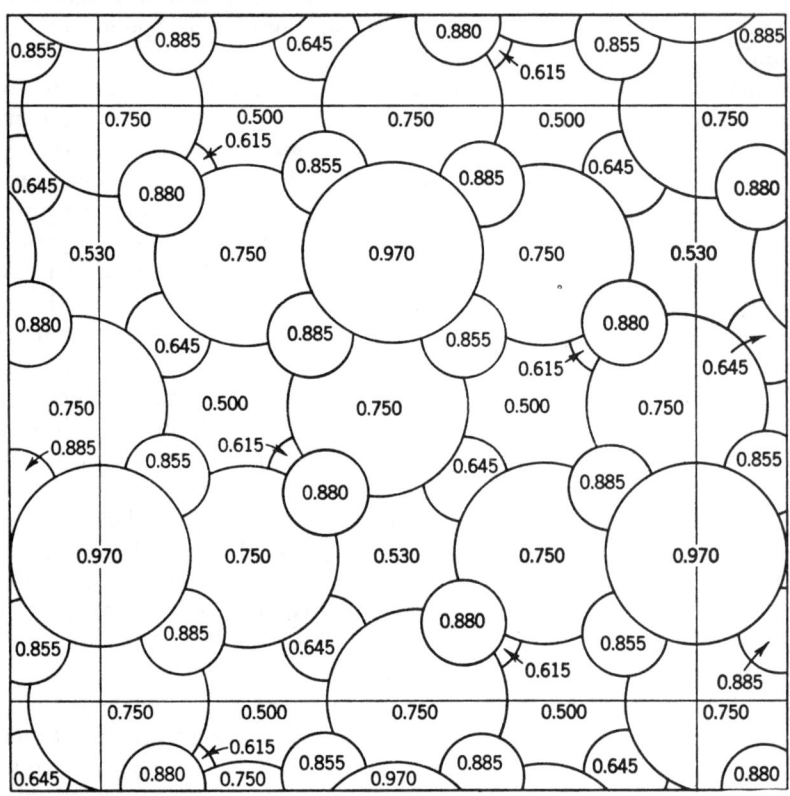

Fig. 3-9. The Tl_2O_3 structure. Heights of atoms, as fractions of the side of the cubic unit cell, are indicated in the figure. The square shown inside the figure is the cubic unit cell. Large spheres are Tl, smaller are O. The regions whose height is 0.500 or 0.530 are thallium atoms which can be seen between the other atoms, though their spheres do not show. There are also thallium atoms, which we have omitted in the figure, above these, at heights of 1.000 or 1.030, respectively.

that the parameters have been measured more accurately for this case than for any other crystal with the same structure. We have chosen the radius of the sphere representing the metallic atom to be 1.40 A, the value given for both Mn and Fe in Table 3-1.

The space group, T_h^7 ($Ia3$), can be discussed by comparison with

T_h^6 ($Pa3$), which we have met in our discussion of the pyrite crystal in Sec. 3-3. The present space group has a body-centered cubic Bravais lattice, rather than the simple cubic lattice of T_h^6, and it has different nonprimitive translations. If we refer back to the discussion of the point group T_h in Sec. 3-3, we can easily state the nonprimitive translations of T_h^7 ($Ia3$). Associated with the operations R_3, R_6, R_{12}, R_3', R_6', R_{12}' of the point group, we have the nonprimitive translation $(a/2)\mathbf{i}$; associated with R_4, R_7, R_{10}, R_4', R_7', R_{10}' we have $(a/2)\mathbf{j}$; associated with R_2, R_8, R_{11}, R_2', R_8', R_{11}' we have $(a/2)\mathbf{k}$; and associated with R_1, R_5, R_9, R_1', R_5', R_9' we have no nonprimitive translation. We then have the atoms in the following positions of the unit cell, described in terms of the unit vectors $a\mathbf{i}$, $a\mathbf{j}$, $a\mathbf{k}$:

Tl: ¼, ¼, ¼; ¼, ¾, ¾; ¾, ¼, ¾; ¾, ¾, ¼
Tl: ±(u,0,¼; ¼,u,0; 0,¼,u; −u,½,¼; ¼,−u,½; ½,¼,−u)
O: ±($\xi\eta\zeta$; ξ, −η, ½ − ζ; ½ − ξ, η, −ζ; −ξ, ½ − η, ζ;   (3-2)
    $\zeta\xi\eta$; ½ − ζ, ξ, −η; −ζ, ½ − ξ, η; ζ, −ξ, ½ − η;
    $\eta\zeta\xi$; −η, ½ − ζ, ξ; η, −ζ, ½ − ξ; ½ − η, ζ, −ξ)

The positions we have listed are to be supplemented by those in the body-centered positions, obtained from those listed by adding the displacement ½, ½, ½. In the case of (Fe,Mn)$_2$O$_3$, the parameters are given by the values $u = -0.030$, $\xi = 0.385$, $\eta = 0.145$, $\zeta = 0.380$, and the length of the side of the cubic cell, a, is 9.37 A.

It is clear that this structure, with four numerical parameters, is a good deal more complicated than the other cubic ones we have discussed. Metallic atoms are of two sorts, the first ones at special positions without parameters, the second ones at positions with a single parameter, while the oxygens are at general positions, so that the number of oxygens per primitive unit cell is 24, the number of operations in the point group; in a cubic unit cell we have 48 oxygens. There are in all 4 metallic atoms of the first sort, and 12 of the second sort, or 16 in all, in the primitive unit cell, with 32 in the cubic unit cell, leading to the correct chemical composition.

The structure is generally discussed as a much modified version of the fluorite structure. If the parameter u were equal to zero, instead of -0.030, the metallic atoms would be on a face-centered cubic lattice, with a cubic unit cell of $a/2$. This would be like the calciums in fluorite, but the unit cell of Tl$_2$O$_3$ would be twice as long on a side, or would hold eight times as many atoms, as in fluorite. If the parameters x, y, z were equal, respectively, to 0.375, 0.125, and 0.375, instead of 0.385, 0.145, and 0.380, the oxygen atoms would be in the positions of fluorines in fluorite. There are, however, only three-quarters as many oxygens as fluorines, so that a quarter of the positions occupied by fluorines in fluorite would be empty here.

We can see the resemblance to fluorite by examining the structure shown in Fig. 3-9. In fluorite, each calcium atom is at the center of a cube of fluorines, with eight fluorine neighbors, while each fluorine is surrounded tetrahedrally by four calciums. In the Tl_2O_3 structure, each metallic atom has only six oxygen neighbors, two of the eight corners of the cubic environment being empty. We see an example of this situation in the lower right corner of the cube in Fig. 3-9, where the metallic atom at height 0.750 is surrounded by six oxygens at heights 0.885, 0.880, 0.855, 0.645, 0.620 (hidden by the atom at 0.885), and $0.615a$. These heights would be 0.875, 0.625 in the fluorite case. If all possible positions were occupied by the oxygens, we should have an additional oxygen above that with height 0.645, and one below that with height 0.855. On the other hand, each oxygen has four metallic neighbors, approximately tetrahedrally arranged, as in fluorite. We see an example in the oxygen at height 0.645, which we have just been discussing, in the lower right-hand corner of the figure. It is surrounded by two metallic atoms below it, one at height 0.500 and the other at 0.530, and two above, both at height 0.750.

The last cubic structure which we shall discuss is the garnet structure, found in a considerable class of minerals and synthetic materials, some of them of great importance on account of their magnetic properties. An example, as we mentioned earlier, is the calcium aluminum orthosilicate, $Ca_3Al_2(SiO_4)_3$. A layer of atoms from the unit cell is shown in Fig. 3-10. The space group is O_h^{10} ($Ia3d$), as we mentioned earlier, the Bravais lattice being body-centered cubic. In Fig. 3-10 we show a unit cube of the substance, equal to two primitive unit cells. This unit cube holds eight molecular constituents; that is, 24 Ca's, 16 Al's, 24 Si's, and 96 O's. The layer which we have drawn shows only something less than a quarter of the contents of the cubic cell. We see that the tetrahedral SiO_4 complexes are packed between the metallic atoms, in a rather complicated way. The Ca atoms, the largest ones in the figure, have eight neighboring oxygens (one above and one below each atom are not shown in the figure). The aluminums, the next in size, have six neighbors (again one above and one below each atom are not shown). The silicons have four neighboring oxygens.

Many substitutions are possible for the Ca, Al, and Si atoms, and as we stated earlier, some of these substituted garnets have very important magnetic properties. There are the iron garnets, in which iron is substituted for both aluminum and silicon. We have, for instance, yttrium iron garnet, often abbreviated YIG, $Y_3Fe_2(FeO_4)_3$, whose general appearance is very similar to that shown in Fig. 3-10, except that the iron atoms are somewhat larger than the aluminum and silicon. The importance of these iron garnets is that the two types of iron atoms have magnetic moments which can be coupled to each other in complicated and interesting ways. An even more complicated case is ytterbium iron garnet, $Yb_3Fe_2(FeO_4)_3$, in which the rare earth ion Yb also has a magnetic moment, which interacts with the moments on the two sets of iron atoms.

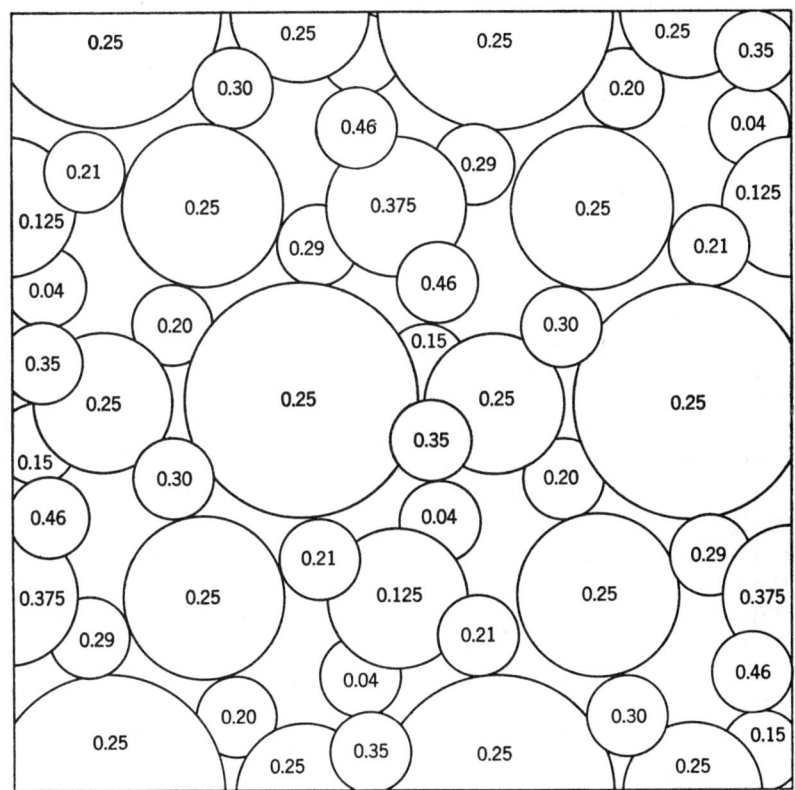

FIG. 3-10. Cubic unit cell of the garnet structure. The parameters ξ, η, ζ are taken to be 0.04, 0.05, and 0.65. The size of the unit cell and the sizes of the atoms are chosen to represent $Ca_3Al_2(SiO_4)_3$. The figure looks very much the same for all the garnets. The calcium atoms are the largest, then the aluminum, silicon, and oxygen. Each of the calcium and aluminum atoms has another oxygen above it, and another below it, which are not shown in the figure. The heights of the various atoms are shown as multiples of a, the side of the cubic unit cell.

Here, as with the magnetic properties of the ferrites, we shall postpone discussion until a later volume.

Now let us consider the details of the space group. The Bravais lattice, as we have stated, is body-centered cubic. The point group is O_h. The operations are associated with nonprimitive translations, as follows, where we are giving the x, y, z components of the translation, in terms of a, the side of the cubic cell:

$$\begin{array}{ll}
\text{No translation} & R_1, R_5, R_9, R'_1, R'_5, R'_9 \\
\tfrac{1}{2}, \tfrac{1}{2}, 0 & R_2, R_8, R_{11}, R'_2, R'_8, R'_{11} \\
0, \tfrac{1}{2}, \tfrac{1}{2} & R_3, R_6, R_{12}, R'_3, R'_6, R'_{12} \\
\tfrac{1}{2}, 0, \tfrac{1}{2} & R_4, R_7, R_{10}, R'_4, R'_7, R'_{10} \\
\tfrac{1}{4}, \tfrac{1}{4}, \tfrac{1}{4} & R_{19}, R_{21}, R_{23}, R'_{19}, R'_{21}, R'_{23} \\
\tfrac{3}{4}, \tfrac{3}{4}, \tfrac{1}{4} & R_{13}, R_{18}, R_{22}, R'_{13}, R'_{18}, R'_{22} \\
\tfrac{1}{4}, \tfrac{3}{4}, \tfrac{3}{4} & R_{14}, R_{15}, R_{24}, R'_{14}, R'_{15}, R'_{24} \\
\tfrac{3}{4}, \tfrac{1}{4}, \tfrac{3}{4} & R_{16}, R_{17}, R_{20}, R'_{16}, R'_{17}, R'_{20}
\end{array} \qquad (3\text{-}3)$$

In $Ca_3Al_2(SiO_4)_3$, the Ca atoms are located at the special positions $\pm(\tfrac{1}{8}, 0, \tfrac{1}{4}; \tfrac{1}{4}, \tfrac{1}{8}, 0; 0, \tfrac{1}{4}, \tfrac{1}{8}; \tfrac{5}{8}, 0, \tfrac{1}{4}; \tfrac{1}{4}, \tfrac{5}{8}, 0; 0, \tfrac{1}{4}, \tfrac{5}{8})$, together with the other positions required by the body-centered nature of the lattice, namely, the values above plus $\tfrac{1}{2}, \tfrac{1}{2}, \tfrac{1}{2}$. The Al atoms are at the positions $0, 0, 0; \tfrac{1}{2}, \tfrac{1}{2}, 0; 0, \tfrac{1}{2}, \tfrac{1}{2}; \tfrac{1}{2}, 0, \tfrac{1}{2}; \tfrac{1}{4}, \tfrac{1}{4}, \tfrac{1}{4}; \tfrac{3}{4}, \tfrac{3}{4}, \tfrac{1}{4}; \tfrac{1}{4}, \tfrac{3}{4}, \tfrac{3}{4}; \tfrac{3}{4}, \tfrac{1}{4}, \tfrac{3}{4}$, and these values plus $\tfrac{1}{2}, \tfrac{1}{2}, \tfrac{1}{2}$. The Si atoms are at the positions $\pm(\tfrac{3}{8}, 0, \tfrac{1}{4}; \tfrac{1}{4}, \tfrac{3}{8}, 0; 0, \tfrac{1}{4}, \tfrac{3}{8}; \tfrac{7}{8}, 0, \tfrac{1}{4}; \tfrac{1}{4}, \tfrac{7}{8}, 0; 0, \tfrac{1}{4}, \tfrac{7}{8})$; and these values plus $\tfrac{1}{2}, \tfrac{1}{2}, \tfrac{1}{2}$. The oxygens are at general positions, derived from an initial set of values $a(\xi \mathbf{i} + \eta \mathbf{j} + \zeta \mathbf{k})$ by all 96 operations, the 48 of the space group, and the other 48 arising from these by adding $\tfrac{1}{2}, \tfrac{1}{2}, \tfrac{1}{2}$. In most garnets, the values of ξ, η, ζ are found to be very close to 0.04, 0.05, and 0.65, respectively. The value a of the side of the cubic cell is close to 12 A in all the cases observed. For example, in $Y_3Fe_2(FeO_4)_3$, a equals 12.38 A, and ξ, η, ζ can be taken, respectively, to be 0.0274, 0.0572, and 0.6492.[1]

The garnet structure is the only silicate which we shall describe in detail. However, the silicates form a very large and important group of solids, including a large fraction of the known minerals. The tetrahedral silicate group, four oxygens tetrahedrally arranged around a central silicon atom, is their characteristic feature. In some cases, the separate groups are independent of each other, as in garnet. However, there are many compounds in which adjacent silicate groups share an oxygen, so that they are linked together by Si—O—Si bonds; generally there is approximately a tetrahedral angle between the two bonds attached to an oxygen, just as in water. These linkages can extend throughout the crystal. If we have one-dimensional linkages, we have chains or fibers of atoms; asbestos is an example of such a structure. If we have two-dimensional linkages, we have sheets, of which mica is a familiar case.

---

[1] In the paper of S. Geller and M. A. Gilleo, *J. Phys. Chem. Solids*, **3**:30 (1957), in which the crystal structure of this substance is reported, quite different values of ξ, η, and ζ are given. However, the values quoted above represent one of the 96 sets of values found from the reported values by application of the operations of the space group.

Three-dimensional linkages lead to substances held together very tightly, of which the various forms of silica, SiO_2, are examples.

Silica exists in many crystalline forms, of which the simplest to describe is that known as high cristobalite, found at high temperature. We can visualize this by starting with a diamond structure, replacing the carbons by silicons, and inserting an oxygen in the middle of the bond between neighboring silicons. Since the Si—O—Si bonds are in a straight line, this is a structure of high energy; it is thought to exist only at high temperatures because temperature agitation is required to straighten out the Si—O—Si bonds, which are normally bent. At lower temperatures the substance SiO_2 takes on much more complicated forms, in which the Si—O—Si bonds show the angle which we have mentioned earlier. One of these structures, of very considerable complication, is the familiar quartz.

The silicates are not the only substances showing tetrahedral groups like SiO_4. Other examples of the same geometrical structures are sulfates, selenates, phosphates, arsenates, germanates, containing groups SO_4, SeO_4, PO_4, AsO_4, GeO_4. In the case of some of these groups, one can build up crystal structures similar to those of the silicates. For example, one can start with one of the structures of silica, and replace alternate silicons by aluminum and phosphorus, respectively, one having one electron less, the other having one electron more, than silicon. Similarly, in a germanate, analogous to a silicate, but formed from germanium, one can replace alternate germaniums by gallium and arsenic. These compounds, phosphates and arsenates, have a certain analogy to the 3-5 compounds formed by replacing germanium in its elementary form by gallium and arsenic.

With these remarks we merely suggest the great variety of important compounds, met both in mineralogy and in the study of synthetically prepared inorganic compounds, which we are not able to touch on in this text. Most of the structures of these compounds are complicated, in many cases more complicated than any of those which we are taking up in this volume. The reader should realize that, even though we are not discussing them, they still have their great importance, and should be studied by anyone who really wishes to understand the solid state.

3-5. The Calcite, Corundum, and Ilmenite Structures. A great many important compounds crystallize in the hexagonal and trigonal systems, which we can conveniently group together. The first two which we shall consider are calcite, one of the forms of $CaCO_3$, and corundum, one of the forms of Al_2O_3. These are representatives of important groups of crystals. They have different structures, but both have the same space group, D_{3d}^6 ($R\bar{3}c$), belonging to the trigonal system, with the rhombohedral Bravais lattice. Unlike the other space groups with the point group D_{3d} which we have encountered so far, namely, D_{3d}^3 ($P\bar{3}m1$) and

D_{3d}^5 ($R\bar{3}m$), the present one is nonsymmorphic, so that we must specify the nonprimitive translations going with some of the operations.

For this purpose we remind the reader of the operations of the point group D_{3d}. These are given in Eq. (A12-64), in Volume 1, Appendix 12. They are

$$X_q\psi(r,\phi,z) = \psi\left[r,\, \phi + \frac{\pi q}{3},\, (-1)^q z\right]$$
$$Y_q\psi(r,\phi,z) = \psi\left[r,\, -\phi + \frac{\pi q}{3},\, (-1)^q z\right] \quad (3\text{-}4)$$

with $q = 0, \pm 1, \ldots, \pm(n-1), n$, which in the present case reduce to

$$X_0\psi(r,\phi,z) = \psi(r,\phi,z) \qquad Y_0\psi(r,\phi,z) = \psi(r,-\phi,z)$$
$$X_{\pm 1}\psi(r,\phi,z) = \psi\left(r,\, \phi \pm \frac{\pi}{3},\, -z\right) \qquad Y_{\pm 1}\psi(r,\phi,z)$$
$$= \psi\left(r,\, -\phi \pm \frac{\pi}{3},\, -z\right)$$
$$X_{\pm 2}\psi(r,\phi,z) = \psi\left(r,\, \phi \pm \frac{2\pi}{3},\, z\right) \qquad Y_{\pm 2}\psi(r,\phi,z) \quad (3\text{-}5)$$
$$= \psi\left(r,\, -\phi \pm \frac{2\pi}{3},\, z\right)$$
$$X_3\psi(r,\phi,z) = \psi(r,\, \phi + \pi,\, -z) \qquad Y_3\psi(r,\phi,z)$$
$$= \psi(r,\, -\phi + \pi,\, -z)$$

The nonprimitive translations are then easily stated: there are none

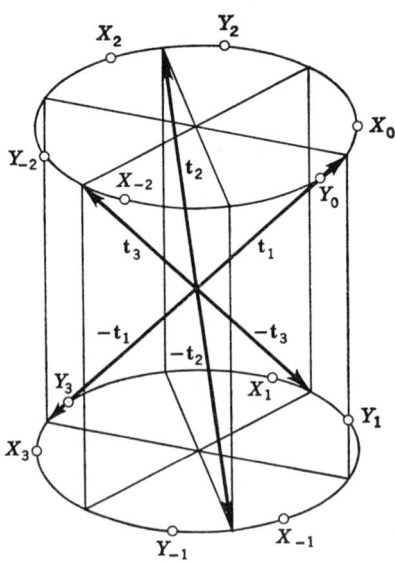

Fig. 3-11. Diagram to illustrate symmetry operations of the point group D_{3d}. The origin is at the center of the figure. The various operations of the group transform the point X_0 to the points indicated.

Sec. 3-5] SPACE GROUPS FOR STRUCTURES OF COMPOUNDS

associated with the X operators, while all the Y operators are associated with the nonprimitive translation $\frac{1}{2}(t_1 + t_2 + t_3)$, where we remember that t_1, t_2, t_3 all have the magnitude a, and make angles α with each other, as described in Sec. 1-5.

It is convenient to rewrite Eq. (3-5) in terms of ξ, η, ζ, where the radius vector $\mathbf{r}$ is given by $\xi t_1 + \eta t_2 + \zeta t_3$. To do so we refer to Fig. 3-11, similar to Fig. A12-2 of Volume 1, showing the symmetry operations of the point group D_{3d} graphically. The points marked X_0, X_1, etc., in the figure are the points to which X_0 is transformed by the operations enumerated in Eq. (3-5). We now see that the point X_1 stands in the same relation to the vectors $-t_3$, $-t_1$, $-t_2$ which X_0 does with respect to t_1, t_2, t_3. In other words,

$$X_1\psi(\xi t_1 + \eta t_2 + \zeta t_3) = \psi[\xi(-t_3) + \eta(-t_1) + \zeta(-t_2)] = \psi(-\eta t_1 - \zeta t_2 - \xi t_3)$$

so that X_1 transforms ξ, η, ζ into $-\eta$, $-\zeta$, $-\xi$. Similarly, Y_0 stands in the same relation to t_1, t_3, t_2 that X_0 does to t_1, t_2, t_3. Hence

$$Y_0\psi(\xi t_1 + \eta t_2 + \zeta t_3) = \psi(\xi t_1 + \zeta t_2 + \eta t_3)$$

We can investigate the other operations in the same way.

We can then combine this information with the nonprimitive translations, and the translations $\mathbf{R}_n$, to write the operations of the space group in a form analogous to what we have done for the group D_{6h}^4 ($P6_3/mmc$) in Eq. (2-13). We have

$$\begin{aligned}
\{X_0|R_n\}\psi(\mathbf{r}) &= \psi[(\xi + n_1)t_1 + (\eta + n_2)t_2 + (\zeta + n_3)t_3] \\
\{X_1|R_n\}\psi(\mathbf{r}) &= \psi[(-\eta + n_1)t_1 + (-\zeta + n_2)t_2 + (-\xi + n_3)t_3] \\
\{X_{-1}|R_n\}\psi(\mathbf{r}) &= \psi[(-\zeta + n_1)t_1 + (-\xi + n_2)t_2 + (-\eta + n_3)t_3] \\
\{X_2|R_n\}\psi(\mathbf{r}) &= \psi[(\zeta + n_1)t_1 + (\xi + n_2)t_2 + (\eta + n_3)t_3] \\
\{X_{-2}|R_n\}\psi(\mathbf{r}) &= \psi[(\eta + n_1)t_1 + (\zeta + n_2)t_2 + (\xi + n_3)t_3] \\
\{X_3|R_n\}\psi(\mathbf{r}) &= \psi[(-\xi + n_1)t_1 + (-\eta + n_2)t_2 + (-\zeta + n_3)t_3] \\
\{Y_0|R_n\}\psi(\mathbf{r}) &= \psi[(\xi + n_1 + \tfrac{1}{2})t_1 + (\zeta + n_2 + \tfrac{1}{2})t_2 \\
&\qquad + (\eta + n_3 + \tfrac{1}{2})t_3] \\
\{Y_1|R_n\}\psi(\mathbf{r}) &= \psi[(-\zeta + n_1 + \tfrac{1}{2})t_1 + (-\eta + n_2 + \tfrac{1}{2})t_2 \\
&\qquad + (-\xi + n_3 + \tfrac{1}{2})t_3] \\
\{Y_{-1}|R_n\}\psi(\mathbf{r}) &= \psi[(-\eta + n_1 + \tfrac{1}{2})t_1 + (-\xi + n_2 + \tfrac{1}{2})t_2 \\
&\qquad + (-\zeta + n_3 + \tfrac{1}{2})t_3] \\
\{Y_2|R_n\}\psi(\mathbf{r}) &= \psi[(\eta + n_1 + \tfrac{1}{2})t_1 + (\xi + n_2 + \tfrac{1}{2})t_2 \\
&\qquad + (\zeta + n_3 + \tfrac{1}{2})t_3] \\
\{Y_{-2}|R_n\}\psi(\mathbf{r}) &= \psi[(\zeta + n_1 + \tfrac{1}{2})t_1 + (\eta + n_2 + \tfrac{1}{2})t_2 \\
&\qquad + (\xi + n_3 + \tfrac{1}{2})t_3] \\
\{Y_3|R_n\}\psi(\mathbf{r}) &= \psi[(-\xi + n_1 + \tfrac{1}{2})t_1 + (-\zeta + n_2 + \tfrac{1}{2})t_2 \\
&\qquad + (-\eta + n_3 + \tfrac{1}{2})t_3]
\end{aligned} \quad (3\text{-}6)$$

as the operations of the space group D_{3d}^6 ($R\bar{3}c$).

We are now ready to proceed with the calcite and corundum structures. A view of the calcite structure is shown in Fig. 3-12. We see a layer of calcium atoms, and above them a layer of triangular planar carbonate groups. The next-higher layer of calciums would fall over the open spaces in the figure. Thus each calcium is bonded to six oxygens, three below and three above it, and each oxygen is bonded to the carbon in its own carbonate group, and to two calciums, one above and one below.

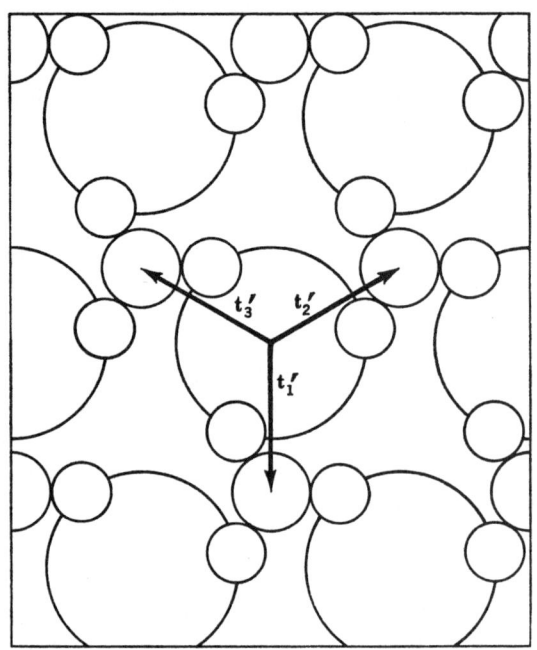

FIG. 3-12. The calcite structure, $CaCO_3$. The large spheres representing the calcium atoms are at height 0, while the CO_3 groups are at height $r/4$, as indicated in the text and in Table 3-3. The vectors t_1', t_2', t_3' are the components of the vectors t_1, t_2, t_3 in the xy plane.

The calciums are in the special positions for which ξ, η, ζ are 0, 0, 0 and $\frac{1}{2}$, $\frac{1}{2}$, $\frac{1}{2}$. The carbons are at $\pm(\frac{1}{4}, \frac{1}{4}, \frac{1}{4})$. The oxygens are at $\pm(\frac{1}{4} - u, \frac{1}{4} + u, \frac{1}{4}; \frac{1}{4} + u, \frac{1}{4}, \frac{1}{4} - u; \frac{1}{4}, \frac{1}{4} - u, \frac{1}{4} + u)$. The parameter u equals 0.243, the length a of the vectors t is 6.36 A, and the angle α is 46°6′, so that $s = 2.87$ A, $r = 5.67$ A.

As in other cases with the rhombohedral Bravais lattice, it is a little complicated to see which atoms lie in each plane. First, we have a calcium atom at the origin, and above the origin at heights $\frac{3}{2}r$, $3r$, . . . , where we remember that r is the z component of any one of the three vectors t_1, t_2, t_3. If we displace from the origin by any vector such as

Sec. 3-5]   SPACE GROUPS FOR STRUCTURES OF COMPOUNDS   81

$t_1 - t_2$, for which the sum of the vertical components adds to zero, we again have a calcium atom in the plane $z = 0$. It is these calcium atoms that are shown in Fig. 3-12.

Next we have carbon atoms at $z = \pm \frac{3}{4}r$, directly above the origin. It is not these carbons, however, which appear in Fig. 3-12. Rather, we have the carbon displaced from the point $-\frac{1}{4}(t_1 + t_2 + t_3)$ by the vector t_1, and the other vectors derived from this by adding such vectors as $t_1 - t_2$. The height of these carbons is $\frac{1}{4}r$, and they are the ones appearing in the figure. From each carbon in this plane, we have oxygens displaced by the amounts $u(t_1 - t_2)$, $u(t_2 - t_3)$, $u(t_3 - t_1)$, vectors in the plane of constant z. From the carbons in the other type of plane, starting with the position $\frac{1}{4}(t_1 + t_2 + t_3)$, the oxygen displacements are the negatives of these values, so that the carbonate group in the next-higher plane is rotated through 180° with respect to those shown in the figure. The complete repeating range in the crystal comprises six layers similar to that shown in Fig. 3-12, each consisting of a calcium and a carbonate layer. In half of them, the carbonates are rotated. The arrangement of these layers is as shown in Table 3-3.

Table 3-3
Arrangement of successive layers in calcite crystal

Height

0, Ca at origin, shown in Fig. 3-12
$(-\frac{3}{4} + 1)r = \frac{1}{4}r$, CO_3, displaced from origin by t_1', shown in Fig. 3-12
$(\frac{3}{2} - 1)r = \frac{1}{2}r$, Ca, displaced from origin by $-t_1'$
$\frac{3}{4}r$, CO_3, C above origin, rotated 180° from plane shown in Fig. 3-12
$(0 + 1)r = r$, Ca, displaced from origin by t_1'
$(-\frac{3}{4} + 2)r = \frac{5}{4}r$, CO_3, like plane shown, displaced from origin by $2t_1'$
$\frac{3}{2}r$, Ca, over origin
$(\frac{3}{4} + 1)r = \frac{7}{4}r$, CO_3, rotated, displaced from origin by t_1'
$2r$, Ca, displaced from origin by $2t_1'$
$(-\frac{3}{4} + 3)r = \frac{9}{4}r$, CO_3, like plane shown in Fig. 3-12, over origin
$(\frac{3}{2} + 1)r = \frac{5}{2}r$, Ca, displaced from origin by t_1'
$(\frac{3}{4} + 2)r = 1\frac{1}{4}r$, CO_3, rotated, displaced from origin by $2t_1'$
$3r$, like 0

There are a considerable number of compounds with the calcite structure. These include a number of carbonates, lithium and sodium nitrate, and several borates. In all of them the carbonate, nitrate, or borate group has the same form of an equilateral triangle of oxygens, with the other atom at the center, in the same plane.

Next we consider the structure of corundum, $\alpha = Al_2O_3$. We show a projection of the structure along the threefold axis in Fig. 3-13. We see equilateral triangles of oxygens, similar to those in the calcite structure, but there is no atom at the center of the triangle in the same plane. Furthermore, the aluminum atoms are not all in a plane: in the figure we show three different layers of aluminums. We can understand the situa-

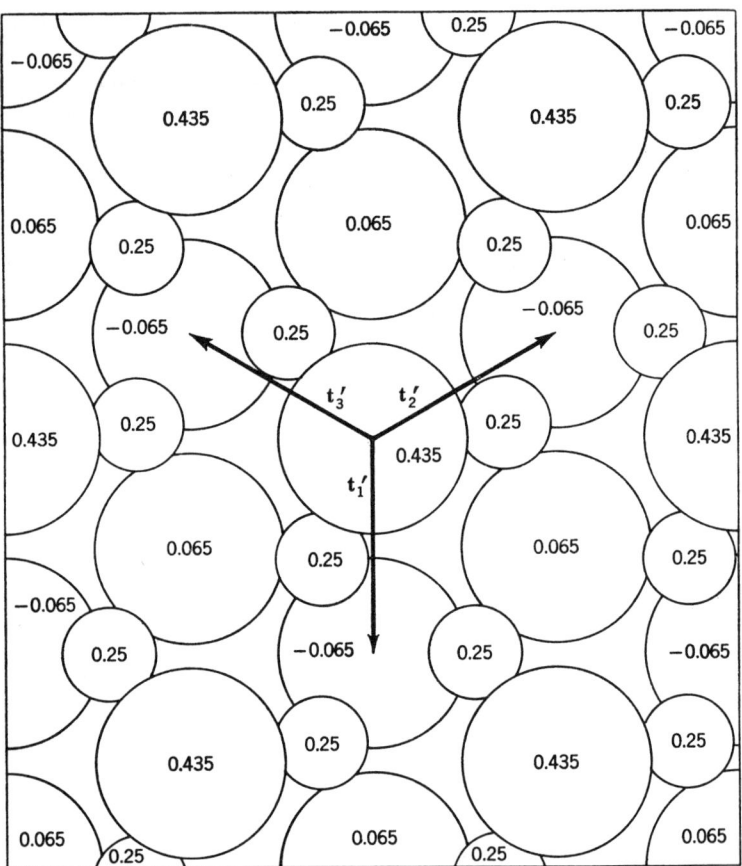

FIG. 3-13. The corundum structure, Al_2O_3. Larger spheres, Al; smaller, O. The heights of the atoms are shown as multiples of r, the component of the vectors t_1, t_2, t_3 along the z axis. The vectors t_1', t_2', t_3' are the components of the vectors t_1, t_2, t_3 in the xy plane.

tion better when we state the positions of the atoms in the structure. Aluminums are in the positions $\xi t_1 + \eta t_2 + \zeta t_3$, whose ξ, η, ζ are given by $\pm(\frac{1}{4} + u, \frac{1}{4} + u, \frac{1}{4} + u; \frac{1}{4} - u, \frac{1}{4} - u, \frac{1}{4} - u)$, with $u = 0.105$. The oxygens are in the same type of positions as in calcite, namely, $\pm(\frac{1}{4} - v, \frac{1}{4} + v, \frac{1}{4}; \frac{1}{4} + v, \frac{1}{4}, \frac{1}{4} - v; \frac{1}{4}, \frac{1}{4} - v, \frac{1}{4} + v)$, with

Sec. 3-5]   SPACE GROUPS FOR STRUCTURES OF COMPOUNDS   83

$v = 0.303$. The length a of the vectors $\mathbf{t}_1$, $\mathbf{t}_2$, $\mathbf{t}_3$ is 5.13 A, the angle α is 55°6', giving $s = 2.74$ A, $r = 4.33$ A.

We can now make a table, like Table 3-3, describing the atoms to be found in each layer at constant z. This is given in Table 3-4. The

Table 3-4
Arrangement of successive layers in corundum crystal

Height

$(-\frac{3}{4} - 3u + 1)r = -0.065r$, Al, displaced from origin by $\mathbf{t}_1'$ (shown in figure)
$(\frac{3}{4} + 3u - 1)r = 0.065r$, Al, displaced from origin by $-\mathbf{t}_1'$ (shown)
$(-\frac{3}{4} + 1)r = 0.25r$, O$_3$, displaced from origin by $\mathbf{t}_1'$ (shown)
$(\frac{3}{4} - 3u)r = 0.435r$, Al, over origin (shown)
$(-\frac{3}{4} + 3u + 1)r = 0.565r$, Al, displaced from origin by $\mathbf{t}_1'$
$\frac{3}{4}r = 0.75r$, O$_3$, rotated by 180°, above origin
$(-\frac{3}{4} - 3u + 2)r = 0.965r$, Al, displaced from origin by $2\mathbf{t}_1'$
$(\frac{3}{4} + 3u)r = 1.065r$, Al, over origin
$(-\frac{3}{4} + 2)r = 1.25r$, O$_3$, like plane shown, displaced from origin by $2\mathbf{t}_1'$
$(\frac{3}{4} - 3u + 1)r = 1.435r$, Al, displaced from origin by $\mathbf{t}_1'$
$(-\frac{3}{4} + 3u + 2)r = 1.565r$, Al, displaced from origin by $2\mathbf{t}_1'$
$(\frac{3}{4} + 1)r = 1.75r$, O$_3$, rotated, displaced from origin by $\mathbf{t}_1'$
$(-\frac{3}{4} - 3u + 3)r = 1.965r$, Al, over origin
$(\frac{3}{4} + 3u + 1)r = 2.065r$, Al, displaced from origin by $\mathbf{t}_1'$
$(-\frac{3}{4} + 3)r = 2.25r$, O$_3$, like plane shown, over origin
$(\frac{3}{4} - 3u + 2)r = 2.435r$, Al, displaced from origin by $2\mathbf{t}_1'$
$(-\frac{3}{4} + 3u + 3)r = 2.565r$, Al, over origin
$(\frac{3}{4} + 2)r = 2.75r$, O$_3$, rotated, displaced from origin by $2\mathbf{t}_1'$
$(-\frac{3}{4} - 3u + 4)r = 2.965r$, Al, like first plane listed

planes in succession are O$_3$, two planes of aluminums, then another O$_3$, and so on. This succession is better shown in Fig. 3-14, in which we look at the crystal along the direction $-\mathbf{t}_1'$, perpendicular to the threefold axis. In the repeating length of the crystal along the z direction, which is the ordinate in Fig. 3-14, we have six layers of oxygens, twelve of aluminums. We see from the figure that two adjacent aluminum layers are only slightly separated. The way in which the oxygens hold together the successive layers of aluminums is well shown in Fig. 3-14. Each aluminum atom is surrounded by six oxygens, each oxygen is bonded to three aluminums.

A number of important oxides crystallize in the corundum structure, most of them oxides of transition metals, such as Fe$_2$O$_3$, hematite. A particularly interesting case is Cr$_2$O$_3$, for it mixes in all proportions with Al$_2$O$_3$, the Al and Cr atoms freely taking each other's place in the lattice. This is not surprising, since the atomic radii, 1.25 A for Al and 1.40 A for Cr, are not very different. These mixed crystals are ruby; the more chromium they contain, the redder they are. The particular importance

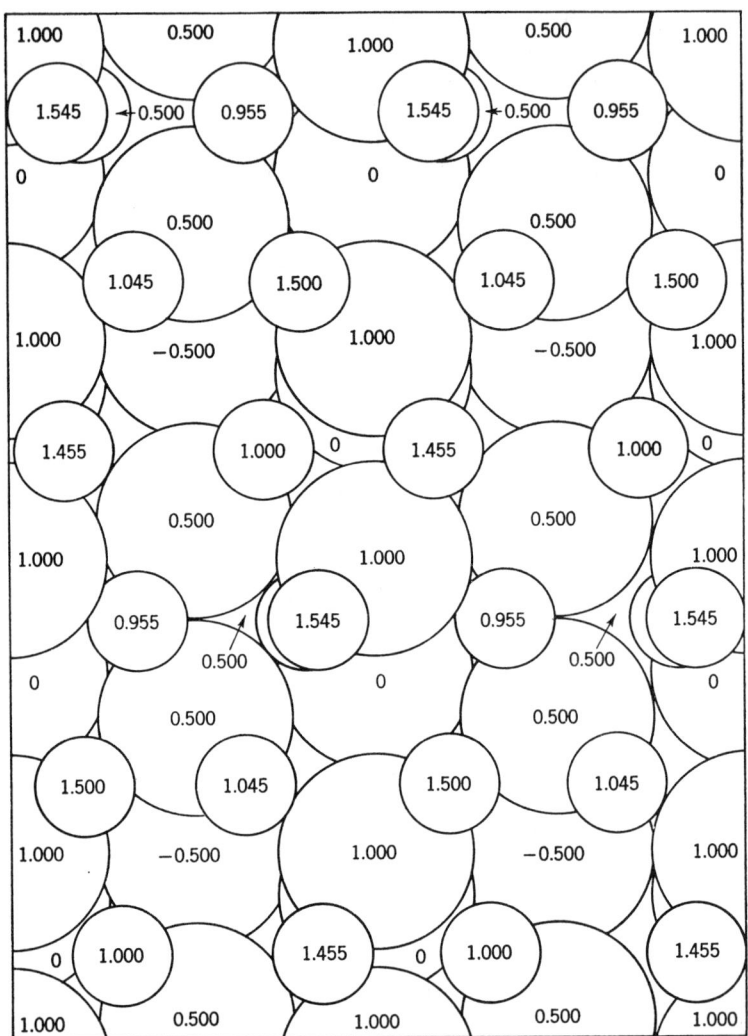

FIG. 3-14. The corundum structure, Al_2O_3, viewed along the direction $-t_1'$, perpendicular to the trigonal axis. The height of the figure is the repeating length along z, or $3r$. The heights of the atoms are shown as multiples of s, the component of the vectors t_1, t_2, t_3 perpendicular to the z axis. Larger spheres, Al; smaller, O.

of the ruby does not lie in its use as a gem (for at present large synthetic crystals can be prepared), but rather in the fact that it has found wide application in the optical maser, a device which we shall take up in a later volume.

We shall next discuss the ilmenite structure, named for ilmenite,

FeTiO$_3$. This structure has a close resemblance to corundum, half the aluminums being replaced by Fe, half by Ti. The projection along the trigonal axis looks very much like Fig. 3-13. As for the arrangement of successive layers, the situation is similar to that of corundum, given in Table 3-4. First there are two iron layers, then O$_3$, then two titaniums, then another O$_3$, and so on. On account of the lower symmetry resulting from the two types of metallic atoms, however, the space group is different. It is C_{3i}^2 ($R\bar{3}$). This has the same rhombohedral Bravais lattice as the group D_{3d}^6 ($R\bar{3}c$) which we have just been discussing, but its point group includes only rotations through 120° and 240°, and the identity, plus these operations combined with an inversion. In terms of ξ, η, ζ, the operations are those which transform ξ, η, ζ into itself; η, ζ, ξ; ζ, ξ, η; $-\xi$, $-\eta$, $-\zeta$; $-\eta$, $-\zeta$, $-\xi$; $-\zeta$, $-\xi$, $-\eta$. The space group is made up out of the operations of this point group and the translation group, without nonprimitive translations. The atomic positions in ilmenite, in terms of ξ, η ζ, are given by the values u, u, u; $-u$, $-u$, $-u$, for Fe, where $u = 0.358$; v, v, v; $-v$, $-v$, $-v$ for Ti, where $v = 0.142$; and general positions $\pm(\xi,\eta,\zeta; \eta,\zeta,\xi; \zeta,\xi,\eta)$ for 0, where $\xi = 0.555$, $\eta = -0.055$, $\zeta = 0.250$.

A number of important oxides have the ilmenite structure. One of the particularly interesting features of them is that two of them, namely, LiNbO$_3$ and LiTaO$_3$, show ferroelectric properties, having permanent electric dipole moments, resembling in this respect BaTiO$_3$.

3-6. Additional Trigonal, Hexagonal, and Orthorhombic Structures. In this section we shall group together several additional trigonal and hexagonal structures, of moderate importance, as well as one orthorhombic structure. We shall start with the structure of AlF$_3$, found also with the halides of a number of trivalent metals. The space group is D_3^7 ($R32$), the Bravais lattice being rhombohedral, the point group D_3, with no nonprimitive translations. The six operations of the point group are those in which the quantities ξ, η, ζ are transformed, respectively, into ξ, η, ζ; ζ, ξ, η; η, ζ, ξ; $-\eta$, $-\xi$, $-\zeta$; $-\zeta$, $-\eta$, $-\xi$; $-\xi$, $-\zeta$, $-\eta$. The aluminum atoms are found at positions u, u, u; $-u$, $-u$, $-u$, where for AlF$_3$ $u = 0.237$. There are two types of fluorines, the first being found at 0, v, $-v$; $-v$, 0, v; v, $-v$, 0, where $v = 0.430$; and the second being found at ½, w, $-w$; $-w$, ½, w; w, $-w$, ½, where $w = 0.070$. For AlF$_3$, $a = 5.03$ A, $\alpha = 58°31'$. The structure is shown in Fig. 3-15. As we can see, each aluminum atom is surrounded by six fluorines, in approximate octahedral arrangement. As in the structures we have been taking up in the preceding section, there are successive vertical layers, which in this case alternate between layers of aluminums and layers with three times as many fluorines. There are twelve such layers, six of aluminum and six of fluorine, in the repeating length along the z axis. Half of these layers are shown in the figure.

Next we shall take up two types of rare earth compounds, of which LaOF and La_2O_3 are examples. The structure of LaOF is shown in Fig. 3-16, and that of La_2O_3 in Fig. 3-17. In each of them one sees the large lanthanum atoms, approximately closely packed in a plane, with the oxygen and fluorine atoms in the interstices. When we examine the

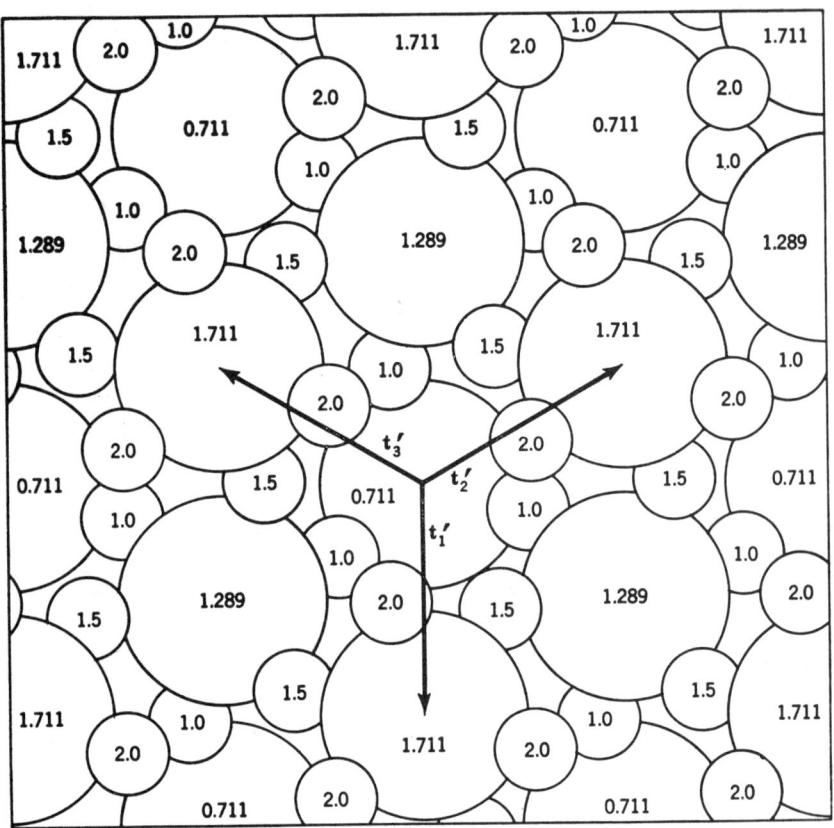

FIG. 3-15. The AlF_3 structure, looking down along the threefold axis. Larger spheres, Al; smaller, F. Heights are represented as multiples of r, the z component of t_1, t_2, t_3.

details of the structures, however, we see that they are quite different. In the first place, the space groups are different. LaOF has the space group D_{3d}^5 ($R\bar{3}m$), which we have already met with the arsenic structure in Sec. 2-6. La_2O_3 has the space group D_{3d}^3 ($P\bar{3}m1$), which we have met in our discussion of the CdI_2 structure in Sec. 3-3. The greatest difference between the two cases is that the first has the rhombohedral Bravais lattice, while the second has the hexagonal lattice. This results in a much more complicated succession of planes of atoms in LaOF than in La_2O_3.

In the LaOF structure, all atoms are in special positions

$$\pm u(\mathbf{t}_1 + \mathbf{t}_2 + \mathbf{t}_3)$$

where $u = 0.242$ for La, 0.122 for F, and 0.370 for O. The length of the

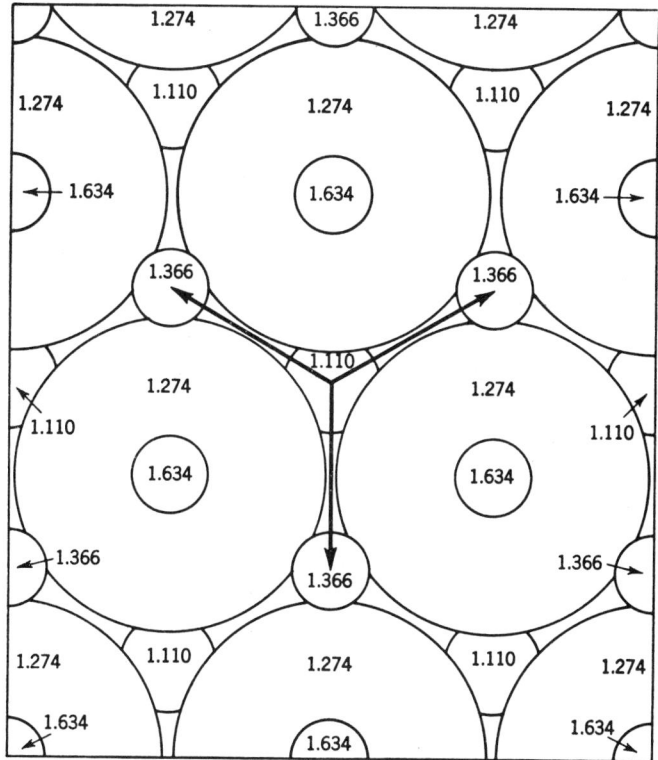

Fig. 3-16. The LaOF structure, looking down along the threefold axis. Largest spheres, La; next smaller, O; smallest, F. The heights are represented as multiples of r, the z component of $\mathbf{t}_1$, $\mathbf{t}_2$, and $\mathbf{t}_3$. The slightly larger spheres at height 1.110 represent oxygen, and the smaller ones at 1.366 and 1.634 are fluorines.

vectors, a, is 7.13 A, the angle α is $33°0'$, so that $s = 2.34$ A, $r = 6.75$ A. As in the other cases we have met with the rhombohedral Bravais lattice, it is convenient to make a table showing the arrangement of successive layers; this is given in Table 3-5. We find, as in the case of the corundum structure given in Table 3-3, that there are eighteen different layers in the repeating distance. Each lanthanum atom in the plane shown in Fig. 3-16 has a fluorine directly above it, distant by the amount 2.41 A,

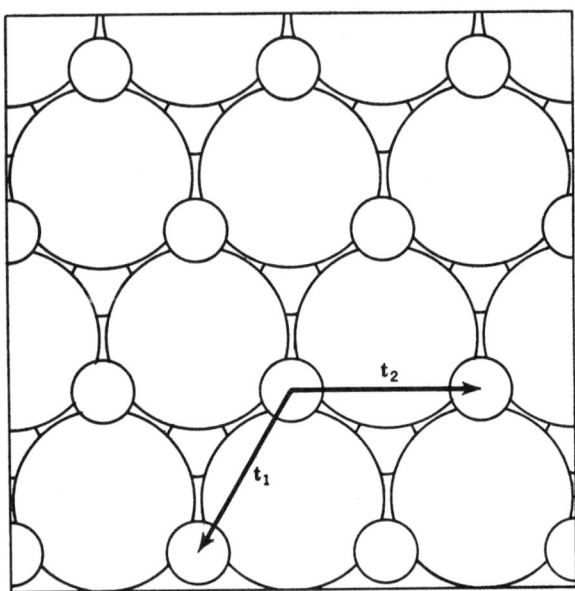

Fig. 3-17. The La_2O_3 structure, looking down along the hexagonal axis. Larger spheres, La; smaller, O. The unit vectors t_1 and t_2 of the hexagonal Bravais lattice are shown. The heights of the atoms as multiples of c, the length of the vector t_3, are as follows: the lanthanums, 0.77; the lower oxygens, 0.63; the upper oxygens, 1.00.

Table 3-5
Arrangement of successive layers in LaOF crystal.
$3u_{La} = 0.726, 3u_F = 0.360, 3u_O = 1.110$

Height

$(1.110 - 1)r = 0.110r$, O, displaced by $-t_1'$
$(-0.726 + 1)r = 0.274r$, La, displaced by t_1'
$0.360r$, F, over origin
$(-0.366 + 1)r = 0.634r$, F, displaced by t_1'
$0.726r$, La, over origin
$(-1.110 + 2)r = 0.890r$, O, displaced by $2t_1'$
$1.110r$, O, over origin
$(-0.726 + 2)r = 1.274r$, La, displaced by $2t_1'$
$(0.366 + 1)r = 1.366r$, F, displaced by t_1'
$(-0.366 + 2)r = 1.634r$, F, displaced by $2t_1'$
$(0.726 + 1)r = 1.726r$, La, displaced by t_1'
$(-1.110 + 3)r = 1.890r$, O, displaced by $3t_1'$ (that is, over origin)
$(1.110 + 1)r = 2.110r$, O, displaced by t_1'
$(-0.726 + 3)r = 2.274r$, La, displaced by $3t_1'$ (over origin)
$(0.366 + 2) = 2.366r$, F, displaced by $2t_1'$
$(-0.366 + 3)r = 2.634r$, F, displaced by $3t_1'$ (over origin)
$(0.726 + 2)r = 2.726r$, La, displaced by $2t_1'$
$(-1.110 + 4)r = 2.890r$, O, displaced by $4t_1'$ (that is, by t_1')

Sec. 3-6] SPACE GROUPS FOR STRUCTURES OF COMPOUNDS

and three fluorines surrounding it, slightly above it, distant by the amount 2.42 A (as compared with the sum of atomic radii, 2.40 A). Similarly, it has three oxygens slightly below, at distance 2.59 A, and one directly below, also at distance 2.59 A (as compared with the sum of radii, 2.50 A). Half the lanthanum layers are like this; the other half have the oxygens above and the fluorines below. Each lanthanum atom, then, has eight electronegative neighbors. Also we notice that the lanthanum atoms themselves are practically close enough to be bonding.

In the La_2O_3 structure, the situation is described in terms of hexagonal vectors t_1, t_2, t_3, instead of rhombohedral. The lanthanums are at special positions $\frac{1}{3}$, $\frac{1}{3}$, u, and $\frac{2}{3}$, $\frac{1}{3}$, $-u$, where $u = 0.23$, and there are two types of oxygens, one at the origin, the other at the positions $\frac{1}{3}$, $\frac{2}{3}$, v, and $\frac{2}{3}$, $\frac{1}{3}$, $-v$, where $v = 0.63$. The lengths of the vectors t_1 and t_2, namely, a and c, are 3.95 A and 6.15 A, respectively. There are now only five layers of atoms in the repeating distance. We have a layer of oxygens at the origin, a layer of lanthanums at $0.23c$, a layer of oxygens at $(1.00 - 0.63)c = 0.37c$, another layer of oxygens at $0.63c$, and a layer of lanthanums at $(1.00 - 0.23)c = 0.77c$.

Each of the lanthanum atoms has seven oxygen neighbors: if we examine the figure, we have three at height $1.00c$, whose distance from the lanthanum is 2.68 A; three at $0.63c$, at a distance of 2.45 A; and one directly below it, not visible in the figure, at distance 2.46 A. These values are to be compared with the sum of atomic radii, 2.50 A. Each of the oxygens like that shown in the figure at $1.00c$ has three lanthanum atoms bonded to it, as shown, and each of those like that at $0.63c$ has four lanthanums, three being shown in the figure, and the fourth being directly below it.

The next structure which we shall consider is that of AlB_2, which has the same space group D_{3d}^3 ($P\bar{3}m1$) which we have just met for La_2O_3, as well as earlier for CdI_2. This is shown in Fig. 3-18, where we show two layers, the lower one just showing borons, the upper one showing aluminums. This is a very simple structure, with the aluminums at the origin of the hexagonal lattice, the borons at the special positions $\frac{1}{3}$, $\frac{2}{3}$, u and $\frac{2}{3}$, $\frac{1}{3}$, $-u$, where $u = \frac{1}{2}$. As we can see, each aluminum is surrounded closely by twelve borons, six in a regular hexagon below it, six more above it. The borons are bonded into a hexagonal structure like a layer of graphite. It is not surprising that, with this bonding between borons, this crystal resembles a metal or alloy, rather than a salt.

The last structure of the trigonal or hexagonal system which we shall take up is that of Na_3As, which has the space group D_{6h}^4 ($P6/mmc$), the same one as the hexagonal close-packed structure of elements. The atomic arrangement, however, shown in Fig. 3-19, is quite different. In this crystal, we have sodiums at two types of sites, the first being 0, 0, $\frac{1}{4}$ and 0, 0, $\frac{3}{4}$, and the second being $\pm(\frac{1}{3}, \frac{2}{3}, u; \frac{2}{3}, \frac{1}{3}, u + \frac{1}{2})$, where

$u = 0.583$. The arsenics are at the sites $\frac{1}{3}, \frac{2}{3}, \frac{1}{4}$ and $\frac{2}{3}, \frac{1}{3}, \frac{3}{4}$. The values of a and c are 5.09 A and 8.98 A, respectively. We see that in planes perpendicular to the threefold axis we have a structure of sodiums and arsenics, each arsenic being bonded to three sodiums in this plane, and each sodium to three arsenics. The bonding out of the plane is more complicated, however. It is better shown in Fig. 3-20, in which we make a cut perpendicular to the plane of the paper in Fig. 3-19, passing through the origin, and cutting through the row of atoms which appears

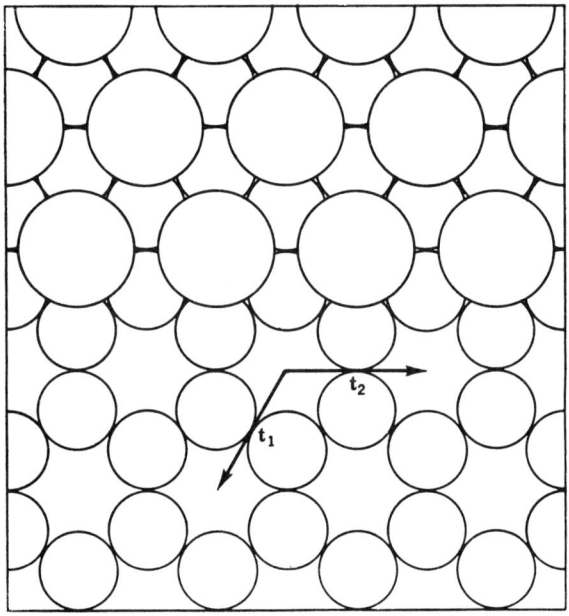

FIG. 3-18. The AlB_2 structure. The layer of borons shown, the smaller spheres, are at height $c/2$, where c is the length of the vector $\mathbf{t}_3$ in the hexagonal Bravais lattice. The aluminums, shown only in the upper half of the diagram, are at height c.

vertically in Fig. 3-19. When we examine Fig. 3-20, we find that each arsenic atom is really bonded not merely to three sodiums in the same horizontal plane, but also to a sodium directly above and another directly below, making a total of five. In addition, there are six sodiums, three in a plane somewhat above and three in a plane somewhat below, at only slightly greater distance from the arsenics, making a total of eleven sodium neighbors for each arsenic. The distance of the three in the same horizontal plane is 2.93 A; of the ones directly above and directly below,

2.97 A; and of the more distant neighbors, 3.30 A. The sum of atomic radii is 2.95 A, showing that the last six sodiums are rather far off to be considered to be bonding.

One remarkable feature of this structure, which is shown in Fig. 3-20, is the close approach to each other of some of the sodium atoms. This is only 2.98 A, whereas the sum of the radii is 3.60 A. The value of 2.98 A is slightly less than the interatomic distance in the Na_2 molecule, considerably less than is found in other crystals; in the sodium crystal

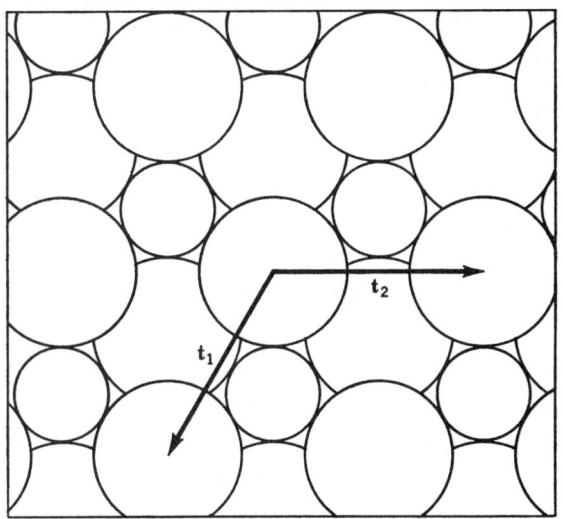

FIG. 3-19. The Na_3As structure, looking down along the threefold axis. The upper layer of sodiums (the larger spheres) and the arsenics are at height $0.75c$, where c is the length of the vector t_3, and the lower layer of sodiums is at height $0.583c$.

the interatomic distance is 3.72 A. We shall comment in the next chapter on some other cases in which alkali atoms are surprisingly close to each other in some types of crystals.

We shall describe only one additional structure, from the orthorhombic system. The structure is that of aragonite, a form of $CaCO_3$, which has the space group D_{2h}^{16} (*Pnma*), with a simple orthorhombic Bravais lattice. We recall that the eight operations of the point group D_{2h} are those in which x, y, or z can change sign. Most of these operations are associated with nonprimitive translations in this space group. We may set up the operations of the space group in the same sort of way which we used in Eq. (2-22) for the similar space group D_{2h}^{18} (*Cmca*), which we met with the

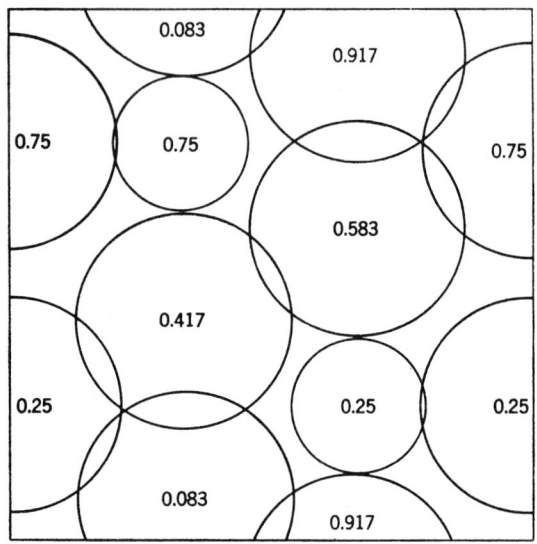

FIG. 3-20. Section of the Na$_3$As structure, along plane normal to the paper, and to the vector t_2, in Fig. 3-19, cutting through the central atom in that figure. The values of z for the various atoms, as multiples of c, are indicated in the figure; the height of the figure is c, extending from $z = 0$ to $z = c$. The sodium atoms (larger spheres) are represented by circles which are allowed to overlap, to indicate that the sodium-sodium distance in this crystal is less than the sum of the atomic radii.

iodine structure. We have

$$\begin{aligned}
\{R_1|\mathbf{R}_n\}\psi(x,y,z) &= \psi(x + n_1 a,\ y + n_2 b,\ z + n_3 c) \\
\{R_2|\mathbf{R}_n\}\psi(x,y,z) &= \psi[x + (n_1 + \tfrac{1}{2})a, \\
&\qquad -y + (n_2 + \tfrac{1}{2})b,\ -z + (n_3 + \tfrac{1}{2})c] \\
\{R_3|\mathbf{R}_n\}\psi(x,y,z) &= \psi[-x + n_1 a,\ y + (n_2 + \tfrac{1}{2})b,\ -z + n_3 c] \\
\{R_4|\mathbf{R}_n\}\psi(x,y,z) &= \psi[-x + (n_1 + \tfrac{1}{2})a,\ -y + n_2 b,\ z + (n_3 + \tfrac{1}{2})c] \\
\{R_5|\mathbf{R}_n\}\psi(x,y,z) &= \psi(-x + n_1 a,\ -y + n_2 b,\ -z + n_3 c) \qquad (3\text{-}7)\\
\{R_6|\mathbf{R}_n\}\psi(x,y,z) &= \psi[-x + (n_1 + \tfrac{1}{2})a, \\
&\qquad y + (n_2 + \tfrac{1}{2})b,\ z + (n_3 + \tfrac{1}{2})c] \\
\{R_7|\mathbf{R}_n\}\psi(x,y,z) &= \psi[x + n_1 a,\ -y + (n_2 + \tfrac{1}{2})b,\ z + n_3 c] \\
\{R_8|\mathbf{R}_n\}\psi(x,y,z) &= \psi[x + (n_1 + \tfrac{1}{2})a,\ y + n_2 b,\ -z + (n_3 + \tfrac{1}{2})c]
\end{aligned}$$

The atoms are then found at the following positions. The Ca atoms are at special positions

$$\pm \left[a\xi \mathbf{i} + \frac{b}{4}\mathbf{j} + c\zeta \mathbf{k},\ (\tfrac{1}{2} + \xi)a\mathbf{i} + \frac{b}{4}\mathbf{j} + (\tfrac{1}{2} - \zeta)c\mathbf{k} \right] \qquad (3\text{-}8)$$

where $\xi = \tfrac{3}{4}$, $\zeta = 0.417$. The C atoms are at special positions like those of Eq. (3-8), with $\xi = 0.417$, $\zeta = \tfrac{3}{4}$. One set of oxygens is at special positions like those of Eq. (3-8), with $\xi = 0.417$, $\zeta = 0.583$. The remaining eight oxygens are at general positions, one of them being $a\xi\mathbf{i} + b\eta\mathbf{j} + c\zeta\mathbf{k}$, with $\xi = 0.417$, $\eta = 0.48$, $\zeta = 0.83$, and the others derived from this position by the operations of Eq. (3-7). We have $a = 5.72$ A, $b = 4.94$ A, $c = 7.94$ A for the lengths of the three vectors $\mathbf{t}_1$, $\mathbf{t}_2$, $\mathbf{t}_3$ along x, y, z, respectively.

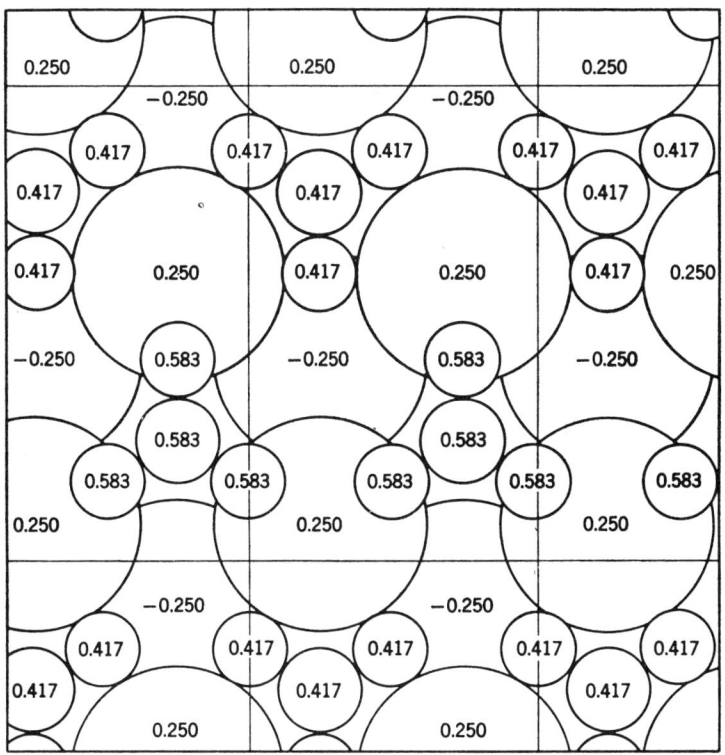

Fig. 3-21. The aragonite structure, $CaCO_3$. Large spheres, Ca; next smaller, C; smallest, O. The yz plane is shown, and the rectangle included within the diagram is the section of the orthorhombic unit cell in the yz plane. The heights of the atoms are given as multiples of a, the magnitude of the vector $\mathbf{t}_1$, along x.

We note from these positions that the carbon and oxygen atoms all have the same values of x. In other words, as the structure is described, the planes of the triangular carbonate groups are planes $x = $ constant. Hence it is useful to show a diagram of the structure looking down along the x direction, and this is done in Fig. 3-21. It is interesting to compare this figure with Fig. 3-12, showing the alternative calcite structure for the same substance $CaCO_3$. The arrangement of calcium atoms in the plane is essentially the same in both structures. The carbonate groups are also essentially identical in the two cases, but they are arranged more symmetrically with respect to the calciums in the aragonite structure. However, there are two types of carbonates, oppositely oriented, in a layer of aragonite, while in calcite all carbonates in a plane are equivalent, but there are two sorts of planes with different orientations. There is also a definite difference in the stacking of the layers. In aragonite, there are only two sorts of calcium layers, which alternate as shown in the figure, while in calcite there are three sorts of calcium layers, and two sorts of carbonate layers, leading to twelve layers in all, six of calcium and six of carbonate, in the repeating distance in the crystal.

4
Atomic Radii and the Chemical Bond

4-1. Ionic and Atomic Radii. The reader who is familiar with the chemical or crystallographic literature has probably been very much upset at seeing the diagrams of crystal structure in Chap. 3, for they show electropositive atoms like the alkalies as large spheres, electronegative atoms like oxygen or the halogens as small spheres, directly opposite to most crystal-structure diagrams, including those in the works of Wyckoff and Wells, quoted earlier. In the present chapter we shall explain the significance of this apparent discrepancy, and we shall see that our radii, and those commonly used, are not really in conflict, but refer to different things. We do point out one fact at the outset, however. Our radii, in which in particular oxygen and halogen atoms have small radii, allow us to give diagrams of structures containing oxygen and the halogens which are rather easier to understand than the conventional ones, in which the large spheres representing oxygen and the halogens tend to obscure the other atoms. This practical advantage of the present radii in drawing structural diagrams is a very different matter, however, from understanding the physical significance of the radii.

We have already mentioned that as soon as x-ray data on crystal structures began to accumulate, it became clear that atomic distances could be approximately regarded as sums of radii. This was pointed out in 1920 by W. L. Bragg,[1] who derived a set of radii in rather close agreement with our values of Table 3-1 and compared the sums of radii with observed interatomic distances in a good many crystals, similar to those included in Appendix 1, but much less extensive. To indicate the closeness of agreement between our radii of Table 3-1 and Bragg's values, we give a comparison of the two sets of radii in Table 4-1. The differences come partly from the fact that we are fitting many more compounds, partly from revised distances which have been determined since the time of Bragg's paper. Since, as we have mentioned earlier, the average dis-

[1] W. L. Bragg, *Phil. Mag.*, **40**:169 (1920).

crepancy between the sum of radii and the observed distance is about 0.1 A, it has not seemed worthwhile to set up our radii more closely than in steps of 0.05 A, rather than 0.01 A, as the other writers whom we shall mention later have done.

Soon after Bragg's suggested radii of 1920, Wyckoff[1] and others examined the data then available, showed that many measurements could be explained in terms of the additivity of radii, but that many others could

Table 4-1
Comparison of Bragg's atomic radii with those of Table 3-1

Element	Bragg	Table 3-1	Element	Bragg	Table 3-1
Li	1.50	1.45	Ni	1.35	1.35
Be	1.15	1.05	Cu	1.37	1.35
C	0.77	0.70	Zn	1.32	1.35
N	0.65	0.65	As	1.26	1.15
O	0.65	0.60	Se	1.17	1.15
F	0.67	0.50	Br	1.19	1.15
Na	1.77	1.80	Rb	2.25	2.35
Mg	1.42	1.50	Sr	1.95	2.00
Al	1.35	1.25	Ag	1.77	1.60
Si	1.17	1.10	Cd	1.60	1.55
S	1.02	1.00	Sn	1.40	1.45
Cl	1.05	1.00	Sb	1.40	1.45
K	2.07	2.20	Te	1.33	1.40
Ca	1.70	1.80	I	1.40	1.40
Ti	1.40	1.40	Cs	2.37	2.60
Cr	1.40	1.40	Ba	2.10	2.15
Mn	1.47	1.40	Tl	2.25	1.90
Fe	1.40	1.40	Pb	1.90	1.80
Co	1.37	1.35	Bi	1.48	1.60

Note: Bragg gives two radii for Cr and Mn, giving the value 1.17 for compounds which he designates as "electronegative," in contrast to the values given in the table for other compounds.

not. A long history has followed in which successively more elaborate modifications of the radii have been made, with the aim of bringing about closer agreement between the experiment and some interpretation of atomic radii. This history is well outlined by Pauling,[2] who has played a leading part in the development. The main result of this development has been to set up different sorts of radii to describe different sorts of crystals, with rules for passing from one sort of radius to another, thus

[1] R. W. G. Wyckoff, "The Structure of Crystals," Chemical Catalog Company, Inc., New York, 1924, p. 399. M. L. Huggins, *Phys. Rev.*, **28**:1086 (1926).

[2] L. Pauling, "The Nature of the Chemical Bond," 3d ed., Cornell University Press, Ithaca, N.Y., 1960. See particularly chaps. 7, 11, and 13.

introducing a great deal of complication, but in return getting closer agreement with experiment. With all this elaboration, the simple fact has rather been lost sight of, that by the straightforward use of a single set of radii, one can do fairly well, as the extensive data of Appendix 1 show.

The first step which led to the elaboration of the concept of atomic radii was the introduction of a different set of radii for use with ionic crystals. There had been before 1920 a large amount of development of the theory of ionic crystals, particularly the alkali halides, and to some extent of the corresponding oxides, sulfides, selenides, and tellurides of the alkaline earths, by Born and many coworkers. It seemed natural to ask if one could not get better agreement by using different radii for the ionic crystals from those used for metals. Landé, Fajans, Grimm, and Herzfeld[1] by 1920 had investigated the additivity of ionic radii in these materials, showing that to a first approximation one could set up radii for positive and negative ions, such that their sums would reproduce as accurately as possible the interionic distances in the appropriate ionic compounds, and had considered the experimental departures from additivity and their probable explanations.

It is of course trivial, given a set of experimental interatomic distances such as we have in Appendix 1, which were available for these ionic compounds by 1920, to subtract one distance from another, such as subtracting all the chlorides from the corresponding bromides, to find if the radius differences between chlorine and bromine are constant. Thus, we have $LiBr - LiCl = 2.75 - 2.57 = 0.18$ A, and similarly for the Na, K, Rb, and Cs salts, we find values of 0.16, 0.15, 0.14, and 0.14 A, for the size of the bromine ion or atom, minus the size of the chlorine (in our Table 3-1, we have struck an average by using a difference of 0.15 in this case). One finds equally good agreement in each case, and in this way one can find the radii of all the alkali and halogen ions, with only one additive arbitrary constant: we can increase all the alkalies, and decrease all the halogens, by the same amount, without changing the sums. Similarly, one can find the radii of all the divalent positive and negative ions, with another arbitrary constant.

Wasastjerne,[2] in 1923, attempted to determine this additive constant in an ingenious way. In the study of the dielectric constants and refractive indices of crystals, it is possible to assign polarizabilities to each of the ions, such that the sum of the polarizabilities of the various ions leads to the polarizability, and hence the dielectric constant and refractive index, of the crystal as a whole. Study of the empirical material relating

---

[1] A. Landé, *Z. Physik*, **1**:191 (1920). K. Fajans and H. Grimm, *Z. Physik*, **2**:299 (1920). K. Fajans and K. F. Herzfeld, *Z. Physik*, **2**:309 (1920).

[2] J. A. Wasastjerne, *Soc. Sci. Fennica, Commentatienes Phys. Math.*, **38**:1 (1923).

to the dielectric constants of a set of compounds such as the alkali halides, or the divalent ionic crystals, has allowed the determination of the separate polarizabilities of the various ions. But a very simple type of theory suggests that the polarizability of an ion should be approximately proportional to its volume. One of the simplest models of a polarizable ion is a perfectly conducting sphere: if placed in an external electric field, it acquires a dipole moment proportional to the field, and the constant of proportionality, the polarizability, is a universal constant times the volume of the sphere. Wasastjerne assumed this simple relation, and that allowed him to find, quite independent of the x-ray evidence, quantities proportional to the volumes, and hence by taking cube roots, quantities proportional to the radii, of the ions.

He then examined these values, and found that if he chose the constant of proportionality properly, he could arrange it so that the ionic radii, found from the polarizabilities, would add to give the interatomic separations. Of course, a little adjustment was necessary to get agreement, but by examination of the available data, both from refractive indices and from x rays, he arrived at a satisfactory set of radii. These radii, together with several other similar sets, are given in Table 4-2. Before discussing Wasastjerne's radii, we shall introduce other sets of similar radii suggested shortly after him by Goldschmidt,[1] Pauling,[2] and Zachariasen.[3]

Goldschmidt was a geologist who was interested in correlating data on interatomic distances in a large number of compounds whose structures had been determined by x-ray methods. He started with Wasastjerne's radii, and examined them to see if small changes would not improve the agreement between measured distances and the sums of radii. He found that this was the case, and his radii are not different in principle from those of Wasastjerne, but are adjusted to fit many more crystals.

Pauling pointed out particularly the fact that a given atomic or ionic radius must not be assumed to be a constant, but actually must be a function of the coordination number (the number of ions of the opposite sign surrounding a given ion), the radius ratio (ratio of radius of positive to negative ion, or vice versa), and the charge on the ion. He worked out so-called univalent radii, radii which would hypothetically be found if ions actually with different charges were artificially assumed to have a standard charge of one unit. Thus, he considered the sequence of ions S^{-2}, Cl^{-1}, K^+, Ca^{+2}, all of which have the same number of electrons, 18. With the increasing nuclear charge, we should expect the radii to shrink,

---

[1] V. M. Goldschmidt, *Skrifter Norske Videnskaps-Akad. Oslo*, I, *Mat-Naturv. Kl.*, 1926; *Z. Tech. Phys.*, **8**(7):251 (1927); *Trans. Faraday Soc.*, **25**:253 (1929).

[2] L. Pauling, *Proc. Roy. Soc. (London)*, **A114**:181 (1927). L. Pauling and J. Sherman, *Z. Krist.*, **81**:1 (1932).

[3] W. H. Zachariasen, *Z. Krist.*, **80**:137 (1931).

as we go from S^{-2} to Ca^{+2}. He estimated this shrinkage, by use of simple arguments based on screening constants. This gave him univalent radii, which for the four ions just mentioned he assumed to be 2.19, 1.81, 1.33, and 1.18 A, respectively. In the real crystals, however, the S^{-2} and Ca^{+2} ions will be attracted more strongly than if they had unit negative and positive charges, respectively, and by use of an argument based on this

Table 4-2
Ionic radii of Wasastjerne, Goldschmidt, Pauling, and Zachariasen, compared with atomic radii of Table 3-1. Numbers in parentheses are atomic minus ionic radii

	Atomic	Wasastjerne	Goldschmidt	Pauling	Zachariasen
Li^+	1.45		0.78(0.67)	0.60(0.85)	0.68(0.77)
Na^+	1.80	1.01(0.79)	0.98(0.82)	0.95(0.85)	0.98(0.82)
K^+	2.20	1.30(0.90)	1.33(0.87)	1.33(0.87)	1.33(0.87)
Rb^+	2.35	1.50(0.85)	1.49(0.86)	1.48(0.87)	1.48(0.87)
Cs^+	2.60	1.75(0.85)	1.65(0.95)	1.69(0.91)	1.67(0.93)
Be^{+2}	1.05		0.34(0.71)	0.31(0.74)	0.39(0.66)
Mg^{+2}	1.50	0.75(0.75)	0.78(0.72)	0.65(0.85)	0.71(0.79)
Ca^{+2}	1.80	1.02(0.78)	1.06(0.74)	0.99(0.81)	0.98(0.82)
Sr^{+2}	2.00	1.20(0.80)	1.27(0.73)	1.13(0.87)	1.15(0.85)
Ba^{+2}	2.15	1.40(0.75)	1.43(0.72)	1.35(0.80)	1.31(0.84)
F^-	0.50	1.33(−0.83)	1.33(−0.83)	1.36(−0.86)	1.33(−0.83)
Cl^-	1.00	1.72(−0.72)	1.81(−0.81)	1.81(−0.81)	1.81(−0.81)
Br^-	1.15	1.92(−0.77)	1.96(−0.81)	1.95(−0.80)	1.96(−0.81)
I^-	1.40	2.19(−0.79)	2.20(−0.80)	2.16(−0.76)	2.19(−0.79)
O^{-2}	0.60	1.32(−0.72)	1.32(−0.72)	1.40(−0.80)	1.40(−0.80)
S^{-2}	1.00	1.69(−0.69)	1.74(−0.74)	1.84(−0.84)	1.85(−0.85)
Se^{-2}	1.15	1.77(−0.62)	1.91(−0.76)	1.98(−0.83)	1.96(−0.81)
Te^{-2}	1.40	1.91(−0.51)	2.11(−0.71)	2.21(−0.81)	2.18(−0.78)

fact, Pauling assumed that the real radii of S^{-2} and Ca^{+2} should be diminished to 1.84 and 0.99 A, respectively, as given in Table 4-2.

In the matter of coordination number, it is an easily determined fact that the more neighbors an ion has, the larger is its distance from the neighbors. This is seen clearly in Appendix 1 in the examples of MnS and MnSe, which exist in three forms each, the NaCl, zinc-blende, and wurtzite structures. In the NaCl structure each ion has six nearest neighbors, while in the other two it has four. We see that the interatomic distance shrinks from 2.61 A for MnS in the NaCl structure to 2.43 or 2.41 A in the other two structures, and similarly the distance in MnSe shrinks from 2.72 A in the NaCl structure to 2.52 A in the other two. It is obvious that under these circumstances no single set of radii can hope to reproduce all the distances. Pauling proposed analytical methods of approximating for this effect, and his radii are set up for the coordination number of 6, as found in the NaCl structure. On the contrary, in our atomic radii, we have disregarded this effect, but in each case have tried to make a compromise in determining the radii. For the compounds just cited, a glance at the numbers will show that we have chosen the Mn radius to fit the tetrahedral coordination, rather than the sixfold case.

Zachariasen proceeded along much the same lines as Pauling, but made use of somewhat more experimental material, and followed Goldschmidt's empirical radii wherever possible. Zachariasen's radii, from Table 4-2, like Pauling's, are meant to be used with sixfold coordination. This has one specific effect: it means that when we are considering the crystals having the CsCl structure, with eightfold coordination, we must not use the radius of Cs^+ tabulated, but a larger value, corrected according to the theories of Pauling or Zachariasen for the larger coordination number.

We may now examine the ionic radii of Table 4-2, and see how they compare with our atomic radii of Table 3-1, and what sort of agreement with experiment they lead to for the sums of radii. To aid in this comparison, we have given in Table 4-2 the differences between atomic radii and the various sets of ionic radii. We find that the ionic radii of the positive ions are much smaller than the atomic radii, while the ionic radii of the negative ions are correspondingly larger than the atomic radii. As a rough approximation, we can say that the ionic radii of the electropositive elements from Table 4-2 are about 0.85 A smaller than the atomic radii, while the ionic radii of the electronegative elements are about 0.85 A larger than the atomic radii. If these differences were exactly constant through the table, the sums of ionic radii would exactly equal the sums of atomic radii, for ionic compounds, and either set of radii would give equally good agreement with experiment. The deviations from this situation are of the order of a few hundredths of an Angstrom.

An entirely different approach to a set of radii arose, not from the study of the alkali halides, but from the crystals such as GaAs, ZnSe, and CuBr, having the zinc-blende or wurtzite structures, with tetrahedral coordination. It was obvious very early that if we set up atomic radii by using half the interatomic distances in the elements, and added them for these compounds, the result gave a good agreement with experiment. In this way, by modifying the radii so derived to fit a maximum number of compounds, Huggins and Pauling[1] derived a set of so-called tetrahedral covalent radii, which would give by their sums the interatomic distances in a large number of tetrahedrally coordinated compounds. These principles are the same ones which we have used in setting up the radii of Table 3-1,

[1] M. L. Huggins, *Phys. Rev.*, **28**:1086 (1926). L. Pauling and M. L. Huggins, *Z. Krist.*, **87**:205 (1934).

and which Bragg had earlier used, and as a matter of fact almost all the tetrahedral covalent radii of Huggins and Pauling agree within a few hundredths of an Angstrom with the radii given in Table 3-1. These writers have also extended similar radii to other coordinations, and Pauling in his book lists in addition metallic radii for many metals, which give good sums when used for intermetallic compounds. These radii have close resemblance to our radii of Table 3-1.

With this discussion of some of the principal sets of radii in use, we can see the relation between them and our radii of Table 3-1. For the univalent and bivalent positive and negative ions, our radii differ from the ionic radii by fixed amounts, approximately ± 0.85 A, such that the sums are essentially the same, and we get as good sums of radii as can be obtained from ionic radii. For the other types of compounds, with tetrahedral coordination, we essentially duplicate Pauling's tetrahedral covalent radii. Our radii differ from the existing sets, however, aside from Bragg's radii, in that we can use them with tolerable accuracy for the elements, intermetallic compounds, and other nonionic compounds, as well as for ionic compounds. Since we are not including corrections for coordination number and the other effects considered by Pauling and Zachariasen, we are subject to larger errors in some cases, and cannot hope to reproduce all the distances as accurately as they do. It is worthwhile to point out, however, that the worst errors in our agreement of Appendix 1 come for compounds involving a few elements, among which are Ag and Tl, which seem to be peculiar in their behavior as far as interatomic distances are concerned, and which Pauling and Zachariasen are not able to handle significantly better than we are.

4-2. The Physical Significance of Atomic and Ionic Radii. In the preceding section we have discussed the various sets of atomic and ionic radii from the purely empirical point of view of their success in reproducing the observed interatomic distances. Let us now ask the broader question as to what these radii mean, why the atomic and ionic radii differ as they do, and what connection they have with the wave functions of the atoms and ions, which in the last analysis must determine the radii.

As a first step, we must presumably expect a correlation between the atomic radius and the radius of maximum charge density in the outermost electron shell of the atom. Huggins (*loc. cit.*) investigated such a correlation between his atomic radii, which approximate ours, and the information available regarding dimensions of atomic wave functions when he wrote in 1926, which naturally was very inadequate, and found evidence of correlation. To get up-to-date information regarding such correlation, we give in Table 4-3 a comparison of the atomic radii of Table 3-1, and the radii of the maximum radial charge density in the outermost shells of the atoms, the latter coming from unpublished calculations of D.

Liberman, J. M. Waber, and D. T. Cromer, of the Los Alamos Scientific Laboratory, who have computed self-consistent wave functions for all the atoms, with relativistic corrections, using the approximate treatment of exchange suggested by the present writer, which we shall comment on in Sec. 4-5. The writer is greatly indebted to Dr. Waber for the use of this unpublished information, which for the first time has made it possible to study the dimensions of the heavier elements. Similar very recent calculations of Herman and Skillman[1] also make such study possible, and

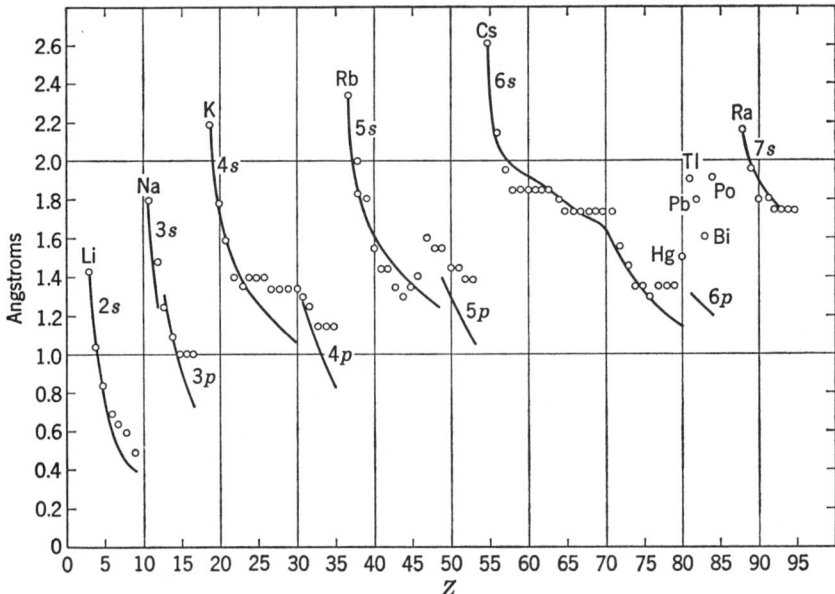

FIG. 4-1. Atomic radii of Table 3-1, indicated by small circles, compared with computed radii of maximum radial charge density in the outermost shells of the atoms, as found by J. M. Waber et al. (unpublished). The computed radii are given by the curves, and the quantum numbers of the outermost atomic shell are indicated.

agree well with those of Waber and collaborators. The results of Table 4-3 are shown graphically in Fig. 4-1.

From Table 4-3, and particularly from Fig. 4-1, we see that there is in fact a very good correlation, and in many cases practically an equality, between the atomic radii of Table 3-1, which were found purely from study of the empirical interatomic distances, before the results of Waber's calculations were available, and the radius of maximum radial charge density in the outermost shell of the atom. This correlation is particularly strik-

[1] F. Herman and S. Skillman, "Atomic Structure Calculations," Prentice-Hall, Inc., Englewood Cliffs, N.J., 1963.

Table 4-3
Comparison of atomic radii of Table 3-1 with radius of maximum radial charge density in the outermost shell in atoms, as computed by Waber et al. Radii are in Angstroms

	Atomic radius from Table 3-1	Waber's radius of electronic shell	Electronic shell		Atomic radius from Table 3-1	Waber's radius of electronic shell	Electronic shell
Li	1.45	1.59	$2s$	In	1.55	1.38	$5p$
Be	1.05	1.04	$2s$	Sn	1.45	1.24	$5p$
B	0.85	0.78	$2p$	Sb	1.45	1.19	$5p$
C	0.70	0.62	$2s$	Te	1.40	1.11	$5p$
N	0.65	0.52	$2s$	I	1.40	1.04	$5p$
O	0.60	0.45	$2s$	Cs	2.60	2.52	$6s$
F	0.50	0.40	$2s$	Ba	2.15	2.06	$6s$
Na	1.80	1.71	$3s$	La	1.95	1.92	$6s$
Mg	1.50	1.28	$3s$	Ce	1.85	1.98	$6s$
Al	1.25	1.31	$3p$	Pr	1.85	1.94	$6s$
Si	1.10	1.07	$3p$	Nd	1.85	1.91	$6s$
P	1.00	0.92	$3p$	Pm	1.85	1.88	$6s$
S	1.00	0.81	$3p$	Sm	1.85	1.85	$6s$
Cl	1.00	0.73	$3p$	Eu	1.85	1.83	$6s$
K	2.20	2.16	$4s$	Gd	1.80	1.71	$6s$
Ca	1.80	1.69	$4s$	Tb	1.75	1.78	$6s$
Sc	1.60	1.57	$4s$	Dy	1.75	1.75	$6s$
Ti	1.40	1.48	$4s$	Ho	1.75	1.73	$6s$
V	1.35	1.40	$4s$	Er	1.75	1.70	$6s$
Cr	1.40	1.45	$4s$	Tu	1.75	1.68	$6s$
Mn	1.40	1.28	$4s$	Yb	1.75	1.66	$6s$
Fe	1.40	1.23	$4s$	Lu	1.75	1.55	$6s$
Co	1.35	1.18	$4s$	Hf	1.55	1.48	$6s$
Ni	1.35	1.14	$4s$	Ta	1.45	1.41	$6s$
Cu	1.35	1.19	$4s$	W	1.35	1.36	$6s$
Zn	1.35	1.07	$4s$	Re	1.35	1.31	$6s$
Ga	1.30	1.25	$4p$	Os	1.30	1.27	$6s$
Ge	1.25	1.09	$4p$	Ir	1.35	1.23	$6s$
As	1.15	1.00	$4p$	Pt	1.35	1.22	$6s$
Se	1.15	0.92	$4p$	Au	1.35	1.19	$6s$
Br	1.15	0.85	$4p$	Hg	1.50	1.13	$6s$
Rb	2.35	2.29	$5s$	Tl	1.90	1.32	$6p$
Sr	2.00	1.84	$5s$	Pb	1.80	1.22	$6p$
Y	1.80	1.69	$5s$	Bi	1.60	1.30	$6p$
Zr	1.55	1.59	$5s$	Po	1.90	1.21	$6p$
Nb	1.45	1.59	$5s$	Ra	2.15	2.04	$7s$
Mo	1.45	1.52	$5s$	Ac	1.95	1.90	$7s$
Tc	1.35	1.39	$5s$	Th	1.80	1.79	$7s$
Ru	1.30	1.41	$5s$	Pa	1.80	1.80	$7s$
Rh	1.35	1.36	$5s$	U	1.75	1.78	$7s$
Pd	1.40			Np	1.75	1.74	$7s$
Ag	1.60	1.29	$5s$	Pu	1.75	1.78	$7s$
Cd	1.55	1.18	$5s$	Am	1.75	1.76	$7s$

ing through the whole series of elements from cesium to platinum, in which the $6s$ electron is the outer electron. The only set of elements in which there is particularly bad correlation is the group from mercury to polonium, in which the $6p$ orbital is coming in as the outermost orbital. It is perhaps significant that some of these elements are among those to which it is very difficult on the basis of empirical evidence to assign a unique radius from the crystal structures.

Table 4-4
Pauling's ionic radii, from Table 4-3, compared with radius of maximum radial charge density in outermost shell of electrons in ions, and ratio of ionic radius to radius of shell

Ion	Ionic radius	Radius of shell	Ratio
Li^+	0.60 A	0.186 A $(1s)$	3.2
Na^+	0.95	0.316 $(2s)$	3.0
K^+	1.33	0.59 $(3p)$	2.2
Rb^+	1.48	0.74 $(4p)$	2.0
Cs^+	1.90	0.92 $(5p)$	2.1
Be^{+2}	0.31	0.138 $(1s)$	2.2
Mg^{+2}	0.65	0.285 $(2s)$	2.2
Ca^{+2}	0.99	0.54 $(3p)$	1.8
Sr^{+2}	1.13	0.69 $(4p)$	1.6
Ba^{+2}	1.35	0.87 $(5p)$	1.6
F^-	1.36	0.40 $(2s)$	3.4
Cl^-	1.81	0.73 $(3p)$	2.5
Br^-	1.95	0.85 $(4p)$	2.3
I^-	2.16	1.04 $(5p)$	2.1
O^{-2}	1.40	0.45 $(2s)$	3.1
S^{-2}	1.84	0.81 $(3p)$	2.3
Se^{-2}	1.98	0.92 $(4p)$	2.2
Te^{-2}	2.21	1.11 $(5p)$	2.0

Next we may consider the ionic radii from Table 4-2. Here we naturally compare these with the radii of the outermost shell of the ion, lacking the outer electrons. When we make this comparison, we find that the ionic radii are of the order of magnitude of twice to three times the radii of the outermost shell of the ion. This comparison is given in Table 4-4. The information regarding the calculated radii is less satisfactory than in Table 4-3, since for most of the ions we do not have the recent calculations which we have from Waber and from Herman and Skillman for the neutral atoms.

The fact, seen in Table 4-4, that the ratio of ionic radius to the radius of the electronic shell is approximately the same for such a sequence as S^{-2}, Cl^{-1}, K^+, and Ca^{+2} is to be expected from the way in which Pauling derived the ionic radii from the radii of the shells. He used an argument based on screening constants, because the wave functions calculated by the self-consistent field were not all available at the time he made his comparison, but we see that the values of Table 4-4, based on self-consistent-field calculations, are in agreement with his general argument.

Now we must correlate the results of these two tables, and find the interpretation of the difference between atomic and ionic radii. The atomic radii can be used in cases where two atoms are held to each other by a covalent bond, or a metallic bond, which is essentially of the same nature. Such a bond, as we know from the molecular structures described in Volume 1, depends on the overlapping of charge in the outer shells of the two atoms being bonded together. We expect this overlapping to be a maximum when the maximum charge densities of the outer shells of the two atoms coincide, that is, when the atoms approach to such a distance that the atomic radii, which as we see from Table 4-3 and Fig. 4-1 are approximately the distances out to the radius of maximum radial charge density, add to give an interatomic distance. This is the general significance of the atomic radii.

The conventional ionic picture of an ionic compound is very different from this. We start, in such a case as KCl, by removing the outer electron from the electropositive element, in this case by removing the $4s$ electron from the potassium atom, and using it to fill in the gap in the outer shell of the electronegative element, in this case the chlorine. The $4s$ electron in potassium, as we see from Table 4-3, has a radius of maximum radial charge density of 2.16 A, whereas the outer shell of the positive ion, which is all that is left when the $4s$ electron is removed, has a radius of only 0.59 A, as we see from Table 4-4. In other words, the potassium positive ion is a very much smaller structure than the neutral potassium atom. On the other hand, the addition of the single electron to the $3p$ shell of chlorine, to complete this shell and convert this to the Cl^- ion, makes only a small increase in the radius of this shell, perhaps of the order of 0.01 A, not given in Tables 4-3 and 4-4 because it is not accurately known.

Once we have made these ions, we then allow them to be bonded together in a crystal by the Coulomb attractions between the oppositely charged ions. It is this attraction that was discussed by Born and others in the early development of the theory of ionic crystals, which we have mentioned earlier. The ions, consisting of closed shells, will have a repulsion for each other at small distances, just like inert gas atoms, in this case argon atoms. From Volume 1, we have seen the nature of this repulsion, identical with that met with two helium atoms. The equilibrium in the crystal is a result of the Coulomb attraction working against this repulsion.

With the amount of attraction produced by the Coulomb forces, the inert gas shells still are quite far apart when the repulsion is great enough to balance the attraction. We see this from Table 4-4, where, in the sequence we are considering, the interatomic distance at equilibrium is something more than twice the sum of the radii of maximum radial charge density in the two ions. In other words, it is necessary only for the tails of the wave functions of the outer electrons to overlap in order to produce enough repulsion of the inert gas shells to balance the Coulomb attractions. This is very much less overlapping than would be found in the case of a covalent bond, where, as we have seen previously, the radii of maximum radial charge density in the two atoms approximately coincide. The difference between the two cases, of course, is that in the case of the covalent bond there are vacancies in the outer shells of the bonding atoms, while with the inert gas shells the outer shells are filled, and the exclusion principle prevents their overlapping.

We can now understand the difference between the two sorts of radii, in such cases as the sequence of atoms from sulfur to calcium. For the electropositive elements, the atomic radius is approximately the actual radius of the valence electron shell. The ionic radius is something like twice the radius of the next inner shell, the outermost one in the inner core of the ion. It is this latter quantity which is about 0.85 A smaller than the radius of the valence electron wave function. For the electronegative element, the added electron or electrons in the outer shell, required to produce the negative ion, cause a slight increase in the size of the shell, but this is a minor effect. The main reason why the ionic radius of the electronegative element is about 0.85 A larger than the atomic radius is the factor of approximately 2 found in Table 4-4: the ionic radius extends out into the tail of the wave function of the outer electrons, whereas the atomic radius extends out only to the maximum charge density in the shell. We see, in other words, that the statement made in the first paragraph of this chapter is justified, namely, that the atomic and ionic radii measure quite different things, and there is no real conflict between them.

There remains, however, the question as to the possibility of using the atomic radii in discussing ionic crystals, which, as we see from Appendix 1, is possible. Here we gain a rather new insight into the nature of ionic bonds, if we consider the case of such a crystal as KCl from an atomic point of view. Instead of considering the crystal from an ionic point of view, as we did in an earlier paragraph, we can start with the separated neutral atoms at infinite separation, which of course will have a much lower energy than the separated ions. We bring them together, and the energy will not begin to decrease until the charge distributions begin to overlap. We shall start to build up overlap charge between the atoms, as in any case of covalent binding. This charge will appear where the outer wave functions of both atoms are large, that is, in the region between the atoms, but near the Cl. Each K atom, with its one outer electron, will form overlap charge near each of its six neighboring Cl's. Since the total charge arising from the K valence electron cannot be greater than one unit, the overlap can contribute a maximum of one-sixth of an electron to each of the six neighboring Cl's. Each of the Cl's can acquire up to one-sixth of an electron from each of its six neighbors, however, so that the net result is that enough charge can build up on each Cl, arising from the 4s electrons of its neighboring K's, to give the Cl one extra negative charge, or produce a Cl⁻ ion. According to this covalent view of the process of binding, no charge need actually move from the K to the Cl atom to produce the ions: the radius of the outer shell of the K atom is such that the charge will automatically have moved over to the Cl, merely by the fact that the atoms have moved to the observed equilibrium distance.

The true wave function, of course, can be built up as a linear combination of many different types of functions, of which we have the ionic and the covalent types we have just indicated. The situation is not unlike that of the diatomic molecule LiH, studied in detail in Volume 1, Chap. 7, in which it was shown how such ionic and covalent basis functions actually combined by configuration interaction (or, according to other language, by resonance) to produce the final wave function in such a case. We may expect, however, that in the case of KCl the final result will be close to the ionic case. This is shown by the calculation of Löwdin,[1] on the binding energy of the KCl crystal, in which he has started with an ionic model of the crystal, has calculated its energy by proper wave-mechanical methods, and gets very acceptable values for the binding energy.

We see by this example that even in a typically ionic compound we can make a very plausible interpretation of a covalent sort, indicating that such components enter into the description of the wave function by means of configuration interaction. There are many more cases in which an ionic picture can be given, but in which presumably the covalent parts of the wave function are much more important. For instance, we may consider the sequence CuBr, ZnSe, GaAs, and the element Ge, which we have mentioned earlier. We could interpret these in an ionic way, as made from Cu^+ and Br^-, Zn^{+2} and Se^{-2}, Ga^{+3} and As^{-3}, and presumably this interpretation is not unreasonable for CuBr, which is not entirely unlike an alkali halide. By the time we get to the others, however, all

[1] P.-O. Löwdin, *Arkiv Mat. Astron. Fys.*, vol. A35, no. 9 (1947), no. 30 (1948); *Advan. Phys.*, **5**:1 (1956).

experimental indications tell us that there is very little ionic character. By the time we get to GaAs, its properties are very similar to those of germanium. Both are semiconductors, with almost identical energy-band structures, and a bonding mechanism like diamond, definitely covalent. In other words, as we go along such a series, the covalent part of the wave function gets more and more important.

The real significance of the fact that we can get the interatomic distances by adding atomic radii, as in Appendix 1, seems then to be that even in the typical ionic compounds, the covalent contribution to the wave function is large enough to be determining in fixing the interatomic distances. This does not mean at all that ionic elements do not enter into the wave function, but they represent far from the whole wave function. Atoms, in other words, tend to be much more nearly neutral in a crystal than a straight ionic interpretation would indicate.

4-3. Interaction of Negative Ions in Crystals. Our emphasis in the preceding sections has been on atoms, or ions, which were forming bonds with each other. A very important feature of the structure of crystals, however, is the interaction between ions which are presumably repelling each other. The most striking examples of this are seen in crystals containing oxygen. In almost no cases are the oxygens bonded together, and yet their interactions are a feature of fundamental importance in the crystal structure.

Let us consider, for instance, the structure of spinel, as shown in Fig. 3-8, and discussed in Sec. 3-4. We have put our emphasis on the binding between the oxygens and the metallic atoms. We must not overlook the fact, however, that the crystal contains a great many oxygens. If we compute the oxygen-oxygen distances, which we have not considered earlier, we find that they are from about 2.50 A to about 3.18 A, averaging about 2.85 A. These are very far apart compared with the sum of atomic radii, which is 1.20 A. On the other hand, the ionic radius of oxygen is given by either Pauling or Zachariasen, in Table 4-2, as 1.40 A, so that the sum of ionic radii is 2.80 A. In other words, the oxygens in spinel are arranged so that, if we interpret the oxygen sizes in terms of the ionic rather than the atomic radii, they are approximately in contact. In fact, on the ionic picture, the spinel structure is described as an approximately close-packed structure of oxygen ions, with the metallic ions, represented by very small spheres (0.65 A for Mg^{+2}, from Table 4-2, and 0.50 A for Al^{+3}), filling certain interstices. This is shown in Fig. 4-2, where we draw the spinel structure using spheres with the ionic radii.

The distance of approximately 2.80 A between oxygen atoms is found in a great many crystals, and this fact has been used as an important verification of the ionic radius of 1.40 A for oxygen, and indirectly of the whole table of ionic radii. It is not hard to see why this ionic radius should come in, in this way. The oxygen ions in such a crystal as spinel do not directly attract each other; there is in fact a net Coulomb repulsion between them. But the crystal as a whole is held together by Coulomb attractions, and two oxygens will tend to be pushed into contact, until

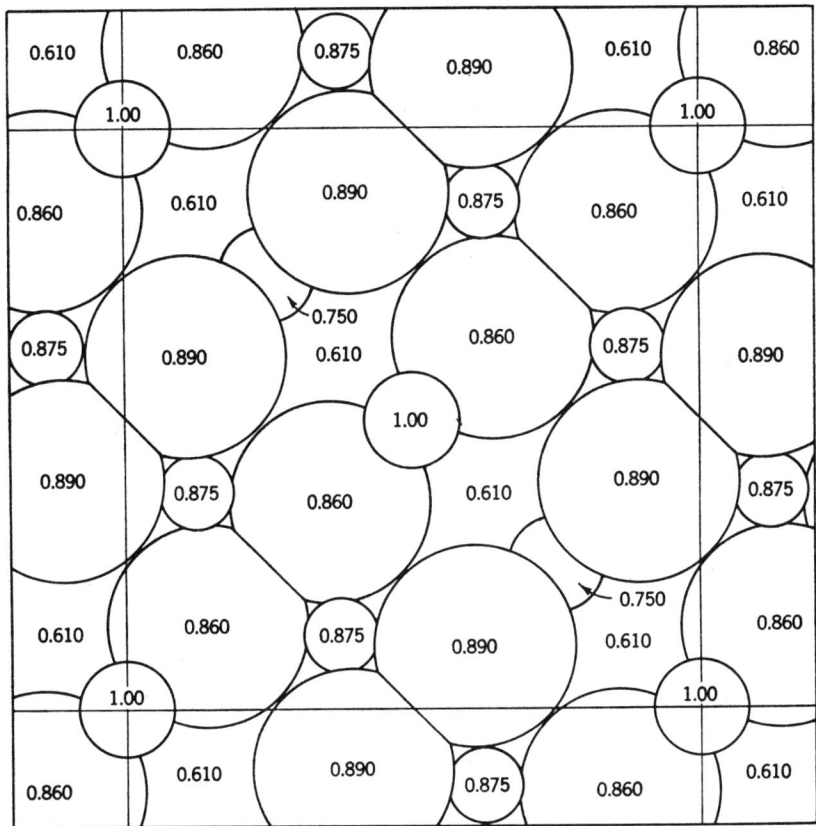

Fig. 4-2. Diagram of spinel structure, Al_2MgO_4, indicating ions by spheres with Pauling's ionic radii. Largest spheres are O^{-2}, next are Mg^{+2}, smallest are Al^{+3}. Heights are indicated in fractions of the side of the cubic unit cell. Note that oxygens at height 1.110, present in Fig. 3-8, are removed, but magnesiums at height 1.000, absent in Fig. 3-8, are shown.

they reach such a distance that the repulsion arising from the interaction between their inert gas shells will balance the net forces pushing them together. We have seen that the ionic radii are such that when two ions are at a distance equal to the sum of their ionic radii, the repulsive forces will be sufficient to balance the Coulomb attractions found in the crystal. These attractions are of the same order of magnitude in different crystals, so that the interionic distances will be found to be roughly the same in each case, whether the ions are held together by direct attraction or by being pushed together by the action of other ions.

One way in which the ionic radius of approximately 1.40 A for oxygen is important is in determining how many oxygen ions can be packed around a metallic ion. It is obvious that we cannot pack an indefinitely large number of oxygens around a small metallic ion, retaining the correct distance from metal to oxygen, without having the oxygens come into contact with each other. There is a tendency to have more oxygens coordinated around a large metallic ion, fewer around a smaller ion, as this would suggest. For example, we found in Sec. 3-6 that in LaOF, each lanthanum has eight oxygen or fluorine neighbors; lanthanum, of course, is a large atom or ion. At the other extreme we have the spinel case, with some ions in octahedral sites surrounded by six oxygens, others in tetrahedral sites surrounded by four; and we have such groups as the sulfate and phosphate, with four oxygens surrounding a sulfur or phosphorus.

Oxygen is not the only negative ion in which we must consider the contact of similar ions in the crystal. We run into the same situation with some of the alkali halides. For instance, in LiI, we see from Appendix 1 that the Li-I distance is 3.00 A, appreciably larger than the sum of atomic radii, which is 1.45 for Li, plus 1.40 for I, adding to 2.85 A. We have the same discrepancy using ionic radii. If we use Pauling's radii, from Table 4-2, we have 0.60 A for Li^+, 2.16 A for I^-, adding to 2.76 A, not very far from the sum of the atomic radii. We note, however, that the distance between adjacent iodines in this crystal is $3.00 \sqrt{2} = 4.25$ A, slightly less than 4.32 A, the sum of two ionic radii for I^-. In other words, on the ionic picture, the iodine is so large that the dimensions of the LiI crystal are set by the iodines coming into contact, leaving a hole for the lithium which is slightly too large for it. A number of cases in Appendix 1 where the distances are not determined correctly by the sum of atomic (or ionic) radii arise from this sort of situation. It is partly by a study of such situations that the ionic radii have been established. These cases, of contact of ions of the same sign, which are not forming bonds, are situations where we must use the ionic radii for a proper discussion. Our atomic radii give information only about the contact of atoms or ions between which bonds are formed.

When we use atomic radii, we also meet some cases of contact of atoms or ions of the same sign, which should be discussed. We have noted in Sec. 3-6 that in the Na_3As structure the sodium atoms approached to a distance of 2.98 A, less than twice the atomic radius of sodium. This is not the only case where we have such situations. We meet examples in the alkali halides; not, as in LiI, with large halogens, but in the opposite extreme of large alkalies. For example, in CsF, having the NaCl structure, the Cs-F distance is 3.00 A, slightly less than the sum 3.10 A of the atomic radii 2.60 A for Cs and 0.50 A for F, or the sum 3.05 A of Pauling's ionic radii 1.69 A for Cs^+ and 1.36 A for F^-. The particular point here is not the slight discrepancy, which is within the probable error of the additivity relations, but rather the fact that the Cs-Cs distance of 4.25 A is much less than twice the atomic radius, which is 5.20 A. In other words, it seems that in such cases alkali atoms can approach considerably closer than the sum of their atomic radii. The explanation here is presumably simple. No one would doubt that CsF is essentially an ionic crystal. The valence electron of Cs, the 6s, surely is absorbed into the structures of the neighboring fluorine ions, and there is no charge left in the Cs 6s wave function to oppose two cesiums approaching closer than the atomic radius would suggest. There would surely not be much sense in proposing that there is any Cs—Cs bonding or repulsion in such a crystal. No doubt something of the same explanation holds for the case of Na_3As, which we pointed out earlier.

4-4. The Charge Distribution and Potential in a Crystal. In our discussion in Sec. 4-2, we concluded that atoms in crystals tend to be more nearly neutral than a straight ionic interpretation would suggest. When we start to consider just what we mean by electrical neutrality, however, or by the amount of charge on an ion in a crystal, we find that the situation is more involved than one would think at first sight. It is important to understand these questions, for they are significant when we come to apply the self-consistent-field method to a crystal. We have to find the potential in which an electron moves, and this in turn means that we must know what the charge density is, in order to get the electrostatic potential by Poisson's equation. We therefore shall now consider just what is likely to be the charge distribution in a crystal, both of an ionic compound and of a substance held by covalent or metallic bonds.

If we had a good self-consistent-field solution for problems of solids, we should be able to give a straightforward theoretical answer to the question as to the charge distribution within a crystal. Unfortunately we do not; we are still in the early stages of being able to apply the self-consistent-field method to solids. There is also an experimental method of finding the charge distribution. In x-ray diffraction, the scattering is done by the electrons, not by the nuclei, and in principle the x-ray method is capable of giving the distribution of electronic charge. In practice, though attempts have been made to derive accurate charge distributions from x-ray data, the errors in the outer part of the atoms, where the only real questions arise, are unfortunately great enough so that no very decisive information has been obtained by this method. We are forced, then, to examine the problem in the light of our general information, checking the results where possible by other theoretical or experimental methods.

Let us start our discussion with an alkali halide; we may as well choose KCl again. A very natural way to make a first approximation to the charge distribution in this crystal is to start with the known self-consistent-field solutions of the K^+ and Cl^- ions, and add their charge distributions. We have already seen that the ions do not overlap very much, so that surely any corrections to this charge distribution will be minor. In the early days of calculations of self-consistent fields, this calculation

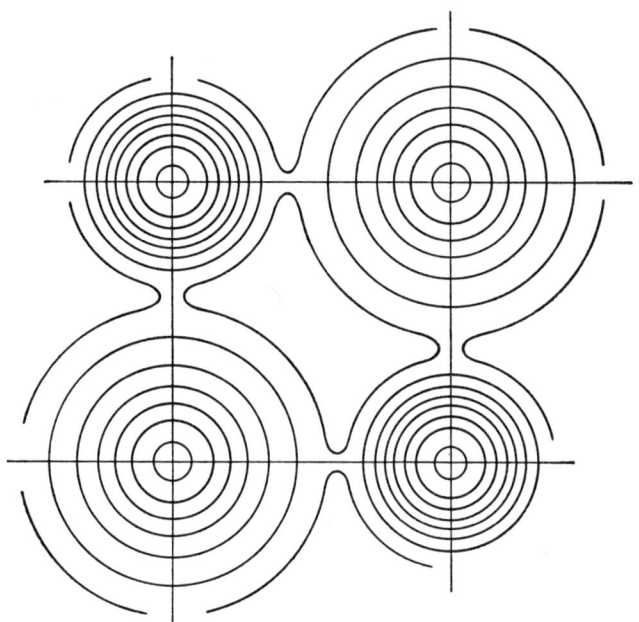

FIG. 4-3. Density contours in NaCl crystal, following Witte and Wölfel. The smaller ion is sodium. For KCl the results would be similar, but the K would be larger than Na.

was made[1] for NaCl, the x-ray scattering was calculated from the resulting charge distribution, and it agreed with experiment within the limit of error of the experiment and the calculation. Hence we may expect that this method will give us a good approximation to the correct charge for an alkali halide.

In Fig. 4-3 we give contours of constant charge density for the resulting charge. This figure was given by Witte and Wölfel[2] as an idealized result of x-ray measurements on this crystal, but as a matter of fact it essentially gives the calculation from the self-consistent-field distribution. We may

[1] R. W. James, I. Waller, and D. R. Hartree, *Proc. Roy. Soc. (London)*, **A118**:334 (1928).
[2] H. Witte and E. Wölfel, *Z. Physik. Chem. (Frankfurt)*, **3**:296 (1955).

assume that it is very close to the correct value for this crystal. We then ask ourselves, does the figure itself indicate that the potassium and chlorine are in the form of atoms or ions, and what values does it tell us to use for the radii of the two atoms or ions? When we ask ourselves these questions, we see that there is no entirely clear-cut answer. There is charge density extending continually from one atom to the next; in the region between them, the tails of the wave functions of the two overlap, with a resulting charge density which does not go to zero. If we arbitrarily divide space into volumes surrounding the potassiums, and other volumes surrounding the chlorines, something like Wigner-Seitz cells, and count the charge enclosed in one type of volume as belonging to the potassium, that enclosed in the other to the chlorine, then it is obvious that the amount of electronic charge which we associate with each atom or ion will depend on just how we draw the volumes. If we make the potassium volume too small, the chlorine volume too large, the potassium will appear too highly positively charged, the chlorine too negative, and vice versa. If the potassium volume extended out to include some of the tails of the chlorine distribution, we could make the potassiums neutral, and likewise the chlorines. In other words, it is partly a matter of choice whether we conclude that the atoms are neutral or ionized.

This conclusion is reinforced by asking how the charge distribution would have looked if we had started with the self-consistent-field densities for neutral potassium and neutral chlorine, instead of the ions, and had superposed these. The $4s$ wave function of the potassium would have been located a long way away from the potassium nucleus. In fact, from our discussion of atomic radii, we know that its maximum radial density would have come in a shell which just cuts through the region of maximum density of the outer chlorine shell. The superposition, at each chlorine, of this $4s$ charge from the six neighboring potassiums, would give an extra amount of electronic charge near the chlorine which would be hard to distinguish from the extra electron found in the Cl^- ion, but missing in the neutral Cl atom. In other words, the difference in charge distribution between the superposition of the ionic charges and the superposition of the atomic charges would be small and subtle, and very hard to determine by examination of the total charge density, as shown in Fig. 4-3.

In Volume 1, we had a simple answer as to whether a diatomic molecule was ionic or not. We merely computed the dipole moment, and compared it with the value which we should have if we had charges $\pm e$ located at the two nuclei. In a symmetrical crystal like KCl, however, there is no such simple way of answering the question. We have just seen how hard it is to answer it by direct inspection of the charge density. There is no dipole moment, on account of symmetry. In other words, the question as to whether a given crystal is ionic or not is almost unanswerable, or meaningless, when considered from the operational point of view of starting with the charge density and proceeding from there.

There is, however, a fairly simple answer in such a case as KCl as to the reasonable radii to use. In Fig. 4-3 it is perfectly obvious that the charge density goes down to a minimum value between the potassiums and chlorines. The contours of constant density indicate that each atom or ion is pretty close to spherical symmetry, out to a radius equal to the distance to the point of minimum density, and in the region between these spherical charge distributions, the charge density is low. It is very natural to use these radii as ionic radii. If we use spheres for the potassium and chlorine, touching each other at the point of minimum density between them, we could define the charge on each ion as the charge within the sphere so constructed. This would, however, leave out of account the small but by no means negligible charge located in the regions between the spheres.

The radii defined in this way do not agree with the values of Pauling and Zachariasen. In the case of KCl, we find that the minimum of charge density comes at a distance of about 1.45 A from the potassium, 1.70 A from the chlorine, suggesting these values for the radii. They compare with 1.33 A for the ionic radius of K^+, and 1.81 A for Cl^-, in Table 4-2. A similar study of MgO and CaO gives a radius of 1.30 A or even less for oxygen, compared with the value of 1.40 A in Table 4-2.

The method we have just outlined is one which can be applied in any given case, but it does not lead to fixed radii for the various ions. The radius which we assign for such an ion as Cl^- by this method is strongly dependent on the ion adjacent to it. Thus, we can consider the crystal of CuCl, which has the zinc-blende structure. The Cu^+ ion, on account of its $3d$ electrons, has much more charge density near its outer boundary than the K^+ ion does. When we superpose the charge distributions of Cu^+ and Cl^- ions, this effect throws the minimum of charge density between the ions closer to the chlorines than in KCl. One finds, in fact, that the minimum of charge density comes about 1.10 A from the Cu, 1.25 A from the Cl. If we examined this crystal rather than KCl, we should then deduce an ionic radius of only 1.25 A for Cl^-, rather than 1.70 A as we found in KCl. Or again in an oxide, if we consider NiO, we find the minimum of charge density coming at about 0.94 A from the nickel, 1.15 A from the oxygen. We should conclude from this case that the ionic radius of oxygen was only 1.15 A, rather than 1.30 as determined from MgO.

In an obviously nonionic crystal of an element, such as a metal, or a crystal like diamond, silicon, or germanium, the minimum of density will come halfway between like atoms, by symmetry. If we carry out a

Wigner-Seitz construction for a body-centered or face-centered cubic metal, the minimum will come at the surface of the Wigner-Seitz cell, and we should derive as the radius of the atom simply half the distance between nearest neighbors, agreeing approximately with our atomic radii. Similarly, for germanium we should find half the interatomic distance.

It is then interesting to consider the sequence CuBr, ZnSe, GaAs, and Ge. In CuBr, we have a situation similar to CuCl, mentioned a short time before, with the minimum about 1.10 A from the Cu, 1.36 A from the Br. As we go through the sequence to germanium, we shall obviously approach the germanium case continuously, in which the minimum will be halfway between the atoms, or at a distance of 1.22 A from each atom. This would give us a sequence of radii for Cu, Zn, Ga, and Ge, which would go from some sort of ionic radius for Cu to a definitely atomic radius for Ge, with a similar situation for the atoms Br, Se, As, and Ge. But these radii would not be found in other compounds of the same elements.

In other words, by considerations based on the position of the minimum charge density, we can derive sensible radii for individual crystals, but they do not agree with either our atomic or ionic radii in general, and they do not show the properties, which those radii do, of being approximately constant, independent of what atoms are being combined with the atom in question. The only conclusion which is reasonable from these facts is that the ionic radii of the type used by Pauling, Zachariasen, and others do not have a very definite meaning in terms of easily defined properties of the electronic charge distribution within the crystal. This situation does not hold for our atomic radii, for we have seen from Table 4-3 that they are fairly closely correlated with a definite atomic property, namely, the radius of maximum radial charge density in the outermost shell of the neutral atom.

4-5. Potentials for the Self-consistent Field in Crystals. In the preceding section we have seen that the familiar ionic radii have no very close connection with the actual electronic charge density in crystals. Furthermore, in a definitely ionic crystal such as KCl, the atomic radii also have no close connection with the charge distribution, since the K^+ ion has no $4s$ valence electron, though they are useful in predicting the experimental internuclear separations. Our major problem in the rest of this volume is to discuss the solution of Schrödinger's equation for the electrons in the crystal, which we shall approach by means of the self-consistent-field method. As we have mentioned earlier, we are not yet in a position to do much more than make the best guess we can as to the potential in which each electron moves, and then solve Schrödinger's equation for an electron moving in that potential. We are now ready, on the basis of our discussion of the preceding chapters, to describe this potential.

It is clearly going to be derived by Poisson's equation from a charge distribution similar to that shown in Fig. 4-3 for the case of KCl. For an ionic crystal we can superpose the charge distributions of the ions, as determined by the self-consistent-field method for individual ions; for a definitely covalent crystal we superpose the charge distributions of the atoms. We have given reasons for thinking that in a doubtful case the final charge distribution will not be very different whether we use the ionic or atomic starting point. In either case, we may expect to find essentially the situation shown in Fig. 4-3, namely, a rather closely spherically symmetrical charge distribution for each atom or ion, with rather small charge density in the space between the spheres. Poisson's equation, of course, is a linear equation, so that if we find the potentials arising from each of these atomic or ionic distributions, and add them, we shall get the potential arising from the total charge distribution. If we are using ionic charge distributions, there is a technical problem associated with adding these potentials: the sum of the potentials of all positive ions will diverge, on account of the divergence of a sum of $1/r$-like terms for a lattice of ions, and we must properly take account of this divergence by balancing the divergences of the positive ions against those from the negative ions. We shall discuss this problem in a later volume, where we shall show that it can be handled in several alternative ways, all quite legitimate.

The radii of the spherical charge distributions representing the various atoms or ions must be determined as in Fig. 4-3, that is, as the distances from the nuclei out to the minimum of charge densities between the atoms. These will not in general be the same as either the atomic or ionic radii, but must be determined separately for each crystal.

We shall see in later chapters that there are now available quite accurate methods of solving Schrödinger's equation in a region in which there are spherically symmetrical potentials within certain spheres, which can touch each other, with a constant potential in the region between the spheres. Work currently under way, which will be discussed later, shows that the energy levels of the one-electron problem set up in this way have fairly good correlation with the observed energy levels of the electrons in the solid. Thus these energy levels are similar to the one-electron energies found in the discussion of atoms or molecules, which we know, by Koopmans' theorem, approximate observed ionization potentials. Calculations by this method, however, as we realize, represent only the first step toward a self-consistent-field calculation. When the one-electron Schrödinger equation has been solved, we should next find the charge density arising from the electrons in the occupied wave functions, compare this with the charge density initially assumed, and use the potential derived from it as the starting point of a next step in the iterative procedure for solving the self-consistent-field problem. A beginning

Sec. 4-5]   ATOMIC RADII AND THE CHEMICAL BOND   117

has been made in this direction, but the problem is a great deal more difficult than with atoms or molecules, and it is likely to be some time before completely satisfactory self-consistent calculations are possible for solids.

So far in this discussion we have disregarded the exchange correction, the fact that an electron should move in the field of all electrons, minus a correction to account for the fact that it does not act on itself. Ideally, this should be handled by using the Hartree-Fock method. In practice, this is so difficult that no straightforward calculations by the method have been carried through. As a simplification, the procedure suggested by the author,[1] involving an exchange potential proportional to the cube root of the total charge density, is practicable, and is finding considerable use at the present time. It seems to represent about the maximum refinement which is possible with present methods. There is a good deal of work going on with the aim of finding whether a superposition of the atomic or ionic potentials, supplemented by this exchange correction, is close enough to the truth so that the one-electron solutions of Schrödinger's equation in this potential form an adequate approximation to the solution of the self-consistent-field problem. The indications are that this method is good enough to be a valid approximation, but that small changes in the assumptions made about the exchange potential can have effects on the energy bands which are great enough to affect seriously the agreement with experiment. We shall come to this question in a later volume, but for the present we shall assume that the potential to be used in applying the self-consistent-field method to solid-state problems can be set up as we have described in the present section. We now go on in the next few chapters to the methods of approximating to the solution of a one-electron Schrödinger equation in such a potential.

[1] J. C. Slater, *Phys. Rev.*, **81**:385 (1951); "Quantum Theory of Atomic Structure," vol. 2, McGraw-Hill Book Company, New York, 1960, sec. 17-3.

5
The Symmetry of Electronic Wave Functions in Crystals

5-1. The Reciprocal Lattice. In the preceding chapters we have been studying the symmetry properties of the charge density, and hence the electrostatic potential, in various types of crystals. The Hamiltonian, of course, will have the same type of symmetry. We shall now take up the related problem: what types of symmetry must the electronic wave function have in crystals whose symmetry is determined by various space groups? It is desirable to investigate this question before we start asking how the wave function is to be computed. We remember that throughout this discussion we are working in the framework of the self-consistent-field method, using a potential determined, for instance, by methods described in Sec. 4-5, where we discussed a potential determined by Poisson's equation from the total electronic charge density plus the nuclear charges, corrected for exchange by an approximate exchange potential proportional to the cube root of the charge density. Our problem will be to solve Schrödinger's equation for a single electron in such a potential. From general methods of group theory, we know that a wave function solving this problem will have the symmetry of one of the irreducible representations of the space group. Our problem in the present chapter is to consider these symmetries.[1]

We have already found, in Sec. 1-2, that the translational symmetry of the crystal leads to the possibility of writing the wave function according to Eq. (1-5), namely, as $u(\mathbf{r}) = e^{i\mathbf{k}\cdot\mathbf{r}} w(\mathbf{r})$, where $\mathbf{k}$ is the wave vector, and $w(\mathbf{r})$ is a periodic function repeating at corresponding points of each unit cell. This suggests the fundamental importance of having an analytic method of describing such a periodic function as $w(\mathbf{r})$, or alternatively, the potential energy function. The general solution of the problem of expressing such a periodic function leads to a three-dimensional Fourier

---
[1] See F. Seitz, Z. Krist., **88**:433 (1934); **90**:289 (1935); **91**:336 (1935); **94**:100 (1936). G. F. Koster, Space Groups and Their Representations, *Solid State Phys.*, **5**:173, 1957.

Sec. 5-1] SYMMETRY OF ELECTRONIC WAVE FUNCTIONS IN CRYSTALS

expansion, which amounts to a superposition of plane waves of definitely determined wave vectors. The study of the wave vectors which will lead to the proper periodicity properties brings us to the idea of the reciprocal lattice, an important concept which was first introduced into vector analysis by Gibbs,[1] and whose application to crystallography was first pointed out by Ewald.[2]

The first step in this process is to set up three vectors $\mathbf{b}_1$, $\mathbf{b}_2$, $\mathbf{b}_3$, which are described as vectors reciprocal to $\mathbf{t}_1$, $\mathbf{t}_2$, $\mathbf{t}_3$. They satisfy the equations

$$\mathbf{t}_i \cdot \mathbf{b}_j = \delta_{ij} \tag{5-1}$$

which say that one of the $\mathbf{b}$'s is perpendicular to the two $\mathbf{t}$'s with different indices, and that the vectors $\mathbf{t}_i$ and $\mathbf{b}_i$ are in a sense reciprocal to each other, in that their scalar product is unity. In terms of these vectors $\mathbf{b}_1$, $\mathbf{b}_2$, $\mathbf{b}_3$, we set up vectors

$$\mathbf{K}_h = 2\pi(h_1 \mathbf{b}_1 + h_2 \mathbf{b}_2 + h_3 \mathbf{b}_3) \tag{5-2}$$

where h_1, h_2, h_3 are integers. Then we find at once that the exponential $e^{i\mathbf{K}_h \cdot \mathbf{r}}$ repeats with identical values at corresponding points of each unit cell, so that it represents a very simple form of the periodic function we are looking for. The reason is that

$$\begin{aligned} e^{i\mathbf{K}_h \cdot (\mathbf{r} + \mathbf{R}_n)} &= e^{i\mathbf{K}_h \cdot \mathbf{R}_n} e^{i\mathbf{K}_h \cdot \mathbf{r}} \\ &= e^{2\pi i (h_1 n_1 + h_2 n_2 + h_3 n_3)} e^{i\mathbf{K}_h \cdot \mathbf{r}} \\ &= e^{i\mathbf{K}_h \cdot \mathbf{r}} \end{aligned} \tag{5-3}$$

where we have used Eq. (1-1), which states that $\mathbf{R}_n = n_1 \mathbf{t}_1 + n_2 \mathbf{t}_2 + n_3 \mathbf{t}_3$, as well as Eqs. (5-1) and (5-2). We can now set up all possible functions $e^{i\mathbf{K}_h \cdot \mathbf{r}}$, where $\mathbf{K}_h$ takes on all values given by Eq. (5-2), and the linear combination of these functions will give a three-dimensional Fourier series, which represents the most general function which repeats in each unit cell. This type of function is represented by a series

$$F(\mathbf{r}) = \Sigma(\mathbf{K}_h) G(\mathbf{K}_h) e^{i\mathbf{K}_h \cdot \mathbf{r}} \tag{5-4}$$

We shall later set up the method of deriving the Fourier coefficients $G(\mathbf{K}_h)$.

The reciprocal space is defined as a space in which the vectors $\mathbf{b}_1$, $\mathbf{b}_2$, $\mathbf{b}_3$ form primitive vectors in the same way that $\mathbf{t}_1$, $\mathbf{t}_2$, $\mathbf{t}_3$ do in ordinary space, so that the vectors $h_1 \mathbf{b}_1 + h_2 \mathbf{b}_2 + h_3 \mathbf{b}_3$ reach from the origin to a lattice of points, called the reciprocal lattice, just as the points $n_1 \mathbf{t}_1 + n_2 \mathbf{t}_2 + n_3 \mathbf{t}_3$ reach from the origin to a lattice of points in ordinary space. Frequently it is more convenient to use a space in which the $\mathbf{K}$'s are plotted, so that we have a lattice of points in $\mathbf{k}$ space given by all vectors indicated in

---

[1] J. W. Gibbs and E. B. Wilson, "Vector Analysis," Yale University Press, New Haven, Conn., 1902, p. 111.
[2] P. P. Ewald, Z. Krist., **56**:129 (1921); **93**:396 (1936).

Eq. (5-2), with all values of the integers h_1, h_2, h_3. We can set up unit cells, in reciprocal space or in k space, just as we set up unit cells in ordinary space. We shall find many uses in the future for the reciprocal space or the k space. One use is obvious at once from our discussion of the preceding paragraph: we may imagine one of the amplitudes $G(\mathbf{K}_h)$ to be associated with each lattice point of the reciprocal lattice, so that this gives a convenient way to classify or order these amplitudes.

Let us now, as an illustration of the use of the reciprocal lattice, set up the function of Eq. (1-5), representing a basis function for an irreducible representation of the translation group, by use of Eq. (5-4). We let the function $w(\mathbf{r})$ of Eq. (1-5) be given by a sum of the type of Eq. (5-4). Then we find that the function $u(\mathbf{r})$ can be written as a sum of exponentials $e^{i(\mathbf{k}+\mathbf{K}_h)\cdot\mathbf{r}}$, where $\mathbf{K}_h$ takes on all possible values. We may write this function as

$$u(\mathbf{k},\mathbf{r}) = \Sigma(\mathbf{K}_h)v(\mathbf{k} + \mathbf{K}_h)e^{i(\mathbf{k}+\mathbf{K}_h)\cdot\mathbf{r}} \qquad (5\text{-}5)$$

which is an obvious generalization of Volume 1, Eq. (8-21). We may now associate one of the amplitudes $v(\mathbf{k} + \mathbf{K}_h)$ with each of the points $\mathbf{k} + \mathbf{K}_h$ in the k space; these lie in corresponding points of each unit cell in k space.

The function of Eq. (5-5) represents a perfectly general basis function for the irreducible representation characterized by the wave vector k. We note from Eq. (1-2) that this function diagonalizes all the translation operators, with the diagonal matrix elements $e^{i\mathbf{k}\cdot\mathbf{R}_n}$. We note further that any two k's which differ by one of the vectors $\mathbf{K}_h$ of the reciprocal lattice will give the same value to this diagonal matrix element. Another way of saying this is that a given irreducible representation is distinguished, not by a specific wave vector k, but by the whole lattice of points $\mathbf{k} + \mathbf{K}_h$, each of which corresponds to the same diagonal matrix element, and which represents the wave vectors of all the plane waves or sinusoidal functions $e^{i(\mathbf{k}+\mathbf{K}_h)\cdot\mathbf{r}}$ which have the same symmetry type. We may, if we choose, describe each irreducible representation by specifying that vector $\mathbf{k} + \mathbf{K}_h$ for it which has the smallest absolute magnitude. Generally we shall choose k such that k itself has a smaller absolute magnitude than any of the other vectors $\mathbf{k} + \mathbf{K}_h$. In this case we refer to it as the reduced wave vector.

We shall wish to consider the nature and symmetry properties of the reciprocal lattice, and of the unit cells in it. We find that it has properties very similar to those of the unit cell and lattice in ordinary space. The reciprocal lattice of a crystal must be one of the 14 Bravais lattices, just as the space lattice is. For any given space group, the reciprocal lattice belongs to the same one of the seven systems as the space lattice,

Sec. 5-2] SYMMETRY OF ELECTRONIC WAVE FUNCTIONS IN CRYSTALS

but it is not necessarily the same Bravais lattice. We may enumerate the 14 cases without trouble; we do so in Table 5-1. In the next section we shall take up the 8 out of the 14 cases which we meet among the space groups which we are discussing in detail.

Table 5-1
Symmetry of reciprocal lattice corresponding to each Bravais space lattice

Crystal system	Space lattice	Reciprocal lattice
Triclinic	Simple triclinic	Simple triclinic
Monoclinic	Simple monoclinic	Simple monoclinic
	One-face-centered monoclinic	One-face-centered monoclinic
Orthorhombic	Simple orthorhombic	Simple orthorhombic
	Face-centered orthorhombic	Body-centered orthorhombic
	Body-centered orthorhombic	Face-centered orthorhombic
	One-face-centered orthorhombic	One-face-centered orthorhombic
Tetragonal	Simple tetragonal	Simple tetragonal
	Body-centered tetragonal	Body-centered tetragonal
Trigonal and	Simple hexagonal	Simple hexagonal
hexagonal	Rhombohedral	Rhombohedral
Cubic	Simple cubic	Simple cubic
	Face-centered cubic	Body-centered cubic
	Body-centered cubic	Face-centered cubic

5-2. Reciprocal Lattices and Brillouin Zones for Eight Bravais Lattices. Before we go on to discuss these eight examples, we should point out that we can set up unit cells in the reciprocal space, just as in ordinary space. Furthermore, we find the same situation discussed in Chap. 1, namely, that if we use the parallelopiped determined by the three vectors b_1, b_2, b_3, it will not in general have the symmetry characteristic of the point group of the crystal. Instead, we can use the equivalent of the Wigner-Seitz construction, to set up unit cells having this symmetry. That is, we set up the vectors from the origin of reciprocal space to all points of the reciprocal lattice; we set up the planes which are the perpendicular bisectors of each of these vectors. The volume enclosed by these planes will be that part of reciprocal space closer to the origin than to any other lattice point. It is the volume within which all reduced wave vectors (in the sense defined in Sec. 5-1) are included. This volume can be used as a unit cell in reciprocal space; similar volumes surrounding each of the points of reciprocal space would just fill space. This volume surrounding the origin of reciprocal space is called the central Brillouin zone. It was introduced into the theory of solids by Brillouin,[1] who also, for reasons

---

[1] L. Brillouin, *J. Phys. Radium*, **1**:377 (1930).

which we shall discuss in Sec. 7-3, introduced further polyhedra, larger than this unit cell, called the second, third, etc., Brillouin zones.

We shall now indicate more of the analytical methods used in handling the reciprocal lattice, and as we have stated, we shall illustrate by means of the 8 Bravais lattices which occur in the 20 space groups which we are discussing. In the first place, from Eq. (5-1), we see that the b_i's can be defined from the equations

$$b_1 = \frac{t_2 \times t_3}{t_1 \cdot (t_2 \times t_3)} \qquad b_2 = \frac{t_3 \times t_1}{t_1 \cdot (t_2 \times t_3)} \qquad b_3 = \frac{t_1 \times t_2}{t_1 \cdot (t_2 \times t_3)} \quad (5\text{-}6)$$

These equations are immediately verified by substitution in Eq. (5-1).

Before we proceed further, we shall prove a simple theorem: the volume of unit cell in the reciprocal space, namely, $b_1 \cdot (b_2 \times b_3)$, is the reciprocal of the volume of unit cell in ordinary space, namely, $t_1 \cdot (t_2 \times t_3)$. To prove this, we start with the vector identity

$$\mathbf{A} \times (\mathbf{B} \times \mathbf{C}) = \mathbf{B}(\mathbf{A} \cdot \mathbf{C}) - \mathbf{C}(\mathbf{A} \cdot \mathbf{B})$$

where $\mathbf{A}, \mathbf{B}, \mathbf{C}$ are any three vectors. We let $\mathbf{A} = t_3 \times t_1$, $\mathbf{B} = t_1$, $\mathbf{C} = t_2$. Then from Eq. (5-6) we have

$$\begin{aligned} b_2 \times b_3 &= \frac{1}{[t_1 \cdot (t_2 \times t_3)]^2} [(t_3 \times t_1) \times (t_1 \times t_2)] \\ &= \frac{1}{[t_1 \cdot (t_2 \times t_3)]^2} \{t_1[t_2 \cdot (t_3 \times t_1)] - t_2[t_1 \cdot (t_3 \times t_1)]\} \\ &= \frac{t_1}{t_1 \cdot (t_2 \times t_3)} \end{aligned} \quad (5\text{-}7)$$

where in the last step we have used the fact that $t_1 \cdot (t_3 \times t_1) = 0$, and that $t_2 \cdot (t_3 \times t_1) = t_1 \cdot (t_2 \times t_3)$. We now take the scalar product of Eq. (5-7) with b_1, use the fact that $b_1 \cdot t_1 = 1$ on account of Eq. (5-1), and the result is

$$b_1 \cdot (b_2 \times b_3) = \frac{1}{t_1 \cdot (t_2 \times t_3)} \quad (5\text{-}8)$$

which was to be proved.

Let us now use Eq. (5-6), and apply it to finding the reciprocal vectors for each of our eight Bravais lattices. For the simple cubic lattice, we see from Eq. (1-17) that $t_1 = ia$, $t_2 = ja$, $t_3 = ka$. The volume of the unit cell, $t_1 \cdot (t_2 \times t_3)$, of course is a^3. From Eq. (5-6), the reciprocal vectors are $b_1 = i/a$, $b_2 = j/a$, $b_3 = k/a$, showing that the unit cell in reciprocal space is simple cubic, like that in real space, agreeing with Table 5-1, and furthermore that its volume is $1/a^3$, as it should be.

For the body-centered cubic space lattice, the vectors t_i are given in Eq. (1-18), namely, $t_1 = (a/2)(-i + j + k)$, $t_2 = (a/2)(i - j + k)$,

Sec. 5-2] SYMMETRY OF ELECTRONIC WAVE FUNCTIONS IN CRYSTALS 123

$t_3 = (a/2)(i + j - k)$. From Eq. (5-6) we find

$$b_1 = \frac{1}{a}(j + k) \qquad b_2 = \frac{1}{a}(i + k) \qquad b_3 = \frac{1}{a}(i + j) \qquad (5\text{-}9)$$

which, as we see from Eq. (1-19), are proportional to t_1, t_2, t_3 for the face-centered cubic space lattice, hence verifying the result that the reciprocal lattice to a body-centered cubic space lattice is face-centered cubic. The volume $b_1 \cdot (b_2 \times b_3)$ equals $2/a^3$, which is the reciprocal of the volume $a^3/2$ of the unit cell in ordinary space for the body-centered cubic structure. Similarly, if we start with the values of t_1, t_2, t_3 for the face-centered cubic space lattice, as given in Eq. (1-19), and use Eq. (5-6), we find that the reciprocal vectors for the face-centered cubic space lattice are

$$b_1 = \frac{1}{a}(-i + j + k) \qquad b_2 = \frac{1}{a}(i - j + k) \qquad b_3 = \frac{1}{a}(i + j - k) \qquad (5\text{-}10)$$

which are proportional to the vectors t_i for the body-centered cubic space lattice, as given in Eq. (1-18). Hence we verify that the reciprocal lattice to a face-centered cubic space lattice is body-centered cubic. The volume $b_1 \cdot (b_2 \times b_3)$ equals $4/a^3$, the reciprocal of the volume $a^3/4$ of the space lattice for the face-centered cubic case.

For the hexagonal Bravais space lattice, we have found in Eq. (1-20) that the t_i's are given by $t_1 = (a/2)(\sqrt{3}\,i - j)$, $t_2 = aj$, $t_3 = ck$. From Eq. (5-6), we then find that

$$b_1 = \frac{2}{\sqrt{3}\,a} i \qquad b_2 = \frac{1}{\sqrt{3}\,a}(i + \sqrt{3}\,j) \qquad b_3 = \frac{1}{c}k \qquad (5\text{-}11)$$

These vectors are not immediately recognizable by comparison with the t_i's, but we can easily see that they also lead to a hexagonal lattice in reciprocal space. To see this, we show in Fig. 5-1 the vectors b_1 and b_2 in the xy plane. Instead of being 120° apart, as t_1 and t_2 are, they are only 60° apart. However, we show in Fig. 5-1 the vectors to the six closest lattice points to the origin, in the xy plane of reciprocal space. We see that these are six vectors making angles of 60° with each other, like the six vectors shown in Fig. 1-5. When we carry out a Wigner-Seitz construction to get the section of the central Brillouin zone in the xy plane, we find as in Fig. 1-5 that it is again a regular hexagon, but rotated by 90° with respect to the hexagon of Fig. 1-5, so that a flat face, instead of an edge, faces along the x direction. We verify, in other words, the fact that the reciprocal lattice to the hexagonal space lattice is again hexagonal, but rotated through 90° about the z axis. The volume of the unit cell in ordinary space is $(\sqrt{3}/2)a^2c$, as we mentioned in Sec. 1-5, and the volume of unit cell in reciprocal space is $2/(\sqrt{3}\,a^2c)$.

Next we consider the rhombohedral space lattice. The t_i's are given in Eq. (1-21) as

$$t_1 = s\mathbf{i} + r\mathbf{k} \qquad t_2 = (s/2)(-\mathbf{i} + \sqrt{3}\,\mathbf{j}) + r\mathbf{k}$$
$$t_3 = (s/2)(-\mathbf{i} - \sqrt{3}\,\mathbf{j}) + r\mathbf{k}$$

From Eq. (5-6) we find

$$\mathbf{b}_1 = \frac{2}{3s}\mathbf{i} + \frac{1}{3r}\mathbf{k} \qquad \mathbf{b}_2 = \frac{1}{3s}(-\mathbf{i} + \sqrt{3}\,\mathbf{j}) + \frac{1}{3r}\mathbf{k}$$
$$\mathbf{b}_3 = \frac{1}{3s}(-\mathbf{i} - \sqrt{3}\,\mathbf{j}) + \frac{1}{3r}\mathbf{k} \tag{5-12}$$

These are derived from the t_i's of the rhombohedral space lattice by substituting $s' = 2/(3s)$ in place of s, and $r' = 1/(3r)$ in place of r. Hence

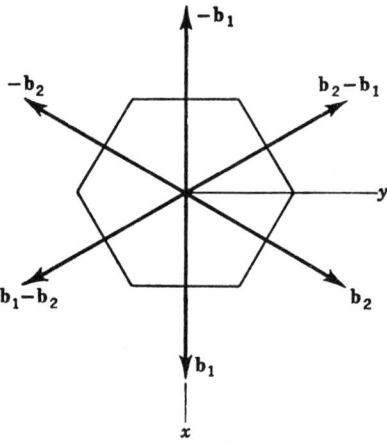

Fig. 5-1. The central Brillouin zone for the hexagonal Bravais space lattice, projected on the xy plane.

the reciprocal lattice to the rhombohedral space lattice is itself rhombohedral, verifying the statement in Table 5-1. The volume of the unit cell in ordinary space is $(3\sqrt{3}/2)s^2 r$, that in reciprocal space is $2/(3\sqrt{3}\,s^2 r)$.

It is interesting to find the relationship between the ratio r/s for the space lattice and the corresponding ratio r'/s' for the reciprocal lattice; these ratios determine the angles between the three fundamental vectors, by Eq. (1-22). From the values given in the last paragraph, we have

$$\frac{r'}{s'} = \frac{1}{2}\frac{s}{r} \tag{5-13}$$

so that the two ratios are reciprocal to each other. In other words, when the angle α between the vectors $\mathbf{t}_i$ is small, the angle α' between the $\mathbf{b}_i$'s is

Sec. 5-2] SYMMETRY OF ELECTRONIC WAVE FUNCTIONS IN CRYSTALS 125

large, and vice versa. We pointed out in Sec. 1-5 that when $r/s = 1/\sqrt{2}$, the rhombohedral lattice reduces to the simple cubic case. In this case, from Eq. (5-13), we see that r'/s' also equals $1/\sqrt{2}$, so that the reciprocal lattice is also simple cubic. Again, we showed that the body-centered cubic lattice is a special case of the rhombohedral lattice with $r/s = 1/2\sqrt{2}$, and the face-centered cubic is a special case with $r/s = \sqrt{2}$. From Eq. (5-13) we see that if r/s has one of these two values, r'/s' will have the other, thereby verifying the statements that the reciprocal lattice to a body-centered cubic lattice is face-centered cubic, and vice versa, as special cases of the rhombohedral lattice.

In addition to the five reciprocal lattices which we have set up so far, the remaining three Bravais lattices which we meet in our detailed examples are the simple tetragonal, simple orthorhombic, and one-face-centered orthorhombic. For the simple tetragonal space lattice, the t_i's are given in Eq. (1-23), namely, $t_1 = a\mathbf{i}$, $t_2 = a\mathbf{j}$, $t_3 = c\mathbf{k}$. The reciprocal vectors are $\mathbf{b}_1 = \mathbf{i}/a$, $\mathbf{b}_2 = \mathbf{j}/a$, $\mathbf{b}_3 = \mathbf{k}/c$, showing that the reciprocal lattice again is simple tetragonal. The volume of the unit cell in ordinary space is a^2c, that in reciprocal space is $1/a^2c$. Similarly, for the simple orthorhombic space lattice the t_i's are $t_1 = a\mathbf{i}$, $t_2 = b\mathbf{j}$, $t_3 = c\mathbf{k}$, and the reciprocal vectors are $\mathbf{b}_1 = \mathbf{i}/a$, $\mathbf{b}_2 = \mathbf{j}/b$, $\mathbf{b}_3 = \mathbf{k}/c$. The volume of the unit cell in ordinary space is abc, and in reciprocal space it is $1/abc$.

We have met the one-face-centered orthorhombic space lattice in connection with the iodine and gallium structure in Sec. 2-8. In Eq. (2-20), we have given the t_i's as $t_1 = \mathbf{i}a/2 + \mathbf{j}b/2$, $t_2 = -\mathbf{i}a/2 + \mathbf{j}b/2$, $t_3 = \mathbf{k}c$. Then we find that

$$\mathbf{b}_1 = \frac{\mathbf{i}}{a} + \frac{\mathbf{j}}{b} \qquad \mathbf{b}_2 = -\frac{\mathbf{i}}{a} + \frac{\mathbf{j}}{b} \qquad \mathbf{b}_3 = \frac{\mathbf{k}}{c} \qquad (5\text{-}14)$$

which again is of the one-face-centered tetragonal type. The unit cell in ordinary space has the volume $abc/2$, and in reciprocal space $2/abc$.

The remaining six Bravais lattices which we have not discussed are the simple triclinic, simple monoclinic, one-face-centered monoclinic, body-centered orthorhombic, face-centered orthorhombic, and body-centered tetragonal. The reader can easily work out these cases by methods similar to those we have used, and should have no trouble doing so, except perhaps in the case of the body-centered tetragonal lattice. The straightforward discussion of the reciprocal lattice to a body-centered tetragonal lattice would suggest that it should be a face-centered tetragonal lattice, and yet that type does not appear in the list of 14 Bravais lattices. The reason is that a face-centered tetragonal lattice can also be considered as body-centered tetragonal, by rotation through 45° about the z axis. We show this in Fig. 5-2. Here we have indicated a hypothetical face-centered tetragonal lattice, as observed looking down along the $-z$ direction. The circles indicate the lattice points in one plane, the crosses in a higher or lower plane. In the lower half of the diagram we draw cells at 45° to the original ones, exhibiting the way in which these describe the same lattice points as body-centered tetragonal, but with a lattice whose side, in the xy plane, is only $1/\sqrt{2}$ as large as for the face-centered case. We

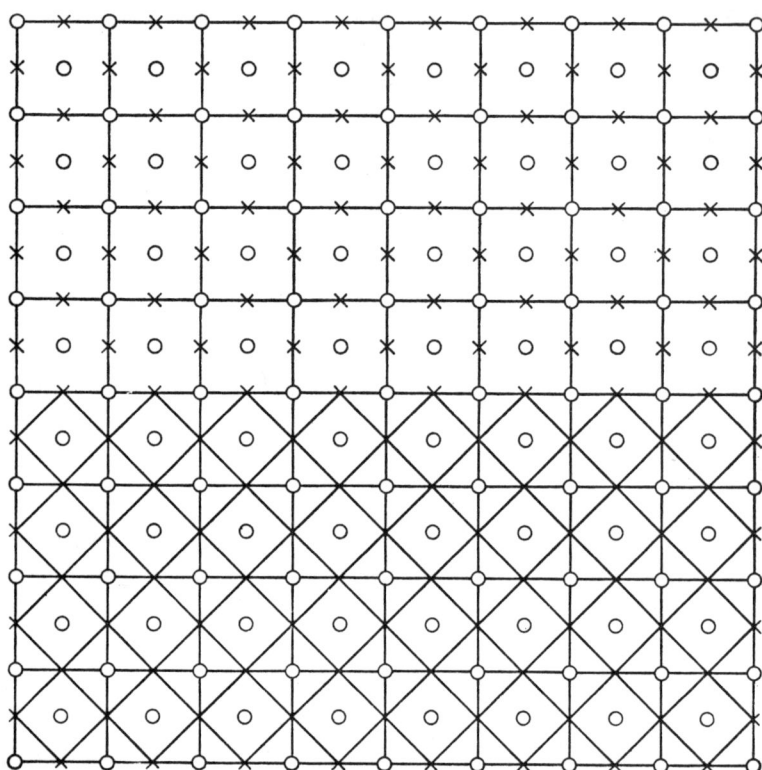

Fig. 5-2. Diagram to indicate how the face-centered tetragonal lattice can also be described as body-centered tetragonal, rotated through 45°, as described in text.

note from this that a face-centered cubic lattice, a special case of that shown, could be described as a body-centered tetragonal lattice, whose height c is the same as the side a of the cubic cell for the face-centered cubic case, but whose tetragonal cell has dimensions of only $a/\sqrt{2}$ in the xy plane, the axis being 45° from the x and y axes used in describing the cubic cell.

5-3. The Effect of Rotational Operations on a Plane Wave. We have now investigated the effect of the translational operations on a plane wave $e^{i(\mathbf{k}+\mathbf{K}_h)\cdot\mathbf{r}}$, and the Brillouin zones, or unit cells in the reciprocal lattice, which are related to these translational operations. Now we shall find the effect on the plane waves of the rotational operations, which are associated in the operations of the space groups. We can do this straightforwardly from Eq. (1-10), in which we have defined the effect of one of the operators of a space group on an arbitrary function ψ. Let us replace $\psi(\mathbf{r})$ by the special value $e^{i(\mathbf{k}+\mathbf{K}_h)\cdot\mathbf{r}}$ It is convenient to write this in the form

$$\psi(x_1,x_2,x_3) = e^{i(\mathbf{k}+\mathbf{K}_h)\cdot\mathbf{r}} = \exp\left[i\Sigma(p)(\mathbf{k}+\mathbf{K}_h)_p x_p\right] \quad (5\text{-}15)$$

where we have denoted x, y, z by x_1, x_2, x_3, and where $(\mathbf{k}+\mathbf{K}_h)_1$ represents the x component of the vector $\mathbf{k}+\mathbf{K}_h$, and so on. Then we apply Eq. (1-10), supplemented by Eqs. (1-7) and (1-9), and find

$$\begin{aligned} \{R_i|\mathbf{R}_n\}e^{i(\mathbf{k}+\mathbf{K}_h)\cdot\mathbf{r}} &= \exp\left\{i\Sigma(p)(\mathbf{k}+\mathbf{K}_h)_p[\Sigma(q)\alpha^i_{pq}x_q + \tau^i_p + \mathbf{R}_{np}]\right\} \\ &= \exp\left\{i\Sigma(q)[\Sigma(p)\alpha^i_{pq}(\mathbf{k}+\mathbf{K}_h)_p]x_q\right\}\exp\left[i\Sigma(p)(\mathbf{k}+\mathbf{K}_h)_p(\tau^i+\mathbf{R}_n)_p\right] \\ &= \exp\left[i\Sigma(q)(\mathbf{k}+\mathbf{K}_h)'_q x_q\right]\exp\left[i\Sigma(p)(\mathbf{k}+\mathbf{K}_h)_p(\tau^i+\mathbf{R}_n)_p\right] \quad (5\text{-}16) \end{aligned}$$

where

$$(\mathbf{k}+\mathbf{K}_h)'_q = \Sigma(p)\alpha^i_{pq}(\mathbf{k}+\mathbf{K}_h)_p \quad (5\text{-}17)$$

Thus we see that the effect of performing the operation $\{R_i|\mathbf{R}_n\}$ on the plane wave can be described in two parts. First, it produces a new plane wave, with a transformed wave vector whose rectangular components are given in Eq. (5-17). Second, it multiplies by the factor $\exp\left[i\Sigma(p)(\mathbf{k}+\mathbf{K}_h)_p(\tau^i+\mathbf{R}_n)_p\right]$, which depends on the translations.

In the transformation of Eq. (5-17) for finding the transformed wave vectors, the coefficients α^i_{pq} do not appear as they do in the expression of Eq. (1-7) for the transformed coordinates. Instead of summing over the second subscript q of α^i_{pq}, as in Eq. (1-7), we sum over the first, p. We shall now investigate the significance of this fact, and shall show that the operators of the point group operating on the wave vectors are the inverse operators of those operating on the coordinates of a point in ordinary space. To show this, we may use the general theorem of group theory,

$$\Gamma(R_k^{-1})_{ij} = \Gamma(R_k)^*_{ji} \quad (5\text{-}18)$$

holding for an operator R_k and its inverse R_k^{-1}, stated in Volume 1, Eq. (A12-11). We have here the case where the matrix elements are real, so that the complex conjugate is not required, and we see that interchange of the order of the subscripts in the matrix representing the operator carries us from the operation to its inverse. Without using this general theorem, however, we can at once give a proof in our case, from the orthogonality relations, Eq. (1-8). We substitute $\alpha^i_{pq} = \alpha^j_{qp}$ in the first equation of Eq. (1-8), which then becomes

$$\Sigma(p)\alpha^j_{qp}\alpha^i_{pr} = \delta_{qr} \quad (5\text{-}19)$$

This shows that the matrix product of the matrices α_{qp}^j and α_{pr}^i is the delta function, or the matrix of unity. That is, the product of these two matrices is unity, which shows that they are inverse to each other.

We have verified our statement, then, that the transformation of the wave vector, described by Eq. (5-17), proceeds by the inverse operations of the point group to the transformations of the coordinates. At this point we shall call attention to a fact which was earlier passed over without comment. In Eq. (1-13), in which we found the matrices α_{pu}^k arising from the successive application of the operators R_iR_j of the point group, we found that this quantity was given by $\Sigma(q)\alpha_{pq}^j\alpha_{qu}^i$. We pointed out at the time that this was the matrix product of the matrices representing the two operations, so that it represented the matrix for an operation of the point group. We did not comment on the fact, however, that the two matrices appeared in the opposite order to what we should have expected, α^j appearing before α^i, even though the operators appeared in the order R_iR_j. We wish to point out now, however, that if we rewrite Eq. (1-13) in terms of the inverse operators, which we may denote by $(\alpha^j)^{-1}$, and so on, the situation is different. We take the inverse of Eq. (1-13), and find

$$(\alpha^k)_{pu}^{-1} = \alpha_{up}^k = \Sigma(q)\alpha_{uq}^j\alpha_{qp}^i$$
$$= \Sigma(q)(\alpha^i)_{pq}^{-1}(\alpha^j)_{qu}^{-1} \qquad (5\text{-}20)$$

which shows that the inverse operators combine in the expected order. In other words, the operators which operate on the wave vectors combine according to the ordinary multiplication rule, whereas those operating on the coordinates do not. This has a bearing on the remark which we made at the end of Sec. 1-3, in discussing Eq. (1-16), to the effect that many writers on group theory define the operators by an inverse process to what we do, each operator according to their definition being the inverse of the operator as we have defined it. If we adopt that more usual convention, we find that the matrices representing the operations operating on the coordinates combine according to the usual multiplication rule, while those operating on the wave vector do not. We cannot have both at the same time. In treating wave mechanics, the transformation applied to wave vectors is of more importance than that applied to the coordinates, and it is for this reason that the choice of definitions of the operators of the space group made in this volume seems more convenient than the standard one.

We can appreciate the advantage of our definition when we come to consider the translations associated with the operations, and in particular the translations involved in the successive application of two operations, as given in Eq. (1-14). This expression seems very complicated, and it is not obvious at first what physical meaning it has. However, if we

Sec. 5-3]   SYMMETRY OF ELECTRONIC WAVE FUNCTIONS IN CRYSTALS   129

apply the operators directly to plane waves, it becomes very simple and straightforward. Let us go through the equivalent of the derivation of Eq. (1-12), describing the successive application of the two operators of the space group, applying the operations directly to a function ψ in the form of a plane wave. From Eqs. (5-16) and (5-17), and the discussion we have just given, we see that the effect of the application of one operator of the space group to a plane wave is to create a new plane wave whose wave vector $(\mathbf{k} + \mathbf{K}_h)'$ is derived from the original wave vector $(\mathbf{k} + \mathbf{K}_h)$ by the operator R_i^{-1} of the point group, inverse to R_i, and to multiply this by the constant factor

$$\exp\left[i\Sigma(p)(\mathbf{k} + \mathbf{K}_h)_p(\boldsymbol{\tau}^i + \mathbf{R}_n)_p\right] = \exp\left[i(\mathbf{k} + \mathbf{K}_h) \cdot (\boldsymbol{\tau}^i + \mathbf{R}_n)\right]$$

which is merely the value of the plane wave $e^{i(\mathbf{k}+\mathbf{K}_h)\cdot\mathbf{r}}$ at the displaced point $\mathbf{r} = \boldsymbol{\tau}^i + \mathbf{R}_n$. Now let us operate on this function with another operator of the space group. This second operator has no effect whatever on the constant factor, which does not involve the coordinates that are being transformed. It merely makes the same sort of change on the transformed plane wave which the original operator made on the original plane wave. Let us write this down analytically, and show that the result is the same which we have already discussed.

We let ψ be given by Eq. (5-15), and follow Eq. (5-16) in finding the effect of $\{R_j|\mathbf{R}_m\}$ on it. We find

$$\{R_j|\mathbf{R}_m\}e^{i(\mathbf{k}+\mathbf{K}_h)\cdot\mathbf{r}} = e^{i(\mathbf{k}+\mathbf{K}_h)'\cdot\mathbf{r}}e^{i(\mathbf{k}+\mathbf{K}_h)\cdot(\boldsymbol{\tau}^j+\mathbf{R}_m)} \qquad (5\text{-}21)$$

where

$$(\mathbf{k} + \mathbf{K}_h)'_q = \Sigma(p)(\alpha^j)_{qp}^{-1}(\mathbf{k} + \mathbf{K}_h)_p = \Sigma(p)\alpha_{pq}^j(\mathbf{k} + \mathbf{K}_h)_p \qquad (5\text{-}22)$$

We now wish to perform the operation $\{R_i|\mathbf{R}_n\}$ on this function. We have

$$\{R_i|\mathbf{R}_n\}\{R_j|\mathbf{R}_m\}e^{i(\mathbf{k}+\mathbf{K}_h)\cdot\mathbf{r}} = e^{i(\mathbf{k}+\mathbf{K}_h)''\cdot\mathbf{r}}e^{i(\mathbf{k}+\mathbf{K}_h)'\cdot(\boldsymbol{\tau}^i+\mathbf{R}_n)}e^{i(\mathbf{k}+\mathbf{K}_h)\cdot(\boldsymbol{\tau}^j+\mathbf{R}_m)} \qquad (5\text{-}23)$$

where

$$(\mathbf{k} + \mathbf{K}_h)''_r = \Sigma(s)(\alpha^i)_{rs}^{-1}(\mathbf{k} + \mathbf{K}_h)'_s = \Sigma(s,t)(\alpha^i)_{rs}^{-1}(\alpha^j)_{st}^{-1}(\mathbf{k} + \mathbf{K}_h)_t$$
$$= \Sigma(s,t)\alpha_{ts}^j\alpha_{sr}^i(\mathbf{k} + \mathbf{K}_h)_t \qquad (5\text{-}24)$$

In Eq. (5-24) we see the effect of the two successive operations of the point group on the wave vector of the transformed wave, agreeing with our previous discussion. To investigate the translations, we have

$$e^{i(\mathbf{k}+\mathbf{K}_h)'\cdot(\boldsymbol{\tau}^i+\mathbf{R}_n)}e^{i(\mathbf{k}+\mathbf{K}_h)\cdot(\boldsymbol{\tau}^j+\mathbf{R}_m)}$$
$$= \exp\left[i\Sigma(q)(\mathbf{k} + \mathbf{K}_h)'_q(\boldsymbol{\tau}^i + \mathbf{R}_n)_q + i\Sigma(p)(\mathbf{k} + \mathbf{K}_h)_p(\boldsymbol{\tau}^j + \mathbf{R}_m)_p\right]$$
$$= \exp\left\{i\Sigma(p)(\mathbf{k} + \mathbf{K}_h)_p[\Sigma(q)\alpha_{pq}^j(\boldsymbol{\tau}^i + \mathbf{R}_n)_q + (\boldsymbol{\tau}^j + \mathbf{R}_m)_p]\right\} \qquad (5\text{-}25)$$

which is the exponential factor that would follow from the general definition of Eq. (5-16), if we used the translation given in Eq. (1-14) for the combined operation.

We thus verify that this straightforward procedure involving the plane waves leads to the same result for the successive application of two operators of the space group which we have found earlier in Sec. 1-3 by direct operation on the coordinates: the rotation of coordinates given by Eq. (5-24) is the same one given in Eq. (1-13), except for a difference in notation, and the translation involved in Eq. (5-23) is the same one given in Eq. (1-14). This procedure of applying successive operators to a plane wave is so simple and natural that we can easily fall into the error of thinking that it must obviously hold. However, if we had adopted the definition of Eq. (1-16) rather than Eq. (1-10) for the operations of the space group, it would not go through in any such simple way.

5-4. Stars of Wave Vectors. From the preceding section, we see that there is a close correspondence between the action of an operation of the space group in ordinary space and in reciprocal space. The operation induces rotations and reflections of the wave vector in reciprocal space, as it does with the position vector in ordinary space; the rotations and reflections produced in reciprocal space are the same as those produced in ordinary space by the inverse operation of the point group. There is nothing in reciprocal space corresponding to the translations, primitive or nonprimitive, associated with the operations in ordinary space; the magnitude of the wave vector is never affected by the operations of the space group, so that all operations resemble those of a symmorphic space group. The effect of the translations is felt only in the constant factor multiplying the transformed wave function or plane wave, that is, in the values of the matrix elements associated with the various operators.

As we have seen in preceding chapters, one of the features of space groups is the existence of general positions and special positions within the unit cell. We recall that a general position is one which transforms to a different point under each of the operations of the space group, so that if an atom is located at such a general position, there must be as many similar atoms in the unit cell as there are operations in the point group. On the other hand, with a special position, more than one of the operations of the point group will transform this position into the same point, so that the number of separate atoms which we must find in the unit cell is less than the number of operations of the point group, if they are located at special positions. The limiting case is that found, for instance, at the origin of some space groups, a special position which in some groups transforms into itself under all the operations of the point group, so that we can have a crystal with only one atom in the unit cell.

Similar situations must exist in the transformations of the wave vectors. If a wave vector terminates at a general position in reciprocal space, application of all operations of the space group to the plane wave described by this wave vector will result in as many plane waves as there are operations in the point group, with a corresponding number of wave vectors, which as we have mentioned above are all of the same length. Such a set of wave vectors is said to form a star. We note that there is a real difference between the situation in ordinary space and in reciprocal space, in that in ordinary space the translations R_n associated with the operations of the space group produce transformed positions equivalent to a given point in each unit cell. In reciprocal space, where the effect of the translations R_n is only to multiply the plane waves by a factor, they do not result in increasing the number of independent wave vectors, which is still equal only to the number of operations in the point group.

Next let us consider the situation if the wave vector terminates at a special position. Then in general the number of independent wave vectors is less than the number of operations in the point group; we have a star with fewer vectors. The limiting case is, as in some space groups with the origin as a special position, that in which the wave vector is zero. Then it forms a star by itself, with one member; any operation of the space group transforms it into itself. The existence of these stars is of primary importance in studying the symmetry properties and dimensionality of the irreducible representations of the wave functions, as was first pointed out by Bouckaert, Smoluchowski, and Wigner.[1]

It will simplify our discussion if we have a special case to think about, and for that purpose we shall consider the simple cubic structure, with the space group O_h^1 ($Pm3m$). For this space group, as for all of them, the special positions are listed in the International Tables, quoted earlier. In Table 5-2 we list the special positions given there. The first column, with the listing $a \cdot \cdot \cdot m$, is Wyckoff's notation for these special positions. The next column, with numbers from 1 to 24, gives the number of equivalent points. Next we have coordinates of a typical point, reduced to a dimensionless form by giving x/a, etc. Then we have the notation introduced by Bouckaert, Smoluchowski, and Wigner for most of the special positions. These positions are indicated in Fig. 5-3, in which we show the Brillouin zone for this case, with the notations of Bouckaert, Smoluchowski, and Wigner for the various symmetry points and directions. Finally, we give the point group representing the symmetry at this point. We recall that the side of the unit cell in reciprocal space, for this case, is $1/a$, and the side of the unit cell in $\mathbf{k}$ space is $2\pi/a$. Hence the coordinate $\frac{1}{2}$, in Table 5-2, refers to a distance π/a in $\mathbf{k}$ space, which we shall use in describing the situation.

We start with the origin, the point Γ in the notation of Bouckaert, Smoluchowski, and Wigner (which we shall sometimes abbreviate BSW). We have pointed out earlier that any operation of the space group applied

[1] L. P. Bouckaert, R. Smoluchowski, and E. Wigner, *Phys. Rev.*, **50**:58 (1936).

Table 5-2
Special positions for the space group O_h^1 ($Pm3m$). Coordinates given in the form x/a, y/a, z/a, where a is the side of the cubic cell

Wyckoff's notation	No. of equivalent positions	Coordinates of typical point in special position	Notation of BSW for corresponding point in reciprocal space	Type of symmetry at corresponding point in reciprocal space
a	1	0, 0, 0	Γ	O_h
b	1	½, ½, ½	R	O_h
c	3	½, ½, 0	M	D_{4h}
d	3	½, 0, 0	X	D_{4h}
e	6	u, 0, 0	Δ	C_{4v}
f	6	½, ½, u	T	C_{4v}
g	8	u, u, u	Λ	C_{3v}
h	12	½, u, 0	Z	C_{2v}
i	12	u, u, 0	Σ	C_{2v}
j	12	½, u, u	S	C_{2v}
k	24	0, u, v		C_s
l	24	½, u, v		C_s
m	24	u, u, v		C_s

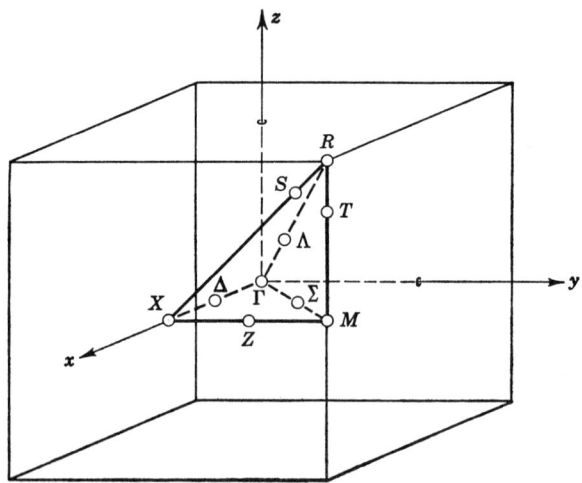

FIG. 5-3. Brillouin zone for the simple cubic space group, O_h^1 ($Pm3m$), showing notation of Bouckaert, Smoluchowski, and Wigner for the various symmetry points and directions. The side of the cube equals $1/a$.

Sec. 5-5] SYMMETRY OF ELECTRONIC WAVE FUNCTIONS IN CRYSTALS 133

to the wave vector zero leads to zero, so that this star has only one member. Next we shall consider the point e in Wyckoff's notation, or Δ in the BSW notation, namely, the case where $k_x \neq 0$, but $k_y = k_z = 0$. The wave vector is along the x direction. If we operate on this wave vector with all operations of the point group, we obtain only wave vectors of identical magnitude along $\pm x$, $\pm y$, $\pm z$, or six vectors in all. The extremity of this vector along the x direction is $k_x = \pi/a$, $k_y = k_z = 0$. This is the point d, or X, in Table 5-2. In this case the vectors $k_x = \pi/a$ and $-\pi/a$ differ by a vector of magnitude $2\pi/a$ along the x direction. This is one of the vectors $\mathbf{K}_h$: it equals $2\pi\mathbf{b}_1$. Hence these two vectors $\pm\pi/a$ both correspond to the same reduced wave vector, and they are not considered as being separate members of the star. Similarly, $k_y = \pm\pi/a$, $k_x = k_z = 0$ are equivalent, and $k_z = \pm\pi/a$, $k_x = k_y = 0$ are equivalent, leaving only three separate vectors in this star, in agreement with Table 5-2. Again, the vector g, or Λ, with equal components k_x, k_y, k_z, is one of eight members of the star, pointing toward the eight corners of the cube. If we are at the end of this vector, however, the point b or R, we can get from any one of the eight corners to any other corner by vectors $\pm 2\pi\mathbf{b}_1$ or $\pm 2\pi\mathbf{b}_2$ or $\pm 2\pi\mathbf{b}_3$, so that all eight vectors are equivalent, and there is only one separate vector in the star. These examples are enough to show the way in which the entries of Table 5-2 are arrived at. A general point, in the notation of Table 5-2, would have coordinates u, v, w, all different, and there would be 48 separate vectors in the star.

5-5. Symmetrized Plane Waves and Irreducible Representations. We now understand enough about the geometry of the reciprocal space so that we can go ahead with our primary task: to construct wave functions having the symmetry of various irreducible representations of the space group. We know from Eq. (5-5) that a sum of plane waves of the form $e^{i(\mathbf{k}+\mathbf{K}_h)\cdot\mathbf{r}}$, where $\mathbf{k}$ is fixed but $\mathbf{K}_h$ takes on all values $2\pi(h_1\mathbf{b}_1 + h_2\mathbf{b}_2 + h_3\mathbf{b}_3)$, where the h's are integers, will diagonalize the translational operations. Obviously, only plane waves associated with a single reduced wave vector $\mathbf{k}$ can enter into the expansion of a given wave function. Let us now see if we cannot build up a smaller sum than this, whose coefficients can be determined by general principles of symmetry, which can be used as symmetry orbitals, that is, as in the molecular problem, as functions having the correct symmetry properties of the final wave function. The final wave function then can be built up as a linear combination of symmetry orbitals of the proper symmetry, with coefficients to be determined. We shall call these symmetry orbitals set up as linear combinations of plane waves symmetrized plane waves.

We start with a single plane wave and let all operations of the space group operate on it. We know from the preceding section that we shall generate in this way a star of wave vectors and the corresponding plane waves, the dimensionality of the star in the simple cubic case being given by Table 5-2. On the other hand, from general principles of group theory, as described in Volume 1, Appendix 12, we know that if we start with a general function and operate on it with all operations of a group, the resulting set of functions forms a basis set for a representation of the group, called a regular representation. We know, furthermore, that by well-defined methods of group theory we can make linear combinations of these basis functions which will serve as bases for the various irreducible representations of the group. This suggests a method of procedure in the present case. We need merely make linear combinations of the wave functions arising by operating on a single plane wave with all the operations of the space group, to set up symmetrized plane waves.

Let the single plane wave have a wave vector $\mathbf{k} + \mathbf{K}_h$, which for generality we assume is not in the central Brillouin zone, so that $\mathbf{K}_h \neq 0$. Even if $\mathbf{k}$ is at a special position in the central Brillouin zone, as given in Table 5-2, we can easily choose a value of $\mathbf{K}_h$ such that the star of wave vectors $\mathbf{k} + \mathbf{K}_h$ will include the complete set of 48 wave vectors (for the cubic case). Let us suppose that $\mathbf{k}$ is at a special position. For instance, let $\mathbf{k}$ be along the x direction, so that it is at the special position Δ, in the BSW notation, but let h_1, h_2, h_3 have three different nonvanishing values. In this case the star of $\mathbf{k}$ vectors will have only six members, but we have 48 plane waves. Hence eight of these plane waves must correspond to each of the six values of $\mathbf{k}$. Out of the operations of the cubic point group, there are eight which leave x invariant; similarly, there are eight which transform it into $-x$, eight which transform it into y, and so on. The operations which leave x invariant form a subgroup of the complete point group. They are called the group of the wave vector along x.

We now see that it is only the plane waves generated by the group of the wave vector which we may combine in building up symmetrized plane waves; for if we combined wave vectors corresponding to different $\mathbf{k}$'s, the translational operations would no longer be diagonal with the same diagonal matrix element for all plane waves. Hence in the particular example we are considering, we must combine the eight plane waves leaving x invariant, to build up symmetrized plane waves. The group of the wave vector in this case is C_{4v}. We can see this by looking along the x axis, realizing that the crystal has a square type of symmetry, and remembering that the group C_{4v} will carry a square into itself. This group has eight operations, as it should.

We now have a familiar situation. We have eight functions, formed by applying the eight operations of C_{4v} to the original plane wave. These serve as basis functions for the regular representation of the group C_{4v}. We recall that this group has four one-dimensional irreducible representa-

Sec. 5-5] SYMMETRY OF ELECTRONIC WAVE FUNCTIONS IN CRYSTALS 135

tions and one two-dimensional irreducible representation. We furthermore recall that in the reduction of the regular representation, we shall find basis functions for each of the irreducible representations, the number of different sets of basis functions for each irreducible representation being equal to the dimensionality of that representation. In the present case, we find one basis function for each of the one-dimensional irreducible representations, and two sets of two basis functions each for the two-dimensional irreducible representation. That is, we find in all $4 + 2 \times 2 = 8$ basis functions, which is the correct number, since we start with eight plane waves.

We recall that the general method of setting up the basis functions for these irreducible representations is the method of projection operators. This method is formulated in Volume 1, Sec. A12-3, and is based on Eq. (A12-26) of that section, which is

$$f_{jl}^{(p)} = \Sigma(R_i)\Gamma_p(R_i)_{jl}^* R_i \psi \tag{5-26}$$

The symbols in this equation have the following meanings. The operators R_i are the operators of a group. The quantities $\Gamma_p(R_i)_{jl}$ are the matrix elements of this operator between the jth and lth basis functions for an irreducible representation denoted by the subscript p. In Eq. (5-26) we are directed to form all the functions $R_i\psi$, formed by allowing all operators of the group to operate on an arbitrary function ψ. We then make a linear combination of these functions, the coefficients being the complex conjugates of the quantities $\Gamma_p(R_i)_{jl}$. The statement of the equation is that the quantity so formed, called $f_{jl}^{(p)}$, is the jth partner in a set of basis functions for the pth irreducible representation of the group. We can choose l to be any one of the indices of the matrix elements $\Gamma_p(R_i)_{jl}$, so that we have as many values of l as the dimensionality of the irreducible representation. Each l gives a different set of basis functions. It is for this reason that we can form from the regular representation as many sets of basis functions for a given irreducible representation as the dimensionality of that representation.

To apply this method, we must then first have the table of matrix elements $\Gamma_p(R_i)_{jl}$. We shall now give this table for the group C_{4v}, which we are using to illustrate the method. The operations of the group C_{4v} are given in Volume 1, Eq. (8-3), which is stated in polar coordinates. If instead we express them in terms of rectangular coordinates y, z, taking x as the rotation axis, we have

$$\begin{aligned}
X_0\psi(y,z) &= R_1\psi = \psi(y,z) & Y_0\psi(y,z) &= R_4'\psi = \psi(y,-z) \\
X_1\psi(y,z) &= R_{13}'\psi = \psi(-z,y) & Y_1\psi(y,z) &= R_{19}\psi = \psi(z,y) \\
X_{-1}\psi(y,z) &= R_{14}'\psi = \psi(z,-y) & Y_{-1}\psi(y,z) &= R_{20}\psi = \psi(-z,-y) \\
X_2\psi(y,z) &= R_2\psi = \psi(-y,-z) & Y_2\psi(y,z) &= R_3'\psi = \psi(-y,z)
\end{aligned} \tag{5-27}$$

The first notation given in Eq. (5-27) is that of Volume 1, Eq. (8-3), and the second is that of Table 2-2 of this volume.

Next we need simple basis functions for the four one-dimensional irreducible representations and the two-dimensional representation, so as to find the desired matrix elements. Simple basis functions for these representations were given in Volume 1, Sec. 8-5. For the one-dimensional representations we have functions 1, $\sin 4\phi$, $\cos 2\phi$, $\sin 2\phi$, in polar coordinates. These are for the representations which in the notation of that section were denoted as Σ^+, Σ^-, Δ^+, Δ^-. For the two-dimensional representation Π we may choose basis functions $e^{i\phi}$, $e^{-i\phi}$, as in Volume 1, Sec. 8-5. We may equally well choose $\cos \phi$ and $\sin \phi$, which for the present purpose are more convenient. In rectangular coordinates, corresponding basis functions are 1, $yz(y^2 - z^2)$, $y^2 - z^2$, yz, for the one-dimensional representations, and y, z, for the two-dimensional representation. These simple polynomials have the same transformation properties as the trigonometric functions.

We may then use Eq. (5-27) to find the effect of operating with each of the operators on each of these basis functions. We can then set up a table, similar to Volume 1, Tables A12-3 and A12-4, for C_{3v}, giving the matrix elements for the group C_{4v}. This is given in Table 5-3. We can now use Eq. (5-27) to set up the suitable linear combinations of the eight plane waves corresponding to the same reduced wave vector, having the correct symmetries to represent the various irreducible representations. As a first step, let us write down these plane waves.

The original plane wave with which we start is

$$\psi = e^{i(k_x + 2\pi h_1/a)x} e^{2\pi i(h_2 y + h_3 z)/a} \qquad (5\text{-}28)$$

corresponding to having the reduced wave vector **k** along the x direction. When we allow the operations of Eq. (5-27) to act on this function, they change only the second factor. Hence the first factor, depending on x, is constant throughout, and we may omit it from our discussion, writing ψ as $e^{2\pi i(h_2 y + h_3 z)/a}$. Then we have

$$\begin{aligned}
X_0 \psi &= e^{2\pi i(h_2 y + h_3 z)/a} \\
X_1 \psi &= e^{2\pi i(-h_2 z + h_3 y)/a} = e^{2\pi i(h_3 y - h_2 z)/a} \\
X_{-1} \psi &= e^{2\pi i(-h_3 y + h_2 z)/a} \\
X_2 \psi &= e^{2\pi i(-h_2 y - h_3 z)/a} \\
Y_0 \psi &= e^{2\pi i(h_2 y - h_3 z)/a} \\
Y_1 \psi &= e^{2\pi i(h_3 y + h_2 z)/a} \\
Y_{-1} \psi &= e^{2\pi i(-h_3 y - h_2 z)/a} \\
Y_2 \psi &= e^{2\pi i(-h_2 y + h_3 z)/a}
\end{aligned} \qquad (5\text{-}29)$$

We note that we have in this table an example of the general principle described earlier, namely, that the application of the operators of the

Table 5-3

Matrix elements $\Gamma_p(R_i)_{jl}$ for the group C_{4v}, using basis functions 1, $yz(y^2 - z^2)$, $y^2 - z^2$, yz, and y and z, for the irreducible representations Σ^+, Σ^-, Δ^+, Δ^-, Π, which for purposes of convenience we can also denote as $\Gamma_1, \Gamma_2, \Gamma_3, \Gamma_4, \Gamma_5$ in order. We give also the symbols for these irreducible representations introduced by BSW. Indices j and l for the two-dimensional representation Γ_5 or Π or Δ_5 are indicated immediately after the symbols of the representation.

		BSW	$X_0 = R_1$	$X_1 = R'_{13}$	$X_{-1} = R'_{14}$	$X_2 = R_2$	$Y_0 = R'_4$	$Y_1 = R_{19}$	$Y_{-1} = R_{20}$	$Y_2 = R'_3$
Σ^+	Γ_1	Δ_1	1	1	1	1	1	1	1	1
Σ^-	Γ_2	Δ'_1	1	1	1	1	−1	−1	−1	−1
Δ^+	Γ_3	Δ_2	1	−1	−1	1	1	−1	−1	1
Δ^-	Γ_4	Δ'_2	1	−1	−1	1	−1	1	1	−1
Π_{11}	$(\Gamma_5)_{11}$	$(\Delta_5)_{11}$	1	0	0	−1	1	0	0	−1
Π_{21}	$(\Gamma_5)_{21}$	$(\Delta_5)_{21}$	0	−1	1	0	0	1	−1	0
Π_{12}	$(\Gamma_5)_{12}$	$(\Delta_5)_{12}$	0	1	−1	0	0	1	−1	0
Π_{22}	$(\Gamma_5)_{22}$	$(\Delta_5)_{22}$	1	0	0	−1	−1	0	0	1
$\chi(\Pi)$	$\chi(\Gamma_5)$	$\chi(\Delta_5)$	2	0	0	−2	0	0	0	0

point group to the plane wave produces the same effect as the application of the inverse operators to the wave vector. Thus, if we started with the wave vector $2\pi(h_2\mathbf{b}_2 + h_3\mathbf{b}_3)$, which equals $2\pi(\mathbf{j}h_2 + \mathbf{k}h_3)/a$, and operated on it with the inverse operator to X_1, which as we recall is X_{-1}, we should change the y component to z, the z component to $-y$, according to Eq. (5-27). That is, we should change the wave vector to $2\pi(\mathbf{j}h_3 - \mathbf{k}h_2)/a$, which would lead to the result of Eq. (5-29) for $X_1\psi$. We can verify the other entries in Eq. (5-29) in a similar way, when we recall that X_1 and X_{-1} are inverse to each other, while the other operators of the group C_{4v} form their own inverses.

We now use Eq. (5-26) to set up linear combinations of the eight functions of Eq. (5-29) which will form basis functions for irreducible representations of the group C_{4v}. For the representation Δ_1, in the BSW notation, we have the sum of all eight functions of Eq. (5-29). If we add these functions, and combine terms, the result can be expressed as 4 times

$$\cos\frac{2\pi h_2 y}{a}\cos\frac{2\pi h_3 z}{a} + \cos\frac{2\pi h_3 y}{a}\cos\frac{2\pi h_2 z}{a} \qquad (5\text{-}30)$$

This function is unchanged by any of the operations of the group, so that it transforms the same way that a constant would. A linear combination of an infinite number of such terms, with all integral values of h_2 and h_3, each multiplied by a factor $e^{i(k_x + 2\pi h_1/a)x}$, where we also use all integral h_1's, would represent a general basis function for a representation of this type.

Next we take the representation Δ_1', for which from Eq. (5-26) and Table 5-3 the basis function is the sum of the first four functions of Eq. (5-29), minus the last four functions. The resulting function can be rearranged to be -4 times

$$\sin\frac{2\pi h_2 y}{a}\sin\frac{2\pi h_3 z}{a} - \sin\frac{2\pi h_3 y}{a}\sin\frac{2\pi h_2 z}{a} \qquad (5\text{-}31)$$

In this case it is not obvious by inspection how this basis function is related to the basis function $yz(y^2 - z^2)$, which we have quoted earlier for this irreducible representation. However, if we expand the function of Eq. (5-31) in power series in y and z, we find that the first nonvanishing term is

$$\frac{1}{3!}\left(\frac{2\pi}{a}\right)^4 h_2 h_3(h_3^2 - h_2^2)yz(y^2 - z^2) \qquad (5\text{-}32)$$

which has the required form. We note both from Eq. (5-31) and (5-32) that we cannot set up such a function in case h_2 or h_3 is zero, or in case $h_3 = \pm h_2$. We note that the function of Eq. (5-31), like that of Eq. (5-30), repeats in each unit cell, unlike the polynomial of Eq. (5-32);

Sec. 5-5] SYMMETRY OF ELECTRONIC WAVE FUNCTIONS IN CRYSTALS 139

increase of either y or z by a leaves it unchanged. This is required by the fact that our function, which is multiplied by the factor $e^{i(k_x+2\pi h_1/a)x}$, has the characteristics of the function $w(r)$ from Eq. (1-5) and must repeat in each unit cell. The function will then have the type of behavior indicated by Eq. (5-32), around corresponding points in each unit cell of the crystal.

When we proceed in the same way, we find for the representation Δ_2 the function

$$\cos \frac{2\pi h_2 y}{a} \cos \frac{2\pi h_3 z}{a} - \cos \frac{2\pi h_3 y}{a} \cos \frac{2\pi h_2 z}{a} \qquad (5\text{-}33)$$

whose power-series expansion starts with the term

$$\frac{1}{2!}\left(\frac{2\pi}{a}\right)^2 (h_3^2 - h_2^2)(y^2 - z^2) \qquad (5\text{-}34)$$

from which we see that we do not obtain such a function if $h_2 = \pm h_3$. For Δ_2' we have a function

$$\sin \frac{2\pi h_2 y}{a} \sin \frac{2\pi h_3 z}{a} + \sin \frac{2\pi h_3 y}{a} \sin \frac{2\pi h_2 z}{a} \qquad (5\text{-}35)$$

whose expansion starts as

$$\left(\frac{2\pi}{a}\right)^2 h_2 h_3 yz \qquad (5\text{-}36)$$

so that we have no function of this type if either h_2 or h_3 is zero.

For the two-dimensional representation Δ_5 we can obtain two sets of basis functions, using either the sets of matrix elements with indices 11 and 21 or those with indices 12 and 22, as the coefficients in the expansions of Eq. (5-26). For the first set we have

$$\sin \frac{2\pi h_2 y}{a} \cos \frac{2\pi h_3 z}{a} \quad \text{and} \quad \cos \frac{2\pi h_3 y}{a} \sin \frac{2\pi h_2 z}{a} \qquad (5\text{-}37)$$

whose expansions obviously start off like y and z, as they should. For the second set we have

$$\sin \frac{2\pi h_3 y}{a} \cos \frac{2\pi h_2 z}{a} \quad \text{and} \quad \cos \frac{2\pi h_2 y}{a} \sin \frac{2\pi h_3 z}{a} \qquad (5\text{-}38)$$

whose expansions start off in the same way. Obviously, these two sets of basis functions are of the same type, differing only in the interchange of h_2 and h_3. We note that if either h_2 or h_3 is zero, one of the two functions (5-37) or (5-38) vanishes, and if $h_2 = \pm h_3$, the two become equal to each other, so that in any one of these cases we have only one set of basis functions.

We have seen from this example how to set up the symmetrized plane waves in a simple case. The general case for all 20 space groups which we are considering in detail is treated in Appendix 3, in which we give the necessary tables of matrix elements for the irreducible representations of the groups of the wave vectors, for all symmetry points. Our Table 5-3 is equivalent to Table A3-21, which is for the same case. In the next section we shall give further general information about the nature of the results.

5-6. General Discussion of Symmetry of Wave Functions. From our discussion of propagation along the direction Δ in the simple cubic crystal, given in the preceding section, we understand the general method of setting up symmetry orbitals out of plane waves. In the next chapter we shall find how to make linear combinations of these symmetrized plane

Table 5-4
Basis functions for the irreducible representations of the group O_h, in the form of polynomials, with the notation of Bouckaert, Smoluchowski, and Wigner for the irreducible representations

Γ_1:	1
Γ_1':	$xyz[x^4(y^2 - z^2) + y^4(z^2 - x^2) + z^4(x^2 - y^2)]$
Γ_2:	$x^4(y^2 - z^2) + y^4(z^2 - x^2) + z^4(x^2 - y^2)$
Γ_2':	xyz
Γ_{12}:	$3x^2 - r^2, \sqrt{3}\,(y^2 - z^2)$
Γ_{12}':	$xyz(3x^2 - r^2), \sqrt{3}\,xyz(y^2 - z^2)$
Γ_{15}:	x, y, z
Γ_{15}':	$yz(y^2 - z^2), zx(z^2 - x^2), xy(x^2 - y^2)$
Γ_{25}:	$x(y^2 - z^2), y(z^2 - x^2), z(x^2 - y^2)$
Γ_{25}':	yz, zx, xy

waves so as to solve Schrödinger's equation. There are still a number of general points, however, which we can bring out by further discussion of the simple cubic case.

One of these points relates to the so-called compatibility relations. Let us consider the point Γ, the center of the Brillouin zone. This is a point, as we see from Table 5-2, or as we can see from general considerations, where the symmetry is the full O_h symmetry of the cubic group. In other words, the group of the wave vector is the complete point group. We then know that there are 10 irreducible representations of this group. In Table 5-4 we give basis functions for these irreducible representations, in the form of polynomials, together with the notation of BSW for these irreducible representations. This information is taken from Volume 1, Table A12-14.

We now notice that the line Δ terminates in the point Γ. Hence there must be some sort of continuity between the symmetry orbitals at Δ and at Γ. We can examine this continuity by comparing the basis functions of Table 5-4 with the polynomial basis functions for the propagation along Δ, which as we have seen are 1 for Δ_1, $yz(y^2 - z^2)$ for Δ_1', $y^2 - z^2$ for Δ_2, yz for Δ_2', and y and z for Δ_5. The operations of the group of the wave vector Δ do not involve x, so that it can be treated as a constant as far as those operations are concerned. We may now find which wave functions along Δ are compatible with each type of function at Γ, by seeing which ones have the same type of behavior under the operations of the group of Δ.

It is obvious that Γ_1 and Δ_1 have the same type of symmetry, so that a state of symmetry Δ_1 can join smoothly to a state of symmetry Γ_1. More interesting is the case of Γ_1'. Here the wave function from Table 5-4 can be rewritten in the form $x[x^4 - x^2(y^2 + z^2) + y^2z^2]yz(y^2 - z^2)$. Now the function $x[x^4 - x^2(y^2 + z^2) + y^2z^2]$ is unchanged under all the operations of the group C_{4v}, as given in Eq. (5-27). Hence the function for Γ_1' has the same dependence on the operations of C_{4v} as the factor $yz(y^2 - z^2)$, which is the basis function for Δ_1', showing that there can be continuity between Γ_1' and Δ_1'. In a similar way the function for Γ_2, from Table 5-4, is of the same form as $y^2 - z^2$, the basis function for Δ_2, and xyz, the function for Γ_2', is of the same form as yz, the function for Δ_2'. In other words, as far as these one-dimensional basis functions are concerned, they have been labeled in the notation of BSW in such a way that a function at Δ with a given subscript has the same symmetry as the function at Γ with the same subscript.

A similar principle of labeling has been used for the other cases. Thus, let us consider the two-dimensional irreducible representation Γ_{12}, with basis functions $3x^2 - r^2$, $\sqrt{3}\,(y^2 - z^2)$. This reduces to two types of symmetry along Δ. The first function is unchanged under any of the operations of C_{4v}, and hence is of type Δ_1, while the second is of type Δ_2. The subscript 12 on Γ_{12} indicates compatibility between this two-dimensional representation of the group O_h and the two one-dimensional representations Δ_1 and Δ_2 of C_{4v}. Similarly, the basis functions of Γ_{12}' are of the form of basis functions for Δ_2' [namely, $xyz(3x^2 - r^2)$, which has the same symmetry as yz] and Δ_1' [namely, $xyz(y^2 - z^2)$, which has the same symmetry as $yz(y^2 - z^2)$], so that Γ_{12}' is compatible with Δ_1' and Δ_2'. As for the three-dimensional representations of Γ, we have compatibility between one of these and a one- and a two-dimensional irreducible representation at Δ. Thus the function x of the three basis functions for Γ_{15} is of the symmetry of Δ_1, and the remaining two, y and z, form basis functions for Δ_5. Skipping over to Γ_{25}', the function yz is a basis function for Δ_2', and xy and xz for Δ_5, so that Γ_{25}' is compatible with Δ_2' and Δ_5. The cases of Γ_{15}' and Γ_{25} are a little more complicated, in that we must make linear combinations of the basis functions for Γ, to get those for Δ, but the result is the same, that Γ'_{15} is compatible with Δ'_1 and Δ_5, and Γ_{25} with Δ_2 and Δ_5.

In Appendix 3 we not only discuss the symmetrized plane waves for all the space groups we are considering, but we also give compatibility relations, of the type just discussed. The cases just considered are included in Table A3-31. There are much more powerful methods, based on group theory, for getting at these compatibility relations, without the need of considering the analytic form of the basis functions. Unfortunately, the notation is not chosen in such a way as to indicate the compatibility automatically, except in the one case we have mentioned, that between Γ and Δ.

Now let us take a somewhat broader view of what we have been accomplishing by this study of symmetry. We continue to take the simple cubic case, and now consider a hypothetical crystal in which we have just one atom per unit cell in such a lattice, at the origin of the unit cell. We have found that for the reduced wave vector equal to zero, at point Γ of the Brillouin zone, the wave function must have the symmetry of one of the irreducible representations of the point group O_h. But we recall that at Γ, the wave function has identical values in each unit cell of the crystal, so that it has a real cubic type of symmetry. These irreducible representations of O_h, as we know from Volume 1, Sec. A12-8, give the types of symmetry arising from an atomic wave function perturbed by a cubic electrostatic field. We have such a field in this case; consequently, each atomic energy level will be split into one of these types of symmetry, either nondegenerate, doubly degenerate, or triply degenerate. The notation of BSW is used to distinguish between these different types of symmetry. Thus, an atomic s state is unaffected by the field, and appears as Γ_1 symmetry. An atomic p state retains its threefold degeneracy, and has symmetry Γ_{15}. An atomic d state is split into a twofold degenerate state Γ_{12} and a threefold degenerate Γ'_{25}, and so on.

Now let us go away from the origin of the Brillouin zone along the x direction, the direction Δ. The Γ_1 state, arising from the atomic s state, will be renamed Δ_1, but will change the properties of its wave function continuously as k_x increases. The threefold degenerate Γ_{15}, arising from the atomic p state, will split into the nondegenerate Δ_1 and the doubly degenerate Δ_5, the first coming from the x-like wave function, the second from the functions y and z. It is obvious that for propagation along x, the x axis is a preferred axis, explaining why the x-like function has a different energy from the other two. Similar splittings happen to the other symmetry types at Γ.

As we go to points of lower symmetry, there will be successively more splitting, until finally at a general point of the Brillouin zone there will be no degeneracy at all. The star of the wave vector here has 48 members,

Sec. 5-6] SYMMETRY OF ELECTRONIC WAVE FUNCTIONS IN CRYSTALS 143

and only one function out of the 48 arising by operating on a plane wave with the operations of the space group will have each of the 48 wave vectors. Consequently, there is no possibility of making linear combinations of plane waves to represent functions of different symmetry types. All these facts are of great service when it comes to calculating the energy bands. For if we are at a special symmetry position in the Brillouin zone, we can deal from the outset with functions of the proper symmetry for the irreducible representation we are considering, thereby simplifying the problem of solving Schrödinger's equation. In later chapters, when we consider various methods of approximate solution of this problem, we shall see specifically how a consideration of symmetry simplifies the solution. In the early days of calculation of energy bands, this simplification was so important that it was only at points of high symmetry that satisfactory calculations of the wave function and energy could be made. This is no longer the case; modern methods of attack on the problem allow us to calculate general positions in the Brillouin zone. But even so, consideration of symmetry is a help. Furthermore, there are many important problems which concern the behavior of the wave functions and energy bands in the neighborhood of symmetry points, and the type of discussion we have given in this chapter, and which is continued in detail in Appendix 3, is of vital importance for these problems.

We shall now go on, in Chap. 6, to a study of the first and most obvious method of approximately solving Schrödinger's equation for the periodic potential problem: the expansion of the wave function in plane waves. This is the method to which our discussion of the present chapter has been leading up. It is a very valuable method for gaining a general insight into the nature of the problem and for proving general theorems. Unfortunately, it is a very poor method for practical computation, on account of its poor convergence. We shall therefore in later chapters have to consider more rapidly convergent procedures. They can hardly be understood, however, without first understanding the nature of the plane-wave expansion of the wave function.

6

Plane-wave Expansions of Wave Functions in Crystals; The One-dimensional Case

6-1. Fourier Expansions of Periodic Functions. In this chapter and the following one we shall explore further the technique of expanding the wave function of an electron in a crystal as a linear combination of plane waves, as given in Eq. (5-5):

$$u(\mathbf{k},\mathbf{r}) = \Sigma(\mathbf{K}_h)v(\mathbf{k} + \mathbf{K}_h)e^{i(\mathbf{k}+\mathbf{K}_h)\cdot\mathbf{r}} \tag{6-1}$$

As a first step in doing this, we must explore a number of simple points of technique. First we must carry further the general method of expanding a function which repeats periodically at corresponding points of each unit cell of the crystal. This method was stated in Eq. (5-4), which is

$$F(\mathbf{r}) = \Sigma(\mathbf{K}_h)G(\mathbf{K}_h)e^{i\mathbf{K}_h\cdot\mathbf{r}} \tag{6-2}$$

where $F(\mathbf{r})$ is any periodic function satisfying the condition that

$$F(\mathbf{r} + \mathbf{R}_n) = F(\mathbf{r})$$

We recall that in these expressions

$$\mathbf{R}_n = n_1\mathbf{t}_1 + n_2\mathbf{t}_2 + n_3\mathbf{t}_3$$

and

$$\mathbf{K}_h = 2\pi(h_1\mathbf{b}_1 + h_2\mathbf{b}_2 + h_3\mathbf{b}_3)$$

where the relation between the t's and the b's is given in Sec. 5-1.

Our first problem, which we have postponed so far, is to find the Fourier coefficients $G(\mathbf{K}_h)$. To find them, we expect that we shall have orthogonality properties of the plane waves, and that if we multiply $F(\mathbf{r})$ by the plane wave $e^{-i\mathbf{K}_h'\cdot\mathbf{r}}$, and integrate over a unit cell, we should be able to find $G(\mathbf{K}_h')$. For this purpose we shall use the form of unit cell given by the parallelopiped determined by the vectors $\mathbf{t}_1$, $\mathbf{t}_2$, $\mathbf{t}_3$; on account of the periodicity of all functions concerned, the integral will be the same as over a unit cell of the Wigner-Seitz type. In integrating, it will be convenient to represent the vector $\mathbf{r}$ by the expression $\xi \mathbf{t}_1 + \eta \mathbf{t}_2 + \zeta \mathbf{t}_3$, where ξ, η, ζ are allowed to range from 0 to 1. If now we let $\mathbf{K}'_h$ stand for the combination $2\pi(h'_1\mathbf{b}_1 + h'_2\mathbf{b}_2 + h'_3\mathbf{b}_3)$, we find

$$\int F(\mathbf{r})e^{-i\mathbf{K}_h'\cdot\mathbf{r}}\,dv = \Sigma(h_1 h_2 h_3)G(\mathbf{K}_h)$$
$$\times \int \exp\{2\pi i[(h_1 - h'_1)\mathbf{b}_1 + (h_2 - h'_2)\mathbf{b}_2 + (h_3 - h'_3)\mathbf{b}_3]\}\cdot\mathbf{r}\,dx\,dy\,dz \quad (6\text{-}3)$$

where the integration is to be extended over the unit cell. We may now replace the integration with respect to $dx\,dy\,dz$ by an integration with respect to $d\xi\,d\eta\,d\zeta$, as stated above. In doing this, we use the relation

$$dx\,dy\,dz = J\,d\xi\,d\eta\,d\zeta$$

where the Jacobian J is equal to

$$J = \begin{vmatrix} \dfrac{\partial x}{\partial \xi} & \dfrac{\partial x}{\partial \eta} & \dfrac{\partial x}{\partial \zeta} \\ \dfrac{\partial y}{\partial \xi} & \dfrac{\partial y}{\partial \eta} & \dfrac{\partial y}{\partial \zeta} \\ \dfrac{\partial z}{\partial \xi} & \dfrac{\partial z}{\partial \eta} & \dfrac{\partial z}{\partial \zeta} \end{vmatrix} = \begin{vmatrix} t_{1x} & t_{2x} & t_{3x} \\ t_{1y} & t_{2y} & t_{3y} \\ t_{1z} & t_{2z} & t_{3z} \end{vmatrix}$$
$$= \Omega \quad (6\text{-}4)$$

where Ω is the volume of the unit cell. Then the integral in Eq. (6-3) becomes Ω times the product of the integral

$$\int_0^1 e^{2\pi i(h_1 - h_1')\xi}\,d\xi$$

and similar integrals over η and ζ. Since each of these integrals is zero if the corresponding $h - h'$ is an integer different from zero, but equal to unity if $h - h' = 0$, we have finally from Eq. (6-3)

$$G(\mathbf{K}_h) = \Omega^{-1}\int F(\mathbf{r})e^{-i\mathbf{K}_h\cdot\mathbf{r}}\,dv \quad (6\text{-}5)$$

where the integration is extended over the unit cell.

There are two simple examples of the application of this formula. In the first place, if the potential energy of the electron is $V(\mathbf{r})$, a periodic function of position, then we may write

$$V(\mathbf{r}) = \Sigma(\mathbf{K}_h)W(\mathbf{K}_h)e^{i\mathbf{K}_h\cdot\mathbf{r}}$$

where

$$W(\mathbf{K}_h) = \Omega^{-1}\int V(\mathbf{r})e^{-i\mathbf{K}_h\cdot\mathbf{r}}\,dv \quad (6\text{-}6)$$

Similarly, if the wave function is given by Eq. (6-1), we may write

$$v(\mathbf{k} + \mathbf{K}_h) = \Omega^{-1}\int u(\mathbf{k},\mathbf{r})e^{-i(\mathbf{k}+\mathbf{K}_h)\cdot\mathbf{r}}\,dv \quad (6\text{-}7)$$

6-2. Periodic Boundary Conditions. If we regard a crystal in a mathematically abstract way, we may assume it to extend indefinitely in each direction and be infinitely large. In this case, there will be an infinite number of translation vectors $\mathbf{R}_n$ which will carry it into itself, and hence we shall be dealing with an infinite group of operators. For many mathematical reasons it is more convenient to have a finite crystal to deal with; and yet we do not wish to become involved in the problems, which are really physical rather than mathematical, that would arise if we had a finite crystal bounded by exterior surfaces, outside which we had empty space. The surface layer of a crystal is a very complicated thing, to be treated separately, and deserving of very careful discussion. To avoid this difficulty, and yet to have the mathematical advantages of a finite crystal, it is customary to use what are called periodic boundary conditions. We can introduce the idea behind this method by considering the case of a ring of N hydrogen atoms, such as the case of H_6 discussed in Volume 1, Appendix 13.

Let us suppose that we are dealing with a ring or chain of N hydrogen atoms, where N is so great that the circumference of the ring is of ordinary dimensions, say, a centimeter. It would make no essential difference in this case whether the chain was in the form of a ring, closing on itself, or formed part of a long straight line of atoms. Any effect of the ends will hardly extend inward more than a few atoms' distance, and by far the larger part of the chain or ring will be independent of the ends, and independent of whether it is closed on itself or not. If, however, we had a straight line of atoms, we could get a situation mathematically like the ring by assuming that the wave function had to come back to its initial value after going along the line a distance of N atoms, just as with the ring the wave function must come back to its initial value after going once around the ring. Thus we arrive at the idea that an infinite line of atoms, with periodic boundary conditions such that we have a repetition after N atoms, will have the mathematical advantages of a finite ring, and yet will correspond to a linear array of atoms. If we investigate the properties of the chain of atoms, and allow N to be so great that the bulk properties become independent of N, then we may be confident that we have not interfered with the applicability of the mathematics to any aspect of the problem except that of surface behavior, which we shall specifically disregard.

By analogy with this, we may assume in three dimensions that the wave function will repeat periodically if we go along the direction of the vector $\mathbf{t}_1$ by the amount $N_1\mathbf{t}_1$, along the vector $\mathbf{t}_2$ by the amount $N_2\mathbf{t}_2$, and along the vector $\mathbf{t}_3$ by $N_3\mathbf{t}_3$, where N_1, N_2, N_3 are large integers. We shall then have a fundamental volume containing $N_1N_2N_3 = N$ unit cells. If this volume is so large in each direction that it has ordinary macroscopic

Sec. 6-2] PLANE-WAVE EXPANSIONS—ONE DIMENSION 147

dimensions, rather than being on an atomic scale, we may be quite safe in assuming that any physical results which we shall derive from the model will be independent of N. This is the method of periodic boundary conditions.

If N_1 successive applications of the displacement $\mathbf{t}_1$ bring the wave function back to its initial value, it is clear that $e^{i\mathbf{k}\cdot N_1\mathbf{t}_1}$ must be equal to unity. If we write $\mathbf{k} = 2\pi(p_1\mathbf{b}_1 + p_2\mathbf{b}_2 + p_3\mathbf{b}_3)$, this condition demands that $N_1 p_1$ should be an integer, or p_1 should be an integer divided by N_1. Since the interior of the unit cell in reciprocal space is contained in the range $0 \leq p_1 < 1$, if for the moment we use the parallelopiped form of unit cell, we see that this means that p_1 can take on the values $0, 1/N_1, 2/N_1, \ldots, (1 - 1/N_1)$. The next value, 1, is equivalent to the value 0 already indicated, so that we have N_1 different values of p_1, spaced a distance $1/N_1$ apart. Similarly, the quantity p_2 can take on values spaced a distance $1/N_2$ apart, and p_3 can take on values spaced a distance $1/N_3$ apart. We have, in other words, a closely spaced lattice of allowable values of $\mathbf{k}$, in the $\mathbf{k}$ space, so closely spaced that the unit cell contains $N_1 N_2 N_3 = N$ such points, distributed with uniform density in $\mathbf{k}$ space. Ordinarily, when we choose N to be a very large number, the spacing is practically continuous, but nevertheless it is very convenient to be able to treat $\mathbf{k}$ as a quantity taking on only discrete values, rather than being continuously variable.

The situation which we then have is very similar to what we are already familiar with in the group C_N, which we have taken up in detail in Volume 1. We have only N independent translation vectors $\mathbf{R}_n$ (including zero) which will carry us from one unit cell to all unit cells inside our fundamental region of N unit cells; any larger value of $\mathbf{R}_n$, which would carry us outside this fundamental region, can be brought back into the fundamental region by adding or subtracting integral multiples of the vectors $N_1\mathbf{t}_1, N_2\mathbf{t}_2, N_3\mathbf{t}_3$. Thus we have N operators in our translation group, instead of an infinite number. This is entirely analogous to the way in which, say, in the group C_6, we can add or subtract integral multiples of 6 to the index q of the operators X_q discussed in Volume 1, Chap. 8, without changing the operators. Furthermore, as in the case of C_N, we have the same number of irreducible representations of the translation group as we have of operators, since the translation group is Abelian. Each irreducible representation is characterized by the value of $\mathbf{k}$, and we have just seen that there are N values of $\mathbf{k}$ inside the central unit cell in reciprocal space. If we prefer to use the central Brillouin zone rather than the central unit cell, we again have N values of $\mathbf{k}$ inside it. This is analogous to the six values of m, representing irreducible representations in the case C_6. The use of the Brillouin zone corresponds, in the case of C_6, to the use of $m = -2, -1, 0, 1, 2, 3$, instead of to the use of $m = 0, 1, 2, 3, 4, 5$, as we should have if we used the analogue of the unit cell determined by the vectors b_1, b_2, b_3. The reader is referred to Volume 1, Sec. 12-2, where we discuss the benzene molecule, for an example of the group C_6, worked out in close analogy to the methods which we are preparing to use for discussing energy bands.

If we use periodic boundary conditions, then it is natural to normalize the wave function over this fundamental repeating volume of space. If we write the wave function in the form of Eq. (6-1), we can then check the normalization and orthogonality properties of the wave functions. By methods similar to those used in deriving Eq. (6-5), we can show that

$$\int e^{-i(\mathbf{k}+\mathbf{K}_h)\cdot\mathbf{r}} e^{i(\mathbf{k}'+\mathbf{K}_h')\cdot\mathbf{r}} \, dv \qquad (6\text{-}8)$$

integrated over the fundamental volume of N unit cells is zero if $\mathbf{k} + \mathbf{K}_h$ and $\mathbf{k}' + \mathbf{K}_h'$ represent any two different wave vectors in $\mathbf{k}$ space satisfying the periodic boundary conditions. If they represent the same wave vector, the integral obviously equals the volume of the N unit cells. Hence we find that

$$\int u^*(\mathbf{k},\mathbf{r}) u(\mathbf{k},\mathbf{r}) \, dv = 1 = N\Omega \Sigma (\mathbf{K}_h) v^*(\mathbf{k} + \mathbf{K}_h) v(\mathbf{k} + \mathbf{K}_h) \qquad (6\text{-}9)$$

where the integration is over the N unit cells. In analogy with this result we note that in Eq. (6-5), if we integrate over N unit cells, the right side should contain a factor $N^{-1}\Omega^{-1}$ in place of Ω^{-1} as found in that equation. Furthermore, there will be an extra factor N^{-1} in Eqs. (6-6) and (6-7), if the integration is over N unit cells rather than a single unit cell.

If we consider the orthogonality integral, $\int u^*(\mathbf{k},\mathbf{r}) u(\mathbf{k}',\mathbf{r}) \, dv$, where $\mathbf{k}$ and $\mathbf{k}'$ are different allowed reduced wave vectors, we must get zero, for we have no nonvanishing integrals of the type of Eq. (6-8). Hence two functions associated with different $\mathbf{k}$'s are orthogonal. This is an explicit proof, in this simple case, of the orthogonality of wave functions associated with different irreducible representations of the group of symmetry operators.

6-3. Schrödinger's Equation for the Periodic-potential Problem. In Eq. (6-1) we have set up a general basis function for the irreducible representation of the group of translations R_n, defined by the reduced wave vector $\mathbf{k}$. By general principles we know that the solutions of Schrödinger's equation for the periodic-potential problem can be expressed in this form. Hence we can convert Schrödinger's equation into a set of equations for the coefficients $v(\mathbf{k} + \mathbf{K}_h)$ appearing in Eq. (6-1). Since the expansion of Eq. (6-1) is of the form of a linear combination of orthogonal basis functions $e^{i(\mathbf{k}+\mathbf{K}_h)\cdot\mathbf{r}}$, the problem of finding the coefficients is like any problem of expanding a solution of Schrödinger's equation as a linear combination of orthogonal functions. As is familiar from

Sec. 6-3] PLANE-WAVE EXPANSIONS—ONE DIMENSION 149

all such cases, we can get at the equations for the coefficients by substituting the expression of Eq. (6-1) into Schrödinger's equation.

The Hamiltonian operator for our one-electron periodic-potential problem is

$$(H)_{\text{op}} = -\frac{\hbar^2}{2m}\nabla^2 + V(r) = -\frac{\hbar^2}{2m}\nabla^2 + \sum(\mathbf{K}_l)W(\mathbf{K}_l)e^{i\mathbf{K}_l \cdot \mathbf{r}} \quad (6\text{-}10)$$

where we have used the expansion of Eq. (6-6) for the periodic potential. We let $(H)_{\text{op}}$ in this form operate on the wave function expressed in the form of Eq. (6-1). We set this equal to $E(\mathbf{k})u(\mathbf{k},\mathbf{r})$, or to $E(\mathbf{k})\Sigma(\mathbf{K}_h)v(\mathbf{k} + \mathbf{K}_h)e^{i(\mathbf{k}+\mathbf{K}_h)\cdot\mathbf{r}}$, multiply by the conjugate of one of the plane waves, namely, $e^{-i(\mathbf{k}'+\mathbf{K}_j)\cdot\mathbf{r}}$, where $\mathbf{k}'$ is an allowed reduced wave vector, and integrate over the volume. We then obtain

$$\int e^{-i(\mathbf{k}'+\mathbf{K}_j)\cdot\mathbf{r}}\sum(\mathbf{K}_h)v(\mathbf{k}+\mathbf{K}_h)\left[(\mathbf{k}+\mathbf{K}_h)^2\frac{\hbar^2}{2m}\right.$$

$$\left. + \sum(\mathbf{K}_l)W(\mathbf{K}_l)e^{i\mathbf{K}_l\cdot\mathbf{r}}\right]e^{i(\mathbf{k}+\mathbf{K}_h)\cdot\mathbf{r}}\,dv$$

$$= \int e^{-i(\mathbf{k}'+\mathbf{K}_j)\cdot\mathbf{r}}E(\mathbf{k})\sum(\mathbf{K}_h)v(\mathbf{k}+\mathbf{K}_h)e^{i(\mathbf{k}+\mathbf{K}_h)\cdot\mathbf{r}}\,dv \quad (6\text{-}11)$$

We shall get nothing from the integration, as a result of orthogonality, unless the wave vector $\mathbf{k}' + \mathbf{K}_j$ equals the wave vector of the exponential function by which it is multiplied. This implies first that $\mathbf{k}'$ must equal $\mathbf{k}$; that is, the Hamiltonian has no nondiagonal matrix elements between functions of different $\mathbf{k}$'s. Furthermore, in the first term on the left side, we have nothing unless $\mathbf{K}_h = \mathbf{K}_j$, or in the second term unless

$$\mathbf{K}_l + \mathbf{K}_h = \mathbf{K}_j$$

Similarly, on the right we have nothing unless $\mathbf{K}_h = \mathbf{K}_j$. Thus we are left with

$$(\mathbf{k}+\mathbf{K}_j)^2\frac{\hbar^2}{2m}v(\mathbf{k}+\mathbf{K}_j) + \sum(\mathbf{K}_l)W(\mathbf{K}_l)v(\mathbf{k}+\mathbf{K}_j-\mathbf{K}_l)$$

$$= E(\mathbf{k})v(\mathbf{k}+\mathbf{K}_j) \quad (6\text{-}12)$$

In Eq. (6-12) we have a set of simultaneous linear algebraic equations for the quantities $v(\mathbf{k} + \mathbf{K}_j)$. As usual, they cannot be satisfied unless the determinant of coefficients vanishes, which provides a secular equation for the energy $E(\mathbf{k})$. For a given value of $\mathbf{k}$, the equation will have an infinite number of eigenvalues, which we may denote as $E_n(\mathbf{k})$, numbering them in order of ascending energies. Associated with each eigenvalue $E_n(\mathbf{k})$ we shall find sets of coefficients $v_n(\mathbf{k} + \mathbf{K}_j)$, which will define the eigenfunction according to Eq. (6-1). Of course, we cannot solve these equations in a practical way if we are dealing with an infinite set of basis functions $e^{i(\mathbf{k}+\mathbf{K}_j)\cdot\mathbf{r}}$ However, if we restrict ourselves to a finite but large set, the problem becomes one in algebra, of the type which can be handled by digital computers, and in some simple cases, as we shall discuss later, we are able to handle a sufficiently large number of plane waves in this way to get an acceptable approximate solution of Schrödinger's equation. Whether we can do this or not as a practical matter, nevertheless this method in principle is very straightforward, and is useful for the insight it gives us into the solutions of the periodic-potential problem.

We now take together the lowest eigenvalues, $E_1(\mathbf{k})$, associated with each $\mathbf{k}$ value. Since the values of $\mathbf{k}$ are very close together in $\mathbf{k}$ space, this will give us very nearly a continuous function of $\mathbf{k}$. This set of energies forms the lowest energy band. We take the next lowest energy, $E_2(\mathbf{k})$, and treat these values as the second energy band, and so on. We shall have many examples before we are through, of energy bands, as functions of $\mathbf{k}$. For a value of $\mathbf{k}$ having special symmetry properties, such as we have been discussing in the preceding chapter, the wave functions determined directly by solution of Eq. (6-12) will automatically take on the symmetry properties of one or another of the irreducible representations of the group of the appropriate wave vector. If the symmetry demands a degenerate state, a suitable number of energy bands, whose energies are different away from the symmetry point, will automatically draw together in energy, leading to an exactly degenerate energy level at the symmetry value of $\mathbf{k}$.

We recall that there are N allowed values of $\mathbf{k}$, in the central Brillouin zone. Hence a single energy band comprises N different states, each capable of holding two electrons, one of each spin. We then may have in the crystal $2N$ electrons in each energy band, as far as the exclusion principle is concerned. Since there are N unit cells in the repeating volume of the crystal, this means that each energy band can hold two electrons, one of each spin, per unit cell. If we are dealing with the very simple case where there is one atom per unit cell, as in a metal having the body-centered or face-centered cubic structure (and where we are using primitive unit cells), this means that we have two electrons, one of each spin, per atom. Since each of the energy levels has arisen from a single atomic level, we have accommodations for just as many electrons in the crystal energy levels as in the atomic levels from which they arise. As a general thing, the energy values associated with an energy band will approach the common atomic energy value as the internuclear separation becomes infinite, and we approach the problem of the isolated atoms.

In the self-consistent-field model, we expect energy levels to be filled up, starting at the bottom, until all the electrons are accounted for. We are then interested, as was pointed out in Volume 1, Sec. 9-4, in the question as to whether we have partially filled energy bands, or entirely filled bands, with energy gaps above. In the first case we have a conductor, in the second case a semiconductor or insulator. In some cases we can answer this question in the most elementary fashion. For instance, metallic sodium crystallizes in the body-centered cubic structure, one atom per unit cell. We know that in the isolated atom, we have enough electrons to fill the $1s$, $2s$, and $2p$ shells, and there is one electron per atom left over for the $3s$ shell. Correspondingly, in the crystal, there are enough electrons to fill the energy bands arising from the $1s$, $2s$, and $2p$ shells, with one electron per atom left over for the band arising from the $3s$. But this band can hold two electrons per atom. Hence it will be half filled, and sodium must be a conductor. In other cases, the question as to whether a given substance is a conductor or an insulator is not so simple; we shall be able to answer the question, in each case, when we have worked out its energy bands.

6-4. Momentum Eigenfunctions. The set of quantities $v(\mathbf{k} + \mathbf{K}_h)$, introduced in Eq. (6-1), has a very interesting physical significance. The quantity $N\Omega v^*(\mathbf{k} + \mathbf{K}_h)v(\mathbf{k} + \mathbf{K}_h)$, occurring in Eq. (6-9), measures the probability that the electron, whose wave function is $u(\mathbf{k},\mathbf{r})$, should have the momentum $(\mathbf{k} + \mathbf{K}_h)\hbar$. This is very similar to the significance of the ordinary wave function, for which $u^*(\mathbf{r})u(\mathbf{r})\, dv$ measures the probability that the coordinates of the particle lie in the range dv. For this reason the set of quantities $v(\mathbf{k} + \mathbf{K}_h)$ is sometimes called the momentum eigenfunction. For a crystal, we have only a discrete set of quantities to represent the momentum eigenfunction, on account of the periodicity properties of the wave functions, but in a nonrepeating structure, such as an individual atom or molecule, the momentum eigenfunction is a continuous function of the momentum (or of $\mathbf{k}$), just as the ordinary eigenfunction $u(\mathbf{r})$ is a continuous function of $\mathbf{r}$.

Let us now inquire how we can verify this interpretation of $v(\mathbf{k} + \mathbf{K}_h)$. The physical reasoning underlying the interpretation is that if the wave function is a plane wave, it represents a particle of precisely determined momentum. If it is a superposition of plane waves, the square of the amplitude of each such plane wave measures the probability that the particle is in that wave, or has that momentum. To get a somewhat more formal proof, let us see how to find the average of any function of the momentum $\mathbf{p}$, averaged over the wave function $u(\mathbf{r})$ of Eq. (6-1).

Let us take an operator $F(\mathbf{p})$, which we assume can be expanded as a power series in $\mathbf{p}$. If $\mathbf{p}$, regarded as the operator $-i\hbar\nabla$, operates on a plane wave $e^{i\mathbf{k}\cdot\mathbf{r}}$ it simply multiplies the wave function by the vector quantity $\mathbf{k}\hbar$. Hence, if the operator $F(\mathbf{p})$ operates on the same plane wave function, it multiplies that function by the factor $F(\mathbf{k}\hbar)$. We see then that the effect of allowing such an operator $F(\mathbf{p})$ to operate on $u(\mathbf{r})$, expressed as in Eq. (6-1), is to give the result

$$F(\mathbf{p})u(\mathbf{r}) = \Sigma(\mathbf{K}_h)F[(\mathbf{k}+\mathbf{K}_h)\hbar]v(\mathbf{k}+\mathbf{K}_h)e^{i(\mathbf{k}+\mathbf{K}_h)\cdot\mathbf{r}} \qquad (6\text{-}13)$$

If we then multiply by $u^*(\mathbf{r})$, and integrate over the fundamental volume of N unit cells, we find for the expectation value of $F(\mathbf{p})$

$$[F(\mathbf{p})]_{av} = \int u^*(\mathbf{r})F(\mathbf{p})u(\mathbf{r})\,dv$$
$$= N\Omega\Sigma(\mathbf{K}_h)v^*(\mathbf{k}+\mathbf{K}_h)F[(\mathbf{k}+\mathbf{K}_h)\hbar]v(\mathbf{k}+\mathbf{K}_h) \qquad (6\text{-}14)$$

In other words, Eq. (6-14) expresses the expectation value of a function of a momentum as a weighted mean of the value of the function of momentum, calculated at values $(\mathbf{k}+\mathbf{K}_h)\hbar$ of momentum, weighted by the quantities $N\Omega v^*(\mathbf{k}+\mathbf{K}_h)v(\mathbf{k}+\mathbf{K}_h)$.

It is clear, then, that our interpretation of v as a momentum eigenfunction is correct, and that furthermore we can introduce a momentum space, proportional to the reciprocal space or the $\mathbf{k}$ space, the momentum being given by $\mathbf{k}\hbar$. We have unit cells, Brillouin zones, etc., in the momentum space, just as in the reciprocal space. We have translation vectors $\mathbf{P}_h$, equal to $\mathbf{K}_h\hbar$. In terms of this momentum space, we may rewrite Eq. (6-1) as

$$u(\mathbf{p},\mathbf{r}) = \Sigma(\mathbf{P}_h)v(\mathbf{p}+\mathbf{P}_h)e^{i(\mathbf{p}+\mathbf{P}_h)\cdot\mathbf{r}/\hbar} \qquad (6\text{-}15)$$

where $v(\mathbf{p}+\mathbf{P}_h)$ is the same quantity as $v(\mathbf{k}+\mathbf{K}_h)$ defined before, but merely carrying the index $\mathbf{p}+\mathbf{P}_h$ rather than $\mathbf{k}+\mathbf{K}_h$.

It is possible to set up a very interesting parallel symbolism between the ordinary coordinate eigenfunction, and the method of constructing operators from functions of the coordinates and momenta to operate on it, and a similar method of constructing operators to operate on the momentum eigenfunction. It proves to be the case that, if we start with any function $F(x,y,z,p_x,p_y,p_z)$, then in dealing with momentum eigenfunctions we construct an operator in which the momentum components are unchanged, but the coordinates x, y, z, wherever they appear, are to be replaced by the operators $i\hbar\partial/\partial p_x$, $i\hbar\partial/\partial p_y$, $i\hbar\partial/\partial p_z$. Then we can get the average value of a function by use of this operator and the momentum eigenfunction, and we can set up the analogue of Schrödinger's equation by use of this operator. We shall illustrate this possibility by showing that Eq. (6-12) can be derived in this way.

The Hamiltonian which we are assuming for the periodic potential problem is $p^2/2m + V(r)$. We expand $V(r)$ in Fourier expansion, which in terms of the momentum notation is

$$V(\mathbf{r}) = \Sigma(\mathbf{P}_h)W(\mathbf{P}_h)e^{i\mathbf{P}_h\cdot\mathbf{r}/\hbar} \qquad (6\text{-}16)$$
$$W(\mathbf{P}_h) = \Omega^{-1}\int V(\mathbf{r})e^{-i\mathbf{P}_h\cdot\mathbf{r}/\hbar}\,dv$$

which follows at once from Eq. (6-6), where in the integration we are integrating over the unit cell. In setting up the operator form of the Hamiltonian, we leave the kinetic energy, $\mathbf{p}^2/2m = (p_x^2 + p_y^2 + p_z^2)/2m$, unchanged, but in the expansion of the potential energy, we are to replace x, y, z in the exponential by differential operators, as we have just stated. The exponential $e^{i\mathbf{P}_h \cdot \mathbf{r}/\hbar}$ factors into three factors, of which one is $e^{iP_{hx}x/\hbar}$, which is to be transformed according to the rule into $e^{-P_{hx}\partial/\partial p_x}$, so that the potential-energy operator becomes

$$(V)_{\mathrm{op}} = \sum (\mathbf{P}_h) W(\mathbf{P}_h) \exp\left[-\left(P_{hx}\frac{\partial}{\partial p_x} + P_{hy}\frac{\partial}{\partial p_y} + P_{hz}\frac{\partial}{\partial p_z}\right)\right] \quad (6\text{-}17)$$

When this operates on a function of momenta, we can get the effect by expanding the exponential in power series, letting each term operate on the function of momenta, in which case it reduces to an ordinary three-dimensional Taylor expansion. If the function of momenta is $v(\mathbf{p})$, we find

$$(V)_{\mathrm{op}} v(\mathbf{p}) = \Sigma(\mathbf{P}_h) W(\mathbf{P}_h) v(\mathbf{p} - \mathbf{P}_h) \quad (6\text{-}18)$$

If we now substitute the Hamiltonian operator arising in this way into Schrödinger's equation for the momentum eigenfunction $v(\mathbf{p})$, we find

$$\frac{p^2}{2m} v(\mathbf{p}) + \sum (\mathbf{P}_h) W(\mathbf{P}_h) v(\mathbf{p} - \mathbf{P}_h) = E v(\mathbf{p}) \quad (6\text{-}19)$$

which aside from a change of notation is identical with Eq. (6-12). The quantity $v(\mathbf{p})$ written here is equivalent to $v(\mathbf{k} + \mathbf{K}_h)$ written in Eq. (6-12), and p^2 is equivalent to $(\mathbf{k} + \mathbf{K}_j)^2 \hbar^2$. Hence we verify that the formalism of momentum eigenfunctions leads to the same result as the ordinary formalism of coordinate eigenfunctions, in this case. We note that, on account of dealing here with a periodic potential, the Fourier expansion of the potential involves only Fourier components $W(\mathbf{P}_h)$ lying at lattice points of the reciprocal lattice. However, if we had been treating a nonperiodic potential, this Fourier expansion would have become a Fourier integral, and the summation over $\mathbf{P}_h$ in Eq. (6-19) would have been replaced by a continuous integration over a continuous variable $\mathbf{P}$. In this case the set of difference equations of Eq. (6-19), relating the values of $v(\mathbf{p})$ at equivalent points of all unit cells in reciprocal space, would have been replaced by equations relating the values of $v(\mathbf{p})$ at all points of a continuous momentum space. In other words, we should have found nondiagonal matrix elements of the Hamiltonian between functions which correspond to different irreducible representations of the translation group of the crystal, as we should expect if we were dealing with a nonperiodic potential.

For a periodic potential, we have seen that the momentum eigenfunction associated with a given value of $\mathbf{p}$ is defined only on a lattice of points

$\mathbf{p} + \mathbf{P}_h$, of which one point, which we may take to be $\mathbf{p}$, is in the central Brillouin zone, and one point is found in each unit cell of reciprocal or momentum space. We have already pointed out that the eigenvalues E associated with different $\mathbf{p}$'s, or different $\mathbf{k}$'s, in the central Brillouin zone, will form practically a continuous function of $\mathbf{p}$, or $\mathbf{k}$, which take on N closely spaced values within this zone, and we have pointed out how we can get a continuous function $E_n(\mathbf{p})$ or $E_n(\mathbf{k})$ of position in this zone, representing the energy. In a similar way, the functions $v(\mathbf{p} + \mathbf{P}_h)$, which like the energy should carry a subscript n to indicate the nth energy band, will be essentially continuous functions of the momentum, when we consider together all the wave functions corresponding to an energy band. As N, the number of unit cells in the repeating region, becomes infinite, both these quantities $E_n(\mathbf{p})$ and $v_n(\mathbf{p} + \mathbf{P}_h)$ become really continuous functions.

We may expect that there will be some sort of continuity between the values of $v_n(\mathbf{p})$ in the central Brillouin zone and the values $v_n(\mathbf{p} + \mathbf{P}_h)$ in other unit cells.[1] This function, regarded as a continuous function throughout momentum space, will give complete information about all wave functions in an energy band. We shall show in the next section that it has very interesting relations to the Wannier functions, whose significance we have already found in Volume 1, Sec. 9-5, and which are useful in considering the localization of electrons in bonds. It is a function which does not have the properties of periodicity in the momentum space. In contrast, the energy $E_n(\mathbf{p})$ repeats in each unit cell of momentum space; the energy is determined only by the lattice of points on which $v(\mathbf{p} + \mathbf{P}_h)$ is defined, and all lattice points given by the same $\mathbf{p}$, but different $\mathbf{P}_h$'s, correspond to the same lattice, and hence the same energy.

6-5. The Wannier Functions. We can set up Wannier functions[2] from the wave functions of each energy band. If $u_n(\mathbf{k},\mathbf{r})$ represents the wave function of the nth band, corresponding to the wave vector $\mathbf{k}$, we may write by analogy with Volume 1, Eq. (9-29),

$$a_n(\mathbf{r} - \mathbf{R}_j) = N^{-\frac{1}{2}} \Sigma(\mathbf{k}) u_n(\mathbf{k},\mathbf{r}) e^{-i\mathbf{k}\cdot\mathbf{R}_j} \qquad (6\text{-}20)$$

which represents the Wannier function in the unit cell located at vector position $\mathbf{R}_j$. As in the proof of Volume 1, Eq. (9-30), we can show that these Wannier functions are normalized and orthogonal. Furthermore, as in Volume 1, Eq. (9-31), we may prove that the wave function $u_n(\mathbf{k},\mathbf{r})$ can be written as a Bloch sum of Wannier functions:

$$u_n(\mathbf{k},\mathbf{r}) = N^{-\frac{1}{2}} \Sigma(\mathbf{R}_j) a_n(\mathbf{r} - \mathbf{R}_j) e^{i\mathbf{k}\cdot\mathbf{R}_j} \qquad (6\text{-}21)$$

[1] For exceptions to this statement, see Secs. 7-4 and 7-5.
[2] G. H. Wannier, *Phys. Rev.*, **52**:191 (1937).

Sec. 6-5] PLANE-WAVE EXPANSIONS—ONE DIMENSION 155

There is one point to note concerning the Wannier functions, which arises from our use of periodic boundary conditions. Since our functions $u_n(\mathbf{k},\mathbf{r})$ repeat periodically in each fundamental region of N unit cells, the Wannier function of Eq. (6-20) will also repeat in each fundamental region, though it is large just near one particular unit cell in each such region. In the summation of Eq. (6-21), we sum over only the vectors $\mathbf{R}_j$ in the fundamental region including the origin; the periodicity of the Wannier functions will automatically bring about the periodicity of $u_n(\mathbf{k},\mathbf{r})$ demanded by the periodic boundary conditions.

There are important relations between the Wannier functions and the momentum eigenfunctions. Let us start with Eq. (6-20) for the Wannier function, and use Eq. (6-1) for the wave function, in terms of the momentum eigenfunction. We have

$$\begin{aligned}a_n(\mathbf{r} - \mathbf{R}_i) &= N^{-\frac{1}{2}}\Sigma(\mathbf{k})u_n(\mathbf{k},\mathbf{r})e^{-i\mathbf{k}\cdot\mathbf{R}_i} \\ &= N^{-\frac{1}{2}}\Sigma(\mathbf{k},\mathbf{K}_j)v_n(\mathbf{k} + \mathbf{K}_j)e^{i(\mathbf{k}+\mathbf{K}_j)\cdot\mathbf{r}}e^{-i\mathbf{k}\cdot\mathbf{R}_i} \\ &= N^{-\frac{1}{2}}\Sigma(\mathbf{k},\mathbf{K}_j)v_n(\mathbf{k} + \mathbf{K}_j)e^{i(\mathbf{k}+\mathbf{K}_j)\cdot(\mathbf{r}-\mathbf{R}_i)}\end{aligned} \quad (6\text{-}22)$$

In the summation over $\mathbf{k}$ and $\mathbf{K}_j$, in Eq. (6-22), we really have a summation over all $\mathbf{k}$ space; the summation over $\mathbf{k}$ carries us over the central zone in $\mathbf{k}$ space, and the summation over $\mathbf{K}_j$ carries us over each other zone or unit cell. We may now convert this summation, over the very closely spaced values of $\mathbf{k} + \mathbf{K}_j$, into an integration over $dk_x\,dk_y\,dk_z$; the summation over $\mathbf{k}$ goes into the integration over $dk_x\,dk_y\,dk_z$, divided by the volume of $\mathbf{k}$ space associated with each term in the summation. This volume is $1/N$ times the volume of unit cell in $\mathbf{k}$ space. The latter is $(2\pi)^3$ times the volume of unit cell in reciprocal space, which is the reciprocal of the volume Ω in real space. Hence the summation over $\mathbf{k}$ goes into $N\Omega/(2\pi)^3$ times the integral over $dk_x\,dk_y\,dk_z$. Finally, then, we have

$$a_n(\mathbf{r} - \mathbf{R}_i) = \frac{N^{\frac{1}{2}}\Omega}{(2\pi)^3}\int v_n(\mathbf{k})e^{i\mathbf{k}\cdot(\mathbf{r}-\mathbf{R}_i)}\,dk_x\,dk_y\,dk_z \quad (6\text{-}23)$$

where the integration is to be extended over all values of $\mathbf{k}$ space. We see, then, that the Wannier function $a_n(\mathbf{r} - \mathbf{R}_i)$ is in a sense a Fourier transform of the momentum eigenfunction $v_n(\mathbf{k})$, regarded as a continuous function of $\mathbf{k}$.

In line with this fact, we can also express $v_n(\mathbf{k})$ as a Fourier transform of a_n. We start with Eq. (6-7), in which we express $v_n(\mathbf{k} + \mathbf{K}_h)$ in terms of an integral over the wave function $u_n(\mathbf{k},\mathbf{r})$. We recall that the integration in Eq. (6-7) is over a unit cell; the integral has identical values over each unit cell, and if we integrate over our fundamental region of N unit cells, as we wish to do here, we find N times as large an integral, so that we must multiply by N^{-1} to have the equation satisfied. We then have

$$v_n(\mathbf{k} + \mathbf{K}_h) = N^{-1}\Omega^{-1}\int u_n(\mathbf{k},\mathbf{r})e^{-i(\mathbf{k}+\mathbf{K}_h)\cdot\mathbf{r}}\,dv \quad (6\text{-}24)$$

where the integration is over the fundamental region. In this integral we then substitute for $u_n(\mathbf{k},\mathbf{r})$ from Eq. (6-21), and find

$$v_n(\mathbf{k} + \mathbf{K}_h) = N^{-3/2}\Omega^{-1}\Sigma(\mathbf{R}_j)\int a_n(\mathbf{r}-\mathbf{R}_j)e^{-i(\mathbf{k}+\mathbf{K}_h)\cdot(\mathbf{r}-\mathbf{R}_j)}\,dv \quad (6\text{-}25)$$

where again the integration is over the fundamental region. But now the integrand depends only on $\mathbf{r} - \mathbf{R}_j$, and if we were to change the value of $\mathbf{R}_j$, it would amount only to shifting the Wannier function from one unit cell to another, without changing the whole integral. Hence each term of the summation over $\mathbf{R}_j$ is equal, and is equal to the term for which $\mathbf{R}_j = 0$. Thus we have

$$v_n(\mathbf{k} + \mathbf{K}_h) = N^{-1/2}\Omega^{-1}\int a_n(\mathbf{r})e^{-i(\mathbf{k}+\mathbf{K}_h)\cdot\mathbf{r}}\,dv \quad (6\text{-}26)$$

In Eq. (6-26) we have the analogue of Eq. (6-23), the two equations together expressing the fact that a_n and v_n are Fourier transforms of each other.

We see, then, that these two functions, the Wannier function $a_n(\mathbf{r})$ and the momentum eigenfunction $v_n(\mathbf{k})$, are closely related, and are very powerful functions, in that either one allows us, through Eq. (6-1) or Eq. (6-21), to write the wave functions of all states in an energy band. We have such functions for each energy band. It is not always easy to determine their form, but it is important to know that they exist, and to realize the amount of information which they carry with them. We shall later find examples of both types of functions, in Sec. 6-6 and Chap. 7.

Since the Wannier functions are not solutions of Schrödinger's equation, the Hamiltonian operator will have nondiagonal matrix elements between them. Let us find what these components are. We let $(H)_{op}$ operate on $a_n(\mathbf{r} - \mathbf{R}_j)$, as given in Eq. (6-20). We then require $(H)_{op}u_n(\mathbf{k},\mathbf{r})$, and since $u_n(\mathbf{k},\mathbf{r})$ is by hypothesis a solution of Schrödinger's equation, this quantity must equal $E_n(\mathbf{k})u_n(\mathbf{k},\mathbf{r})$. Thus we find

$$(H)_{op}a_n(\mathbf{r}-\mathbf{R}_j) = N^{-1/2}\Sigma(\mathbf{k})E_n(\mathbf{k})u_n(\mathbf{k},\mathbf{r})e^{-i\mathbf{k}\cdot\mathbf{R}_j} \quad (6\text{-}27)$$

We then substitute for $u_n(\mathbf{k},\mathbf{r})$ from Eq. (6-21), changing the index of summation in that equation to $\mathbf{R}_i$, and find

$$(H)_{op}a_n(\mathbf{r}-\mathbf{R}_j) = N^{-1}\Sigma(\mathbf{R}_i,\mathbf{k})E_n(\mathbf{k})e^{i\mathbf{k}\cdot(\mathbf{R}_i-\mathbf{R}_j)}a_n(\mathbf{r}-\mathbf{R}_i) \quad (6\text{-}28)$$

We let

$$N^{-1}\Sigma(\mathbf{k})E_n(\mathbf{k})e^{i\mathbf{k}\cdot(\mathbf{R}_i-\mathbf{R}_j)} = \mathcal{E}_n(\mathbf{R}_i-\mathbf{R}_j) \quad (6\text{-}29)$$

Then we have

$$(H)_{op}a_n(\mathbf{r}-\mathbf{R}_j) = \Sigma(\mathbf{R}_i)\mathcal{E}_n(\mathbf{R}_i-\mathbf{R}_j)a_n(\mathbf{r}-\mathbf{R}_i) \quad (6\text{-}30)$$

If we multiply Eq. (6-30) on the left by the complex conjugate of one of the a_n's, and integrate over the volume, then on account of the orthogonality of the Wannier functions we have only one nonvanishing term on the right, and we find

$$\int a_n^*(\mathbf{r} - \mathbf{R}_i)(H)_{op} a_n(\mathbf{r} - \mathbf{R}_j)\, dv = \mathcal{E}_n(\mathbf{R}_i - \mathbf{R}_j) \tag{6-31}$$

These quantities $\mathcal{E}_n(\mathbf{R}_i - \mathbf{R}_j)$, which as we see are the matrix elements of the Hamiltonian with respect to the Wannier functions, have a simple significance which we can find from Eq. (6-29). Let us rewrite this equation in the form

$$N^{-1}\Sigma(\mathbf{k})E_n(\mathbf{k})e^{i\mathbf{k}\cdot\mathbf{R}_s} = \mathcal{E}_n(\mathbf{R}_s) \tag{6-32}$$

multiply by $e^{-i\mathbf{k}'\cdot\mathbf{R}_s}$, and sum over all vectors $\mathbf{R}_s$ in the fundamental volume of space, equal to N unit cells. This involves $\Sigma(\mathbf{R}_s)e^{i(\mathbf{k}-\mathbf{k}')\cdot\mathbf{R}_s}$, which can easily be shown to vanish if $\mathbf{k} \neq \mathbf{k}'$, but to equal N if $\mathbf{k} = \mathbf{k}'$. Thus we find

$$E_n(\mathbf{k}) = \Sigma(\mathbf{R}_s)\mathcal{E}_n(\mathbf{R}_s)e^{-i\mathbf{k}\cdot\mathbf{R}_s} \tag{6-33}$$

In this important formula, we express $E_n(\mathbf{k})$, the energy in the nth energy band, as a function of the reduced wave vector $\mathbf{k}$, in the form of a Fourier series in $\mathbf{k}$, the coefficients being the matrix elements of the Hamiltonian between Wannier functions, as given in Eq. (6-31). Each of the terms of this Fourier series will be periodic in the $\mathbf{k}$ space, repeating if we add any of the fundamental vectors $\mathbf{K}_j$ to $\mathbf{k}$; but this is in agreement with the fact, which we have pointed out several times, that the energy is a periodic function in $\mathbf{k}$ space, repeating in each unit cell.

In Eq. (6-33) we have the generalization, for the case of the crystal, of the expression for the energy given in Volume 1, Eq. (9-22), which is

$$H_{mm} = \frac{H(0) + 2\cos(2\pi m/N)H(1) + 2\cos(4\pi m/N)H(2) + \cdots}{1 + 2\cos(2\pi m/N)S(1) + 2\cos(4\pi m/N)S(2) + \cdots}$$

We recall that in that case the quantities $H(0)$, $H(1)$, etc., are the matrix elements of the Hamiltonian between atomic orbitals located on the various atomic sites, which reduce in the present case, where we are using Wannier functions in place of atomic orbitals, to the expressions of Eq. (6-31). We also recall that the quantities $S(1)$, $S(2)$, etc., in the denominator of Volume 1, Eq. (9-22), are overlap integrals, which vanish in the case of the orthogonal Wannier functions. We now see the close relationship between our present treatment and that of Bloch sums. In Eq. (6-21) we have expressed the wave function as a Bloch sum of Wannier functions; in Eq. (6-33) we have shown that the energy as a function of $\mathbf{k}$ is given by a formula of the type of Volume 1, Eq. (9-22), which is appropriate for finding the energy bands in the case of Bloch sums. Here, however, in contrast to Volume 1, Eq. (9-22), we are using the Wannier functions derived from exact solutions of the periodic potential problem, so that our relations are exact, rather than approximate as they would be if we used atomic orbitals instead, and if the Bloch sum were a linear combination of atomic orbitals.

There is one simple and important qualitative result of Eq. (6-33). As the interatomic distance increases to infinity, the Wannier functions will approach atomic orbitals, which will not overlap, and which will not have any nondiagonal matrix elements of the Hamiltonian between them, as defined in Eq. (6-31); the only nonvanishing elements of this sort will be the diagonal ones, for $\mathbf{R}_i = \mathbf{R}_j$, the case where both orbitals are on the same atomic site. In this case, as we see from Eq. (6-33), all Fourier components in the expansion of $E(\mathbf{k})$ except the constant term will vanish, and the energy will be independent of $\mathbf{k}$, or the energy band will have zero width. Thus we see in a very general way that the energy bands shrink to zero width as the interatomic distances become infinite, and furthermore that the energy values approach those of the isolated atoms.[1]

6-6. Energy Bands in the One-dimensional Case. We have now outlined the main features of the energy-band problem, and it is time to have examples which will illustrate the general principles. We shall not take up the practical methods of solving Schrödinger's equation for the cases actually met in real crystals until Chaps. 8 and 9, but first we wish to study simplified examples which resemble the true problem. We shall now take up the one-dimensional case, and then go on in the next chapter to discuss two- and three-dimensional crystals. The one-dimensional case is of course a generalization of the case of C_{Nv} which we have taken up in Volume 1, Chaps. 8 and 9, but we shall handle it by the more powerful methods which we have been outlining in this chapter.

We shall assume that the potential is periodic with period a along the x direction. In the reciprocal space, we shall have periodicity with a period $b = 1/a$; in $\mathbf{k}$ space, periodicity with a period $2\pi/a$; and in momentum space periodicity with period $2\pi\hbar/a = h/a$. We shall measure x in units of a, so that the symbol x will stand for the distance divided by a; energies in units of $h^2/2ma^2$, so that E stands for the energy divided by $h^2/2ma^2$; and momentum in units of h/a, so that p stands for momentum divided by h/a. The potential energy $V(x)$, following Eq. (6-6), will be written in the form

$$V(x) = \Sigma(l)W(l)e^{2\pi i l x} \tag{6-34}$$

where in place of K_l we use its value $2\pi l/a$. The quantity l takes on all

---

[1] This statement should be qualified by saying that in a two- or three-dimensional case we must use modified momentum eigenfunctions as described in Sec. 7-5 to achieve this situation.

integral values. The wave function $u(p,x)$ will be written, in accordance with Eq. (6-1), in the form

$$u(p,x) = \Sigma(h)v(p + h)e^{2\pi i(p+h)x} \qquad (6\text{-}35)$$

where h is an integer (not to be confused with Planck's constant). It is most convenient to handle Schrödinger's equation in the form of Eq. (6-12), where we are dealing with the momentum eigenfunction; we have

$$(p + h)^2 v(p + h) + \Sigma(l)W(l)v(p + h - l) = E(p)v(p + h) \qquad (6\text{-}36)$$

where the kinetic energy term takes this simple form on account of our units of energy and momentum. The central Brillouin zone in k space extends from $-\pi/a$ to π/a, having a length equal to the fundamental period $2\pi/a$, and being centered on the origin. In terms of p, the central zone extends from $p = -\frac{1}{2}$ to $\frac{1}{2}$. We shall find that $E(p)$ is periodic in p with a period of unity.

As a first example,[1] we shall take the case where the potential is a sinusoidal function of x; that is, $W(l)$ is zero except for $l = \pm 1$. It is convenient to let $W(1) = W(-1) = -W$, in which case, from Eq. (6-34), the potential energy will be $V(x) = -2W \cos 2\pi x$; by using the negative sign for W, we have a minimum of potential at the origin. Schrödinger's equation for this problem is a well-known form called Mathieu's equation, which has been extensively studied; the method of treatment is based on the use of momentum eigenfunctions. For this special case, Eq. (6-36) reduces to

$$(p + h)^2 v(p + h) - Wv(p + h - 1) - Wv(p + h + 1) = E(p)v(p + h) \qquad (6\text{-}37)$$

To solve this equation we note that it is a recursion formula for the quantities $v(p + h)$. Thus we can rewrite it in the form

$$Wv(p + h + 1) = [(p + h)^2 - E(p)]v(p + h) - Wv(p + h - 1) \qquad (6\text{-}38)$$

We see from this equation that if we are constructing a table of values of the quantities $v(p + h)$ as functions of h, we can find each entry in the table from the two preceding entries. This is a convenient method for actually computing the momentum eigenfunction in this problem, provided we know two initial entries in the table (the initial conditions) and the energy.

This situation is similar to that found in solving Schrödinger's equation for a one-dimensional problem: we can integrate the equation, provided we know initial conditions (usually the value of the function and its derivative at a given point) and the energy. As with Schrödinger's equation, if we choose the boundary conditions and the energy in an arbitrary

[1] For this case, see J. C. Slater, *Phys. Rev.*, **87**:807 (1952).

way, we generally find that the wave function goes infinite for infinite values of the argument, both at plus and minus infinity. Similarly, in Eq. (6-38), if we carry out the procedure just sketched, $v(p + h)$ will generally become infinite as h becomes positively or negatively infinite. By choosing the initial conditions properly, we can make v go asymptotically to zero at one limit, either plus or minus infinity, but not in general at both. For certain special energies, however, the eigenvalues of the problem, v will go to zero in both limits. It is only such solutions which

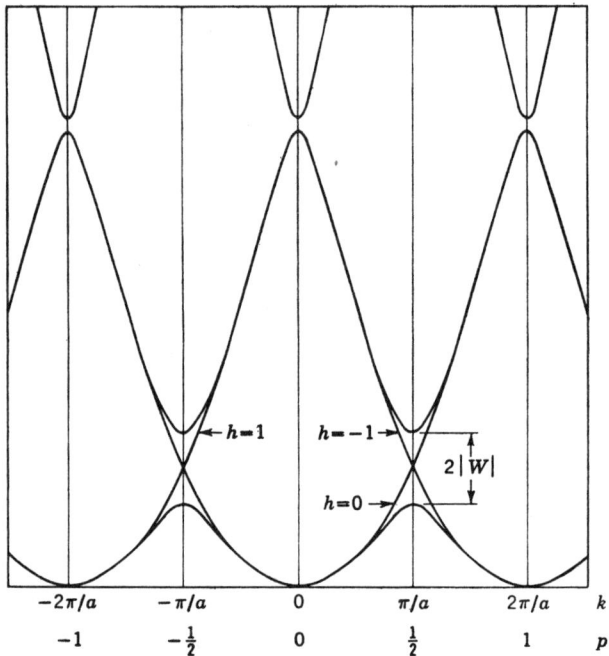

Fig. 6-1. Energy bands for one-dimensional periodic potential.

we can use. It is not hard to work out practical methods for carrying through this calculation, and finding the eigenvalues and eigenfunctions of the problem.

When this is done, we get curves for the energy E_p, and the momentum eigenfunction $v(p)$, of which typical examples are given in Figs. 6-1 and 6-2. These are given for a relatively small amplitude W of the potential energy, namely, $W = 0.0625$; on an energy scale, as we shall prove in a moment, the value of W is approximately half the separation between the energies of the two lowest bands at $k = \pi/a$. In Fig. 6-1 we see the periodicity of the energy in the k space, with period $2\pi/a$, which we have expected. We see that for certain ranges of energy, the energy bands, we have energy levels, while between them we have energy gaps in which there are no energy levels.

It is easy to understand the general form of the curves for energy bands, in Fig. 6-1. The equation for the momentum eigenfunction, Eq. (6-37), is of the type always found in quantum mechanics when we express a wave function as a linear combination of unperturbed functions, in this case exponentials, with coefficients, in this case the v's, to be determined. The diagonal matrix element of the Hamiltonian is $(p + h)^2$ for the hth function. The nondiagonal matrix elements are nonvanishing only between functions with adjacent values of h, and for those cases are equal

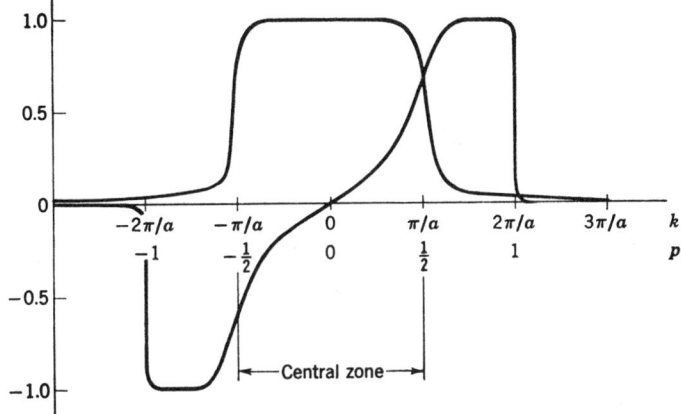

Fig. 6-2. Momentum eigenfunctions for two lowest energy bands. Even function, lowest band; odd function, second band.

to $-W$. In Fig. 6-1, we have in the first place drawn parabolas corresponding to the diagonal matrix elements for several of the functions, namely, those for $h = 0, \pm 1$. These cross at $p = \pm \frac{1}{2}$; at $p = \frac{1}{2}$, the two cases with $h = 0$ and $h = -1$ have identical energies. Near these crossings, the nondiagonal matrix elements of the Hamiltonian become important, and just at the crossover points, where the problem would be degenerate if it were not for the nondiagonal matrix elements, the unperturbed energies, represented by the intersecting parabolas, are pushed apart by amounts $\pm W$, producing the energy gap. Much smaller gaps are found between the second and third bands, and at higher crossovers; examination of the problem shows that the gaps here are proportional to W^2, W^3, etc., as we go higher in energy.

The physical phenomenon occurring to produce the gaps is of interest, particularly in connection with the interpretation which we shall later give to the three-dimensional case. We may imagine an incident wave, whose space part is given by e^{ikx}, where we now use ordinary units, to be traveling down the crystal. It will be scattered by each of the peaks of the sinusoidal potential. The scattered waves traveling backward, along $-x$, will be able to reinforce each other if they are in phase with each other, which means that the path difference from one peak to the next and back must be a wavelength or integral multiple of a wavelength. The de Broglie wavelength is here $h/p = 2\pi/k$, where p is now the momentum in ordinary units, so that the edge of the Brillouin zone, $k = \pi/a$, corresponds to a wavelength equal to $2a$. Since the wave scattered from each successive peak must travel a distance of $2a$ greater than from the preceding peak, we see that we have just this reinforcement of scattered waves which is necessary for reflection, and for a standing wave. We expect, then, that at this value of k, there will be a superposition of the direct and scattered or reflected wave, resulting in a standing wave. The standing wave can have its maxima either at the minima of the potential, in which case the energy will be depressed, or at the maxima of potential, in which case it will be raised. These two cases correspond to the two limits of the energy gap. We shall see, when we come to the three-dimensional case in Sec. 7-6, that a similar argument will show us that Bragg scattering of the electron waves by the atoms is responsible for the boundaries of the Brillouin zones, and for the energy gaps which will be found to appear there.

The momentum eigenfunction, shown in Fig. 6-2, fits in with this interpretation. To find the momentum eigenfunction associated with a given energy band, or the quantities $v(p + h)$, where p is now in a dimensionless form, we must take a set of points p, $p \pm 1$, $p \pm 2$, ... and read off the value of the curve at these points. Suppose that we start with p near zero. Then from the figure we see that the value of the momentum eigenfunction of the lowest energy band associated with $h = 0$ is practically unity (we have normalized the momentum eigenfunction in this figure so as to omit the factor depending on $N\Omega$), and all other h values give practically zero. In other words, the wave function is practically a single sinusoidal function, $e^{2\pi i p x}$. As p increases, however, approaching $\frac{1}{2}$, we see that not only the point p, but also $p - 1$, correspond to values of the abscissa at which we have considerable values of the ordinate. Just at $p = \frac{1}{2}$, these two values of v are equal, and each is approximately $1/\sqrt{2}$, showing that we have here the superposition of the direct and reflected waves, giving a cosine function, with a maximum of charge density at the minimum of the potential. On the other hand, for the second energy band, for $p = \frac{1}{2}$, the two values of v are of opposite sign. To make the wave function real, we must here multiply the value of v shown in the figure by i, so as to produce a sine function for the eigen-

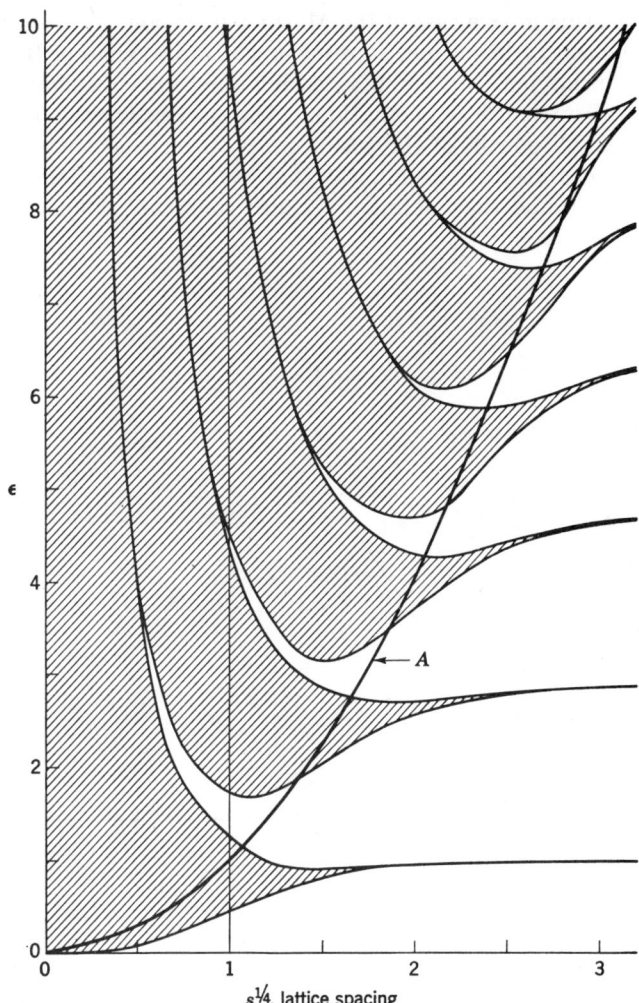

Fig. 6-3. Energy bands as function of lattice-spacing, one-dimensional Mathieu problem. Curve A represents the height of the potential barrier between atoms, for each lattice spacing. Line at $s = 1$ corresponds to case of Figs. 6-1 and 6-2; $s^{1/4} = 3.16$, to Fig. 6-4.

function, with the maximum charge density at the maximum of potential. For a value of p somewhat greater than $\frac{1}{2}$, the largest Fourier component for the second band comes for the value lying at this p; the values of p smaller by 1, 2, . . . have practically no amplitude, so that again the wave function is practically a single sinusoidal function.

It is interesting in this case of the Mathieu equation to investigate the nature of the energy bands and gaps for much larger values of W than we have been considering. In Fig. 6-3 we show the bands of energy levels as a function of internuclear separation. This is constructed as follows. We let the amplitude W of the cosine function increase with increasing internuclear separation, in such a way that the curvature of the potential energy curve around the minimum remains constant, independent of a.

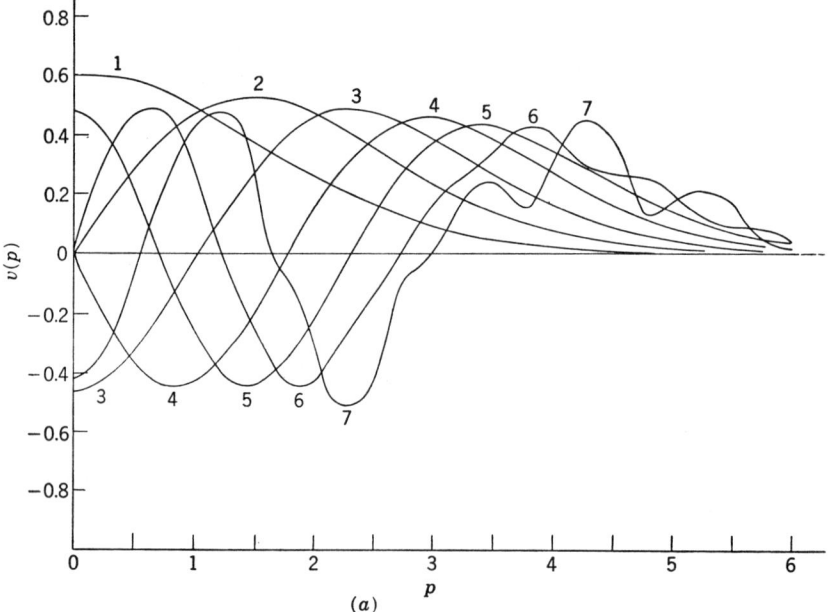

Fig. 6-4. Momentum eigenfunctions $v(p)$, for $s = 100$ (as indicated in Fig. 6-3). (a) First seven bands. (b) Seventh to eleventh bands.

We furthermore adjust the zero of energy to lie at the bottom of the potential energy well. Thus, in the limit of large internuclear separation, the potential energy about the minimum is given approximately by a parabola, and the electron moving at the bottom of one of the wells approaches a linear oscillator problem. The energy levels then approach the values $(n + \frac{1}{2})h\nu$ of a linear oscillator of appropriate frequency, at large values of a. Each of these energy levels broadens into an energy band with decreasing internuclear separation, with gaps between them which grow narrower as a decreases to zero. By a vertical line in Fig. 6-3, we show the value of a chosen for the case of Figs. 6-1 and 6-2. The parabolic line in Fig. 6-3 shows the top of the potential barrier, at a distance $4W$ above

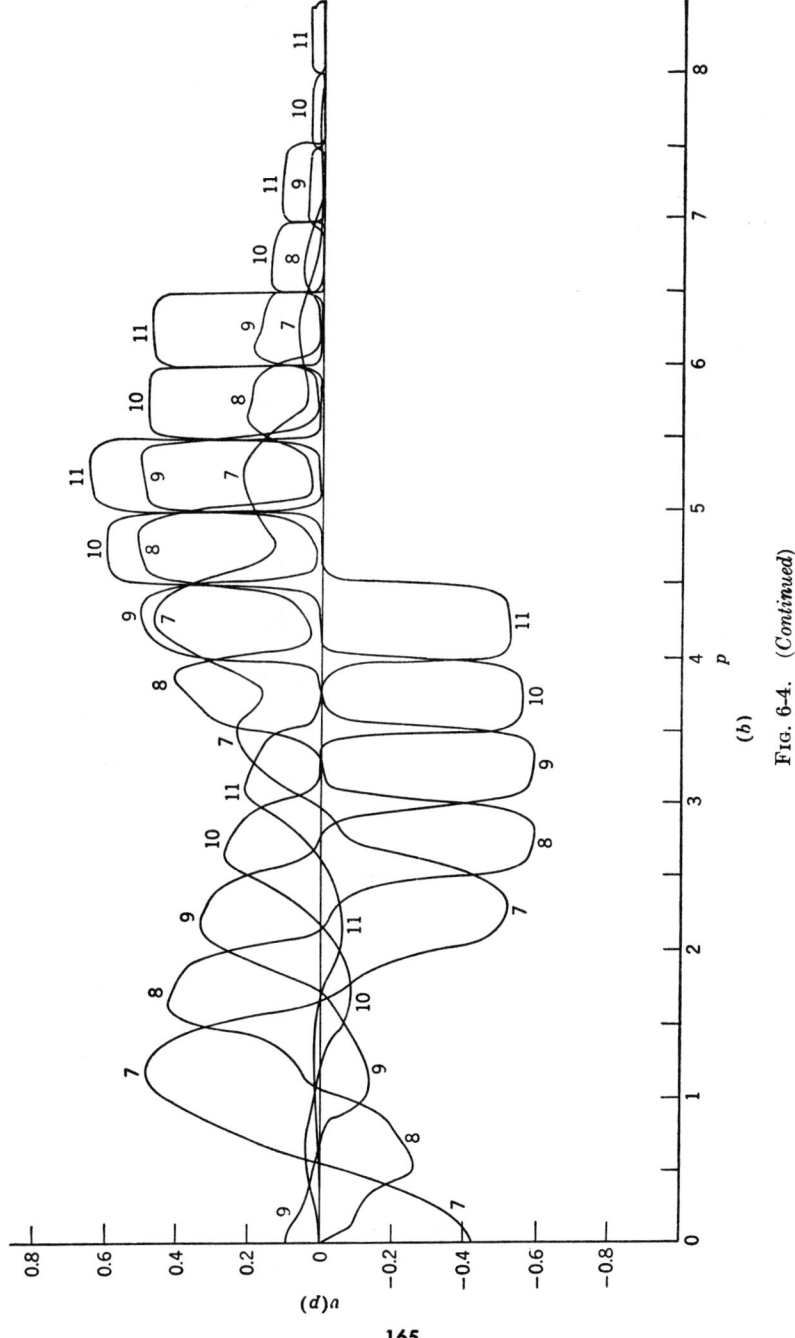

Fig. 6-4. (Continued)

the energy minimum (arising from the potential energy $-2W\cos 2\pi x$). We observe that the energy levels below this potential barrier are only moderately broadened, but that those above the barrier are almost continuous, with only narrow gaps.

For the low bands in a case with large internuclear distances, or with a high barrier, the wave functions are almost entirely concentrated within the potential wells, becoming very small in the barriers. These cases correspond to the inner electrons of an atom, in a crystal, in which these electrons from inner shells are confined almost entirely to the immediate neighborhood of the nucleus. In such a concentrated wave function, the momentum eigenfunction resembles that of an isolated atom, and spreads far out in momentum space; for the dimensions of the momentum eigenfunction vary inversely as the size of the orbit in ordinary space. Furthermore, in the case of the Mathieu-function problem, these momentum eigenfunctions take on a very simple form; for it can be shown that the Fourier transform of the type of wave function we have for a linear oscillator is the same type of function of k. In Fig. 6-4 we show the momentum eigenfunctions for a value of a equal to about the maximum value shown in Fig. 6-3. For the lowest energy bands, as we see, these have forms similar to those of the lowest eigenfunctions of the linear-oscillator problem in coordinate space, and extending outward over many unit cells in k space. As we get to higher energy bands, however, the momentum eigenfunctions become of more and more peculiar forms, eventually reaching a behavior somewhat similar to that shown in Fig. 6-2, for high energies. Careful analysis of the functions, in these cases, shows that for the energies far above the top of the potential barriers, the wave functions resemble plane waves.

The case of the Mathieu functions, which we have been describing, is important, on account of the fact that it can be solved exactly; however, it is a very special case. If we have more than one nonvanishing Fourier component of the potential, the qualitative situation is not changed. However, in a curve of energy versus k, as in Fig. 6-1, we find that each of the energy gaps begins to broaden out as the values of the $W(l)$'s increase from zero, and the gap at the value $k = \pi l/a$ is equal to a first order to $2W(l)$. In other words, all energy gaps vary linearly with the magnitude of the perturbing potential, rather than having the second, third, etc., gaps going as the second, third, etc., powers of the perturbing potential, as with the Mathieu case. Aside from this, the results have a great deal of similarity with the case we have been taking up.

7

Plane-wave Expansion in the Two- and Three-dimensional Cases

7-1. The Two-dimensional Square Lattice. In Chap. 5 we have given a general discussion of symmetrized combinations of plane waves. In Chap. 6 we have considered how to solve Schrödinger's equation, in one dimension, by using a linear combination of plane waves. We shall now combine these two types of discussion, going to two- and three-dimensional cases in which we not only are interested in solving Schrödinger's equation, but in which the simplification resulting from the use of symmetrized plane waves is of great importance. We shall postpone until Sec. 7-7 the question as to whether this is a practical method of computing solutions of Schrödinger's equation in a crystal. At that time we shall find that it is not sufficiently rapidly convergent to give adequate results for real crystals, but nevertheless the calculation points the path to the true solution of Schrödinger's equation, and a thorough understanding of it is necessary for a study of the more adequate methods of approximation.

A great deal of the information we wish can be found from the two-dimensional square lattice, and we shall first discuss it in detail, later sketching the extension to more complicated cases. We assume identical potential wells at the origin and points displaced from it by integral multiples of a along either x or y. As in Sec. 6-6, we shall measure distances in units of a, and energies in units of $h^2/2ma^2$. Then, as in Eq. (6-34), the potential energy can be written in the form

$$V(x,y) = \Sigma(h_1,h_2)W(h_1h_2)e^{2\pi i(h_1 x + h_2 y)} \qquad (7\text{-}1)$$

where h_1 and h_2 are integers. The wave function can be written as a sum of exponential functions,

$$u(p_1,p_2; x,y) = \Sigma(h_1,h_2)v(p_1 + h_1, p_2 + h_2)e^{2\pi i[(p_1+h_1)x + (p_2+h_2)y]} \qquad (7\text{-}2)$$

where the reduced wave vector has components $2\pi p_1/a$, $2\pi p_2/a$. The diagonal matrix element of Schrödinger's equation, for one of the exponential functions in Eq. (7-2), is $(p_1 + h_1)^2 + (p_2 + h_2)^2$, and the nondiagonal matrix element between two such functions is $W(h_1 - h_1',\ h_2 - h_2')$, where $h_1,\ h_2$ and $h_1',\ h_2'$ characterize the two plane waves. Thus Schrödinger's equation, in the form of Eq. (6-12), becomes

$$[(p_1 + h_1)^2 + (p_2 + h_2)^2]v(p_1 + h_1, p_2 + h_2)$$
$$+ \Sigma(h_1',h_2')W(h_1',h_2')v(p_1 + h_1 - h_1', p_2 + h_2 - h_2')$$
$$= E(p_1,p_2)v(p_1 + h_1, p_2 + h_2) \quad (7\text{-}3)$$

To have a specific case to describe, calculations have been carried out by Prof. J. H. Wood for an example in which the potential energy is given as follows:

$$W(00) = 0$$
$$W(10) = W(-10) = W(01) = W(0 - 1) = -0.040$$
$$W(11) = W(-11) = W(1 - 1) = W(-1 - 1) = -0.025$$
$$W(20) = W(-20) = W(02) = W(0 - 2) = -0.010$$
$$W(21) = W(-21) = W(2 - 1) = W(-2 - 1) = W(12)$$
$$\qquad = W(-12) = W(1 - 2) = W(-1 - 2) = -0.006$$
$$W(22) = W(-22) = W(2 - 2) = W(-2 - 2) = -0.0015$$
Other W's $= 0$
$\quad(7\text{-}4)$

The form of this potential energy, along the x axis, for $y = 0$, is shown in Fig. 7-1. It has a potential well of approximately Gaussian form at the

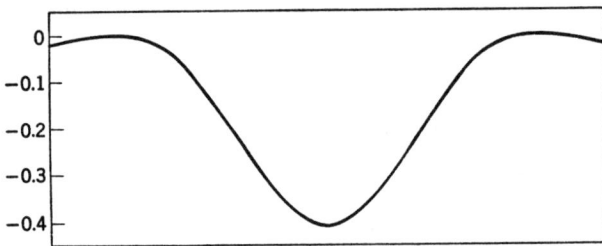

FIG. 7-1. Potential of Eq. (7-4), as function of x, for $y = 0$, within unit cell.

center of each unit cell. To carry out an approximate solution, we have used plane waves corresponding to h_1 and h_2 going from -2 to 2, or 25 plane waves in all. The secular problem was solved on a digital computer, producing both the energy and the coefficients $v(p_1 + h_1, p_2 + h_2)$, as functions of p_1 and p_2. Computations were made for values of p_1 and p_2 differing by 0.1 unit, as if we had periodic boundary conditions with $N_1 = N_2 = 10$, $N = 100$. This involves 21 separate calculations, for which p_1, p_2 are given by the values 00, 10, 20, 30, 40, 50, 11, 21, 31, 41,

Sec. 7-1]   PLANE-WAVE EXPANSION—TWO AND THREE DIMENSIONS   169

51, 22, 32, 42, 52, 33, 43, 53, 44, 54, 55, in which such a symbol as 21 indicates that $p_1 = 0.2$, $p_2 = 0.1$. From these, the other points in the Brillouin zone, at intervals of 0.1, can be obtained by symmetry. The Brillouin zone, of course, extends from $-\frac{1}{2}$ to $\frac{1}{2}$, for p_1 and p_2.

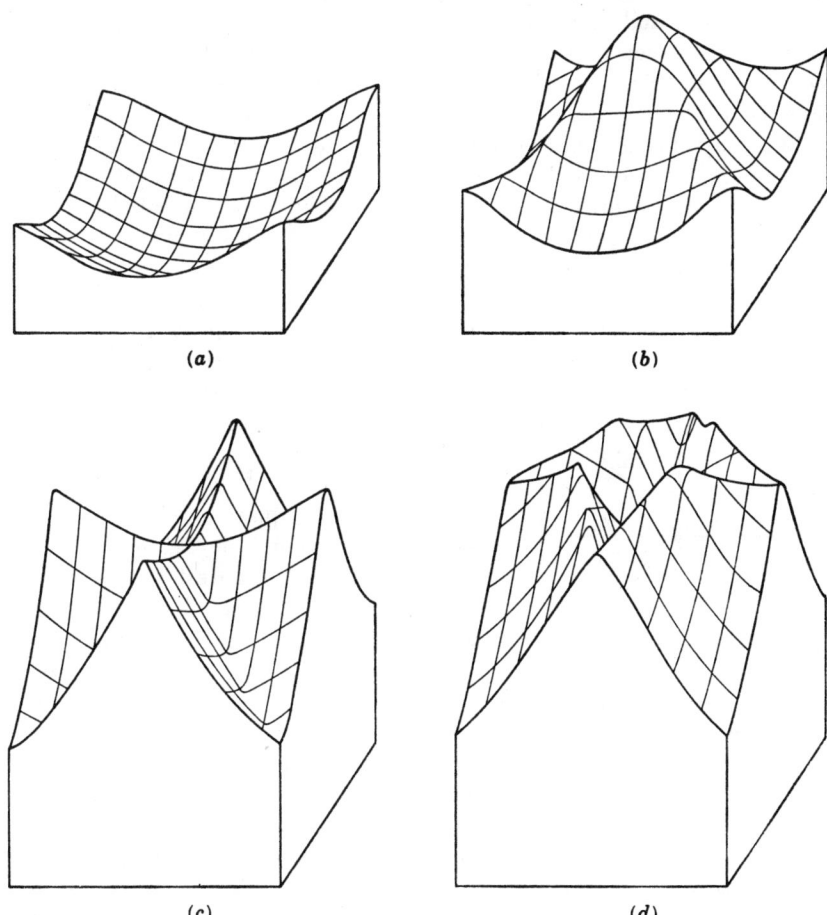

FIG. 7-2. (a) Perspective of lowest energy surface, two-dimensional square lattice, as function of wave vector within the central Brillouin zone. (b) Second energy surface. (c) Third energy surface. (d) Fourth energy surface.

The results of this calculation are rather hard to present, and this is typical of the results of energy-band calculations in general. First let us consider the energy. For each value of p_1, p_2, we have 25 energy values, as a result of solving the 25×25 determinantal equation. We may expect that the lower energies are fairly good approximations to the lower energy levels of the problem, though the higher ones will be quite inaccurate. (For discussion of this point, see J. C. Slater, "Quantum Theory of Atomic Structure," vol. 1, sec. 5-2.) As always, the lowest energies will combine to form the lowest energy band, the second lowest the second energy band, and so on. These can be plotted as surfaces, the energy being plotted vertically upward, above a plane in which p_1 and p_2 are coordinates. This is carried out for the four lowest energy bands in Fig. 7-2a–d, where the surfaces are shown in perspective.

We may discuss these energy surfaces by comparison with Fig. 6-1. In that figure, we drew a parabola $E = (p + h)^2$ at each lattice point, where p is the coordinate, h an integer. The lowest parabolas in each unit cell of the reciprocal space combined to form a curve which, when slightly rounded off at the sharp cusp on account of the secular problem, produced the energy curve for the lowest energy band. Similarly, the next-lowest parabolas in each unit cell combined to produce the second-lowest energy band. In the two-dimensional case of Fig. 7-2, we have in a similar way a paraboloid of revolution, given by

$$E = (p_1 + h_1)^2 + (p_2 + h_2)^2$$

centered in each square unit cell. The lowest paraboloids combine, again with rounding off, to form the lowest energy surface, the second-lowest paraboloids produce the second energy surface, and so on.

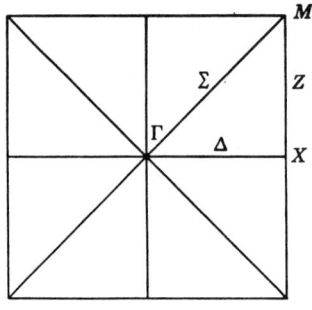

Fig. 7-3. Symmetry directions in two-dimensional square Brillouin zone.

This is not too good a way to visualize the bands, however. It is hard to understand the perspective, and furthermore, this is not a method which can be extended to the three-dimensional case, where the energy forms a fourth variable. We then look for other methods of displaying the energy. There are two schemes, which have proved to be very useful in the three-dimensional case, which can be illustrated by our two-dimensional example. The first of them is to consider only symmetry directions, and plot energy as a one-dimensional problem, for wave vectors along these symmetry directions. Let us illustrate specifically by our square lattice. In Fig. 7-3 we show the Brillouin zone for this case, with the symmetry directions indicated, denoted by symbols suggested by the simple cubic lattice of Fig. 5-3. That is, we let the origin be denoted by Γ; the x axis, for which $p_2 = 0$, by Δ; its extremity, given by $p_1 = \frac{1}{2}$, $p_2 = 0$, by X; the direction for which $x = y$, or $p_1 = p_2$, by Σ; its extremity, given by $p_1 = p_2 = \frac{1}{2}$, by M; and the line $p_1 = \frac{1}{2}$, p_2

arbitrary, by Z. Out of the 21 points for which numerical calculations have been made in the present case, 15 lie in these various symmetry directions. We can then give the energy as a function of p_1 along Δ, from Γ to X; next as a function of p_2, p_1 being $\frac{1}{2}$, along Z, from X to M; then as a function of the magnitude of p along the line Σ, from M back to Γ. This is done in Fig. 7-4, where we give in Fig. 7-4a the free-electron energy, or $(p_1 + h_1)^2 + (p_2 + h_2)^2$, and in Fig. 7-4b the calculated energy found by solving the secular problem. We see that the resemblance between the two cases is similar to that between the parabolas and the actual energy curves in Fig. 6-1.

The second method of indicating the energy surface is to draw contours of constant energy, in the reciprocal space. For the free-electron case, these will be made up of circles, of appropriate radius, centered on the lattice points of the reciprocal lattice. For the actual solutions of Schrödinger's equation, the contours will become distorted. Energy contours, for the four lowest energy bands, are given in Fig. 7-5. The reader should compare these contour maps with the perspectives of Fig. 7-2, as well as with Fig. 7-4.

There is one important point regarding these energy contours: a contour of constant energy can often give information about several bands simultaneously. This is clear from Fig. 7-4, in which it is obvious that a line at constant height, or constant energy, will often cut a number of energy surfaces. Wherever we have an energy contour in Fig. 7-5, we may visualize the situation like a contour map in geography. We think of a plane of constant height, cutting through such a surface as is shown in Fig. 7-2. The latter surface has peaks and valleys. Within a closed energy contour, there must be either a peak or a valley, the energy either rising above the value characteristic of the contour or falling below it. Hence, as we go through an energy contour in the reciprocal space, there will be a change by one unit in the number of energy bands lying below the energy for which the contour is drawn. We can, then, take the reciprocal space, take a contour like one of those given in Fig. 7-5, and indicate in each of the regions of this space the number of energy bands lying below this energy, for the particular values of p_1 and p_2 lying in this region. This is shown for a number of cases in Fig. 7-6. We see in this figure that there can well be regions, in the same diagram, in which there are several different numbers of energy bands lying below the energy characteristic of the contour.

This situation is important when we come to consider the Fermi level. This is the energy representing the topmost occupied energy level in a crystal. If we are dealing with a metal, it lies within an energy band. If then we take an energy contour whose energy is the Fermi energy, we shall find a situation like those shown in Fig. 7-6. There will be certain

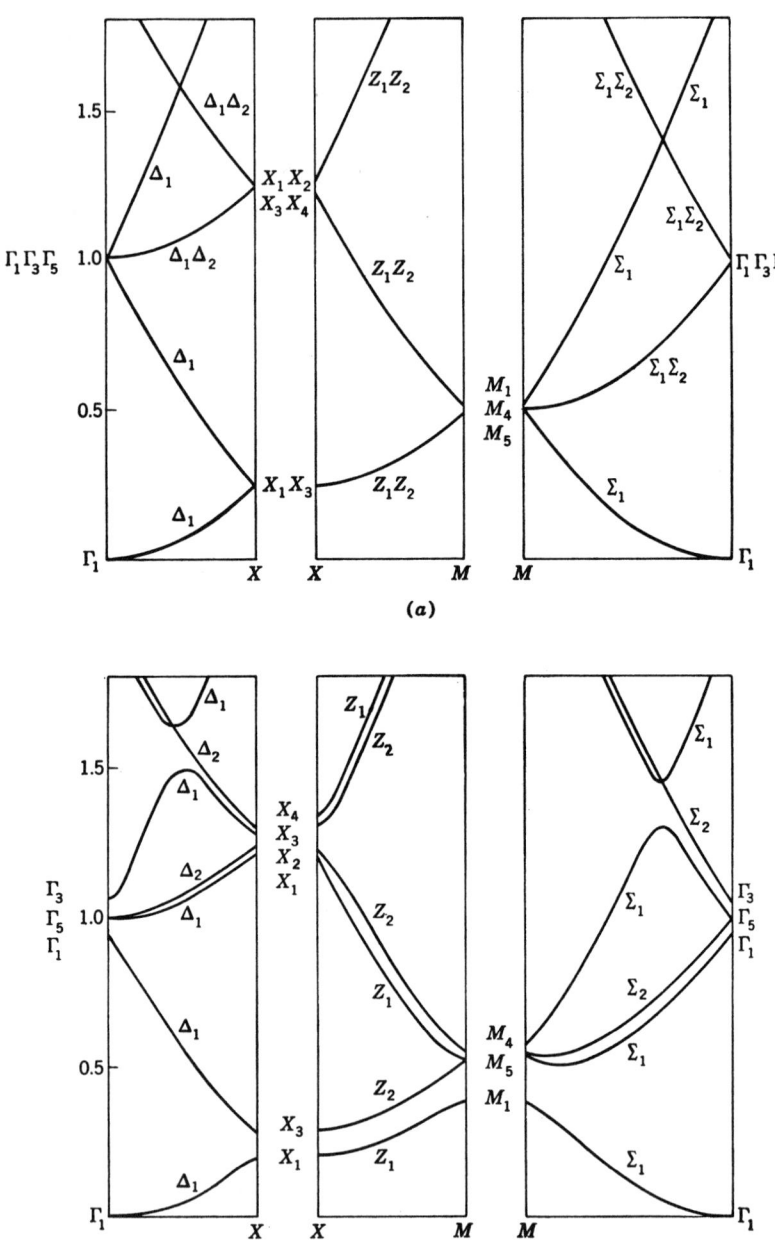

Fig. 7-4. Energy bands of two-dimensional square lattice, as function of wave vector, along symmetry directions in Brillouin zone. (a) Free-electron approximation. (b) Calculated energy for case worked out in text.

regions of the Brillouin zone with each of several integral numbers of energy bands whose energy lies below the Fermi level. Each point in this region (that is, each set of p_1, p_2 allowed by the periodic boundary conditions, and lying within this region) can accommodate two electrons, one of each spin, per band. Since the number of points per unit area is fixed, equal to $1/N$, where N is the number of unit cells in the repeating region

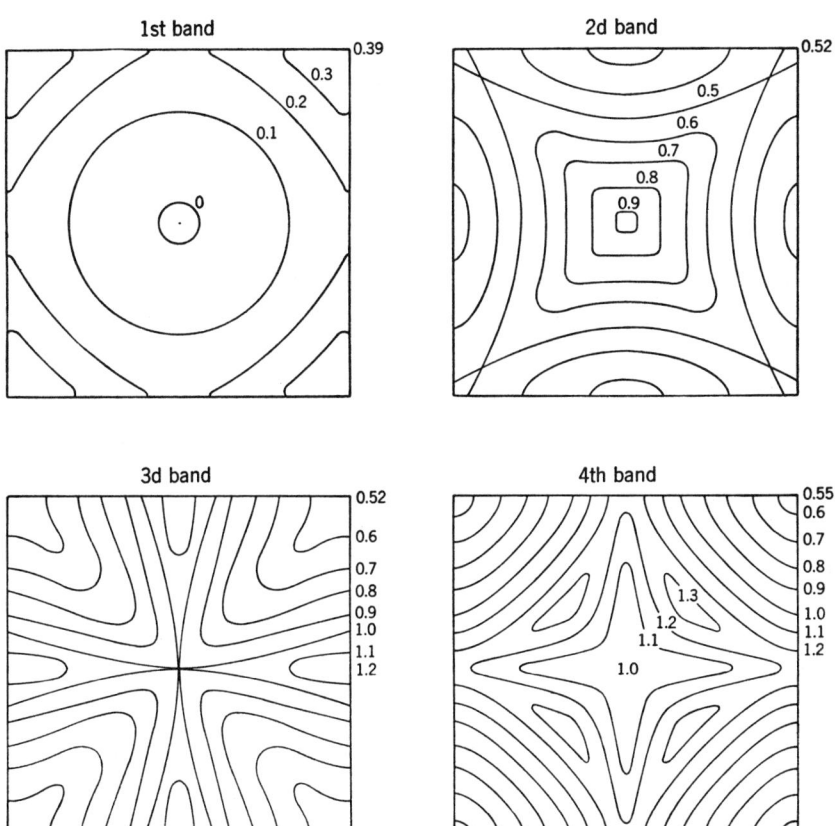

Fig. 7-5. Energy contours for first four energy bands of two-dimensional square lattice. These are for the same four bands shown in perspective in Fig. 7-2.

of the crystal, we see that the number of electrons accommodated in each of the regions of the Brillouin zone must equal N times the area of the region, multiplied by the number of bands in this region, and multiplied by 2, on account of the spin. If we sum these quantities for the whole Brillouin zone, we must get the number of electrons in the crystal. Stated otherwise, it is this condition which determines the Fermi level, or defines which energy contour corresponds to the Fermi energy.

7-2. Symmetry Properties of the Wave Functions. In the preceding section, we have considered the energy bands of our square two-dimensional crystal, but we have not considered the symmetry properties of the wave functions. In this particular example, symmetry was not used to

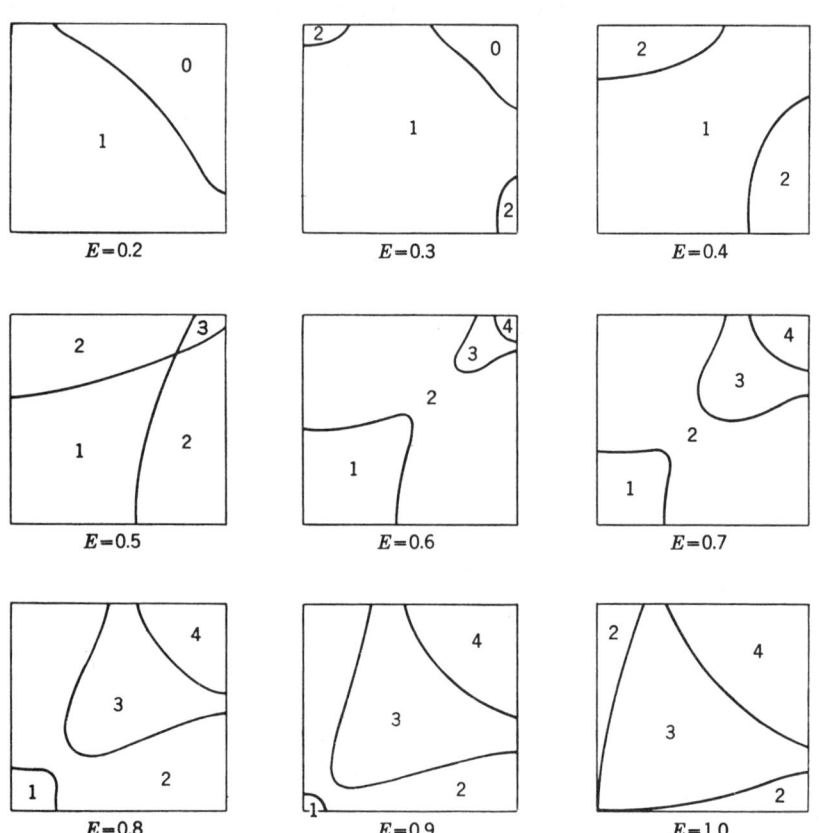

FIG. 7-6. Diagram indicating number of occupied energy bands below Fermi energy E, throughout Brillouin zone, for various values of E, two-dimensional square lattice. Only one-fourth of the Brillouin zone is shown in each case, the origin being in the lower left-hand corner.

simplify the solution of the secular problem; each of the 21 secular problems was solved as if there were no symmetry involved. When the results are examined, however, it automatically turns out that each of the combinations of plane waves, of the form given in Eq. (7-2), shows the symmetry of one of the appropriate irreducible representations for the group of the

Sec. 7-2] PLANE-WAVE EXPANSION—TWO AND THREE DIMENSIONS 175

wave vector.[1] In the present section we shall state what these irreducible representations are, and show how to build up symmetrized plane waves, from which the wave function could have been built up in the first place.

The point group associated with the simple square space group is C_{4v}, which we have already discussed in Sec. 5-5. At the point Γ, the center of the Brillouin zone, the group of the wave vector is the complete group C_{4v}, so that the irreducible representations are as given in Table 5-3. From the 25 plane waves which we are using, we can form 6 stars. The first one consists of the single wave for h_1, h_2 equal to 00. The second consists of the four waves 10, 01, -10, $0 - 1$, the third of the four waves 11, -11, $1 - 1$, $-1 - 1$, the fourth of the four waves 20, 02, -20, $0 - 2$, the fifth of the eight waves 21, -21, $2 - 1$, $-2 - 1$, 12, -12, $1 - 2$, $-1 - 2$, and the sixth of the four waves 22, -22, $2 - 2$, $-2 - 2$. From each of these stars we can construct basis functions for several of the irreducible representations of Table 5-3, using the method of projection operators.

The single wave 00, which is a constant, is a basis function for Γ_1, or Σ^+ or Δ_1, in the other two notations used in Table 5-3. We shall use the Γ notation in the present section. If we consider the next four functions, 10, 01, -10, $0 - 1$, we find that they can be found from the first function 10 by the following operations:

$$\begin{aligned} X_0\psi(10) &= \psi(10) & Y_0\psi(10) &= \psi(10) \\ X_1\psi(10) &= \psi(0-1) & Y_1\psi(10) &= \psi(01) \\ X_{-1}\psi(10) &= \psi(01) & Y_{-1}\psi(10) &= \psi(0-1) \\ X_2\psi(10) &= \psi(-10) & Y_2\psi(10) &= \psi(-10) \end{aligned} \quad (7\text{-}5)$$

In our derivation of these equations we have used Eq. (5-29), replacing the coordinates y and z which appear there by x and y. Then we find from Eq. (5-26) and Table 5-3 that we have the following symmetrized plane waves formed from this star:

$$\begin{aligned} \Gamma_1 &: \psi(10) + \psi(0-1) + \psi(01) + \psi(-10) \\ \Gamma_3 &: \psi(10) - \psi(0-1) - \psi(01) + \psi(-10) \\ \Gamma_5 &: \psi(10) - \psi(-10) - \psi(0-1) + \psi(01) \end{aligned} \quad (7\text{-}6)$$

Basis functions for Γ_2 and Γ_4 cannot be formed from the functions of this star, as we saw in the discussion of Eqs. (5-32) and (5-36). Also only one set of basis functions for Γ_5 can be formed, as we saw in the discussion of Eq. (5-38).

---

[1] This is not strictly true. If we have two or more almost degenerate levels, the computer may produce linear combinations of the true wave functions, since the program has a finite tolerance which will not distinguish between exactly degenerate and almost degenerate cases.

In a similar way we can discuss the other cases of stars at Γ. We find that the four functions formed from ± 1, ± 1 form basis functions for representations Γ_1, Γ_4, Γ_5; the four formed from ± 2, 0, 0 ± 2 lead to Γ_1, Γ_3, Γ_5, while those formed from $\pm 2 \pm 1$, $\pm 1 \pm 2$ form a regular representation, with Γ_1, Γ_2, Γ_3, Γ_4, and two different functions for Γ_5. Finally the four functions formed from $\pm 2 \pm 2$ lead to representations Γ_1, Γ_4, Γ_5. We see, then, that we have basis functions for six symmetrized plane waves corresponding to Γ_1, one for Γ_2, three for Γ_3, three for Γ_4, six for Γ_5. These add to 25 basis functions in all, when we recall that each Γ_5 comprises two basis functions. It is of course necessary that the total equal the number of plane waves with which we started. It now proves to be

Table 7-1
Irreducible representations for symmetry directions, two-dimensional square symmetry

	X_0	Y_0		X_0	Y_1		X_0	Y_2
Δ_1	1	1	Σ_1	1	1	Z_1	1	1
Δ_2	1	−1	Σ_2	1	−1	Z_2	1	−1

	X_0	X_2	Y_0	Y_2
X_1	1	1	1	1
X_2	1	1	−1	−1
X_3	1	−1	1	−1
X_4	1	−1	−1	1

For M, the group of the wave vector is the complete group. We have representations, M_1, M_2, M_3, M_4, M_5, M_6 being two-dimensional and the others one-dimensional. The matrix elements are the same as those for the point Γ, as given in Table 5-3.

the case that each of the solutions of the secular equation for the point Γ, or $p_1 = p_2 = 0$, is of one of these types, as we have indicated for the lower bands in Fig. 7-4. If we had started from the beginning with symmetrized plane waves, of the type shown in Eq. (7-6), we should have had only a 6×6 secular equation for Γ_1 and Γ_5, no secular equation at all for Γ_2, cubics for Γ_3 and Γ_4.

We can proceed in similar ways for the other symmetries. In Table 7-1 we give the irreducible representations for the various symmetry points aside from Γ. The reader can easily verify these from the methods described in Chap. 5. The compatibility relations between the representations are given in Table 7-2. There are many examples of the use of these compatibility relations in Fig. 7-4.

7-3. Brillouin Zones.
When we compare the free-electron energy bands, and the correctly calculated ones, in Fig. 7-4, we see that the effect of the nondiagonal matrix elements of the Hamiltonian is to push two curves of energy as a function of **k** apart, where they would cross each other in the unperturbed, or free-electron, case. This, of course, is characteristic of any degenerate perturbation problem. Brillouin[1] considered this situation in the reciprocal space, asking where in this space we should find such degeneracies. Specifically, he started with the plane wave centered on the origin, with energy $p_1^2 + p_2^2$ in our present notation, whose energy as a function of p_1 and p_2 is given by a paraboloid of revolution centered on the origin, and asked where this plane wave would be degenerate with another plane wave, corresponding to integers h_1 and h_2, with energy $(p_1 + h_1)^2 + (p_2 + h_2)^2$. This energy is represented by a

Table 7-2
Compatibility relations for the two-dimensional square lattice

Γ_1	Γ_2	Γ_3	Γ_4	Γ_5		M_1	M_2	M_3	M_4	M_5
Δ_1	Δ_2	Δ_1	Δ_2	$\Delta_1\Delta_2$		Z_1	Z_2	Z_1	Z_2	Z_1Z_2
Σ_1	Σ_2	Σ_2	Σ_1	$\Sigma_1\Sigma_2$		Σ_1	Σ_2	Σ_2	Σ_1	$\Sigma_1\Sigma_2$

X_1	X_2	X_3	X_4
Δ_1	Δ_2	Δ_1	Δ_2
Z_1	Z_2	Z_2	Z_1

paraboloid centered on the position given by components $-h_1$, $-h_2$. If we now set up the perpendicular bisector of the line joining the origin and the point $-h_1$, $-h_2$, every point of this bisector will be equally far from the centers of the two paraboloids, and hence the two energies will be equal at all points along it.

The central Brillouin zone is formed by the lines of this sort, forming the perpendicular bisectors of the lines joining the origin and the points 10, 01, -10, $0-1$, which are the closest points of the reciprocal lattice to the origin. Brillouin, in addition, drew the bisectors of lines to all close lattice points, arriving at the result shown in Fig. 7-7. The central Brillouin zone is the square centered on the origin, shown unshaded in the figure. In addition, Brillouin defined second, third, fourth, etc., zones, as indicated by shading in the figure. The nth Brillouin zone

[1] L. Brillouin, *J. Phys. Radium*, **1**:377 (1930), and other references given in the Bibliography.

includes all parts of the reciprocal space in which the paraboloid centered on the origin has an energy which is the nth from the lowest. Thus in the triangle immediately to the right of the central Brillouin zone, we are nearer the point with coordinates $h_1 = 1$, $h_2 = 0$ than we are to the origin. Thus the paraboloid centered at this point, with energy $(p_1 - 1)^2 + p_2^2$, has a lower energy than that centered on the origin, with energy $p_1^2 + p_2^2$,

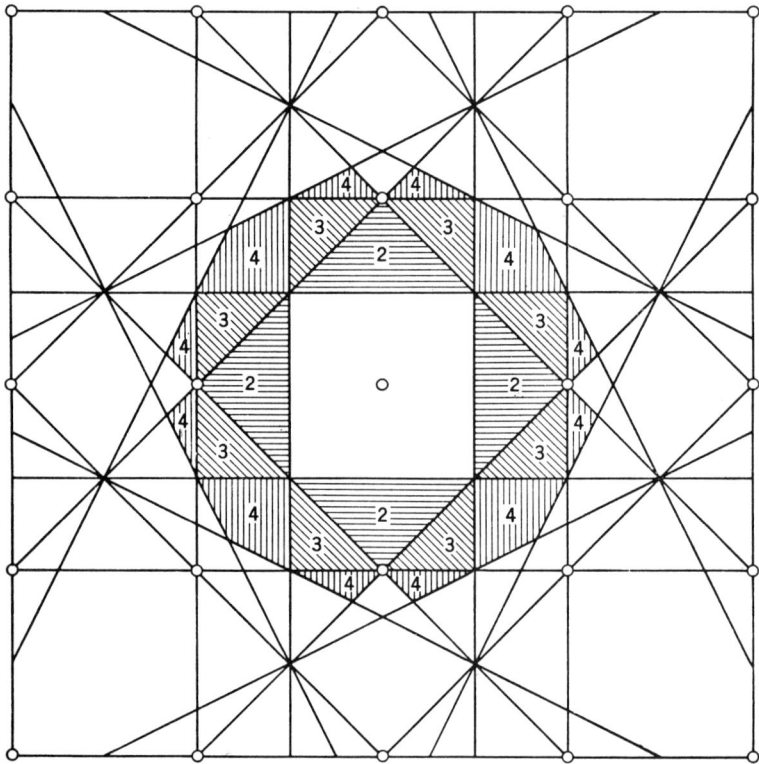

Fig. 7-7. Brillouin's construction for Brillouin zones, square lattice, second, third, and fourth zones indicated. Circles indicate lattice points of reciprocal lattice.

though all other paraboloids have a higher energy. Hence the points in this triangle form part of the second zone. Similar arguments apply to all the various segments of the reciprocal space indicated in Brillouin's construction.

The reader can verify by study of Fig. 7-7 that the segments of the nth Brillouin zone can be translated bodily, by integral distances along both axes, so that they just combine to fill the central Brillouin zone. This is illustrated for the fourth zone, in Fig. 7-8. Hence each zone has an area equal to that of the central Brillouin zone. Let us furthermore set up circles on which the energy $p_1^2 + p_2^2$ is constant. If we take these energy contours as they pass through the nth zone, in Brillouin's construction, and then translate them as we have just described to the central Brillouin zone, they form the contours of energy of the nth energy band, in the free-electron approximation. In other words, Brillouin's construction may be considered to be a way of setting up these energy contours. We see why the central Brillouin zone is often referred to as the reduced zone: we start with the extended segments as shown in Fig. 7-7 or 7-8, and reduce them to the shape of the central zone by the translations we have described.

In a case, such as the one discussed in the present chapter, in which the energy bands are not very far from the free-electron values, we can profitably reverse this process. Thus, we can take the energy contours for the nth energy band, as shown in Fig. 7-5. We can cut up the central Brillouin zone into the segments suggested for the nth zone, as we have shown for the fourth zone in Fig. 7-8. Then we can translate the resulting segments back to the positions shown for the nth zone in Fig. 7-7. In this

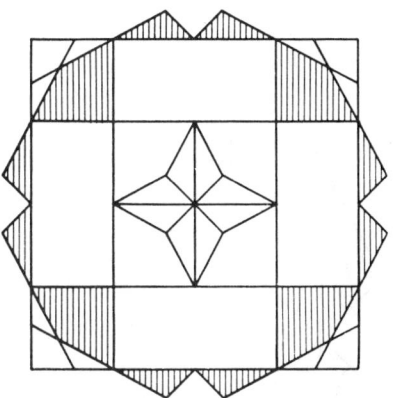

FIG. 7-8. Fourth Brillouin zone (shaded) as shown in Fig. 7-7, together with construction to show how it can be fitted into central zone.

way we set up what is called the extended-zone scheme for describing the energy bands. We do this in Figs. 7-9 and 7-10 for the case we are considering. In Fig. 7-9 we show a set of energy contours describing all the energy bands, the nth energy band being shown in the nth Brillouin zone. The contours differ from the concentric circles with center at the origin, which would be the energy contours for the unperturbed plane wave with energy $p_1^2 + p_2^2$, only by the effects of the nondiagonal matrix elements of the Hamiltonian. If these nondiagonal matrix elements are small, as they are in the case we are considering, the departure from concentric circles will not be great enough to destroy the usefulness of the comparison. In Fig. 7-10 we show the same thing as a perspective of the energy surface, in the extended-zone scheme.

We can now understand the significance of Brillouin's construction. We start with the energy of the unperturbed plane wave, $p_1^2 + p_2^2$, in which for the moment we are allowing p_1 and p_2 to take on any values, not necessarily within the central zone. We then apply the perturbation, consisting of the nondiagonal matrix elements of the Hamiltonian. The effect is to introduce energy gaps across the lines shown in Brillouin's construction. On such a line, we recall that two plane waves, that corresponding to a given h_1 and h_2 and that for which h_1 and h_2 are zero, are degenerate. Suppose we were attempting to solve our secular problem

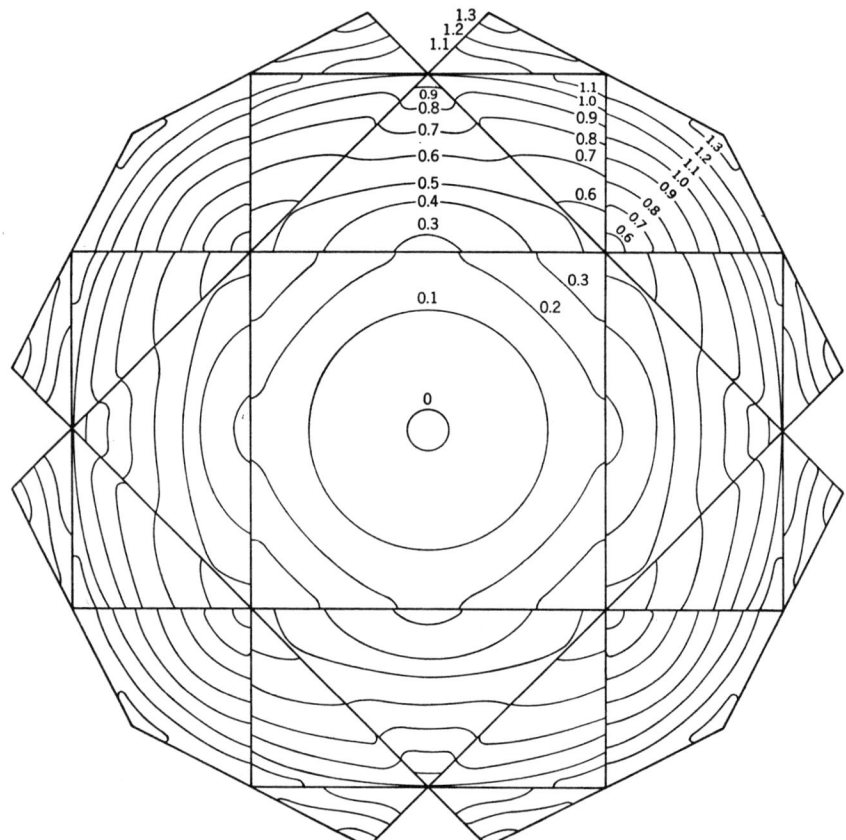

FIG. 7-9. Energy contours of first four energy bands in the extended-zone scheme, square lattice.

by perturbation methods. The first step would be to remove the degeneracy of the plane waves which have identical energies, by solving a secular equation involving only these plane waves; after that, we could proceed by second-order perturbation theory. In the removal of this degeneracy, the energies of the two perturbed states will be pushed apart, with respect to the unperturbed degenerate energy, by plus or minus the nondiagonal matrix elements of the energy between the two unperturbed functions. In the case considered, this matrix element is $W(h_1,h_2)$. Hence we conclude that to the first order of approximation, an energy gap of amount $2W(h_1,h_2)$ will be introduced across each such line in the Brillouin construction. This is like the one-dimensional case which we have discussed in connection with Fig. 6-1.

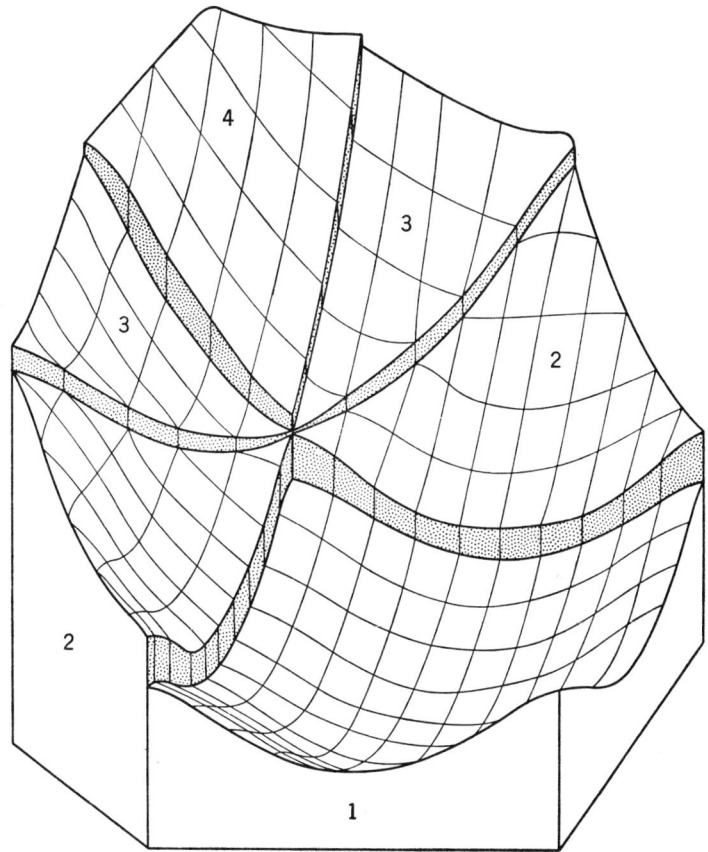

Fig. 7-10. Energy bands in the extended-zone scheme, perspective.

This very simple conclusion was used a great deal in deriving qualitative properties of energy bands in the early days of energy-band theory.[1] We can see from our present example that it is hardly accurate enough to be of any great quantitative value. For example, the boundary

[1] See for instance H. Jones, *Proc. Roy. Soc. (London)*, **A144**:225 (1934); **A147**:396 (1934).

between zones 2 and 3, when we translate to the central zone to use the reduced-zone scheme, falls along the directions Σ. In Table 7-3 we give the correct separations at points along this line, as found from the solution of the secular equation. When we compare with the separation given by Brillouin's simple scheme, namely, 2W(11), which is 0.050 unit, we see that the agreement, though qualitatively satisfactory, is not really quantitative. It breaks down completely as we approach the point M, where a number of plane waves are degenerate.

7-4. The Momentum Eigenfunction for the Two-dimensional Case. The calculation which we have been discussing for the two-dimensional case yields the momentum eigenfunction, as well as the energy values. We shall now discuss these results, and shall show that in a two- or three-dimensional problem the situation is a good deal more complicated than in the one-dimensional example of Chap. 6.

Table 7-3
Energy separation between second and third bands, along direction Σ. Values of p_1 and p_2 are given along line Σ

Point	2d band	3d band	Difference
$\Gamma(0,0)$	0.9376	1.0067	0.0691
$\Sigma(0.1,0.1)$	0.7897	0.8426	0.0529
$\Sigma(0.2,0.2)$	0.6517	0.7031	0.0514
$\Sigma(0.3,0.3)$	0.5534	0.6033	0.0499
$\Sigma(0.4,0.4)$	0.4991	0.5433	0.0442
$M(0.5,0.5)$	0.5234	0.5234	0.0000

When we assemble the results of the calculations, for the various points within the Brillouin zone for which calculations are made, we note in the first place that the program evaluating the eigenvectors of the problem of course cannot determine the sign of the whole function. It is set up to produce a normalized function, but we still can change the sign of all components $v_n(\mathbf{k} + \mathbf{K}_i)$ corresponding to a given band n, and a given value of $\mathbf{k}$. When we inspect the results, we find that by suitable choice of these signs, the computed values appear in fact to form entries in a two-dimensional table of values for a continuous function. Thus in Table 7-4 we give the entries for a quadrant of the Brillouin zone for the lowest band. To save space in writing the table, we have given entries only to three figures, though they are computed to greater accuracy, and we do not carry the table out to values of $p + h = 2.5$, the extreme values for which the calculations have been made, but rather stop the tabulation when the numbers have sunk to a small value.

For this lowest band, the situation is very simple. The table repeats

Sec. 7-4] PLANE-WAVE EXPANSION—TWO AND THREE DIMENSIONS 183

symmetrically in each quadrant, and represents a perfectly continuous function. We note that this function has the Γ_1 type of symmetry, in the momentum space. That is, it is unchanged when we make rotations of 90°, 180°, 270°, or when we reflect in the horizontal axis and follow by one of these rotations. Alternatively, the function is even when we perform a reflection in the x or y axis, or in one of the 45° lines. These are the characteristics which the wave function itself has. As we see from Fig. 7-4, for the lowest band we have the symmetry Γ_1 at the origin of the Brillouin zone, Δ_1 along the x axis, and Σ_1 along the 45° line, each of the latter two symmetries indicating that the function is even with respect to the corresponding reflection.

Table 7-4
Momentum eigenfunction for lowest band, two-dimensional case. Horizontally we give $p_1 + h_1$, vertically $p_2 + h_2$. Only a quadrant is tabulated

	0	0.1	0.2	0.3	0.4	0.5	0.6	0.7	0.8	0.9	1.0	1.1	1.2
1.2	0.031	0.031	0.032	0.032	0.033	0.034	0.022	0.018	0.015	0.013	0.012	0.011	0.010
1.1	0.036	0.037	0.037	0.037	0.039	0.040	0.025	0.020	0.017	0.015	0.013	0.012	0.011
1.0	0.044	0.044	0.044	0.045	0.047	0.047	0.030	0.023	0.019	0.017	0.015	0.013	0.012
0.9	0.054	0.055	0.055	0.056	0.058	0.059	0.036	0.027	0.022	0.019	0.017	0.015	0.013
0.8	0.073	0.073	0.074	0.075	0.078	0.078	0.046	0.033	0.026	0.022	0.019	0.017	0.015
0.7	0.108	0.105	0.110	0.112	0.116	0.114	0.063	0.043	0.033	0.027	0.023	0.020	0.018
0.6	0.207	0.205	0.210	0.213	0.222	0.206	0.101	0.063	0.046	0.036	0.030	0.025	0.022
0.5	0.704	0.703	0.703	0.697	0.675	0.499	0.206	0.114	0.078	0.059	0.047	0.040	0.034
0.4	0.975	0.974	0.972	0.967	0.943	0.675	0.222	0.116	0.078	0.058	0.047	0.039	0.033
0.3	0.991	0.991	0.989	0.985	0.967	0.697	0.213	0.112	0.075	0.056	0.045	0.037	0.032
0.2	0.994	0.994	0.993	0.989	0.972	0.703	0.210	0.110	0.074	0.055	0.044	0.037	0.032
0.1	0.995	0.995	0.994	0.991	0.974	0.703	0.205	0.105	0.073	0.055	0.044	0.037	0.031
0.0	0.996	0.995	0.994	0.991	0.975	0.704	0.207	0.108	0.073	0.054	0.044	0.036	0.031

It is interesting to look at the actual values of the entries in Table 7-4, as well as at their symmetry. We observe that we have entries only slightly smaller than unity, for points within the central Brillouin zone (that is, for $|p_1 + h_1|$ and $|p_2 + h_2|$ each smaller than $\frac{1}{2}$), while for points on the surface of the zone the entries are approximately $0.707 = 1/\sqrt{2}$, except at the corner of the zone, the point M, where the entry is approximately $0.500 = 1/\sqrt{4}$. Outside the central Brillouin zone the entries are very small, their squares being of the order of magnitude of a few hundredths of a unit. We can interpret this momentum eigenfunction completely in terms of the one-dimensional case which we have taken up in Chap. 6.

We note that if only the energy component

$$W(10) = W(01) = W(-10) = W(0-1)$$

which in Eq. (7-4) is given the value -0.040, were different from zero, and the other components of Eq. (7-4) vanished, we should have a problem whose potential energy was $W(10)$ $(2 \cos 2\pi x + 2 \cos 2\pi y)$, as we see from Eq. (7-1). In this case, we could separate variables in Schrödinger's equation, and could write the wave function as a product of a function of x and a function of y. Each of these functions would satisfy a one-dimensional Schrödinger equation of the type discussed in Sec. 6-6. As we see from Eq. (6-35), the wave function, being a product of the functions of x and y, could be written in the form

$$u(p_1,p_2;x,y) = \Sigma(h_1,h_2) v_1(p_1 + h_1) v_2(p_2 + h_2) e^{2\pi i[(h_1+p_1)x+(h_2+p_2)y]} \quad (7\text{-}7)$$

where v_1, v_2 are the momentum eigenfunctions of the one-dimensional problems associated with x and y, respectively. In other words, the function $v(p_1 + h_1, p_2 + h_2)$ of Eq. (7-2) can be written in this case as a product of a function of p_x and a function of p_y, where $p_x = p_1 + h_1$, $p_y = p_2 + h_2$.

Each wave function of the two-dimensional case can then, to this approximation, be considered to be made up of two one-dimensional functions, and to be characterized by the quantum numbers of these two functions. In the case of the lowest energy band, both the functions v_1 and v_2 correspond to the lowest band of the one-dimensional case, whose momentum eigenfunction is given in Fig. 6-2. We expect, then, that to a certain approximation the momentum eigenfunction of the two-dimensional problem, as given in Table 7-4, could be written as a product of two such functions. This expectation is entirely borne out by the facts. Over the range of p from $-\frac{1}{2}$ to $\frac{1}{2}$, where the one-dimensional function is very nearly unity, the two-dimensional function is likewise nearly unity. Along the boundaries of the Brillouin zone, we have noted in the discussion of Fig. 6-2 that the one-dimensional function is approximately 0.707, and the two-dimensional function also has this value. At the corner of the zone, where both $p_1 + h_1$ and $p_2 + h_2$ equal $\frac{1}{2}$, the function is the product of two factors each equal to 0.707, or 0.500. Outside the Brillouin zone the function rapidly falls off to zero. In other words, the approximation of separability is well justified in this case. This assumption was discussed at length in the paper by the present author cited in Sec. 6-6.

When we go to the higher energy bands, however, the situation becomes much more involved, as was discussed in the reference just cited. The situation is illustrated in Table 7-5, where we give the actually calculated values of the momentum eigenfunction, for the second band. As in Table 7-4, we have cut off the table, in the interests of brevity, before we have gone to very large values of $p + h$. Furthermore, the entries corresponding to the point M, such as for $p_1 = p_2 = 0.5$, are omitted.

Table 7-5
Momentum eigenfunction for second band, two-dimensional case

	−1.0	−0.9	−0.8	−0.7	−0.6	−0.5	0.0	0.1	0.2	0.3	0.4	0.5	1.0	1.1	1.2	1.3	1.4	1.5
1.2			0.023	0.028	0.026	0.008			0.028	0.023	0.015	−0.008			0.004	0.002	0.001	−0.004
1.1		0.028	0.034	0.033	0.030	0.009		0.058	0.042	0.029	0.018	−0.009		0.007	0.004	0.002	0.001	−0.005
1.0	0.043	0.044	0.042	0.040	0.036	0.011		0.128	0.062	0.038	0.003	−0.011		0.009	0.004	0.003	0.001	−0.005
0.4					0.651	0.705					−0.330	−0.705					0.007	−0.011
0.3				0.691	0.962	0.706				−0.164	−0.232	−0.706				0.016	0.001	−0.011
0.2			0.699	0.982	0.972	0.707			−0.109	−0.122	−0.217	−0.707			0.028	0.008	0.002	−0.011
0.1		0.700	0.986	0.989	0.974	0.707	−0.090	−0.083	−0.083	−0.115	−0.213	−0.707		0.058	0.014	0.007	0.002	−0.011
0.0	0.496	0.978	0.991	0.990	0.975	0.707		−0.068	−0.080	−0.114	−0.212	−0.707	0.496	0.036	0.013	0.007	0.002	−0.011
−0.6					0.208	0.055					0.651	−0.055					0.014	−0.007
−0.7				0.125	0.096	0.028				0.691	0.098	−0.028				0.019	0.003	−0.006
−0.8			0.088	0.071	0.062	0.018			0.699	0.117	0.048	−0.018			0.023	0.005	0.001	−0.006
−0.9		0.065	0.054	0.051	0.046	0.014		0.700	0.123	0.058	0.031	−0.013		0.028	0.006	0.003	0.001	−0.005
−1.0	0.043	0.044	0.042	0.040	0.036	0.011	0.496	0.128	0.062	0.038	0.023	−0.011	0.043	0.009	0.004	0.003	0.002	−0.004

The reason is that at this point the second and third bands are degenerate with each other, so that the wave functions are linear combinations of two degenerate functions, and no unique function is determined from the calculation.

We must now ask ourselves how to extend the results, by symmetry, to the rest of the Brillouin zone; examination of Table 7-5 will show that only one-eighth of the central Brillouin zone is filled in, in the table. For the lowest band the procedure was obvious: we extended the function by using its symmetry on reflection in the coordinate axes and the 45° lines, and the result was a continuous function everywhere. We can do the same thing here, and secure continuity within the central Brillouin zone. When we do this, however, we find a function which shows a discontinuity around the boundary of this central zone. We can see this from Table 7-5, by looking at the entries corresponding to $p_2 + h_2 = 0$. From Table 7-5, the entries for $p_1 + h_1$ equal to $-1.0, -0.9, \ldots, -0.5$ are, respectively, 0.496, 0.978, 0.991, 0.990, 0.975, and 0.707. However, if we fill out the central zone in the manner described, the entries for -0.5, $-0.4, -0.3, -0.2, \ldots, 0.4, 0.5$ would be, respectively, $-0.707, -0.212$, $-0.114, -0.080, -0.068, -0.090, -0.068, -0.080, -0.114, -0.212$, -0.707. Next we come to the entries from 0.5 to 1.0, which would automatically be determined by symmetry to be 0.707, 0.975, 0.990, 0.991, 0.978, 0.496. In other words, at ± 0.5, we have discontinuities, the function approaching 0.707 from one side, and -0.707 from the other.

There is no way to avoid this discontinuity, which is a result of a type of internal inconsistency in the symmetry properties of this band. We note from Fig. 7-4 that this second band, though it has the same symmetries $\Gamma_1, \Delta_1, \Sigma_1$ as the first band, has the symmetry Z_2 rather than Z_1, as well as M_5 rather than M_1. From Table 7-1 we see that the symmetry Z_2 implies that the function changes sign under the operation Y_2, whereas with Z_1 there is no change of sign. We can see from Eq. (5-29) that this operation Y_2 is one in which the sign of the quantity $p_1 + h_1$ is changed, while $p_2 + h_2$ remains unchanged. In other words, when we go from $p_1 + h_1 = 0.5$ to -0.5, keeping $p_2 + h_2$ fixed, the Z_2 symmetry demands that the function change sign, whereas the Γ_1 type of symmetry, which we are trying to build up from the momentum eigenfunction as a whole, demands that the function be unchanged by the same operation. It is for this reason that the two conflicting values 0.707 and -0.707 are found as we approach the values 0.5 or -0.5 from opposite sides. These inconsistent symmetry requirements are not in contradiction to the ordinary compatibility relations, which determine the behavior of the function only in a microscopic region near a symmetry point. They come, rather, by comparing the behavior of the function at points distant from each other in reciprocal space, such as at 0.5 and -0.5.

Similar inconsistencies of symmetry properties are found in most energy bands, though not in all, with the result that the momentum eigenfunction, though continuous over limited regions of reciprocal space, shows lines over which there are discontinuities. We shall show in the next section how we can modify the definitions of the energy bands and momentum eigenfunctions so as to minimize these discontinuities, and we shall see why it is desirable to minimize them. Before we go on to this point, however, we should examine the general nature of the momentum eigenfunctions for the various energy bands, and investigate the relation of these energy bands to the Brillouin zones.

In both Tables 7-4 and 7-5, there are some entries which are almost equal to unity, whereas most others are very small. In other words, the eigenfunction in most cases is nearly a single plane wave, with only small contributions from plane waves with other wave vectors. The reason of course is that our example is only slightly removed from the free-electron case. The part of reciprocal space where the values of almost unity occur is the same as the Brillouin zone of the corresponding energy band. Thus, in Table 7-4, we have pointed out for the first band that the contributions of unity are found for p_1 and p_2 each equal to values between -0.5 and 0.5, which is just the central Brillouin zone. In Table 7-5, for the second band, we find the large contributions in the left-hand central triangle, or for the region bounded by the horizontal axis $p_2 + h_2 = 0$, the vertical line $p_1 + h_1 = -0.5$, and the 45° line running from $p_1 + h_1 = -1.0$, $p_2 + h_2 = 0$, up to $p_1 + h_1 = -0.5$, $p_2 + h_2 = 0.5$. As we see from Fig. 7-7, this triangle forms one-eighth of the second Brillouin zone, and when we extend the momentum eigenfunction as we have described, the large entries will fill the second zone. Similarly, if we work out the momentum eigenfunctions of the third and fourth energy bands, we find that the contributions are large just in the third and fourth zones, as shown in Fig. 7-7. Thus we see the significance of Brillouin's construction, in terms of momentum eigenfunctions: the large contributions to the momentum eigenfunction, in the nth band, for the almost free electron case, come in the nth Brillouin zone. Contributions in other regions represent the amplitudes of other waves which must be mixed with the primary plane wave to take account of the departure of the wave function from a single plane wave.

7-5. Modified Momentum Eigenfunctions and Wannier Functions.
We can gain a great deal of insight into what is going on in the second, and higher, energy bands of the two-dimensional problem by considering them as perturbations of the separable problem arising from the one-dimensional case of Sec. 6-6. We have already pointed out that the lowest energy band of the two-dimensional case has a wave function which is very nearly the product of the wave functions of the one-dimensional case,

each corresponding to the lowest state of the one-dimensional problem, with the symmetrical momentum eigenfunction given in Fig. 6-2. If the problem in two dimensions were separable, the next-lowest stationary state of the two-dimensional case would correspond to having different quantum numbers for the x and y parts of the problem. Thus, in Eq. (7-7), we could have a momentum eigenfunction $v_1(p_1 + h_1)v_2(p_2 + h_2)$, where one of the functions v_1 or v_2 would correspond to the lowest state, the other to the next-lowest state, of the one-dimensional case. Obviously, there are two ways to build up this function, one arising from the other by a rotation of axes through 90°, so that one of the functions has a node at $p_1 + h_1 = 0$, the other at $p_2 + h_2 = 0$. Such a pair of momentum eigenfunctions would show Γ_5 symmetry.

The energy of one of these functions would be the sum of the energy of the lowest one-dimensional state, as a function of p_1, plus the second

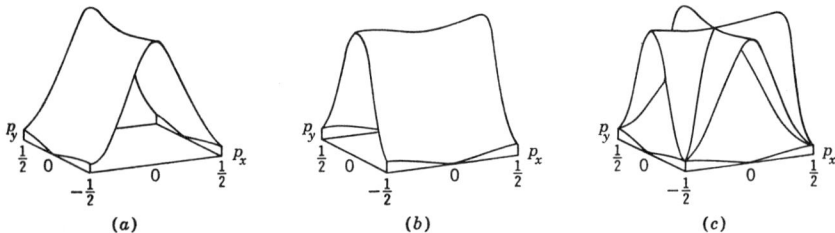

FIG. 7-11. Energy as a function of momentum, p-like states, two-dimensional Mathieu problem. Energy surface for (a) case $n_x = 1$, $n_y = 0$; (b) case $n_x = 0$, $n_y = 1$; (c) combination of cases (a) and (b), showing degeneracy, or intersection of surfaces, along diagonals $k_y = \pm k_x$.

lowest as a function of p_2, or vice versa. In Fig. 7-11 we show the resulting energy surfaces in perspective. In (a) we show the case where n_x, the quantum number associated with the x direction, is 1, while n_y is 0, 0 being the lowest state, 1 the next lowest. In (b) we show the other case, where $n_x = 0$, $n_y = 1$. In (c) we combine these two cases, showing the two energy surfaces intersecting each other along the 45° lines in the central Brillouin zone, at which the unperturbed problem is degenerate. We shall show shortly that by action of the perturbative terms, those present in our two-dimensional case but not in the separable problem, the two degenerate states are split apart along this 45° line, except at the origin of reciprocal space, the point Γ, and at the corner, the point M. When this happens, it is obvious by comparison of Fig. 7-11 with Fig. 7-2b and c that the two resulting energy surfaces approach those for the second and third energy bands of the two-dimensional case.

Let us now verify our statement that there is a nondiagonal matrix

Sec. 7-5]   PLANE-WAVE EXPANSION—TWO AND THREE DIMENSIONS   189

element of the Hamiltonian introduced between these two unperturbed functions, except at the points Γ and M. If we denote the momentum eigenfunction of the lowest one-dimensional state by $v_1(p + h)$, and of the second state by $v_2(p + h)$, then the two functions in question are that given in Eq. (7-7) and that with the two subscripts on v_1 and v_2 interchanged. It is then easy to show that the nondiagonal matrix element between the two functions, introduced by a Fourier component $W(l_1,l_2)$ of the potential energy, is

$$\Sigma(h_1,h_2)v_1^*(p_1 + h_1)v_2(p_1 + h_1 - l_1)$$
$$v_2^*(p_2 + h_2)v_1(p_2 + h_2 - l_2)W(l_1,l_2) \quad (7\text{-}8)$$

As a specific example, we shall consider the Fourier components

$$W(\pm 1, \pm 1)$$

the largest ones disregarded in the separable case, as we see from Eq. (7-4), which we have seen in Table 7-3 to be responsible for the energy-band separation along Σ. We sum the expression of Eq. (7-8) over the four such cases, and compute for the particular case $p_1 = p_2$, or the direction Σ, where we have the degeneracy. Then the expression can be rewritten in the form

$$W(11)|\Sigma(h_1)[v_1^*(p_1 + h_1 - 1) + v_1^*(p_1 + h_1 + 1)]v_2(p_1 + h_1)|^2 \quad (7\text{-}9)$$

In Table 7-6 we give the quantities $v_1(p_1 + h_1 - 1) + v_1(p_1 + h_1 + 1)$ and $v_2(p_1 + h_1)$ as functions of $p_1 + h_1$. We see from the table that the function $v_1(p_1 + h_1 - 1) + v_1(p_1 + h_1 + 1)$ is an even function of $p_1 + h_1$, while $v_2(p_1 + h_1)$ is an odd function. We are to multiply the entries in the two tables together at a set of points p_1, $p_1 \pm 1$, $p_1 \pm 2$, and add them. The product is an odd function. If $p_1 = 0$, or $p_1 = \frac{1}{2}$, namely, at the points Γ and M, we are computing this odd function at a sequence of points symmetrically placed with respect to the origin, so that the sum is necessarily zero. This is the reason for our statement **that** there is no nondiagonal matrix element of the Hamiltonian between the two wave functions at Γ and M. On the other hand, at the points $p_1 = 0.1, 0.2, 0.3, 0.4$, there is one large entry in the sum, coming from $h_1 = -1$, and very nearly equal to -1. In fact, the sums at these four points are found to be -1.008, -0.996, -0.980, and -0.916, respectively. The squares, then, which appear in Eq. (7-9), are approximately unity, so that the magnitude of the nondiagonal matrix element, given in Eq. (7-9), is approximately $W(11)$, which would be the value found from the free-electron approximation. It was this value which we compared with the actual separations along the line Σ in Table 7-3.

We see, then, that by and large the second and third bands may be derived by perturbations from these two unperturbed functions arising

Table 7-6

Functions $v_1(p_1 + h_1)$, $v_1(p_1 + h_1 - 1) + v_1(p_1 + h_1 + 1)$, and $v_2(p_1 + h_1)$, as functions of $p_1 + h_1$, for one-dimensional problem approximating our two-dimensional case

$p_1 + h_1$	$v_1(p_1 + h_1)$	$v_1(p_1 + h_1 - 1) + v_1(p_1 + h_1 + 1)$	$v_2(p_1 + h_1)$
−2.0	0.003	0.044	−0.009
−1.9	0.004	0.054	−0.015
−1.8	0.005	0.073	−0.016
−1.7	0.006	0.108	−0.017
−1.6	0.009	0.207	−0.018
−1.5	0.019	0.704	−0.011
−1.4	0.025	0.975	0.002
−1.3	0.028	0.991	0.007
−1.2	0.031	0.994	0.013
−1.1	0.036	0.995	0.036
−1.0	0.044	0.999	−0.707
−0.9	0.054	0.999	−0.978
−0.8	0.073	0.999	−0.991
−0.7	0.108	0.997	−0.990
−0.6	0.207	0.984	−0.975
−0.5	0.704	0.723	−0.707
−0.4	0.975	0.232	−0.212
−0.3	0.991	0.136	−0.114
−0.2	0.994	0.104	−0.080
−0.1	0.995	0.090	−0.068
0.0	0.996	0.088	0.000
0.1	0.995	0.090	0.068
0.2	0.994	0.104	0.080
0.3	0.991	0.136	0.114
0.4	0.975	0.232	0.212
0.5	0.704	0.723	0.707
0.6	0.207	0.984	0.975
0.7	0.108	0.997	0.990
0.8	0.073	0.999	0.991
0.9	0.054	0.999	0.978
1.0	0.044	0.999	0.707
1.1	0.036	0.995	−0.036
1.2	0.031	0.994	−0.013
1.3	0.028	0.991	−0.007
1.4	0.025	0.975	−0.002
1.5	0.019	0.704	0.011
1.6	0.009	0.207	0.018
1.7	0.006	0.108	0.017
1.8	0.005	0.073	0.016
1.9	0.004	0.054	0.015
2.0	0.003	0.044	0.009

from the separable free-electron case. This is not true in all details. At Γ, in particular, we see from Fig. 7-4 that it is the third and fourth bands, not the second and third, which approach the doubly degenerate Γ_5 state. The reason for this is that in this neighborhood there are really four, not two, states arising from the almost free electron case which interact. The two which we have been neglecting arise from the cases $n_x = 2$, $n_y = 0$, and $n_x = 0$, $n_y = 2$, of the separable problem. When they are considered, we must solve a problem involving all four functions to get the proper values at Γ, and we find that a state Γ_1 arising from these higher functions is actually pushed below the Γ_5, so that the second band in fact approaches Γ_1, rather than Γ_5, as we go to $p_1 = p_2 = 0$.

The unperturbed momentum eigenfunction, of the nature of

$$v_1(p_1 + h_1)v_2(p_2 + h_2) \quad \text{or} \quad v_2(p_1 + h_1)v_1(p_2 + h_2)$$

which we have just been discussing, is a much simpler quantity than the one which we should build up out of the second band alone, and it shows no discontinuities of any large amount. For many purposes it is desirable to build up such a function. We can secure it by using, at different values of p_1 and p_2, contributions from different bands, either the second, third, or fourth, rather than building the momentum eigenfunction only from a single band. Thus we can indicate in Fig. 7-12 the different regions of momentum space, and we show in this figure the band whose momentum eigenfunction we are to use in this region to build up the continuous eigenfunction. At Γ, we use one or the other of the two degenerate functions arising from the Γ_5 state, belonging to the third and fourth bands. At M similarly we use one or the other of the two degenerate functions arising from the M_5, from the second and third bands. Along Σ we use $1/\sqrt{2}$ times the sum or difference of the functions arising from the second and third bands. In the remaining parts of momentum space we use the momentum eigenfunction of either the second or third band, with appropriately chosen sign. When we combine these entries, we have a fairly continuous function, of the required properties.

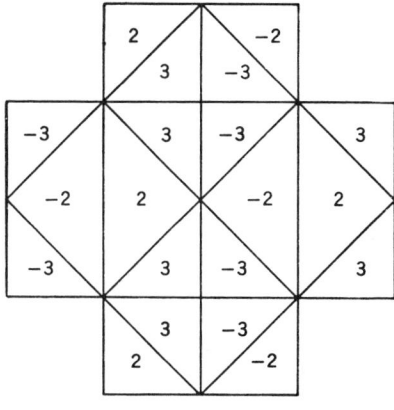

Fig. 7-12. Functions to be used in different areas of k space, to produce p_x-like momentum eigenfunction, square lattice.

The function we are setting up, of course, is defined only at a grid of points in momentum space, since it is only for values of p differing by 0.1 unit that calculations have been made. As we have noted earlier, this is just what we should have if we were imposing periodic boundary conditions, with a periodic length of 10 lattice points along each of the two axes, or with 100 unit cells in the repeating region. From this example, we have a graphic picture of the results of the fact that the periodic boundary conditions limit us to a discrete lattice of points in momentum space. A consequence of this is that there is always some arbitrariness in drawing a surface of energy versus p_1 and p_2, and there is no strict way of defining whether the resulting surface is continuous or not. However, by the

Table 7-7
Momentum eigenfunction of p_x-like type. Only one-quarter of the momentum space is indicated, $p_1 + h_1$ being horizontal, $p_2 + h_2$ vertical. The function is an odd function of $p_1 + h_1$, an even function of $p_2 + h_2$

	0	0.1	0.2	0.3	0.4	0.5	0.6	0.7	0.8	0.9	1.0
1.2	0.000	−0.040	−0.030	−0.023	−0.015	0.008	0.026	0.028	0.029	0.029	0.015
1.1	0.000	−0.061	−0.042	−0.029	−0.018	0.009	0.030	0.033	0.034	0.035	0.021
1.0	0.000	−0.128	−0.062	−0.038	−0.023	0.011	0.036	0.040	0.042	0.044	0.025
0.9	0.000	0.004	−0.123	−0.058	−0.031	0.013	0.046	0.051	0.054	0.048	0.031
0.8	0.000	0.132	0.006	−0.117	−0.048	0.018	0.062	0.071	0.062	0.057	0.041
0.7	0.000	0.070	0.132	0.011	−0.098	0.028	0.096	0.080	0.084	0.092	0.061
0.6	0.000	0.053	0.076	0.207	0.039	0.055	0.147	0.702	0.175	0.182	0.118
0.5	0.000	0.061	0.079	0.115	0.207	0.500	0.675	0.697	0.702	0.703	0.500
0.4	0.000	0.056	0.072	0.098	0.233	0.705	0.959	0.972	0.978	0.979	0.697
0.3	0.000	0.053	0.061	0.116	0.232	0.706	0.962	0.981	0.988	0.991	0.704
0.2	0.000	0.049	0.077	0.122	0.217	0.707	0.972	0.982	0.991	0.987	0.706
0.1	0.000	0.059	0.083	0.115	0.213	0.707	0.974	0.989	0.986	0.991	0.707
0.0	0.000	0.068	0.080	0.114	0.212	0.707	0.975	0.990	0.991	0.978	0.707

present procedure, we certainly get a much closer approach to continuity than by extending the entries of Table 7-5 throughout the momentum space according to the Γ_1 symmetry. The second and third bands are fundamentally a result of the twofold degeneracy of p-type bands, and the present scheme, using two momentum eigenfunctions of Γ_5 symmetry, one being derived from the other by a 90° rotation, exhibits this fact in the most natural way possible. We show the resulting momentum eigenfunction in Table 7-7. It is from this table that the function $v_2(p_1 + h_1)$ of Table 7-6 is derived.

In this discussion, we have been proceeding in a very free manner to build up a momentum eigenfunction in quite a different manner from the one we have discussed earlier. Instead of choosing all the entries from the same energy band, we have taken different entries from different bands. The possibility of doing this has been discussed by Koster.[1] As he has shown or suggested in his paper, we really have much more freedom in setting up momentum eigenfunctions, and consequently Wannier functions, than we have indicated in our discussion up to this point. Thus, in our discussions of Secs. 6-4 and 6-5, we have assumed **that** the nth energy band, out of which we were constructing a Wannier function, consisted of the nth energy levels measured up from the bottom, at each value of $\mathbf{k}$. In our derivations, however, we have nowhere used this specific assumption. We could, if we chose, define the nth band as consisting of one energy level at one $\mathbf{k}$ value, another at another $\mathbf{k}$, and so on, entirely discarding the definition of the nth band as being the nth level measured from the bottom. As long as we used each momentum function $v_n(\mathbf{k} + \mathbf{K}_h)$ only in one energy band, and had a value of $v_n(\mathbf{k} + \mathbf{K}_h)$ defined for each $\mathbf{k} + \mathbf{K}_h$ in our network of points, all proofs of Secs. 6-4 and 6-5 would go through without alteration. In particular, these functions would still be orthogonal, they would still lead to orthogonal Wannier functions, and the theorems concerning Fourier resolution, as well as the Fourier components of the Hamiltonian between Wannier functions, would go through just as before. In other words, we have been unnecessarily restrictive in our definitions.

The only respect in which we have gone beyond this in the definition of the momentum eigenfunction of Table 7-7 is along the direction Σ, where we have used a linear combination of momentum eigenfunctions from two different bands, the second and third, in building up our composite function. It is easy to show that the effect of this is to introduce nondiagonal matrix elements of the Hamiltonian between Wannier functions derived from our two momentum eigenfunctions forming the two-dimensional Γ_5 representation. It is the corresponding nondiagonal matrix elements of the Hamiltonian between the corresponding Bloch functions which we have evaluated in Eq. (7-9) and Table 7-6. This situation is different from that of Eq. (6-30), which indicates no nondiagonal matrix elements of the Hamiltonian between Wannier functions connected with different bands, if we use the conventional definition.

Let us now ask why it is desirable to use the type of continuous momentum eigenfunction which we have just been discussing, rather than the discontinuous function of Γ_1 symmetry for the second band, which we discussed in connection with Table 7-5. The reason comes in the definition of the Wannier functions. The Fourier transform of a function showing discontinuities falls off much more slowly with distance than that of a continuous function. The object of setting up Wannier functions is to get as concentrated orbitals as possible. We can use them in discussing the localization of chemical bonds, just as we did in connection

[1] G. F. Koster, *Phys. Rev.*, **89**:67 (1953).

with bonds in such molecules as H_2O, NH_3, CH_4, and benzene, in Volume 1. They have no point if they extend outward too far.

Even with the approximate discontinuity shown by such functions as those given in Fig. 6-2, the Wannier functions extend out rather far in space, and if further discontinuities were introduced, they would extend even further. In Fig. 7-13 we show the Wannier functions found by Fourier transformation from the one-dimensional functions v_1 and v_2 of Table 7-6, which represent functions very similar to those of Fig. 6-2. We see that while the largest amplitude of these functions is located in the

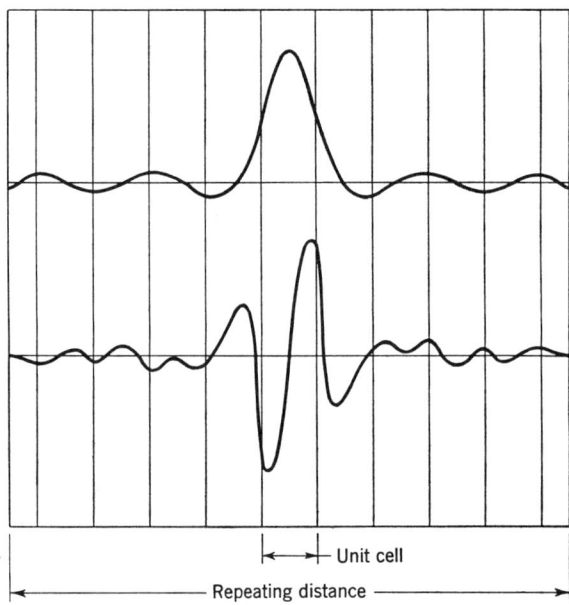

FIG. 7-13. Wannier functions related to first and second one-dimensional momentum eigenfunctions v_1 and v_2.

central unit cell of ordinary space, for which we have set up the Wannier functions, still they extend out into other unit cells with quite large amplitudes. These functions represent the best we can do, for this particular case, in getting concentrated functions.[1]

It is because we are dealing with an almost free electron case that the Wannier functions spread out so far. If we were dealing with a quite different type of problem, in which the momentum eigenfunction was

[1] W. Kohn, *Phys. Rev.*, **115**:809 (1959), has discussed the continuity of momentum eigenfunctions, and the rate at which the Wannier functions fall off, in the one-dimensional case, coming to this same conclusion.

spread out over many unit cells in the reciprocal space, as in Fig. 6-4, and the ordinary wave function was concentrated into the central part of the unit cell in ordinary space, we should find that the Wannier function could also be made to be concentrated in the same way. This could only be done, however, for the p-like functions, if we used composite momentum eigenfunctions, of the sort described in the present section, having p-like or Γ_5 symmetry. If we tried to use momentum eigenfunctions, each constructed from one energy band, by the ordinary definition, the discontinuities in the momentum eigenfunction would result in Wannier functions spread out in space.

This will be strikingly the case if the energy bands are of a complicated form, such for instance as the fourth band along the direction Σ, as shown in Fig. 7-4. We recall that according to Eq. (6-33), the quantities $\mathcal{E}_n(\mathbf{R}_s)$ form the coefficients in a Fourier expansion of the energy bands $E_n(\mathbf{k})$ as functions of $\mathbf{k}$. For an energy band of very peculiar shape, it will require many Fourier components to give an adequate description of the energy band. But in addition, according to Eq. (6-31), these same quantities $\mathcal{E}_n(\mathbf{R}_s)$ are the nondiagonal matrix elements of the Hamiltonian between Wannier functions distant from each other by the amount $\mathbf{R}_s$. If such quantities are large, for large $\mathbf{R}_s$, it must be that Wannier functions even at a considerable distance apart will overlap each other, which means that they are very spread out. This is what we want to avoid. Hence we wish to build up a composite energy band, out of parts of other bands, which will be as smooth a function of $\mathbf{k}$ as possible, which is what we have done in building up our momentum eigenfunction of Γ_5 symmetry for the second and third bands. In this way we shall get as concentrated Wannier functions as is possible.

7-6. The Relation of Brillouin Zones to Bragg's Law. Let us now return to the general significance of Brillouin zones and the interpretation of Fig. 7-7. The lines forming the perpendicular bisectors of the lines joining the origin and the points h_1, h_2, which are used in Brillouin's construction, have a relationship to the Bragg law of x-ray diffraction, as we pointed out in Sec. 6-6 in treating the one-dimensional problem. In the present section we shall explain this relationship.

We start with the simplest description of Bragg's law. In Fig. 7-14 we show the atoms in a two-dimensional square lattice, in ordinary space. We can draw sets of planes through these atoms; the set drawn makes angles of 45° with the coordinate axes. According to Bragg, we may regard x-ray diffraction as reflection in the planes of the atoms, in such a way that the reflected rays from successive planes of atoms interfere constructively. We show the incident and reflected beams, and note that, if θ is the glancing angle of incidence, the path difference between rays reflected from successive planes is $2d \sin \theta$, where d is the perpendicular distance between adjacent planes. If we require that this equal $n\lambda$, where n is an integer, so that the path difference will be a whole number of wavelengths, we have

$$n\lambda = 2d \sin \theta \qquad (7\text{-}10)$$

which is Bragg's law.

A more sophisticated extension of Bragg's ideas correlates the x-ray scattering with the Fourier components of the charge density, rather than with the planes of atoms themselves. Thus, for the planes of atoms shown in Fig. 7-14, we have an infinite set of plane waves, with the same

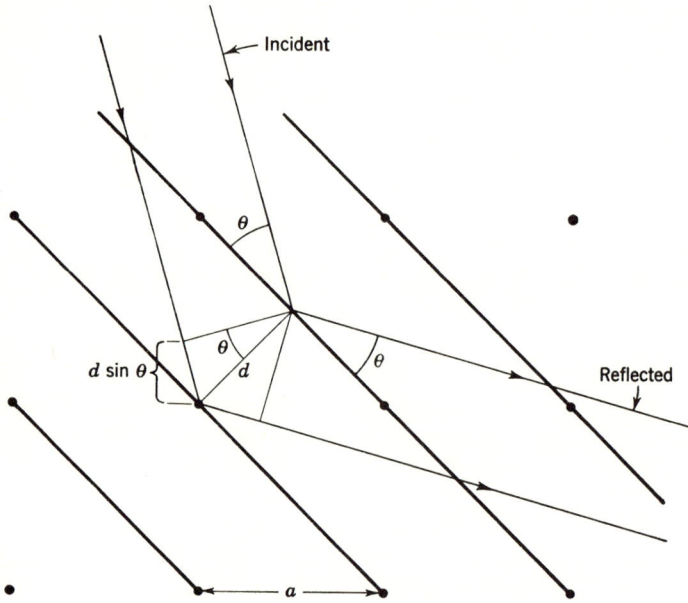

Fig. 7-14. Diagram to illustrate Bragg's law in ordinary space.

wave normal. They are those represented by the exponential functions $e^{i\mathbf{K}_l \cdot \mathbf{r}}$, where $\mathbf{K}_l$ is a vector of the reciprocal lattice, normal to the planes. For the case shown, this vector is $2\pi(\mathbf{i} + \mathbf{j})n/a$, where n is an integer, equal to 1 if the spacing between wave fronts is d, as shown in the figure, but equal to a greater integer if the spacing is d/n. More generally, the magnitude of $\mathbf{K}_l$ is $2\pi n/d$. It is then shown in x-ray scattering theory that the intensity of the nth-order scattered wave is determined by the corresponding nth Fourier component of the charge density. If we consider the scattering of the incident wave by the sinusoidal component of charge density of spacing d/n, we need use only a path difference of a single wavelength, and rewrite Eq. (7-10) in the form $\lambda = 2(d/n) \sin \theta$, which is equivalent to Bragg's equation. This law makes use only of the wave nature of the diffracted beam, so that it holds for electron diffraction as well as for x-ray diffraction, and is equally applicable whether the electrons enter the crystal from outside or form plane waves within the crystal, helping to make up the solution of Schrödinger's equation for the periodic potential problem. The only difference between the x-ray and electron case is that for electron scattering it is the Fourier component of the potential, which we have denoted

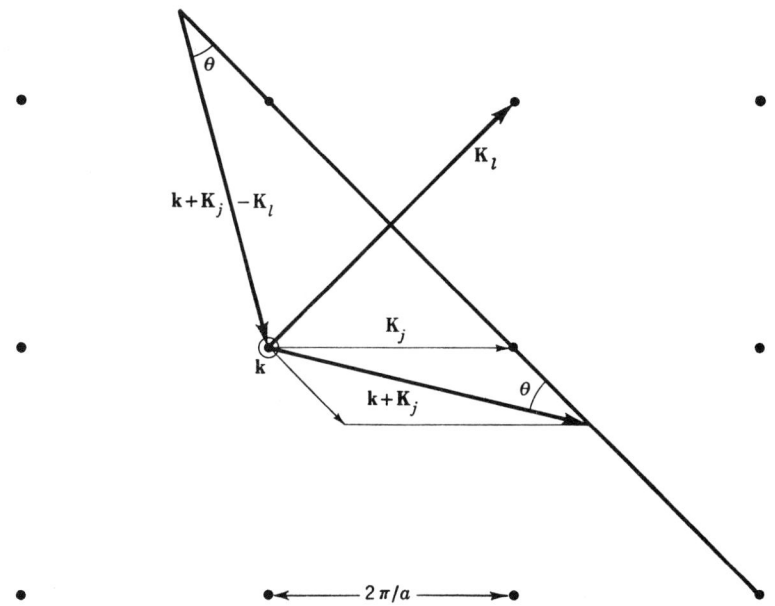

FIG. 7-15. Diagram to illustrate Bragg's law in momentum space.

by $W(\mathbf{K}_l)$, which determines the amplitude of electron scattering, rather than the Fourier component of charge density.

In any such scattering problem, we can carry through a discussion in the momentum space, or $\mathbf{k}$ space, which supplements the treatment in ordinary space which we have shown in Fig. 7-14. We give the required diagram in Fig. 7-15. The points now indicate the reciprocal lattice, rather than the space lattice, and our vector $\mathbf{K}_l$, which is along the direction of the wave normal, is the wave vector involved in the Fourier component $W(\mathbf{K}_l)e^{i\mathbf{K}_l \cdot \mathbf{r}}$ concerned in the scattering in question. We have drawn in Fig. 7-15 the plane (or, in two dimensions, the line) forming the perpendicular bisector of this vector; it is necessarily parallel to the wave fronts in Fig. 7-14. We also draw wave vectors in the directions of the wave normals of the incident and diffracted waves which were shown in Fig. 7-14. These vectors are drawn so that that of the incident wave starts on the perpendicular bisector and terminates at the origin, while that of the diffracted wave starts at the origin and terminates on the bisecting plane. From the construction, the magnitudes of these vectors are equal, and equal to $|\mathbf{K}_l|/(2 \sin \theta)$. Since $|\mathbf{K}_l| = 2\pi n/d$, and from Eq. (7-10) we know that $\sin \theta = n\lambda/2d$, we see that the magnitude of these vectors equals

$$\frac{2\pi n}{d} \frac{1}{2} \frac{2d}{n\lambda} = \frac{2\pi}{\lambda} \tag{7-11}$$

which is the correct value for the wave vector of the incident or diffracted wave.

We have shown, then, that the vectors in Fig. 7-15, in the direction of the wave normals of the incident and diffracted waves, equal both in direction and magnitude the wave vectors of these two waves, which then by the construction differ by the amount $\mathbf{K}_l$. If the wave vector of the incident wave is written as $\mathbf{k} + \mathbf{K}_j - \mathbf{K}_l$, that of the diffracted wave is $\mathbf{k} + \mathbf{K}_j$. These are the two waves between which the Fourier component $W(\mathbf{K}_l)e^{i\mathbf{K}_l \cdot \mathbf{r}}$ of the potential will have a nonvanishing matrix element, given by

$$\int e^{-i(\mathbf{k}+\mathbf{K}_j) \cdot \mathbf{r}} W(\mathbf{K}_l) e^{i\mathbf{K}_l \cdot \mathbf{r}} e^{i(\mathbf{k}+\mathbf{K}_j-\mathbf{K}_l) \cdot \mathbf{r}} \, dv = W(\mathbf{K}_l)$$

if the integration is over unit volume. We see, then, that we can state the equivalent of the Bragg law in the following simple form: the wave vectors of the incident and diffracted waves have equal magnitude, and differ vectorially by the vector $\mathbf{K}_l$.

In Fig. 7-15, the perpendicular bisector of the vector $\mathbf{K}_l$, on which the interacting wave vectors terminate, is the boundary of one of the Brillouin zones, according to Brillouin's construction. It is not an accident that we have this relation between the periodic potential problem on the one hand, with the energy gaps which appear at the boundaries of the Brillouin zone, and the problem of electron diffraction on the other hand. As a matter of fact, the present analysis was first carried through by Bethe,[1] in a very early study of the theory of electron diffraction. He was dealing with a beam of fast electrons striking a crystal and being diffracted. He treated the periodic potential as a perturbation, and considered the interaction between the incident wave and a scattered wave. The matrix element of the periodic potential between the incident and scattered wave was found to be as in the preceding paragraph, equal to zero unless the two waves satisfied the relation that their wave vectors

[1] H. Bethe, *Ann. Physik*, **87**:55 (1928).

differed by $\mathbf{K}_l$, in which case the matrix element was $W(\mathbf{K}_l)$. The scattered wave had the same energy as the incident wave. Hence the two principles, the equality of energy and the wave vectors differing by $\mathbf{K}_l$, which define the boundaries between the Brillouin zones, are the same principles which Bethe found to define the wave vectors of plane waves which could be scattered, or which satisfied the Bragg law.

Bethe went further than this, and deduced the existence of the energy gaps. These have an important relation to the theory of electron diffraction. A beam of electrons striking a crystal with just the energy and direction suitable to satisfy the Bragg conditions will have an energy falling within the energy gap; for it is in this way that the energy gaps are set up, when we use the free-electron approximation. Hence the electrons cannot form a propagated beam within the crystal; one can show that the wave function, for these energies, will have a complex or imaginary wave vector, corresponding to damping along the direction of the wave vector, rather than real propagation. This leads to total reflection, and is one way of describing the Bragg scattering of the electron beam. A similar theory of x-ray diffraction, called the dynamical theory of x-ray diffraction, predicts forbidden gaps of propagation for x-rays, in much the same way, coming just at the Bragg angles.

7-7. The Three-dimensional Case, and the Plane-wave Expansion for Real Crystals. Most of our discussion so far in the present chapter, except for the derivation of Bragg's law which we have just taken up, has been based on the two-dimensional square lattice. However, the changes involved in going to three dimensions are not large. Of course, we can no longer plot the energy as a function of the three rectangular components of the wave vector, as we did for the two-dimensional case in Fig. 7-2. The Brillouin zone must be shown in three-dimensional space, but we have become familiar with that from Chap. 5. The method used in Fig. 7-4, in which we give the energy as a function of distance in the reciprocal space, along various symmetry directions, can be carried out in three dimensions as well as in two, and we shall find many examples in later chapters.

The energy contours now become surfaces in the three-dimensional reciprocal space. We can indicate them by giving sections of an energy surface cut out by an intersecting plane. For instance, we could indicate the energy contours by giving diagrams, like those of Fig. 7-5, for a number of equally spaced values of k_z, or p_3. Alternatively, we can give perspective drawings of the three-dimensional energy surfaces. Such methods have been used a great deal for showing Fermi surfaces, and we shall encounter examples in later volumes. Brillouin's construction for the higher Brillouin zones has been carried through for many three-dimensional cases, but it obviously becomes very complicated geometrically. As an example, we show in Fig. 7-16 the shapes of the surfaces bounding the first, second, and third Brillouin zones for the face-centered cubic Bravais lattice. When we try to plot the energy bands in the extended-zone scheme, showing the nth energy band in the nth Brillouin zone, it is hard to indicate the result graphically in any way except by taking sections through the figure, corresponding to different constant values of k_z, or p_3.

The problem of symmetry of wave functions, and the method of denoting the energy bands along various symmetry directions, is not essentially more complicated in three than in two dimensions. In Chap. 5 we have already indicated how to handle the problem for the simple cubic Bravais lattice, and in Appendix 3 we discuss all the various space groups which we are working out in the present text, in great detail.

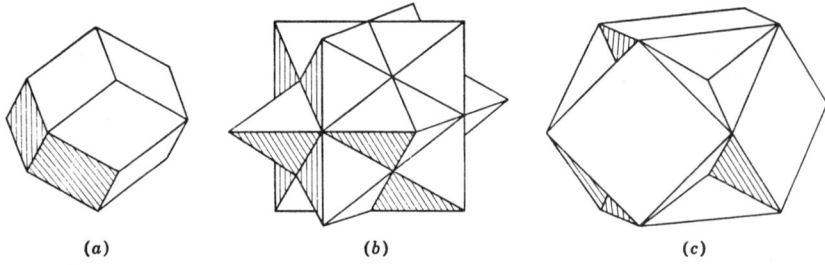

FIG. 7-16. Zones in momentum space, for body-centered cubic space lattice, face-centered momentum lattice. (a) First solid represents central Brillouin zone, bounded by 12 faces parallel to 110 planes. (b) Second zone includes volume between first and second solids, also bounded by 110. (c) Third zone is included between second and third, bounded by 110 and 100.

There will be examples in later chapters of the use of these general methods.

Formally, then, there is no particular difficulty in going to three dimensions, and in extending the methods we have been describing to real three-dimensional crystals. We now ask ourselves, is this method of expansion in plane waves practical for the actual solution of the energy-band problem in crystals? The answer is twofold, involving a paradox which it will take us a considerable time to explain. We shall first show that the method is certainly not adequate for real crystals. The reason is that the matrix elements of the potential energy are much larger in proportion than in the simple example we have taken up in this chapter, with elements of appreciable size extending out to large values of h_1, h_2, h_3. This brings with it the consequence that a very large number of plane waves is required to get an adequate expansion of the wave function, as in the one-dimensional case described in Fig. 6-4. The reason for this is that the real wave functions are so concentrated in space that their Fourier transforms, the momentum eigenfunctions, extend far out in momentum space, requiring many nonvanishing Fourier coefficients. This is a situation which gets more serious the heavier the atoms concerned, for then the inner parts of the wave function are more concentrated toward small values of the radius. Hence we might expect that this method might be practicable, if at all, only for the lightest atoms.

This is actually the case. Schlosser[1] has carried through treatments of the lithium and sodium crystals, using straightforward numerical solutions of Eq. (6-12), and has found that for lithium it is practicable to use the method at the center of the Brillouin zone, taking advantage of the simplification introduced by using symmetrized plane waves, but that for sodium the convergence is very poor. Even in the case of lithium, he was forced to carry out a form of extrapolation to get reasonable results. He solved Eq. (6-12), using successively more and more symmetrized plane waves, with wave vectors extending out further and further in $\mathbf{k}$ space. He solved for the eigenvalues of each of these problems, and plotted the resulting energy levels as functions of the reciprocal of the number of symmetrized plane waves considered. Extrapolation of the curve to an infinite number of values, or zero value of the reciprocal of this number, gave fairly reasonable energy values. Even the largest number which he was able to use, however, still gave energies rather far from the limiting values. We conclude, therefore, that this direct solution of Eq. (6-12), though at first sight attractive, is not actually of much practical value.

Nevertheless, there is a remarkable fact which has been uncovered by many calculations, carried out by much more powerful methods which we shall discuss in later chapters. This is that when the energy bands corresponding to the outer electrons of a crystal are studied, they prove to resemble very closely the results of a plane-wave expansion with small nondiagonal matrix elements of the Hamiltonian, even though the true nondiagonal matrix elements are very much larger. This resemblance is so close that the results of the sort of expansion discussed in the present chapter have been used successfully in many cases to discuss actual energy bands of metals, with fairly satisfactory agreement with experiment. This fact is so remarkable that it will deserve detailed discussion, which we shall give in Chap. 9. It is impossible to understand the reasons for it, however, without much more complete understanding than we have achieved so far of the actual nature of energy bands in real crystals. Consequently we shall go ahead, in the next two chapters, with the dis-

[1] H. C. Schlosser, Ph.D. thesis, Carnegie Institute of Technology, Pittsburgh, Pa., 1960 (unpublished). See also H. C. Schlosser and P. M. Marcus, *Phys. Rev.*, **131**:2529 (1963).

cussion of much more powerful methods of approximating to the solution of Schrödinger's equation in a periodic potential than the simple expansion in plane waves, and the application of these methods to the calculation of the energy bands in actual crystals. When we have carried through this discussion, and have demonstrated by actual examples the resemblance of the final results to those of the simple plane-wave expansion with small nondiagonal matrix components of the Hamiltonian, we shall be able to point out the features in the actual problem which lead to the comparative success of the plane-wave, or free-electron, method.

8

The Tight-binding and Orthogonalized Plane-wave Methods

8-1. The General Idea of the LCAO and OPW Methods. In the preceding chapter, we have stated that the most obvious method of approximating to the energy bands in a crystal, namely, by expanding the wave function in plane waves, is so slowly convergent as to be impracticable. The difficulty with the method is that it takes a very large number of plane waves to expand the inner parts of the wave functions, with their rapid oscillations near the nuclei. The next method which would naturally occur to us is the extension of the method of linear combinations of atomic orbitals, or the LCAO method, which we have described extensively in Volume 1 in its application to molecules. This method has had a good deal of success, and was one of the first to be suggested historically. Bloch,[1] in his first paper on energy bands, made use of it, and for that reason it is often referred to as the Bloch method. It is often also referred to as the tight-binding method, since it works particularly well for electrons which are bound tightly to their nuclei.

We have already laid the foundation for the study of this method in Volume 1, Chap. 9, where we were describing its application to the case of a ring or linear array of atoms. The extension to the three-dimensional case follows very easily. We start with an atomic orbital $a(\mathbf{r})$, centered on a nucleus at the origin. For an orbital centered on the corresponding site in the unit cell displaced by a vector $\mathbf{R}_j$ from the origin, we have $a(\mathbf{r} - \mathbf{R}_j)$. We then form a Bloch sum

$$b(\mathbf{k},\mathbf{r}) = \Sigma(\mathbf{R}_j) e^{i\mathbf{k}\cdot\mathbf{R}_j} a(\mathbf{r} - \mathbf{R}_j) \qquad (8\text{-}1)$$

where $\mathbf{k}$ is the reduced wave vector, by a method entirely analogous to Volume 1, Eq. (9-1). In a later section we shall show how to compute the matrix elements of unity, and of the Hamiltonian, between different Bloch

---

[1] F. Bloch, *Z. Physik*, **52**:555 (1928).

sums. We find that there are no nondiagonal matrix elements of unity or the Hamiltonian between Bloch sums of different reduced wave vectors **k**, but there are such nondiagonal matrix elements between sums of the same **k** value, formed from different atomic orbitals.

Let us then consider how the method would be used in an actual calculation of energy bands, in a predetermined periodic potential. A first, but oversimplified, method would be to set up a single Bloch function, from a single atomic orbital, and to use the diagonal matrix element of this Bloch function. This would correspond to the type of treatment used in Volume 1, Appendix 13, for the problem of H_6, where we used merely linear combinations of 1s atomic orbitals on the various hydrogen atoms. To get more satisfactory results, we should set up Bloch sums of a number of atomic orbitals, and use linear combinations of these, determined by solving secular equations. This is more nearly the method used in Volume 1, in the molecular-orbital method of handling molecules, particularly benzene, except that here we are assuming a periodic potential as a starting point, whereas there, dealing with a simpler problem, we were able to carry through a Hartree-Fock calculation, so as to determine the potential by a self-consistent-field method. As in the earlier case, by making linear combinations of different Bloch functions with the same **k**, but different atomic orbitals, we can attain the benefits of hybridization, and achieve some of the improvements which could otherwise result from configuration interaction. For instance, in the sodium crystal, we could make Bloch sums of the atomic 1s, 2s, $2p_x$, $2p_y$, $2p_z$, 3s, $3p_x$, $3p_y$, $3p_z$ atomic orbitals. We should expect that the inner shells of the atoms would arise in the crystal as in the isolated atoms from the atomic 1s, 2s, and 2p orbitals, and that the conduction band would be formed from the Bloch sums of the 3s and 3p atomic orbitals.

One can obtain in this way good qualitative results, and a usable first approximation toward quantitative results. The method gives very good representations of the functions corresponding to the inner electrons of the atoms. However, it does not form by itself a good starting point for making a more accurate calculation. If we could build up an infinite set of basis functions for the periodic potential problem, or a large but finite set, we might hope by linear combinations to get as good approximations to the true solution as we pleased. The possibility of doing this straightforwardly, by using Bloch sums corresponding to more and more highly excited atomic orbitals, was investigated by Parmenter,[1] who studied the lithium problem by this method. He found that the addition of Bloch sums for higher atomic orbitals than the 1s, 2s, $2p_x$, $2p_y$, and $2p_z$ was very nearly useless for improving the results found from those Bloch sums alone, which correspond to the set described above for sodium.

[1] R. H. Parmenter, *Phys. Rev.*, **86**:552 (1952).

The difficulty was that, for instance, the Bloch sums formed from $3s, 4s, \ldots$ atomic orbitals in this case were practically indistinguishable from each other. The reason is that at the ordinary internuclear separation distance in the crystal, the atomic orbitals corresponding to the higher principal quantum numbers extend very far out from the atom as compared with the internuclear separation. At a given point in space the Bloch sum is composed in this case of contributions from the atomic orbitals on a great many distant atoms, overlapping and combining, some with plus, some minus signs, so that these contributions practically cancel, leaving only the contributions from the nearby atoms. We remember, however, that all the various s atomic orbitals behave similarly near the nucleus. Hence these similarities persist, the differences are ironed out, and as we stated earlier, the Bloch sums are practically identical for all principal quantum numbers.

To get around this difficulty, Parmenter was led to the conclusion that the best way to supplement the Bloch sums of the occupied states in the atom was not to use Bloch sums of excited atomic orbitals, but rather to use a set of plane waves, suitably orthogonalized to the inner Bloch sums. This method was not new; it was the method of orthogonalized plane waves, abbreviated OPW, which had been suggested in 1940 by Herring,[1] and which had already shown itself capable of giving valuable results.

As we have just stated, the fundamental idea of this method is to use as basis functions for an expansion of the solution of the periodic potential problem a composite set, consisting first of Bloch sums of atomic orbitals of the inner electrons of the atom, and second of a limited number of plane waves. If the inner atomic orbitals do not overlap appreciably, the Bloch sums formed from them will form very good basis functions for the wave functions of these inner electrons. The addition of a fairly small number of plane waves, with the same reduced wave vector as the Bloch sums, will then make it possible to carry out a fairly good expansion of the wave functions in the higher energy bands. For sodium, as an example, in the OPW method, we should use Bloch sums only of the $1s$, $2s$, $2p_x$, $2p_y$, and $2p_z$ atomic orbitals, and should rely on combinations of these sums and a limited number of plane waves to approximate the conduction band and higher energy bands. This method proves, as we should expect, to give much more rapid convergence than an attempt to expand the complete wave function in plane waves. Since the plane waves do not have to take care of the inner features of the wave function, we do not need many of them to give a satisfactory approximation.

We then assume that we can approximate the wave functions of the

[1] C. Herring, *Phys. Rev.*, **57**:1169 (1940); C. Herring and A. G. Hill, *Phys. Rev.*, **58**:132 (1940); For a general review, see T. O. Woodruff, *Solid State Phys.*, **4**:367 (1957).

outer electrons by using linear combinations of the plane waves, and the Bloch sums formed from atomic orbitals of the inner electrons. Since we know that the true wave functions will be orthogonal to each other, it is convenient to set up linear combinations of a single plane wave and the Bloch sums, with coefficients chosen so that the combination is orthogonal to all the Bloch sums used. Such a combination is called an orthogonalized plane wave (OPW). The process of adding some of the Bloch sums to the plane wave, to produce orthogonalization, automatically introduces a behavior near the nuclei of the atoms very similar to what is found in the true wave functions.

In order to understand what we are accomplishing in this way, it is a great help if we have a clear idea of the energy bands and wave functions encountered in real crystals. Accordingly, before going further with the discussion of the LCAO and OPW methods, we shall go on in the next two sections to a discussion of the energy bands, wave functions, and momentum eigenfunctions of a real crystal, choosing sodium for the purpose. This problem has been discussed by many methods,[1] all leading to substantially the same results, and we shall take information from several sources in the description of the energy bands and wave functions which we shall give.

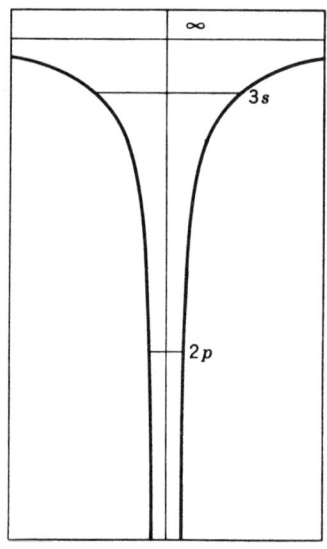

FIG. 8-1. Potential within a sodium atom in a crystal.

8-2. Energy Bands of Sodium. First, in Fig. 8-1, we show the potential energy of an electron in a sodium crystal, as a function of distance along a line in the 111 direction, pointing from one atom to its nearest neighbor in the body-centered cubic crystal. In Fig. 8-2 we give the Fourier components $W(\mathbf{K}_h)$ of this potential, as a function of the magnitude of the wave vector $\mathbf{K}_h$. In carrying out the Fourier resolution, we have assumed that the potential function is spherically symmetrical within a sphere of such a radius that neighboring spheres touch, and is constant over the region between spheres, a moderately good approximation, as mentioned in Sec. 4-5. The method of carrying out the Fourier resolution is discussed in Appendix 4. For this case, $W(\mathbf{K}_h)$ is a function only of the magnitude of $\mathbf{K}_h$, not of its direction.

[1] See bibliography at the end of Chap. 10 for a listing. Most of the results of the next two sections are from J. C. Slater, *Rev. Mod. Phys.*, **6**:209 (1934).

Sec. 8-2] THE TIGHT-BINDING AND OPW METHODS

In Fig. 8-2, we indicate the values of the magnitudes of the vectors $\mathbf{K}_h$, equal to $(h_1^2 + h_2^2 + h_3^2)^{1/2}$ in the units used. The energy is given in units of $h^2/2ma^2$, used in our discussions of Chaps. 6 and 7. All the Fourier components of the potential energy indicated in Fig. 8-2 are greater than 0.10, in these units, the largest nondiagonal components being around 0.30, as compared with the example used in Sec. 7-1, an almost free electron case, in which the largest component considered had the magnitude 0.040, and the components very rapidly decreased to much smaller values. From this fact we see that in the case of sodium we are dealing with a much larger potential energy than in these almost free electron cases. This is one piece of information leading to the result, stated in the preceding chapter, that the plane-wave expansion is poorly convergent in such a problem.

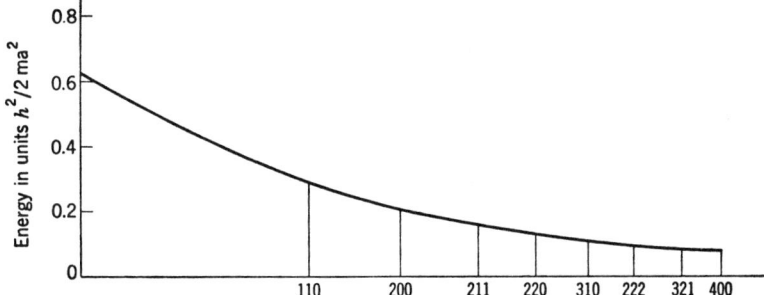

FIG. 8-2. Fourier components $W(\mathbf{K}_j)$ of sodium potential, for various $\mathbf{K}_j$'s.

We can see in another way that the case of the sodium energy bands is much further from the free-electron case than that treated in Sec. 7-1. In Fig. 8-3 we show the boundaries of the energy bands as functions of the internuclear separation. At the equilibrium distance of separation, the 1s, 2s, and 2p bands are hardly broadened at all, compared with the atomic energy levels, but the band formed from the atomic 3s level has broadened, merged with that formed from the 3p, and formed a composite conduction band. It is interesting to compare Fig. 8-3 with the similar Fig. 6-3, in which we showed the energy bands in the one-dimensional case as a function of the magnitude W of the nondiagonal matrix element of the Hamiltonian. The case of almost free electrons corresponds to a small value of W; the case shown in Fig. 6-1 is drawn for $W = 0.0625$. The sodium case obviously corresponds to a much larger value of W, large enough so that at least the two lowest energy bands are hardly broadened at all. (Of course it must be recalled that the one-dimensional sinusoidal potential, discussed in Fig. 6-1, is not really comparable with the three-dimensional sodium potential.)

The conduction band shown in Fig. 8-3 will hold one electron per atom, which will then half fill the band arising from the atomic 3s states. The maximum energy of the occupied levels is the Fermi level, which will lie in the middle of the 3s band. We notice then from Fig. 8-3 that the occupied conduction levels will have an energy at the equilibrium separation which is considerably below the energy in the isolated atom; in a qualitative way, this is connected with the binding energy of the crystal.

Even though the energy-band structure of the sodium crystal is far from the free-electron case, nevertheless there is one way of looking at it in

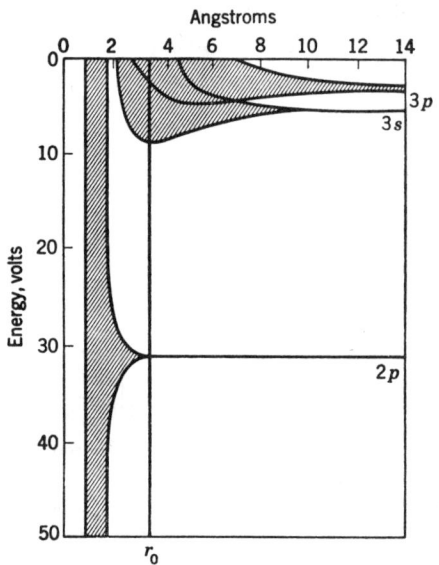

Fig. 8-3. Energy bands of sodium as function of internuclear separation. Equilibrium distance indicated by r_0.

which there is a striking resemblance to that case. This is the reason why, as we have mentioned earlier, a free-electron approximation can in fact go a long way toward describing actual metals. We must fix our attention only on the conduction band and higher energy bands, disregarding the inner energy levels of the atom, arising from the atomic 1s, 2s, and 2p states. In Fig. 8-4 we show the energy of the conduction band and higher energy bands, along symmetry directions, using the method of plotting illustrated for the two-dimensional case in Fig. 7-4. Here, as there, we give both the computed energy bands of sodium and corresponding free-electron energies. The resemblance is striking, and it is the type that would be found if the problem were only a very slight modification of a free-electron problem, being in fact much closer to the free-electron case than our two-dimensional example of Sec. 7-1. It is this close agreement which we have mentioned in the last paragraph of Chap. 7. We shall come later to the explanation of this agreement, which proves to hold for many crystals.

8-3. Wave Functions and Momentum Eigenfunctions for Sodium.
Now let us consider the wave functions. In Fig. 8-5 we show wave functions for the inner electrons, $1s$, $2s$, and $2p$, as functions of distance along a line in the 111 direction through the crystal, passing through successive atoms. The first curve of Fig. 8-5 shows the $1s$ energy band for $\mathbf{k} = 0$. This is a superposition of atomic $1s$ electrons on the various atomic sites, if we use the LCAO method; the result will be practically identical with any reasonable method of calculation. The wave function near a nucleus falls off as e^{-ar}, where a is an effective nuclear charge, r the distance from the nucleus; we see the function falling off very rapidly as we go away from each nucleus. The next curves, (b) and (c), show the wave functions for the $1s$ energy band for a value of $\mathbf{k}$ such that the wavelength is 10 atomic distances. The Bloch sum is a complex quantity, but in Fig. 8-5b and c we show the real and imaginary parts of this function, respectively, behaving in an overall fashion like a cosine and sine curve, though these sinusoidal curves merely indicate the height to which the peaks of the $1s$ atomic orbitals rise. In Fig. 8-5d we show the case where the $1s$ functions have opposite signs on adjacent atoms, so that the interatomic distance corresponds to a half wavelength. This is the boundary of the Brillouin zone in this direction.

We note that the function shown in Fig. 8-5d has a cosinelike behavior with respect to each atom: the functions are even on reflection. If we take a one-dimensional case, they correspond in symmetry properties to the lowest energy level at $k = \pi/a$ in Fig. 6-1, representing the top of the lowest energy band. Next, in Fig. 8-5e, we show the wave function for the same $\mathbf{k}$ value, describing the bottom of the next-higher energy band. These two bands we are discussing are of the nondegenerate type Λ_1, described in Table A3-22; the twofold degenerate bands of type Λ_3 have wave functions which vanish along the line passing through adjacent atoms, and cannot be shown on a diagram like that of Fig. 8-5.

This band shown in Fig. 8-5e is formed from atomic $2p$ functions. We make the linear combination $2p_x + 2p_y + 2p_z$, which points along the direction of propagation, and combine the functions with opposite signs at adjacent atoms, since we are dealing with the boundary of the Brillouin zone. We see that this function is antisymmetric with respect to each of the atomic sites; it stands with respect to the functions of Fig. 8-5d as a sine function to a cosine. As in Fig. 6-1, this second energy band has its lowest energy at the boundary of the Brillouin zone, and its highest energy

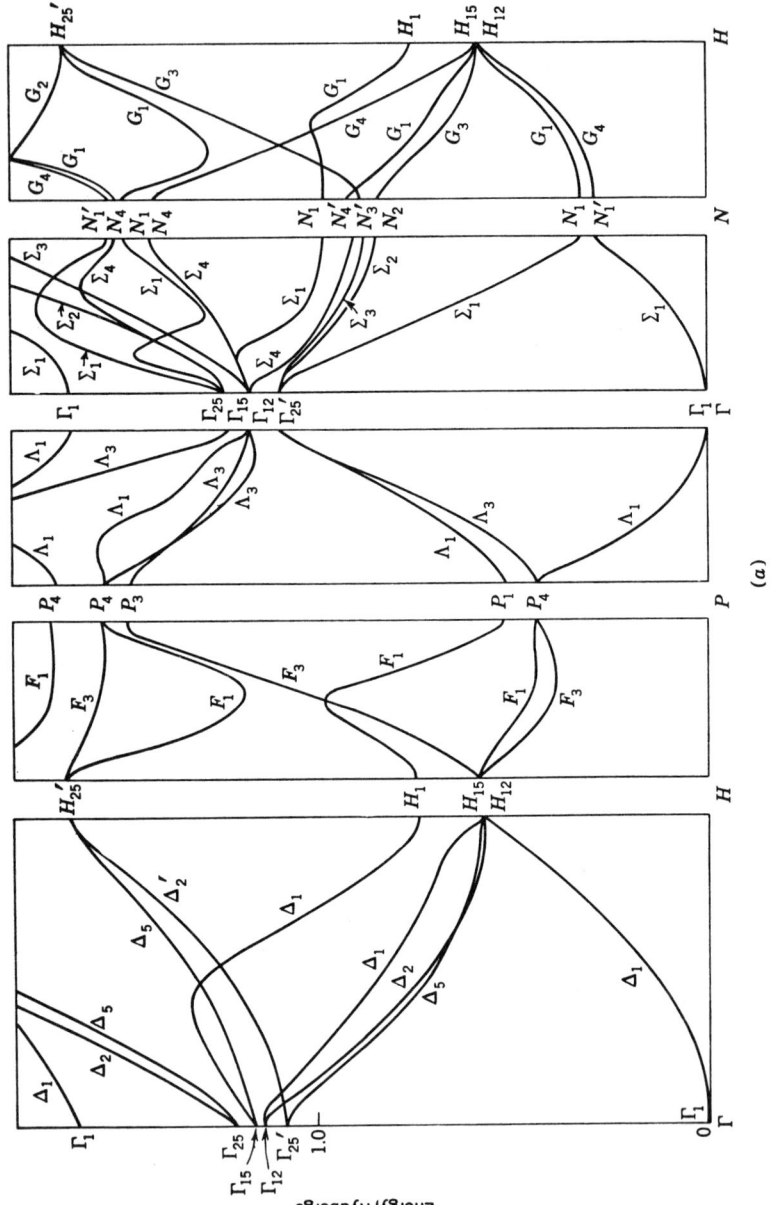

(a)

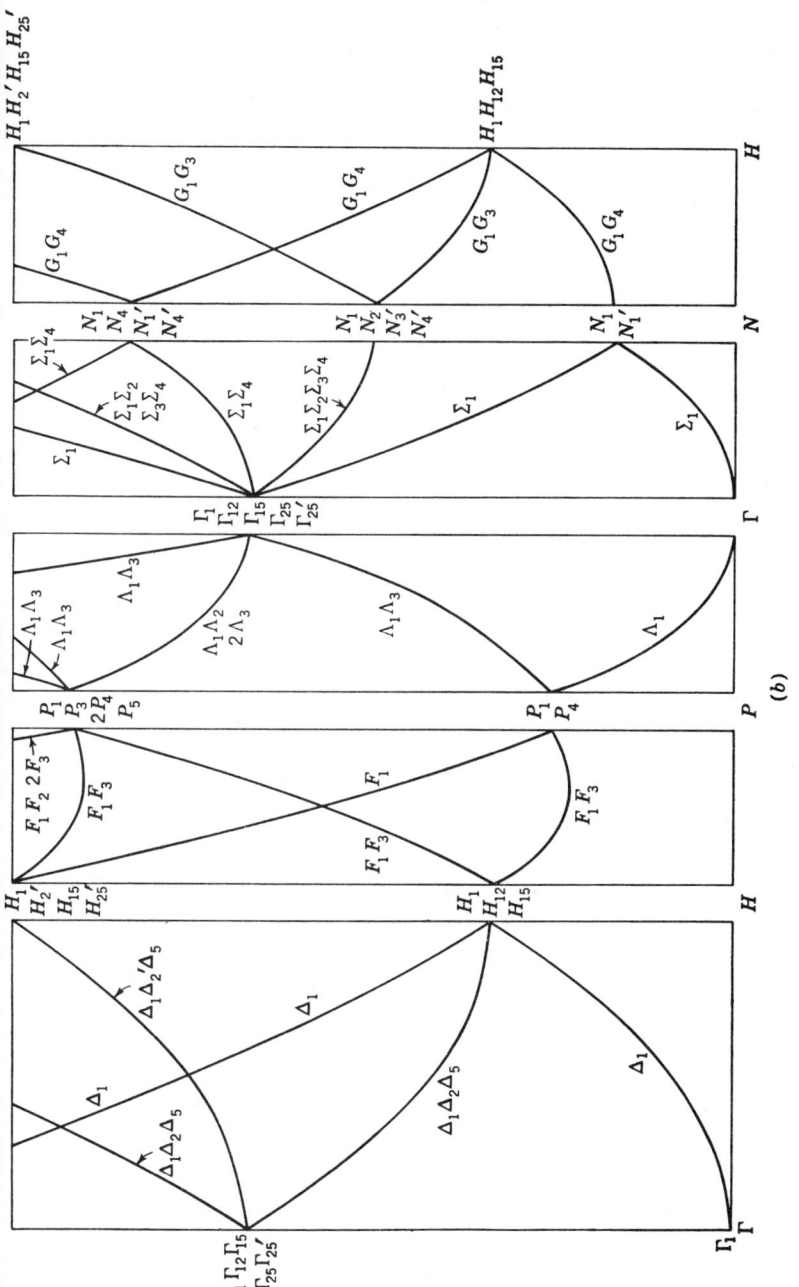

FIG. 8-4. Energy of conduction and higher energy bands of sodium along symmetry directions, from Kenney. (a) Actual bands. (b) Free-electron bands.

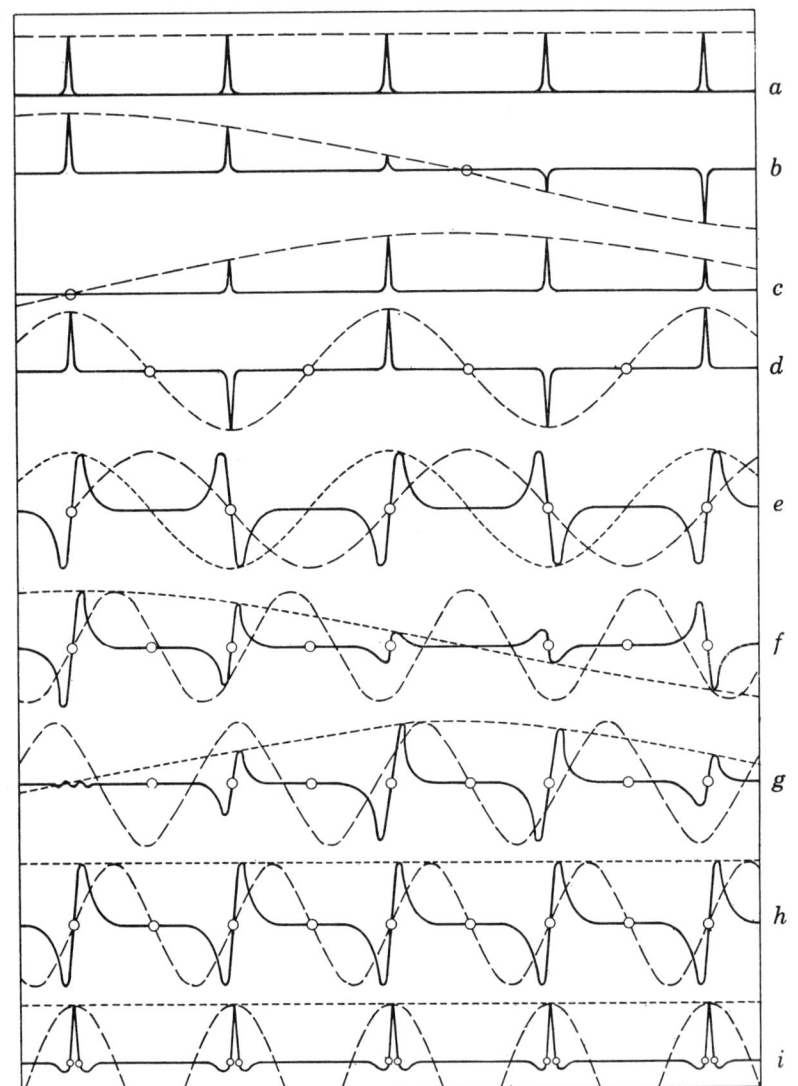

Fig. 8-5. Wave functions for bound electrons for sodium, as function of distance along a line in 111 direction.

at the center. We go through a series of higher and higher energies in Fig. 8-5e through h, showing the wave functions of this band. For each of the cases, we show two sinusoidal curves, a dotted and a dashed curve. The dotted curve corresponds to the cosine curve derived from $e^{i\mathbf{k}\cdot\mathbf{r}}$, in which we use the value of $\mathbf{k}$ in the central Brillouin zone, going to $\mathbf{k} = 0$

for curve (h), the center of the reduced zone. The dashed curve corresponds to values of **k** in the second Brillouin zone (corresponding to values going from π/a to $2\pi/a$ in Fig. 6-1), so that the wavelength gets shorter, rather than longer, as we get to the top of the band, curve (h). We see that the dashed curve has the same number of nodes as the true wave function, and that this number of nodes increases as the energy rises; this is a usual situation, though except in one-dimensional problems there is no general proof that increasing energy goes with increasing number of nodes. Figure 8-5h corresponds to the center of the reduced zone, and a p-like function. Finally, Fig. 8-5i, also corresponding to the value $\mathbf{k} = 0$, shows a $2s$-like function, corresponding to the irreducible representation Γ_1 of the group O_h, as the case of (h) corresponds to Γ_{15}.

Next, in Fig. 8-6, we show a similar set of curves for the conduction band of sodium. Curve (a) shows the s-like wave function for $\mathbf{k} = 0$, the basis function for the irreducible representation Γ_1. In (b) and (c) we show real and imaginary parts of the wave function for the same band, for the case where the wavelength is 8 atomic distances. Here we can see very clearly how, if we build up the wave function by the LCAO method, we must combine $3s$ and $3p$ functions. In case (b), for instance, the wave function is $3s$-like at the atomic site at the left of the figure, but two atoms to the right it is entirely $3p$-like, having a node at the atom. Curve (c), the imaginary part of the function for which (b) is the real part, has its p-like function on the atom at the left. In cases (d) and (e) we have the situation at the edge of the Brillouin zone, (d) corresponding to a $3s$-like combination, which is found to correspond to the higher energy, and (e) corresponding to a $3p$-like combination. Functions (f) and (g) correspond to real and imaginary parts of a function mostly constructed from the $3p$ functions, in an intermediate position in the band.

From these diagrams we get a general idea of the type of wave function which we must describe mathematically, by some approximate method. Near the nuclei, the functions go through violent fluctuations, such as we can get by superposing the atomic orbitals, as in the LCAO method. On the other hand, between the atoms, as we see particularly clearly from Fig. 8-6, the wave functions of the conduction band approach quite closely to a simple sinusoidal function, with a wave vector in the central Brillouin zone. We can now see why the LCAO method works fairly well for the conduction band, but not for excited energy levels. The $3s$ atomic orbital has approximately the same behavior near a nucleus that the function of Fig. 8-6 has, so that a Bloch sum of $3s$ functions behaves suitably at the nucleus. The outermost maxima of the atomic wave functions come just about at the midpoint between adjacent atoms, as we pointed out in Chap. 4, so that the atomic orbitals can superpose to give a fairly adequate representation of the wave function in the crystal. However, if

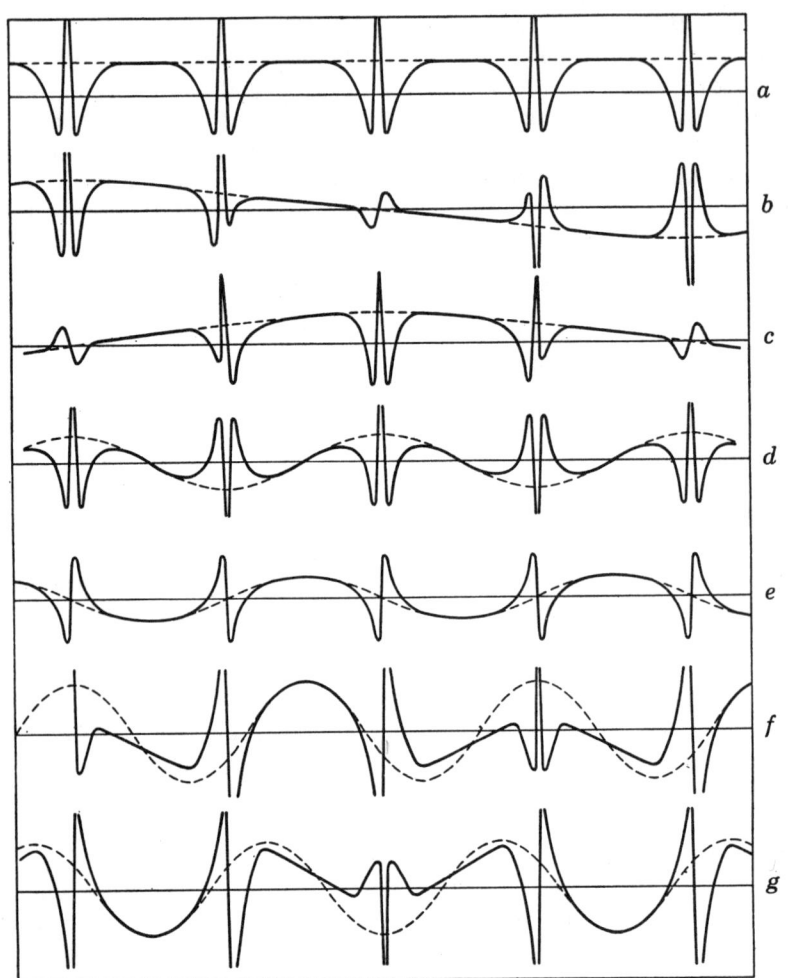

Fig. 8-6. Wave functions for free electrons, for sodium, as function of distance along a line in 111 direction.

we use $4s$ or $5s$ wave functions, the outermost maxima lie at distances of several atomic units from the atom, and if we superpose them, as we have mentioned earlier, they practically cancel, and give nothing new.

We can understand this situation better if we consider the Fourier representation of the wave function, or the momentum eigenfunction. Let us start our discussion of this question by finding the momentum eigenfunction arising from a Bloch sum of atomic orbitals. We start with Eq. (6-24), giving the general expression for the momentum eigenfunction, and Eq. (8-1) for an unnormalized Bloch sum. We shall later work out the normalization integral, but for the present purpose the unnormalized function will be satisfactory. We combine Eqs. (6-24) and (8-1), and find for the unnormalized momentum eigenfunction related to the Bloch function b_n

$$v_n(\mathbf{k} + \mathbf{K}_h) = N^{-1}\Omega^{-1}\int b_n(\mathbf{k},\mathbf{r})e^{-i(\mathbf{k}+\mathbf{K}_h)\cdot\mathbf{r}}\,dv$$
$$= N^{-1}\Omega^{-1}\Sigma(\mathbf{R}_j)\int e^{i\mathbf{k}\cdot\mathbf{R}_j}a_n(\mathbf{r}-\mathbf{R}_j)e^{-i(\mathbf{k}+\mathbf{K}_h)\cdot\mathbf{r}}\,dv \quad (8\text{-}2)$$

Since $e^{i\mathbf{K}_h\cdot\mathbf{R}_j} = 1$, we may rewrite this as

$$v_n(\mathbf{k} + \mathbf{K}_h) = N^{-1}\Omega^{-1}\Sigma(\mathbf{R}_j)\int a_n(\mathbf{r}-\mathbf{R}_j)e^{-i(\mathbf{k}+\mathbf{K}_h)\cdot(\mathbf{r}-\mathbf{R}_j)}\,dv \quad (8\text{-}3)$$

Each term in the summation over $\mathbf{R}_j$ is equal, since the integrand depends on $\mathbf{r} - \mathbf{R}_j$, and hence we have

$$v_n(\mathbf{k} + \mathbf{K}_h) = \Omega^{-1}\int a_n(\mathbf{r})e^{-i(\mathbf{k}+\mathbf{K}_h)\cdot\mathbf{r}}\,dv \quad (8\text{-}4)$$

The relation of this equation to Eq. (6-26), which holds specifically for the Fourier expansion of the Wannier functions, is obvious (the constant factor being different because our present functions are not normalized).

We see, then, that the momentum eigenfunction of a Bloch sum can be found from the Fourier resolution of the atomic function itself, or from the momentum eigenfunction of the isolated atom. That is a continuous function of the wave vector, or of the momentum; we are, however, to pick out of this continuous function the discrete values which lie at the points in $\mathbf{k}$ space given by $\mathbf{k} + \mathbf{K}_h$, to get the Fourier expansion of a particular Bloch function. But this momentum-eigenfunction expansion of an atomic function is a straightforward problem, which has been handled for hydrogen functions by Podolsky and Pauling.[1] It is easy to show[2] that for an atomic orbital depending in the usual way on angles, the momentum eigenfunction depends in the same way on angles in the $\mathbf{k}$ space, and the radial part of the momentum eigenfunction is related to the radial part of the atomic orbital, $R_{nl}(r)$, by an equation which, apart from normalization constants, is

$$v_{nl}(|\mathbf{k}|) = \int_0^\infty r^2 R_{nl}(r) j_l(|\mathbf{k}|r)\,dr \quad (8\text{-}5)$$

where j_l is a spherical Bessel function.

We show the radial dependence of the momentum eigenfunctions for hydrogen wave functions in Fig. 8-7. We see that each momentum eigenfunction shows the same number of nodes as the corresponding coordinate eigenfunction. There is an inverse type of relation between the two functions. Thus for s states the number of nodes in either func-

---

[1] B. Podolsky and L. Pauling, *Phys. Rev.*, **34**:109 (1929).
[2] See Appendix 4 for a discussion of the momentum eigenfunction.

tion is one less than the principal quantum number n. In the coordinate eigenfunction, as n increases, more and more maxima are added to the function for large values of r. The inner maxima, for small r, remain approximately fixed in position, though they decrease in magnitude, while the size of the orbit as a whole increases rapidly with n. Now the inner maxima of the coordinate eigenfunction correspond to the outer maxima of the momentum eigenfunction, and vice versa, for it is when the electron is far out, corresponding to large r, that it is moving slowly, in a

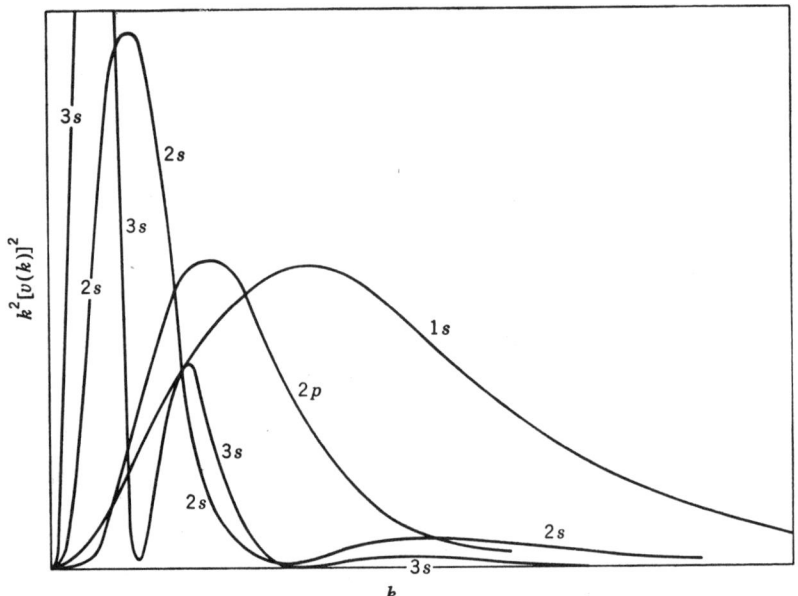

FIG. 8-7. Momentum eigenfunctions for atomic hydrogen problem, from Podolsky and Pauling. Quantity plotted is k^2 times the square of the radial momentum eigenfunction.

region of small momentum and wave vector. Thus as n increases, the new maxima of v are added inside the old ones, rather than outside. The outer maxima retain their approximate positions, but become reduced in amplitude, the innermost maximum having by far the greatest amplitude. Thus the effect is that the average magnitude of momentum, and consequently the kinetic energy, decrease rapidly as n increases. This of course is just what we should expect. The total energy decreases rapidly in magnitude as n increases, and for an inverse square field the kinetic energy is numerically equal to the total energy. Qualitatively similar results are found for other azimuthal quantum numbers, and the inner maxima agree roughly with those for s states, while the outer maxima, corresponding to the inner part of the orbit, are absent.

It is now important to understand the scale of our momentum eigenfunctions: how large are the vectors $\mathbf{K}_j$ of the reciprocal lattice, in a diagram like that of Fig. 8-7? Of course, that figure refers to hydrogen, and for another atom, the momentum eigenfunctions associated with the various principal quantum numbers will resemble those of Fig. 8-7, but with different scales: the larger the effective nuclear charge, and the smaller the wave function in ordinary space, the larger it will be in momentum space. As a rough rule of thumb, we can easily verify the following statement, proved in Appendix 4: the ratio of the total extension of the momentum eigenfunction (say, out to the outermost maximum, or somewhat beyond it) to the size of a unit cell or central Brillouin zone, in momentum space, is comparable with the ratio of the size of a unit cell in ordinary space to the extension of the atomic orbital in question. Thus, we have seen in Fig. 8-5 that the $1s$ and $2s$ atomic orbitals in sodium are small compared with the size of the unit cell. Therefore their momentum eigenfunctions will extend over a large distance compared with the central Brillouin zone. This has a very simple meaning in terms of the Fourier expansion which is represented by the momentum eigenfunction. We are trying to make an expansion of functions such as those of Fig. 8-5, with details very small compared with the unit cell. In building up the Fourier expansion, we must use sinusoidal functions whose wavelengths are comparable in size with the details we are trying to describe. This means that the wavelengths of these important Fourier components must be small compared with the dimensions of the unit cell, which in turn demands that the wave vectors associated with them must be large compared with the Brillouin zone.

We have in this fact the reason why the momentum-eigenfunction expansion is so slowly convergent that it does not form a practical method of solving the periodic-potential problem. Not only to expand the $1s$, $2s$, and $2p$ eigenfunctions, but to expand the inner details of the $3s$ and $3p$, as we see them in Fig. 8-5, we need appreciable values of the Fourier components out to a large value of $\mathbf{k} + \mathbf{K}_j$, which means that we must have a great many nonvanishing Fourier components, ordinarily many thousands of them of appreciable magnitude, compared with the 25 plane waves which we used in a nearly free-electron two-dimensional case in Sec. 7-1. It is the large number of such components which makes the method impractical. The reason why lithium is almost the only case where it can be used is that there the inner details of the wave function are larger, on account of the small atomic number, than in almost any other crystal.

Our interpretation of the momentum eigenfunction also shows us why

Parmenter found that the Bloch sums of excited atomic orbitals were not useful basis functions for an expansion of the wave function. Just as the expansion of the inner details of the wave functions comes at large values of $\mathbf{k} + \mathbf{K}_j$, the expansion of the outer part comes at very small $\mathbf{k}$, within the central Brillouin zone. An atomic orbital of large principal quantum number will have a momentum eigenfunction, like those of Fig. 8-7, with many nodes at very small values of $\mathbf{k}$, corresponding to the outer nodes in coordinate space. The only difference between one value of n and another will come in this inner part of the momentum eigenfunction, located in the interior of the central Brillouin zone. But in the Fourier expansion of a Bloch wave, we are directed to take only one point $\mathbf{k}$ of our lattice of points $\mathbf{k} + \mathbf{K}_j$ from the central Brillouin zone. The value of the momentum eigenfunction at this one value $\mathbf{k}$ represents all the information used in building up the Bloch sum, out of all the detail found inside the central Brillouin zone in the Fourier expansion of the atomic orbital. Thus almost all the information which distinguishes one value of n from another will be lost when we construct the Bloch sum.

This gives us, then, a clear picture of the difficulties with both the free-electron and the LCAO methods. The free-electron method has trouble introducing enough plane waves to describe the part of the wave function near the atomic nucleus, though it might be able to handle the situation well in the region between the atoms. The LCAO method, on the contrary, though it describes the part of the wave function near the nucleus well, is unable to do well with the part of the wave function between the atoms. It seems as if we should use Bloch sums of atomic orbitals for the regions near the nuclei, and plane waves in the region between. This is just what we accomplish with the orthogonalized-plane-wave method.

8-4. Matrix Elements for the LCAO and OPW Methods. We have outlined in the preceding sections the general idea behind the LCAO and OPW methods, but have not given the technical details as to how to find the matrix elements of unity and the Hamiltonian between the various wave functions concerned. We shall work out these matrix elements in the present section. First we handle the LCAO method; the OPW method follows simply from it. We start with a Bloch sum,

$$b_n(\mathbf{k},\mathbf{r}) = \Sigma(\mathbf{R}_j) e^{i\mathbf{k}\cdot\mathbf{R}_j} a_n(\mathbf{r} - \mathbf{R}_j)$$

formed according to Eq. (8-1) from atomic orbitals a_n in the various unit cells. The wave vector $\mathbf{k}$ is the reduced wave vector; we gain nothing new if we set up a Bloch sum with vectors $\mathbf{k} + \mathbf{K}_h$, since in the resulting exponential $e^{i(\mathbf{k}+\mathbf{K}_h)\cdot\mathbf{R}_j}$ the factor $e^{i\mathbf{K}_h\cdot\mathbf{R}_j}$ is unity. We assume periodic boundary conditions, so that $\mathbf{k}$ is limited to one of the values described in Sec. 6-2. Since the Bloch sum can be expanded in plane waves, as in Eq. (6-1), we can use the proofs given in Secs. 6-2 and 6-4, to the effect

Sec. 8-4] THE TIGHT-BINDING AND OPW METHODS 219

that wave functions with different **k** values are orthogonal, and there is no nondiagonal matrix element of the Hamiltonian between two Bloch sums with different **k** values. More generally, we can use the same method to prove that no symmetric operator—that is, no operator which repeats periodically in each unit cell—can have a nonvanishing nondiagonal matrix element between two Bloch sums with different **k** values.

Let us then take a symmetric operator $(F)_{op}$, which is a periodic function of the coordinates, and may depend also on the momenta (so that it could be the sum of potential and kinetic energies), and let us find its matrix element between two Bloch functions,

$$b_n(\mathbf{k},\mathbf{r}) = \Sigma(\mathbf{R}_j)e^{i\mathbf{k}\cdot\mathbf{R}_j}a_n(\mathbf{r} - \mathbf{R}_j) \tag{8-6}$$

where the subscript n refers to the quantum numbers of the atomic orbital, and a similar function $b_m(\mathbf{k},\mathbf{r})$, connected with different atomic quantum numbers; in this way we shall arrive at nondiagonal as well as diagonal matrix elements. The integral equals

$$\int b_n^*(\mathbf{k},\mathbf{r})(F)_{op}b_m(\mathbf{k},\mathbf{r})\,dv$$
$$= \Sigma(\mathbf{R}_i,\mathbf{R}_j)e^{i\mathbf{k}\cdot(-\mathbf{R}_i+\mathbf{R}_j)}\int a_n^*(\mathbf{r} - \mathbf{R}_i)(F)_{op}a_m(\mathbf{r} - \mathbf{R}_j)\,dv \tag{8-7}$$

As in the derivation of Volume 1, Eq. (9-12), the integral depends only on the vector distance between the two atomic orbitals. Thus we let $\mathbf{R}_i - \mathbf{R}_j = \mathbf{R}_l$. Then we may transform Eq. (8-7) into

$$\int b_n^*(\mathbf{k},\mathbf{r})(F)_{op}b_m(\mathbf{k},\mathbf{r})\,dv = N\Sigma(\mathbf{R}_l)e^{-i\mathbf{k}\cdot\mathbf{R}_l}F_{nm}(\mathbf{R}_l) \tag{8-8}$$

where

$$F_{nm}(\mathbf{R}_l) = \int a_n^*(\mathbf{r})(F)_{op}a_m(\mathbf{r} + \mathbf{R}_l)\,dv \tag{8-9}$$

In carrying out the integrations over the volume, we must observe several facts. First, the integration is supposed to be over the repeating volume of N unit cells. The integrand, $a_n^*(\mathbf{r} - \mathbf{R}_i)(F)_{op}a_m(\mathbf{r} - \mathbf{R}_j)$, will be of appreciable size only if $\mathbf{R}_i$ and $\mathbf{R}_j$ are nearly the same, or $\mathbf{R}_l$ is small, since the atomic functions will fall off rapidly with distance, and they must overlap to give an appreciable integrand. Furthermore, the integrand will be only of appreciable size within the repeating volume if the atoms lie within the volume. Hence there are N atomic sites, or values for $\mathbf{R}_i$, for which the integrand is appreciable, explaining the factor N which has appeared in Eq. (8-8).

There will, however, be tails of the functions associated with atomic sites within the repeating volume which will extend outside, and tails of those associated with atomic sites outside the repeating volume which will extend inside. The summation over $\mathbf{R}_j$ in Eq. (8-6), and the summations over $\mathbf{R}_i$ and $\mathbf{R}_j$ in Eq. (8-7), extend over all space, so that these atomic sites outside the repeating volume contribute to the integrand of Eq. (8-7). However, on account of the periodicity, the tails extending in over one side of the repeating volume will just balance those extending out over the opposite side. One can readily prove from this that the use of the factor N in Eq. (8-8) is rigorous, and that the integration of Eq. (8-9) should be carried out over all space, though of course the integrand will rapidly become small as r becomes large.

We may now use the result of Eq. (8-8) to discuss the normalization, orthogonality, and matrix elements of the Hamiltonian, for the Bloch functions. First, for normalization, we replace $(F)_{op}$ by unity, and let $n = m$. If we let

$$\int a_n^*(\mathbf{r})a_n(\mathbf{r} + \mathbf{R}_l) \, dv = S_{nn}(\mathbf{R}_l) \tag{8-10}$$

so that this quantity is an overlap integral between orbitals a_n on atomic sites displaced by $\mathbf{R}_l$, we have

$$\int b_n^*(\mathbf{k},\mathbf{r})b_n(\mathbf{k},\mathbf{r}) \, dv = N\Sigma(\mathbf{R}_l)e^{-i\mathbf{k}\cdot\mathbf{R}_l}S_{nn}(\mathbf{R}_l) \tag{8-11}$$

in analogy with Volume 1, Eq. (9-21). To get a normalized Bloch function we divide the Bloch sum of Eq. (8-1) or (8-6) by the square root of this quantity of Eq. (8-11). We note that to normalize Eq. (8-4), which was derived in an unnormalized form, we must divide it by this square root.

Next we consider the orthogonality of Bloch sums formed from different atomic orbitals. If we let

$$\int a_n^*(\mathbf{r})a_m(\mathbf{r} + \mathbf{R}_l) \, dv = S_{nm}(\mathbf{R}_l) \tag{8-12}$$

we have

$$\int b_n^*(\mathbf{k},\mathbf{r})b_m(\mathbf{k},\mathbf{r}) \, dv = N\Sigma(\mathbf{R}_l)e^{-i\mathbf{k}\cdot\mathbf{R}_l}S_{nm}(\mathbf{R}_l) \tag{8-13}$$

Since there is no reason why the overlap integrals of Eq. (8-12), between atomic orbitals of different quantum numbers on different sites, should vanish, we see that two Bloch functions for the same $\mathbf{k}$, but formed from atomic orbitals of different atomic quantum numbers, will not be orthogonal. Hence we must either orthogonalize them by the Schmidt process or deal with secular equations involving nonorthogonal basis functions. One ordinarily does not deal with very large secular equations between Bloch functions, so that ordinarily one retains the nonorthogonality and proceeds by the use of secular equations between nonorthogonal basis sets.

Next we consider the matrix elements of the Hamiltonian. If

$$(F)_{op} = (H)_{op}$$

we may use Eq. (8-8) to give both the diagonal and nondiagonal matrix elements of the Hamiltonian. If we define

$$H_{nm}(\mathbf{R}_l) = \int a_n^*(\mathbf{r})(H)_{op}a_m(\mathbf{r} + \mathbf{R}_l) \, dv \tag{8-14}$$

we find

$$\int b_n^*(\mathbf{k},\mathbf{r})(H)_{op}b_m(\mathbf{k},\mathbf{r}) \, dv = N\Sigma(\mathbf{R}_l)e^{-i\mathbf{k}\cdot\mathbf{R}_l}H_{nm}(\mathbf{R}_l) \tag{8-15}$$

in close analogy with Eqs. (6-31) and (6-33). We recall that to get the matrix elements of the Hamiltonian we must use normalized Bloch functions, which requires that we divide the quantities of Eq. (8-15) by the product of the square roots of the normalization integrals of the two functions, as given in Eq. (8-13).

In Eqs. (8-11) and (8-15) we have set up the matrix elements of unity and the Hamiltonian between different Bloch sums, giving all that is required to formulate the secular equation for the LCAO method. A good deal of use has been made of simplified versions of this method, such for instance as neglecting overlap integrals between orbitals on different atomic sites, and neglecting all matrix elements $H_{nm}(\mathbf{R}_l)$ except those between rather close neighbors. Formulas for computing the required integrals have been set up by the author and Koster[1] for this simplified case. The result of a discussion by such simplified methods can often give useful qualitative information, but of course it cannot be trusted quantitatively. Since we are more interested in this discussion in the rigorous and quantitative aspect of the method, it must be emphasized that the complete matrix elements, as defined in Eqs. (8-11) and (8-15), are required for such a treatment.

We note that the integrals defined in Eq. (8-14) are of the general sort met in molecular work, discussed in Volume 1. Thus, $(H)_{op}$ involves a potential energy which is ordinarily approximated as a sum of spherically symmetrical potentials located on the various atoms. In this case the integrand of Eq. (8-14) has two atomic orbitals on different sites, and a potential-energy term which can be on still another site, leading then to three-center integrals, of the type met for molecules. If a more accurate treatment is desired, the potential energy will involve that arising from overlap charges, each such term involving two centers rather than one, and when we combine these with the remaining two centers met with the two atomic orbitals found in the integrand, we find four-center integrals. Treatments by the LCAO method, including evaluation of these integrals, have been carried out for a few substances, and will be reported later in Chap. 10. The bibliography at the end of that chapter tells which energy bands have been calculated by this method. Two conspicuous examples are graphite and the alkali halides.

We now pass to the OPW method, which as we have described it consists merely of the use of a composite basis function containing some Bloch sums and some plane waves. In finding the matrix elements of unity and the Hamiltonian, we find elements between two Bloch sums, between two plane waves, and between a Bloch sum and a plane wave. The elements of the first type are those which we have just been treating by the LCAO method, and those between two plane waves are of the type

[1] J. C. Slater and G. F. Koster, *Phys. Rev.*, **94**:1498 (1954).

already met in discussing the plane-wave method of expansion in Chaps. 6 and 7. Thus the only new integrals which we require are those between a Bloch sum and a plane wave. These are handled by expanding the Bloch sum in plane waves, and then using the technique of Chaps. 6 and 7 for finding the matrix elements between plane waves. The expansion of the Bloch sum in plane waves is the ordinary expansion in momentum eigenfunctions. We combine Eqs. (6-1), (8-4), and (8-11), and have for the normalized Bloch function

$$u_n(\mathbf{k},\mathbf{r}) = b_n(\mathbf{k},\mathbf{r})[N\Sigma(\mathbf{R}_l)e^{-i\mathbf{k}\cdot\mathbf{R}_l}S_{nn}(\mathbf{R}_l)]^{-\frac{1}{2}}$$
$$= \Sigma(\mathbf{k} + \mathbf{K}_h)v(\mathbf{k} + \mathbf{K}_h)e^{i(\mathbf{k}+\mathbf{K}_h)\cdot\mathbf{r}} \quad (8\text{-}16)$$

where

$$v_n(\mathbf{k} + \mathbf{K}_h) = N^{-\frac{1}{2}}\Omega^{-1}\int a_n^*(\mathbf{r})e^{-i(\mathbf{k}+\mathbf{K}_h)\cdot\mathbf{r}}\,dv[\Sigma(\mathbf{R}_l)e^{-i\mathbf{k}\cdot\mathbf{R}_l}S_{nn}(\mathbf{R}_l)]^{-\frac{1}{2}} \quad (8\text{-}17)$$

Next we must find the overlap integral between such a function and a plane wave. For this purpose we multiply Eq. (8-16) by the complex conjugate of the plane wave, whose normalized wave function is $(N\Omega)^{-\frac{1}{2}}e^{i(\mathbf{k}+\mathbf{K}_i)\cdot\mathbf{r}}$, and integrate over the volume of the repeating region. We find without trouble that this overlap integral is

$$(N\Omega)^{\frac{1}{2}}v_n(\mathbf{k} + \mathbf{K}_i) \quad (8\text{-}18)$$

where $v_n(\mathbf{k} + \mathbf{K}_i)$ is given in Eq. (8-17). For the Hamiltonian, we start by operating with the Hamiltonian on the wave function in the form of Eq. (8-16). We multiply by the complex conjugate of the plane wave, integrate, and find for the nondiagonal matrix element of the Hamiltonian between the plane wave and the Bloch function the quantity

$$(N\Omega)^{\frac{1}{2}}\left[(\mathbf{k} + \mathbf{K}_i)^2\frac{\hbar^2}{2m}v_n(\mathbf{k} + \mathbf{K}_i) + \sum(\mathbf{K}_l)W(\mathbf{K}_l)v_n(\mathbf{k} + \mathbf{K}_i - \mathbf{K}_l)\right] \quad (8\text{-}19)$$

as in the derivation of Eq. (6-12).

It is now a simple problem, which we do not have to discuss in detail, to construct the orthogonalized plane waves, and find the matrix elements of the Hamiltonian between them. For each $\mathbf{k}$ value in the reduced zone we have a secular equation to solve. At a symmetry point, we can construct linear combinations of orthogonalized plane waves having the appropriate symmetry, and use these for basis functions, as will be described in the next section. In this case, the secular equation will have fewer rows and columns than the number of orthogonalized plane waves used, since a number of orthogonalized plane waves, forming one of the symmetry combinations, will have to have the same coefficients. At a general point of the Brillouin zone, this simplification will not occur.

For this reason it is far easier to get the solutions at symmetry points than at general points, and in many of the calculations using the method, only symmetry points have been used, and the energy bands at general values of k have been estimated by interpolation. Now that more powerful computational methods are becoming available, however, it is becoming practical to determine the energy bands at general points of the Brillouin zone.

A number of calculations using the orthogonalized-plane-wave method have been carried through, as we shall describe in Chap. 10, and list in the bibliography at the end of that chapter, and it proves to be one of the very valuable and accurate methods of approximating the true solutions of Schrödinger's equation in a periodic potential. If the number of plane waves can be extended without limit, of course it becomes an exact solution; but even with a limited number of plane waves, it proves to converge moderately well, in the sense that addition of further plane waves is making a fairly small change in the results.

Unfortunately, many of the calculations by the method make an approximation whose justification is doubtful, and which may throw doubt on the accuracy of the results. These calculations use a set of basis functions including only the orthogonalized plane waves, but neglect the nondiagonal matrix elements of the Hamiltonian between these functions and the Bloch sums of the inner atomic orbitals. These nondiagonal matrix elements cannot be assumed to vanish, though they will be small if the Bloch sums of the inner orbitals form good approximations to the true solutions of the periodic potential problem. They should properly be included, the Bloch sums of the inner orbitals being included in the basis set. This feature of the method seems to have been overlooked by many of its users, though it is well known to others.

These misgivings about the accuracy of current calculations with the OPW method extend even more to those using a method suggested by Phillips and Kleinman,[1] sometimes denoted by the name of the pseudopotential method. This starts with an assumption similar to that of the OPW method, but then throws the effect of the orthogonalization of the plane waves with the core functions into a form suggesting a repulsive potential. Such a treatment is in principle possible, and had been in use long before the suggestion of Phillips and Kleinman,[2] but the latter workers have made so many approximations in carrying it out that the correctness of their results is questionable.

[1] J. C. Phillips and L. Kleinman, *Phys. Rev.*, **116**:287 (1959). L. Kleinman and J. C. Phillips, *Phys. Rev.*, **116**:880 (1959); **117**:460 (1960); **118**:1153 (1960); **125**:819 (1962).
[2] See for instance H. Hellmann, *Acta Physicochimica URSS*, **1**:913 (1935); **4**:225 (1936), and later papers; also various papers by P. Gombas, found in the Bibliography.

The OPW method is more satisfactory in some cases than in others; the doubtful cases are mostly those in which it is not very obvious which electrons to regard as inner ones, which as conduction electrons. A typical example is that of an element in a transition group, such as the iron group, where the $3d$ electrons are being added to the periodic table. Should we include a Bloch sum of atomic $3d$ orbitals in the set of basis functions or not? They are in a sense outer electrons, and the general scheme of procedure with the orthogonalized-plane-wave method has generally been to handle the outer electrons by plane waves, using the Bloch sums only for the definitely inner electrons. Yet if we were to try to handle them by plane waves, examination of the nature of the $3d$ wave functions shows that the maximum of the charge density lies rather far inside the atom, and a great many plane waves would be required to get good convergence.

The method could certainly be extended to such questionable cases, though so far its main successes have been in such cases as diamond, silicon, and germanium, where the four outer electrons are handled by plane waves, and the inner ones ($1s$ in diamond, or $1s$, $2s$, and $2p$ in silicon, and $1s$, $2s$, $2p$, $3s$, $3p$, $3d$ in germanium) by Bloch sums. In any case, it remains a method of series expansion, in which it is not certain that it is practicable to retain enough terms to get adequate approximations. If the Bloch sums form a really good representation of the wave functions of the inner electrons, the method is likely to converge well; but if they are less adequate, the plane waves will have to remedy their deficiencies, and the convergence can be poor. This suggests that other methods of approximation also have their uses, in which we can be more sure of getting the real solution of Schrödinger's equation. In the next chapter we shall start the discussion of several related methods which are capable, at least in principle, and probably in practice, of giving results of very satisfactory accuracy.

8-5. Crystal Symmetry and the LCAO and OPW Methods. In Chap. 5 and Sec. 7-2, we have discussed the effect of the rotational and reflection symmetry in the crystal on the wave functions, and have shown how to set up symmetrized plane waves, linear combinations of plane waves, corresponding to the various irreducible representations of the group of the wave vector, for special positions in the Brillouin zone. Now we shall ask how these general methods apply when we are using the LCAO and OPW methods. Our discussion will apply specifically to the LCAO method. If we are using the OPW method, we must merely use symmetrized combinations of Bloch sums, as in the LCAO method, and at the same time symmetrized plane waves of the same symmetry type, constructed as we have described in Sec. 7-2. For the sake of definiteness, we shall take as before the example of the crystal of sodium, body-centered cubic with one atom per unit cell. As far as the point Γ and propagation along the directions Δ, Λ, and Σ are concerned (that is, the center of the Brillouin zone, and propagation along the x axis, along the line $x = y = z$, and along the line $x = y$, $z = 0$), the body-centered cubic case is just like the simple cubic case treated in Chap. 5, so that we can use the results of that chapter in discussing this example.

Let us start with the center of the Brillouin zone, the point Γ. Bloch sums of atomic orbitals, for $\mathbf{k} = 0$, are simply the superpositions of the atomic orbitals at corresponding points in each unit cell. The symmetry of each Bloch sum can then be studied by examining the dependence on x, y, and z, the coordinates measured from the origin of space, where the atom is assumed to be located. A Bloch sum of atomic s functions will then be found, following the discussion in Sec. 5-6, to be a symmetry orbital of the Γ_1 type, since it has no angular dependence at the atomic site. Bloch sums of the p_x, p_y, and p_z-like atomic orbitals will behave like x, y, z near the nucleus, and hence, according to Table 5-4, will furnish symmetry orbitals for the three-dimensional representation Γ_{15}. If we make Bloch sums of d-like atomic orbitals, the sums of the xy, yz, and zx-like d orbitals will have the symmetry of $\Gamma_{25'}$, while the sums of the $3x^2 - r^2$ and $\sqrt{3}\,(y^2 - z^2)$ types of d orbitals will have the symmetry Γ_{12}. In other words, by studying the behavior of the atomic orbitals near the nucleus, and comparing with the behavior of the symmetrized plane waves at the same point, as given in Table 5-4, we can find the symmetry type of each Bloch sum.

When now we start solving the Schrödinger equation by finding linear combinations of Bloch sums, we see that if we are looking for a state Γ_1, we should use atomic s orbitals; for a state Γ_{15}, atomic p orbitals; for a state $\Gamma_{25'}$, atomic d orbitals; and so on. We shall find no nondiagonal matrix elements of the Hamiltonian between Bloch sums corresponding to different symmetry types. If we are using Bloch sums only of the occupied atomic orbitals, namely, in sodium the $1s$, $2s$, $3s$, $2p$, it is clear that we can make up the states Γ_1 as three linear combinations of the three Bloch sums formed from the $1s$, $2s$, and $3s$ atomic orbitals; we can make up Γ_{15} from the atomic $2p$'s. In the OPW method, as we have mentioned earlier, we should not use the Bloch sum of $3s$ atomic orbitals, but should supplement the Bloch sums we do use by corresponding symmetrized plane waves.

As we go along the direction Δ, propagation in the x direction, we have seen from Sec. 5-6 that a wave function with the symmetry type Δ_1 can have s-like, or p_x-like, behavior around the origin. In other words, the Bloch sums of the atomic s and p_x orbitals will all have Δ_1 symmetry. The Bloch sums of the p_y and p_z orbitals will provide the two components for basis functions of the Δ_5 type. We can then solve a secular problem involving the Bloch sums of the atomic $1s$, $2s$, $3s$, and $2p_x$ orbitals to get the Δ_1 states, in the LCAO method for sodium; or, in the OPW method, we omit the $3s$ Bloch sums, as before, and include suitably symmetrized plane waves. The $2p_y$ and $2p_z$ Bloch sums will provide as they stand a pair of basis functions for the Δ_5 state, to be supplemented in the OPW method by symmetrized plane waves. In similar ways we can examine the Bloch sums which will contribute toward the wave functions for each symmetry type for propagation along Λ or Σ.

In the preceding section, we have described how to find the matrix elements of the Hamiltonian between Bloch sums and plane waves, and from these we can easily get the matrix elements between the symmetry orbitals formed from Bloch sums and from plane waves. It is interesting to study the behavior of the matrix elements, and the solutions of the secular equation determining the energy, as we start at Γ, and go along such a direction as Δ. At Γ, as we have mentioned earlier, we shall find Bloch sums of either s-like or p-like atomic orbitals, with suitably symmetrized plane waves. As we go along Δ, however, we find nondiagonal matrix elements of the Hamiltonian between the s-like and p_x-like Bloch sums, which prove to be proportional to k_x. When we set up the secular equation for the Δ_1 type of crystalline state, we must include all these s and p_x Bloch sums. We shall then find that the Δ_1 state starting out from a state Γ_1 will have a p_x-like contribution proportional to k_x, and similarly the Δ_1 state starting out from a state Γ_{15} will have an s-like contribution proportional to k_x. There will be corresponding corrections to the energy proportional to k_x^2 in each case, and with the states reducing to Γ_{15}, this results in a separation between the Δ_1 and Δ_5 states proportional to k_x^2. Once we get away from the origin in **k** space, in other words, we can no longer speak of an energy band as being exclusively s-like or exclusively p-like.

As we go along the direction Δ toward the outer surface of the Brillouin zone, we find that the relative amounts of s-like and p_x-like contributions to each of the Δ_1 states, those arising from Γ_1 and Γ_{15}, can change in quite arbitrary ways. At the surface of the zone, it proves to be the case that again a symmetry orbital must be entirely s-like or entirely p-like, but there is no guarantee that a state which starts at Γ as an s-like Γ_1 may not change its character, and at the outer surface of the zone become entirely p-like. This is a phenomenon in energy bands similar to one pointed out in Volume 1, Sec. 12-2, in the discussion of the benzene molecule, where we had the analogue of energy bands. We pointed out in that section the existence of an energy band formed at the center of the Brillouin zone (denoted there as $m = 0$) from carbon $2s$ and $2p_r$ orbitals, and at the edge of the zone (denoted there by $m = 3$) from carbon $2p_\phi$ orbitals, representing a change of symmetry type similar to the one discussed here for sodium. Our two-dimensional example of Fig. 7-4 provides many examples of the same phenomenon. We shall later discuss such situations in actual calculations of energy bands for metals.

If we had gone away from the point $\mathbf{k} = 0$ in a different direction from the x axis, we should have found different behavior. Along the directions Λ and Σ, we should again find a splitting into a nondegenerate state and a twofold degenerate state. On the other hand, in an arbitrary direction different from these, all three states would split apart, since at a general point of $\mathbf{k}$ space there is no degeneracy. The behavior of the three energy bands arising from a threefold degenerate state Γ_{15} at $\mathbf{k} = 0$ is quite involved, but can be investigated by much more general methods, which allow us to set up secular equations for any state degenerate at $\mathbf{k} = 0$, in terms of a power-series expansion in $\mathbf{k}$. This general method, called the $\mathbf{k} \cdot \mathbf{p}$ method, is taken up in Appendix 5. In the case of the behavior of the states arising from a Γ_{15} state as we go away from the origin, it would lead to a cubic secular equation for the energy. To the approximation described there, the energy will prove to be proportional to the square of the magnitude of $\mathbf{k}$, but multiplied by a complicated function of the angle, such that we automatically get twofold degeneracy and a third nondegenerate state along the directions Δ, Λ, and Σ, but three distinct energies in all other directions.

9
The Cellular and Augmented-plane-wave Methods

9-1. The Wigner-Seitz and Cellular Methods. In the preceding chapters we have been taking up the methods of solving for the energy bands in a solid which were derived directly from the two oldest methods: the free-electron method, or expansion in plane waves, and the LCAO or tight-binding method, or expansion in Bloch sums of atomic orbitals. The orthogonalized-plane-wave method, as we have seen, is a combination of these two, expansion in a composite basis set including both Bloch sums and plane waves.

Wigner and Seitz,[1] in 1933, suggested an approach which broke away in a new direction from these two original methods. Their work was applied to the sodium crystal, which we recall is a body-centered cubic structure. In their paper they first suggested the use of the Wigner-Seitz cell, which was shown in Fig. 1-2 for this case, and noted that, though it is a polyhedron for the body-centered structure, nevertheless it does not depart very far from being spherical. If one starts thinking about the crystal in this way, one asks if the potential energy felt by an electron, inside a Wigner-Seitz cell, is really very far from spherical symmetry. The charge within the sphere is approximately spherical, and a spherical charge distribution produces a spherical potential. The atoms in the other cells are uncharged, and if they were also spherical, they would produce no field at exterior points. It seems like a promising first approximation to treat the potential within a cell as being spherically symmetrical, treating the departure from this symmetry later as a perturbation, if necessary. But if we have a spherically symmetrical potential, we should be able to get an exact solution of Schrödinger's equation; we have the same symmetry situation as in an isolated atom. Thus the possibility opens up of getting what amounts to an exact solution of Schrödinger's equation in the type of potential we have in a crystal. To

[1] E. Wigner and F. Seitz, *Phys. Rev.*, **43**:804 (1933); **46**:509 (1934).

achieve this, however, involves a good many steps, which have taken a long time to work out.

The primary aim of Wigner and Seitz was rather limited: they wished to find the energy of the bottom of the conduction band of sodium, as a function of internuclear separation; that is, the one-electron energy of the wave function with $\mathbf{k} = 0$, formed essentially from atomic $3s$ orbitals, but without assuming that the $3s$ has the same form in the crystal that it would have in the isolated atom. For a state with $\mathbf{k} = 0$, the wave function must repeat in each unit cell of the crystal. For this type of symmetry, the Γ_1 symmetry in group-theoretical language, the wave function near the nucleus has first a term independent of angle, as we see from Volume 1, Table A12-17; then a term corresponding to $l = 4$, depending on coordinates like $x^4 + y^4 + z^4 - \frac{3}{5}r^4$, times a function of r, a quantity which would presumably be negligible if the potential were approximately spherical; and higher terms. Wigner and Seitz disregarded these dependences on angle, assuming a spherically symmetrical wave function. They had, then, to find a spherically symmetrical, or s-type, solution of Schrödinger's equation inside the Wigner-Seitz cell, which could join smoothly onto an identical wave function within the next cell.

From Fig. 8-6 we have a general idea of the type of wave function to expect: it will have a high peak at the nucleus, go through two nodes, and become approximately constant as we approach the boundary of the Wigner-Seitz cell. As a result of the periodicity, the normal derivative must be zero around the boundary of the cell, so that the function and its derivative may join smoothly from one cell to the next. Wigner and Seitz approached the problem of computing this function in the following way. They replaced the polyhedral cell by a sphere of the same volume, and set up a spherically symmetrical potential within this sphere, approximating the true potential as well as they could. Then they solved the Schrödinger equation for this spherical potential, taking an s-type solution, and looked for the particular energy for which the wave function would have zero slope at the boundary of the sphere. The potential which they assumed was essentially that felt by an electron in the field of the inner shells of the sodium atom; such a potential had been set up semiempirically by Prokofjew,[1] in a study of the wave functions of valence electrons in the sodium atom, and Wigner and Seitz adopted Prokofjew's potential, which is known to give eigenvalues in good agreement with the energy levels of the sodium atom.

The radial wave equation for a spherically symmetrical potential, and for a given azimuthal quantum number (in this case 0, for an s state), of course has solutions behaving regularly at the origin for any energy value.

[1] W. Prokofjew, *Z. Physik*, **58**:255 (1929).

Only for the eigenvalues of the problem of the isolated atom does the same solution behave properly at infinity, but in the present case, where we are interested only in the solution within a sphere, this does not concern us. If we integrate outward numerically from the nucleus, with an arbitrary energy, the solution may have a form such as is given in Fig. 9-1. This solution goes infinite as r becomes infinite, but there are two fairly large values of r for which its derivative is zero. If the radius of the Wigner-Seitz sphere happened to have either of these values, the solution would be an appropriate one to use. Wigner and Seitz carried out such integrations for a variety of energies, and found the radii of the spheres on which there was zero slope. They then computed the internuclear spacing which would lead to these radii (the two quantities of course are proportional to each other), and this allowed them to plot for each energy the internuclear spacing, which then gives the energy as a function of internuclear spacing, for the bottom of the conduction band.

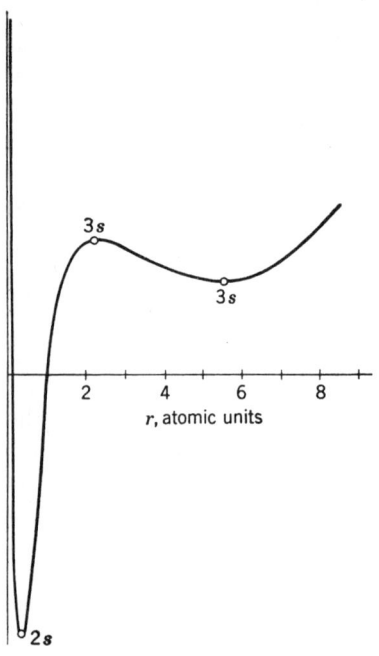

Fig. 9-1. Radial wave function in sodium potential, as function of r, showing points where derivative is zero.

The resulting curves are those representing the bottoms of the bands in Fig. 8-3. The case shown in Fig. 9-1 would correspond to an energy somewhat above the minimum of the 3s band in Fig. 8-3; the two points where the derivative of the wave function is zero correspond to the two intersections which a horizontal line of constant energy would have in this figure with the curve representing the bottom of the band. There is another place where the derivative is zero, at much smaller value of r; this would correspond to the bottom of the 2s band, at the same energy. We can inquire how Fig. 9-1 would be modified, if the energy either increased or decreased, and thereby understand the form of Fig. 8-3. If the energy were decreased, the two outer points where the derivative of the wave function was zero would move together, coalescing at the energy corresponding to the bottom of the 3s band. For energies below this, there is no such point where the derivative is zero; we have only the much smaller value, at the bottom of the 2s band. If we proceed in the opposite

Sec. 9-1] THE CELLULAR AND AUGMENTED-PLANE-WAVE METHODS 231

direction, increasing the energy, the outermost maximum in Fig. 9-1 moves somewhat inward, and the outermost minimum moves rapidly outward toward infinity. The energy where it has reached infinity is that corresponding to the eigenfunction of the isolated atom; the minimum not only comes for infinite r, but corresponds to the wave function being zero there. This explains how the energy representing the bottom of the $3s$ band, in Fig. 8-3, merges with the energy of the isolated atom at infinite internuclear separation.

Wigner and Seitz in their first paper worked out only this energy of the bottom of the $3s$ band. They then used a free-electron approximation for the energy of the electrons with $\mathbf{k}$ greater than zero, assuming that a kinetic energy $k^2\hbar^2/2m$ could be added to the energy of the state with $\mathbf{k} = 0$, to get the higher energy levels. They used an average excitation energy, added this to the energy at the bottom of the band, and by various approximations estimated the total energy of the sodium crystal as a function of internuclear separation, and hence the binding energy of the crystal.

The author[1] extended the method of Wigner and Seitz to obtain estimates of the energies of states of other $\mathbf{k}$ values as functions of the internuclear separation; Wigner and Seitz in their second paper used equivalent methods. Instead of using merely an s-like solution of the Schrödinger equation in the spherical potential within the Wigner-Seitz cell, it is necessary for other $\mathbf{k}$ values to use a mixture of solutions of different l values. Thus, we have seen in Sec. 8-5 that for propagation along the Δ direction, or along the x direction, with symmetry Δ_1, we can use functions which have the symmetry of atomic s orbitals, or of p_x; we can also use functions of type $3x^2 - r^2$, and so on. In fact, if we take the x axis as the axis of quantization of the atomic orbitals, we can combine orbitals of any l values, so long as they have m_l, the component of angular momentum along that axis, equal to zero. For this reason, for a wave vector along x, it was assumed that the wave function could be built up out of a linear combination of solutions of Schrödinger's equation for the spherical potential, using functions of the types just described, with coefficients to be determined; all these solutions were determined for a single energy, to be found as a function of the assumed reduced wave vector $\mathbf{k}$ in the following manner.

The actual Wigner-Seitz cell was used, rather than the sphere of the same volume as Wigner and Seitz had used. Then an effort was made to have the solution in one cell join smoothly onto that in the adjacent cell. The condition for this can be stated rigorously, in terms of the wave function within a single cell. We know that in going a distance $\mathbf{R}_i$, the wave function must be multiplied by $e^{i\mathbf{k}\cdot\mathbf{R}_i}$. But now if we go from a point on

[1] J. C. Slater, *Phys. Rev.*, **45**:794 (1934); *Rev. Mod. Phys.*, **6**:209 (1934).

one face of the Wigner-Seitz cell to a point perpendicularly opposite on the other face, as shown in Fig. 9-2, the displacement vector is one of the vectors $\mathbf{R}_i$ of the lattice. Hence we conclude that the wave function on the right-hand face of the cell shown in Fig. 9-2 must be $e^{i\mathbf{k}\cdot\mathbf{R}_i}$ times the value on the left-hand face. Also, in order to secure continuity of the derivative of the function as well as of the function itself, the derivative must be multiplied by this same factor in going from one face to the other.

To get continuity all over the faces, we should have to apply these conditions an infinite number of times, at all points of the faces. This would require an infinite number of disposable constants, and could be arranged only if we used a sum of functions of all l values. This is impracticable, and to make it unnecessary, only a few values of l were used, and the matching conditions were applied only at the centers of the faces of the Wigner-Seitz cell. This led to manageable equations for determining the coefficients of the wave functions of various l values entering into the solution, and for finding the energy as a function of $\mathbf{k}$. This method was used originally for finding the sodium energy bands discussed in Sec. 8-3. The same method has been applied to a good many other crystals, as enumerated in the bibliography of energy-band calculations at the end of Chap. 10.

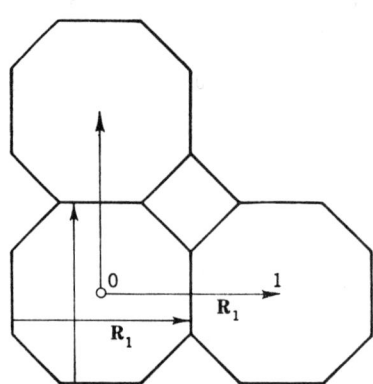

FIG. 9-2. Perpendicular to face of polyhedral cell. Illustrated with section of body-centered cell taken through nuclei. Vector $\mathbf{R}_1$ from nucleus 0 to nucleus 1 equals vector from one face of cell to opposite face.

This method is clearly only approximate, in that the series of l values is not carried very far. This was pointed out by Shockley,[1] in a paper in which he investigated to see how well the method could reproduce the solution in case there really is only a constant potential throughout the lattice. In this case, the so-called empty-lattice test, one is trying to reproduce the true solution of the problem, a plane wave, by superposing a quite small number of solutions of the spherical problem, with different l values. As one would expect, the resulting wave function, and energy, were quite far from the true values, and it was obvious from this example that the cellular method as used in the early papers was subject to at least equally large errors. Attempts to improve the method have been made

[1] W. Shockley, *Phys. Rev.*, **52**:866 (1937).

since that time, by a number of writers,[1] who have used more terms, and fitted the boundary conditions at more points around the boundaries of the Wigner-Seitz cell. No completely unambiguous method of carrying this out has been devised, however, and while this method, generally called the cellular method, has given useful results, it is generally felt at present that it is not susceptible of the highest accuracy. The mathematical difficulty with it has been pointed out by Ham[2] and by Saffren.[3] They have shown that a series of the type that is used in the method, a sum of functions all of which are solutions of the same Schrödinger problem of spherical symmetry, with different l values, must not be expected to converge to the true solution of Schrödinger's equation for the periodic-potential problem, except within a sphere inscribed within the Wigner-Seitz cell. Outside that sphere, in the corners of the cell, some other form of expansion must be used.

9-2. The Augmented-plane-wave Method. The difficulty which we have just pointed out in the cellular method comes in the corners of the cell, outside the inscribed sphere. To get around this difficulty, the author, in 1937,[4] suggested an alternative method, which has become known as the augmented-plane-wave method. This starts by assuming that we are dealing with a potential energy function which is spherically symmetrical within the inscribed spheres in the Wigner-Seitz cells, and which is constant outside these spheres. This is not a perfect representation of the potential actually found in a crystal, but it is not a bad approximation, and it has the advantage that we can set up a really rigorous solution of Schrödinger's equation in this case. Having found these solutions, the small departures of the potential from the assumed form can be treated as perturbations.

The idea of the method is that in the spheres we use the same sort of expansion as in the cellular method, namely, an expansion in l, using functions of all l values derived by solving Schrödinger's equation for the spherical potential, for an energy value which later is adjusted to equal the true energy of the stationary state. Outside the spheres, we use an expansion in plane waves. This gives the result which we have indicated as being desirable: an expansion of the function near the nucleus in terms of solutions of the atomic problem, and an expansion outside in plane

---

[1] See for example F. von der Lage and H. Bethe, *Phys. Rev.*, **71**:612 (1947); D. J. Howarth and H. Jones, *Proc. Phys. Soc. (London)*, **A65**:355 (1952); S. L. Altmann, *Proc. Roy. Soc. (London)*, **A244**:141, 153 (1958).

[2] F. S. Ham, Ph.D. thesis, Harvard University, Cambridge, Mass., 1954; *Phys. Rev.*, **128**:82 (1962).

[3] M. Saffren, Ph.D. thesis, Massachusetts Institute of Technology, Cambridge, Mass., 1959.

[4] J. C. Slater, *Phys. Rev.*, **51**:846 (1937); **92**:603 (1953). M. M. Saffren and J. C. Slater, *Phys. Rev.*, **92**:1126 (1953).

waves. Here, however, instead of superposing solutions of the two sorts, as in the orthogonalized-plane-wave method, we have a dividing surface, the surface of the sphere, where we change from the one type of expansion to the other. The essential point of the method, of course, is to get continuity across the boundary of the sphere. The reason why this method is so much more manageable than the cellular method is that it is easy to fit boundary conditions on the surface of a sphere, but hard over the odd-shaped surface of the polyhedral Wigner-Seitz cell.

The first step in the method is to take a single plane wave, of wave vector $\mathbf{k}$, which for the moment represents the whole wave vector ordinarily described as $\mathbf{k} + \mathbf{K}_j$, in the region between spheres, and then to fit solutions of the spherically symmetrical Schrödinger equation within the spheres to it so as to make the function continuous (though we cannot simultaneously make the normal derivative continuous). There is a familiar expansion of a plane wave, in terms of functions around a given center. This is

$$e^{i\mathbf{k}\cdot\mathbf{r}} = \Sigma(l)(2l + 1)i^l P_l(\cos\theta) j_l(|\mathbf{k}|r) \qquad (9\text{-}1)$$

Here the axis of spherical polar coordinates is taken along the direction of the wave vector $\mathbf{k}$. The quantities $j_l(|\mathbf{k}|r)$ are spherical Bessel functions, which we discuss later, in Eqs. (9-4) and (9-5). If then the radius of the sphere inscribed in the Wigner-Seitz cell is R, we find the value of the plane wave around the boundary by substituting $r = R$ in Eq. (9-1). We now must match this to a solution of Schrödinger's equation inside the sphere. To do this, we merely must make the amplitude of the function of each l value within the sphere have such a value that at $r = R$ it will match the corresponding function from Eq. (9-1).

We call such a plane wave, joined to a suitable solution of the spherical problem within each of the spheres, an augmented plane wave. It has a good many of the same properties as an orthogonalized plane wave. Now we use a set of such augmented plane waves, with wave vectors $\mathbf{k} + \mathbf{K}_j$, where $\mathbf{k}$ is here the reduced wave vector, as basis functions for the solution of the periodic-potential problem, making linear combinations with coefficients to be determined. In Appendix 6 we show how to set up the matrix elements of the Hamiltonian, and of unity, with respect to these functions, and to set up the secular equation between them. The secular equation is formed by setting the determinant of quantities $(H - E)_{ij}$ equal to zero, where H is the Hamiltonian, E the energy, and i and j refer to augmented plane waves with wave vectors $\mathbf{k}_i$ and $\mathbf{k}_j$, respectively, which as in Eq. (9-1) stand for the complete wave vectors $\mathbf{k} + \mathbf{K}_i, \mathbf{k} + \mathbf{K}_j$.

We find that we have

$$(H - E)_{ij} = (\mathbf{k}_i \cdot \mathbf{k}_j - E)\delta_{ij} + \frac{1}{\Omega} \sum (n) e^{i(\mathbf{k}_j - \mathbf{k}_i)\cdot \mathbf{r}_n} F_{nij} \qquad (9\text{-}2)$$

Sec. 9-2] THE CELLULAR AND AUGMENTED-PLANE-WAVE METHODS 235

where

$$F_{nij} = 4\pi R_n^2 \left[-(\mathbf{k}_i \cdot \mathbf{k}_j - E) \frac{j_1(|\mathbf{k}_j - \mathbf{k}_i| R_n)}{|\mathbf{k}_j - \mathbf{k}_i|} \right.$$
$$\left. + \sum_{l=0}^{\infty} (2l + 1) P_l(\cos \theta_{ij}) j_l(|\mathbf{k}_i| R_n) j_l(|\mathbf{k}_j| R_n) \frac{u'_{nl}(R_n)}{u_{nl}(R_n)} \right] \quad (9\text{-}3)$$

In these equations, the symbols have the following significance. The quantity Ω is the volume of the unit cell. The summation over n is over all atoms in the unit cell, the nth atom being located at vector position $\mathbf{r}_n$. The quantity R_n is the radius of the sphere representing the nth atom, within which the potential has spherical symmetry. The potential outside all the spheres is assumed to be zero. The angle θ_{ij} is that between the wave vectors $\mathbf{k}_i$ and $\mathbf{k}_j$. The quantities j_1, j_l, are spherical Bessel functions. The quantities $u_{nl}(R_n)$ and $u'_{nl}(R_n)$ are the radial wave function, and its derivative with respect to r, for the nth atom with angular quantum number l, computed at the distance R_n. As we have mentioned earlier, these wave functions are to be computed by numerical integration outward from $r = 0$, and are the solutions of the radial wave equation which are regular at the origin. The integration is to be carried out with an energy parameter E which is held as a disposable parameter during the calculation, so that $(H - E)_{ij}$ involves E both explicitly and indirectly through its occurrence in u'_{nl}/u_{nl}. On account of this occurrence of E in two ways in the secular equation, it is not practical to solve it in the ordinary way. Instead, the value of the determinant is computed as a function of E, and the zeros of this function are found. Each one of these corresponds to an energy value for the assumed value of $\mathbf{k}$, the reduced wave vector.

This is a somewhat difficult process to carry through, but it has been programmed by J. H. Wood for a digital computer, and can be handled successfully. When the problem has been solved, the wave function within the sphere will be an exact solution of Schrödinger's equation in that region. The function outside will be as good a solution as we can get with the number of plane waves used. The effect of solving the secular equation is to remove, or greatly minimize, the discontinuity of first derivative which there would be at the surface of the spheres if we had used only a single augmented plane wave, which has a continuous function, but not a continuous first derivative, at the surface of the sphere. It is found that in fact it is possible to handle a linear combination of enough augmented plane waves so that good convergence is secured, even in such a problem as a crystal formed from atoms of the iron group. There is every reason to think that the resulting function is a very accurate solution of Schrödinger's equation for the assumed potential. In the expansion over different l values, it is necessary to go up to something like $l = 12$ for an iron-group atom, and it is necessary to use something like 40 plane waves to get an adequate approximation,[1] at a point of no symmetry. Of course, symmetrized waves are used at symmetry points, greatly reducing the size of the secular equations.

It has been pointed out by Saffren,[2] who made a good deal of the early development of the method, that there are various different ways in which it can be regarded. One is as a modification of the orthogonalized-plane-wave method. Instead of stopping the plane wave at the surface of the sphere, and using a different solution inside, we could look at the same solution by saying that the plane wave extended throughout the crystal, and inside the spheres we had a function which equaled the true function minus the plane wave, plus the plane wave. We could use the expansion of Eq. (9-1) for the plane wave within the sphere, so that from the function of spherical type corresponding to each l value, we should subtract the corresponding term of Eq. (9-1) to get the quantities which would be added to the plane wave to get the complete solution. These resulting functions within the spheres would be identical in each sphere, except for being multiplied by factors $e^{i(\mathbf{k}+\mathbf{K}_j)\cdot\mathbf{R}_i}$. But this is just like a Bloch sum of these functions within the spheres. The result is like an orthogonalized plane wave, formed from a real plane wave, and a Bloch sum of these functions which are found within the spheres, but which do not extend outside.

As a result of this formal similarity between the methods of the augmented plane wave and the orthogonalized plane wave, many of the technical methods of calculation used for one can be used for the other. In particular, there is no reason why the augmented plane waves could not be used as basis functions for solving a problem in which the potential energy did not have precisely the prescribed behavior of being spherical within the spheres, constant outside. One would start with a potential within the sphere which was a spherical average of the true potential, and outside the spheres had a value equal to the average of the true potential over the region outside the spheres. Since the augmented plane waves form such an accurate solution of the unperturbed problem, the expansion in terms of them would be expected to show rapid convergence. However, this method has not so far been used in a real case.

9-3. The Scattered-wave Method. There are several other methods which can be shown to be mathematically equivalent to the augmented-plane-wave method, but which proceed along quite different lines. They all use the same potential, however, spherically symmetrical within

[1] J. H. Wood, *Phys. Rev.*, **126**:517 (1962).
[2] M. M. Saffren, Ph.D. thesis, Massachusetts Institute of Technology, Cambridge, Mass., 1959.

Sec. 9-3] THE CELLULAR AND AUGMENTED-PLANE-WAVE METHODS 237

spheres representing the atoms, and constant outside. One of these investigates the scattering of an impinging plane wave on such a spherical atom.[1] This method was suggested by a treatment originally used by Ewald[2] for the problem of x-ray scattering. We have already pointed out, in connection with our discussion of Bragg scattering, that x-ray-diffraction theory and energy-band theory have a close relationship.

Ewald's discussion was carried out in the very early days of x-ray research, and was started before the discovery of x-ray diffraction, as a study of the refraction of light by crystals. It was known, from much earlier work on the optical properties of solids, that when a wave of light falls on an atom, it induces an oscillating electric dipole in the atom, which emits a spherical electromagnetic wave. These emitted spherical waves interfere with each other, and with the incident wave, to produce the scattered light wave. If the atoms are arranged on a regular lattice, this scattered wave is the ordinary refracted wave, traveling with a different velocity from that found in free space (thus leading to the index of refraction), and having a diminished intensity (the effect of absorption). If the atoms have a random arrangement, there will also be light scattered to the side, as is observed when light passes through a liquid or a gas. The well-known Rayleigh scattering of sunlight by the atmosphere, which results in the blue appearance of the sky, is an example of this random scattering.

The theories of this effect which were available at the time of Ewald's work were of a macroscopic sort, taking no real account of the atomic nature of matter. They merely assumed a certain definite number of scattering atoms per unit volume. Ewald, on the contrary, assumed a regular arrangement of the atoms on a lattice. He proceeded in the following manner. He assumed that the total electromagnetic field inside the crystal consisted of a sum of the incident wave, plus the scattered waves emitted by all atoms. He then investigated the balance of radiation in the neighborhood of any atom. The atom emitted a spherical wave. It also had an incoming wave, consisting of the wave incident on the crystal from outside, plus the resultant of the scattered waves from all other atoms of the crystal. He analyzed this incoming field into an infinite set of plane waves, considered the scattered wave produced by each of these plane-wave components, and applied the condition that the sum of all these scattered waves had to equal the wave actually emitted by the atom. Since this atom was typical of all in the crystal, there was a sort of self-consistent condition: the wave emitted by a given atom has to be such that, if we assumed similar waves to be emitted by all atoms, and

[1] J. Korringa, *Physica*, **13**:392 (1947). P. M. Morse, *Proc. Natl. Acad. Sci.*, **42**:276 (1956).
[2] P. P. Ewald, *Ann. Physik*, **49**:1, 117 (1916); **54**:519 (1917).

superposed their effect as an incident wave on a typical atom, they would produce just the scattered wave which had been originally postulated.

Ewald was able to carry through this requirement mathematically. One of the important parts of his analysis was the superposition of the waves scattered by all atoms of the crystal. This problem of the evaluation of spherical waves emitted by atoms arranged on a lattice resulted in the famous Ewald summation method, an ingenious mathematical device which has since been applied to many different problems involving similar sums. By use of these mathematical methods he arrived at a rigorous theory of the index of refraction and absorption of a crystal. He was furthermore in position to derive a theory of x-ray diffraction. Once it was clear, through the work of von Laue and his collaborators in 1912, that x rays could be diffracted by a crystal, leading to diffracted beams determined by the Bragg law, Ewald could consider such diffraction, without modification of his analysis, merely applying it to much shorter wavelengths than the optical wavelengths for which it was designed.

The scattered-wave treatment of the electrons in a periodic potential is the application of this same method to the electron-diffraction problem. The idea behind the methods of Korringa and of Morse is the same as Ewald's: to consider the scattering of electron waves by a spherical atom, and to synthesize the scattered wave actually emitted by an atom out of the scattered waves produced by the incident waves. The first part of the problem, in this method, is to investigate the scattering of an electron wave by a single atom. This is a problem which had already been considered by Morse,[1] and his solution is directly adaptable to the present case.

Morse was interested in the scattering of a beam of electrons by a gas. He wished specifically to explain the Ramsauer effect.[2] This was an effect which had been discovered experimentally some time before. Ramsauer was investigating the passage of a beam of electrons through a gas, as a function of the electron velocity or energy. For most gases, each atom acts as if it had a considerable cross section. The beam of electrons is weakened by passing through the gas as if all those electrons striking an atomic cross section were scattered out of the beam, being scattered to the side, either elastically, without loss of energy (except recoil energy of the atom), or inelastically (with loss of excitation energy to the atom). For very slow electrons, below the excitation potentials of the atoms, the scattering would have to be entirely elastic. In most

---

[1] W. P. Allis and P. M. Morse, *Z. Physik*, **70**:567 (1931). P. M. Morse, *Rev. Mod. Phys.*, **4**:577 (1932).

[2] C. Ramsauer, *Ann. Physik*, **64**:513 (1921); **66**:545 (1921); **72**:345 (1923). For review of the subject, see R. B. Brode, *Rev. Mod. Phys.*, **5**:257 (1933).

Sec. 9-3] THE CELLULAR AND AUGMENTED-PLANE-WAVE METHODS

cases the resulting cross section varied only slowly with electron energy, and was comparable in size with the atomic dimensions. Ramsauer found, however, that for some gases, notably the inert gases, the cross section decreased rapidly as the electron energy decreased, becoming exceedingly small for very slow electrons. This surprising effect could not be explained classically. Morse showed, however, that by treating the scattering of the electron waves by the atoms by a wave-mechanical method, the effect could be understood. He replaced the atom by a spherically symmetrical potential, and assumed that it was surrounded by a region in which the potential was constant. Thus he was dealing with the same sort of potential which we assume in the APW method. However, he was treating the scattering by a single atom, rather than the behavior of a crystal lattice of atoms.

Morse's treatment proceeded essentially in the following manner. For incident electrons of a given energy, he first solved the wave equation inside the sphere representing the atom, for the assumed energy. As we have seen in our discussion of the cellular and the APW methods, we can find such a solution for each l value, regular at the origin, and we can find the value of the function, or more properly of the logarithmic derivative $u_l'(R)/u_l(R)$, at the surface of the sphere. We must then join this solution continuously, with continuous derivative, to a solution outside the sphere.

In this outside region, where the potential is constant, and may be taken to be zero, Schrödinger's equation becomes $-\nabla^2 u = Eu$, which is the time-independent form of the wave equation. Radial solutions are known to be the spherical Bessel functions, and the spherical Neumann functions. These solutions have the following properties. At large values of r, the asymptotic forms are

$$j_l(x) = \frac{1}{x} \cos\left(x - \frac{l+1}{2}\pi\right)$$
$$n_l(x) = \frac{1}{x} \sin\left(x - \frac{l+1}{2}\pi\right) \qquad (9\text{-}4)$$

where j_l and n_l are the spherical Bessel and Neumann functions, respectively. At small values, the asymptotic forms are

$$j_l(x) = \frac{x^l}{1 \cdot 3 \cdot 5 \cdot \cdots \cdot (2l+1)}$$
$$n_l(x) = -\frac{1 \cdot 1 \cdot 3 \cdot 5 \cdot \cdots \cdot (2l-1)}{x^{l+1}} \qquad (9\text{-}5)$$

The solutions are given by these functions, where $x = |\mathbf{k}|r$, $\mathbf{k}$ being the wave vector of the incoming plane wave.

We see that the Bessel function behaves regularly at the origin, while the Neumann function does not. Hence if we are investigating the solution of the wave equation in a restricted region including the origin, we use only the Bessel function, since the Neumann function diverges. However, in the present case, we are interested in the solution outside the atomic sphere, where both solutions are acceptable. By using appropriate linear combinations of the Bessel and Neumann functions, we can get any desired value of the logarithmic derivative, and hence can match the logarithmic derivative found from the solution of Schrödinger's equation inside the atomic sphere.

We must ask how the resulting solution outside the sphere is to be interpreted. We wish this solution to consist of two parts: a plane wave, expressed as in Eq. (9-1), in terms of spherical Bessel functions, and a set of outgoing spherical waves, one for each l value. Such a spherical outgoing wave is not expressed either by a Bessel or a Neumann function, but by a linear combination called a spherical Hankel function, $j_l(x) + in_l(x)$, which behaves asymptotically at large x like

$$\frac{1}{x}\left[\cos\left(x - \frac{l+1}{2}\pi\right) + i\sin\left(x - \frac{l+1}{2}\pi\right)\right]$$
$$= \frac{1}{x}\exp i\left(x - \frac{l+1}{2}\pi\right) \quad (9\text{-}6)$$

which obviously represents the outgoing wave, with amplitude falling off as $1/r$, and intensity as $1/r^2$. We see that if we superpose the spherical Bessel function $j_l(|\mathbf{k}|r)$ from Eq. (9-1), and a constant times the spherical Hankel function of Eq. (9-6), the ratio of the coefficient of the Neumann function to that of the Bessel function can be adjusted at will, by proper choice of the coefficient multiplying the Hankel function. In particular, we can adjust it to give the same logarithmic derivative at the surface of the sphere required to match the value found inside the sphere. Thus the amplitude of the spherical wavelet is determined.

We notice that this amplitude is greater, the greater the deviation of the logarithmic derivative found inside the sphere, from that which would be found for a spherical Bessel function $j_l(|\mathbf{k}|r)$ alone. If by chance the value of the logarithmic derivative for the spherical Bessel function matched the value found from the solution inside the sphere, at the sphere boundary, no mixture of the Neumann function would be required, and hence the amplitude of the scattered wavelet would be zero. This is the accident which Morse found to occur if he used a spherical potential appropriate for an inert gas atom. The matching occurred only for a range of energies of the electron near zero, and led to values for the amplitudes of the scattered wavelets, which, when interpreted in terms of atomic cross sections, agreed well with Ramsauer's observations. We shall come back to this important fact in the next section.

This sketch of the treatment of scattering by a single atom shows how one part of the calculation of energy bands goes, when one uses the scattered-wave method. The rest of the analysis proceeds essentially as in Ewald's theory. It is this method which was suggested by Korringa, and carried through by Morse. It has not been used for calculations of very high accuracy, but there is no doubt that it would reduce to the same final results as the augmented-plane-wave method if it were carried to equal accuracy. The reason for this statement is that it is using the same form of expansion: radial solutions of the spherical problem, multiplied by spherical harmonics, within the spherical atoms, joined smoothly onto a superposition of plane waves outside the atoms.

9-4. The Free-electron Approximation and the Scattered-wave Method. The greatest importance of the scattered-wave method lies not in its particular convenience for calculation, but in the light which it sheds on the physically important fact that an almost free electron calculation is capable of giving energy bands in very close agreement with experiment, in a good many cases. We have mentioned this fact before, but without trying to indicate an explanation. Now we are in position to provide the explanation, and to show that it is closely tied in with the phenomenon of the Ramsauer effect.

The metallic crystals for which the free-electron approximation is most accurate are the alkali metals. For these, the conduction electron is moving in the field of the inner shells, which have an inert-gas structure. Thus the potential which the electron sees is not very different from that seen by an electron in the Ramsauer experiment, approaching an inert gas atom. It is those particular atoms which have the property that the logarithmic derivative of the wave function, at the boundary of an atomic sphere, is approximately equal to that found from a spherical Bessel function, so that a single plane wave will join practically continuously to the solution inside the sphere, without the need of scattered waves. For the alkaline earth elements, and the following trivalent metals, the agreement between the free-electron approximation and the actual bands gets poorer, but is still good enough to be valuable. For the alkali metals, we can see in a striking way how close the agreement is, by plotting the logarithmic derivative $u_l'(R)/u_l(R)$ as a function of energy, and comparing with the corresponding plot for j_l. This is done in Fig. 9-3, for sodium. In contrast to this, we show in Fig. 9-4 the corresponding curves for copper. In this case, for the d electrons, $l = 2$, there is no agreement at all between the actual curve and that for the free electrons, found from the spherical Bessel function. As a consequence of this, the free-electron approximation is entirely erroneous for copper, as for other transition elements, in which a band composed largely of d-like wave functions introduces itself into the middle of a band composed of s- and p-like functions.

We have seen that as the logarithmic derivative of the true wave function u approaches that of the spherical Bessel function, the free-electron approximation becomes more nearly applicable. This fact is not at all obvious from the matrix elements given in Eq. (9-3), for the APW

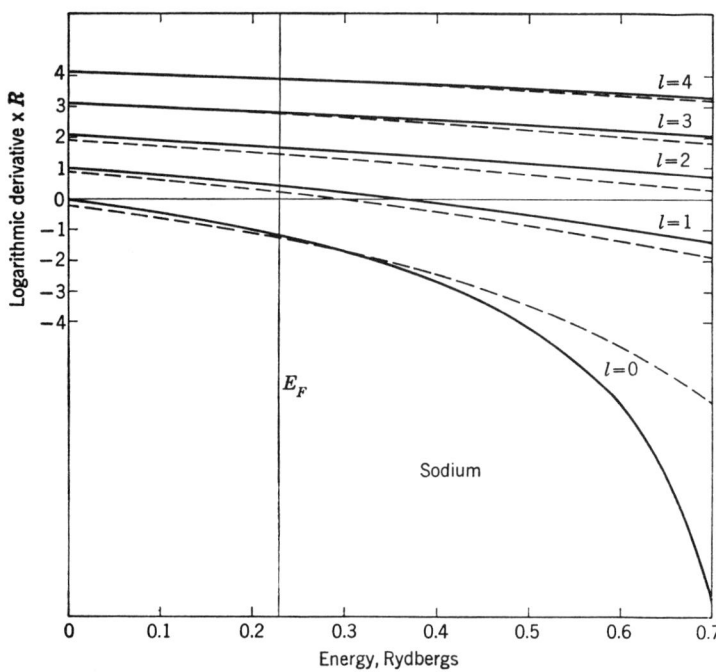

FIG. 9-3. Logarithmic derivatives $u_l'(R)/u_l(R)$ as function of energy, for sodium, compared with corresponding free-electron function $j_l'(kR)/j_l(kR)$, where $k^2 = E$. Free-electron case, full lines; sodium, dashed lines. The writer is indebted to Dr. A. C. Switendick, Prof. J. H. Wood, and J. F. Kenney for the calculations resulting in these curves.

method. However, as shown in Appendix 6, the expression of Eq. (9-3) can be transformed to the form

$$F_{nij} = 4\pi R_n^2 \left\{ -(k_j^2 - E) \frac{j_1(|\mathbf{k}_j - \mathbf{k}_i|R_n)}{|\mathbf{k}_j - \mathbf{k}_i|} \right.$$
$$+ \sum_{l=0}^{\infty} (2l+1) P_l(\cos \theta_{ij}) j_l(|\mathbf{k}_i|R_n) j_l(|\mathbf{k}_j|R_n)$$
$$\left. \left[\frac{u_{nl}'(R_n)}{u_{nl}(R_n)} - \frac{j_l'(|\mathbf{k}_j|R_n)}{j_l(|\mathbf{k}_j|R_n)} \right] \right\} \quad (9\text{-}7)$$

This form, which is not symmetrical in i and j, is less convenient for computation than that of Eq. (9-3), but it is better for discussing the free-electron limit.

To consider this limit, let us first take a diagonal matrix element, $i = j$. One can show, from Eq. (9-5), that the limiting value of $j_1(|\mathbf{k}_j - \mathbf{k}_i|R_n)/|\mathbf{k}_j - \mathbf{k}_i|$, as $\mathbf{k}_j - \mathbf{k}_i$ approaches zero, is $R_n/3$. If we

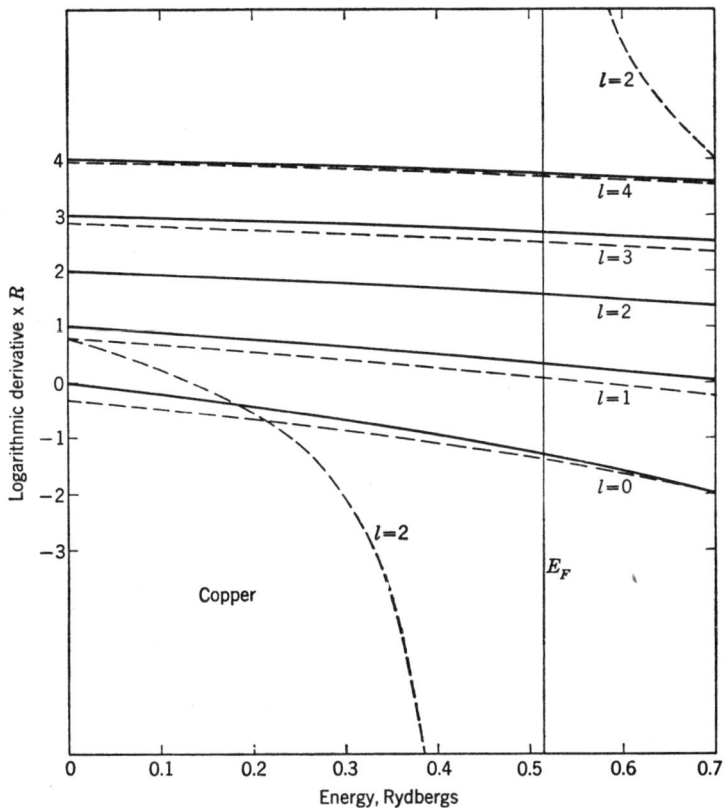

FIG. 9-4. Logarithmic derivatives as in Fig. 9-3, for copper.

substitute this value in Eq. (9-7), and insert the resulting quantity into Eq. (9-2), we find that

$$(H - E)_{jj} = (k_j^2 - E)\left[1 - \sum (n)\frac{4\pi}{3}\frac{R_n^3}{\Omega}\right]$$
$$+ \sum (n)\frac{4\pi R_n^2}{\Omega}\sum_{l=0}^{\infty}(2l+1)[j_l(|\mathbf{k}_j|R_n)]^2$$
$$\left[\frac{u'_{nl}(R_n)}{u_{nl}(R_n)} - \frac{j'_l(|\mathbf{k}_j|R_n)}{j_l(|\mathbf{k}_j|R_n)}\right] \quad (9\text{-}8)$$

We shall show shortly that in the free-electron limit the nondiagonal matrix elements are very small. Then it is legitimate to use first-order perturbation theory, which amounts to replacing the secular determinant by the product of its diagonal terms. If we set this equal to zero, we can set one term such as $(H - E)_{jj}$ equal to zero. In this way from Eq. (9-8) we find

$$E = k_j^2 + \sum (n) \frac{4\pi R_n^2}{\Omega - \Sigma(m)\frac{4}{3}\pi R_m^3} \sum_{l=0}^{\infty} (2l + 1)[j_l(|\mathbf{k}_j|R_n)]^2$$
$$\times \left[\frac{u'_{nl}(R_n)}{u_{nl}(R_n)} - \frac{j'_l(|\mathbf{k}_j|R_n)}{j_l(|\mathbf{k}_j|R_n)} \right] \quad (9\text{-}9)$$

If we have the situation we have described earlier, we shall almost find for a particular value of $\mathbf{k}_j$, that

$$\frac{u'_{nl}(R_n)}{u_{nl}(R_n)} = \frac{j'_l(|\mathbf{k}_j|R_n)}{j_l(|\mathbf{k}_j|R_n)} \quad (9\text{-}10)$$

in which each side is a function of the energy, which in the spherical Bessel functions on the right is equated to k_j^2. This is the case shown in Fig. 9-3, where our condition is almost fulfilled for the case of sodium, over a considerable range of energies. To the approximation to which this condition holds, Eq. (9-9) leads to the free-electron energy, $E = k_j^2$. In this case, we may next look at the nondiagonal matrix elements $(H - E)_{ij}$, as given in Eqs. (9-2) and (9-7), for the case in which the jth state is that for which Eq. (9-10) is approximately fulfilled. We may approximately substitute the expression given by Eq. (9-9) for $k_j^2 - E$ into Eq. (9-7). When we have done this, we see that every term of the nondiagonal matrix element involves a term like $u'_{nl}(R_n)/u_{nl}(R_n) - j'_l(|\mathbf{k}_j|R_n)/j_l(|\mathbf{k}_j|R_n)$, which is very small when we have approximately the condition of Eq. (9-10).

Thus we verify our statement that in the case where the atoms have small scattering cross section, or the Ramsauer effect holds, the problem of solving Schrödinger's equation in the periodic potential will approach the free-electron case, in that the nondiagonal matrix elements between augmented plane waves will be small, and the energy will be given by the free-electron value k_j^2 except for small terms. Near a point in the Brillouin zone where two plane waves have the same energy, there will of course be a case of degeneracy, and one must consider a secular equation between the almost degenerate augmented plane waves, so that their energies will be pushed apart as in the nearly free electron case.

The arguments which we have been bringing up in this section to show that one can have energy bands approaching those of free electrons have been carried further by Fues and Statz.[1] These writers have examined, both in a one-dimensional model and in the three-dimensional case, the conditions required to have the quantities

$$\frac{u'_{nl}(R_n)}{u_{nl}(R_n)} - \frac{j'_l(|\mathbf{k}_j|R_n)}{j_l(|\mathbf{k}_j|R_n)}$$

approximately vanish, which as we have seen must occur if the free-electron model is to be approximately correct. They have shown that if the potential energy has very deep potential wells, of small cross section, such as one has near the nuclei of an alkali metal, one automatically sets up the required conditions for these quantities to be small. Hence it is no accident that we have the nearly free electron case for the alkali metals.

We see from the discussion of this section that this free-electron-like behavior of the energy bands in many cases can be explained in a very straightforward manner in terms of the scattered-wave method of treating energy bands, which is mathematically equivalent to the APW method. In contrast, it is a very involved matter to discuss the same problem in the language of the OPW method. There is a considerable literature devoted to this effort, the principal papers being by Phillips and Kleinman[2] and by Cohen and Heine.[3] In contrast to the very straightforward treatment following from the APW type of discussion, this treatment on the basis of the OPW method, leading to a so-called pseudopotential, seems not as satisfying physically as is the interpretation based on scattering theory.

One remark regarding the success of the nearly free electron approximation for the energy of energy bands is of great importance. We must not forget that though the energies are close to free-electron values, the wave functions are very far from the plane waves representing free electrons. The point is that the wave function is represented with good accuracy by a single plane wave between the spheres representing the atoms, but inside such a sphere, the wave functions go through all the oscillations characteristic of an atomic wave function. However, we have the special case in which a single plane wave outside the atom can join smoothly onto a solution of the spherically symmetrical Schrödinger equation within each atomic sphere, the energy of this solution being the same as that of the plane wave.

[1] E. Fues and H. Statz, *Z. Naturforsch.*, **7a**:2 (1952).
[2] J. C. Phillips, *Phys. Rev.*, **112**:685 (1958). J. C. Phillips and L. Kleinman, *Phys. Rev.*, **116**:287 (1959). L. Kleinman and J. C. Phillips, *Phys. Rev.*, **116**:880 (1959); **117**:460 (1960); **118**:1153 (1960). J. C. Phillips and L. Kleinman, *Phys. Rev.*, **128**:2098 (1962).
[3] M. H. Cohen and V. Heine, *Phys. Rev.*, **122**:1821 (1961).

9-5. Other Related Methods. Kohn and Rostoker[1] have suggested a method for solving the periodic-potential problem which uses the same type of potential as the APW and scattered-wave methods, namely, a spherically symmetrical potential in a sphere representing each atom, and a constant outside these spheres, but which proceeds by different methods. Like the APW method, it has been programmed for digital computers, and has been used, principally by Segall and by Ham (see references in the bibliography at the end of Chap. 10) for calculations on a number of metals. It is an open question which method, that of Kohn and Rostoker or the APW method, is more convenient for practical calculations. Fortunately, the two methods have been intercompared by calculations on identical potentials for copper, by Segall,[2] by the Kohn-Rostoker method, and by Burdick[3] by the APW method, and the results agree within the numerical accuracy of the calculation. It thus appears that both methods provide valid solutions of Schrödinger's equation for the type of potential used.

The Kohn-Rostoker method proceeds by the use of Green's function, and it is often called by that name. A good account of the present procedure for using this method has been given by Ham and Segall.[4] The idea of Green's function is of course well known in mathematical physics. We are interested in solving Schrödinger's equation,

$$(-\nabla^2 + V)\psi = E\psi \tag{9-11}$$

where V is the periodic potential. We write this equation in the form

$$(\nabla^2 + E)\psi = V\psi \tag{9-12}$$

This has the general form of the inhomogeneous wave equation,

$$(\nabla^2 + E)\psi = f \tag{9-13}$$

where f is a function of position. Since this is a linear differential equation, one can subdivide the right-hand side into elements one of which has the value $V\psi$ inside a small volume dv, with zero everywhere else. The general solution of Eq. (9-12) will be the sum of the solutions of all such equations, plus a general solution of the homogeneous equation $(\nabla^2 + E)\psi = 0$.

The solution of Eq. (9-12), with $V\psi$ replaced by a quantity which is zero except within dv, is known to be a spherical wave, of wave vector **k** deter-

---

[1] W. Kohn and N. Rostoker, *Phys. Rev.*, **94**:1111 (1954).
[2] B. Segall, *Phys. Rev. Letters*, **7**:154 (1961); *Phys. Rev.*, **125**:109 (1962).
[3] G. A. Burdick, *Phys. Rev. Letters*, **7**:156 (1961); *Phys. Rev.* **129**:138 (1963).
[4] F. S. Ham and B. Segall, *Phys. Rev.*, **124**:1786 (1961).

mined by $k^2 = E$, and amplitude inversely proportional to the distance from the volume element dv. This function is Green's function for the problem. More precisely, we have

$$d\psi = -\frac{1}{4\pi}(V\psi\,dv)\frac{e^{-ikr}}{r} \quad (9\text{-}14)$$

where the $d\psi$ given on the left of Eq. (9-14) is the wave emitted by the volume element dv, within which $V\psi$ will have the value indicated by the expression $V\psi\,dv$. The distance r is that measured from the volume element to the point where ψ is being computed. To get the complete solution ψ, we must then sum or integrate contributions like that of Eq. (9-14) over all space, obtaining

$$\psi = -\frac{1}{4\pi}\int\frac{e^{-ikr}}{r}V\psi\,dv \quad (9\text{-}15)$$

If we add a general solution of the equation $(\nabla^2 + E)\psi = 0$, we have a general solution of the Schrödinger equation.

We observe that this is an integral equation for ψ: it appears both on the left side of Eq. (9-15) and inside the integral sign. This can be interpreted in a way entirely analogous to that which we have described in discussing the scattered-wave method. The wave function at a given point of space equals the incident wave [the general solution of $(\nabla^2 + E)\psi = 0$], plus scattered wavelets (the functions $e^{-ikr/r}$) originating in all volume elements. The integral equation of Eq. (9-15) is a self-consistency equation: the sum of all scattered wavelets, plus the incident wave, add to the wave function itself. The difference between the Green's function method and the scattered-wave method of Ewald, Korringa, and Morse is that in the latter method we handle together the scattering from all volume elements of a spherical atom, then add up the contributions of each atom, while in the Green's function method we directly sum the scattered wavelets from each volume element, though in the course of the subsequent calculation this is converted into the scattered waves from individual atoms.

In the Green's function method, as in the Korringa-Morse method, the summation of the scattered wavelets is the principal task, and in each case it is carried out by Ewald's method, which we have mentioned earlier. Kohn, Rostoker, and Segall (*loc. cit.*) have shown how to reduce the problem to the form of a determinantal equation, not unlike that met with the APW method. That is, the matrix elements found in the determinant involve the wave vector **k**, the energy E, and the logarithmic derivative u'/u of the solution of the spherically symmetrical problem at the radius R of the sphere representing the atom. They have put these matrix elements in such a form that they carry out once for all the calculation required for each k and E value, for a given crystal structure, and it is then a simple task to insert the logarithmic derivative, compute the value of the determinant, and vary either E or k (they choose the latter) to make the determinant equal to zero. Thus the general outline of the calculation is the same by either the Green's function or the APW method, and each one involves the same type of information about the atoms. Since each method gives a correct solution of the same problem, one would expect that the determinant used in the Green's function method could be transformed into that used in the APW method.

There is one device which the users of the Green's function method have employed, though it could equally well be used in the APW method. This is the so-called quantum-defect method, first used by Kuhn and Van Vleck,[1] and recently improved by Brooks and Ham.[2] To introduce this method, let us consider an alkali metal, such as sodium, for which the method is particularly well adapted. We know that for the APW or Green's function method we require the logarithmic derivative of the radial wave function, for each l value, at the radius R, as a function of energy. We note, however, that essentially the same potential would be found for the isolated sodium atom. The eigenfunctions of the atom will have definite values of u'/u at the radius R. These eigenfunctions are known, as are their eigenvalues. In fact, for large values of r, the potential surrounding a sodium atom is a simple Coulomb potential, so that the eigenfunctions are like those for hydrogen, only computed for the observed eigenvalues. They are functions which reduce to zero at infinity, and join at smaller values of r onto the correct sodium eigenfunctions. From these facts it is possible to derive the values of u'/u at radius R, knowing only the observed eigenvalues for the atom.

It is also possible to derive a general formula for the value of u'/u for each l value, as a function of energy. This formula is determined from general principles, except for a phase factor, and this latter factor can be found as described in the preceding paragraph, from the observed eigenvalues. In other words, we find the logarithmic derivative needed for solving the scattering problem, by use only of the observed spectrum. These values may well be more accurate than those calculated by numerical integration of assumed potentials, for they automatically contain corrections required to get agreement with experiment for the optical spectrum of the isolated atom. Such corrections include, for instance, the effect of polarization of the inner core by the valence electron, an effect disregarded in the self-consistent-field method. The outer electron polarizes the core, this results in a dipole moment, which in turn exerts a

[1] T. S. Kuhn and J. H. Van Vleck, *Phys. Rev.*, **79**:382 (1950).
[2] H. Brooks and F. S. Ham, *Phys. Rev.*, **112**:344 (1958).

Sec. 9-5] THE CELLULAR AND AUGMENTED-PLANE-WAVE METHODS

field on the outer electron, which appears as a correction to the self-consistent potential. It could be derived from the self-consistent approach in terms of configuration interaction, or second-order perturbation theory. Use of the quantum-defect method makes this type of calculation unnecessary. The method has been used with good success by Ham[1] and others, principally for studies of energy bands in alkali metals.

[1] F. S. Ham, *Phys. Rev.*, **128**:82, 2524 (1962).

10

Calculations of Energy Bands in Crystals

10-1. The Free-electron Approximation. We have been studying the methods of calculating energy bands in crystals, as well as the empirically known crystal structures of elements and simple binary compounds. Now we come to a discussion of calculations which have been made for these elements and compounds, with the results they lead to regarding the nature of the energy bands. We shall put off until a later volume the discussion of the experimental methods which have been used to verify and extend these calculations. It would hardly have been possible to have deduced anything regarding the nature of energy bands directly from experiment, without the guidance which the theory has given us. With the existence of the theory, however, it has proved possible to devise a variety of very ingenious experimental methods to investigate many details of the band structure. The general result of these experiments is that the broad outlines of the theory are verified.

We shall not try in this chapter to give an exhaustive discussion of all the energy-band calculations which have been made, though a fairly complete bibliography is given at the end of the chapter. Rather, we shall present a number of typical calculations, which will serve to illustrate the general situation. Some of those we shall choose are among the most elaborate and accurate calculations which have yet been made, in some cases amplified by incorporation of experimental results. In other cases, we use simpler calculations, where they illustrate important points on which no more accurate work has been done. In tying these calculations together, we may seek some unifying principle to understand the really very complicated results which we shall find, and this principle is the resemblance of the energy bands to those which would be found from slightly perturbed free electrons.

In the earliest calculations of energy bands, such as those for sodium discussed in Sec. 8-3, it was observed that the energy in the conduction band as a function of the wave vector had a close resemblance to what would be found from slightly perturbed plane waves. We illustrated this

Sec. 10-1]   CALCULATIONS OF ENERGY BANDS IN CRYSTALS   251

in Fig. 8-4, where the calculations of energy versus **k** for sodium showed less deviation from the free-electron parabolic curves than we illustrated for a one-dimensional case in Fig. 6-1. This was interpreted by some writers as meaning that the free-electron approximation was really applicable with good accuracy to the problem of energy bands in real crystals. It is abundantly clear from our discussion of Sec. 9-4 that this is not the case, and we must look more deeply for an explanation of the success of the free-electron approximation. The interpretation, as we have shown in Sec. 9-4, is that a single augmented plane wave forms a good approximation to a solution of such a crystal as an alkali metal, except in the immediate neighborhood of the surface of the Brillouin zone, where we must take into account the interaction of waves with approximately equal energy. As we showed in Eq. (9-9), the energy of a single augmented plane wave for an alkali metal depends on **k** very much like the energy of a plane wave. Hence the final secular problem which we must solve, combining augmented plane waves in linear combinations to form the best possible approximations to the wave function, is not only formally similar to the secular problem of the free-electron method, but also, the nondiagonal matrix elements being small, it corresponds to the case of rather free electrons. It is on account of the small size of the nondiagonal matrix elements that this method converges so rapidly, in contrast to the true free-electron method of expanding the whole wave function in plane waves. Similar arguments apply if we are dealing with orthogonalized plane waves, though the arguments are not so straightforward.

Since these early calculations, a great variety of evidence has indicated that the similarity to free electrons is found in many cases. In the first place, by making the hypothesis that the free-electron method could be used, energy bands were deduced for a variety of materials, and good correlations were found between these bands and experiment, during the period before it was practicable to make really good energy-band calculations. Second, more recent very accurate calculations for other types of materials than alkali metals have verified the closeness of the free-electron picture. As we have already indicated, this resemblance has been found to hold for the monovalent, divalent, and trivalent elements. Even the tetravalent elements, carbon, silicon, and germanium, with their diamond structure, show a moderate resemblance to the free-electron picture, and useful deductions have been made for the pentavalent element bismuth, on a free-electron basis, with explanation of some of the puzzling properties of that element. Alloys as well as elements have been found to agree fairly well with a free-electron picture.[1]

On the other hand, there are other cases of great importance where it

[1] The following papers, arranged chronologically, include some of the most important

fails completely. The most conspicuous of these examples are found in the transition groups, where iron, for instance, shows a very different energy-band structure from what the free-electron picture would suggest. The interpretation of this is simple: the d electrons, which are in the process of being bound into the core of the atom at this part of the periodic table, are behaving neither like inner electrons nor like free electrons, but midway between them. This fact is of the greatest significance in the study of the transition elements, and their rather anomalous energy-band structure contributes a great deal to the interpretation of their properties, particularly their magnetic behavior.

Even though we have these exceptions to the agreement between the free-electron picture and the actual energy bands, still the agreement is striking enough so that we can use it as a guide in our study of the energy bands of metals and binary compounds. We shall therefore go on in the next section to a discussion of the methods of extending the kind of calculation which was made for a one-dimensional problem in Fig. 6-1 to a three-dimensional crystal of the symmetry types which we meet.

---

references dealing with nearly free electron calculations of energy bands:

H. Jones, *Proc. Roy. Soc. (London)*, **A144**:225 (1934); alloys in γ-phase; **A147**:396 (1934); bismuth.
H. Hellmann, *Acta Physicochimica URSS*, **1**:913 (1935), **4**:225 (1936); general method.
N. F. Mott and H. Jones, "The Theory of the Properties of Metals and Alloys," Oxford University Press, Fair Lawn, N.J., 1936; many examples.
P. Gombas, *Z. Physik*, **113**:150 (1939); alkali metals.
H. Jones, *Physica*, **15**:13 (1949); structural and elastic properties of metals.
H. Jones, *Phil. Mag.*, **41**:663 (1950); magnesium solid solutions.
E. Fues and H. Statz, *Z. Naturforsch.*, **7a**:2 (1952); general discussion.
F. Herman, *Rev. Mod. Phys.*, **30**:102 (1958); examples of free-electron energy bands.
J. Callaway, *Phys. Rev.*, **112**:322 (1958); sodium.
J. C. Phillips, *J. Phys. Chem. Solids*, **8**:369, 379 (1959); silicon and germanium.
J. C. Phillips and L. Kleinman, *Phys. Rev.*, **116**:287 (1959); general discussion.
L. Kleinman and J. C. Phillips, *Phys. Rev.*, **116**:880 (1959); diamond; **117**:460 (1960); BN; **118**:1153 (1960); silicon.
W. A. Harrison, *Phys. Rev.*, **116**:555 (1959); **118**:1182 (1960); aluminum; **118**:1190 (1960); general method. *J. Phys. Chem. Solids*, **17**:171 (1960); bismuth.
W. A. Harrison, "The Fermi Surface," John Wiley & Sons, Inc., New York, 1960, p. 28; general method.
H. Jones, "The Theory of Brillouin Zones and Electronic States in Crystals," Interscience Publishers, Inc., New York, 1960; general methods.
J. F. Cornwell and E. P. Wohlfarth, *Nature*, **186**:379 (1960); lithium.
J. F. Cornwell, *Proc. Roy. Soc. (London)*, **A261**:551 (1961); alkalies and beryllium, *Phil. Mag.*, **6**:727 (1961); noble metals.
M. Cohen and V. Heine, *Phys. Rev.*, **122**:1821 (1961); general method.
L. M. Falicov, *Phil. Trans. Roy. Soc. (London)*, **A255**:55 (1962); magnesium.
W. A. Harrison, *Phys. Rev.*, **126**:497 (1962); **129**:2503, 2512 (1963); zinc.

10-2. Free-electron Energy Bands in Crystals. We have seen in our earlier work that for free electrons, the energy surfaces will be found by setting up the functions $(\mathbf{k} + \mathbf{K}_j)^2 \hbar^2/2m$, where $\mathbf{k}$ is the wave vector, and $\mathbf{K}_j$ is one of the vectors of the reciprocal lattice. Since we have an infinite number of vectors $\mathbf{K}_j$, we have an infinite number of such functions. If we consider the values of these functions in the central Brillouin zone, the lowest energy value will correspond to the lowest energy band, the second lowest to the second energy band, and so on. These energy bands will be separated from each other if we are not at special symmetry points within the zone, but at the symmetry points two or more of the functions will have identical energies. At these points, where we then have degeneracy in the unperturbed problem, we can make linear combinations of the plane waves $e^{i(\mathbf{k}+\mathbf{K}_j)\cdot\mathbf{r}}$ whose energy is $(\mathbf{k} + \mathbf{K}_j)^2 \hbar^2/2m$, having appropriate symmetry for the various irreducible representations of the symmetry group of the crystal. In this way we find the energy as a function of $\mathbf{k}$, and the particular symmetry behavior, for the free-electron approximation. When perturbative interactions are introduced, they will push apart energy levels which accidentally agree in the free-electron case, so that the only degeneracy remaining will be that required by symmetry.

We now wish to discuss such calculations for the actual cases we shall meet. We shall show the energy as a function of the magnitude of $\mathbf{k}$, along various symmetry directions, terminating at points of special symmetry, as we did in Fig. 7-4 for the two-dimensional case, and in Fig. 8-4 for the body-centered cubic structure found in sodium. We give in Figs. 10-1 and 10-2 the free-electron energy bands for the body-centered and face-centered cubic cases, respectively. The details of the method of constructing these figures, by means of the group theory, are given in Appendix 7. We have already given a simplified version of Fig. 10-1 in Fig. 8-4.

To interpret these figures, we give the Brillouin zones for these cases, with the various symmetry directions and points shown, in Fig. 10-3; the same information is given in Figs. A3-3 and A3-2, respectively. In Fig. 10-1, for the body-centered cubic case, we start at the left side of the figure with a diagram of energy versus $\mathbf{k}$, along the direction 100, or the x axis, which in the notation of Bouckaert, Smoluchowski, and Wigner is the direction Δ. At the extremity of the Brillouin zone we come to the point H. Next we have a diagram showing energy versus distance along the surface of the Brillouin zone, from point H to point P, along the direction F in Fig. 10-3. The point P is the extremity of the direction Λ, or the 111 direction, and in the third diagram we come back from P to Γ along that direction, returning to the original values. Next we go outward again along the 110 direction, or Σ, to the point N, the extremity of the Brillouin zone in that direction; and finally from N to H, along G.

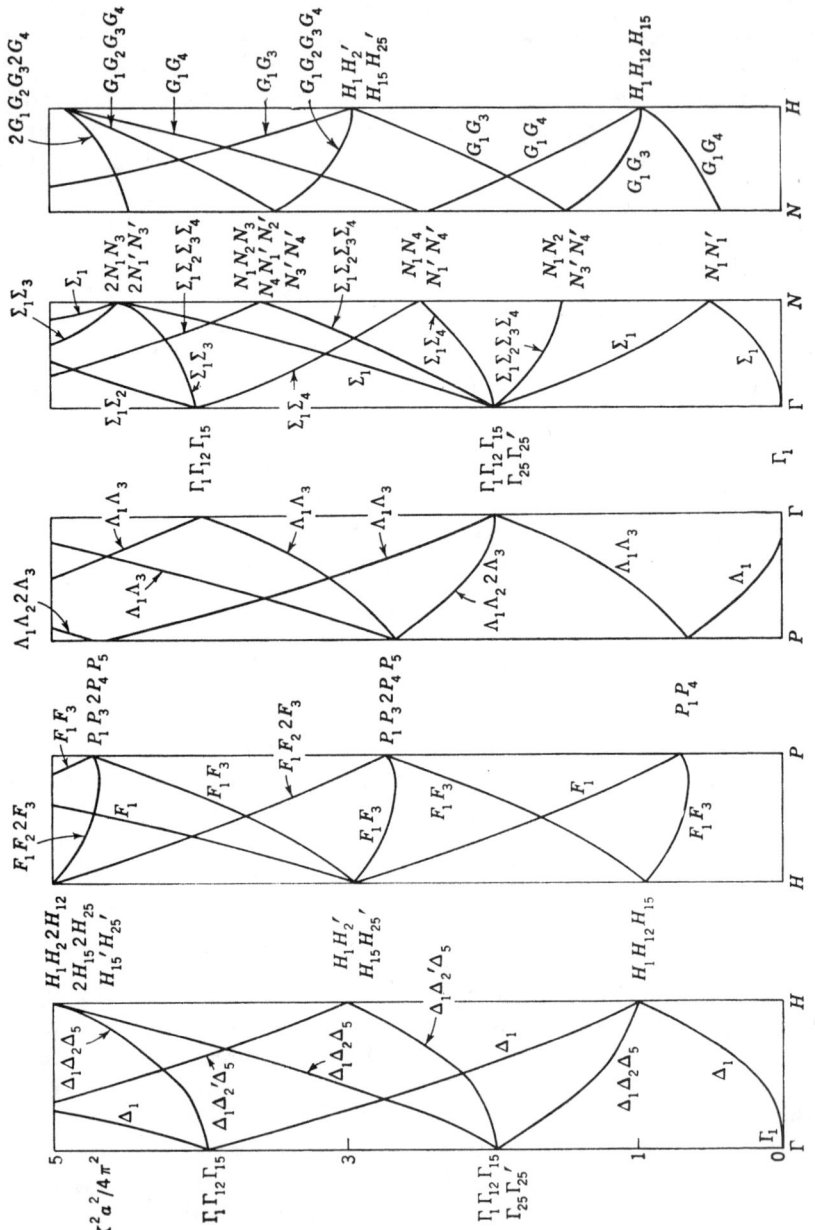

Fig. 10-1. Free-electron energy bands, body-centered cubic structure.

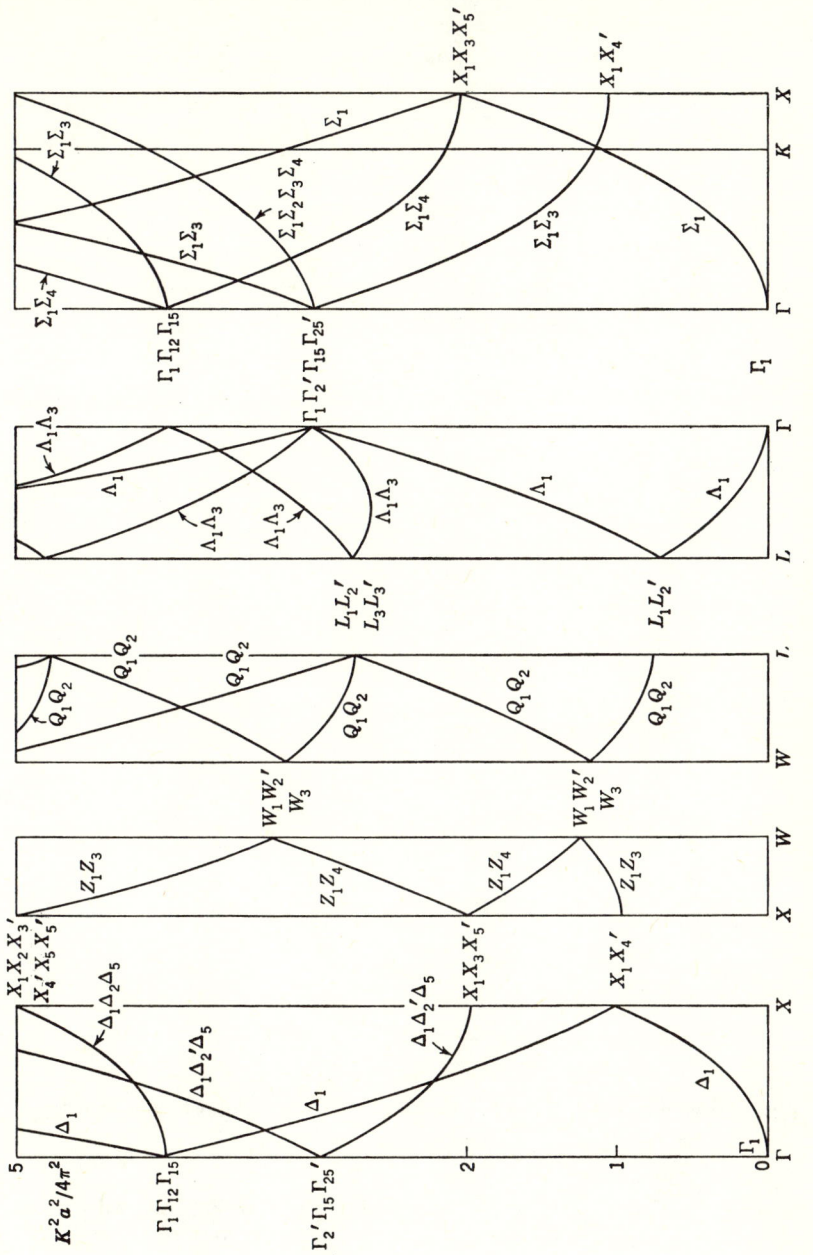

Fig. 10-2. Free-electron energy bands, face-centered cubic structure.

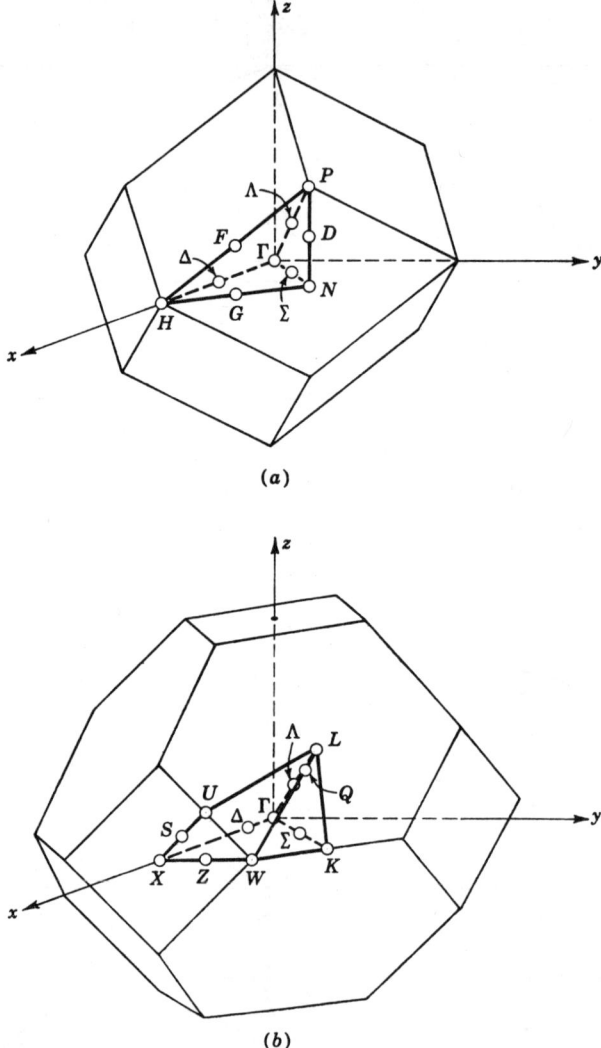

Fig. 10-3. Brillouin zones for the body-centered and face-centered cubic structures (a and b, respectively).

As we see from Fig. 10-3, this includes all the special directions and points of the Brillouin zone except for the line D, joining N and P.

In Fig. 10-2 we go through a similar cycle for the face-centered cubic case, again going out along the x axis to X, along the direction Z to W, along Q to L, the extremity of the 111 direction, and back along Λ to Γ;

then out Σ to its extremity K, and beyond this point to the equivalent of X. In Fig. 10-4 we show, by indicating a projection of several unit cells, how it can be that a prolongation of the line Σ can carry us to the point X. We see, then, that diagrams like those of Figs. 10-1 and 10-2 can give almost complete information about the energy along symmetry directions. For many crystals this is the only information we have, and even if we have further data, it is hard to show it graphically.

10-3. Energy Bands of Typical Elements: Monovalent, Divalent, and Trivalent. We are now ready to consider the actual energy bands of some typical elements and compounds. We shall make no distinction

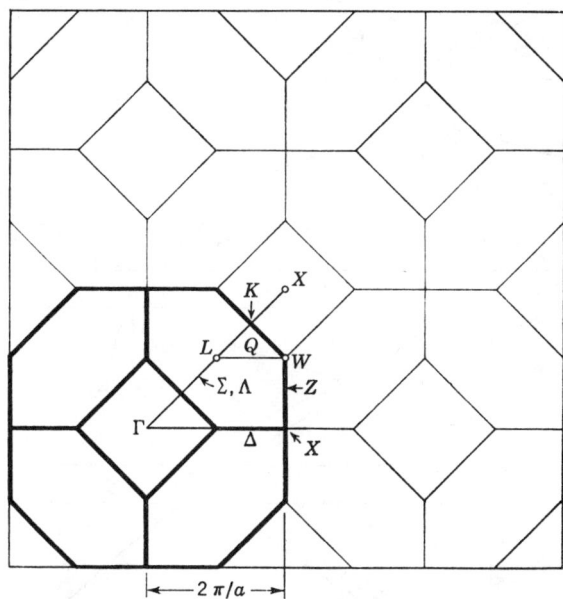

Fig. 10-4. Projection of several unit cells in xy plane, face-centered cubic structure, showing symmetry points and directions. Central Brillouin zone in lower left corner.

between the methods which have been used to calculate the energy bands; the results of different methods are so similar that the differences are not significant. Also we shall make no distinction in this first survey between different crystal structures; we shall see later that for the nonmetallic elements the crystal structure is very significant, but for the metallic elements we find similar results independent of the crystal structure. We shall rather arrange our discussion on the basis of the valence of the element. We start with elements not falling in the transition groups, and proceed from monovalent to divalent, trivalent, etc., elements. The bibliography at the end of this chapter is arranged in the same way.

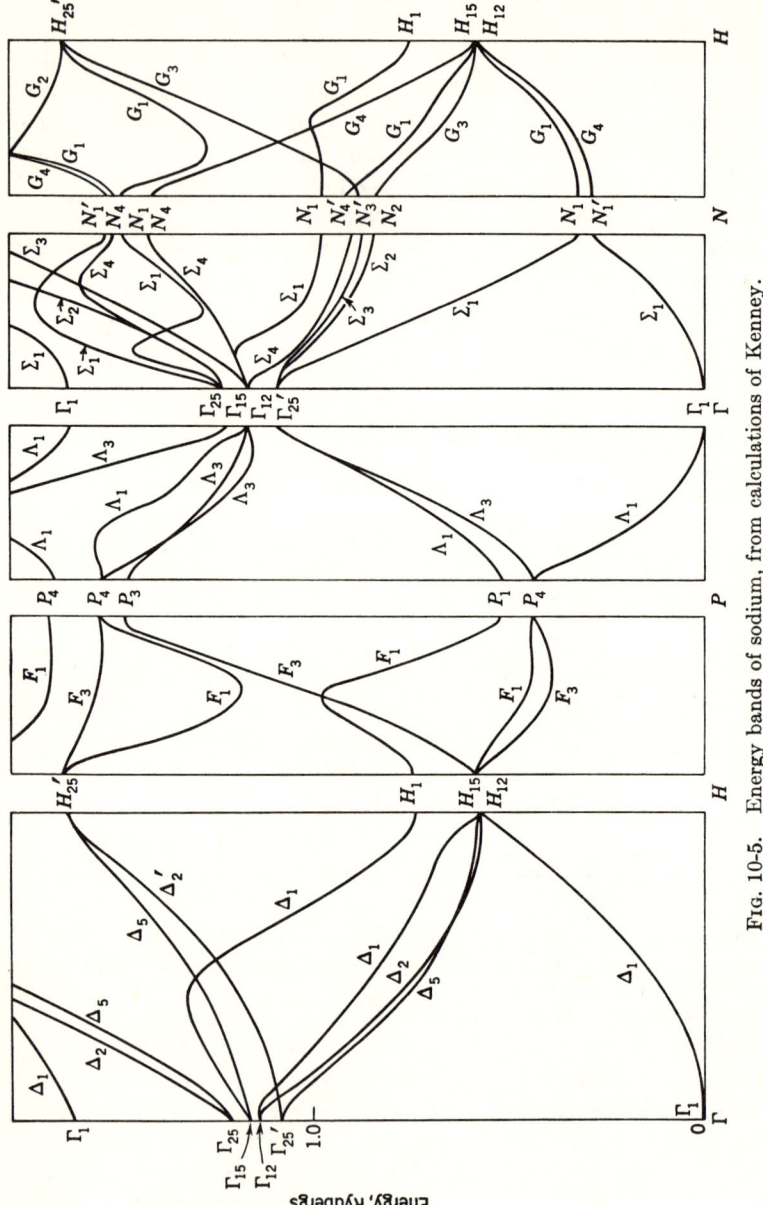

Fig. 10-5. Energy bands of sodium, from calculations of Kenney.

The alkali metals, though they were the first investigated, have not been as well worked out as some others, until very recently. We have used the early cellular work on sodium as an example for most of the discussion in Chap. 8. The most complete published calculation by modern methods is that of Ham[1] carried out by the Green's function method, for all the alkalies. Much more complete calculation by the APW method, not yet published, has been carried out by J. F. Kenney at M.I.T. In Fig. 10-5, which is identical with the earlier Fig. 8-4a, we give the energy bands of sodium as found by Kenney; they agree well with Ham. The energy, up to the Fermi energy, is so closely free-electron-like that the deviations would not show on a diagram on the scale of this figure. However, as we see by comparison with Fig. 10-1, there are considerable deviations from free-electron behavior as we get up to the top of the first band. We see that there is a considerable energy separation between states H_1, H_{12}, and H_{15}, which would coincide in the free-electron picture. Likewise, there is considerable separation between P_1 and P_4, though the separation between N_1 and N_1' is much smaller.

Ham has done one thing which few modern calculations have done: he has found energy bands as functions of internuclear separation, though over a rather limited range. In Fig. 10-6 we show a number of the features of the energy bands as functions of this distance. For comparison, we indicate on the same figure the bottom and top of the 3s and 3p bands, and the bottom of the 3d band, as found from the old cellular calculation shown in Fig. 8-3, which was carried out over wide ranges of internuclear distance. If the two calculations agreed exactly, the state Γ_1 would coincide with the bottom of the 3s band; they differ slightly, an indication of the inaccuracy of the cellular calculation. The state N_1', which has p-like characteristics, forms essentially the bottom of the p-like states, and we see that it falls close to the bottom of the 3p band in the cellular calculation. In Fig. 10-6 we indicate with each of these bands whether it is of s, p, or d character, as we can find from the tables of Appendix 3. It is interesting to see the s-like states tending downward more rapidly than the p-like states, with increasing interatomic distances, the atomic 3s state lying below the 3p. We see that the part of the figure occupied by both 3s and 3p bands according to the cellular calculation has in fact states of both sorts, according to the more accurate Green's function calculation. From this figure we get a clearer understanding of the significance of the energy bands shown in Fig. 8-3.

The other alkali metals are considerably less free-electron-like than sodium, according to all calculations. For comparison with sodium, we give in Fig. 10-7 the curves for lithium, potassium, rubidium, and cesium, as found by Kenney. For these elements, there are wide separations

[1] F. S. Ham, *Phys. Rev.*, **128**:82 (1962).

between the energies of states which would coincide in the free-electron case. Furthermore, even below the Fermi level the energy is less accurately given by the free-electron parabola than for sodium. It was an interesting accident of history that the first energy bands carefully calculated, those of sodium, were at the same time the most nearly free-electron-like of any which have been investigated. This undoubtedly had a great deal to do with the importance which the free-electron model

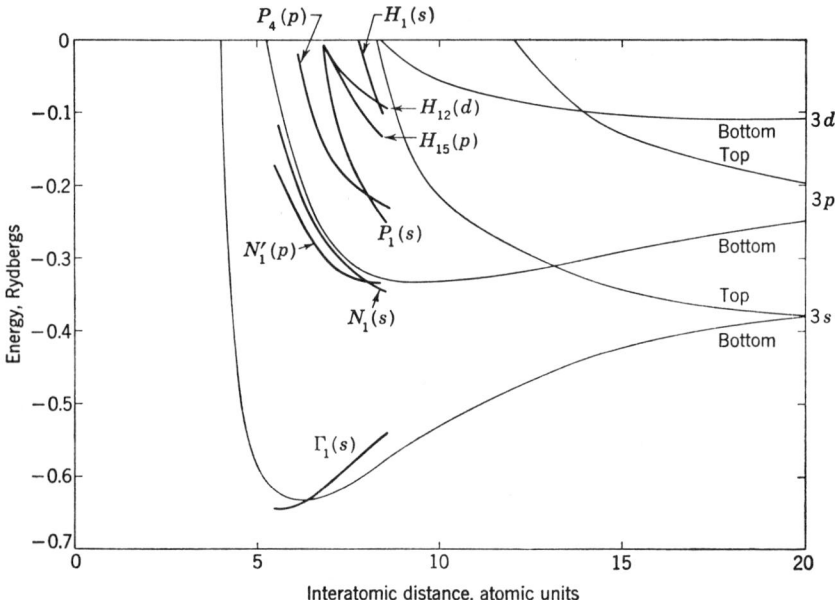

FIG. 10-6. Features of the sodium energy bands as functions of interatomic distance, according to Ham. Energy bands found by the cellular method, from Fig. 8-3, are shown for comparison. Symbols (s), (p), (d) indicate the type of atomic symmetry encountered in each case.

has had in the subsequent development of the theory, an importance which was perhaps greater than was strictly warranted.

Next we come to divalent elements, the alkaline earths. From Table 2-1 we see that here we have hexagonal structures as the only structures in beryllium and magnesium, and as alternative structures in calcium and strontium, the latter two having face-centered cubic structures as well. We do not have calculations of the face-centered cases for alkaline earths carried out by very modern and accurate methods; an early cellular calculation for calcium by Manning and Krutter[1] is probably not very

[1] M. F. Manning and H. M. Krutter, *Phys. Rev.*, **51**:761 (1937).

Sec. 10-3]   CALCULATIONS OF ENERGY BANDS IN CRYSTALS   261

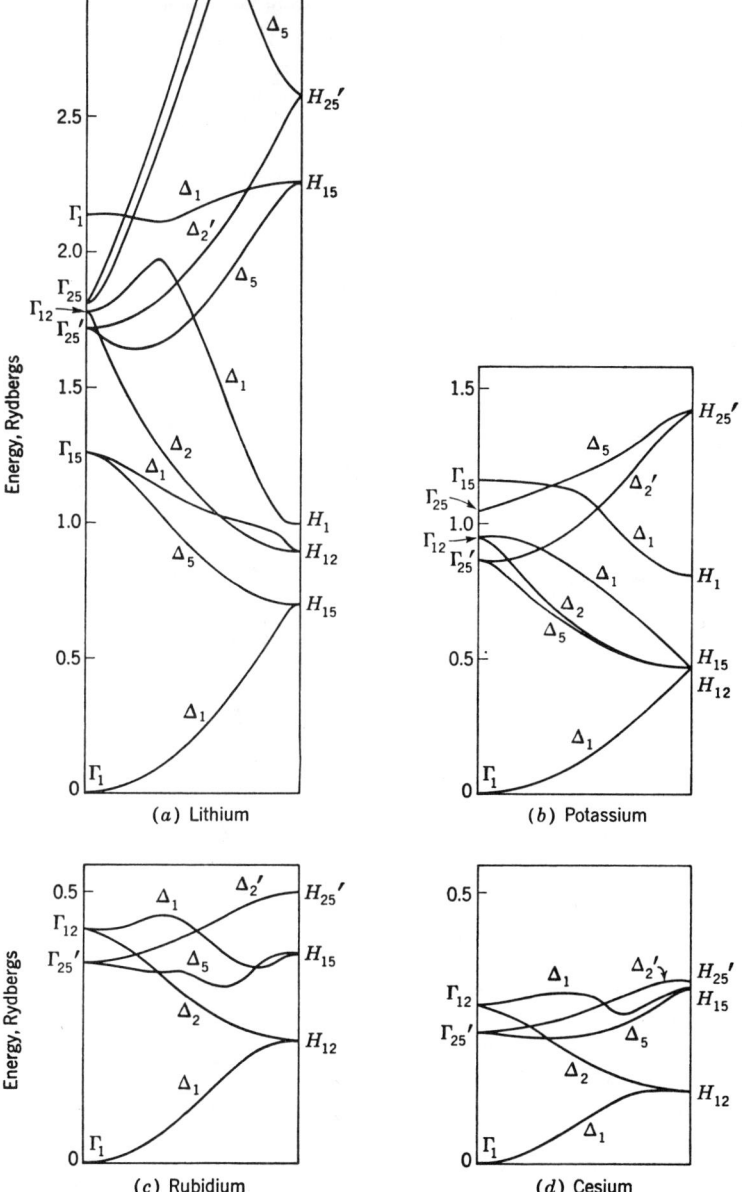

FIG. 10-7. Energy bands of (a) lithium, (b) potassium, (c) rubidium, and (d) cesium from calculations of Kenney.

accurate. For hexagonal beryllium, however, we have the calculation of Herring and Hill[1] by the orthogonalized-plane-wave method, the first application of that method, and a very carefully worked out case. We also have a very recent OPW calculation by Loucks and Cutler,[2] and unpublished calculations by J. R. Terrall, by the APW method, agreeing

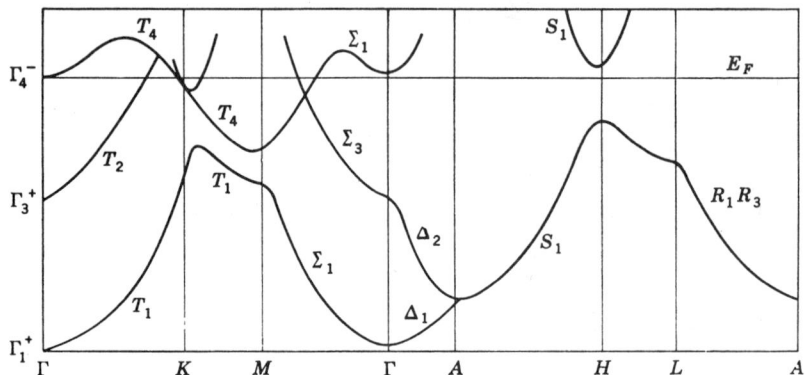

FIG. 10-8. Energy bands of beryllium, according to Herring and Hill.

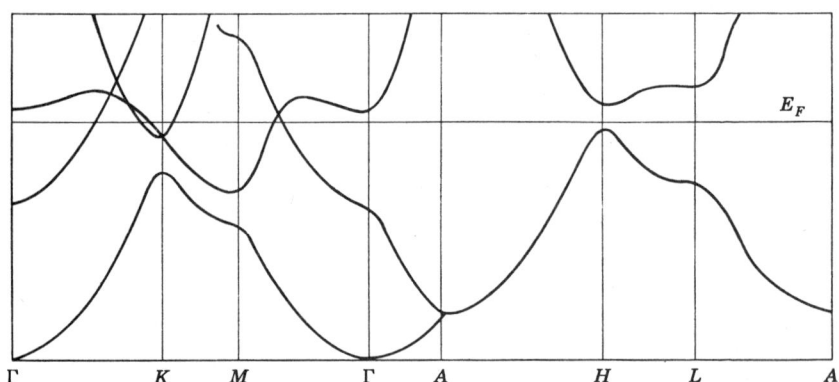

FIG. 10-9. Energy bands of beryllium, according to Loucks and Cutler.

closely with each other. In Fig. 10-8 we show the calculations of Herring and Hill, in Fig. 10-9 those of Loucks and Cutler, and in Fig. 10-10 the corresponding free-electron calculation for a hexagonal close-packed crystal, as computed by methods discussed in Appendix 7. The Brillouin zone and symmetry points are shown in Fig. 10-11, identical with Fig.

[1] C. Herring and A. G. Hill, *Phys. Rev.*, **58**:132 (1940).
[2] T. L. Loucks and P. H. Cutler, *Phys. Rev.*, **133**:A819 (1964).

A3-1. As one will see by careful examination of that figure, the sequence of points shown in Figs. 10-8, 10-9, and 10-10 is a rather intricate one on the surface of the Brillouin zone, chosen in such a way as to include all the symmetry directions for which calculations have been made.

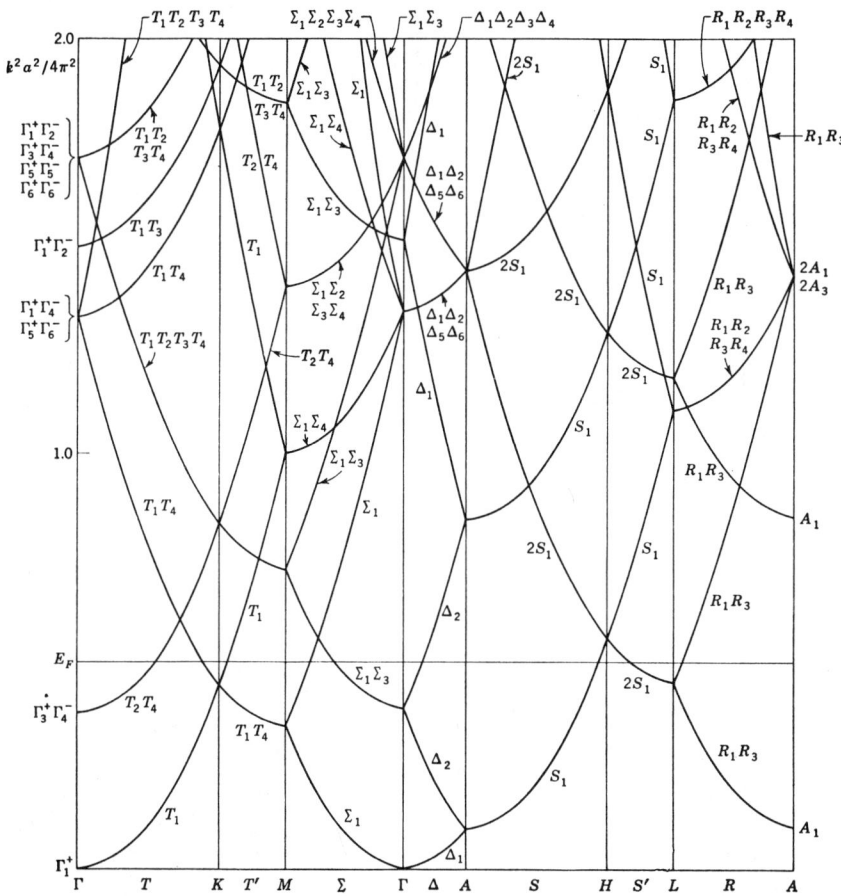

FIG. 10-10. Free-electron energy bands, hexagonal close-packed structure.

The striking feature of Figs. 10-8 to 10-10 is the close resemblance between the early calculations of Herring and Hill, which are unusually complete for energy-band calculations, the recent OPW calculation of Loucks and Cutler, and the free-electron case of Fig. 10-10. Both the calculations and the free-electron case are very complicated, and it should be obvious to the reader why it is very difficult to get any sensible interpretation of the calculations without the free-electron case for comparison.

The resemblance is obviously close enough to bear out our statement that the free-electron approximation forms a valid first approximation to the actual energy bands.

To get further understanding of the nature of the energy bands, we should interpret them in terms of the behavior of the wave functions around the atoms, the type of behavior which we should find if we set up the wave functions as Bloch sums of atomic orbitals. We have seen in Sec. 2-3 that the positions of the atoms in the hexagonal close-packed structure are special positions, with the symmetry D_{3h}. Hence the wave functions in the neighborhood of an atom must have the symmetry of one

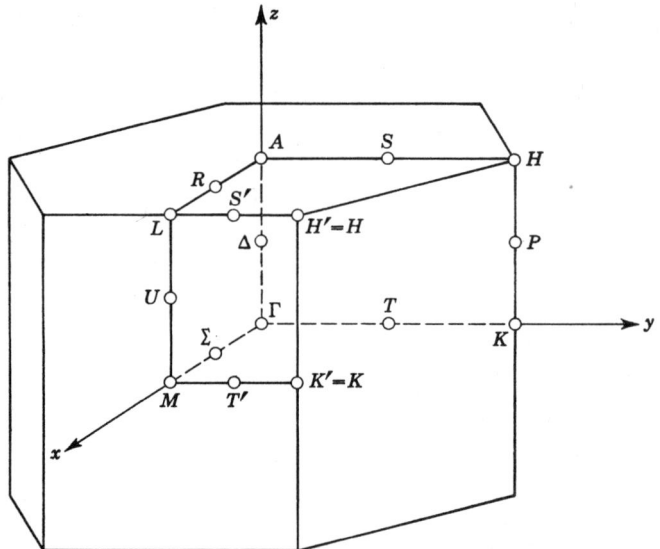

FIG. 10-11. Brillouin zone for the hexagonal Bravais lattice.

or another of the irreducible representations of this point group, for the states at Γ, in which we have periodicity from cell to cell. In Appendix 3 we show how to find the symmetry behavior of the individual irreducible representations of the space group at Γ, in terms of their behavior about the atom, and the relative behavior on the two atoms in a unit cell. We proceed essentially as we did in Sec. 5-5, where we discussed a simple two-dimensional example.

From Volume 1, Chap. 8 and Appendix 12, we know that there are six irreducible representations for the group D_{3h}. In the first place, for the group C_{3v}, we have two one-dimensional irreducible representations, which we denoted as Σ^+ and Σ^-, and a two-dimensional irreducible representation, which we denoted as Π. Simple basis functions for these representations, in polar coordinates, were a constant, sin 3ϕ, and the two functions cos ϕ, sin ϕ for the two-dimensional representation. In D_{3h}, each of these representations of C_{3v} goes into two, one even and the other odd in z, the direction along the axis of rotation. If we express these basis functions in rectangular coordinates, which is more convenient for the present purpose, the functions even in z are a constant, the function $y(3x^2 - y^2)$, and x, y, while those odd in z have an extra factor z. Thus an atomic s state goes into the Σ^+ even in z, while an atomic p is split into the Σ^+ odd in z (the p_z) and the Π even in z (the p_x, p_y).

We show in Sec. A3-1 how to find what behavior about an atom each of the irreducible representations at the point Γ has. In Table 10-1 we

Table 10-1

Irreducible representations at the point Γ, in hexagonal close-packed structure. Information taken from Table A3-2. Functions behave around each atom like the various irreducible representations of the point group D_{3h}, for operations about that atom as a center. There are two representations of each type, one bonding, the other antibonding, between the two atoms of the unit cell. Atomic states leading to the various representations are shown.

	Atomic state	Bonding	Antibonding
Even in z:			
Σ^+	s	Γ_1^+	Γ_4^-
Σ^-	f	Γ_2^+	Γ_3^-
Π	p_x, p_y	Γ_5^+	Γ_6^-
Odd in z:			
Σ^+	p_z	Γ_3^+	Γ_2^-
Σ^-	g	Γ_4^+	Γ_1^-
Π	d_{xz}, d_{yz}	Γ_6^+	Γ_5^-

give this information. For each of the six types of irreducible representation of D_{3h}, we have two irreducible representations at the point Γ for the hexagonal structure. One of these irreducible representations is bonding, while the other is antibonding, and this is indicated in the table.

We can now use the information in Table 10-1 to get further understanding about the nature of the energy bands in Fig. 10-8. The lowest level, of symmetry Γ_1^+, is a Bloch sum of s wave functions on all atoms, going to Σ^+ type in the D_{3h} symmetry. Since we are dealing with beryllium, we could make this up by superposing atomic $2s$ orbitals at all atomic sites, resulting in overlapping charge between the atoms, and bonding. The next-higher level which we find here, Γ_3^+, is a bonding combination of p_z orbitals on all atomic sites. That is, we have the p_z's on one layer of atoms with the $+$ part of the wave function pointing up, the $-$ part pointing down. On the next-higher layer, the signs are reversed, so that the + part of the wave functions on this next layer points down, and reinforces the + part of the functions from the next-lower layer. In other words, these are arranged so as to have overlap charge, or to produce bonding, just as the 2s orbitals are in the state Γ_1^+. From a molecular orbital point of view, it is this overlapping which produces the bonding holding the crystal together. The next-higher energy level at $\mathbf{k} = 0$, the Γ_4^-, can be regarded as a Bloch sum of 2s orbitals, with opposite signs on adjacent layers, so as to produce an antibonding behavior. These levels are not occupied; as we see from Figs. 10-8 and 10-9, they lie just above the Fermi surface. Higher Γ levels, not shown in the figure, would arise from p_x and p_y orbitals, largely nonbonding, and from antibonding combinations of the p_z orbitals, as well as from orbitals with higher principal quantum numbers.

From the lowest level Γ_1^+, we see in Fig. 10-8 that we can go along the direction T to the point K with a function of type T_1, that on the other hand we can go along the direction Σ to the point M with a function of type Σ_1, and that these levels at K and M can join. These directions indicate propagation in the basal plane of the hexagonal prism forming the Brillouin zone, shown in Fig. 10-11. The direction T is that pointing out to one of the corners of the hexagon, the direction Σ points to the center of a face, and the line from K to M is along the outer surface of the Brillouin zone. The symmetry of T_1 is that which would be produced from a mixture of atomic s orbitals, and p orbitals pointing in the direction of propagation. We have a similar situation for Σ_1. In other words, these have the same type of behavior which we described for the case of sodium and the body-centered cubic case in Sec. 8-2. From the next higher level, Γ_3^+, we have a level T_2, which is made up of p_z functions, mixed with d orbitals of a yz type, if the propagation is along the y axis. That is, the behavior as we proceed along the direction of propagation is like that for T_1, but we have a nodal plane in the basal plane, $z = 0$. On the other hand, the level T_4, arising from Γ_4^-, is made up as Γ_1^+ is from atomic s orbitals and from p orbitals like p_y pointing along the direction of propagation, but just as with the case Γ_4^-, we have opposite signs for the orbitals on the atoms in adjacent planes, or a nodal plane halfway between adjacent planes, and antibonding. The situation of these higher levels along the direction Σ is similar.

We can also go from $\mathbf{k} = 0$ along the z axis, the axis of the hexagon, or the direction Δ. Here the function Δ_1 is made up of a combination of s and p_z functions, the amplitudes and phases being determined by a Bloch factor $e^{i\mathbf{k}\cdot\mathbf{R}}$, where we are to use an appropriate value of R for each atom, no matter which of the two in the unit cell it is. The function Δ_2 differs from Δ_1 in having the opposite sign for the wave function at the second atom in the unit cell. At the boundary of the Brillouin zone, the point A, there is degeneracy between the functions Δ_1 and Δ_2. This arises because the phase factors multiplying the wave function at the two atoms in the unit cell, which are ± 1 for Δ_1 and Δ_2 at the center of the Brillouin zone, have changed to $\pm i$ at the edge of the zone, so that one wave function is the complex conjugate of the other, resulting in the degeneracy, and showing how there can be a continuous change in properties from the one symmetry type to the other. In the case shown, we have a continuous change in properties from the state Γ_1^+, an s-like state, up through Δ_1 and A_1 to Δ_2, and then to Γ_3^+, a p_z-like function. This is similar to the continuous change in properties from an s-like to a p-like band which we have commented on earlier in the case of sodium.

When we notice the position of the Fermi level in Fig. 10-8, we see that the energy bands lying below it are of the bonding variety, while those above it are antibonding, leading to the same qualitative explanation of the bonding in this crystal which we have found in studying diatomic molecules. In this case of the divalent elements, we have a particularly interesting situation regarding bonding and antibonding orbitals. If we had widely separated beryllium atoms, each with its two valence electrons, they would both be in $2s$ atomic orbitals, since this orbital lies considerably below the $2p$. These would be closed shells, and the atoms would repel, just as two helium atoms do. As the atoms approach, the $2s$ and $2p$ bands will broaden and eventually overlap, as the $3s$ and $3p$ bands did in sodium, illustrated in Fig. 8-3. It is only when they overlap that we can begin to get bonding. The $2s$ band itself consists of bonding and antibonding orbitals, and if they were both occupied, they would result in a net repulsion. When we have the overlapping of the bands, however, we can have hybridizing of s and p orbitals, forming several bands of bonding characteristics, and it is these bands, which naturally have the lower energy, which are shown as being occupied in Fig. 10-8.

Let us ask in terms of that figure just what changes would occur as the internuclear separation was increased, and as the $2s$ band became disentangled from the $2p$ band. The calculations have not been made as a function of internuclear distance, but we can easily figure out what would happen. If we had a band composed only of $2s$ atomic orbitals, its lower edge would come from the Γ_1^+ representation, consisting of a bonding combination of $2s$ orbitals, and its upper edge would come from the Γ_4^-, the antibonding combination of $2s$ orbitals. The lower edge of a band composed of $2p$ orbitals would be the Γ_3^+, the bonding combination of $2p_z$ orbitals, which are so oriented as to lead to bonding. The overlapping of the bands is seen in Fig. 10-8 in the fact that the Γ_3^+ lies below the Γ_4^-. However, as the internuclear separation was increased, these two would change places, the Γ_4^- moving down closer and closer to the Γ_1^+ as the separation became infinite. At the point where the Γ_3^+ and Γ_4^- changed places, the bands terminating in them would change their properties. The Δ_2 would terminate in Γ_4^-, with which it is compatible, rather than in Γ_3^+; this would then join to the Σ_1 as in Fig. 10-8, and the s-like band would lie entirely below the p-like band. At the same time, the T_2 band would move above the T_4. We should then have a complete disengagement of the two bands, and as the internuclear separation became infinite, the $2s$ and $2p$ bands would shrink to zero in width.

At the same time that these changes were occurring, there would be a change in the conductivity properties of the crystal. At the observed separation, as shown in Fig. 10-8, the Fermi level cuts a number of the energy bands, and we have each of these bands partly occupied with electrons, partly empty, leading to a conductor. The Fermi surface, the surface in the Brillouin zone on which the energy equaled the Fermi energy, would be a very complicated surface. On the other hand, as the internuclear separation increased enough so that the $2s$ and $2p$ bands became separated from each other, the $2s$ band would have just enough levels to accommodate two electrons per atom, as in the isolated atoms, so that we should have an insulator. We can see this from Fig. 10-8, in which we note that in this case there are two energy bands which would be occupied, the ones which we have been describing as the bonding and antibonding $2s$ bands. Each of these bands holds two electrons, one of each spin, per unit cell, so that we have four electrons per unit cell in the $2s$ bands. Since there are two atoms per unit cell, this corresponds to two electrons per atom.

We see from this discussion that a divalent element like beryllium has a special feature as far as its energy bands are concerned, in that its metallic properties depend on the overlapping between the s and the p bands. This overlapping, however, is not very great, as we see from Fig. 10-8. This fact is also brought out in the density-of-states curve, in which we plot the number of energy levels per unit energy range, as a function of the energy. The density of states of course goes to zero in the gaps between energy bands, and rises to a peak where the levels are close together. It is interesting to consider the nature of this curve for beryllium. It is shown in Fig. 10-12. We see that it falls to a sharp minimum very close to the Fermi level. This shows in a very graphic way how the overlapping of the $2s$ and $2p$ bands is not really very great at the equilibrium distance in the crystal. If the distance were somewhat greater, so that we had no overlapping of the bands, the density of states would fall to zero between the two bands. This deep minimum of the density-of-states curve for divalent metals at the Fermi level is not dependent on their having the hexagonal structure. In Fig. 10-12 we show also the curve for the face-centered cubic modification of calcium, as computed by Manning and Krutter in the calculation mentioned earlier. It has the same feature shown by hexagonal beryllium.

Next we come to the trivalent elements. A very careful study of face-centered aluminum has been recently made by Segall,[1] using the Green's function method. His energy bands are shown in Fig. 10-13. Earlier calculations of Heine[2] agree well with Segall's results. These energy bands

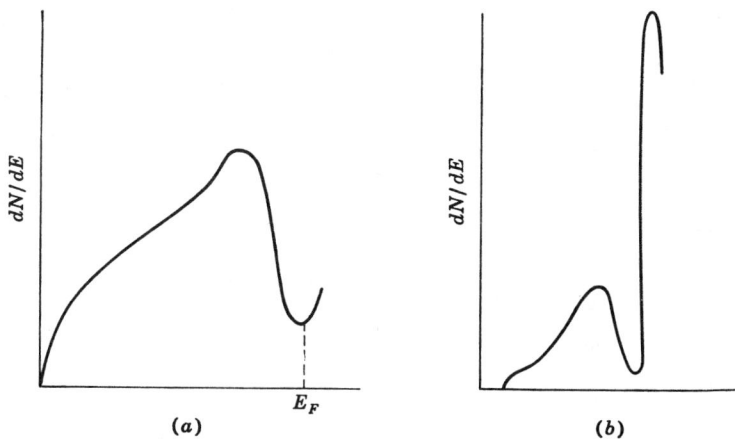

FIG. 10-12. Density of states for (a) beryllium, from the calculation of Herring and Hill, and for (b) calcium, from Manning and Krutter.

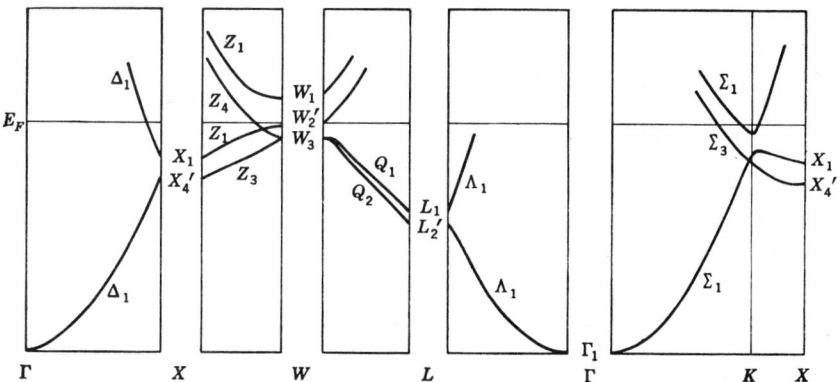

FIG. 10-13. Energy bands of aluminum, from calculations of Segall.

should be compared with the free-electron bands for the face-centered cubic structure shown in Fig. 10-2. We see a striking similarity between the two cases, verifying the free-electron-like nature of the energy bands of aluminum. Let us now consider the nature of the various energy bands, in

[1] B. Segall, *Phys. Rev.*, **124**:1797 (1961).
[2] V. Heine, *Proc. Roy. Soc. (London)*, **A240**:340, 354, 361 (1957).

terms of linear combinations of atomic orbitals, as we did with beryllium, so that we can understand the nature of the occupied energy levels.

First we start with propagation along the direction Δ, the 100 direction, or the x axis. The bottom of the band, the point of symmetry Γ_1 at $\mathbf{k} = 0$, is like a Bloch sum of $3s$ electrons on all atoms, and the band Δ_1 has wave functions made up of linear combinations of $3s$ and $3p$ orbitals, the latter pointing along the x axis or the direction of propagation, in the way with which we are now familiar. At the edge of the Brillouin zone, the point X, the representation Δ_1 is compatible with two irreducible representations, X_1 and X_4'. These differ in their behavior under the inversion, which is a member of the group of the wave vector at X, but not at a general point along the x axis; the inversion changes the wave vector $\mathbf{k}$ into $-\mathbf{k}$, and this changes it into an identical reduced wave vector at X, but not otherwise. The representation X_1 gives functions which are even on inversion, and therefore are of s-like type, while X_4' has basis functions which are odd on inversion, and therefore p-like. The calculations of Segall show that X_4', the p-like state, lies slightly lower than X_1, the s-like state; they would be coincident in the free-electron approximation.

From the state X_1, which lies slightly higher than X_4', we can proceed backward along the direction Δ, this time again with a function of symmetry type Δ_1, which is compatible with both X_1 and X_4'. Segall's calculations do not follow this state back to Γ, but from Fig. 10-2 we expect that, since bands of the same symmetry type cannot cross, the Δ_1 band will follow along to the crossover point between the parabola heading for the energy of four units at Γ in the figure, and the other parabola heading for the energy of three units. It will follow the latter parabola, and will terminate at point Γ in either a Γ_1 or a Γ_{15} state, whichever lies lower. As we see from Fig. 10-13, the Fermi level comes not very far above the points X_1 and X_4', so that the higher part of the level in question, and its behavior at Γ, would correspond to unoccupied parts of the band.

Next let us start again at Γ, and go along the direction Λ to L, the extremity of the band in the 111 direction. Here, as we have mentioned earlier in Sec. 8-2, we have the symmetry type Λ_1, made up as Δ_1 is from a linear combination of s and p orbitals, in this case $3s$ and $3p$. At the boundary of the Brillouin zone we have two states, L_1 and L_2', compatible with Λ_1, as we can show from Table A3-31, their relation being just like that of X_1 and X_4' above: L_1 has basis functions even on inversion, and which therefore can be constructed from $3s$ orbitals, while L_2' has basis functions odd on inversion, which can be built up from $3p$ functions. Here again, as in the 100 direction, the calculations of Segall indicate that the p-like functions have a slightly lower energy. Here again, also, there is another Λ_1 level arising from the L_1 point, rising as we go back along the Λ direction, and also terminating presumably in either the state Γ_1 or Γ_{15}.

Next we may consider the joining of the points X and L, through the path X-Z-W-Q-L. We see from Fig. 10-3 that X is at the center of the square face of the Brillouin zone, L at the center of the hexagonal face, W is a common corner of a square and a hexagonal face, and Z and Q are straight lines in the outer surface of the zone connecting X and W, W and L, respectively. The method of joining is indicated in Fig. 10-13, on the basis of Segall's calculations. To indicate the type of basis functions, in the form of linear combinations of atomic orbitals, we give in Fig. 10-14 a concise description of the basis functions for each of the symmetry types, arranged schematically to agree with the energy levels as given in Fig 10-13. We see that at the point W we have an s-like state, W_1, which again lies slightly above the p-like states W'_2 and W_3, which are split on account of symmetry into the nondegenerate W'_2 and the doubly degenerate W_3.

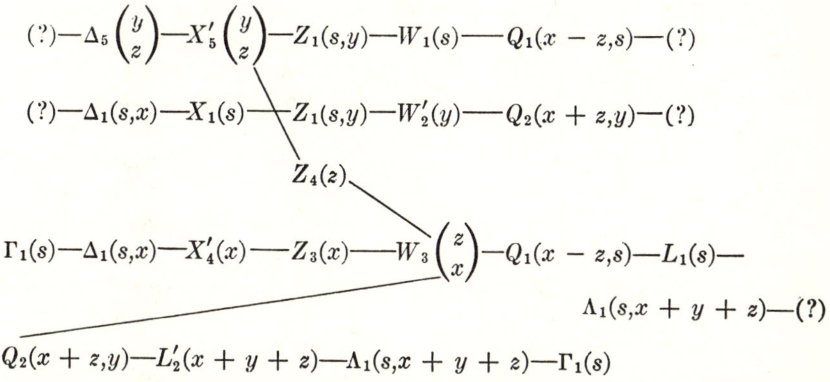

$$(?)-\Delta_5 \binom{y}{z}-X'_5 \binom{y}{z}-Z_1(s,y)-W_1(s)-Q_1(x-z,s)-(?)$$

$$(?)-\Delta_1(s,x)-X_1(s)-Z_1(s,y)-W'_2(y)-Q_2(x+z,y)-(?)$$

$$Z_4(z)$$

$$\Gamma_1(s)-\Delta_1(s,x)-X'_4(x)-Z_3(x)-W_3\binom{z}{x}-Q_1(x-z,s)-L_1(s)-$$

$$\Lambda_1(s,x+y+z)-(?)$$

$$Q_2(x+z,y)-L'_2(x+y+z)-\Lambda_1(s,x+y+z)-\Gamma_1(s)$$

FIG. 10-14. Lower sequences of symmetry states for aluminum, as given in Fig. 10-13, together with symbols such as s, x, y, etc., indicating type of atomic orbital which must be combined on the various lattice sites to produce the wave function of the appropriate symmetry.

This discussion is enough to give the general nature of the wave functions and their symmetry properties. As for the energy levels, Segall finds values very closely agreeing with the free-electron values. The free-electron parabolas reproduce almost exactly the lowest Δ_1, Λ_1, and Σ_1 levels, and come between the Z_1 and Z_3, the Q_1 and Q_2, and the states between K and X terminating in X_1 and X'_4. Segall finds that matrix elements of a nearly free electron model can be chosen which very accurately reproduce the curves of Fig. 10-13, as a result of solving a secular equation describing interaction of plane waves with almost degenerate energies.

We may lastly consider the position of the Fermi level. For aluminum we must accommodate three electrons per atom, so that the Fermi level will lie higher than in the monovalent and divalent elements. As we see from Fig. 10-13, the computed Fermi level lies high enough so that the lowest energy band is almost completely occupied with two electrons per atom, and in addition we have enough electrons in the second energy band to account for the remaining electron. In the case of aluminum, unlike the divalent elements, there is no striking change in conductivity properties as the internuclear separation increases; at large distances, the lowest band, having s-like behavior, would become separated from the next lowest, having p-like behavior. The lowest band would hold two electrons per atom, the next lowest one, and consequently there would be conductivity arising from this next-lowest band.

10-4. Energy Bands of Tetravalent Elements and 3-5 Compounds.
The elements carbon, silicon, germanium, and tin in its so-called gray modification have the diamond structure, and are insulators or semiconductors, in total contrast to the elements which we have just been taking up, which are metals. We shall now consider these elements, using germanium as our main example, and shall also consider the 3-5 compounds, such as GaAs, which have the similar zinc-blende structure. Let us first take up the free-electron case for these crystals. We recall that as far as the lattice is concerned, they correspond to a face-centered cubic lattice, but the point group is different from that for the face-centered cubic structure. We have the same Brillouin zone as for the face-centered cubic structure, and can use the same plane waves as for the face-centered cubic case. However, on account of the nonprimitive translations associated with some of the operators in the diamond structure, the effect of operating on the plane waves with the various operators is not the same as with the face-centered cubic structure. Hence we do not find in every case the same symmetries associated with each of the energy bands which we have for the face-centered cubic structure, as shown in Fig. 10-2. It is no more difficult to find the symmetry types than for the face-centered cubic crystal; we must use the information about the symmetry operations for the diamond structure, as found in Sec. A3-2, and use the methods of Appendix 7. When we carry through this discussion, we find that the free-electron energy bands have just the same appearance as in Fig. 10-2, but the designations of the energy bands are as given in Fig. 10-15.

We now give, in Fig. 10-16, a set of energy bands for germanium, as given by Herman.[1] These are partly a result of calculation, by the orthogonalized-plane-wave method, and partly a result of experiment; more extensive experimental investigation of the energy bands has been given for germanium than for any other element. The calculations, however, are not as complete as for aluminum, and some of the curves are estimated. Energies at the point Γ calculated by Segall[2] by the Green's function method are in agreement with Herman. More recent OPW cal-

---
[1] F. Herman, *Phys. Rev.*, **93**:1214 (1954); *Physica*, **20**:801 (1954); *J. Electronics*, **1**:103 (1955).
[2] B. Segall, *J. Phys. Chem. Solids*, **8**:371, 379 (1959).

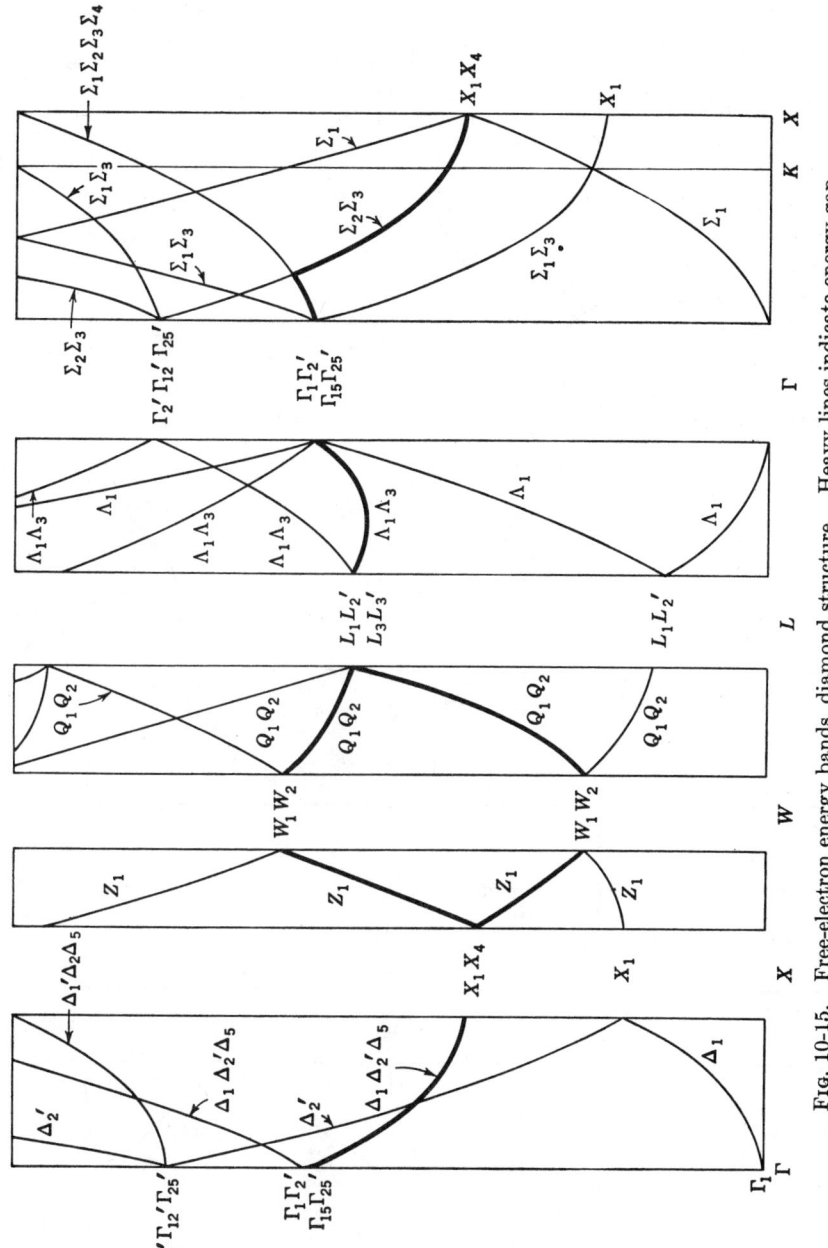

Fig. 10-15. Free-electron energy bands, diamond structure. Heavy lines indicate energy gap.

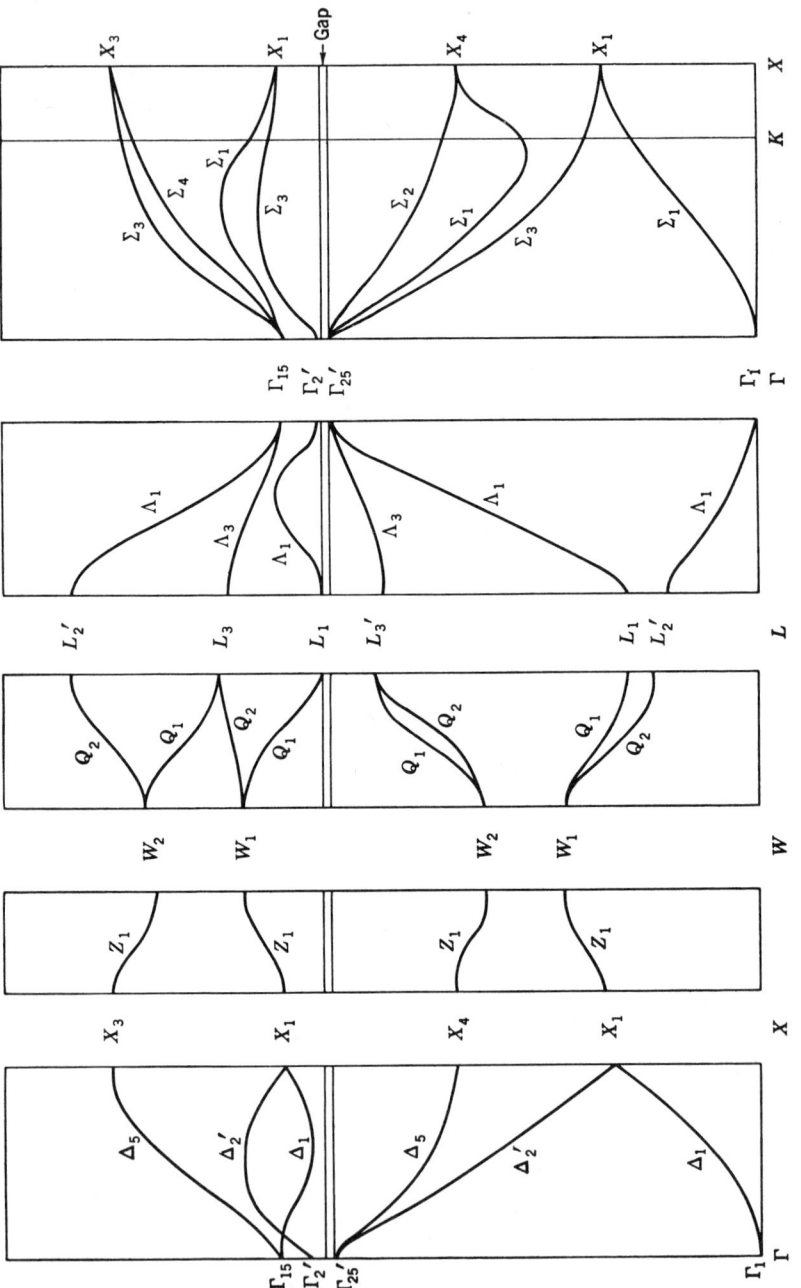

Fig. 10-16. Energy bands of germanium, as estimated by Herman.

culations by Bassani and Yoshimine,[1] less complete, agree in most respects with Herman's estimates, though there are some significant discrepancies.

In Fig. 10-16, there is a region, bounded on the bottom by the highest occupied energy band, and on the top by the lowest empty energy band, representing the energy gap, within which the Fermi level must lie. We can verify that this gap comes at the right place. We know that we must have four outer electrons per atom, or eight per unit cell, which contains two atoms in the diamond structure. When we recall the degeneracy of the various bands, we see that in each case we have four occupied energy bands, each holding two electrons per unit cell. For instance, along the direction Δ, we have states Δ_1, Δ_2', each nondegenerate, and Δ_5, doubly degenerate, among the occupied levels.

The semiconducting properties of germanium arise because we have these four lower bands completely filled, then a gap, then the upper bands completely empty. For this to occur, we must have the situation that the lowest point of the conduction band, which comes at the level L_1, must lie above the highest point of the valence band, which comes at Γ_{25}'. This is the case, but the rather small energy difference between these two energy values, which gives the band gap, is much smaller than the energy separation between the valence and conduction bands in some other directions, as for instance W. For help in visualizing the position of the gap, we have followed Herman, and have indicated the position of the top of the valence band and the bottom of the conduction band in the free-electron case of Fig. 10-15. For the directions Z and Q, with W between them, there would be a band gap in the free-electron case, explaining therefore the wide gap found in the actual case of germanium at W. Along the other directions, however, there would be degeneracy in the free-electron case, which is removed, and the bands separate to form a gap, in the case of germanium. We shall be able to understand this separation better when we investigate the nature of the wave functions in the various bands.

The wave function of type Γ_1, at the bottom of the valence band, can be considered to be made up of a Bloch sum of atomic s functions on all atoms of the crystal, all with the same coefficients. The function Γ_2', like Γ_1, is made up of atomic s functions, but here the signs on the two atoms in the unit cell are opposite, so that the wave function is of an antibonding rather than a bonding type. The functions Γ_{25}' and Γ_{15} are triply degenerate, and can be considered to be made up of p_x, p_y, or p_z functions on the atoms. As with Γ_1 and Γ_2', the difference between the two types of function comes in the signs of the atomic functions. In Γ_{25}' the functions on the two atoms in the unit cell have opposite signs, so that their wave functions, being of the p type, add in the region of overlapping between the two atoms, leading to a bonding type of interaction, while in Γ_{15} the signs are the same, there

[1] F. Bassani and M. Yoshimine, *Phys. Rev.*, **130**:20 (1963).

are nodes of the wave function between the neighboring atoms, and we have antibonding. We can now see the feature which pushes the energy bands apart, which would be degenerate in the free-electron case, and which thus produces the energy gap: it is the bonding nature of the lower energy bands, the valence bands, and the antibonding nature of the higher, conduction bands.

For propagation along the 100 or x direction, the function of symmetry Δ_1 is made up on one set of atoms of a Bloch sum of s and p_x functions (also, more generally, of d functions behaving like yz or like $3x^2 - r^2$). On the other set of atoms, we have a similar Bloch sum, the phases being determined by a Bloch factor $e^{i\mathbf{k}\cdot\mathbf{R}}$, where $\mathbf{R}$ is the radius vector to the atom in question (rather than merely to the corner of the unit cell), and $\mathbf{k}$ is the reduced wave vector. The function of symmetry Δ_2' is made up of the same functions on the atoms, but with an additional change of sign between the two atoms in the unit cell. There is an interesting degeneracy between the states of these two types at the edge of the unit cell, the point X, similar to that found for the hexagonal close-packed case at point A, which is commented on in Sec. 10-3. This arises because the ratio of the wave functions on the two atoms in unit cell is in one case i, in the other case $-i$, at this point, so that one wave function is the complex conjugate of the other, and they are necessarily degenerate. The function of symmetry Δ_5 is twofold degenerate, one of the functions being a linear combination of p functions like $p_y + p_z$ on one set of atoms with a similar combination of p functions on the other set, while the other function is an identical combination of functions $p_y - p_z$ on the two sets of atoms. For propagation along the 111 direction, Λ_1 is a combination of an s function and p's pointing along the 111 direction, while Λ_3 is a twofold degenerate set of functions, made up from two types of p functions at right angles to the 111 direction, and to each other. In other directions we have much the same situation as in those we have discussed, the valence-band wave functions being made up of s and p functions, in essentially bonding combinations, while the conduction-band wave functions are also made of s and p functions, but in antibonding combinations.

An interesting feature of the energy bands of the tetravalent elements appears when we consider the dependence of the energy bands on interatomic distance. Here the situation in a sense is the opposite of what we have found for the divalent elements. At small distances, the elements with the diamond structure are semiconductors or insulators, but at larger distances they would become conductors. The reason is that with four electrons per atom, two of these could be accommodated in a band arising from the atomic s electrons, leaving two more which would partially fill the band arising from the atomic p electrons. Only when the two bands overlap, as they do at the equilibrium interatomic distance, does the situation change so that we have just enough valence-band levels to accommodate the electrons, with a gap above.

No calculations of the dependence of the germanium bands on interatomic distance have been made, but the properties which we have just been discussing are illustrated by early cellular calculations of diamond, by Kimball, and by Hund and Mrowka.[1] They have given curves showing the energy bands of diamond as functions of the interatomic distance, shown in Fig. 10-17. The shaded areas of this figure show energy levels, the unshaded areas being gaps. We observe that at the point where the $2s$ and $2p$ bands touch, there is a change of properties, a gap appearing above a lower band which now becomes capable of holding four electrons per atom, rather than two per atom as in the $2s$ band at large internuclear

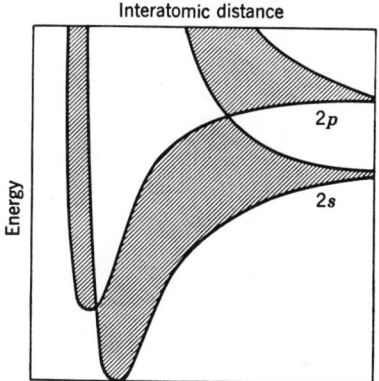

Fig. 10-17. Boundaries of energy bands of diamond, as functions of interatomic distance, from Kimball.

distances. The detailed nature of the energy bands as given by these earlier calculations is somewhat different from that found by more recent and accurate work, but the general situation illustrated seems to be that actually found in diamond-type crystals.

We can now ask in terms of the energy bands of Fig. 10-16 just what changes are expected to occur as a function of interatomic distances, to lead to the results of Fig. 10-17. At large distances, the states Γ_1 and Γ'_2 will lie below Γ'_{25} and Γ_{15}, since the s state from which Γ_1 and Γ'_2 are formed will lie below the atomic p state from which Γ'_{25} and Γ_{15} are formed, and the bands are only slightly broadened. Then at large distances there will be a band of character Δ_1 along the 100 direction, changing to Δ'_2 at the

[1] G. E. Kimball, *J. Chem. Phys.*, **3**:560 (1935). F. Hund and B. Mrowka, *Sächsische Akad. Wiss. Leipzig*, **87**:185 (1935).

edge of the band, and joining the state Γ_2' at the center of the zone, as indicated schematically in Fig. 10-18. This band would hold two electrons per atom. This will lie entirely below the higher band, of p character, as shown in Fig. 10-18. As the bands broaden, the level Γ_2' will be pushed above Γ_{25}', and the levels will join as indicated in Fig. 10-16, so that we shall get the situation found in germanium. With still broader bands, the level Γ_2' can rise above the Γ_{15}, and this seems to be the situation actually found in diamond and silicon, though not in germanium. It makes changes in the arrangements of the conduction bands, but does not change the fundamental properties of the system of energy bands.

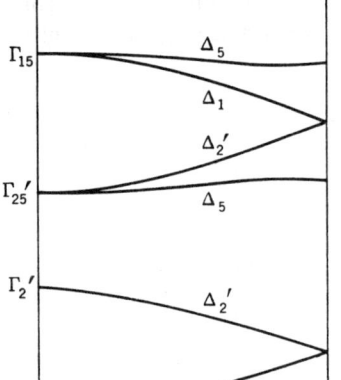

Fig. 10-18. Energy as a function of k, along the 100 direction, for diamond, at large internuclear distance (schematic).

An important feature of the structure of germanium and the other similar elements is that it is a covalently bonded substance, unlike all others which we have been discussing. It should be possible, then, to describe the four tetrahedral bonds connecting each atom with its neighbors in the same sort of language which we have used for polyatomic molecules in Volume 1. This can be done, and we take it up in Appendix 8. In place of the energy-band wave functions, which have the same nature as molecular orbitals in a molecular problem, we can set up linear combinations of the valence-band functions which have the properties of the equivalent orbitals, in the language of Lennard-Jones and his colleagues, as was discussed in Volume 1. One of these equivalent orbitals will be located in each bond, and will be a symmetrical function, on account of the fact that the two atoms being bonded together are identical. Similarly, one can make up antibonding equivalent orbitals, antisymmetric functions, formed in a similar way from antibonding conduction-band functions. From these bonding and antibonding equivalent orbitals, one can set up linear combinations A and B, like those discussed in Volume 1, Secs. 10-5 and 10-6, and can use these to set up a valence-bond function, treating it by the method of Hurley, Lennard-Jones, and Pople. An essentially equivalent treatment of the binding energy of diamond, by Schmid,[1] is described in Appendix 8, and illustrates the method which could be used in handling the binding energy of a covalently bonded

[1] L. A. Schmid, *Phys. Rev.*, **92**:1373 (1953).

crystal by the valence-bond method. Similar methods have been applied to the 3-5 compounds, by Coulson, Redei, and Stocker.[1]

The 3-5 compounds belong to the zinc-blende rather than to the diamond structure. This, however, has no profound effect on their energy bands. The free-electron approximation is not altered, except in the notations for the various bands, which we shall not go into in detail. Herman[2] has made estimates of the energy bands of the 3-5 compounds, based on such theoretical and experimental information as was available at the time, and more recently Bassani and Yoshimine[3] have made OPW

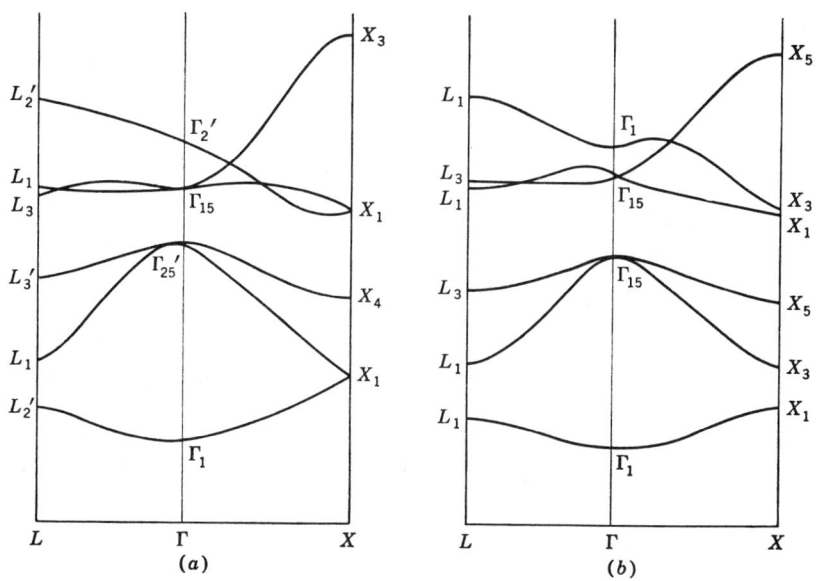

FIG. 10-19. Energy bands of Ge compared with GaAs, from Bassani and Yoshimine. (a) Ge. (b) GaAs.

calculations of a number of these compounds, as well as of tetravalent elements, agreeing in general with the estimates of Herman. In Fig. 10-19 we show the energy bands along the 100 and 111 directions, for GaAs, as compared with Ge, according to the calculations of Bassani and Yoshimine. We see that there is no striking difference in the energy bands in the two cases. Semiempirical studies have been made of a number of 3-5 compounds, and also of alloys of two tetravalent elements, as for instance

[1] C. A. Coulson, L. B. Redei, and D. Stocker, *Proc. Roy. Soc. (London)*, **A270**:357 (1962). L. B. Redei, *Proc. Roy. Soc. (London)*, **A270**:373, 383 (1962). D. Stocker, *Proc. Roy. Soc. (London)*, **A270**:397 (1962).

[2] F. Herman, *J. Electronics*, **1**:103 (1955).

[3] F. Bassani and M. Yoshimine, *Phys. Rev.*, **130**:20 (1963).

mixtures of silicon and germanium. There have been a good many detailed studies of the relative heights of the various bands, as determined by experimental methods, and of the effect of chemical composition on these heights. We shall not go into these studies, but shall refer the reader to the original papers for further details.

There is one feature of the energy bands of the crystals with diamond and zinc-blende structure (and with other structures as well) which is of considerable practical as well as theoretical importance, and which has been studied in a great deal of detail. This is the effect of spin-orbit interaction on the energy levels. We know from atomic theory that, for instance, in an atomic p state, for a single-electron atom, the degeneracy of the state will be removed by the spin-orbit interaction. This results in two energy levels, $^2P_{1/2}$ and $^2P_{3/2}$, of which the first is doubly degenerate (on account of the two orientations of the angular-momentum vector $\mathbf{J} = 1/2$ in space), while the second is fourfold degenerate (on account of the four orientations of the vector $\mathbf{J} = 3/2$). A similar situation occurs in crystals, at the symmetry points and in their neighborhood. For instance, the level Γ'_{25} of germanium, at the top of the valence band, would be threefold degenerate according to our treatment, which neglects spin. If we included spin, instead of having sixfold degeneracy, it is found that there is a splitting as in the atomic case, into a twofold and a fourfold degenerate state. This splitting is important near the symmetry point, resulting in quite different detailed relations of the energy bands, but becomes of less importance as we go away from points of special symmetry. The general question of spin-orbit interaction, and the so-called double groups to which it leads, is taken up in Appendix 9.

10-5. Energy Bands of Transition Elements. Next we shall consider the energy bands of transition elements of the iron group. These show resemblances, and also great differences, as compared with those which we have taken up so far. For a first survey of the group, we show in Fig. 10-20 a set of calculations made by Mattheiss[1] for elements from argon to zinc, using the APW method. For purposes of comparison, we give each set of energy bands only for one direction of propagation in each case. Later we shall take up some of the cases in more detail.

The characteristic of this series of elements of course is the presence of the d electrons. At the beginning of the iron group the $3d$ electrons have orbitals outside the atom, and their energy levels are empty in the crystal or atom. This is shown in argon, the first element shown in Fig. 10-20. In this case the energy bands shown, arising from the atomic $4s$, $4p$, and $3d$ atomic levels, are entirely excited levels. The occupied band, the valence band, lies far below the levels shown. This is indicated in Fig.

[1] L. F. Mattheiss, *Bull. Am. Phys. Soc.*, ser. II, **8**:222 (1963); *Phys. Rev.*, **133**:A1399 (1964).

10-21, in which Mattheiss has calculated the valence band as well as the conduction band of argon, again by the APW method. The valence band is made up out of 3s and 3p atomic levels. The wide gap between valence

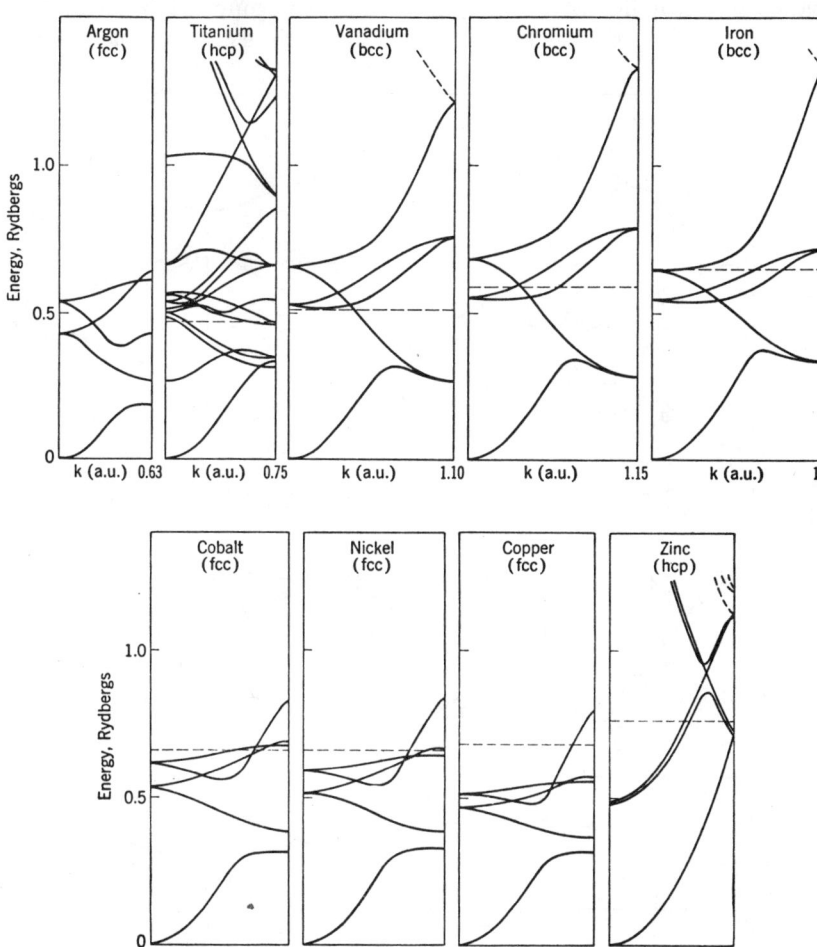

FIG. 10-20. Energy bands of 3d transition elements, along a single direction, from Mattheiss.

and conduction bands in this case leads to the fact that solid argon is an insulator.

As we go through the series of transition elements, we see from Fig. 10-20 that the levels become filled up, as the Fermi level moves higher and higher with addition of electrons to the atoms, so that the 3d-like bands, which appear midway up the band formed from 4s and 4p, become filled up, being entirely filled in copper. The wave functions corresponding to these bands at the same time are becoming more concentrated, and the 3d bands are becoming narrower. As we go beyond copper, to zinc, gallium, and germanium, the 3d band moves below the 4s. This has already

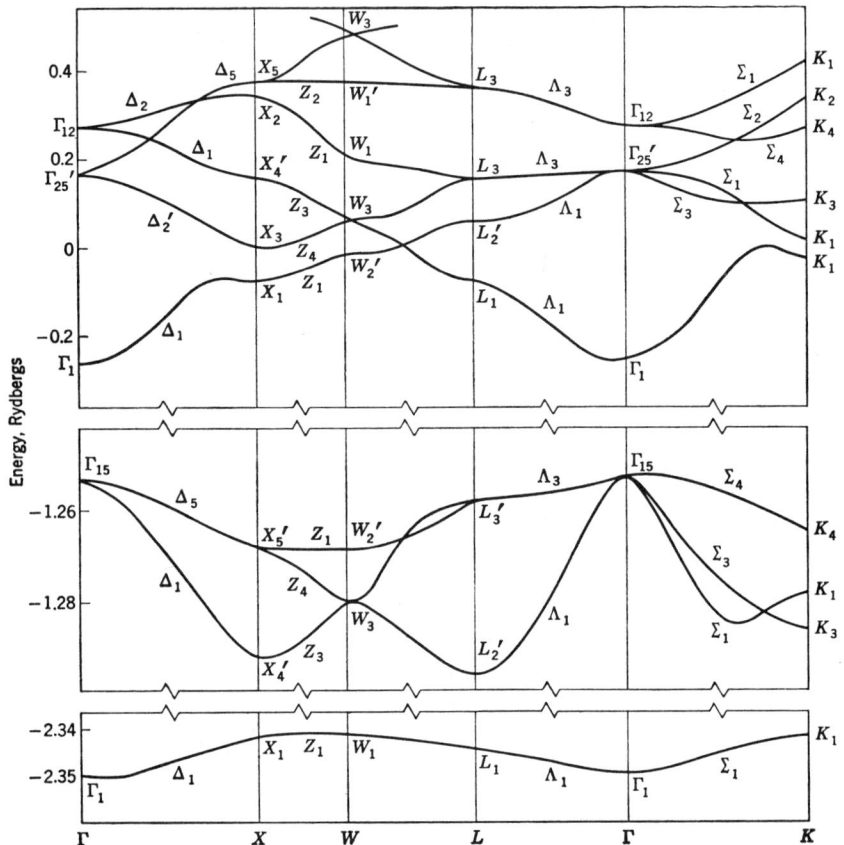

FIG. 10-21. Energy bands of argon, as calculated by Mattheiss. Note energy gaps and changes of scale.

occurred in zinc, in which the 3d band does not appear in Fig. 10-20. We shall give energy bands for gallium in Sec. 10-6, and there too the 3d bands lie below the valence band. In germanium this is also the case, and in Fig. 10-16, showing the energy bands in germanium, the 3d band is far below the bottom of the figure. As we go along in the periodic table beyond these elements, the 3d band goes down to form an x-ray level, its breadth decreases practically to zero, and the wave functions are located close to the nucleus. In contrast to these cases, it is the partially filled d bands which are responsible for the interesting features of the transition elements, from scandium to nickel.

Let us now start a more detailed discussion with copper, in which the Fermi level has already moved above the top of the $3d$ bands. Here we show the energy bands in Fig. 10-22, showing calculations by the APW method by Burdick,[1] and by the Green's function method by Segall.[2] This is the case mentioned earlier in Sec. 9-5, in which the excellent agreement between the calculations has convinced us of the accuracy of these energy-band calculations.

The crystal is face-centered cubic, so that we may compare the bands with the free-electron case of Fig. 10-2, and with the case of aluminum, as given in Fig. 10-13. We see a profound difference between the two cases. Slightly above the bottom of the valence band, at Γ_1, we find levels Γ_{12} and Γ'_{25}, the characteristic d-band wave functions, which in the free-electron case and in aluminum are found far above the bottom of the band. We have a collection of bands associated with these limiting values, concentrated within a width of two- or three-tenths of a Rydberg, and constituting essentially a d band capable of holding 10 electrons per atom. Rising above these bands, we have a Δ_1 state leading to X'_4, from there through a state Z_3 to W_3, down by a state Q_2 to L'_2, and thence to Λ_1; and a state Σ_1 leading up to K_1 and X_1, with K_3 and X'_4 in the neighborhood. These states have a striking resemblance to the bands found around the Fermi level in aluminum, as shown in Fig. 10-13. It is rather as if we had the d band inserted into the middle of the free-electron-like set of bands. The Fermi level has a height such that the d bands are entirely occupied, the one remaining electron being in the free-electron-like band, similar to an alkali.

One notices that following out of the lowest level, Γ_1, there is a state Δ_1 in the 100 direction, and a state Λ_1 in the 111 direction, as there would be for aluminum. Instead of rising to the X'_4 and L'_2 states, as they would in aluminum, they change their character, and terminate in states X_1 and L_1 instead; along these directions we have wave functions which start as s-like functions at $\mathbf{k} = 0$, but turn into d-like functions by the time we get to the edges of the Brillouin zone. Other levels Δ_1 and Λ_1 arising from Γ_{12} and Γ'_{25}, respectively, take their place, and go up to the free-electron-like higher levels of the energy-band scheme. This phenomenon is sometimes described by saying that the levels Δ_1 and Λ_1, which for the free-electron case would extend right through the region occupied by the d band, become perturbed, since two levels of the same symmetry type cannot cross each

[1] G. A. Burdick, *Phys. Rev.*, **129**:138 (1963).
[2] B. Segall, *Phys. Rev.*, **125**:109 (1962).

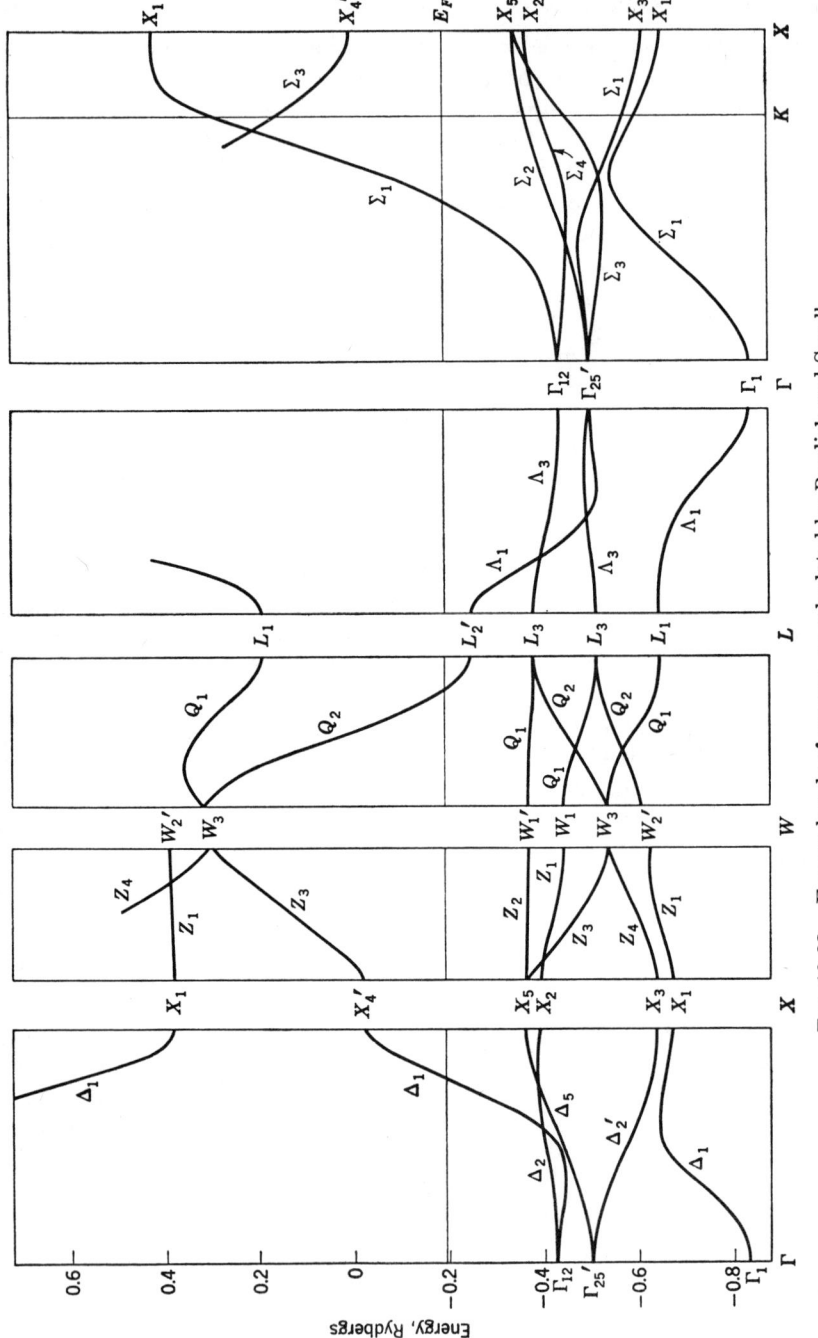

FIG. 10-22. Energy bands of copper, as calculated by Burdick and Segall.

other. We may regard the true analogue of the state Δ_1 forming the lowest energy band of aluminum in the 100 direction as consisting partly of the lower part of the lowest Δ_1 state of copper, and the upper part of the next higher Δ_1, with similar results in the other directions.

Since the Fermi level for copper lies well above the d band, the electrons at the Fermi surface are rather free-electron-like, though there are appreciable deviations from the free-electron behavior. For instance, if we had the free-electron case, we should find the levels X'_4 and X_1, which come at energies of about 0 and 0.4 Rydberg, coinciding, and similarly the levels L'_2 and L_1, at about -0.2 and 0.2 Rydberg, would coincide. Their wide separation shows the considerable deviation from the free-electron case. But nevertheless this deviation is small compared with the profound modification of the energy levels arising from the d band. The conduction electron of copper, in other words, is rather like that in an alkali metal, though considerably more modified by the deviations from the free-electron model, but slightly below the Fermi surface we have the d band.

As we go back to lighter elements than copper, the Fermi level, instead of lying above the d band as in copper, will lie within it. This is shown in Fig. 10-23, where we show the energy bands for face-centered and body-centered cubic iron, as calculated by Wood[1] by the APW method. We see that for face-centered cubic iron (not the usual modification at room temperature) the d-band structure is very similar to that of copper, but it is somewhat broadened, and the Fermi level comes toward the upper part of the d band. The electrons around the Fermi surface, then, have a very different behavior from free electrons. The body-centered cubic modification of iron, also shown in Fig. 10-23, though different in details from the face-centered cubic case, has a strong general resemblance to it. It is the body-centered cubic case that is shown in Fig. 10-20.

There are several important consequences of the partial occupation of the d band in such an element as iron. First, on account of the deviation of the conduction electrons from a free-electron-like behavior, the conductivity is much less than for copper or for an akali metal. Second, the partial occupation of the d band results in the magnetic behavior of these elements. We know that if we have an individual atom with some of its d states empty, we shall have a multiplet structure, with orbital and spin angular momentum. The situation becomes much more complicated in the metallic crystal, but nevertheless it is this same effect which results in the magnetic moments of the atoms in a ferromagnetic metal. These magnetic effects will be treated in a later volume of this series. Third, the upper wave functions for the d band, as for any band, tend to be antibonding, while the lower ones are bonding. Since the upper ones are empty, the lower ones occupied, in the crystal, this means that there is

[1] J. H. Wood, *Phys. Rev.*, **126**:517 (1962).

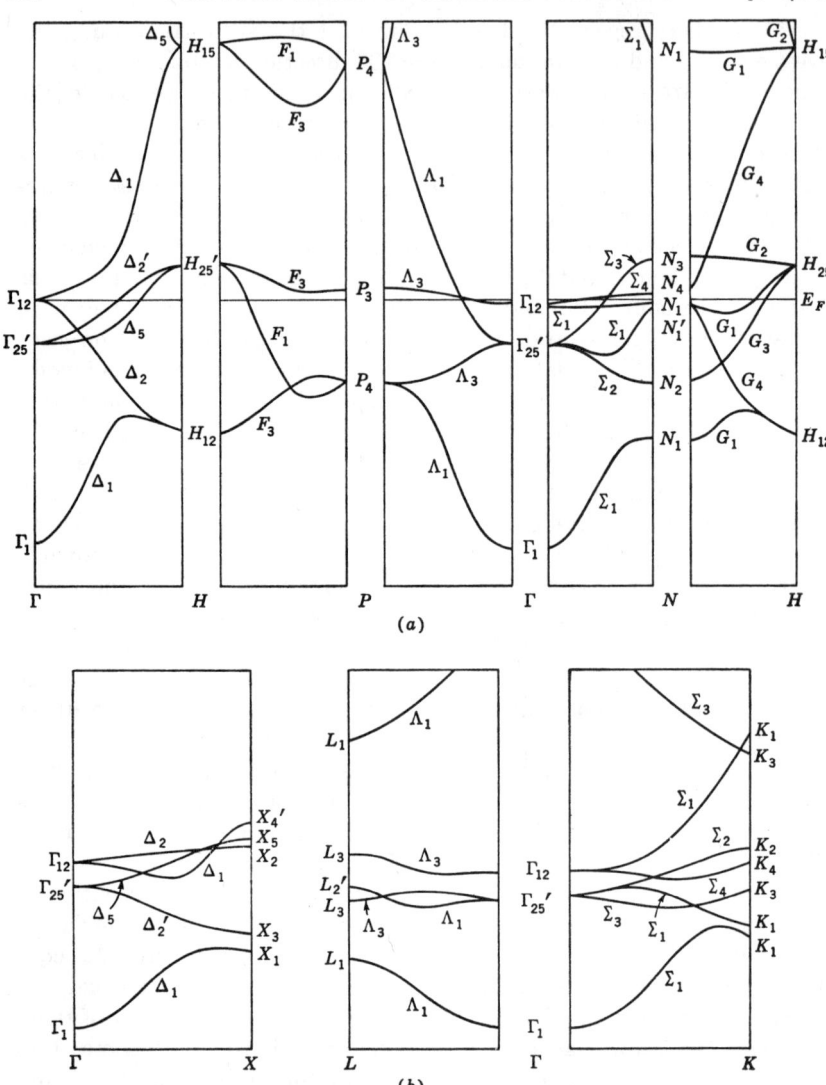

FIG. 10-23. Energy bands of body-centered and face-centered cubic iron, as calculated by Wood. (a) Body-centered. (b) Face-centered.

net binding arising from the d electrons, as well as from the s electrons, and this adds to the strength of binding in the transition elements.

The existence of the d bands superposed on the ordinary free-electron-like bands of the transition elements has a great effect on the density-of-states curve. We have enough energy levels concentrated within a narrow energy range to accommodate 10 electrons per atom, resulting then in a very high peak in the density-of-states curve. In Fig. 10-24 we show the density-of-states curve for body-centered iron, as computed by Wood from the calculations represented in Fig. 10-23, including also calculations in nonsymmetry directions, which are not shown in Fig. 10-23. We not only see the high peak corresponding to the d band, but also a good deal of structure in the peak. In particular, we have a specially high density of states toward the top of the d band, then a minimum, then a broader but lower maximum of density further down in the band. This general character has been found in all density-of-states curves which have been computed for transition elements, and it has an important bearing on a number of properties of these elements, particularly the magnetic behavior.

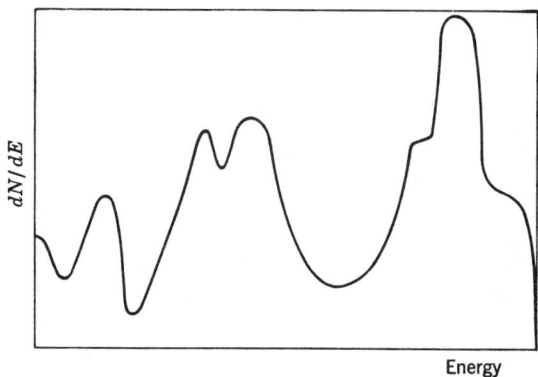

FIG. 10-24. Density of states for body-centered cubic iron, as calculated by Wood.

A number of writers have suggested that the d band should really be double, consisting of an upper part capable of holding four electrons per atom, and a lower, broader part holding six electrons per atom. This suggestion arose from the assumption that the bands arising from the state Γ_{12} would not become mixed with those arising from Γ'_{25}. Examination of Fig. 10-23 shows that this assumption is not justified. However, the density of states from Fig. 10-24 is not very different from what we should have if the band were split into two parts, and this is all that is necessary to explain the experimental phenomena.

10-6. Energy Bands of Elements with Special Crystal Structures. We do not propose to go in detail into the work that has been done on the structure of the energy bands for the elements with peculiar crystal structures. We shall, however, indicate enough of the results to give some idea as to the type of energy bands arising in these cases. The cases which have been investigated are the pentavalent element bismuth; the hexavalent elements selenium and tellurium; the graphite form of carbon; and one or two others. These all form special cases, though they have some features of resemblance. We shall start with some discussion of the case of graphite.

The graphite crystal structure consists of sheets of carbon atoms arranged at the corners of an array of regular hexagons, as we know from Sec. 2-5, illustrated in Fig. 2-8. The successive sheets are so far apart that they are held together only by Van der Waals forces. As a result of this large distance between sheets, one can get a fairly adequate idea of some of the properties of the energy bands of graphite by considering only a single sheet, forming a two-dimensional problem. Studies of the two-dimensional energy bands have been made by a number of writers, listed in the bibliography at the end of the chapter, including Wallace, Coulson and his students, Corbato, and a number of others, almost all of whom have used the Bloch or tight-binding method. The symmetry of the two-dimensional case is different from the real three-dimensional crystal, as we can see from Fig. 2-8. The two types of atoms in the basal plane are equivalent in the two-dimensional case, but not in the three-dimensional case, where half the atoms in one plane have neighbors in the next plane above and below, while the others do not. We shall first consider the two-dimensional case, but shall not go into the details of its symmetry properties and irreducible representations. We shall merely indicate the type of wave functions concerned in the various low-lying energy bands, using for this purpose the calculations of Corbato.[1]

The two-dimensional Brillouin zone is a hexagon, like the basal plane of the three-dimensional Brillouin zone of Fig. 10-11. As in that case, we shall designate the origin, $\mathbf{k} = 0$, as Γ; the center of one of the faces of the hexagon as M; and a corner as K. We now show, in Fig. 10-25, the energy bands around the circuit Γ-K-M-Γ, as found by Corbato. We notice in the first place that the bands are denoted as σ or π bands. This is in accordance with the notation used for plane molecules, in which the wave functions even in z, where z is the coordinate normal to the plane, are denoted as σ, and those odd in z are denoted as π. In the case of graphite, where the atomic orbitals (aside from the $1s$ orbitals) are the $2s$, $2p_x$, $2p_y$, $2p_z$, we see that only the $2p_z$ is of the π type, all others being of the σ type. We have energy bands of both types in both the valence and conduction bands.

The lowest band at point Γ is made up, according to the tight-binding scheme, which Corbato used, as a linear combination of atomic $2s$ orbitals on all atoms, which on account of overlap gives a bonding molecular orbital. As we go along the direction either toward K or toward M, we

[1] F. Corbato, Ph.D. thesis, Massachusetts Institute of Technology, Cambridge, Mass., p. 195; *Proc.* 1957 *Carbon Conference*, Pergamon Press, New York, 1957, p. 173.

have a mixture of $2s$ and the suitable combination of $2p_x$ and $2p_y$ to point along the propagation direction. At K a degeneracy is demanded by the symmetry, and as we go back along the upper of the two degenerate bands toward the point Γ again, the wave function takes on more and more of the p type, until at Γ it joins to a twofold degenerate representation, the two basis functions being $2p_x$ and $2p_y$, or linear combinations of these. These basis functions have opposite sign on adjacent atoms, so that the $2p$ functions will overlap and produce bonding orbitals.

The remaining band coming from this twofold degenerate Γ state follows along much the same value, toward the point K, as the one we have

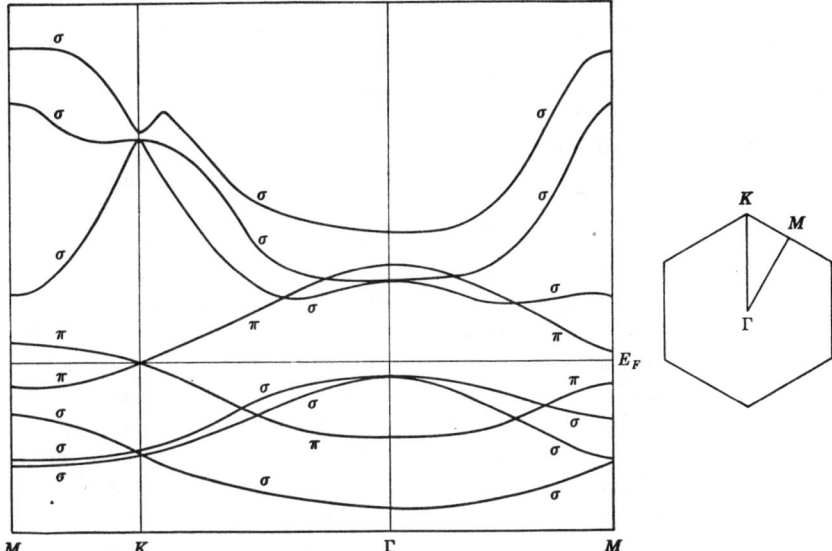

FIG. 10-25. Energy bands of graphite, as calculated by Corbato (two-dimensional model).

already discussed. This behavior of bands is quite similar to that of the lowest bands of germanium, Δ_1, Δ_2', and Δ_5, which we have shown in Fig. 10-16, except that in place of the doubly degenerate Δ_5, we have only a single level. The third p state at $\mathbf{k} = 0$ is displaced from the p_x and p_y states, since z, the direction normal to the plane, is not equivalent to x and y as in a cubic crystal. This third p function forms an additional Γ state, of the π type, between the two we have described. The energy band arising from it rises as we go to K, has a twofold degenerate representation there, and then as we go back to Γ, the second π band continues to rise, above the energy gap.

Corresponding to each of the bonding types of energy bands, which we have described, there is an antibonding type, lying above the energy gap. At $\mathbf{k} = 0$, the antibonding s-type wave function has opposite signs on adjacent atoms, and the antibonding p-type functions for the σ bands have the same signs on adjacent atoms, while the p_z functions for the π band have opposite signs on adjacent atoms for the antibonding wave function. These antibonding energy bands lie above the energy gap, except for the π case, in which an account of the degenerate representation at the point K, the gap shrinks to zero. There are just enough energy bands below the gap to accommodate all the electrons, four per atom (aside from the $1s$ electrons). Thus, we have three σ bands and one π band, each holding two electrons per unit cell, or eight electrons per unit cell in all, which gives four per atom, since there are two atoms per unit cell. We see, then, that if graphite consisted of completely separated two-dimensional sheets, there would be a gap of zero width, at the point K, between the valence band and the conduction band.

When the three-dimensional nature of the crystal is considered, several changes result. First, each of the energy bands found in the two-dimensional case becomes doubled, with a separation between the two resulting bands which is very small, since the layers are far apart. The two bands result from the two possible behaviors of the wave function when we go from one layer of atoms to the other. Just as in the hexagonal close-packed structure, there is a symmetry operation in the graphite structure carrying an atom of one layer into an atom of the other, consisting of a nonprimitive translation plus a 60° rotation. The wave function will be either even or odd when this operation is performed, a distinction which would have no effect on the energy if the planes were infinitely far apart, but which produces a separation of the energy levels if they are close enough together to affect each other. This does not change the general nature of the bands, as shown in Fig. 10-25. The doubling of the number of bands is what is needed to accommodate twice as many electrons per unit cell, since the unit cell in the three-dimensional problem holds four atoms, while that in the two-dimensional case holds only two.

The only place where there is appreciable change on account of going to three dimensions is in the neighborhood of the point K. Here it is found that there is quite an involved modification of the simple case of Fig. 10-25, the net result being that graphite behaves definitely like a metal, rather than like the peculiar intermediate case found in the two-dimensional approximation. There is one band which has a few electrons in it, another one with a few vacancies. We refer the reader to the original papers for the discussion of this complicated behavior.[1]

We have mentioned that in addition to graphite, some work has been

---

[1] See F. Herman, *Rev. Mod. Phys.*, **30**:102 (1958), as well as papers on graphite listed in the bibliography at the end of this chapter.

done on the energy bands of bismuth, sometimes called a semimetal, and on selenium and tellurium. We shall not discuss the cases of selenium and tellurium; some discussion of these elements is given in the paper of Herman just quoted. As for bismuth, there is an interesting paper by Jones,[1] one of the early ones on the band theory of metals, in which he set up the free-electron approximation for bismuth, and investigated where the Fermi level would come, and the nature of the Fermi surface. He found that there would be a situation something like that which we have just discussed for graphite: there would be one band, almost filled, and another one almost empty, so that one would contain a few vacancies, the other a few electrons. There have been more recent calculations, tight-binding calculations by Morita, Mase, and Behrens, and a pseudopotential calculation by Harrison, all listed in the bibliography at the end of the chapter, which lead to the same general type of result, though they do not agree in details. It would seem that more refined calculations are needed for this element. There are many peculiar properties which bismuth possesses, more than any other element in the periodic table, which would be explained in a general way by a structure of this type. The existing theories, however, are far from sufficient to describe the many types of experimental data available.

The only element with a complicated crystal structure for which we possess an adequate treatment of the energy bands is gallium, which has been carried out by Wood[2] using the APW method. We give the calculated energy bands in Fig. 10-26. This crystal has four atoms per unit cell, more than in any of the other cases we have so far discussed, with the result that there are a great many energy bands. As we see from the figure, the Fermi energy comes near the top of some energy bands, near the bottom of others, with the result that here again, as in graphite and bismuth, some energy bands will be almost filled, others almost empty. So far, there has not been enough comparison of the calculation with the results of experiment, which are very complicated, to make sure whether the theory agrees with experiment or not. In one qualitative respect there is agreement, namely, in the complicated nature of the bands. It is, however, interesting to see from this calculation that the APW method can be used for a crystal as complicated as this, and that it converges with adequate rapidity to a reliable answer.

10-7. Energy Bands for Alkali Halide Crystals and Other Compounds. The only binary compounds which we have mentioned so far in our discussion of energy bands are the 3-5 compounds, in which the energy bands resemble closely those of the tetravalent element which stands between the elements of valences 3 and 5. In Volume 1, Chap. 7, we have pointed

[1] H. Jones, *Proc. Roy. Soc. (London)*, **A147**:396 (1934).
[2] J. H. Wood, *Bull. Am. Phys. Soc.*, ser. II, **8**:222 (1963).

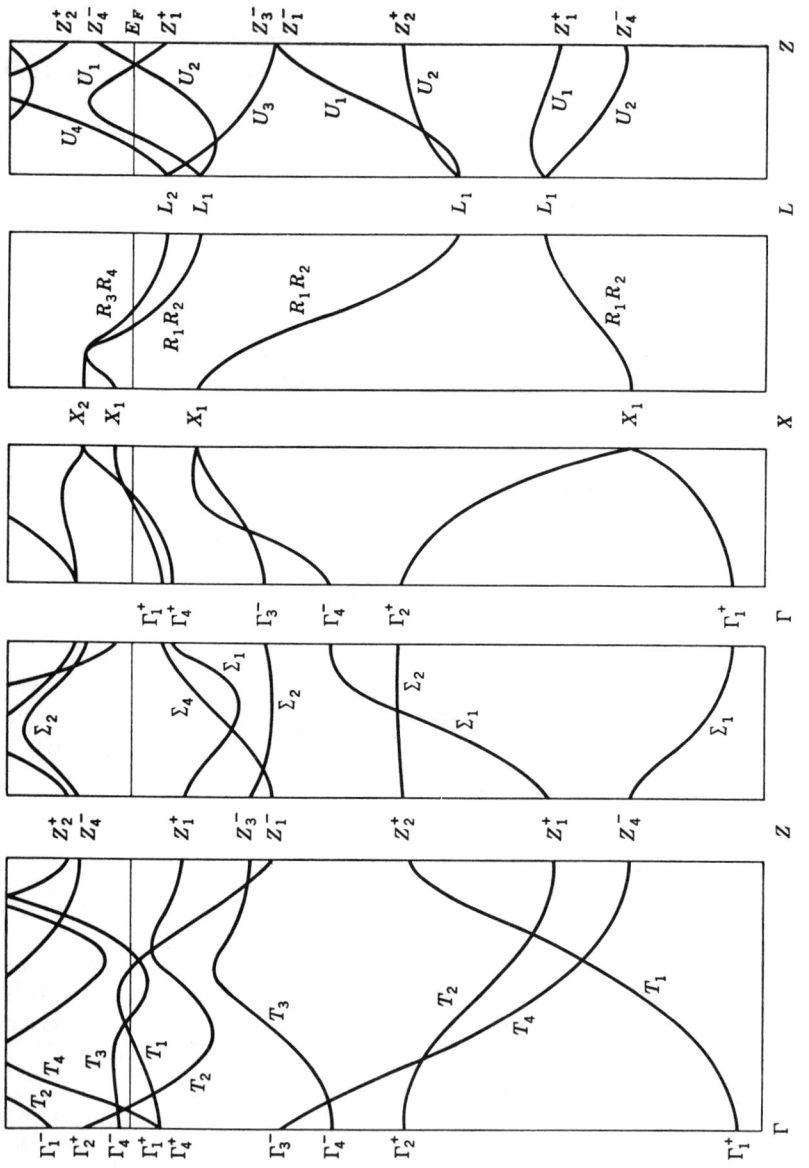

FIG. 10-26. Energy bands of gallium, as calculated by Wood.

out an example in the study of diatomic molecules which resembles this: the molecules CO and BF, composed of two atoms standing equal distances on both sides of the atom N, so that these molecules try to simulate N_2, and resemble it closely in electronic structure. We have a very different situation, however, in the alkali halides, in which the two atoms behave like ions, simulating the inert gases which are adjacent to them. The closest analogue to this case among the diatomic molecules which we have taken up in detail in Volume 1 is LiH, and also LiF, both of which we have discussed carefully. We shall now take up the nature of the energy bands arising when these same substances are formed into

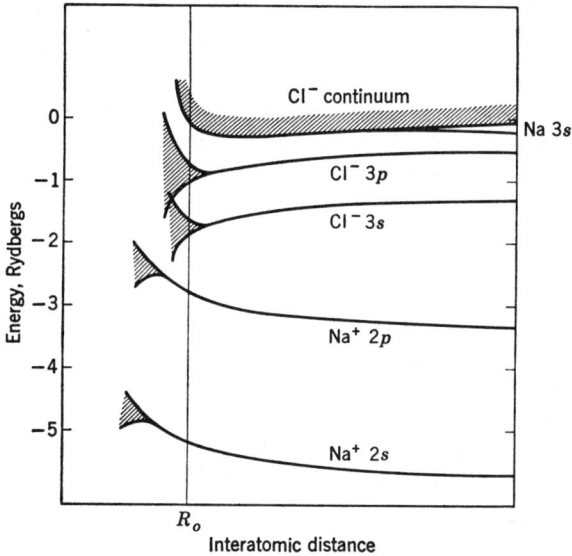

FIG. 10-27. Boundaries of energy bands of NaCl, as functions of interatomic distance, as estimated by Slater and Shockley.

crystals instead of diatomic molecules. We shall take up only compounds having the sodium chloride structure; there has been little study of those crystallizing in the cesium chloride structure, as far as their energy bands are concerned.

As a first step in discussing these compounds, we shall give an estimate made by the author and Shockley,[1] regarding the dependence of the energy bands of NaCl on internuclear separation, since this illustrates points met with ionic crystals of this sort, which we have not encountered so far. In Fig. 10-27 we show these estimated energy levels. This figure is comparable to Fig. 8-3, in which we showed the energy bands for metallic sodium,

[1] J. C. Slater and W. Shockley, *Phys. Rev.*, **50**:705 (1936).

except that Fig. 10-27 is only schematic, not the result of calculations as is Fig. 8-3. This diagram has a number of features which we have not met so far in this chapter. First, we show a number of energy bands; whereas in this chapter, the bands shown from the inner electrons of the atom were not shown. Second, we meet bands associated with either the Na^+ or the Cl^- ion; we have not had such a situation previously, where we had only one type of atom. Third, the energy bands vary strongly with internuclear separation even at large distances. This is a result of the electrostatic long-range forces, an effect which we shall comment on shortly. Fourth, at the equilibrium internuclear distance, the bands are much less spread out than in a metal; for instance, the Cl^- $3s$ and $3p$ orbitals have not spread enough to overlap. The reason is that, as we have noticed in Chap. 4, the interatomic distances in an alkali halide are much larger, in terms of the ionic sizes, than in a metal, a result of the fact that the attractive forces are provided by the electrostatic attraction between the ions, and do not depend on overlap between charge clouds, which is only enough to produce the repulsive interaction between the closed shells.

Let us ask why we have the upward or downward tendency of the energy levels at large distances in Fig. 10-27. An electron located on a negative ion, Cl^-, will be surrounded by six nearest neighbors which will be positive ions, Na^+; the closest negative ions will be farther away. We may expect, then, that this will result in a net decrease of energy, proportional to $1/R$, as the internuclear distance R decreases, and it is this decrease which appears in Fig. 10-27. Similarly, an electron located on a positive ion, Na^+, will be surrounded by six negatively charged neighbors, and its energy will rise. The method of calculating this shift of energy was worked out many years ago by Madelung, for the purpose of studying the cohesive energy of alkali halide crystals, and it is consequently referred to as a Madelung energy. In constructing Fig. 10-27, the energy levels are chosen so as to go to the correct values at infinite separation, and the dependence on distance is computed by Madelung's method. The spreading of the bands at small distance is estimated.

A quantitative calculation of a similar case has been given for KCl by Howland.[1] His calculation in a way is better than most calculations of one-electron energies. He has used the tight-binding method, has set up a determinantal function describing the state of the complete crystal when an electron is removed from one Bloch state of a given wave vector and in a given band, and has found the average energy of the crystal as a function of the wave vector of the removed electron. This automatically takes care of the potential in which the electron is supposed to move, and since a determinantal function is used, it is equivalent to the use of a Hartree-Fock method. The results of Howland for the energy bands of

[1] L. P. Howland, *Phys. Rev.*, **109**:1927 (1958).

KCl are given in Fig. 10-28. The resemblance to Fig. 10-27 is obvious, and shows that the methods used in setting up that estimate were justified.

Howland has not only calculated the boundaries of the various energy bands, as in Fig. 10-28, but has also calculated the energy of an electron in the $3p$ band of Cl^- as a function of the wave vector. The curves are given in Fig. 10-29. The band is triply degenerate at $\mathbf{k} = 0$, of type Γ_{15}, as it should be for the face-centered cubic structure. It splits as we go

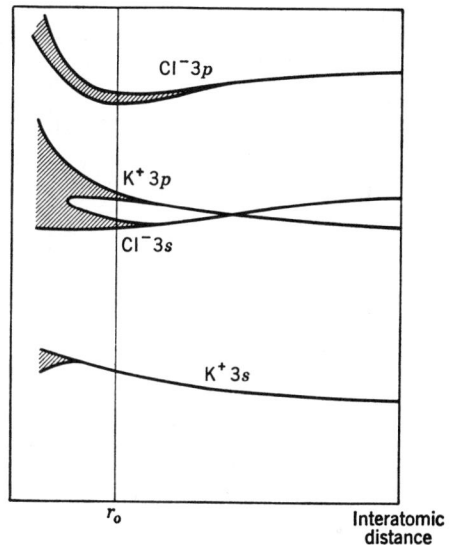

FIG. 10-28. Boundaries of energy bands of KCl, as functions of interatomic distance, as calculated by Howland.

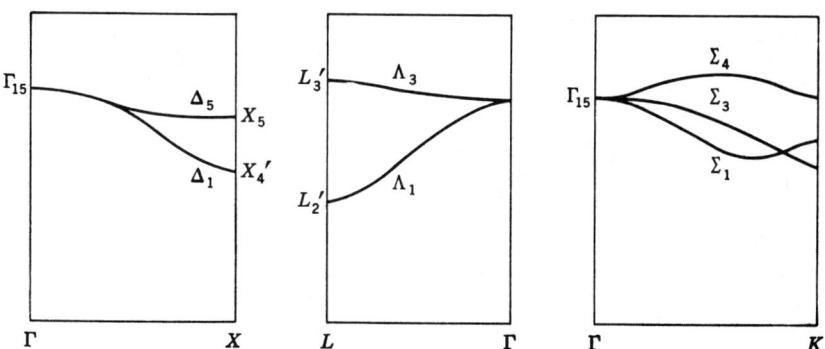

FIG. 10-29. Energy bands of KCl arising from $3p$ atomic orbitals of Cl^-, according to Howland.

away from $\mathbf{k} = 0$ to a nondegenerate band, Δ_1 and Λ_1, respectively, and a doubly degenerate band, Δ_5 and Λ_3, respectively. The total splitting of the band is rather small compared with the distance between the $3s$ and $3p$ bands at the equilibrium distance.

It is very interesting to consider the wave functions as well as the energy bands, and for this purpose we shall use another case, an early

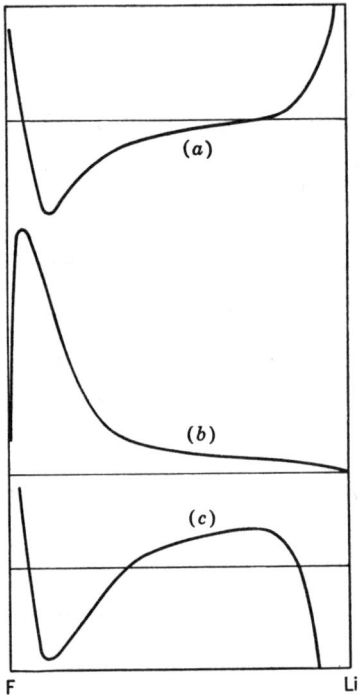

Fig. 10-30. Wave function for LiF, along line joining Li and F atoms, as calculated by Ewing and Seitz. Functions (a) and (b) represent occupied bands, (c) the lowest excited band.

calculation of the energy bands of LiF by the cellular method, by Ewing and Seitz.[1] These writers found energy bands for the F$^-$ $2s$ and $2p$ states which are much like those for the Cl$^-$ $3s$ and $3p$ shown in Fig. 10-29, but also they found an excited state, lying above the energy gap, having the type Γ_1 at $\mathbf{k} = 0$, and Δ_1 and Λ_1. In addition to this, they showed the form of the wave functions, as function of the distance along a line from an ion of one sign to the nearest neighbor ion of the opposite sign. In Fig. 10-30 we show these wave functions, for the three bands we have just

[1] D. H. Ewing and F. Seitz, *Phys. Rev.*, **50**:760 (1936).

mentioned. For the band which we have described as the F^- $2s$, the wave function is shown in Fig. 10-30a. We see that it is by no means all concentrated on the fluorine ion; there is an appreciable concentration on the lithium ion. This shows that the crystal is by no means exactly formed from ions; the wave function concentrates a small amount of charge on the lithium, in addition to its K electrons, so that it does not have a full unit of positive charge. The band which we have described as F^- $2p$ has a wave function as shown in Fig. 10-30b. This has no concentration on the lithium, and hence does not contribute to the neutralization of the lithium positive charge.

The excited-state wave function, shown in Fig. 10-30c, has a strong concentration of charge on the lithium, but a considerable amount as well on the fluorine. Since the main concentration is on the lithium, we see that an electronic excitation in which an electron was removed from the state (b), the fluorine $2p$, to the excited state (c) would be what is called a charge-transfer process: charge is transferred from the fluorine to the lithium, in the process of excitation. On account of the considerable amount of charge on both ions in this excited state, however, we do not by any means have a hundred percent charge transfer; to a considerable extent the excitation results in the electron being raised to a higher level of the fluorine. This mixed situation was illustrated in Fig. 10-27, in which the excited level of NaCl was shown as originating both from the sodium $3s$ level, which would normally be empty in an ionic crystal, and from the continuum of excited states of Cl^-. These two types of excited states would be estimated to lie at about the same energy, and the calculations of Ewing and Seitz show that the true excited state is made up of a mixture of both. These simple examples, then, illustrate two important points in the electronic structure of binary crystals: the way in which the wave functions can be modified so that they are made neither purely of ions nor of neutral atoms; and the way in which an excitation process can be a mixture of excitation on one atom or ion, and transfer of charge from one ion to the other, in such a way ordinarily as to tend to neutralize the charges. Similar phenomena occur regularly in more complicated crystals, for which energy-band calculations are not yet available.

We may make one final remark about the energy bands of the alkali halides: on account of the large internuclear separations in these substances, the energy bands have no resemblance at all to the free-electron case which we have used in describing the energy bands of metallic crystals, and of those like diamond held by covalent bonds. They are much more like the energy levels of the separated atoms, and it is for this reason that the Bloch or tight-binding type of calculation used by Howland, in the paper cited earlier, can be quite accurate for such a substance, and much closer to the truth than any free-electron approximation would be.

In addition to alkali halides, calculations have been made of several other compounds with the sodium chloride structure. One such is the calculation made by Bell et al.[1] on PbS, using the cellular method. This substance is an important semiconductor and infrared-sensitive material, but the cellular method as used in this calculation is hardly accurate enough to give reliable results. More recently, however, Johnson, Conklin, and Pratt[2] have made calculations on the similar material PbTe, using the APW method. This calculation is particularly interesting, in

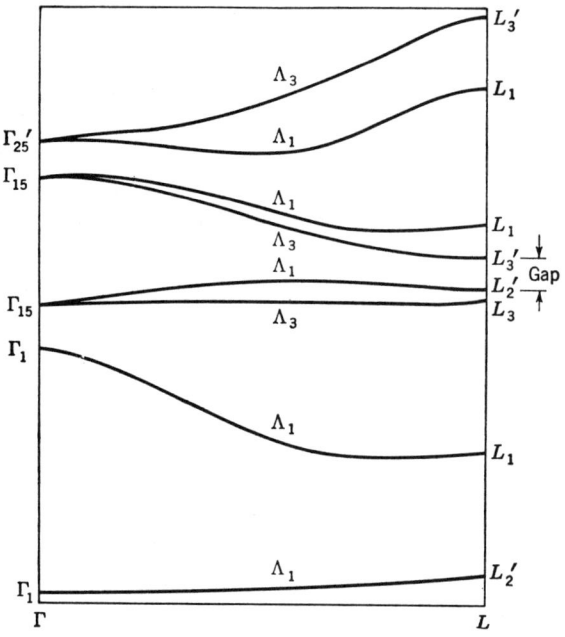

Fig. 10-31. Energy bands of lead telluride, along 111 direction, as calculated by Johnson, Conklin, and Pratt (nonrelativistic).

that the atoms, particularly lead, are so heavy that the spin-orbit and other relativistic corrections are of great importance. Energy bands as calculated by Johnson, Conklin, and Pratt are shown in Fig. 10-31. The energy gap comes at the point L, as indicated in the figure, and the calculated energy gap, found from nonrelativistic calculation, comes out a good deal larger than the experimentally observed value. In Fig. 10-32 we show the gap at this point, along with the other energy levels there,

[1] D. G. Bell, D. M. Hum, L. Pincherle, D. W. Sciama, and P. M. Woodward, *Proc. Roy. Soc. (London)*, **A217**:71 (1953).
[2] L. E. Johnson, J. B. Conklin, and G. W. Pratt, Jr., *Phys. Rev. Letters*, **11**:538 (1963).

first calculated nonrelativistically, as in Fig. 10-31; then with the spin-orbit correction alone, producing a splitting of certain energy bands, as indicated in Sec. 10-4 and Appendix 9. Finally, we give the calculation including the other relativistic terms. These are the terms arising from the change of mass with velocity, and other effects, discussed for instance in Slater, "Quantum Theory of Atomic Structure," vol. 2, chaps. 23 and

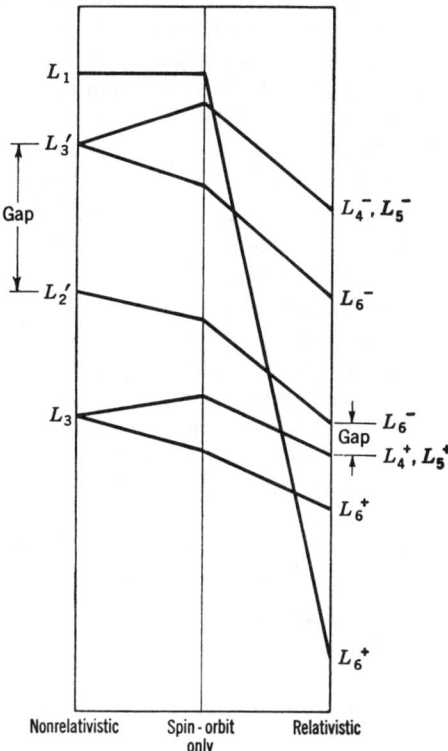

FIG. 10-32. Effects of spin-orbit interaction, and complete relativistic treatment, on energy bands of lead telluride, at point L, as calculated by Johnson, Conklin, and Pratt.

24. We see that the complete relativistic treatment has changed the order of terms, and has narrowed the energy gap, until the final value agrees as far as order of magnitude is concerned with experiment. This is a particularly graphic illustration of the importance of relativistic treatment for energy bands of crystals containing heavy elements. It seems likely that even for fairly low atomic weights the relativistic corrections are important enough so that they must be included for anything approaching a quantitative agreement with experiment.

A number of other energy-band calculations for compounds have been made, with varying degrees of approximation. However, one must realize that the study of energy bands of compounds is only beginning. In Chap. 3 we have discussed a great many types of compounds, and this has been done with the expectation that in the course of time, energy-band studies of many of them will be made. Only through such studies will one be able to answer theoretically even the simple question as to whether a given compound should be expected to be a semiconductor, insulator, or metallic conductor. The methods of calculation which we have described in the preceding chapters are adapted to compounds as well as to metals, as the success of the calculation for PBTe by Johnson, Conklin, and Pratt shows. Partially completed calculations of Mattheiss, using the APW method, on compounds of type V_3X, where X stands for Ga, Ge, As, or similar elements, crystallizing with eight atoms in a unit cell, are pointing the way toward the application of these methods to compounds more complicated than any so far investigated.

In this way one should open up a large field for future work, not only theoretical, but also experimental. For the experimental methods of studying energy bands, which we shall take up in later volumes, are adapted to compounds as well as to elements. We realize from the work which has been done already that inorganic compounds, or even organic ones, in the crystalline form, have very elaborate properties, in the nature of their energy-band behavior, which are now susceptible to quantitative study, and which in the course of time are bound to contribute greatly to the knowledge of the properties, physical and chemical, of these materials.

10-8. Bibliography of Energy-band Calculations. This bibliography contains most of the principal papers dealing with detailed energy-band calculations. The references are arranged chronologically under each type of crystal. The references, arranged alphabetically by authors, and including titles, are also included in the main bibliography at the end of this volume.

Alkali Metals

Wigner, E., and F. Seitz: *Phys. Rev.*, **43**:804 (1933); **46**:509 (1934). Sodium, cellular.
Slater, J. C.: *Phys. Rev.*, **45**:794 (1934). Sodium, cellular.
Millman, J.: *Phys. Rev.*, **47**:286 (1935). Lithium, cellular.
Seitz, F.: *Phys. Rev.*, **47**:400 (1935). Lithium, cellular.
Gombas, P.: *Z. Physik*, **113**:150 (1939), Na, K, Rb, Cs, pseudopotential.
von der Lage, F., and H. Bethe: *Phys. Rev.*, **71**:612 (1947). Sodium, cellular.
Sternheimer, R.: *Phys. Rev.*, **78**:235 (1950). Cesium, cellular.
Silverman, R. A.: *Phys. Rev.*, **85**:227 (1952). Lithium, k · p.
Parmenter, R. H.: *Phys. Rev.*, **86**:552 (1952). Lithium, OPW.
Howarth, D., and H. Jones: *Proc. Phys. Soc. (London)*, **A65**:355 (1952). Sodium, cellular.
Kohn, W., and J. Rostoker: *Phys. Rev.*, **94**:1111 (1954). Lithium, Green's function.

Schiff, B.: *Proc. Phys. Soc. (London)*, **A67**:2 (1954). Lithium, cellular.
Callaway, J.: *Phys. Rev.*, **103**:1219 (1956). Potassium, OPW and cellular.
Miasek, M.: *Bull. Acad. Polon. Sci.*, Cl. III, **4**:453 (1956). Sodium, variation.
Callaway, J., and E. L. Haase: *Phys. Rev.*, **108**:217 (1957). Cesium, OPW and cellular.
Brown, E., and J. A. Krumhansl: *Phys. Rev.*, **109**:30 (1958). Lithium, modified OPW.
Glasser, M. L., and J. Callaway: *Phys. Rev.*, **109**:1541 (1958). Lithium, OPW.
Callaway, J.: *Phys. Rev.*, **112**:322 (1958). Sodium, pseudopotential.
Callaway, J., and D. F. Morgan, Jr.: *Phys. Rev.*, **112**:334 (1958). Rubidium, cellular.
Callaway, J.: *Phys. Rev.*, **112**:1061 (1958). Cesium, cellular, terms in k^2.
Cohen, M. H., and V. Heine: *Advan. Phys.*, **7**:395 (1958). Alkali metals.
Bassani, F.: *J. Phys. Chem. Solids*, **8**:375, 379 (1959). Sodium in diamond lattice, OPW.
Bassani, F., and V. Celli: *Nuovo Cimento*, **11**:805 (1959). Lithium in diamond lattice, OPW.
Callaway, J.: *Phys. Rev.*, **119**:1012 (1960). Potassium, cellular, terms in k^2.
Cornwell, J. F., and E. P. Wohlfarth: *Nature*, **186**:379 (1960). Lithium, pseudopotential.
Callaway, J.: *Phys. Rev.*, **123**:1255 (1961). Sodium, cellular, terms in k^2.
Callaway, J.: *Phys. Rev.*, **124**:1824 (1961). Lithium, OPW, and comparison of various results.
Cornwell, J.: *Proc. Roy. Soc. (London)*, **A261**:551 (1961). Alkali metals and beryllium, pseudopotential.
Callaway, J., and W. Kohn: *Phys. Rev.*, **127**:1913 (1962). Lithium, cellular, terms in k^2.
Ham, F. S.: *Phys. Rev.*, **128**:82 (1962). Alkali metals, Green's function.

Divalent and Trivalent Elements

Manning, M. F., and H. Krutter: *Phys. Rev.*, **51**:761 (1937). Calcium, cellular.
Herring, C., and A. G. Hill: *Phys. Rev.*, **58**:132 (1940). Beryllium, OPW.
Matyas, Z.: *Phil. Mag.*, **30**:429 (1948). Aluminum, tight-binding.
Jones, H.: *Phil. Mag.*, **41**:663 (1950). Magnesium, nearly free electrons.
Donovan, B.: *Phil. Mag.*, **43**:868 (1952). Beryllium, cellular.
Trlifaj, M.: *Czech. J. Phys.*, **1**:110 (1952). Magnesium, APW.
Antoncik, E.: *Czech. J. Phys.*, **2**:18 (1953). Aluminum, APW.
Heine, V.: *Proc. Roy. Soc. (London)*, **A240**:340, 354, 361 (1957). Aluminum, OPW.
Harrison, W. A.: *Phys. Rev.*, **116**:555 (1959); **118**:1182 (1960). Aluminum, nearly free electrons, pseudopotential.
Segall, B.: *Phys. Rev.*, **124**:1797 (1961). Aluminum, Green's function.
Cornwell, J. F.: *Proc. Roy. Soc. (London)*, **A261**:551 (1961). Beryllium, pseudopotential.
Falicov, L. M.: *Phil. Trans. Roy. Soc. (London)*, **A255**:55 (1962). Magnesium, OPW.
Harrison, W. A.: *Phys. Rev.*, **126**:497 (1962); **129**:2503, 2512 (1963). Zinc, pseudopotential.
Segall, B.: *Phys. Rev.*, **131**:121 (1963). Aluminum, Green's function.
Loucks, T. L., and P. H. Cutler: *Phys. Rev.*, **133**:A819 (1964). Beryllium, OPW.

Diamond, Silicon, Germanium, 3-5 Compounds

Kimball, G. E.: *J. Chem. Phys.*, **3**:560 (1935). Diamond, cellular.
Hund, F., and B. Mrowka: *Sächsische Akad. Wiss. Leipzig*, **87**:185, 325 (1935). Diamond, cellular.

Mullaney, J. F.: *Phys. Rev.*, **66**:326 (1944).   Silicon, cellular.
Morita, A.: *Sci. Rept. Tohoku Univ.*, **33**:92 (1949).   Diamond, tight-binding.
Holmes, D. K.: *Phys. Rev.*, **87**:782 (1952).   Silicon, cellular.
Herman, F.: *Phys. Rev.*, **88**:1210 (1952).   Diamond, OPW.
Hall, G. G.: *Phil. Mag.*, **43**:338 (1952); *Phys. Rev.*, **90**:317 (1953).   Diamond, equivalent orbital tight-binding.
Herman, F., and J. Callaway: *Phys. Rev.*, **89**:518 (1953).   Germanium, OPW.
Yamaka, E., and T. Sugita: *Phys. Rev.*, **90**:992 (1953).   Silicon, cellular.
Herman, F.: *Phys. Rev.*, **93**:1214 (1954).   Diamond and germanium, OPW.
Herman, F.: *Physica*, **20**:801 (1954).   Germanium, OPW.
Jenkins, D. P.: *Physica*, **20**:967 (1954).   Silicon, cellular.
Bell, D. G., R. Hensman, D. P. Jenkins, and L. Pincherle: *Proc. Phys. Soc. (London)*, **A67**:562 (1954).   Silicon, cellular.
Herman, F.: *Proc. Inst. Radio Engrs.*, **43**:1703 (1955).   Silicon and germanium, OPW.
Herman, F.: *J. Electronics*, **1**:103 (1955).   General discussion, OPW and experiment.
Woodruff, T. O.: *Phys. Rev.*, **98**:1741 (1955); **103**:1159 (1956).   Silicon, OPW.
Jenkins, D. P.: *Proc. Phys. Soc. (London)*, **A69**:548 (1956).   Silicon, cellular.
Kobayasi, S.: *J. Phys. Soc. Japan*, **11**:175 (1956); **13**:261 (1958).   Carborundum, tight-binding, OPW.
Bassani, F.: *Phys. Rev.*, **108**:263 (1957).   Silicon, OPW, tight-binding, interpolation.
Callaway, J.: *J. Electronics*, **2**:330 (1957).   GaAs, perturbation of Ge.
Kane, E. O.: *J. Phys. Chem. Solids*, **1**:82 (1957).   Germanium, silicon, $k \cdot p$.
Kane, E. O.: *J. Phys. Chem. Solids*, **1**:249 (1957).   InSb, $k \cdot p$.
Hall, G. G.: *Phil. Mag.*, **3**:429 (1958).   Diamond, silicon, germanium, equivalent orbital tight-binding.
Morita, A.: *Progr. Theoret. Phys. (Kyoto)*, **19**:534 (1958).   Diamond type, semilocalized combination of orbitals.
Kleinman, L., and J. C. Phillips: *Phys. Rev.*, **116**:880 (1959).   Diamond, pseudopotential.
Phillips, J. C.: *J. Phys. Chem. Solids*, **8**:369, 379 (1959).   Silicon, germanium, pseudopotential.
Segall, B.: *J. Phys. Chem. Solids*, **8**:371, 379 (1959).   Germanium, Green's function.
Gubanov, A. I., and A. A. Nranyan: *Fiz. Tverd. Tela*, **1**:1044 (1959).   3-5 compounds, tight-binding and equivalent orbitals.
Bassani, F.: *Nuovo Cimento*, **13**:244 (1959).   Silicon, tight-binding.
Kleinman, L., and J. C. Phillips: *Phys. Rev.*, **117**:460 (1960).   BN, pseudopotential.
Kleinman, L., and J. C. Phillips: *Phys. Rev.*, **118**:1153 (1960).   Silicon, pseudopotential.
Nranyan, A. A.: *Fiz. Tverd. Tela*, **2**:1650 (1960).   3-5 compounds, tight-binding equivalent orbitals.
Gashimzade, F. M., and V. E. Khartsiev: *Fiz. Tverd. Tela*, **3**:1453 (1961).   Silicon, germanium, GaAs, OPW.
Phillips, J. C.: *Phys. Rev.*, **125**:1931 (1962).   Silicon and germanium, general discussion.
Braunstein, R., and E. O. Kane: *J. Phys. Chem. Solids*, **23**:1423 (1962).   3-5 compounds.
Coulson, C. A., L. B. Redei, and D. Stocker: *Proc. Roy. Soc. (London)*, **A270**:357 (1962).   3-5 compounds, OPW.
Redei, L. B.: *Proc. Roy. Soc. (London)*, **A270**:373, 383 (1962).   Diamond, OPW.
Stocker, D.: *Proc. Roy. Soc. (London)*, **A270**:397 (1962).   3-5 compounds, OPW.
Bassani, F., and M. Yoshimine: *Phys. Rev.*, **130**:20 (1963).   Group 4 elements and 3-5 compounds, OPW.

Braunstein, R.: *Phys. Rev.*, **130**:869 (1963). Germanium-silicon alloys.
Bassani, F., and L. Liu: *Phys. Rev.*, **132**:2047 (1963). Gray tin.
Cohan, N. V., D. Pugh, and R. H. Tredgold: *Proc. Phys. Soc. (London)*, **82**:65 (1963). Diamond, tight-binding, equivalent orbitals.

Transition and Other Elements with fcc, bcc, or Hexagonal Structure

Krutter, H. M.: *Phys. Rev.*, **48**:664 (1935). Copper, cellular.
Tibbs, S. R.: *Proc. Cambridge Phil. Soc.*, **34**:89 (1938). Copper, silver, cellular.
Chodorow, M. I.: *Phys. Rev.*, **55**:675 (1939). Copper, APW.
Manning, M. F., and M. I. Chodorow: *Phys. Rev.*, **56**:787 (1939). Tungsten, cellular.
Manning, M. F.: *Phys. Rev.*, **63**:190 (1943). Iron, cellular.
Greene, J. B., and M. F. Manning: *Phys. Rev.*, **63**:203 (1943). Iron, fcc, cellular.
Fletcher, G. C., and E. P. Wohlfarth: *Phil. Mag.*, **42**:106 (1951). Nickel, tight-binding.
Fletcher, G. C.: *Proc. Phys. Soc. (London)*, **A65**:192 (1952). Nickel, tight-binding.
Howarth, D. J.: *Proc. Roy. Soc. (London)*, **A220**:513 (1953). Copper, cellular.
Koster, G. F.: *Phys. Rev.*, **98**:901 (1955). Nickel, tight-binding.
Howarth, D. J.: *Phys. Rev.*, **99**:469 (1955). Copper, APW.
Callaway, J.: *Phys. Rev.*, **99**:500 (1955). Iron, OPW.
Schiff, B.: *Proc. Phys. Soc. (London)*, **A68**:686 (1955); **A69**:185 (1956). Titanium, cellular.
Fukuchi, M.: *Progr. Theoret. Phys.*, **16**:222 (1956). Copper, OPW.
Altmann, S. L., and N. V. Cohan: *Proc. Phys. Soc. (London)*, **71**:383 (1958). Titanium, cellular.
Altmann, S. L.: *Proc. Roy. Soc. (London)*, **A244**:141, 153 (1958). Zirconium, cellular.
Stern, F.: *Phys. Rev.*, **116**:1399 (1959). Iron, tight-binding.
Belding, E. F.: *Phil. Mag.*, **4**:1145 (1959). Cr, Fe, Ni, tight-binding.
Wood, J. H.: *Phys. Rev.*, **117**:714 (1960). Iron, APW.
Segall, B.: *Phys. Rev. Letters*, **7**:154 (1961). Copper, Green's function.
Burdick, G. A.: *Phys. Rev. Letters*, **7**:156 (1961). Copper, APW.
Asdente, M., and J. Friedel: *Phys. Rev.*, **124**:384 (1961); **126**:2262 (1962). Chromium, tight-binding, 4s omitted.
Knox, R. S., and F. Bassani: *Phys. Rev.*, **124**:652 (1961). Argon, tight-binding and OPW.
Cornwell, J. F.: *Phil. Mag.*, **6**:727 (1961). Noble metals, pseudopotential.
Segall, B.: *Phys. Rev.*, **125**:109 (1962). Copper, Green's function.
Wood, J. H.: *Phys. Rev.*, **126**:517 (1962). Iron, APW.
Asdente, M.: *Phys. Rev.*, **127**:1949 (1962). Chromium, tight-binding.
Altmann, S. L., and C. J. Bradley: *Phys. Letters*, **1**:336 (1962). Zirconium, cellular.
Cornwell, J. F., and E. P. Wohlfarth: *J. Phys. Soc. Japan*, **17** (suppl. B-1):32 (1962). Iron.
Glasser, M. L.: *Rev. Mexicana Fiz.*, **11**:31 (1962). Silver.
Lomer, W. M.: *Proc. Phys. Soc. (London)*, **80**:489 (1962). Chromium, general discussion.
Burdick, G. A.: *Phys. Rev.*, **129**:138 (1963). Copper, APW.
Yamashita, J., M. Fukuchi, and S. Wakoh: *J. Phys. Soc. Japan*, **18**:999 (1963). Nickel, tight-binding and Green's function.
Mattheiss, L. F.: *Bull. Am. Phys. Soc.*, ser. II, **8**:222 (1962). 3d transition elements to Cu, Zn, APW.
Fowler, W. B.: *Phys. Rev.*, **132**:1594 (1963). Krypton, tight-binding and OPW.
Mattheiss, L. F.: *Phys. Rev.*, **133**:A1399 (1964). Argon, APW.

Graphite

Wallace, P. R.: *Phys. Rev.*, **71**:622 (1947); **72**:258 (1947). Tight-binding.
Coulson, C. A.: *Nature*, **159**:265 (1947). Tight-binding.
Coulson, C. A., and R. Taylor: *Proc. Phys. Soc. (London)*, **A65**:815 (1952). Tight-binding.
Carter, J. L., and J. A. Krumhansl: *J. Chem. Phys.*, **21**:2238 (1953). Tight-binding.
Ariyama, K., and S. Mase: *Progr. Theoret. Phys.*, **12**:244 (1954). Tight-binding.
Lomer, W. M.: *Proc. Roy. Soc. (London)*, **A227**:330 (1955). Tight-binding.
Johnston, D. F.: *Proc. Roy. Soc. (London)*, **A227**:349 (1955); **A237**:48 (1956). Tight-binding.
McClure, J. W.: *Phys. Rev.*, **108**:612 (1957). k · p.
Yamazaki, M.: *J. Chem. Phys.*, **26**:930 (1957). Tight-binding.
Corbato, F.: *Proc. 1957 Carbon Conference*, Pergamon Press, New York, p. 173. Tight-binding.
Slonczewski, J. C., and P. R. Weiss: *Phys. Rev.*, **109**:272 (1958). k · p.
Haering, R. R.: *Can. J. Phys.*, **36**:352 (1958). Tight-binding.
Mase, S.: *J. Phys. Soc. Japan*, **13**:563 (1958). Tight-binding.
Peacock, T. E., and R. McWeeny: *Proc. Phys. Soc. (London)*, **74**:385 (1959). Tight-binding.
Barriol, J.: *J. Chim. Phys.*, **57**:837 (1960); J. Barriol and J. Metzger, *J. Chim. Phys.*, **57**:848 (1960). Tight-binding.
Anno, T., and C. A. Coulson: *Proc. Roy. Soc. (London)*, **A264**:165 (1961). Tight-binding (semiempirical).

Elements with Other Crystal Structures

Jones, H.: *Proc. Roy. Soc. (London)*, **A147**:396 (1934). Bismuth, nearly free electrons.
Morita, A.: *Sci. Rept. Tohoku Univ.*, **33**:144 (1949). Bismuth, tight-binding.
Reitz, J. R.: *Phys. Rev.*, **105**:1233 (1957). Selenium, tellurium, tight-binding.
Gaspar, R.: *Acta Phys. Hung.*, **7**:289 (1957). Selenium and tellurium, tight-binding.
Ridley, E. C.: *Proc. Roy. Soc. (London)*, **A247**:199 (1958). Uranium, cellular.
Mase, S.: *J. Phys. Soc. Japan*, **13**:434 (1958); **14**:584 (1959). Bismuth, tight-binding.
de Carvalho, A. P.: *Compt. rend.*, **248**:778 (1959). Tellurium, tight-binding.
Harrison, W. A.: *J. Phys. Chem. Solids*, **17**:171 (1960). Bismuth, pseudopotential.
Miasek, M.: *Bull. Acad. Polon. Sci., Ser. Sci. Math. Astron. Phys.*, **8**:9 (1960). White tin, OPW.
Bergson, G.: *Arkiv Kemi*, **16**:315 (1960). Sulfur, tight-binding.
Behrens, E.: *Z. Physik*, **161**:279 (1961). Bismuth, tight-binding.
Behrens, E.: *Z. Physik*, **163**:140 (1961). Selenium, tight-binding.
Miasek, M.: *Phys. Rev.*, **130**:11 (1963). White tin, OPW.
Wood, J. H.: *Bull. Am. Phys. Soc.*, ser. II, **8**:222 (1963). Gallium, APW.

Compounds, Other Than 3-5

Jones, H.: *Proc. Roy. Soc. (London)*, **A144**:225 (1934). Alloys, γ-phase, nearly free electrons.
Slater, J. C., and W. Shockley: *Phys. Rev.*, **50**:705 (1936). Sodium chloride, general discussion.
Shockley, W.: *Phys. Rev.*, **50**:754 (1936). Sodium chloride, cellular.
Ewing, D. H., and F. Seitz: *Phys. Rev.*, **50**:760 (1936). LiF and LiH, cellular.
Tibbs, S. R.: *Trans. Faraday Soc.*, **35**:1471 (1939). Sodium chloride, cellular.

Morita, A., and C. Horie: *Sci. Rept. Tohoku Univ.*, **36**:259 (1952). Barium oxide, tight-binding.
Bell, D. G., D. M. Hum, L. Pincherle, D. W. Sciama, and P. M. Woodward: *Proc. Roy. Soc. (London)*, **A217**:71 (1953). PbS, cellular.
Casella, R. C.: *Phys. Rev.*, **104**:1260 (1956). Sodium chloride, tight-binding.
Yamazaki, M.: *J. Chem. Phys.*, **27**:746 (1957). Boron carbide, tight-binding.
Kucher, T. I.: *Zh. eksperim. i Teor. Fiz.*, **34**:394 (1958); **35**:1049 (1958). NaCl, tight-binding.
Birman, J. L.: *Phys. Rev.*, **109**:810 (1958). ZnS, cellular.
Shakin, C., and J. Birman: *Phys. Rev.*, **109**:818 (1958). ZnS, cellular.
Howland, L. P.: *Phys. Rev.*, **109**:1927 (1958). Potassium chloride, tight-binding.
Tolpygo, K. B., and O. F. Tomasevich: *Ukr. Fiz. Zh.*, **3**:145 (1958). Sodium chloride, tight-binding.
Birman, J. L.: *Phys. Rev.*, **115**:1493 (1959). ZnS, tight-binding.
Birman, J. L.: *J. Phys. Chem. Solids*, **8**:35 (1959). ZnS, general discussion.
O'Sullivan, W.: *J. Chem. Phys.*, **30**:379 (1959). BeO, tight-binding.
Kudinov, E. K.: *Fiz. Tverd. Tela*, **1**:1851 (1959). Bi_2Te_3, tight-binding.
Flodmark, S.: *Arkiv Fiz.*, **14**:513 (1959); **18**:49 (1960). Type MB_6, tight-binding.
Balkanski, M., and J. des Cloizeaux: *J. Phys. Radium*, **21**:825 (1960); *Abhandl. Deut. Akad. Wiss. Berlin, Kl. Math. Phys. Tech.*, **76** (1960). CdS, spin-orbit interaction.
Kucher, T. I., and K. B. Tolpygo: *Fiz. Tverd. Tela*, **2**:2301 (1960). Sodium chloride, tight-binding.
Tolpygo, K. B., and O. F. Tomasevich: *Fiz. Tverd. Tela*, **2**:3110 (1960). Sodium chloride, tight-binding.
Kudinov, E. K.: *Fiz. Tverd. Tela*, **3**:317 (1961). Bi_2Te_3, tight-binding.
Zhilich, A. G., and V. P. Makarov: *Fiz. Tverd. Tela*, **3**:585 (1961). Cuprous oxide, Green's function.
Wood, V. E., and J. R. Reitz: *J. Phys. Chem. Solids*, **23**:229 (1962). Cesium gold, cellular.
Gashimzade, F. M., and V. E. Khartsiev: *Fiz. Tverd. Tela*, **4**:434 (1962). SnS-type compounds, OPW.
Evseev, Z. Ya., and K. B. Tolpygo: *Fiz. Tverd. Tela*, **4**:3644 (1962). Sodium chloride, tight-binding.
Johnson, L. E., J. B. Conklin, and G. W. Pratt, Jr.: *Phys. Rev. Letters*, **11**:538 (1963). PbTe, relativistic APW.
Evseev, Z. Ya.: *Fiz. Tverd. Tela*, **5**:2345 (1963). Sodium chloride, tight-binding.
Lee, P. M., and L. Pincherle: *Proc. Phys. Soc. (London)*, **81**:461 (1963). Bismuth telluride, APW.
Yamashita, J.: *J. Phys. Soc. Japan*, **18**:1010 (1963). TiO and NiO. Tight-binding.

Appendix 1
Interatomic Distances and Crystal Structures

In the table which follows, we have endeavored to present the maximum amount of information about the crystals or molecules concerned consistent with brevity. The following information is given with reference to each bond (all distances are in Angstroms):
First: Bonding atoms, arranged in order of atomic number.
Second: Crystal or molecule concerned.
Third: Description of structure. Some 30 well-known structures are indicated by abbreviations or other symbols. Abbreviations are as follows: ar (aragonite), bcc (body-centered cubic), cal (calcite), cor (corundum), di (diamond), fcc (face-centered cubic), fl (fluorite), ga (garnet), gr (graphite), hex (hexagonal close-packed), il (ilmenite), per (perovskite), py (pyrite), ru (rutile), sp (spinel), wu (wurtzite), znb (zinc-blende). The following structures are written out: AlB_2, AlF_3, As, $CdCl_2$, CdI_2, CsCl, I, La_2O_3, LaOF, Na_3As, NaCl, NiAs, Se, Tl_2O_3. Other structures are indicated by asterisks. Molecules are indicated by "mol."
Fourth: Space group for crystals, in notation of International Tables for X-Ray Crystallography, published for the International Union of Crystallography by the Kynoch Press, Birmingham, England, 1952. Schoenflies notation for point group of molecules.
Fifth: Chapter and paragraph reference from Wyckoff, "Crystal Structures," Interscience Publishers, Inc., New York, 1948, for crystals, indicating place where additional information about the crystal structure, references, and so on, may be found. For molecules, (T.I.D.) indicates Tables of Interatomic Distances and Configurations in Molecules and Ions, The Chemical Society, London, 1958, in which molecular information is included.
Sixth: Observed interatomic distance in bond. Where more than one observed distance is given, it indicates bonds of different lengths between the same kinds of atoms in the same compound. These distances, for crystals, are either as given by Wyckoff or computed or estimated from

crystal parameters given in his text, and should not be regarded as primary references. The reader should consult the original work, to which reference is given by Wyckoff, for further details of any crystal in which he is interested. Distances for molecules are taken from Tables of Interatomic Distances.

Seventh: Sum of atomic radii, from Table 3-1.

H—H	H_2 (mol) $D_{\infty h}$ (T.I.D.)	0.74	0.50
H—Li	LiH (mol) $C_{\infty v}$ (T.I.D.)	1.60	1.70
	LiH (NaCl) O_h^5 $(Fm3m)$ (III, al)	2.04	1.70
H—Be	BeH (mol) $C_{\infty v}$ (T.I.D.)	1.34	1.30
	BeB_2H_8 (mol) D_{2d} (T.I.D.)	1.28, 1.63	1.30
H—B	BH (mol) $C_{\infty v}$ (T.I.D.)	1.23	1.15
	$(BH_4)^-$, in $NaBH_4$, KBH_4, $RbBH_4$ (T.I.D.)	1.25	1.15
H—C	CH (mol) $C_{\infty v}$ (T.I.D.)	1.12	0.95
	$CHBr_3$ (mol) C_{3v} (T.I.D.)	1.09	0.95
	CH_4 (mol) T_d (T.I.D.)	1.09	0.95
H—N	NH (mol) $C_{\infty v}$ (T.I.D.)	1.04	0.90
	NH_3 (mol) C_{3v} (T.I.D.)	1.01	0.90
	$(NH_4)^+$, in NH_4Cl, NH_4Br (T.I.D.)	1.03	0.90
H—O	OH (mol) $C_{\infty v}$ (T.I.D.)	0.97	0.85
	H_2O (mol) C_{2v} (T.I.D.)	0.96	0.85
H—F	HF (mol) $C_{\infty v}$ (T.I.D.)	0.92	0.75
H—Na	NaH (mol) $C_{\infty v}$ (T.I.D.)	1.89	2.05
H—Mg	MgH (mol) $C_{\infty v}$ (T.I.D.)	1.73	1.75
H—Al	AlH (mol) $C_{\infty v}$ (T.I.D.)	1.65	1.50
H—Si	SiH (mol) $C_{\infty v}$ (T.I.D.)	1.52	1.35
	SiH_4 (mol) T_d (T.I.D.)	1.48	1.35
H—P	PH (mol) $C_{\infty v}$ (T.I.D.)	1.43	1.25
	PH_3 (mol) C_{3v} (T.I.D.)	1.42	1.25
H—S	SH (mol) $C_{\infty v}$ (T.I.D.)	1.35	1.25
	H_2S (mol) C_{2v} (T.I.D.)	1.33	1.25
H—Cl	HCl (mol) $C_{\infty v}$ (T.I.D.)	1.27	1.25
H—K	KH (mol) $C_{\infty v}$ (T.I.D.)	2.24	2.45
H—Ca	CaH (mol) $C_{\infty v}$ (T.I.D.)	2.00	2.05
H—Mn	MnH (mol) $C_{\infty v}$ (T.I.D.)	1.73	1.65
H—Co	CoH (mol) $C_{\infty v}$ (T.I.D.)	1.54	1.60
H—Ni	NiH (mol) $C_{\infty v}$ (T.I.D.)	1.47	1.60
H—Cu	CuH (mol) $C_{\infty v}$ (T.I.D.)	1.46	1.60
H—Zn	ZnH (mol) $C_{\infty v}$ (T.I.D.)	1.59	1.60
H—Ge	GeH (mol) $C_{\infty v}$ (T.I.D.)	1.59	1.50
	GeH_4 (mol) T_d (T.I.D.)	1.53	1.50
H—As	AsH_3 (mol) C_{3v} (T.I.D.)	1.52	1.40
H—Se	SeH_2 (mol) C_{2v} (T.I.D.)	1.47	1.40
H—Br	HBr (mol) $C_{\infty v}$ (T.I.D.)	1.41	1.40
H—Rb	RbH (mol) $C_{\infty v}$ (T.I.D.)	2.37	2.60
H—Sr	SrH (mol) $C_{\infty v}$ (T.I.D.)	2.15	2.25
H—Ag	AgH (mol) $C_{\infty v}$ (T.I.D.)	1.62	1.85
H—Cd	CdH (mol) $C_{\infty v}$ (T.I.D.)	1.76	1.80

App. 1] INTERATOMIC DISTANCES AND CRYSTAL STRUCTURES

H—In	InH (mol) $C_{\infty v}$ (T.I.D.)	1.84	1.80
H—Sn	SnH (mol) $C_{\infty v}$ (T.I.D.)	1.79	1.70
H—Sb	SbH$_3$ (mol) C_{3v} (T.I.D.)	1.71	1.70
H—I	HI (mol) $C_{\infty v}$ (T.I.D.)	1.60	1.65
H—Cs	CsH (mol) $C_{\infty v}$ (T.I.D.)	2.49	2.85
H—Ba	BaH (mol) $C_{\infty v}$ (T.I.D.)	2.23	2.40
H—Au	AuH (mol) $C_{\infty v}$ (T.I.D.)	1.52	1.60
H—Hg	HgH (mol) $C_{\infty v}$ (T.I.D.)	1.74	1.75
H—Tl	TlH (mol) $C_{\infty v}$ (T.I.D.)	1.87	2.15
H—Pb	PbH (mol) $C_{\infty v}$ (T.I.D.)	1.84	2.05
H—Bi	BiH (mol) $C_{\infty v}$ (T.I.D.)	1.81	1.85
Li—Li	Li$_2$ (mol) $D_{\infty h}$ (T.I.D.)	2.67	2.90
	metal (bcc) O_h^9 ($Im3m$) (II, c)	3.04	2.90
	metal (fcc) O_h^5 ($Fm3m$) (II, a)	3.11	2.90
	metal (hex) D_{6h}^4 ($P6_3/mmc$) (II, b)	3.11	2.90
Li—O	Li$_2$O (fl) O_h^5 ($Fm3m$) (IV, a1)	2.00	2.05
	LiNO$_3$ (cal) D_{3d}^6 ($R\bar{3}c$) (VII, a1)	2.19	2.05
	LiFePO$_4$ * D_{2h}^{16} ($Pnma$) (VIII, b4a)	2.10–2.26	2.05
Li—F	LiF (mol) $C_{\infty v}$ (T.I.D.)	1.51	1.95
	LiF (NaCl) O_h^5 ($Fm3m$) (III, a1)	2.01	1.95
	BaLiF$_3$ (per) O_h^1 ($Pm3m$) (VII, a5)	2.00	1.95
Li—P	Li$_3$P (Na$_3$As) D_{6h}^4 ($P6_3/mmc$) (V, b18)	2.46–2.50	2.45
Li—S	Li$_2$S (fl) O_h^5 ($Fm3m$) (IV, a1)	2.48	2.45
Li—Cl	LiCl (NaCl) O_h^5 ($Fm3m$) (III, a1)	2.57	2.45
Li—As	Li$_3$As (Na$_3$As) D_{6h}^4 ($P6_3/mmc$) (V, b18)	2.54–2.59	2.60
Li—Se	Li$_2$Se (fl) O_h^5 ($Fm3m$) (IV, a1)	2.60	2.60
Li—Br	LiBr (NaCl) O_h^5 ($Fm3m$) (III, a1)	2.75	2.60
Li—Ag	LiAg (CsCl) O_h^1 ($Pm3m$) (III, b1)	2.75	3.05
Li—Sb	Li$_3$Sb (Na$_3$As) D_{6h}^4 ($P6_3/mmc$) (V, b18)	2.72–2.78	2.90
Li—Te	Li$_2$Te (fl) O_h^5 ($Fm3m$) (IV, a1)	2.82	2.85
Li—I	LiI (mol) $C_{\infty v}$ (T.I.D.)	2.39	2.85
	LiI (NaCl) O_h^5 ($Fm3m$) (III, a1)	3.00	2.85
Li—Hg	LiHg (CsCl) O_h^1 ($Pm3m$) (III, b1)	2.85	2.95
Li—Tl	LiTl (CsCl) O_h^1 ($Pm3m$) (III, b1)	2.97	3.35
Be—Be	metal (hex) D_{6h}^4 ($P6_3/mmc$) (II, b)	2.23–2.29	2.10
Be—C	Be$_2$C (fl) O_h^5 ($Fm3m$) (IV, a1)	1.88	1.80
Be—N	Be$_3$N$_2$ (Tl$_2$O$_3$) T_h^7 ($Ia3$) (V, a2)	1.71	1.75
Be—O	BeO (mol) $C_{\infty v}$ (T.I.D.)	1.33	1.65
	BeO (wu) C_{6v}^4 ($P6_3mc$) (III, c2)	1.64	1.65
Be—F	BeF (mol) $C_{\infty v}$ (T.I.D.)	1.36	1.55
	(NH$_4$)$_2$BeF$_4$ * D_{2h}^{16} ($Pnma$) (VIII, b5)	1.61	1.55
Be—P	Be$_3$P$_2$ (Tl$_2$O$_3$) T_h^7 ($Ia3$) (V, a2)	2.16	2.05
Be—S	BeS (znb) T_d^2 ($F\bar{4}3m$) (III, c1)	2.10	2.05
Be—Co	BeCo (CsCl) O_h^1 ($Pm3m$) (III, b1)	2.26	2.45
Be—Cu	BeCu (CsCl) O_h^1 ($Pm3m$) (III, b1)	2.33	2.40
Be—Se	BeSe (znb) T_d^2 ($F\bar{4}3m$) (III, c1)	2.20	2.20
Be—Pd	BePd (CsCl) O_h^1 ($Pm3m$) (III, b1)	2.44	2.45
Be—Te	(znb) T_d^2 ($F\bar{4}3m$) (III, cl)	2.41	2.45

B—B	B_2 (mol) $D_{\infty h}$ (T.I.D.)	1.59	1.70
	metal * D_{2d}^5 ($P\bar{4}n2$) (II, ea)	1.75–1.80	1.70
	B_4H_{10} * C_{2h}^5 ($P2_1/b$) (V, h11)	1.75–1.85	1.70
	B_5H_{12} * C_{2h}^5 ($P2_1/b$) (V, h14)	1.72–1.86	1.70
	B_4C * D_{3d}^5 ($R\bar{3}m$) (V, c3)	1.74–1.80	1.70
	AlB_2 (AlB_2) D_{3d}^3 ($P\bar{3}m1$) (IV, c6)	1.73	1.70
	ZrB_2 (AlB_2) D_{3d}^3 ($P\bar{3}m1$) (IV, c6)	1.83	1.70
	CaB_6 * O_h^1 ($Pm3m$) (V, f1)	1.72	1.70
	ThB_4 * D_{4h}^5 ($P4/mbm$) (V, c7)	1.74–1.80	1.70
B—C	B_4C * D_{3d}^5 ($R\bar{3}m$) (V, c3)	1.64	1.55
B—N	BN (mol) $C_{\infty v}$ (T.I.D.)	1.28	1.50
	BN * D_{6h}^4 ($P6_3/mmc$) (III, d4a)	1.45	1.50
B—O	BO (mol) $C_{\infty v}$ (T.I.D.)	1.20	1.45
	BPO_4 * S_4^2 ($I\bar{4}$) (VIII, a2)	1.44	1.45
	$BAsO_4$ * S_4^2 ($I\bar{4}$) (VIII, a2)	1.44	1.45
	H_3BO_3 * C_i^1 ($P\bar{1}$) (VII, b8)	1.36	1.45
	$Co_3(BO_3)_2$ * D_{2h}^{12} ($Pnnm$) (VII, b14)	1.34–1.42	1.45
	YBO_3 (cal) D_{3d}^6 ($R\bar{3}c$) (VII, a1)	1.27	1.45
	$Ag_3(BO_3)_2$ * D_{2h}^{12} ($Pnnm$) (VII, b14)	1.34–1.42	1.45
	$InBO_3$ (cal) D_{3d}^6 ($R\bar{3}c$) (VII, a1)	1.19	1.45
B—F	BF (mol) $C_{\infty v}$ (T.I.D.)	1.26	1.35
	BF_3 (mol) D_{3h} (T.I.D.)	1.31	1.35
B—Cl	BCl (mol) $C_{\infty v}$ (T.I.D.)	1.72	1.85
	BCl_3 (mol) D_{3h} (T.I.D.)	1.73	1.85
B—Br	BBr (mol) $C_{\infty v}$ (T.I.D.)	1.88	2.00
	BBr_3 (mol) D_{3h} (T.I.D.)	1.87	2.00
B—Th	ThB_4 * D_{4h}^5 ($P4/mbm$)	2.78–3.10	2.60
C—C	C_2 (mol) $D_{\infty h}$ (T.I.D.)	1.31	1.40
	diamond (di) O_h^7 ($Fd3m$) (II, h)	1.54	1.40
	graphite (gr) D_{6h}^4 ($P6_3mc$) (II, i)	1.42	1.40
	C_2H_6, and single-bond organic (T.I.D.)	1.54	1.40
	C_2H_4, and double-bond organic (T.I.D.)	1.34	1.40
	C_2H_2, and triple-bond organic (T.I.D.)	1.20	1.40
	B_4C * D_{3d}^5 ($R\bar{3}m$) (V, c3)	1.39	1.40
C—N	CN (mol) $C_{\infty v}$ (T.I.D.)	1.17	1.35
	CN_4O_8 (mol) S_4 or D_{2d} (T.I.D.)	1.47	1.35
	NH_4SCN * C_{2h}^5 ($P2_1/b$) (VI, a20)	1.25	1.35
	NaOCN * D_{3d}^5 ($R\bar{3}m$) (VI, a7)	1.13	1.35
	$CaCN_2$ * D_{3d}^5 ($R\bar{3}m$) (VI, a7)	1.16	1.35
C—O	CO (mol) $C_{\infty v}$ (T.I.D.)	1.13	1.30
	CH_3OH, alcohols and ethers (T.I.D.)	1.43	1.30
	NaOCN * D_{3d}^5 ($R\bar{3}m$) (VI, a7)	1.21	1.30
	$CaCO_3$ (ar) D_{2h}^{16} ($Pnma$) (VII, a10)	1.30–1.33	1.30
	$CaCO_3$ (cal) D_{3d}^6 ($R\bar{3}c$) (VII, a1)	1.25	1.30
	Other carbonates, calcite structure	1.15–1.23	1.30
	$Ni(CO)_4$ * T_h^6 ($Pa3$) (V, c9)	1.15	1.30
C—F	CF (mol) $C_{\infty v}$ (T.I.D.)	1.27	1.20
	CF_4 (mol) T_d, F-substituted hydrocarbons (T.I.D.)	1.32	1.20
C—Cl	CCl_4 (mol) T_d, Cl-substituted hydrocarbons (T.I.D.)	1.77	1.70

INTERATOMIC DISTANCES AND CRYSTAL STRUCTURES

C—Br,	CBr_4 (mol), T_d, Br-substituted hydrocarbons (T.I.D.)	1.94	1.85
C—I	CI_4 (mol) T_d, I-substituted hydrocarbons (T.I.D.)	2.15	2.10
N—N	N_2 (mol) $D_{\infty h}$ (T.I.D.)	1.10	1.30
	$N_2H_6SO_4$ * D_2^4 ($P2_12_12_1$) (VIII, a22)	1 40	1.30
	$Sr(N_3)_2$ * D_{2h}^{24} ($Fddd$) (VI, a21)	1.12	1.30
N—O	NO (mol) $C_{\infty v}$ (T.I.D.)	1.15	1.25
	N_2O_5 (mol) (T.I.D.)	1.18	1.25
	N_2O_5 * D_{6h}^4 ($P6_3/mmc$) (V, d5)	1.15–1.24	1.25
	$LiNO_3$ (cal) D_{3d}^6 ($R\bar{3}c$) (VII, a1)	1.17	1.25
	NH_4NO_3 * D_{2h}^{16} ($Pnma$) (VII, a7)	1.24–1.26	1.25
	$NH_4NO_3 \cdot 2HNO_3$ * D_2^3 ($P2_12_12_1$) (VII, b23)	1.20–1.33	1.25
	$NaNO_3$ (cal) D_{3d}^6 ($R\bar{3}c$) (VII, a1)	1.27	1.25
	KNO_3 (ar) D_{2h}^{16} ($Pnma$) (VII, a10)	1.20–1.22	1.25
N—F	NF_3 (mol) C_{3v} (T.I.D.)	1.37	1.15
N—Cl	NClO (mol) C_s (T.I.D.)	1.95	1.65
	$NClO_2$ (mol) C_{2v} (T.I.D.)	1.79	1.65
N—Br	NBrO (mol) C_s (T.I.D.)	2.14	1.80
O—O	O_2 (mol) $D_{\infty h}$ (T.I.D.)	1.21	1.20
O—F	F_2O (mol) C_{2v} (T.I.D.)	1.42	1.10
F—F	F_2 (mol) $D_{\infty h}$ (T.I.D.)	1.42	1.00
Na—O	Na_2O (fl) O_h^5 ($Fm3m$) (IV, a1)	2.41	2.40
	$NaNO_3$ (cal) D_{3d}^6 ($R\bar{3}c$) (VII, a1)	2.37	2.40
	Na_2SO_3 * C_{3i}^1 ($P\bar{3}$) (VII, b5)	2.45	2.40
	$NaClO_3$ * T^4 ($P2_13$) (VII, a4)	2.38–2.48	2.40
	$NaBrO_3$ * T^4 ($P2_13$) (VII, a4)	2.38–2.48	2.40
	$NaNiO_2$ * C_{2h}^3 ($B2/m$) (VI, a23)	2.29–2.34	2.40
	$NaNbO_3$ (per) O_h^1 ($Pm3m$) (VII, a5)	2.75	2.40
	$NaTaO_3$ (per) O_h^1 ($Pm3m$) (VII, a5)	2.74	2.40
	$NaWO_3$ (per) O_h^1 ($Pm3m$) (VII, a5)	2.73	2.40
	$NaH(PO_3NH_2)$ * C_6^6 ($P6_3$) or C_6^3 ($P6_5$) (VIII, b23)	2.39–2.43	2.40
	$NaAlFAsO_4$ * C_{2h}^6 ($B2/b$) (VIII, c7)	2.40–2.47	2.40
	$Na_2Ca_2(CO_3)_3$ * C_{2v}^{14} ($Amm2$) (VII, b24)	2.34	2.40
Na—F	NaF (NaCl) O_h^5 ($Fm3m$) (III, a1)	2.31	2.30
	$NaHF_2$ * D_{3d}^5 ($R\bar{3}m$) (VI, 27)	2.30	2.30
Na—Na	Na_2 (mol) $D_{\infty h}$ (T.I.D.)	3.08	3.60
	metal (bcc) O_h^9 ($Im3m$) (II, c)	3.72	3.60
Na—P	Na_3P (Na_3S) D_{6h}^4 ($P6_3/mmc$) (V, b18)	2.87–2.93	2.80
Na—S	Na_2S (fl) O_h^5 ($Fm3m$) (IV, a1)	2.82	2.80
Na—Cl	NaCl (NaCl) O_h^5 ($Fm3m$) (III, a1)	2.82	2.80
Na—As	Na_3As (Na_3As) D_{6h}^4 ($P6_3/mmc$) (V, b18)	2.94–2.99	2.95
Na—Se	Na_2Se (fl) O_h^5 ($Fm3m$) (IV, a1)	2.95	2.95
Na—Br	NaBr (NaCl) O_h^5 ($Fm3m$) (III, a1)	2.98	2.95
Na—Nb	$NaNbO_3$ (per) O_h^1 ($Pm3m$) (VII, a5)	3.37	3.25
Na—Sb	Na_3Sb (Na_3As) D_{6h}^4 ($P6_3/mmc$) (V, b18)	3.10–3.16	3.25
Na—Te	Na_2Te (fl) O_h^5 ($Fm3m$) (IV, a1)	3.17	3.20
Na—I	NaI (NaCl) O_h^5 ($Fm3m$) (III, a1)	3.24	3.20

Na—Ta	NaTaO$_3$ (per) O_h^1 ($Pm3m$) (VII, a5)	3.36	3.25
Na—W	NaWO$_3$ (per) O_h^1 ($Pm3m$) (VII, a5)	3.34	3.15
Na—Bi	Na$_3$Bi (Na$_3$As) D_{6h}^4 ($P6_3/mmc$) (V, b18)	3.16–3.22	3.40
Mg—N	Mg$_3$N$_2$ (Tl$_2$O$_3$) T_h^7 ($Ia3$) (V, a2)	2.12	2.15
Mg—O	MgO (NaCl) O_h^5 ($Fm3m$) (III, a1)	2.10	2.10
	MgCO$_3$ (cal) D_{3d}^6 ($R\bar{3}c$) (VII, a1)	2.15	2.10
	MgTiO$_3$ (il) C_{3i}^2 ($R\bar{3}$) (V, a5)	1.98–2.17	2.10
	Mg$_2$TiO$_4$ (sp) O_h^7 ($Fd3m$) (VIII, b1)	2.00	2.10
	Mg$_2$SnO$_4$ (sp) O_h^7 ($Fd3m$) (VIII, b1)	2.15	2.10
	Mg(UO$_2$)O$_2$ * D_{2h}^{28} ($Imma$) (VIII, a27)	1.98–2.19	2.10
	Mg$_3$(BO$_3$)$_2$ * D_{2h}^{12} ($Pnnm$) (VII, b14)	2.15	2.10
	Al$_2$MgO$_4$ (sp) O_h^7 ($Fd3m$) (VIII, b1)	1.91	2.10
	Cr$_2$MgO$_4$ (sp) O_h^7 ($Fd3m$) (VIII, b1)	1.93	2.10
	Fe$_2$MgO$_4$ (sp) O_h^7 ($Fd3m$) (VIII, b1)	2.02	2.10
	Ga$_2$MgO$_4$ (sp) O_h^7 ($Fd3m$) (VIII, b1)	2.03	2.10
	In$_2$MgO$_4$ (sp) O_h^7 ($Fd3m$) (VIII, b1)	1.86	2.10
Mg—F	MgF$_2$ (ru) D_{4h}^{14} ($P4_2/mnm$) (IV, b1)	1.98–2.05	2.00
	KMgF$_3$ (per) O_h^1 ($Pm3m$) (VII, a5)	1.99	2.00
Mg—Mg	metal (hex) D_{6h}^4 ($P6_3/mmc$) (II, b)	3.20–3.21	3.00
Mg—Si	Mg$_2$Si (fl) O_h^5 ($Fm3m$) (IV, a1)	2.76	2.65
Mg—P	Mg$_3$P$_2$ (Tl$_2$O$_3$) T_h^7 ($Ia3$) (V, a2)	2.55	2.50
Mg—S	MgS (NaCl) O_h^5 ($Fm3m$) (III, a1)	2.60	2.50
	In$_2$MgS$_4$ (sp) O_h^7 ($Fd3m$) (VIII, b1)	2.48	2.50
Mg—Cl	MgCl$_2$ (CdCl$_2$) D_{3d}^5 ($R\bar{3}m$) (IV, c2)	2.54	2.50
Mg—Ge	Mg$_2$Ge (fl) O_h^5 ($Fm3m$) (IV, a1)	2.76	2.75
Mg—As	Mg$_3$As$_2$ (Tl$_2$O$_3$) T_h^7 ($Ia3$) (V, a2)	2.62	2.65
Mg—Se	MgSe (NaCl) O_h^5 ($Fm3m$) (III, a1)	2.73	2.65
Mg—Br	MgBr$_2$ (CdI$_2$) D_{3d}^3 ($P\bar{3}m1$) (IV, c1)	2.70	2.65
Mg—Sr	MgSr (CsCl) O_h^1 ($Pm3m$) (III, b1)	3.38	3.50
Mg—Ag	MgAg (CsCl) O_h^1 ($Pm3m$) (III, b1)	2.85	3.10
Mg—Sn	Mg$_2$Sn (fl) O_h^5 ($Fm3m$) (IV, a1)	2.93	2.95
Mg—Sb	Mg$_3$Sb$_2$ (La$_2$O$_3$) D_{3d}^3 ($P\bar{3}m1$) (V, a1)	2.82–3.12	2.95
Mg—Te	MgTe (wu) C_{6v}^4 ($P6_3mc$) (III, c2)	2.75	2.90
Mg—I	MgI$_2$ (CdI$_2$) D_{3d}^3 ($P\bar{3}m1$) (IV, c1)	2.94	2.90
Mg—La	MgLa (CsCl) O_h^1 ($Pm3m$) (III, b1)	3.43	3.45
Mg—Ce	MgCe (CsCl) O_h^1 ($Pm3m$) (III, b1)	3.38	3.35
Mg—Pr	MgPr (CsCl) O_h^1 ($Pm3m$) (III, b1)	3.36	3.35
Mg—Au	MgAu (CsCl) O_h^1 ($Pm3m$) (III, b1)	2.83	2.85
Mg—Hg	MgHg (CsCl) O_h^1 ($Pm3m$) (III, b1)	2.98	3.00
Mg—Tl	MgTl (CsCl) O_h^1 ($Pm3m$) (III, b1)	3.14	3.40
Mg—Pb	Mg$_2$Pb (fl) O_h^5 ($Fm3m$) (IV, a1)	2.96	3.30
Mg—Bi	Mg$_3$Bi$_2$ (La$_2$O$_3$) D_{3d}^3 ($P\bar{3}m1$) (V, a1)	2.88–3.18	3.10
Al—B	AlB$_2$ (AlB$_2$) D_{3d}^3 ($P\bar{3}m1$) (IV, c6)	2.37	2.10
Al—C	Al$_4$C$_3$ * D_{3d}^5 ($R\bar{3}m$) (V, g4)	1.90–2.22	1.95
Al—N	AlN (wu) C_{6v}^4 ($P6_3mc$) (III, c2)	1.86	1.90
Al—O	AlO (mol) $C_{\infty v}$ (T.I.D.)	1.62	1.85
	Al$_2$O$_3$ (cor) D_{3d}^6 ($R\bar{3}c$) (V, a4)	1.84–1.98	1.85
	AlPO$_4$ * D_3^4 ($P3_121$) or D_3^6 ($P3_221$) (VIII, a3)	1.70	1.85

INTERATOMIC DISTANCES AND CRYSTAL STRUCTURES

	Compound	Observed	Standard
	$Al(PO_3)_3$ * T_d^6 ($I\bar{4}3d$) (VII, b10)	1.80–1.83	1.85
	Al_2MgO_4 (sp) O_h^7 ($Fd3m$) (VIII, b1)	1.93	1.85
	Al_2MnO_4 (sp) O_h^7 ($Fd3m$) (VIII, b1)	1.95	1.85
	Al_2FeO_4 (sp) O_h^7 ($Fd3m$) (VIII, b1)	1.95	1.85
	Al_2CoO_4 (sp) O_h^7 ($Fd3m$) (VIII, b1)	1.91	1.85
	Al_2NiO_4 (sp) O_h^7 ($Fd3m$) (VIII, b1)	1.90	1.85
	Al_2ZnO_4 (sp) O_h^7 ($Fd3m$) (VIII, b1)	1.91	1.85
	$NaAlFAsO_4$ * C_{2h}^6 ($B2/b$) (VIII, c7)	1.80–1.86	1.85
	$KAlO_2$ * O_h^7 ($Fd3m$) (VI, a13)	1.66	1.85
	$Ca_3Al_2(SiO_4)_3$ (ga) O_h^{10} ($Ia3d$) (XII, a1, 3)	1.93	1.85
	$YAlO_3$ (per) O_h^1 ($Pm3m$) (VII, a5)	1.84	1.85
	$Y_3Al_2(AlO_4)_3$ (ga) O_h^{10} ($Ia3d$) (XII, a1, 3)	1.80–1.87	1.85
	$LaAlO_3$ (per) O_h^1 ($Pm3m$) (VII, a5)	1.89	1.85
Al—F	AlF_3 * D_3^7 ($R32$) (V, b2)	1.70–1.89	1.75
	$TlAlF_4$ * D_{4h}^1 ($P4/mmm$) (VIII, a12)	~1.8	1.75
Al—Al	metal (fcc) O_h^5 ($Fm3m$) (II, a)	2.86	2.50
Al—P	AlP (znb) T_d^2 ($F\bar{4}3m$) (III, c1)	2.35	2.25
Al—Cl	$NaAlCl_4$ * D_2^4 ($P2_12_12_1$) (VIII, a23)	2.11–2.16	2.25
Al—Ni	$AlNi$ (CsCl) O_h^1 ($Pm3m$) (III, b1)	2.44	2.60
Al—As	$AsAl$ (znb) T_d^2 ($F\bar{4}3m$) (III, c1)	2.43	2.40
Al—Br	$AlBr_3$ * C_{2h}^5 ($P2_1/b$) (V, b27)	2.23–2.42	2.40
Al—Sb	$AlSb$ (znb) T_d^2 ($F\bar{4}3m$) (III, c1)	2.66	2.70
Al—Nd	$AlNd$ (CsCl) O_h^1 ($Pm3m$) (III, b1)	3.23	3.10
Si—C	SiC (znb) T_d^2 ($F\bar{4}3m$) (III, c1)	1.89	1.80
Si—O	SiO_2 (several structures and silicates, IV, e2)	1.52–1.69	1.70
	$Ca_3Al_2(SiO_4)_3$ (ga) O_h^{10} ($Ia3d$) (XII, a1, 3)	1.70	1.70
Si—F	SiF (mol) $C_{\infty v}$ (T.I.D.)	1.60	1.60
	SiF_4 (mol) T_d (T.I.D.)	1.55	1.60
	SiF_4 * T_d^3 ($I\bar{4}3m$) (V, c11)	1.56	1.60
	$(SiF_6)^{--}$ in fluosilicates, O_h (T.I.D.)	1.71	1.60
Si—Si	Si_2 (mol) $D_{\infty h}$ (T.I.D.)	2.25	2.20
	element (di) O_h^7 ($Fd3m$) (II, h)	2.35	2.20
	$H_3Si—SiH_3$ (mol) D_{3d} (T.I.D.)	2.32	2.20
	$Cl_3Si—SiCl_3$ (mol) D_{3d} (T.I.D.)	2.32	2.20
Si—Cl	$SiCl_4$ (mol) T_d (T.I.D.)	2.01	2.10
Si—Br	$SiBr_4$ (mol) T_d (T.I.D.)	2.15	2.25
Si—I	SiI_4 (mol) T_d (T.I.D.)	2.43	2.50
P—N	$P_3Cl_6N_3$ (mol) D_{3h} (T.I.D.)	1.65	1.65
P—O	P_2O_5 * C_{2v}^{19} ($Fdd2$) (V, d2)	1.40–1.65	1.60
	$(PO_4)^{-3}$, phosphates (T.I.D.)	1.54–1.56	1.60
	$Al(PO_3)_3$ * T_d^6 ($I\bar{4}3d$) (VII, b10)	1.39–1.60	1.60
P—F	PF_5 (mol) $D_{\infty h}$ (T.I.D.)	1.57	1.50
P—P	P_2 (mol) $D_{\infty h}$ (T.I.D.)	1.89	2.00
	element, black (I) D_{2h}^{18} ($Cmca$) (III, m)	2.17–2.20	2.00
	P_4S_3 (mol) C_{3v} (T.I.D.)	2.25	2.00
P—S	P_4S_3 (mol) C_{3v} (T.I.D.)	2.10	2.00
P—Cl	$P_3Cl_6N_3$ (mol) D_{3h} (T.I.D.)	1.97	2.00
	PCl_5 * C_{4h}^3 ($P4/n$) (V, e1)	1.97–2.08	2.00

S—C	NH_4SCN * C_{2h}^5 ($P2_1/b$) (VI, a20)	1.59	1.70
	CS_2 (mol) $D_{\infty h}$ (T.I.D.)	1.56	1.70
S—N	$H_2N.SO_3H$ (mol) (T.I.D.)	1.76	1.65
S—O	SO (mol) $C_{\infty v}$ (T.I.D.)	1.49	1.60
	$CaSO_4$ * D_{2h}^{17} ($Cmcm$) (VIII, a5)	1.65	1.60
	Na_2SO_3 * C_{3i}^1 ($P\bar{3}$) (VII, b5)	1.39	1.60
	SO_3 (mol) D_{3h} (T.I.D.)	1.43	1.60
	SO_2 * C_{2v}^{17} ($Aba2$) (IV, 16)	1.43	1.60
S—F	SF_6 (mol) O_h (T.I.D.)	1.56	1.50
S—S	S_2 (mol) $D_{\infty h}$ (T.I.D.)	1.92	2.00
	element * D_{2h}^{24} ($Fddd$) (II, p)	2.10	2.00
	CH_3S—SCH_3 (mol) (T.I.D.)	2.04	2.00
	FeS_2 (py) T_h^6 ($Pa3$) (IV, g2)	2.13	2.00
	RuS_2 (py) T_h^6 ($Pa3$) (IV, g2)	2.13	2.00
S—Cl	SCl_2 (mol) C_{2v} (T.I.D.)	2.00	2.00
	SCl_2O (mol) C_s (T.I.D.)	2.07	2.00
S—Br	SBr_2O (mol) C_s (T.I.D.)	2.27	2.15
Cl—O	Cl_2O (mol) C_{2v} (T.I.D.)	1.49	1.60
	$NaClO_4$ * D_{2h}^{17} ($Cmcm$) (VIII, a5)	1.56	1.60
Cl—F	ClF (mol) $C_{\infty v}$ (T.I.D.)	1.63	1.50
	ClF_3 (mol) C_{2v} (T.I.D.)	1.60–1.70	1.50
Cl—Cl	Cl_2 (mol) $D_{\infty h}$ (T.I.D.)	2.00	2.00
	crystal (I) D_{2h}^{18} ($Cmca$) (II, 4a)	2.02	2.00
Cl—I	$CsCl_2I$ * D_{3d}^5 ($R\bar{3}m$) (VI, a7)	2.25	2.40
	NH_4IBrCl * D_{2h}^{16} ($Pnma$) (VI, a16)	2.38	2.40
K—O	K_2O (fl) O_h^5 ($Fm3m$) (IV, a1)	2.79	2.80
	KH_2PO_4 * D_{2d}^{12} ($I\bar{4}2d$) (VIII, c2, c2a)	2.82	2.80
	KNO_2 * C_s^3 (Bm) (VI, a10)	2.75–3.01	2.80
	KNO_3 (ar) D_{2h}^{16} ($Pnma$) (VII, a10)	2.82–2.92	2.80
	$KClO_3$ * D_{2h}^2 ($Pnnn$) (VII, a3)	2.94	2.80
	$KBrO_3$ * C_{3v}^5 ($R3m$) (VII, a2)	2.94	2.80
	$KNbO_3$ (per) O_h^1 (VII, a5)	2.83	2.80
	KIO_3 (per) O_h^1 ($Pm3m$) (VII, a5)	3.16	2.80
	$KTaO_3$ (per) O_h^1 ($Pm3m$) (VII, a5)	2.82	2.80
	$KAgCO_3$ * D_{2h}^{27} ($Ibca$) (VII, b3)	2.65–3.00	2.80
	KIO_2F_2 * C_{2v}^5 ($Pca2_1$) (VIII, a9)	~2.6	2.80
	$KAl_3(OH)_6(SO_4)_2$ * C_{3v}^5 ($R3m$) (VIII, c8)	2.80	2.80
K—F	KF (NaCl) O_h^5 ($Fm3m$) (III, a1)	2.67	2.70
	$KMgF_3$ (per) O_h^1 ($Pm3m$) (VII, a5)	2.81	2.70
	$KNiF_3$ (per) O_h^1 ($Pm3m$) (VII, a5)	2.83	2.70
	$KZnF_3$ (per) O_h^1 ($Pm3m$) (VII, a5)	2.86	2.70
	$KCdF_3$ (per) O_h^1 ($Pm3m$) (VII, a5)	3.03	2.70
	KIO_2F_2 * C_{2v}^5 ($Pca2_1$) (VIII, a9)	2.63	2.70
K—Mg	$KMgF_3$ (per) O_h^1 ($Pm3m$) (VII, a5)	3.44	3.70
K—S	K_2S (fl) O_h^5 ($Fm3m$) (IV, a1)	3.20	3.20
	KCu_4S_3 * D_{4h}^1 ($P4/mmm$) (VII, b21)	3.34	3.20
K—Cl	KCl (NaCl) O_h^5 ($Fm3m$) (III, a1)	3.15	3.20
	K_2PtCl_4 * D_{4h}^1 ($P4/mmm$) (VIII, b15)	3.28	3.20
	$KCrO_3Cl$ * C_{2h}^5 ($P2_1/b$) (VIII, a21)	3.29	3.20

App. 1] INTERATOMIC DISTANCES AND CRYSTAL STRUCTURES 315

K—K	K_2 (mol) $D_{\infty h}$ (T.I.D.)	3.92	4.40
	metal (bcc) O_h^9 ($Im3m$) (II, c)	4.51	4.40
K—Ni	$KNiF_3$ (per) O_h^1 ($Pm3m$) (VII, a5)	3.48	3.55
K—Zn	$KZnF_3$ (per) O_h^1 ($Pm3m$) (VII, a5)	3.52	3.55
K—As	$K_3As(Na_3As)$ D_{6h}^4 ($P6_3/mmc$) (V, b18)	3.34–3.40	3.35
K—Se	K_2Se (fl) O_h^5 ($Fm3m$) (IV, a1)	3.33	3.35
K—Br	KBr (NaCl) O_h^5 ($Fm3m$) (III, a1)	3.30	3.35
K—Nb	$KNbO_3$ (per) O_h^1 ($Pm3m$) (VII, a5)	3.48	3.65
K—Cd	$KCdF_3$ (per) O_h^1 ($Pm3m$) (VII, a5)	3.71	3.75
K—Sb	K_3Sb (Na_3As) D_{6h}^4 ($P6_3/mmc$) (V, b18)	3.48–3.55	3.65
K—Te	K_2Te (fl) O_h^5 ($Fm3m$) (IV, a1)	3.54	3.60
K—I	KI (NaCl) O_h^5 ($Fm3m$) (III, a1)	3.53	3.60
K—Ta	$KTaO_3$ (per) O_h^1 ($Pm3m$) (VIII, a5)	3.45	3.65
K—Bi	K_3Bi (Na_3As) D_{6h}^4 ($P6_3/mmc$) (V, b18)	3.56–3.62	3.80
Ca—B	CaB_6 * O_h^1 ($Pm3m$) (V, fl)	~3.0	2.65
Ca—C	CaC_2 * D_{4h}^{17} ($I4/mmm$) (IV, g1)	2.48–2.83	2.50
Ca—N	Ca_3N_2 (Tl_2O_3) T_h^7 ($Ia3$) (V, a2)	~2.2	2.45
	$CaCN_2$ * D_{3d}^5 ($R\bar{3}m$) (VI, a7)	2.49	2.45
Ca—O	CaO (NaCl) O_h^5 ($Fm3m$) (III, a1)	2.40	2.40
	$CaCO_3$ (cal) D_{3d}^6 ($R\bar{3}c$) (VII, a1)	2.38	2.40
	$CaCO_3$ (ar) D_{2h}^{16} ($Pnma$) (VII, a10)	2.32–2.69	2.40
	$Ca_3Al_2(SiO_4)_3$ (ga) O_h^{10} ($Ia3d$) (XII, a1, 3)	2.34–2.60	2.40
	$CaTiO_3$ (per) O_h^1 ($Pm3m$) (VII, a5)	2.72	2.40
	$CaMnO_3$ (per) O_h^1 ($Pm3m$) (VII, a5)	2.65	2.40
	$CaZrO_3$ (per) O_h^1 ($Pm3m$) (VII, a5)	2.81	2.40
	$CaSnO_3$ (per) O_h^1 ($Pm3m$) (VII, a5)	2.77	2.40
	$CaCeO_3$ (per) O_h^1 ($Pm3m$) (VII, a5)	2.72	2.40
	$CaUO_4$ * D_{3d}^5 ($R\bar{3}m$) (VI, a19)	2.44	2.40
	$Na_2Ca_2(CO_3)_3$ * C_{2v}^{14} ($Amm2$) (VII, b24)	2.32	2.40
Ca—F	CaF_2 (fl) O_h^5 ($Fm3m$) (IV, a1)	2.36	2.30
	$RbCaF_3$ (per) O_h^1 ($Pm3m$) (VII, a5)	2.23	2.30
	$CsCaF_3$ (per) O_h^1 ($Pm3m$) (VII, a5)	2.26	2.30
Ca—S	CaS (NaCl) O_h^5 ($Fm3m$) (III, a1)	2.84	2.80
Ca—Cl	$CaCl_2$ * D_{2h}^{12} ($Pnnm$) (IV, b2)	2.70–2.76	2.80
Ca—Ca	metal (fcc) O_h^5 ($Fm3m$) (II, a)	3.92	3.60
Ca—Ti	$CaTiO_3$ (per) O_h^1 ($Pm3m$) (VII, a5)	3.34	3.20
Ca—Mn	$CaMnO_3$ (per) O_h^1 ($Pm3m$) (VII, a5)	3.24	3.20
Ca—Ga	$CaGa_2$ (AlB_2) D_{3d}^3 ($P\bar{3}m1$) (IV, c6)	3.30	3.10
Ca—Se	$CaSe$ (NaCl) O_h^5 ($Fm3m$) (III, a1)	2.96	2.95
Ca—Zr	$CaZrO_3$ (per) O_h^1 ($Pm3m$) (VII, a5)	3.46	3.35
Ca—Sn	$CaSnO_3$ (per) O_h^1 ($Pm3m$) (VII, a5)	3.40	3.25
Ca—Te	$CaTe$ (NaCl) O_h^5 ($Fm3m$) (III, a1)	3.17	3.20
Ca—I	CaI_2 (CdI_2) D_{3d}^3 ($P\bar{3}m1$) (IV, c1)	3.04	3.20
Ca—Ce	$CaCeO_3$ (per) O_h^1 ($Pm3m$) (VII, a5)	3.35	3.65
Ca—Tl	$CaTl$ (CsCl) O_h^1 ($Pm3m$) (III, b1)	3.34	3.70
Sc—N	ScN (NaCl) O_h^5 ($Fm3m$) (III, a1)	2.22	2.25
Sc—O	Sc_2O_3 (Tl_2O_3) T_h^7 ($Ia3$) (V, a2)	~2.1	2.20
	$ScBO_3$ (cal) D_{3d}^6 ($R\bar{3}c$) (VII, a1)	2.22	2.20

Sc—F	ScF_3 * D_3^7 ($R32$) (VI, b2)	2.36	2.10
Sc—Sc	metal (fcc) O_h^5 ($Fm3m$) (II, a)	3.20	3.20
	metal (hex) D_{6h}^4 ($P6_3/mmc$) (II, b)	3.26	3.20
Ti—C	TiC (NaCl) O_h^5 ($Fm3m$) (III, a1)	2.16	2.10
Ti—N	TiN (NaCl) O_h^5 ($Fm3m$) (III, a1)	2.11	2.05
Ti—O	TiO (NaCl) O_h^5 ($Fm3m$) (III, a1)	2.12	2.00
	TiO_2 (ru) D_{4h}^{14} ($P4_2/mnm$) (IV, b1)	1.89–1.97	2.00
	TiO_2, anatase * D_{4h}^{19} ($I4_1/amd$) (IV, b3)	1.91–1.95	2.00
	TiO_2, brookite * D_{2h}^{15} ($Pbca$) (IV, b4)	1.92–1.98	2.00
	Ti_2O_3 (cor) D_{3d}^6 ($R\bar{3}c$) (V, a4)	1.93–2.08	2.00
	$MgTiO_3$ (il) C_{3i}^2 ($R\bar{3}$) (V, a5)	1.97–2.16	2.00
	Mg_2TiO_4 (sp) O_h^7 ($Fd3m$) (VIII, b1)	2.05	2.00
	$CaTiO_3$ (per) O_h^1 ($Pm3m$) (VII, a5)	1.92	2.00
	$MnTiO_3$ (il) C_{3i}^2 ($R\bar{3}$) (V, a5)	1.99–2.19	2.00
	$FeTiO_3$ (il) C_{3i}^2 ($R\bar{3}$) (V, a5)	1.96–2.15	2.00
	Fe_2TiO_4 (sp) O_h^7 ($Fd3m$) (VIII, b1)	2.06	2.00
	$CoTiO_3$ (il) C_{3i}^2 ($R\bar{3}$) (V, a5)	1.94–2.13	2.00
	$NiTiO_3$ (il) C_{3i}^2 ($R\bar{3}$) (V, a5)	1.93–2.12	2.00
	Zn_2TiO_4 (sp) O_h^7 ($Fd3m$) (VIII, b1)	1.91	2.00
	$SrTiO_3$ (per) O_h^1 ($Pm3m$) (VII, a5)	1.95	2.00
	$BaTiO_3$ (per) O_h^1 ($Pm3m$) (VII, a5)	1.98	2.00
	$BaTiO_3$ * D_{6h}^4 ($P6_3/mmc$) (VII, a17)	1.96	2.00
	$PbTiO_3$ (per) O_h^1 ($Pm3m$) (VII, a5)	1.95	2.00
Ti—S	TiS_2 (CdI_2) D_{3d}^3 ($P\bar{3}m1$) (IV, c1)	2.42	2.40
Ti—Ti	metal (hex) D_{6h}^4 ($P6_3/mmc$) (II, b)	2.90–2.95	2.80
	metal (bcc) O_h^9 ($Im3m$) (II, c)	2.86	2.80
Ti—Se	$TiSe_2$ (CdI_2) D_{3d}^3 ($P\bar{3}m1$) (IV, c1)	2.52	2.55
Ti—Te	$TiTe_2$ (CdI_2) D_{3d}^3 ($P\bar{3}m1$) (IV, c1)	2.72	2.80
Ti—I	TiI_2 (CdI_2) D_{3d}^3 ($P\bar{3}m1$) (IV, c1)	2.92	2.80
V—C	VC (NaCl) O_h^5 ($Fm3m$) (III, a1)	2.07	2.05
V—N	VN (NaCl) O_h^5 ($Fm3m$) (III, a1)	2.06	2.00
V—O	VO (NaCl) O_h^5 ($Fm3m$) (III, a1)	2.04	1.95
	V_2O_3 (cor) D_{3d}^6 ($R\bar{3}c$) (V, a4)	1.95–2.11	1.95
	VO_2 (ru) D_{4h}^{14} ($P4_2/mnm$) (IV, b1)	1.89–1.99	1.95
	V_2O_5 * D_{2h}^{13} ($Pmmn$) (V, d3a)	1.54–2.02	1.95
	$LaVO_3$ (per) O_h^1 ($Pm3m$) (VII, a5)	1.96	1.95
	$CeVO_3$ (per) O_h^1 ($Pm3m$) (VII, a5)	1.95	1.95
	$NdVO_3$ (per) O_h^1 ($Pm3m$) (VII, a5)	1.95	1.95
	$PrVO_3$ (per) O_h^1 ($Pm3m$) (VII, a5)	1.95	1.95
	$SmVO_3$ (per) O_h^1 ($Pm3m$) (VII, a5)	1.95	1.95
	$BiVO_4$ * D_{2h}^{14} ($Pbcn$) (VIII, a20)	1.76–1.95	1.95
V—S	VS (NiAs) D_{6h}^4 ($P6_3/mmc$) (III, d1)	2.42	2.35
	Cu_3VS_4 * T_d^1 ($P\bar{4}3m$) (VIII, c5)	2.19	2.35
V—Cl	$VClO_3$ (mol) C_{3v} (T.I.D.)	2.12	2.35
V—V	metal (bcc) O_h^9 ($Im3m$) (II, c)	2.63	2.70
V—Se	VSe (NiAs) D_{6h}^4 ($P6_3/mmc$) (III, d1)	2.55	2.50
V—Br	VBr_2 (CdI_2) D_{3d}^3 ($P\bar{3}m1$) (IV, c1)	2.67	2.50
V—I	VI_2 (CdI_2) D_{3d}^3 ($P\bar{3}m1$) (IV, c1)	2.85	2.75

App. 1] INTERATOMIC DISTANCES AND CRYSTAL STRUCTURES 317

Cr—N	CrN (NaCl) O_h^5 ($Fm3m$) (III, a1)	2.07	2.05
Cr—O	CrO (mol) $C_{\infty v}$ (T.I.D.)	1.63	2.00
	CrO_2 (ru) D_{4h}^{14} ($P4_2/mnm$) (IV, b1)	1.86–1.94	2.00
	Cr_2O_3 (cor) D_{3d}^6 ($R\bar{3}c$) (V, a4)	1.93–2.08	2.00
	CrO_3 * C_{2v}^{16} ($Ama2$) (V, b9a)	1.79–1.81	2.00
	Cr_2MgO_4 (sp) O_h^7 ($Fd3m$) (VIII, b1)	2.00	2.00
	Cr_2NiO_4 (sp) O_h^7 ($Fd3m$) (VIII, b1)	2.00	2.00
	Cr_2ZnO_4 (sp) O_h^7 ($Fd3m$) (VIII, b1)	2.08	2.00
	Cr_2CdO_4 (sp) O_h^7 ($Fd3m$) (VIII, b1)	2.05	2.00
	$VCrO_4$ * D_{2h}^{17} ($Cmcm$) (VIII, a16)	1.95–2.03	2.00
	$LaCrO_3$ (per) O_h^1 ($Pm3m$) (VII, a5)	1.95	2.00
	$CeCrO_3$ (per) O_h^1 ($Pm3m$) (VII, a5)	1.95	2.00
	$PrCrO_3$ (per) O_h^1 ($Pm3m$) (VII, a5)	1.95	2.00
	$NdCrO_3$ (per) O_h^1 ($Pm3m$) (VII, a5)	1.95	2.00
	$SmCrO_3$ (per) O_h^1 ($Pm3m$) (VII, a5)	1.93	2.00
Cr—S	CrS (NiAs) D_{6h}^4 ($P6_3/mmc$) (III, d1)	2.44	2.40
	Cr_2CdS_4 (sp) O_h^7 ($Fd3m$) (VIII, b1)	2.55	2.40
	Cr_2HgS_4 (sp) O_h^7 ($Fd3m$) (VIII, b1)	2.40	2.40
Cr—Cl	$CrCl_3$ * D_3^7 ($R32$) (V, b2)	~2.5	2.40
	$KCrO_3Cl$ * C_{2h}^5 ($P2_1/c$) (VIII, a21)	2.16	2.40
Cr—Cr	metal (bcc) O_h^9 ($Im3m$) (II, c)	2.50	2.80
	metal (hex) D_{6h}^4 ($P6_3/mmc$) (II, b)	2.71–2.72	2.80
	CrS (NiAs) D_{6h}^4 ($P6_3/mmc$) (III, d1)	2.88	2.80
	CrSe (NiAs) D_{6h}^4 ($P6_3/mmc$) (III, d1)	3.00	2.80
	CrSb (NiAs) D_{6h}^4 ($P6_3/mmc$) (III, d1)	2.72	2.80
	CrTe (NiAs) D_{6h}^4 ($P6_3/mmc$) (III, d1)	3.10	2.80
Cr—Se	CrSe (NiAs) D_{6h}^4 ($P6_3/mmc$) (III, d1)	2.54	2.55
	Cr_2ZnSe_4 (sp) O_h^7 ($Fd3m$) (VIII, b1)	2.60	2.55
	Cr_2CdSe_4 (sp) O_h^7 ($Fd3m$) (VIII, b1)	2.60	2.55
Cr—Br	$CrBr_3$ * D_3^7 ($R32$) (V, 2b)	2.58	2.55
Cr—Sb	CrSb (NiAs) D_{6h}^4 ($P6_3/mmc$) (III, d1)	2.79	2.85
Cr—Te	CrTe (NiAs) D_{6h}^4 ($P6_3/mmc$) (III, d1)	2.77	2.80
Mn—O	MnO (NaCl) O_h^5 ($Fm3m$) (III, a1)	2.22	2.00
	MnO_2 (ru) D_{4h}^{14} ($P4_2/mnm$) (IV, b1)	1.88–1.95	2.00
	Mn_2O_3 (Tl_2O_3) T_h^7 ($Ia3$) (V, a2)	2.01	2.00
	$MnCO_3$ (cal) D_{3d}^6 ($R\bar{3}c$) (VII, a1)	2.22	2.00
	$MnTiO_3$ (il) C_{3i}^2 ($R\bar{3}$) (V, a5)	~2.0–2.1	2.00
	$CaMnO_3$ (per) O_h^1 ($Pm3m$) (VII, a5)	1.87	2.00
	$LaMnO_3$ (per) O_h^1 ($Pm3m$) (VII, a5)	1.94	2.00
	$GdMnO_3$ (per) O_h^1 ($Pm3m$) (VII, a5)	1.91	2.00
	$AgMnO_4$ * C_{2h}^5 ($P2_1/b$) (VIII, a11)	1.48–1.86	2.00
	Al_2MnO_4 (sp) O_h^7 ($Fd3m$) (VIII, b1)	2.00	2.00
Mn—F	MnF_2 (ru) D_{4h}^{14} ($P4_2/mnm$) (IV, b1)	2.10–2.13	1.90
Mn—S	MnS (NaCl) O_h^5 ($Fm3m$) (III, a1)	2.61	2.40
	MnS (znb) T_d^2 ($F\bar{4}3m$) (III, c1)	2.43	2.40
	MnS (wu) C_{6v}^4 ($P6_3mc$) (III, c2)	2.41	2.40
	MnS_2 (py) T_h^6 ($Pa3$) (III, g2)	2.54	2.40
Mn—Cl	$MnCl_2$ ($CdCl_2$) D_{3d}^5 ($R\bar{3}m$) (IV, c2)	2.58	2.40
Mn—Mn	metal (α) * T_d^3 ($I\bar{4}3m$) (II, t)	2.24–3.00	2.80

Mn—Mn	metal (β) * O^6 ($P4_332$) or O^7 ($P4_132$) (II, t)	2.36–2.67	2.80
	metal (γ) (fcc) O_h^5 ($Fm3m$) (II, u)	2.67	2.80
	MnAs (NiAs) D_{6h}^4 ($P6_3/mmc$) (III, d1)	2.84	2.80
	MnSb (NiAs) D_{6h}^4 ($P6_3/mmc$) (III, d1)	2.88	2.80
	MnTe (NiAs) D_{6h}^4 ($P6_3/mmc$) (III, d1)	3.34	2.80
	MnBi (NiAs) D_{6h}^4 ($P6_3/mmc$) (III, d1)	3.06	2.80
Mn—As	MnAs (NiAs) D_{6h}^4 ($P6_3/mmc$) (III, d1)	2.58	2.55
Mn—Se	MnSe (NaCl) O_h^5 ($Fm3m$) (III, a1)	2.72	2.55
	MnSe (znb) T_d^2 ($F\bar{4}3m$) (III, c1)	2.52	2.55
	MnSe (wu) C_{6v}^4 ($P6_3mc$) (III, a2)	2.52	2.55
Mn—Br	MnBr$_2$ (CdI$_2$) D_{3d}^3 ($P\bar{3}m1$) (IV, c1)	2.70	2.55
Mn—Sb	MnSb (NiAs) D_{6h}^4 ($P6_3/mmc$) (III, d1)	2.78	2.85
Mn—Te	MnTe (NiAs) D_{6h}^4 ($P6_3/mmc$) (III, d1)	2.91	2.80
	MnTe$_2$ (py) T_h^6 ($Pa3$) (III, g2)	2.89	2.80
Mn—I	MnI$_2$ (CdI$_2$) D_{3d}^3 ($P\bar{3}m1$) (IV, c1)	2.96	2.80
Mn—Bi	MnBi (NiAs) D_{6h}^4 ($P6_3/mmc$) (III, d1)	2.92	3.00
Fe—O	FeO (NaCl) O_h^5 ($Fm3m$) (III, a1)	2.16	2.00
	Fe$_2$O$_3$ (cor) D_{3d}^6 ($R\bar{3}c$) (V, a4)	1.95–2.10	2.00
	Fe$_3$O$_4$ (sp) O_h^7 ($Fd3m$) (VIII, b2)	1.98	2.00
	FeCO$_3$ (cal) D_{3d}^6 ($R\bar{3}c$) (VII, a1)	2.18	2.00
	FeTiO$_3$ (il) C_{3i}^2 ($R\bar{3}$) (V, a5)	1.97–2.16	2.00
	Fe$_2$MgO$_4$ (sp) O_h^7 ($Fd3m$) (VIII, b1)	1.97	2.00
	Fe$_2$TiO$_4$ (sp) O_h^7 ($Fd3m$) (VIII, b1)	2.00	2.00
	Fe$_2$CuO$_4$ (sp) O_h^7 ($Fd3m$) (VIII, b1)	2.07	2.00
	Fe$_2$ZnO$_4$ (sp) O_h^7 ($Fd3m$) (VIII, b1)	1.99	2.00
	LiFePO$_4$ * D_{2h}^{16} ($Pnma$) (VIII, b4a)	2.08–2.28	2.00
	Al$_2$FeO$_4$ (sp) O_h^7 ($Fd3m$) (VIII, b1)	1.97	2.00
	KFeO$_2$ * O_h^7 ($Fd3m$) (VI, a13)	1.73	2.00
	Y$_3$Fe$_2$(FeO$_4$)$_3$ (ga) O_h^{10} ($Ia3d$) (XII, a1, 3)	1.88–2.00	2.00
	LaFeO$_3$ (per) O_h^1 ($Pm3m$) (VII, a5)	1.95	2.00
Fe—F	FeF$_2$ (ru) D_{4h}^{14} ($P4_2/mnm$) (IV, b1)	2.05–2.08	1.90
	FeF$_3$ * D_3^7 ($R32$) (V, b2)	~2.15	1.90
Fe—S	FeS (NiAs) D_{6h}^4 ($P6_3/mmc$) (III, d1)	2.45	2.40
	FeS$_2$ (py) T_h^6 ($Pa3$) (IV, g2)	2.26	2.40
	KFeS$_2$ * C_{2h}^6 ($B2/b$) (VI, a14)	2.20–2.28	2.40
	CuFeS$_2$ * D_{2d}^{12} ($I\bar{4}2d$) (VI, a11)	2.20	2.40
	Cu$_2$FeSnS$_4$ * D_{2d}^{11} ($I\bar{4}2m$) (VIII, c4)	2.36	2.40
	In$_2$FeS$_4$ (sp) O_h^7 ($Fd3m$) (VIII, b1)	2.47	2.40
Fe—Cl	FeCl$_3$ * D_3^7 ($R32$) (V, b2)	2.45	2.40
	FeCl$_2$ (CdCl$_2$) D_{3d}^5 ($R\bar{3}m$) (IV, c2)	2.54	2.40
Fe—Fe	metal (bcc) O_h^9 ($Im3m$) (II, c)	2.48	2.80
	metal (fcc) O_h^5 ($Fm3m$) (II, a)	2.54	2.80
	FeS (NiAs) D_{6h}^4 ($P6_3/mmc$) (III, d1)	2.84	2.80
	FeSe (NiAs) D_{6h}^4 ($P6_3/mmc$) (III, d1)	2.92	2.80
	FeSb (NiAs) D_{6h}^4 ($P6_3/mmc$) (III, d1)	2.56	2.80
	FeTe (NiAs) D_{6h}^4 ($P6_3/mmc$) (III, d1)	2.82	2.80
Fe—Se	FeSe (NiAs) D_{6h}^4 ($P6_3/mmc$) (III, d1)	2.55	2.55
Fe—Br	FeBr$_2$ (CdI$_2$) D_{3d}^3 ($P\bar{3}m1$) (IV, c1)	2.64	2.55
Fe—Sb	FeSb (NiAs) D_{6h}^4 ($P6_3/mmc$) (III, d1)	2.67	2.85

App. 1]   INTERATOMIC DISTANCES AND CRYSTAL STRUCTURES   319

Fe—Te	FeTe (NiAs) D_{6h}^4 $(P6_3/mmc)$ (III, d1)	2.61	2.80
Fe—I	FeI$_2$ (CdI$_2$) D_{3d}^3 $(P\bar{3}m1)$ (IV, c1)	2.88	2.80
Co—O	CoO (NaCl) O_h^5 $(Fm3m)$ (III, a1)	2.13	1.95
	CoCO$_3$ (cal) D_{3d}^6 $(R\bar{3}c)$ (VII, a1)	2.46	1.95
	CoTiO$_3$ (il) C_{3i}^2 $(R\bar{3})$ (V, a5)	~1.9–2.0	1.95
	Co$_2$SnO$_4$ (sp) O_h^7 $(Fd3m)$ (VIII, b1)	2.15	1.95
	Al$_2$CoO$_4$ (sp) O_h^7 $(Fd3m)$ (VIII, b1)	1.96	1.95
	LaCoO$_3$ (per) O_h^1 $(Pm3m)$ (VII, a5)	1.91	1.95
	PrCoO$_3$ (per) O_h^1 $(Pm3m)$ (VII, a5)	1.88	1.95
	NdCoO$_3$ (per) O_h^1 $(Pm3m)$ (VII, a5)	1.89	1.95
	SmCoO$_3$ (per) O_h^1 $(Pm3m)$ (VII, a5)	1.88	1.95
Co—F	CoF$_2$ (ru) D_{4h}^{14} $(P4_2/mnm)$ (IV, b1)	2.04–2.06	1.85
	CoF$_3$ * D_3^7 $(R32)$ (V, h2)	~2.1	1.85
Co—Si	CoSi$_2$ (fl) O_h^5 $(Fm3m)$ (IV, a1)	2.32	2.45
Co—S	CoS (NiAs) D_{6h}^4 $(P6_3/mmc)$ (III, d1)	2.33	2.35
	CoS$_2$ (py) T_h^6 $(Pa3)$ (IV, g2)	2.31	2.35
	Co$_9$S$_8$ * O_h^5 $(Fm3m)$ (V, h3)	2.14–2.48	2.35
	In$_2$CoS$_4$ (sp) O_h^7 $(Fd3m)$ (VIII, b1)	2.46	2.35
Co—Cl	CoCl$_2$ (CdCl$_2$) D_{3d}^5 $(R\bar{3}m)$ (IV, c2)	2.52	2.35
Co—Co	metal (fcc) O_h^5 $(Fm3m)$ (II, a)	2.51	2.70
	metal (hex) D_{6h}^4 $(P6_3/mmc)$ (II, b)	2.49–2.50	2.70
	CoS (NiAs) D_{6h}^4 $(P6_3/mmc)$ (III, d1)	2.60	2.70
	CoSe (NiAs) D_{6h}^4 $(P6_3/mmc)$ (III, d1)	2.64	2.70
	CoSb (NiAs) D_{6h}^4 $(P6_3/mmc)$ (III, 31)	2.58	2.70
	CoTe (NiAs) D_{6h}^4 $(P6_3/mmc)$ (III, d1)	2.68	2.70
Co—As	CoAs$_3$ * T_h^5 $(Im3)$ (V, b16)	2.35	2.50
Co—Se	CoSe (NiAs) D_{6h}^4 $(P6_3/mmc)$ (III, d1)	2.46	2.50
	CoSe$_2$ (py) T_h^6 $(Pa3)$ (IV, g2)	2.44	2.50
Co—Br	CoBr$_2$ (CdI$_2$) D_{3d}^3 $(P\bar{3}m1)$ (IV, c1)	2.62	2.50
Co—Sb	CoSb (NiAs) D_{6h}^4 $(P6_3/mmc)$ (III, d1)	2.58	2.80
Co—Te	CoTe (NiAs) D_{6h}^4 $(P6_3/mmc)$ (III, d1)	2.62	2.75
	CoTe$_2$ (CdI$_2$) D_{3d}^3 $(P\bar{3}m1)$ (IV, c1)	2.75	2.75
Co—I	CoI$_2$ (CdI$_2$) D_{3d}^3 $(P\bar{3}m1)$ (IV, c1)	2.82	2.75
Ni—C	Ni(CO)$_4$ * T_h^6 $(Pa3)$ (V, c9)	1.84	2.05
Ni—O	NiO (NaCl) O_h^5 $(Fm3m)$ (III, a1)	2.09	1.95
	NiO·BaO * D_{2h}^{17} $(Cmcm)$ (VI, a22)	2.01	1.95
	NiO·3BaO * D_{3d}^6 $(R\bar{3}c)$ (VIII, c13)	1.96	1.95
	NiTiO$_3$ (il) C_{3i}^2 $(R\bar{3})$ (V, a5)	~1.9–2.0	1.95
	NaNiO$_2$ * C_{2h}^3 $(B2/m)$ (VI, a23)	1.95–2.17	1.95
	Al$_2$NiO$_4$ (sp) O_h^7 $(Fd3m)$ (VIII, b1)	1.95	1.95
	Cr$_2$NiO$_4$ (sp) O_h^7 $(Fd3m)$ (VIII, b1)	1.93	1.95
Ni—F	NiF$_2$ (ru) D_{4h}^{14} $(P4_2/mnm)$ (IV, b1)	2.02–2.07	1.85
	KNiF$_3$ (per) O_h^1 $(Pm3m)$ (VII, a5)	2.01	1.85
Ni—S	NiS (NiAs) D_{6h}^4 $(P6_3/mmc)$ (III, d1)	2.38	2.35
	NiS$_2$ (py) T_h^6 $(Pa3)$ (IV, g2)	2.39	2.35
	Ni$_3$S$_2$ * D_3^7 $(R32)$ (V, a9)	2.28	2.35
	In$_2$NiS$_4$ (sp) O_h^7 $(Fd3m)$ (VIII, b1)	2.44	2.35
Ni—Cl	NiCl$_2$ (CdCl$_2$) D_{3d}^5 $(R\bar{3}m)$ (IV, c2)	2.51	2.35

Ni—Ni	metal (fcc) O_h^5 ($Fm3m$) (II, a)	2.49	2.70
	metal (hex) D_{6h}^4 ($P6_3/mmc$) (II, b)	2.65	2.70
	NiS (NiAs) D_{6h}^4 ($P6_3/mmc$) (III, d1)	2.66	2.70
	NiAs (NiAs) D_{6h}^4 ($P6_3/mmc$) (III, d1)	2.50	2.70
	NiSe (NiAs) D_{6h}^4 ($P6_3/mmc$) (III, d1)	2.66	2.70
	NiSn (NiAs) D_{6h}^4 ($P6_3/mmc$) (III, d1)	2.58	2.70
	NiSb (NiAs) D_{6h}^4 ($P6_3/mmc$) (III, d1)	2.56	2.70
	NiTe (NiAs) D_{6h}^4 ($P6_3/mmc$) (III, d1)	2.66	2.70
	NiO·BaO * D_{2h}^{17} ($Cmcm$) (IV, a22)	2.36	2.70
Ni—As	NiAs (NiAs) D_{6h}^4 ($P6_3/mmc$) (III, d1)	2.43	2.50
Ni—Se	NiSe (NiAs) D_{6h}^4 ($P6_3/mmc$) (III, d1)	2.50	2.50
	NiSe$_2$ (py) T_h^6 ($Pa3$) (IV, g2)	2.48	2.50
Ni—Br	NiBr$_2$(CdCl$_2$) D_{3d}^5 ($R\bar{3}m$) (IV, c2)	2.64	2.50
Ni—Sb	NiSb (NiAs) D_{6h}^4 ($P6_3/mmc$) (III, d1)	2.60	2.80
Ni—Te	NiTe (NiAs) D_{6h}^4 ($P6_3/mmc$) (III, d1)	2.64	2.75
	NiTe$_2$ (CdI$_2$) D_{3d}^3 ($P\bar{3}m1$) (IV, c1)	2.60	2.75
Ni—I	NiI$_2$ (CdCl$_2$) D_{3d}^5 ($R\bar{3}m$) (IV, c2)	2.78	2.75
Cu—N	Cu$_3$N * O_h^1 ($Pm3m$) (V, b5)	1.95	2.00
Cu—O	CuO * C_{2h}^6 ($B2/b$) (III, f1)	1.95	1.95
	Cu$_2$O * O_h^4 ($Pn3m$) (IV, f1)	1.84	1.95
	Cu$_2$(OH)PO$_4$ * D_{2h}^{12} ($Pnnm$) (VIII, c6)	1.84–2.34	1.95
	Cu$_2$(OH)AsO$_4$ * D_{2h}^{12} ($Pnnm$) (VIII, c6)	1.84–2.34	1.95
	Fe$_2$CuO$_4$ (sp) O_h^7 ($Fd3m$) (VIII, b1)	1.90	1.95
Cu—F	CuF (znb) T_d^2 ($F\bar{4}3m$) (III, c1)	1.85	1.85
	CuF$_2$ (fl) O_h^5 ($Fm3m$) (IV, a1)	2.34	1.85
Cu—S	Cu$_3$VS$_4$ * T_d^1 ($P\bar{4}3m$) (VIII, c5)	2.19	2.35
	CuFeS$_2$ * D_{2d}^{12} ($I\bar{4}2d$) (VI, a11)	2.32	2.35
	Cu$_2$FeSnS$_4$ * D_{2d}^{11} ($I\bar{4}2m$) (VIII, c4)	2.31	2.35
	Cu$_3$AsS$_4$ * C_{2v}^7 ($Pmn2_1$) (VIII, c3)	2.31	2.35
	CuSbS$_2$ * D_{2h}^{16} ($Pnma$) (VI, a12)	2.25–2.33	2.35
	KCu$_4$S$_3$ * D_{4h}^1 ($P4/mmm$) (VII, b21)	2.34–2.45	2.35
	RbCu$_4$S$_3$ * D_{4h}^1 ($P4/mmm$) (VII, b21)	2.34–2.45	2.35
Cu—Cl	CuCl (znb) T_d^2 ($F\bar{4}3m$) (III, c1)	2.35	2.35
	Cs$_2$CuCl$_4$ * D_{2h}^{16} ($Pnma$) (VIII, b19)	2.21	2.35
	(NH$_4$)$_2$CuCl$_3$ * D_{2h}^{19} ($Cmmm$) (VII, b18)	2.34–2.40	2.35
Cu—Cu	metal (fcc) O_h^5 ($Fm3m$) (II, a)	2.54	2.70
	Cu$_3$N * O_h^1 ($Pm3m$) (V, b5)	2.71	2.70
	CuSn (NiAs) D_{6h}^4 ($P6_3/mmc$) (III, d1)	2.54	2.70
Cu—Zn	CuZn (CsCl) O_h^1 ($Pm3m$) (III, b1)	2.55	2.70
Cu—Br	CuBr (znb) T_d^2 ($F\bar{4}3m$) (III, c1)	2.46	2.50
Cu—Pd	CuPd (CsCl) O_h^1 ($Pm3m$) (III, b1)	2.59	2.75
Cu—Sn	CuSn (NiAs) D_{6h}^4 ($P6_3/mmc$) (III, d1)	2.73	2.80
Cu—Sb	Cu$_2$Sb * D_{4h}^7 ($P4/nmm$) (IV, h3)	~2.5–2.7	2.80
Cu—I	CuI (znb) T_d^2 ($F\bar{4}3m$) (III, c1)	2.62	2.75
Zn—N	Zn$_3$N$_2$ (Tl$_2$O$_3$) T_h^7 ($Ia3$) (V, a2)	~2.1	2.00
Zn—O	ZnO (wu) C_{6v}^4 ($P6_3mc$) (III, c2)	1.95	1.95
	ZnCO$_3$ (cal) D_{3d}^6 ($R\bar{3}c$) (VII, a1)	2.16	1.95
	Zn$_2$TiO$_4$ (sp) O_h^7 ($Fd3m$) (VIII, b1)	2.08	1.95

Zn—O	Zn$_2$SnO$_4$ (sp) O_h^7 ($Fd3m$) (VIII, b1)	2.03	1.95
	Al$_2$ZnO$_4$ (sp) O_h^7 ($Fd3m$) (VIII, b1)	1.96	1.95
	Cr$_2$ZnO$_4$ (sp) O_h^7 ($Fd3m$) (VIII, b1)	1.93	1.95
	Fe$_2$ZnO$_4$ (sp) O_h^7 ($Fd3m$) (VIII, b1)	2.04	1.95
	Pb(Zn,Cu)(OH)VO$_4$ * D_{2h}^{16} ($Pnma$) (VIII, c14)	1.88–1.96	1.95
Zn—F	ZnF$_2$ (ru) D_{4h}^{14} ($P4_2/mnm$) (IV, b1)	2.02–2.07	1.85
	KZnF$_3$ (per) O_h^1 ($Pm3m$) (VII, a5)	2.03	1.85
	AgZnF$_3$ (per) O_h^1 ($Pm3m$) (VII, a5)	1.99	1.85
Zn—P	Zn$_3$P$_2$ * D_{4h}^{15} ($P4_2/nmc$) (V, a3)	2.35	2.35
Zn—S	ZnS (znb) T_d^2 ($F\bar{4}3m$) (III, c1)	2.36	2.35
	ZnS (wu) C_{6v}^4 ($P6_3mc$) (III, c2)	2.33	2.35
Zn—Cl	ZnCl$_2$ (CdCl$_2$) D_{3d}^5 ($R\bar{3}m$) (IV, c2)	2.64	2.35
Zn—Zn	metal (hex) D_{6h}^4 ($P6_3/mmc$) (II, b)	2.66–2.91	2.70
	ZnSb * D_{2h}^{15} ($Pbca$) (III, c4)	2.66–2.74	2.70
Zn—As	Zn$_3$As$_2$ * D_{4h}^{15} ($P4_2/nmc$) (V, c4)	2.43	2.50
Zn—Se	ZnSe (znb) T_d^2 ($F\bar{4}3m$) (III, c1)	2.45	2.50
	Cr$_2$ZnSe$_4$ (sp) O_h^7 ($Fd3m$) (VIII, b1)	2.32	2.50
Zn—Ag	ZnAg (CsCl) O_h^1 ($Pm3m$) (III, b1)	2.73	2.95
Zn—Sb	ZnSb * D_{2h}^{15} ($Pbca$) (III, c4)	2.66–2.74	2.80
Zn—Te	ZnTe (znb) T_d^2 ($F\bar{4}3m$) (III, c1)	2.63	2.75
Zn—I	ZnI$_2$ (CdI$_2$) D_{3d}^3 ($P\bar{3}m1$) (IV, c1)	2.95	2.75
Zn—La	ZnLa (CsCl) O_h^1 ($Pm3m$) (III, b1)	3.25	3.30
Zn—Ce	ZnCe (CsCl) O_h^1 ($Pm3m$) (III, b1)	3.21	3.20
Zn—Pr	ZnPr (CsCl) O_h^1 ($Pm3m$) (III, b1)	3.18	3.20
Zn—Au	ZnAu (CsCl) O_h^1 ($Pm3m$) (III, b1)	2.77	2.70
Ga—N	GaN (wu) C_{6v}^4 ($P6_3mc$) (III, c2)	1.94	1.95
Ga—O	Ga$_2$O$_3$ (cor) D_{3d}^6 ($R\bar{3}c$) (V, a4)	1.89–2.05	1.90
	Ga$_2$MgO$_4$ (sp) O_h^7 ($Fd3m$) (III, b1)	1.94	1.90
	LaGaO$_3$ (per) O_h^1 ($Pm3m$) (VII, a5)	1.95	1.90
Ga—P	GaP (znb) T_d^2 ($F\bar{4}3m$) (III, c1)	2.36	2.30
Ga—Ga	metal (I) D_{2h}^{18} ($Cmca$) (II, g)	2.44–2.76	2.60
	CaGa$_2$ (AlB$_2$) D_{3d}^3 ($P\bar{3}m1$) (IV, c6)	2.50	2.60
	LaGa$_2$ (AlB$_2$) D_{3d}^3 ($P\bar{3}m1$) (IV, c6)	2.51	2.60
	CeGa$_2$ (AlB$_2$) D_{3d}^3 ($P\bar{3}m1$) (IV, c6)	2.49	2.60
Ga—As	GaAs (znb) T_d^2 ($F\bar{4}3m$) (III, c1)	2.43	2.45
Ga—Sb	GaSb (znb) T_d^2 ($F\bar{4}3m$) (III, c1)	2.65	2.75
Ge—O	GeO$_2$ (ru) D_{4h}^{14} ($P4_2/mnm$) (IV, b1)	1.85–1.93	1.85
Ge—S	GeS * D_{2h}^{16} ($Pnma$) (III, h3)	2.47–3.00	2.25
	GeS$_2$ * C_{2v}^{19} ($Fdd2$) (IV, e9)	2.07–2.26	2.25
Ge—Ge	element (di) O_h^7 ($Fd3m$) (II, h)	2.45	2.50
	H$_3$Ge—GeH$_2$—GeH$_3$ (mol) (T.I.D.)	2.41	2.50
Ge—I	GeI$_2$ (CdI$_2$) D_{3d}^3 ($P\bar{3}m1$) (IV, c1)	2.92	2.65
	GeI$_4$ * T_h^6 ($Pa3$) (V, c1)	2.54	2.65
As—O	As$_2$O$_3$ * O_h^7 ($Fd3m$) (V, a6)	2.01	1.75
	As$_2$O$_3$ * C_{2h}^5 ($P2_1/b$) (V, a15)	1.74–1.82	1.75
	BAsO$_4$ * S_4^2 ($I\bar{4}$) (VIII, a2)	1.66	1.75
	NaAlFAsO$_4$ * C_{2h}^6 ($B2/b$) (VIII, c7)	1.68	1.75
	AlAsO$_4$ * D_3^4 ($P3_121$) or D_3^6 ($P3_221$) (VIII, a3)	1.62	1.75

As—O	R—H$_2$AsO$_4$ * D_{2d}^{12} ($I\bar{4}2d$) (VIII, c2)	1.75	1.75
	Cu$_2$(OH)AsO$_4$ * D_{2h}^{12} ($Pnnm$) (VIII, c6)	1.49–1.81	1.75
	BiAsO$_4$ * C_{4h}^6 ($I4_1/a$) (VIII, a6)	1.63	1.75
As—S	As$_2$S$_3$ * C_{2h}^5 ($P2_1/b$) (V, a14)	2.15–2.34	2.15
	Cu$_3$AsS$_4$ * C_{2v}^7 ($Pmn2_1$) (VIII, c3)	2.21–2.24	2.15
	Ag$_3$AsS$_3$ * C_{3v}^6 ($R3c$) (VII, b6)	2.25	2.15
As—As	metal (As) D_{3d}^5 ($R\bar{3}m$) (II, n)	2.50	2.30
	PtAs$_2$ (py) T_h^6 ($Pa3$) (IV, g2)	2.26	2.30
As—I	AsI$_3$ * D_3^7 ($R32$) (V, b2)	2.96	2.55
Se—O	SeO$_2$ * D_{4h}^{13} ($P4_2/mbc$) (IV, i1)	1.75–1.79	1.75
	H$_2$SeO$_3$ * D_2^4 ($P2_12_12_1$) (VII, b13)	1.72–1.76	1.75
	H$_2$SeO$_4$ * D_2^3 ($P2_12_12$) (VIII, b21)	1.57–1.66	1.75
Se—F	SeF$_6$ (mol) O_h (T.I.D.)	1.70	1.65
Se—Cl	Cs$_2$SeCl$_6$ * O_h^5 ($Fm3m$) (IX, c1)	2.41	2.15
Se—Se	element (Se) D_3^4 ($P3_121$) or D_3^6 ($P3_221$) (II, q)	2.32	2.30
	CoSe$_2$ (py) T_h^6 ($Pa3$) (IV, g2)	2.49	2.30
Br—O	NaBrO$_3$ * T^4 ($P2_13$) (VII, a4)	1.78	1.75
Br—F	BrF (mol) $C_{\infty v}$ (T.I.D.)	1.76	1.65
Br—Cl	BrCl (mol) $C_{\infty v}$ (T.I.D.)	2.14	2.15
Br—Br	Br$_2$ (mol) $D_{\infty h}$ (T.I.D.)	2.29	2.30
	crystal (I) D_{2h}^{18} ($Cmca$) (II, v)	2.42	2.30
Rb—O	Rb$_2$O (fl) O_h^5 ($Fm3m$) (IV, a1)	2.92	2.95
	RbIO$_3$ (per) O_h^1 ($Pm3m$) (VII, a5)	3.20	2.95
Rb—F	RbF (NaCl) O_h^5 ($Fm3m$) (III, a1)	2.82	2.85
	RbCaF$_3$ (per) O_h^1 ($Pm3m$) (VII, a5)	3.14	2.85
Rb—S	Rb$_2$S (fl) O_h^5 ($Fm3m$) (IV, a1)	3.32	3.35
	RbCu$_4$S$_3$ * D_{4h}^1 ($P4/mmm$) (VII, b21)	3.39	3.35
Rb—Cl	RbCl (NaCl) O_h^5 ($Fm3m$) (III, a1)	3.29	3.35
	RbCl (CsCl) O_h^1 ($Pm3m$) (III, b1)	3.23	3.35
Rb—Ca	RbCaF$_3$ (per) O_h^1 ($Pm3m$) (VII, a5)	3.86	4.15
Rb—Br	RbBr (NaCl) O_h^5 ($Fm3m$) (III, a1)	3.43	3.50
Rb—Rb	metal (bcc) O_h^9 ($Im3m$) (II, c)	4.88	4.70
Rb—I	RbI (NaCl) O_h^5 ($Fm3m$) (III, a1)	3.66	3.75
	RbIO$_3$ (per) O_h^1 ($Pm3m$) (VII, a5)	3.92	3.75
Sr—O	SrO (NaCl) O_h^5 ($Fm3m$) (III, a1)	2.57	2.60
	SrTiO$_3$ (per) O_h^1 ($Pm3m$) (VII, a5)	2.76	2.60
	SrZrO$_3$ (per) O_h^1 ($Pm3m$) (VII, a5)	2.87	2.60
	SrSnO$_3$ (per) O_h^1 ($Pm3m$) (VII, a5)	2.85	2.60
	SrCeO$_3$ (per) O_h^1 ($Pm3m$) (VII, a5)	3.02	2.60
	SrHfO$_3$ (per) O_h^1 ($Pm3m$) (VII, a5)	2.88	2.60
Sr—F	SrF$_2$ (fl) O_h^5 ($Fm3m$) (IV, a1)	2.51	2.50
Sr—S	SrS (NaCl) O_h^5 ($Fm3m$) (III, a1)	2.94	3.00
Sr—Cl	SrCl$_2$ (fl) O_h^5 ($Fm3m$) (IV, a1)	3.02	3.00
Sr—Ti	SrTiO$_3$ (per) O_h^1 ($Pm3m$) (VII, a5)	3.38	3.40
Sr—Se	SrSe (NaCl) O_h^5 ($Fm3m$) (III, a1)	3.12	3.15
Sr—Br	SrBr$_2$ * D_{2h}^{16} ($Pnma$) (IV, d4)	3.16–3.44	3.15

Sr—Sr	metal (fcc) O_h^5 ($Fm3m$) (II, a)	4.30	4.00
	metal (hex) D_{6h}^4 ($P6_3/mmc$) (II, h)	4.32	4.00
	metal (bcc) O_h^9 ($Im3m$) (II, c)	4.20	4.00
Sr—Zr	SrZrO$_3$ (per) O_h^1 ($Pm3m$) (VII, a5)	3.54	3.55
Sr—Sn	SrSnO$_3$ (per) O_h^1 ($Pm3m$) (VII, a5)	3.48	3.45
Sr—Te	SrTe (NaCl) O_h^5 ($Fm3m$) (III, a1)	3.24	3.40
Sr—Ce	SrCeO$_3$ (per) O_h^1 ($Pm3m$) (VII, a5)	3.70	3.85
Sr—Hf	SrHfO$_3$ (per) O_h^1 ($Pm3m$) (VII, a5)	3.53	3.55
Sr—Tl	SrTl (CsCl) O_h^1 ($Pm3m$) (III, b1)	3.48	3.90
Y—O	Y$_2$O$_3$ (Tl$_2$O$_3$) T_h^7 ($Ia3$) (V, a2)	2.25	2.40
	YBO$_3$ (cal) D_{3d}^6 ($R\bar{3}c$) (VII, a1)	2.40	2.40
	YOF (LaOF) D_{3d}^5 ($R\bar{3}m$) (IV, a5)	2.34–2.42	2.40
	YAlO$_3$ (per) O_h^1 ($Pm3m$) (VII, a5)	2.60	2.40
	Y$_3$Al$_2$(AlO$_4$)$_3$ (ga) O_h^{10} ($Ia3d$) (XII, a1, 3)	2.40–2.46	2.40
	Y$_3$Fe$_2$(FeO$_4$)$_3$ (ga) O_h^{10} ($Ia3d$) (XII, a1, 3)	2.37–2.43	2.40
Y—F	YOF (LaOF) D_{3d}^5 ($R\bar{3}m$) (IV, a5)	2.26–2.27	2.30
Y—Al	YAlO$_3$ (per) O_h^1 ($Pm3m$) (VII, a5)	3.18	3.05
Y—Y	metal (hex) D_{6h}^4 ($P6_3/mmc$) (II, b)	3.56–3.64	3.60
Zr—C	ZrC (NaCl) O_h^5 ($Fm3m$) (III, a1)	2.34	2.25
Zr—N	ZrN (NaCl) O_h^5 ($Fm3m$) (III, a1)	2.31	2.20
Zr—O	ZrO (mol) $C_{\infty v}$ (T.I.D.)	1.73	2.15
	ZrO$_2$ (fl) O_h^5 ($Fm3m$) (IV, a1)	2.20	2.15
	ZrSiO$_4$ * D_{4h}^{19} ($I4_1/amd$) (VIII, a4)	2.05	2.15
	CaZrO$_3$ (per) O_h^1 ($Pm3m$) (VII, a5)	2.00	2.15
	SrZrO$_3$ (per) O_h^1 ($Pm3m$) (VII, a5)	2.05	2.15
	BaZrO$_3$ (per) O_h^1 ($Pm3m$) (VII, a5)	2.20	2.15
	PbZrO$_3$ (per) O_h^1 ($Pm3m$) (VII, a5)	2.32	2.15
Zr—P	ZrP (NaCl) O_h^5 ($Fm3m$) (III, a1)	2.63	2.55
Zr—S	ZrS$_2$ (CdI$_2$) D_{3d}^3 ($P\bar{3}m1$) (IV, c1)	2.58	2.55
Zr—Cl	ZrCl$_4$ (mol) T_d (T.I.D.)	2.34	2.55
	Cs$_2$ZrCl$_6$ * O_h^5 ($Fm3m$) (IX, c1)	2.45	2.55
Zr—Se	ZrSe$_2$ (CdI$_2$) D_{3d}^3 ($P\bar{3}m1$) (IV, c1)	2.67	2.70
Zr—Zr	metal (hex) D_{6h}^4 ($P6_3/mmc$) (II, b)	3.18–3.23	3.10
	metal (bcc) O_h^9 ($Im3m$) (II, c)	3.13	3.10
Nb—C	NbC (NaCl) O_h^5 ($Fm3m$) (III, a1)	2.23	2.15
Nb—N	NbN (NaCl) O_h^5 ($Fm3m$) (III, a1)	2.20	2.10
Nb—O	NbO$_2$ (ru) D_{4h}^{14} ($P4_2/mnm$) (IV, b1)	1.97–2.09	2.05
	NaNbO$_3$ (per) O_h^1 ($Pm3m$) (VII, a5)	1.95	2.05
	KNbO$_3$ (per) O_h^1 ($Pm3m$) (VII, a5)	2.01	2.05
Nb—Cl	NbCl$_5$ (mol) D_{3h} (T.I.D.)	2.29	2.45
Nb—Br	NbBr$_5$ (mol) D_{3h} (T.I.D.)	2.46	2.60
Nb—Nb	metal (bcc) O_h^9 ($Im3m$) (II, c)	2.86	2.90
Mo—O	MoO$_2$ (ru) D_{4h}^{14} ($P4_2/mnm$) (IV, b1)	1.92–2.13	2.05
	MoO$_3$ * D_{2h}^{16} ($Pnmn$) (V, b6)	1.88–2.45	2.05
	Ag$_2$MoO$_4$ (sp) O_h^7 ($Fd3m$) (VIII, b1)	1.83	2.05
Mo—Mo	métal (bcc) O_h^9 ($Im3m$) (II, c)	2.73	2.90

Tc—Tc	metal (hex) D_{6h}^4 ($P6_3/mmc$) (T.I.D.)	2.70–2.74	2.70
Ru—O	RuO$_2$ (ru) D_{4h}^{14} ($P4_2/mnm$) (IV, b1)	1.98	1.90
Ru—S	RuS$_2$ (py) T_h^6 ($Pa3$) (II, g2)	2.35	2.30
Ru—Se	RuSe$_2$ (py) T_h^6 ($Pa3$) (II, g2)	2.47	2.45
Ru—Ru	metal (hex) D_{6h}^4 ($P6_3/mmc$) (II, b)	2.65–2.70	2.60
Ru—Te	RuTe$_2$ (py) T_h^6 ($Pa3$) (II, g2)	2.66	2.70
Rh—O	Rh$_2$O$_3$ (cor) D_{3d}^6 ($R\bar{3}c$) (V, a4)	1.96–2.13	1.95
Rh—F	RhF$_3$ * D_3^7 ($R32$) (V, b2)	2.01	1.85
Rh—P	Rh$_2$P (fl) O_h^5 ($Fm3m$) (IV, a1)	2.39	2.35
Rh—S	RhS$_2$ (py) T_h^6 ($Pa3$) (II, g2)	2.33	2.35
Rh—Rh	metal (fcc) O_h^5 ($Fm3m$) (II, a)	2.69	2.70
Pd—F	PdF$_2$ (ru) D_{4h}^{14} ($P4_2/mnm$) (IV, b1)	2.15–2.16	1.90
	PdF$_3$ * D_3^7 ($R32$) (V, b2)	2.09	1.90
Pd—As	PdAs$_2$ (py) T_h^6 ($Pa3$) (IV, g2)	2.49	2.55
Pd—Pd	metal (fcc) O_h^5 ($Fm3m$) (II, a)	2.74	2.80
	PdSb (NiAs) D_{6h}^4 ($P6_3/mmc$) (III, d1)	2.78	2.80
Pd—Sb	PdSb (NiAs) D_{6h}^4 ($P6_3/mmc$) (III, d1)	2.73	2.85
	PdSb$_2$ (py) T_h^6 ($Pa3$) (IV, g2)	2.69	2.85
Pd—Te	PdTe (NiAs) D_{6h}^4 ($P6_3/mmc$) (III, d1)	2.77	2.80
	PdTe$_2$ (CdI$_2$) D_{3d}^3 ($P\bar{3}m1$) (IV, c1)	2.66	2.80
Ag—N	AgNO$_2$ * C_{2v}^{20} ($Imm2$) (VI, a9)	2.06	2.25
Ag—O	Ag$_2$O * O_h^4 ($Pn3m$) (T.I.D.)	2.05	2.20
	AgNO$_2$ * C_{2v}^{20} ($Imm2$) (VI, a9)	2.69	2.20
	AgClO$_3$ * C_{4h}^5 ($I4/m$) (VII, a13)	2.47–2.55	2.20
	AgMnO$_4$ * C_{2h}^5 ($P2_1/b$) (VIII, a11)	~2.21	2.20
	AgBrO$_3$ * C_{4h}^5 ($I4/m$) (VII, a13)	2.47–2.55	2.20
	Ag$_2$MoO$_4$ (sp) O_h^7 ($Fd3m$) (VIII, b1)	2.43	2.20
	Ag$_2$PbO$_2$ * C_{2h}^5 ($B2/b$) (VI, a25)	2.08–2.10	2.20
Ag—F	AgF (NaCl) O_h^5 ($Fm3m$) (III, a1)	2.46	2.10
	Ag$_2$F (CdI$_2$) D_{3d}^3 ($P\bar{3}m1$) (IV, c1)	2.24	2.10
	AgZnF$_3$ (per) O_h^1 ($Pm3m$) (VII, a5)	2.81	2.10
Ag—S	Ag$_3$AsS$_3$ * C_{3v}^6 ($R3c$) (VII, b6)	2.40	2.60
	Ag$_3$SbS$_3$ * C_{3v}^6 ($R3c$) (VII, b6)	2.40	2.60
Ag—Cl	AgCl (NaCl) O_h^5 ($Fm3m$) (III, a1)	2.77	2.60
	Cs$_2$AgAuCl$_2$ * D_{4h}^{17} ($I4/mmm$) (VII, a6)	2.36	2.60
Ag—Zn	AgZn (CsCl) O_h^1 ($Pm3m$) (III, b1)	2.73	2.95
	AgZnF$_3$ (per) O_h^1 ($Pm3m$) (VII, a5)	3.44	2.95
Ag—Br	AgBr (NaCl) O_h^5 ($Fm3m$) (III, a1)	2.89	2.75
Ag—Ag	metal (fcc) O_h^5 ($Fm3m$) (II, a)	2.89	3.20
Ag—Cd	AgCd (CsCl) O_h^1 ($Pm3m$) (III, b1)	2.89	3.15
Ag—I	AgI (znb) T_d^2 ($F\bar{4}3m$) (III, c1)	2.80	3.00
Ag—La	AgLa (CsCl) O_h^1 ($Pm3m$) (III, b1)	3.26	3.55
Ag—Ce	AgCe (CsCl) O_h^1 ($Pm3m$) (III, b1)	3.23	3.45
Cd—N	Cd$_3$N$_2$ (Tl$_2$O$_3$) T_h^7 ($Ia3$) (V, a2)	2.30	2.20
Cd—O	CdO (NaCl) O_h^5 ($Fm3m$) (III, a1)	2.35	2.15

App. 1] INTERATOMIC DISTANCES AND CRYSTAL STRUCTURES 325

Cd—O	$CdCO_3$ (cal) D_{3d}^6 ($R\bar{3}c$) (VIII, a1)	2.32	2.15
	$CdSnO_3$ (per) O_h^1 ($Pm3m$) (VII, a5)	2.68	2.15
	$CdCeO_3$ (per) O_h^1 ($Pm3m$) (VII, a5)	2.71	2.15
	Cr_2CdO_4 (sp) O_h^7 ($Fd3m$) (VII, b1)	2.00	2.15
Cd—F	CdF_2 (fl) O_h^5 ($Fm3m$) (IV, a1)	2.32	2.05
	$KCdF_3$ (per) O_h^1 ($Pm3m$) (VII, a5)	2.15	2.05
Cd—P	Cd_3P_2 * D_{4h}^{15} ($P4_2/nmc$) (V, a3)	2.52	2.55
Cd—S	CdS (znb) T_d^2 ($F\bar{4}3m$) (III, c1)	2.52	2.55
	CdS (wu) C_{6v}^4 ($P6_3mc$) (III, c2)	2.51	2.55
	Cr_2CdS_4 (sp) O_h^7 ($Fd3m$) (VIII, b1)	2.21	2.55
	In_2CdS_4 (sp) O_h^7 ($Fd3m$) (VIII, b1)	2.56	2.55
Cd—Cl	$CdCl_2$ ($CdCl_2$) D_{3d}^5 ($R\bar{3}m$) (IV, c2)	2.65	2.55
	CH_4CdCl_3 * D_{2h}^{16} ($Pnma$) (VII, a15)	2.60–2.72	2.55
	$RbCdCl_3$ * D_{2h}^{16} ($Pnma$) (VII, a15)	2.60–2.72	2.55
	$CsCdCl_3$ (per) O_h^1 ($Pm3m$) (VII, a5)	2.60	2.55
Cd—As	Cd_3As_2 * D_{4h}^{15} ($P4/nmc$) (V, a3)	2.60	2.70
Cd—Se	CdSe (znb) T_d^2 ($F\bar{4}3m$) (III, c1)	2.62	2.70
	CdSe (wu) C_{6v}^4 ($P6_3mc$) (III, c2)	2.63	2.70
	Cr_2CdSe_4 (sp) O_h^7 ($Fd3m$) (VIII, b1)	2.48	2.70
Cd—Br	$CdBr_2$ ($CdCl_2$) D_{3d}^5 ($R\bar{3}m$) (IV, c2)	2.76	2.70
	$CsCdBr_3$ (per) O_h^1 ($Pm3m$) (VII, a5)	2.67	2.70
Cd—Cd	metal (hex) D_{6h}^4 ($P6_3/mmc$) (II, b)	2.98–3.29	3.10
	CdSb * D_{2h}^{15} ($Pbca$) (III, c4)	2.99	3.10
Cd—Sn	$CdSnO_3$ (per) O_h^1 ($Pm3m$) (VII, a5)	3.30	3.00
Cd—Sb	CdSb * D_{2h}^{15} ($Pbca$) (III, c4)	2.80–2.91	3.00
Cd—Te	CdTe (znb) T_d^2 ($F\bar{4}3m$) (III, c1)	2.78	2.95
Cd—I	CdI_2 (CdI_2) D_{3d}^3 ($P\bar{3}m1$) (IV, c1)	2.98	2.95
Cd—La	CdLa (CsCl) O_h^1 ($Pm3m$) (III, b1)	3.38	3.50
Cd—Ce	CdCe (CsCl) O_h^1 ($Pm3m$) (III, b1)	3.35	3.40
	$CdCeO_3$ (per) O_h^1 ($Pm3m$) (VII, a5)	3.32	3.40
Cd—Pr	CdPr (CsCl) O_h^1 ($Pm3m$) (III, b1)	3.31	3.40
In—N	InN (wu) C_{6v}^4 ($P6_3mc$) (III, c2)	2.13	2.20
In—O	In_2O_3 (Tl_2O_3) T_h^7 ($Ia3$) (V, a2)	2.15	2.15
	$InBO_3$ (cal) D_{3d}^6 ($R\bar{3}c$) (VII, a1)	2.23	2.15
	In_2MgO_4 (sp) O_h^7 ($Fd3m$) (VIII, b1)	2.25	2.15
In—P	InP (znb) T_d^2 ($F\bar{4}3m$) (III, c1)	2.54	2.55
In—S	In_2MgS_4 (sp) O_h^7 ($Fd3m$) (VIII, b1)	2.59	2.55
	In_2FeS_4 (sp) O_h^7 ($Fd3m$) (VIII, b1)	2.57	2.55
	In_2CoS_4 (sp) O_h^7 ($Fd3m$) (VIII, b1)	2.55	2.55
	In_2NiS_4 (sp) O_h^7 ($Fd3m$) (VIII, b1)	2.53	2.55
	In_2CdS_4 (sp) O_h^7 ($Fd3m$) (VIII, b1)	2.60	2.55
	In_2HgS_4 (sp) O_h^7 ($Fd3m$) (VIII, b1)	2.58	2.55
In—As	InAs (znb) T_d^2 ($F\bar{4}3m$) (III, c1)	2.62	2.70
In—In	metal * D_{4h}^{17} ($I4/mmm$) (II, f)	3.24–3.36	3.10
In—Sb	InSb (znb) T_d^2 ($F\bar{4}3m$) (III, c1)	2.80	3.00
Sn—O	SnO (mol) $C_{\infty v}$ (T.I.D.)	1.84	2.05
	SnO_2 (ru) D_{4h}^{14} ($P4_2/mnm$) (IV, b1)	2.04–2.07	2.05
	$CaSnO_3$ (per) O_h^1 ($Pm3m$) (VII, a5)	1.96	2.05

Sn—O	SrSnO$_3$ (per) O_h^1 ($Pm3m$) (VII, a5)	2.02	2.05
	CdSnO$_3$ (per) O_h^1 ($Pm3m$) (VII, a5)	1.90	2.05
	BaSnO$_3$ (per) O_h^1 ($Pm3m$) (VII, a5)	2.05	2.05
	Mg$_2$SnO$_4$ (sp) O_h^7 ($Fd3m$) (VIII, b1)	1.85	2.05
	Co$_2$SnO$_4$ (sp) O_h^7 ($Fd3m$) (VIII, b1)	1.86	2.05
	Zn$_2$SnO$_4$ (sp) O_h^7 ($Fd3m$) (VIII, b1)	2.09	2.05
Sn—S	SnS * D_{2h}^{16} ($Pnma$) (III, h3)	2.62–3.40	2.45
	SnS$_2$ (CdI$_2$) D_{3d}^3 ($P\bar{3}m1$) (IV, c1)	2.56	2.45
	Cu$_2$FeSnS$_4$ * D_{2d}^{11} ($I\bar{4}2m$) (VIII, c4)	2.42	2.45
Sn—Cl	SnCl$_2$ (mol) $C_{\infty v}$ (T.I.D.)	2.42	2.45
	SnCl$_4$ (mol) T_d (T.I.D.)	2.33	2.45
	Cs$_2$SnCl$_6$ * O_h^5 ($Fm3m$) (IX, c1; T.I.D.)	2.43	2.45
Sn—As	SnAs (NaCl) O_h^5 ($Fm3m$) (III, a1)	2.84	2.60
Sn—Sn	metal (di) O_h^5 ($Fm3m$) (II, j)	2.80	2.90
	metal (white tin) * D_{4h}^{19} ($I4_1/amd$) (II, j)	3.02–3.17	2.90
Sn—Sb	SnSb (NaCl) O_h^5 ($Fm3m$) (III, a1)	3.06	2.90
Sn—Te	SnTe (NaCl) O_h^5 ($Fm3m$) (III, a1)	3.14	2.85
Sn—I	SnI$_4$ * T_h^6 ($Pa3$) (V, c1)	2.63	2.85
Sb—O	Sb$_2$O$_3$ * O_h^7 ($Fd3m$) (V, a6)	2.22	2.05
	LiSbO$_3$ * D_{2h}^6 ($Pnna$) (VII, a22)	2.00–2.05	2.05
	Sb$_2$ZnO$_4$ * D_{4h}^{13} ($P4_2/mbc$) (V, g3a)	1.87–2.01	2.05
	BiSbO$_4$ * C_{2h}^6 ($B2/b$) (VIII, a19)	2.52–2.93	2.05
Sb—F	SbF$_3$ (mol) C_s (T.I.D.)	2.03	1.95
	NaSbF$_6$ * T_h^6 ($Pa3$) (IV, b2; T.I.D.)	1.95	1.95
	K$_2$SbF$_5$ * D_{2h}^{17} ($Cmcm$) (IX, a6)	2.02–2.08	1.95
Sb—S	CuSbS$_2$ * D_{2h}^{16} ($Pnma$) (VI, a12)	2.44–2.57	2.45
	Ag$_3$SbS$_3$ * C_{3v}^6 ($R3c$) (VII, b6)	2.45	2.45
Sb—Cl	SbCl$_3$ (mol) C_{3v} (T.I.D.)	2.37	2.45
	SbCl$_5$ (mol) D_{3h} (T.I.D.)	2.31–2.43	2.45
Sb—Br	SbBr$_3$ (mol) C_{3v} (T.I.D.)	2.51	2.60
Sb—Sb	metal (As) D_{3d}^5 ($R\bar{3}m$) (II, n)	2.90	2.90
	ZnSb * D_{2h}^{15} ($Pbca$) (III, c4)	2.81	2.90
	CdSb * D_{2h}^{15} ($Pbca$) (III, c4)	2.81	2.90
Sb—I	SbI$_3$ * D_3^7 ($R32$) (V, h2)	3.08	2.85
Te—O	TeO$_2$ (ru) D_{4h}^{14} ($P4_2/mnm$) (IV, b1)	2.10–2.29	2.00
Te—F	TeF$_6$ (mol) O_h (T.I.D.)	1.82	1.90
Te—Cl	TeCl$_2$ (mol) C_{2v} (T.I.D.)	2.36	2.40
	TeCl$_4$ (mol) C_{2v} (T.I.D.)	2.33	2.40
Te—Br	TeBr$_2$ (mol) C_{2v} (T.I.D.)	2.51	2.55
Te—Te	element (Se) D_3^4 ($P3_121$) or D_3^6 ($P3_221$) (II, q)	2.86	2.80
I—O	HIO$_3$ * D_2^4 ($P2_12_12_1$) (VII, a23)	1.81–1.89	2.00
	KIO$_3$ (per) O_h^1 ($Pm3m$) (VII, a5)	2.23	2.00
	KIO$_4$ * C_{4h}^6 ($I4_1/a$) (VIII, a6)	1.80	2.00
	KIO$_2$F$_2$ * C_{2v}^5 ($Pca2_1$) (VIII, a9)	1.92	2.00
	RbIO$_3$ (per) O_h^1 ($Pm3m$) (VII, a5)	2.26	2.00
	CsIO$_3$ (per) O_h^1 ($Pm3m$) (VII, a5)	2.33	2.00
I—F	KIO$_2$F$_2$ * C_{2v}^5 ($Pca2_1$) (VIII, a9)	1.99	1.90
	IF$_7$ (mol) D_{5h} (T.I.D.)	1.83–1.94	1.90

I—Cl	ICl (mol) $C_{\infty v}$ (T.I.D.)	2.30	2.40
	KICl$_4$ * C_{2h}^5 ($P2_1/b$) (VIII, a13; T.I.D.)	2.34	2.40
	NH$_4$BrICl * D_{2h}^{16} ($Pnma$) (VI, a16)	2.38	2.40
I—Br	NH$_4$BrICl * D_{2h}^{16} ($Pnma$) (VI, a16)	2.50	2.55
I—I	I$_2$ (mol) $D_{\infty h}$ (T.I.D.)	2.66	2.80
	element (I) D_{2h}^{18} ($Cmca$) (II, v)	2.68	2.80
	CsI$_3$ * D_{2h}^{16} ($Pnma$) (VI, a16; T.I.D.)	2.83–3.04	2.80
Cs—O	CsIO$_3$ (per) O_h^1 ($Pm3m$) (VII, a5)	3.30	3.20
Cs—F	CsF (mol) $C_{\infty v}$ (T.I.D.)	2.35	3.10
	CsF (NaCl) O_h^5 ($Fm3m$) (III, a1)	3.00	3.10
	CsCaF$_3$ (per) O_h^1 ($Pm3m$) (VII, a5)	3.20	3.10
Cs—S	Cs$_2$S$_6$ * C_2^1 ($P2$) (V, b33)	3.48	3.60
Cs—Cl	CsCl (CsCl) O_h^1 ($Pm3m$) (III, b1)	3.57	3.60
	CsCl (NaCl) O_h^5 ($Fm3m$) (III, a1)	3.51	3.60
	CsCl$_2$I * D_{3d}^5 ($R\bar{3}m$) (VI, a7)	3.66	3.60
	Cs$_2$CuCl$_4$ * D_{2h}^{16} ($Pnma$) (VIII, b19)	3.42	3.60
	CsCdCl$_3$ (per) O_h^1 ($Pm3m$) (VII, a5)	3.67	3.60
	CsHgCl$_3$ (per) O_h^1 ($Pm3m$) (VII, a5)	3.84	3.60
Cs—Ca	CsCaF$_3$ (per) O_h^1 ($Pm3m$) (VII, a5)	3.92	4.40
Cs—Br	CsBr (CsCl) O_h^1 ($Pm3m$) (III, b1)	3.71	3.75
	CsCdBr$_3$ (per) O_h^1 ($Pm3m$) (VII, a5)	3.77	3.75
	CsHgBr$_3$ (per) O_h^1 ($Pm3m$) (VII, a5)	4.07	3.75
Cs—I	CsI (CsCl) O_h^1 ($Pm3m$) (III, b1)	3.95	4.00
	CsIO$_3$ (per) O_h^1 ($Pm3m$) (VII, a5)	4.05	4.00
Cs—Cs	metal (bcc) O_h^9 ($Im3m$) (II, c)	5.24	5.20
Ba—Li	BaLiF$_3$ (per) O_h^1 ($Pm3m$) (VII, a5)	3.46	3.60
Ba—O	BaO (NaCl) O_h^5 ($Fm3m$) (III, a1)	2.76	2.75
	BaO.NiO * D_{2h}^{17} ($Cmcm$) (VI, a22)	2.80–2.84	2.75
	NiO.3BaO * D_{3d}^6 ($R\bar{3}c$) (VIII, c13)	2.82	2.75
	Ba$_3$(PO$_4$)$_2$ * D_{3d}^5 ($R\bar{3}m$) (VIII, c11)	2.71–3.23	2.75
	BaTiO$_3$ * D_{6h}^4 ($P6_3/mmc$) (VII, a17)	2.78–2.96	2.75
	BaTiO$_3$ (per) O_h^1 ($Pm3m$) (VII, a5)	2.80	2.75
	BaZrO$_3$ (per) O_h^1 ($Pm3m$) (VII, a5)	2.95	2.75
	BaSnO$_3$ (per) O_h^1 ($Pm3m$) (VII, a5)	2.89	2.75
	BaCeO$_3$ (per) O_h^1 ($Pm3m$) (VII, a5)	3 09	2.75
	BaPrO$_3$ (per) O_h^1 ($Pm3m$) (VII, a5)	3.07	2.75
	BaThO$_3$ (per) O_h^1 ($Pm3m$) (VII, a5)	3.16	2.75
	BaUO$_3$ (per) O_h^1 ($Pm3m$) (VII, a5)	3.11	2.75
Ba—F	BaF$_2$ (fl) O_h^5 ($Fm3m$) (IV, a1)	2.68	2.65
	BaLiF$_3$ (per) O_h^1 ($Pm3m$) (VII, a5)	2.82	2.65
Ba—S	BaS (NaCl) O_h^5 ($Fm3m$) (III, a1)	3.18	3.15
Ba—Cl	BaCl$_2$ (fl) O_h^5 ($Fm3m$) (IV, a1)	3.18	3.15
Ba—Ti	BaTiO$_3$ (per) O_h^1 ($Pm3m$) (VII, a5)	3.44	3.55
Ba—Se	BaSe (NaCl) O_h^5 ($Fm3m$) (III, a1)	3.31	3.30
Ba—Zr	BaZrO$_3$ (per) O_h^1 ($Pm3m$) (VII, a5)	3.63	3.70
Ba—Sn	BaSnO$_3$ (per) O_h^1 ($Pm3m$) (VII, a5)	3.56	3.60
Ba—Te	BaTe (NaCl) O_h^5 ($Fm3m$) (III, a1)	3.50	3.55
Ba—Ba	metal (bcc) O_h^9 ($Im3m$) (II, c)	4.35	4.30
Ba—Ce	BaCeO$_3$ (per) O_h^1 ($Pm3m$) (VII, a5)	3.80	4.00

Ba—Pr	BaPrO$_3$ (per) O_h^1 ($Pm3m$) (VII, a5)	3.78	4.00
Ba—Th	BaThO$_3$ (per) O_h^1 ($Pm3m$) (VII, a5)	3.88	3.90
Ba—U	BaUO$_3$ (per) O_h^1 ($Pm3m$) (VII, a5)	3.81	3.90
La—N	LaN (NaCl) O_h^5 ($Fm3m$) (III, a1)	2.63	2.60
La—O	La$_2$O$_3$ (Tl$_2$O$_3$) T_h^7 ($Ia3$) (V, a2)	2.42	2.55
	La$_2$O$_3$ (La$_2$O$_3$) D_{3d}^3 ($P\bar{3}m1$) (V, a1)	2.44–2.68	2.55
	LaOF (LaOF) D_{3d}^5 ($R\bar{3}m$) (IV, a5)	2.50–2.59	2.55
	LaVO$_3$ (per) O_h^1 ($Pm3m$) (VII, a5)	2.76	2.55
	LaCrO$_3$ (per) O_h^1 ($Pm3m$) (VII, a5)	2.76	2.55
	LaMnO$_3$ (per) O_h^1 ($Pm3m$) (VII, a5)	2.74	2.55
	LaFeO$_3$ (per) O_h^1 ($Pm3m$) (VII, a5)	2.75	2.55
	LaCoO$_3$ (per) O_h^1 ($Pm3m$) (VII, a5)	2.70	2.55
	LaGeO$_3$ (per) O_h^1 ($Pm3m$) (VII, a5)	2.75	2.55
La—F	LaF$_3$ * D_{6h}^4 ($P6_3/mmc$) (V, b15a)	2.35–2.39	2.45
	LaOF (LaOF) D_{3d}^5 ($R\bar{3}m$) (IV, a5)	2.41–2.42	2.45
La—P	LaP (NaCl) O_h^5 ($Fm3m$) (III, a1)	3.01	2.95
La—V	LaVO$_3$ (per) O_h^1 ($Pm3m$) (VII, a5)	3.39	3.30
La—Cr	LaCrO$_3$ (per) O_h^1 ($Pm3m$) (VII, a5)	3.38	3.35
La—Mn	LaMnO$_3$ (per) O_h^1 ($Pm3m$) (VII, a5)	3.36	3.35
La—Fe	LaFeO$_3$ (per) O_h^1 ($Pm3m$) (VII, a5)	3.37	3.35
La—Co	LaCoO$_3$ (per) O_h^1 ($Pm3m$) (VII, a5)	3.30	3.30
La—Ga	LaGa$_2$ (AlB$_2$) D_{3d}^3 ($P\bar{3}m1$) (IV, c6)	3.34	3.25
	LaGaO$_3$ (per) O_h^1 ($Pm3m$) (VII, a5)	3.37	3.25
La—As	LaAs (NaCl) O_h^5 ($Fm3m$) (III, a1)	3.06	3.10
La—Sb	LaSb (NaCl) O_h^5 ($Fm3m$) (III, a1)	3.24	3.40
La—La	metal (fcc) O_h^5 ($Fm3m$) (II, a)	3.74	3.90
	metal (hex) D_{6h}^4 ($P6_3/mmc$) (II, b)	3.73–3.75	3.90
La—Bi	LaBi (NaCl) O_h^5 ($Fm3m$) (III, a1)	3.28	3.55
Ce—N	CeN (NaCl) O_h^5 ($Fm3m$) (III, a1)	2.50	2.50
Ce—O	CeO$_2$ (fl) O_h^5 ($Fm3m$) (IV, a1)	2.35	2.45
	Ce$_2$O$_3$ (La$_2$O$_3$) D_{3d}^3 ($P\bar{3}m1$) (V, a1)	2.40–2.64	2.45
	CeVO$_3$ (per) O_h^1 ($Pm3m$) (VII, a5)	2.76	2.45
	CeCrO$_3$ (per) O_h^1 ($Pm3m$) (VII, a5)	2.75	2.45
	CaCeO$_3$ (per) O_h^1 ($Pm3m$) (VII, a5)	1.93	2.45
	SrCeO$_3$ (per) O_h^1 ($Pm3m$) (VII, a5)	2.14	2.45
	CdCeO$_3$ (per) O_h^1 ($Pm3m$) (VII, a5)	1.92	2.45
	BaCeO$_3$ (per) O_h^1 ($Pm3m$) (VII, a5)	2.19	2.45
	PbCeO$_3$ (per) O_h^1 ($Pm3m$) (VII, a5)	1.91	2.45
Ce—P	CeP (NaCl) O_h^5 ($Fm3m$) (III, a1)	2.95	2.85
Ce—S	CeS (NaCl) O_h^5 ($Fm3m$) (III, a1)	2.89	2.85
Ce—V	CeVO$_3$ (per) O_h^1 ($Pm3m$) (VII, a5)	3.38	3.20
Ce—Cr	CeCrO$_3$ (per) O_h^1 ($Pm3m$) (VII, a5)	3.37	3.25
Ce—Ga	CeGa$_2$ (AlB$_2$) D_{3d}^3 ($P\bar{3}m1$) (IV, c6)	3.30	3.15
Ce—As	CeAs (NaCl) O_h^5 ($Fm3m$) (III, a1)	3.03	3.00
Ce—Sb	CeSb (NaCl) O_h^5 ($Fm3m$) (III, a1)	3.20	3.30
Ce—Ce	metal (fcc) O_h^5 ($Fm3m$) (II, a)	3.64	3.70
	metal (hex) D_{6h}^4 ($P6_3/mmc$) (II, b)	3.65	3.70
Ce—Bi	CeBi (NaCl) O_h^5 ($Fm3m$) (III, a1)	3.24	3.45

INTERATOMIC DISTANCES AND CRYSTAL STRUCTURES

Pr—N	PrN (NaCl) O_h^5 ($Fm3m$) (III, a1)	2.58	2.50
Pr—O	PrO$_2$ (fl) O_h^5 ($Fm3m$) (IV, a1)	2.33	2.45
	Pr$_2$O$_3$ (La$_2$O$_3$) D_{3d}^3 ($P\bar{3}m1$) (V, a1)	2.38–2.62	2.45
	Pr$_2$O$_3$ (Tl$_2$O$_3$) T_h^7 ($Ia3$) (V, a2)	2.38	2.45
	PrOF (LaOF) D_{3d}^5 ($R\bar{3}m$) (IV, a5)	2.46–2.55	2.45
	PrVO$_3$ (per) O_h^1 ($Pm3m$) (VII, a5)	2.75	2.45
	PrCrO$_3$ (per) O_h^1 ($Pm3m$) (VII, a5)	2.75	2.45
	PrCoO$_3$ (per) O_h^1 ($Pm3m$) (VII, a5)	2.66	2.45
Pr—F	PrOF (LaOF) D_{3d}^5 ($R\bar{3}m$) (IV, a5)	2.37–2.39	2.35
Pr—P	PrP (NaCl) O_h^5 ($Fm3m$) (III, a1)	2.93	2.85
Pr—V	PrVO$_3$ (per) O_h^1 ($Pm3m$) (VII, a5)	3.37	3.20
Pr—Cr	PrCrO$_3$ (per) O_h^1 ($Pm3m$) (VII, a5)	3.37	3.25
Pr—Co	PrCoO$_3$ (per) O_h^1 ($Pm3m$) (VII, a5)	3.26	3.20
Pr—As	PrAs (NaCl) O_h^5 ($Fm3m$) (III, a1)	3.00	3.00
Pr—Sb	PrSb (NaCl) O_h^5 ($Fm3m$) (III, a1)	3.17	3.30
Pr—Pr	metal (fcc) O_h^5 ($Fm3m$) (II, a)	3.65	3.70
	metal (hex) D_{6h}^4 ($P6_3/mmc$) (II, b)	3.64–3.67	3.70
Pr—Bi	PrBi (NaCl) O_h^5 ($Fm3m$) (III, a1)	3.22	3.45
Nd—N	NdN (NaCl) O_h^5 ($Fm3m$) (III, a1)	2.57	2.50
Nd—O	Nd$_2$O$_3$ (La$_2$O$_3$) D_{3d}^3 ($P\bar{3}m1$) (V, a1)	2.38–2.62	2.45
	Nd$_2$O$_3$ (Tl$_2$O$_3$) T_h^7 ($Ia3$) (V, a2)	2.34	2.45
	NdOF (LaOF) D_{3d}^5 ($R\bar{3}m$) (IV, a5)	2.44–2.53	2.45
	NdVO$_3$ (per) O_h^1 ($Pm3m$) (VII, a5)	2.75	2.45
	NdCrO$_3$ (per) O_h^1 ($Pm3m$) (VII, a5)	2.75	2.45
	NdCoO$_3$ (per) O_h^1 ($Pm3m$) (VII, a5)	2.66	2.45
Nd—F	NdOF (LaOF) D_{3d}^5 ($R\bar{3}m$) (IV, a5)	2.36–2.37	2.35
Nd—P	NdP (NaCl) O_h^5 ($Fm3m$) (III, a1)	2.91	2.85
Nd—V	NdVO$_3$ (per) O_h^1 ($Pm3m$) (VII, a5)	3.37	3.20
Nd—Cr	NdCrO$_3$ (per) O_h^1 ($Pm3m$) (VII, a5)	3.37	3.25
Nd—Co	NdCoO$_3$ (per) O_h^1 ($Pm3m$) (VII, a5)	3.26	3.20
Nd—As	NdAs (NaCl) O_h^5 ($Fm3m$) (III, a1)	2.98	3.00
Nd—Sb	NdSb (NaCl) O_h^5 ($Fm3m$) (III, a1)	3.15	3.30
Nd—Nd	metal (hex) D_{6h}^4 ($P6_3/mmc$) (II, b)	3.63–3.66	3.70
Sm—O	SmO (NaCl) O_h^5 ($Fm3m$) (III, a1)	2.51	2.45
	Sm$_2$O$_3$ (Tl$_2$O$_3$) T_h^7 ($Ia3$) (V, a2)	2.31	2.45
	SmOF (LaOF) D_{3d}^5 ($R\bar{3}m$) (IV, a5)	2.40–2.49	2.45
	SmVO$_3$ (per) O_h^1 ($Pm3m$) (VII, a5)	2.75	2.45
	SmCrO$_3$ (per) O_h^1 ($Pm3m$) (VII, a5)	2.73	2.45
	SmCoO$_3$ (per) O_h^1 ($Pm3m$) (VII, a5)	2.65	2.45
Sm—F	SmOF (LaOF) D_{3d}^5 ($R\bar{3}m$) (IV, a5)	2.32–2.33	2.35
Sm—V	SmVO$_3$ (per) O_h^1 ($Pm3m$) (VII, a5)	3.37	3.20
Sm—Cr	SmCrO$_3$ (per) O_h^1 ($Pm3m$) (VII, a5)	3.34	3.25
Sm—Co	SmCoO$_3$ (per) O_h^1 ($Pm3m$) (VII, a5)	3.25	3.20
Eu—O	Eu$_2$O$_3$ (Tl$_2$O$_3$) T_h^7 ($Ia3$) (V, a2)	2.31	2.45
	EuOF (LaOF) D_{3d}^5 ($R\bar{3}m$) (IV, a5)	2.39–2.48	2.45
Eu—F	EuF$_2$ (fl) O_h^5 ($Fm3m$) (IV, a1)	2.51	2.35
	EuOF (LaOF) D_{3d}^5 ($R\bar{3}m$) (IV, a5)	2.31–2.32	2.35

Eu—S	EuS (NaCl) O_h^5 ($Fm3m$) (III, a1)	2.98	2.85
Eu—Se	EuSe (NaCl) O_h^5 ($Fm3m$) (III, a1)	3.08	3.00
Eu—Te	EuTe (NaCl) O_h^5 ($Fm3m$) (III, a1)	3.28	3.25
Gd—N	GdN (NaCl) O_h^5 ($Fm3m$) (III, a1)	2.50	2.45
Gd—O	Gd_2O_3 (Tl_2O_3) T_h^7 ($Ia3$) (V, a2)	2.30	2.40
	GdOF (LaOF) D_{3d}^5 ($R\bar{3}m$) (IV, a5)	2.38–2.47	2.40
	$GdMnO_3$ (per) O_h^1 ($Pm3m$) (VII, a5)	2.70	2.40
Gd—F	GdOF (LaOF) D_{3d}^5 ($R\bar{3}m$) (IV, a5)	2.30–2.31	2.30
Gd—Mn	$GdMnO_3$ (per) O_h^1 ($Pm3m$) (VII, a5)	3.31	3.20
Gd—Gd	metal (hex) D_{6h}^4 ($P6_3/mmc$) (II, b)	3.58–3.63	3.60
Tb—O	Tb_2O_3 (Tl_2O_3) T_h^7 ($Ia3$) (V, a2)	2.25	2.35
	TbOF (LaOF) D_{3d}^5 ($R\bar{3}m$) (IV, a5)	2.37–2.46	2.35
Tb—F	TbOF (LaOF) D_{3d}^5 ($R\bar{3}m$) (IV, a5)	2.29–2.30	2:25
Tb—Tb	metal (hex) D_{6h}^4 ($P6_3/mmc$) (II, b)	3.55–3.59	3.50
Dy—O	Dy_2O_3 (Tl_2O_3) T_h^7 ($Ia3$) (V, a2)	2.25	2.35
Dy—Dy	metal (hex) D_{6h}^4 ($P6_3/mmc$) (II, b)	3.51–3.58	3.50
Ho—O	Ho_2O_3 (Tl_2O_3) T_h^7 ($Ia3$) (V, a2)	2.25	2.35
Ho—Ho	metal (hex) D_{6h}^4 ($P6_3/mmc$) (II, b)	3.49–3.56	3.50
Er—O	Er_2O_3 (Tl_2O_3) T_h^7 ($Ia3$) (V, a2)	2.24	2.35
Er—Er	metal (hex) D_{6h}^4 ($P6_3/mmc$) (II, b)	3.47–3.56	3.50
Tu—O	Tu_2O_3 (Tl_2O_3) T_h^7 ($Ia3$) (V, a2)	2.24	2.35
Tu—Tu	metal (hex) D_{6h}^4 ($P6_3/mmc$) (II, b)	3.45–3.53	3.50
Yb—O	Yb_2O_3 (Tl_2O_3) T_h^7 ($Ia3$) (V, a2)	2.21	2.35
Yb—Se	YbSe (NaCl) O_h^5 ($Fm3m$) (III, a1)	2.93	2.90
Yb—Te	YbTe (NaCl) O_h^5 ($Fm3m$) (III, a1)	3.17	3.15
Yb—I	YbI_2 (CdI_2) D_{3d}^3 ($P\bar{3}m1$) (IV, c1)	3.12	3.15
Yb—Yb	metal (fcc) O_h^5 ($Fm3m$) (II, a)	3.86	3.50
Lu—O	Lu_2O_3 (Tl_2O_3) T_h^7 ($Ia3$) (V, a2)	2.21	2.35
Hf—C	HfC (NaĆl) O_h^5 ($Fm3m$) (III, a1)	2.23	2.25
Hf—O	HfO_2 (fl) O_h^5 ($Fm3m$) (IV, a1)	2.22	2.15
	$SrHfO_3$ (per) O_h^1 ($Pm3m$) (VII, a5)	2.04	2.15
Hf—Hf	metal (hex) D_{6h}^4 ($P6_3/mmc$) (II, b)	3.13–3.20	3.10
Ta—C	TaC (NaCl) O_h^5 ($Fm3m$) (III, a1)	2.22	2.15
Ta—N	TaN (wu) C_{6v}^4 ($P6_3mc$) (III, c2)	1.85	2.10
Ta—O	TaO (NaCl) O_h^5 ($Fm3m$) (III, a1)	2.22	2.05
	$NaTaO_3$ (per) O_h^1 ($Pm3m$) (VII, a5)	1.94	2.05
	$KTaO_3$ (per) O_h^1 ($Pm3m$) (VII, a5)	2.00	2.05
Ta—Ta	metal (bcc) O_h^9 ($Im3m$) (II, c)	2.86	2.90
W—O	WO_2 (ru) D_{4h}^{14} ($P4_2/mnm$) (IV, b1)	1.91–2.14	1.95
	$NaWO_3$ (per) O_h^1 ($Pm3m$) (VII, a5)	1.93	1.95

INTERATOMIC DISTANCES AND CRYSTAL STRUCTURES

W—Cl	WCl$_6$ * C_{3i}^2 ($R\bar{3}$) (V, f4)	2.24	2.35
W—W	metal (bcc) O_h^9 ($Im3m$) (II, c)	2.74	2.70
Re—S	ReS$_2$ (py) T_h^6 ($Pa3$) (IV, g2)	2.32	2.35
Re—Re	metal (hex) D_{6h}^4 ($P6_3/mmc$) (II, b)	2.74–2.76	2.70
Os—O	OsO$_2$ (ru) D_{4h}^{14} ($P4_2/mnm$) (IV, b1)	1.98–2.01	1.90
	OsO$_4$ * C_2^3 (B2) (V, c10)	(?)1.66	1.90
Os—S	OsS$_2$ (py) T_h^6 ($Pa3$) (IV, g2)	2.34	2.30
Os—Se	OsSe$_2$ (py) T_h^6 ($Pa3$) (IV, g2)	2.47	2.45
Os—Te	OsTe$_2$ (py) T_h^6 ($Pa3$) (IV, g2)	2.66	2.70
Os—Os	metal (hex) D_{6h}^4 ($P6_3/mmc$) (II, b)	2.66–2.74	2.60
Ir—O	IrO$_2$ (ru) D_{4h}^{14} ($P4_2/mnm$) (IV, b1)	1.97–1.98	1.95
Ir—P	Ir$_2$P (fl) O_h^5 ($Fm3m$) (IV, a1)	2.40	2.35
Ir—Sn	IrSn$_2$ (fl) O_h^6 ($Fm3m$) (IV, a1)	2.74	2.80
Ir—Ir	metal (fcc) O_h^5 ($Fm3m$) (II, a)	2.71	2.70
Pt—Al	PtAl$_2$ (fl) O_h^5 ($Fm3m$) (IV, a1)	2.56	2.60
Pt—P	PtP$_2$ (py) T_h^6 ($Pa3$) (IV, g2)	2.37	2.35
Pt—S	PtS$_2$ (CdI$_2$) D_{3d}^3 ($P\bar{3}m$) (IV, c1)	2.40	2.35
Pt—Cl	K$_2$PtCl$_4$ * D_{4h}^1 ($P4/mmm$) (VIII, b15)	2.32	2.35
Pt—Ga	PtGa$_2$ (fl) O_h^5 ($Fm3m$) (IV, a1)	2.56	2.65
Pt—As	PtAs$_2$ (py) T_h^6 ($Pa3$) (IV, g2)	2.49	2.50
Pt—Se	PtSe$_2$ (CdI$_2$) D_{3d}^3 ($P\bar{3}m1$) (IV, c1)	2.50	2.50
Pt—In	PtIn$_2$ (fl) O_h^5 ($Fm3m$) (IV, a1)	2.76	2.90
Pt—Sn	PtSn (NiAs) D_{6h}^4 ($P6_3/mmc$) (III, d)	2.73	2.80
	PtSn$_2$ (fl) O_h^5 ($Fm3m$) (IV, a1)	2.78	2.80
Pt—Sb	PtSb (NiAs) D_{6h}^4 ($P6_3/mmc$) (III, d)	2.75	2.80
	PtSb$_2$ (py) T_h^6 ($Pa3$) (IV, g2)	2.69	2.80
Pt—Te	PtTe$_2$ (CdI$_2$) D_{3d}^3 ($P\bar{3}m1$) (IV, c1)	2.66	2.75
Pt—Pt	metal (fcc) O_h^5 ($Fm3m$) (II, a)	2.77	2.70
	PtSb (NiAs) D_{6h}^4 ($P6_3/mmc$) (III, d)	2.72	2.70
	PtSn (NiAs) D_{6h}^4 ($P6_3/mmc$) (III, d)	2.70	2.70
Pt—Bi	PtBi$_2$ (py) T_h^6 ($Pa3$) (IV, g2)	2.79	2.95
Au—Al	AuAl$_2$ (fl) O_h^5 ($Fm3m$) (IV, a1)	2.60	2.60
Au—Cl	Cs$_2$AgAuCl$_6$ * D_{4h}^{17} ($I4/mmm$) (VII, a6)	2.30	2.35
Au—Zn	AuZn (CsCl) O_h^1 ($Pm3m$) (III, b1)	2.75	2.70
Au—Ga	AuGa$_2$ (fl) O_h^5 ($Fm3m$) (IV, a1)	2.62	2.65
Au—Cd	AuCd (NaCl) O_h^5 ($Fm3m$) (III, a1)	2.90	2.90
Au—In	AuIn$_2$ (fl) O_h^5 ($Fm3m$) (IV, a1)	2.81	2.90
Au—Sn	AuSn (NiAs) D_{6h}^4 ($P6_3/mmc$) (III, d)	2.84	2.80
Au—Sb	AuSb$_2$ (py) T_h^6 ($Pa3$) (IV, g2)	2.77	2.80
Au—Au	metal (fcc) O_h^5 ($Im3m$) (II, a)	2.88	2.70
	AuSn (NiAs) D_{6h}^4 ($P6_3/mmc$) (III, d)	2.54	2.70
Hg—F	HgF$_2$ (fl) O_h^5 ($Fm3m$) (IV, a1)	2.40	2.00
Hg—S	HgS (znb) T_d^2 ($F\bar{4}3m$) (III, c1)	2.53	2.50
	Cr$_2$HgS$_4$ (sp) O_h^7 ($Fd3m$) (VIII, b1)	2.50	2.50
	In$_2$HgS$_4$ (sp) O_h^7 ($Fd3m$) (VIII, b1)	2.59	2.50

Hg—Cl	HgCl$_2$ * D_{2h}^{16} (*Pnma*) (IV, d6)	2.23–2.27	2.50
	CsHgCl$_3$ (per) O_h^1 (*Pm3m*) (VII, a5)	2.72	2.50
Hg—Se	HgSe (znb) T_d^2 ($F\bar{4}3m$) (III, c1)	2.63	2.65
Hg—Br	HgBr$_2$ * C_{2v}^{12} (*Cmc2$_1$*) (IV, d7)	2.48	2.65
	CsHgBr$_3$ (per) O_h^1 (*Pm3m*) (VII, a5)	2.88	2.65
Hg—Te	HgTe (znb) T_d^2 ($F\bar{4}3m$) (III, c1)	2.79	2.90
Hg—I	HgI$_2$ * D_{4h}^{15} (*P4$_2$/nmc*) (IV, c1)	2.78	2.90
Hg—Hg	metal * D_{3d}^5 ($R\bar{3}m$) (II, d)	3.00–3.47	3.00
Tl—N	Tl$_2$N$_3$ (La$_2$O$_3$) D_{3d}^3 ($P\bar{3}m1$) (V, a1)	2.42–2.68	2.55
Tl—O	Tl$_2$O$_3$ (Tl$_2$O$_3$) T_h^7 (*Ia3*) (V, a2)	2.25	2.50
Tl—F	TlAlF$_4$ * D_{4h}^1 (*P4/mmm*) (VIII, a12)	2.88	2.40
Tl—Cl	TlCl (CsCl) O_h^1 (*Pm3m*) (III, b1)	3.31	2.90
Tl—Se	TlSe * D_{4h}^{18} (*I4/mcm*) (III, i2)	2.68–3.42	3.05
Tl—Br	TlBr (CsCl) O_h^1 (*Pm3m*) (III, b1)	3.44	3.05
Tl—Sb	TlSb (CsCl) O_h^1 (*Pm3m*) (III, b1)	3.32	3.35
Tl—I	TlI (CsCl) O_h^1 (*Pm3m*) (III, b1)	3.65	3.30
Tl—Tl	metal (bcc) O_h^9 (*Im3m*) (II, c)	3.36	3.80
	metal (hex) D_{6h}^4 (*P6$_3$/mmc*) (II, b)	3.41–3.47	3.80
Tl—Bi	TlBi (CsCl) O_h^1 (*Pm3m*) (III, b1)	3.45	3.50
Pb—O	PbO * C_{2v}^5 (*Pca2$_1$*) (III, c2)	2.18	2.40
	Pb$_2$O$_3$ * C_{2h}^2 (*P2$_1$/m*) (V, a13)	1.94–2.81	2.40
	PbO$_2$ (ru) D_{4h}^{14} (*P4$_2$/mnm*) (IV, b1)	2.15–2.16	2.40
	PbTiO$_3$ (per) O_h^1 (*Pm3m*) (VII, a5)	2.75	2.40
	PbZrO$_3$ (per) O_h^1 (*Pm3m*) (VII, a5)	3.28	2.40
	PbCeO$_3$ (per) O_h^1 (*Pm3m*) (VII, a5)	2.69	2.40
	Ag$_2$PbO$_2$ * C_{2h}^6 (*B2/b*) (VI, a25)	2.28–2.37	2.40
Pb—F	PbF$_2$ (fl) O_h^5 (*Fm3m*) (IV, a1)	2.58	2.30
Pb—S	PbS (NaCl) O_h^5 (*Fm3m*) (III, a1)	2.97	2.80
Pb—Cl	PbCl$_2$ * D_{2h}^{16} (*Pnma*) (IV, d3)	2.67–3.29	2.80
Pb—Ti	PbTiO$_3$ (per) O_h^1 (*Pm3m*) (VII, a5)	3.37	3.20
Pb—Se	PbSe (NaCl) O_h^5 (*Fm3m*) (III, a1)	3.04	2.95
Pb—Zr	PbZrO$_3$ (per) O_h^1 (*Pm3m*) (VII, a5)	4.02	3.35
Pb—Te	PbTe (NaCl) O_h^5 (*Fm3m*) (III, a1)	3.24	3.20
Pb—I	PbI$_2$ (CdI$_2$) D_{3d}^3 ($P\bar{3}m1$) (IV, c1)	3.13	3.20
Pb—Ce	PbCeO$_3$ (per) O_h^1 (*Pm3m*) (VII, a5)	3.30	3.65
Pb—Pb	metal (fcc) O_h^5 (*Fm3m*) (II, a)	3.50	3.60
Bi—O	BiAsO$_4$ * C_{4h}^6 (*I4$_1$/a*) (VIII, a6)	2.49–2.59	2.20
	BiSbO$_4$ * C_{4h}^6 (*B2/b*) (VII, a19)	~2.2–3.0	2.20
Bi—F	BiF$_3$ * O_h^5 (*Fm3m*) (V, b13)	2.56	2.10
Bi—S	Bi$_2$S$_3$ * D_{2h}^{16} (*Pnma*) (V, a10)	3.05	2.60
	Bi$_2$PbS$_4$ * D_{2h}^{16} (*Pnma*) (VIII, b20)	2.64–3.07	2.60
Bi—Te	Bi$_2$Te$_3$ * D_{3d}^5 ($R\bar{3}m$) (V, a12)	3.12	3.00
Bi—I	BiI$_3$ * C_{3i}^2 ($R\bar{3}$) (V, b1)	3.09	3.00
Bi—Bi	metal (As) D_{3d}^5 ($R\bar{3}m$) (II, n)	3.10	3.20
Po—O	PoO$_2$ (fl) O_h^5 (*Fm3m*) (IV, a1)	2.56	2.50
Po—F	PoF$_2$ (fl) O_h^5 (*Fm3m*) (IV, a1)	2.41	2.40

App. 1] INTERATOMIC DISTANCES AND CRYSTAL STRUCTURES

Ac–O	Ac_2O_3 (La_2O_3) D_{3d}^3 ($P\bar{3}m1$) (V, a1)	2.52–2.86	2.55
Th—O	ThO_2 (fl) O_h^5 ($Fm3m$) (IV, a1)	2.42	2.40
	$BaThO_3$ (per) O_h^1 ($Pm3m$) (VII, a5)	2.24	2.40
Th—P	ThP (NaCl) O_h^5 ($Fm3m$) (III, a1)	2.91	2.80
Th—S	ThS (NaCl) O_h^5 ($Fm3m$) (III, a1)	2.84	2.80
Th—As	ThAs (NaCl) O_h^5 ($Fm3m$) (III, a1)	2.98	2.95
Th—Se	ThSe (NaCl) O_h^5 ($Fm3m$) (III, a1)	2.94	2.95
Th—Th	metal (fcc) O_h^5 ($Fm3m$) (II, a)	3.57	3.60
Pa—O	PaO (NaCl) O_h^5 ($Fm3m$) (III, a1)	2.48	2.40
	PaO_2 (fl) O_h^5 ($Fm3m$) (IV, a1)	2.36	2.40
U—C	UC (NaCl) O_h^5 ($Fm3m$) (III, a1)	2.50	2.45
U—N	UN (NaCl) O_h^5 ($Fm3m$) (III, a1)	2.44	2.40
	UN_2 (fl) O_h^5 ($Fm3m$) (IV, a1)	2.30	2.40
	U_2N_3 (Tl_2O_3) T_h^7 ($Ia3$) (V, a2)	2.25	2.40
U—O	UO (NaCl) O_h^5 ($Fm3m$) (III, a1)	2.46	2.35
	UO_2 (fl) O_h^5 ($Fm3m$) (IV, a1)	2.36	2.35
	$BaUO_3$ (per) O_h^1 ($Pm3m$) (VII, a5)	2.20	2.35
U—P	UP (NaCl) O_h^5 ($Fm3m$) (III, a1)	2.80	2.75
U—S	US (NaCl) O_h^5 ($Fm3m$) (III, a1)	2.74	2.75
U—As	UAs (NaCl) O_h^5 ($Fm3m$) (III, a1)	2.88	2.90
U—Se	USe (NaCl) O_h^5 ($Fm3m$) (III, a1)	2.88	2.90
U—Te	UTe (NaCl) O_h^5 ($Fm3m$) (III, a1)	3.08	3.15
U—Bi	UBi (NaCl) O_h^5 ($Fm3m$) (III, a1)	3.18	3.35
U—U	metal * D_{2h}^{17} ($Cmcm$) (II, s)	2.71–3.36	3.50
	metal (bcc) O_h^9 ($Im3m$) (II, c)	3.01	3.50
Np—N	NpN (NaCl) O_h^5 ($Fm3m$) (III, a1)	2.45	2.40
Np—O	NpO (NaCl) O_h^5 ($Fm3m$) (III, a1)	2.50	2.35
	NpO_2 (fl) O_h^5 ($Fm3m$) (IV, a1)	2.35	2.35
Np—Np	metal (bcc) O_h^9 ($Im3m$) (II, c)	3.05	3.50
Pu—C	PuC (NaCl) O_h^5 ($Fm3m$) (III, a1)	2.46	2.45
Pu—N	PuN (NaCl) O_h^5 ($Fm3m$) (III, a1)	2.45	2.40
Pu—O	PuO (NaCl) O_h^5 ($Fm3m$) (III, a1)	2.48	2.35
	PuO_2 (fl) O_h^5 ($Fm3m$) (IV, a1)	2.33	2.35
Pu—S	PuS (NaCl) O_h^5 ($Fm3m$) (III, a1)	2.77	2.75
Pu—Pu	metal (fcc) O_h^5 ($Fm3m$) (II, a)	3.28	3.50
	metal (bcc) O_h^9 ($Im3m$) (II, c)	3.15	3.50
Am—O	AmO (NaCl) O_h^5 ($Fm3m$) (III, a1)	2.48	2.35
	AmO_2 (fl) O_h^5 ($Fm3m$) (IV, a1)	2.32	2.35
	Am_2O_3 (Tl_2O_3) T_h^7 ($Ia3$) (V, a2)	2.33	2.35

Appendix 2
Crystal Parameters

Parameters of crystals crystallizing in structures described in detail in the text. Structures are arranged as in Table 3-2. Notation follows present text rather than Wyckoff, where they differ.

Orthorhombic:

62. D_{2h}^{16} (*Pnma*), aragonite structure. Sec. 3-6, Wyckoff VII, a10, Tables VIIA, 5 and VIIA, 6.

$CaCO_3$, aragonite. $a = 5.72$ A, $b = 4.94$ A, $c = 7.94$ A
$PbCO_3$, $a = 6.15$ A, $b = 5.17$ A, $c = 8.47$ A
KNO_3, $a = 6.45$ A, $b = 5.43$ A, $c = 9.17$ A

		ξ	η	ζ
$CaCO_3$	Ca	0.750		0.417
	C	0.417		0.750
	O	0.417		0.583
	O	0.417	0.48	0.83
$PbCO_3$	Pb	0.750		0.417
	C	0.403		0.764
	O	0.403		0.591
	O	0.403	0.455	0.809
KNO_3	K	0.750		0.416
	N	0.417		0.750
	O	0.417		0.617
	O	0.417	0.444	0.814

64. D_{2h}^{18} (*Cmca*), iodine structure. Sec. 2-8, Wyckoff II, g, II, m, II, ua, II, v, Table II, 5.

	a	b	c	u	v
P (black)	3.31 A	10.50 A	4.38 A	0.098	0.090
Cl	6.29	4.50	8.21	0.130	0.100
Ga	4.49	7.63	4.51	0.1525	0.0785
Br	6.67	4.48	8.72	0.135	0.110
I	7.27	4.79	9.79	0.149	0.116

Tetragonal:

136. D_{4h}^{14} ($P4_2/mnm$), rutile structure. Sec. 3-3, Wyckoff, IV, b1, Table IV, 3.

	a	c	u
MgF_2	4.66 A	3.08 A	0.31
TiO_2 (rutile)	4.49	2.89	0.31
VO_2	4.54	2.88	
CrO_2	4.41	2.86	
MnO_2	4.44	2.89	
MnF_2	4.87	3.28	
FeF_2	4.67	3.30	0.31
CoF_2	4.69	3.19	0.31
NiF_2	4.71	3.12	0.31
ZnF_2	4.72	3.13	0.31
GeO_2	4.39	2.86	0.31
NbO_2	4.77	2.96	
MoO_2	4.86	2.79	
RuO_2	4.51	3.11	
PdF_2	4.93	3.37	
SnO_2	4.72	3.16	0.31
TeO_2	4.79	3.77	
WO_2	4.86	2.77	
OsO_2	4.51	3.19	
IrO_2	4.49	3.14	
PbO_2	4.93	3.37	

Trigonal:

148. C_{3i}^2 ($R\overline{3}$), ilmenite structure. Sec. 3-5, Wyckoff V, a5 and Table VA-6. Parameters given in Sec. 3-5 for ilmenite, $FeTiO_3$, not determined for other crystals. Cell dimensions given below:

	a	α
$LiNbO_3$	5.47 A	55°43′
$LiTaO_3$		
$MgTiO_3$	5.54	54°39′
$MnTiO_3$	5.62	54°16′
$FeTiO_3$	5.52	54°49′
$CoTiO_3$	5.49	54°42′
$NiTiO_3$	5.45	55°8′
$CdTiO_3$	5.82	53°36′

152. D_3^4 ($P3_121$) or 154. D_3^6 ($P3_221$), selenium structure. Sec. 2-7, Wyckoff II, q. Complete information in Sec. 2-7.

155. D_3^7 ($R32$), AlF$_3$ structure. Sec. 3-6, Wyckoff V, b2, Table Vb, 2. Parameters are given in Sec. 3-6 for AlF$_3$. For FeF$_3$, the proposed parameters are $u = \frac{1}{4}$, $v = \frac{1}{3}$, $w = \frac{1}{6}$, leading to a structure which can also be described in terms of the space group D_{3d}^6 ($R\bar{3}c$); see Wyckoff. Only the constants a and α have been measured for other compounds, as follows:

	a	α
AlF$_3$	5.03 A	58°31'
CrBr$_3$	7.05	52°36'
FeF$_3$	5.39	58°0'
FeCl$_3$	6.69	52°30'
CoF$_3$	5.30	57°0'
AsI$_3$	8.25	51°20'
RhF$_3$	5.34	54°20'
PdF$_3$	5.56	54°0'
SbI$_3$	8.18	54°14'
BiI$_3$	8.13	54°50'

164. D_{3d}^3 ($P\bar{3}m1$), CdI$_2$ structure. Sec. 3-3, Wyckoff IV, c1, Table IV-6. The parameter u has been measured for only a few of these crystals, in which it is about $\frac{1}{4}$. Constants a and c for a number of crystals with this structure are given below:

	a	c		a	c
MgBr$_2$	3.81 A	6.26 A	CoI$_2$	3.96 A	6.65 A
MgI$_2$	4.14	6.88	NiTe$_2$	3.86	5.30
CaI$_2$	4.48	6.96	ZnI$_2$	4.25	6.54
TiS$_2$	3.40	5.87	GeI$_2$	4.13	6.79
TiSe$_2$	3.53	6.00	ZrS$_2$	3.68	5.85
TiTe$_2$	3.77	6.54	ZrSe$_2$	3.79	6.18
TiI$_2$	4.11	6.82	PdTe$_2$	4.03	5.12
VBr$_2$	3.77	6.18	CdI$_2$	4.24	6.84
VI$_2$	4.00	6.67	SnS$_2$	3.64	5.87
MnBr$_2$	3.82	6.19	YbI$_2$	4.48	6.96
MnI$_2$	4.16	6.82	PtS$_2$	3.54	5.02
FeBr$_2$	3.74	6.17	PtSe$_2$	3.72	5.06
FeI$_2$	4.04	6.75	PtTe$_2$	4.01	5.20
CoBr$_2$	3.68	6.12	PbI$_2$	4.54	6.86
CoTe$_2$	3.78	5.40			

In addition to these compounds, a considerable number of hydroxides, such as Cd(OH)$_2$, have this structure.

164. D_{3d}^3 ($P\bar{3}m1$), La_2O_3 structure. Sec. 3-6, Wyckoff V, a1, Table VA-2. The values of u and v are supposed to be approximately the values 0.23 and 0.63, respectively, quoted in Sec. 3-6. Parameters for a number of crystals having this structure are given below, including some in which part of the oxygens are replaced by sulfurs.

	a	c	u	v
Mg_3Sb_2	4.57 A	7.23 A		
Mg_3Bi_2	4.67	7.40		
La_2O_3	3.95	6.15		
La_2O_2S	4.04	6.89	0.29	0.64
Ce_2O_3	3.88	6.06		
Ce_2O_2S	4.01	6.83	0.29	0.64
Pr_2O_3	3.85	6.00		
Nd_2O_3	3.84	6.01		
Tl_2N_3	3.88	6.29		
Ac_2O_3	4.08	6.30		
Pu_2O_2S	3.93	6.77	0.29	0.64

164. D_{3d}^3 ($P\bar{3}m1$) AlB_2 structure. Sec. 3-6, Wyckoff IV, c6, Table IV, 9.

	a	c
AlB_2	3.00 A	3.24 A
$CaGa_2$	4.31	4.31
ZrB_2	3.15	3.53
$LaGa_2$	4.32	4.40
$CeGa_2$	4.30	4.31

166. D_{3d}^5 ($R\bar{3}m$), arsenic structure. Sec. 2-6, Wyckoff II, n, Table II, 7.

	a	α	u
β-graphite	3.64 A	39°30′	0.16
As	4.13	54°10′	0.226
Sb	4.51	57°6′	0.233
Bi	4.75	57°14′	0.237

166. D_{3d}^5 ($R\bar{3}m$), CdCl$_2$ structure. Sec. 3-3, Wyckoff IV, c2, Table IV, 7.

	a	α
MgCl$_2$	6.22 A	33°26′
MnCl$_2$	6.20	34°35′
FeCl$_2$	6.20	33°33′
CoCl$_2$	6.16	33°26′
NiCl$_2$	6.13	33°36′
NiBr$_2$	6.47	33°20′
NiI$_2$	6.92	32°40′
ZnCl$_2$	6.31	34°48′
CdCl$_2$	6.23	36°2′
CdBr$_2$	6.63	34°42′

166. D_{3d}^5 ($R\bar{3}m$), LaOF structure. Sec. 3-6, Wyckoff IV, a5, Table IV, 28. The parameters for LaOF are given in Sec. 3-6. Presumably approximately the same parameters apply in the cases below.

	a	α
YOF	6.70 A	33°12′
LaOF	7.13	33°0′
PrOF	7.02	33°2′
NdOF	6.95	33°2′
SmOF	6.87	33°4′
EuOF	6.83	33°3′
GdOF	6.80	33°3′
TbOF	6.76	33°1′

167. D_{3d}^6 ($R\bar{3}c$), calcite structure. Sec. 3-5, Wyckoff VII, a1, Table VIIIA, 2.

	a	α	u
LiNO$_3$	5.74 A	48°3′	0.264
NaNO$_3$	6.32	47°15′	0.25
MgCO$_3$	5.61	48°12′	
CaCO$_3$ calcite	6.36	46°6′	0.243
ScBO$_3$	5.78	48°28′	
MnCO$_3$	5.84	47°45′	0.27
FeCO$_3$	5.75	47°25′	0.27
CoCO$_3$	5.67	48°14′	
ZnCO$_3$	5.67	48°26′	
YBO$_3$	6.44	46°17′	
CdCO$_3$	6.11	47°24′	0.25
InBO$_3$	5.84	48°10′	

167. D_{3d}^6 ($R\bar{3}c$), corundum structure. Sec. 3-5, Wyckoff V, a4, Table Va, 6. The parameters u and v have been determined only for corundum, Al_2O_3. Values of a and α are tabulated below for other crystals with the same structure.

	a	α
Al_2O_3 corundum	5.13 A	55°6′
Ti_2O_3	5.37	56°48′
V_2O_3	5.43	53°53′
Cr_2O_3	5.38	54°50′
α-Fe_2O_3 hematite	5.41	55°17′
Ga_2O_3	5.28	55°35′
Rh_2O_3	5.47	55°40′

Hexagonal:

186. C_{6v}^4 ($P6_3mc$), graphite structure. Sec. 2-5, Wyckoff II, i. $a = 2.46$ A, $c = 6.70$ A, other information given in Sec. 2-5. No other substances with this structure.

186. C_{6v}^4 ($P6_3mc$), wurtzite structure. Sec. 3-2, Wyckoff III, c2, Table III, 11. Information regarding the parameters u is given in Sec. 3-2. Values of a and c for crystals with this structure are tabulated below.

	a	c		a	c
BeO	2.70 A	4.39 A	GaN	3.18 A	5.17 A
MgTe	4.52	7.33	AgI	4.58	7.49
AlN	3.10	4.97	CdS	4.13	6.69
MnS	3.98	6.43	CdSe	4.30	7.02
MnSe	4.12	6.72	InN	3.53	5.69
CuH	2.89	4.61	TaN	3.05	4.94
ZnO	3.24	5.19			
ZnS wurtzite	3.81	6.23			

194. D_{6h}^4 ($P6_3/mmc$), hexagonal close-packed structure. Sec. 2-3, Wyckoff II, b, Table II, 3

	a	c		a	c
He	3.57 A	5.83 A	Cd	2.98 A	5.62 A
Li	3.09	4.83	La	3.75	6.07
Be	2.29	3.58	Ce	3.65	5.96
Mg	3.21	5.21	Pr	3.67	5.92
Ca	3.98	6.52	Nd	3.66	5.90
Sc	3.31	5.26	Gd	3.63	5.80
Ti	2.95	4.69	Tb	3.59	5.67
Cr	2.72	4.43	Dy	3.59	5.65
Co	2.50	4.07	Ho	3.56	5.63
Ni	2.65	4.33	Er	3.56	5.60
Zn	2.66	4.95	Tu	3.53	5.58
Sr	4.32	7.06	Lu	3.52	5.57
Y	3.64	5.76	Hf	3.20	5.06
Zr	3.23	5.25	Re	2.76	4.46
Tc	2.74	4.39	Os	2.74	4.32
Ru	2.70	4.28	Tl	3.46	5.53

194. D_{6h}^4 ($P6_3/mmc$), NiAs structure. Sec. 3-2, Wyckoff, III, d1, Table III, 15.

	a	c		a	c
VS	3.36 A	5.81 A	CoSe	3.61 A	5.28 A
VSe	3.58	5.98	CoSb	3.87	5.19
CrS	3.45	5.75	CoTe	3.89	5.36
CrSe	3.68	6.02	NiS	3.42	5.30
CrTe	3.98	6.21	NiAs	3.60	5.01
MnAs	3.72	5.70	NiSe	3.66	5.33
MnSb	4.12	5.78	NiSn	4.08	5.17
MnTe	4.12	6.70	NiSb	3.94	5.14
MnBi	4.30	6.12	NiTe	3.96	5.35
FeS	3.45	5.67	CuSn	4.19	5.09
FeSe	3.64	5.96	PdSb	4.07	5.58
FeSn	4.23	5.21	PdTe	4.13	5.66
FeSb	4.06	5.13	PtSn	4.10	5.43
FeTe	3.80	5.65	PtSb	4.13	5.47
CoS	3.37	5.16	AuSn	4.31	5.51

194. D_{6h}^4 ($P6_3/mmc$), Na$_3$As structure. Sec. 3-6, Wyckoff V, b18, Table VB, 9. The parameter u has been determined only for Na$_3$As. The quantities a and c for other crystals with this structure are tabulated below.

	a	c		a	c
Li$_3$P	4.26 A	7.58 A	Na$_3$Sb	5.36 A	9.50 A
Li$_3$As	4.39	7.81	Na$_3$Bi	5.45	9.66
Li$_3$Sb	4.70	8.31	K$_3$As	5.78	10.22
Na$_3$P	4.98	8.80	K$_3$Sb	6.03	10.69
Na$_3$As	5.09	8.98	K$_3$Bi	6.18	10.93

Cubic:

205. T_h^6 ($Pa3$), pyrite structure. Sec. 3-3, Wyckoff IV, g2, Table IV, 19.

	a	u		a	u
MnS$_2$	6.10 A	0.401	RuTe$_2$	6.36 A	
MnTe$_2$	6.94		PdAs$_2$	5.97	
FeS$_2$	5.41	0.386	PdSb$_2$	6.44	
pyrite			ReS$_2$	5.57	
CoS$_2$	5.52		OsS$_2$	5.61	
CoSe$_2$	5.85	0.377	OsSe$_2$	5.93	
NiS$_2$	5.74		OsTe$_2$	6.37	
NiSe$_2$	5.95		PtP$_2$	5.68	
NiAsS	5.68		PtAs$_2$	5.96	0.39
NiSbS	5.91		PtSb$_2$	6.43	
RuS$_2$	5.59	0.39	PtBi$_2$	6.68	0.38
RuSe$_2$	5.92				

206. T_h^7 ($Ia3$), Tl$_2$O$_3$ structure. Sec. 3-4, Wyckoff V, a2, Table Va, 3. Parameters are given in Sec. 3-4 for (Fe,Mn)$_2$O$_3$, which is the only crystal with this structure for which they have been accurately measured. The value of a is given below for other crystals with this structure.

	a		a		a
Be$_3$N$_2$	8.13 A	Y$_2$O$_3$	10.60 A	Dy$_2$O$_3$	10.67 A
Be$_3$P$_2$	10.15	Cd$_3$N$_2$	10.79	Ho$_2$O$_3$	10.61
Mg$_3$N$_2$	9.95	In$_2$O$_3$	10.12	Er$_2$O$_3$	10.55
Mg$_3$P$_2$	12.01	La$_2$O$_3$	11.38	Tu$_2$O$_3$	10.49
Mg$_3$As$_2$	12.33	Pr$_2$O$_3$	11.14	Yb$_2$O$_3$	10.44
Ca$_3$N$_2$	10.40	Nd$_2$O$_3$	11.05	Lu$_2$O$_3$	10.39
Sc$_2$O$_3$	9.79	Sm$_2$O$_3$	10.93	Tl$_2$O$_3$	10.54
Mn$_2$O$_3$	9.41	Eu$_2$O$_3$	10.87	U$_2$N$_3$	10.68
(Fe,Mn)$_2$O$_3$	9.37	Gd$_2$O$_3$	10.81	Am$_2$O$_3$	11.03
Zn$_3$N$_2$	9.74	Tb$_2$O$_3$	10.57		

216. T_d^2 ($F\bar{4}3m$), zinc-blende structure.   Sec. 3-2, Wyckoff III, c1, Table III, 10.

	a		a		a
BeS	4.85 A	CuCl	5.41 A	AgI	6.47 A
BeSe	5.07	CuBr	5.68	CdS	5.82
BeTe	5.54	CuI	6.04	CdTe	6.41
AlP	5.42	ZnS	5.41	InP	5.86
AlAs	5.62	ZnSe	5.65	InAs	6.04
AlSb	6.13	ZnTe	6.07	InSb	6.46
SiC	4.35	GaP	5.44	HgS	5.84
MnS	5.60	GaAs	5.64	HgSe	6.07
MnSe	5.82	GaSb	6.12	HgTe	6.36
CuF	4.26				

221. O_h^1 ($Pm3m$), CsCl structure.   Sec. 3-2, Wyckoff III, b1, Table III, 7.

	a		a		a
LiAg	3.17 A	MgTl	3.63 A	CdLa	3.90 A
LiHg	3.29	AlNi	2.88	CdCe	3.86
LiTl	3.42	AlNd	3.73	CdPr	3.82
BeCo	2.61	CaTl	3.85	CsCN	4.25
BeCu	2.70	CuZn	2.95	CsCl	4.11
BePd	2.81	CuPd	2.99	CsBr	4.29
NH_4Cl	3.87	ZnLa	3.75	CsI	4.56
NH_4Br	4.05	ZnCe	3.70	AuZn	3.19
NH_4I	4.37	ZnPr	3.67	AuCd	3.34
MgSr	3.90	RbCl	3.74	TlCN	3.82
MgAg	3.28	SrTl	4.02	TlCl	3.83
MgLa	3.97	AgZn	3.16	TlBr	3.97
MgCe	3.90	AgCd	3.33	TlSb	3.84
MgPr	3.88	AgLa	3.76	TlI	4.20
MgAu	3.26	AgCe	3.73	TlBi	3.98

App. 2] CRYSTAL PARAMETERS 343

221. O_h^1 (Pm3m), perovskite structure. Sec. 3-4, Wyckoff VII, a5, Table VIIA, 4, 4a, 4b.

	a		a		a
NaNbO₃	3.89 A	SrHfO₃	4.07 A	LaCrO₃	3.90 A
NaTaO₃	3.88	YAlO₃	3.67	LaMnO₃	3.88
NaWO₃	3.86	AgZnF₃	3.98	LaFeO₃	3.89
KNbO₃	4.01	CdSnO₃	3.80	LaCoO₃	3.82
KIO₃	4.46	CdCeO₃	3.83	LaGaO₃	3.89
KTaO₃	3.99	CsIO₃	4.66	CeVO₃	3.90
KMgF₃	3.98	CsCaF₃	4.52	CeCrO₃	3.89
KNiF₃	4.01	CsCdCl₃	5.20	PrVO₃	3.89
KZnF₃	4.05	CsCdBr₃	5.33	PrCrO₃	3.89
KCdF₃	4.29	CsHgCl₃	5.44	PrCoO₃	3.76
CaTiO₃	3.84	CsHgBr₃	5.77	NdVO₃	3.89
CaMnO₃	3.74	BaTiO₃	3.97	NdCrO₃	3.89
CaZrO₃	3.99	BaZrO₃	4.17	NdCoO₃	3.77
CaSnO₃	3.92	BaSnO₃	4.10	SmVO₃	3.89
CaCeO₃	3.85	BaCeO₃	4.38	SmCrO₃	3.86
RbIO₃	4.52	BaPrO₃	4.35	SmCoO₃	3.75
RbCaF₃	4.45	BaThO₃	4.48	GdMnO₃	3.82
SrTiO₃	3.90	BaUO₃	4.40	PbTiO₃	3.89
SrZrO₃	4.09	BaLiF₃	4.00	PbZrO₃	4.64
SrSnO₃	4.03	LaAlO₃	3.78	PbCeO₃	3.81
SrCeO₃	4.27	LaVO₃	3.91		

225. O_h^5 (Fm3m), face-centered cubic structure. Sec. 2-2, Wyckoff II, a, Table II, 2.

	a		a		a
Li	4.40 A	Cu	3.61 A	Ce	5.15 A
Ne	4.52	Kr	5.71	Pr	5.16
Al	4.05	Sr	6.08	Yb	5.46
A	5.43	Rh	3.80	Ir	5.84
Ca	5.58	Pd	3.89	Pt	3.92
Sc	4.54	Ag	4.09	Au	4.08
Fe	3.59	Xe	6.25	Pb	4.95
Co	3.55	La	5.30	Th	5.05
Ni	3.52				

225. O_h^5 (Fm3m), sodium chloride structure. Sec. 3-2, Wyckoff III, a1, Table III, 2, 2a, 2b, 4, 5.

	a		a		a
LiH	4.09 A	NaCN	5.89 A	MgO	4.20 A
LiF	4.02	NaSH	6.06	MgS	5.19
LiCl	5.13	NaCl	5.64	MgSe	5.45
LiBr	5.50	NaSeH	6.30	KCN	6.53
LiI	6.00	NaBr	5.97	KOH	5.78
NaF	4.62	NaI	6.47	KF	5.35

225. O_h^5 ($Fm3m$), sodium chloride structure. Sec. 3-2, Wyckoff III, a1, Table III, 2, 2a, 2b, 4, 5. (*Continued*)

	a		a		a
KSH	6.60 A	ZrN	4.61 A	NdSb	6.31 A
KCl	6.29	ZrP	5.27	SmO	5.02
KSeH	6.92	NbC	4.47	EuS	5.96
KBr	6.60	NbN	4.41	EuSe	6.17
KI	7.07	AgF	4.92	EuTe	6.57
CaNH	5.01	AgCl	5.55	GdN	4.99
CaO	4.81	AgBr	5.77	YbSe	5.87
CaS	5.68	CdO	4.69	YbTe	6.34
CaSe	5.91	SnAs	5.68	HfC	4.46
CaTe	6.35	SnSb	6.13	TaC	4.44
ScN	4.44	SnTe	6.29	TaO	4.43
TiC	4.32	CsF	6.01	PbS	5.94
TiN	4.24	CsCl	7.02	PbSe	6.12
TiO	4.24	BaNH	5.84	PbTe	6.34
VC	4.18	BaO	5.52	ThP	5.82
VN	4.13	BaS	6.35	ThS	5.68
VO	4.08	BaSe	6.62	ThAs	5.97
CrN	4.14	BaTe	6.99	ThSe	5.88
MnO	4.44	LaN	5.30	PaO	4.96
MnS	5.22	LaP	6.01	UC	5.00
MnSe	5.45	LaAs	6.13	UN	4.88
FeO	4.33	LaSb	6.48	UO	4.92
CoO	4.27	LaBi	6.57	UP	5.59
NiO	4.17	CeN	5.01	US	5.48
RbCN	6.82	CeP	5.90	UAs	5.77
RbF	5.64	CeS	5.78	USe	5.75
RbSH	6.93	CeAs	6.06	USb	6.19
RbCl	6.58	CeSb	6.40	UTe	6.16
RbSeH	7.21	CeBi	6.49	UBi	6.36
RbBr	6.85	PrN	5.16	NpN	4.90
RbI	7.34	PrP	5.86	NpO	5.01
SrNH	5.45	PrAs	6.00	PuC	4.92
SrO	5.16	PrSb	6.35	PuN	4.91
SrS	5.87	PrBi	6.45	PuO	4.96
SrSe	6.23	NdN	5.14	PuS	5.54
SrTe	6.47	NdP	5.83	AmO	4.96
ZrC	4.67	NdAs	5.96		

225. O_h^5 ($Fm3m$), fluorite and antifluorite structures. Sec. 3-3, Wyckoff IV, a1, Table IV, 2, 2a, 2b.

	a		a		a
Li$_2$NH	5.05 A	Li$_2$Te	6.50 A	Na$_2$Se	6.81 A
Li$_2$O	4.62	Be$_2$C	4.33	Na$_2$Te	7.31
Li$_2$S	5.71	Na$_2$O	5.55	Mg$_2$Si	6.39
Li$_2$Se	6.01	Na$_2$S	6.53	Mg$_2$Ge	6.38

225. O_h^5 (Fm3m), fluorite and antifluorite structures. Sec. 3-3, Wyckoff IV, a1, Table IV, 2, 2a, 2b. (Continued)

	a		a		a
Mg_2Sn	6.77 A	Rh_2P	5.51 A	$AuGa_2$	6.06 A
Mg_2Pb	6.84	CdF_2	5.39	$AuIn_2$	6.50
K_2O	6.44	BaF_2	6.20	$AuSb_2$	6.66
K_2S	7.39	$BaCl_2$	7.34	HgF_2	5.54
K_2Se	7.68	CeO_2	5.41	PbF_2	5.93
K_2Te	8.15	PrO_2	5.36	PoO_2	5.69
CaF_2	5.46	EuF_2	5.80	RaF_2	6.37
fluorite		HfO_2	5.12	ThO_2	5.60
$CoSi_2$	5.36	Ir_2P	5.54	PaO_2	5.52
CuF_2	5.42	$IrSn_2$	6.34	UN_2	5.31
Rb_2O	6.74	$PtAl_2$	5.91	UO_2	5.47
Rb_2S	7.65	$PtGa_2$	5.91	NpO_2	5.44
SrF_2	5.80	$PtIn_2$	6.35	PuO_2	5.40
$SrCl_2$	6.98	$PtSn_2$	6.43	AmO_2	5.38
ZrO_2	5.07	$AuAl_2$	6.00		

227. O_h^7 (Fd3m), diamond structure. Sec. 2-4, Wyckoff II, h, Table II, 6.

	a		a		a
C	3.56 A	Ge	5.62 A	Sn	6.46 A
Si	5.42				

227. O_h^7 (Fd3m), spinel structure. Sec. 3-4, Wyckoff VIII, b1, Table VIIIB, 2, 2a.

	a	u		a	u
Na_2MoO_4	8.99 A		V_2MgO_4	8.39 A	
Na_2WO_4	8.99		V_2FeO_4	8.47	
Mg_2TiO_4	8.44	0.390	V_2ZnO_4	8.39	
Mg_2VO_4	8.39		Cr_2MgO_4	8.31	0.385
Mg_2SnO_4	8.58	0.375	Cr_2MnO_4	8.44	
Al_2MgO_4	8.08	0.387	Cr_2MnS_4	10.05	
spinel			Cr_2FeO_4	8.38	
Al_2CrS_4	9.90		Cr_2FeS_4	9.97	
Al_2MnO_4	8.27	0.390	Cr_2CoO_4	8.32	
Al_2FeO_4	8.12	0.390	Cr_2CoS_4	9.91	
Al_2CoO_4	8.10	0.390	Cr_2NiO_4	8.30	0.385
Al_2NiO_4	8.05	0.390	Cr_2ZnO_4	8.30	0.375
Al_2CuO_4	8.06		Cr_2ZnS_4	9.92	
Al_2ZnO_4	8.08	0.390	Cr_2ZnSe_4	10.44	0.378
Al_2ZnS_4	9.97		Cr_2CdO_4	8.57	0.385
Al_2SnO_4	8.12		Cr_2CdS_4	10.19	0.375
$K_2Zn(CN)_4$	12.54	0.37	Cr_2CdSe_4	10.72	0.383
$K_2Cd(CN)_4$	12.84	0.37	Cr_2HgS_4	10.20	0.392
$K_2Hg(CN)_4$	12.76	0.37	Mn_2TiO_4	8.67	

227. O_h^7 (Fd3m), spinel structure. Sec. 3-4, Wyckoff VIII, b1, Table VIIIB, 2, 2a. (Continued)

	a	u		a	u
Fe_2MgO_4	8.37 A	0.390	Ni_2FeS_4	9.45 A	
Fe_2TiO_4	8.50	0.390	Ni_3S_4	9.48	
Fe_2MnO_4	8.56		Zn_2TiO_4	8.46	0.380
Fe_3O_4	8.40		Zn_2SnO_4	8.61	0.390
magnetite			Ga_2MgO_4	8.28	0.392
Fe_2CoO_4	8.35		Ga_2ZnO_4	8.32	
Fe_2NiO_4	8.32		Ga_2CdO_4	8.57	
Fe_2CuO_4	8.45	0.380	Rh_2MgO_4	8.51	
Fe_2ZnO_4	8.42	0.389	Rh_2ZnO_4	8.52	
Fe_2CdO_4	8.68		Ag_2MoO_4	9.26	0.364
Fe_2PbO_4	7.81		In_2MgO_4	8.81	0.372
Co_2MgO_4	8.11		In_2MgS_4	10.71	0.384
Co_2TiO_4	8.42		In_2CaS_4	10.80	
Co_3O_4	8.11		In_2MnS_4	10.72	
Co_3S_4	9.36		In_2FeS_4	10.62	0.384
Co_2CuO_4	8.04		In_2CoS_4	10.58	0.384
Co_2CuS_4	9.46		In_2NiS_4	10.49	0.384
Co_2ZnO_4	8.11		In_2CdS_4	10.82	0.386
Co_2SnO_4	8.60	0.375	In_2HgS_4	10.83	0.388

229. O_h^9 (Im3m), body-centered cubic structure. Sec. 2-2, Wyckoff II, c, Table II, 4.

	a		a
Li	3.51 A	Nb	3.30 A
Na	4.29	Mo	3.15
K	5.21	Cs	6.05
Ti	3.31	Ba	5.03
V	3.04	Ta	3.31
Cr	2.88	W	3.16
Fe	2.87	Tl	3.88
Rb	5.63	U	3.47
Sr	4.85	Np	3.52
Zr	3.62	Pu	3.64

230. O_h^{10} (Ia3d), garnet structure. Sec. 3-4, Wyckoff XII, aI, 3, Table VIIIC, Ia.

	a	ξ	η	ζ
$Ca_3Al_2(SiO_4)_3$	12.16 A	0.035	0.04	0.65
$Y_3Al_2(AlO_4)_3$	12.01	0.04	0.055	0.64
$Y_3Fe_2(FeO_4)_3$	12.38	0.0274	0.0572	0.6492

Appendix 3
Symmetry Properties and Projection Operators for 20 Space Groups

A3-1. The Space Groups D_{6h}^4 ($P6_3/mmc$) and C_{6v}^4 ($P6_3mc$). In this Appendix we shall carry further the discussion of the properties of space groups given in Chaps. 1 and 5, and in particular shall set up the method of using projection operators for making symmetrized combinations of plane waves. We shall start in this section with the space groups D_{6h}^4 ($P6_3/mmc$) and C_{6v}^4 ($P6_3mc$), which can conveniently be treated together, and which we shall use as illustrations of the general method. These hexagonal cases are enough more complicated than the familiar cubic space groups so that they provide most of the features found in still more complicated space groups; that is, they are nonsymmorphic, and their primitive unit vectors cannot be taken along the coordinate axes. Hence they form good examples to illustrate the general methods. As has been mentioned in Table 3-2, the space group D_{6h}^4 is found in the hexagonal close-packed structure, discussed in Sec. 2-3, and in the NiAs and Na$_3$As structures, discussed in Secs. 3-2 (NiAs) and 3-6 (Na$_3$As). C_{6v}^4 is found in the graphite and wurtzite structures, discussed in Secs. 2-5 (graphite) and 3-2 (wurtzite). The atomic positions in these structures are given in Table A3-1. (See the end of this section for this table, and others referred to in the section.)

We have discussed the unit cell of the hexagonal Bravais lattice, which we have with both these groups, in Fig. 1-5, and have given the primitive vectors in Eq. (1-20). In Eq. (5-11) we have given the vectors of the reciprocal lattice, and have illustrated them in Fig. 5-1. We show the Brillouin zone, used for both cases under discussion, in Fig. A3-1, together with the notation for the special symmetry directions introduced by Herring,[1] who has given the standard treatment of these hexagonal cases. We have discussed the operations of the space group D_{6h}^4 in detail in Sec. 2-3, and have enumerated them in Eq. (2-13). We now wish to proceed from this point. The reason why it is convenient to discuss the

[1] C. Herring, *J. Franklin Inst.*, **233**:525 (1942).

space group C_{6v}^4 along with D_{6h}^4 is that the former contains the same operations as the unprimed operations of D_{6h}^4 enumerated in Eq. (2-13), without the primed operations.

We shall proceed as in Sec. 5-3, finding the effect of the operators of the space group on a plane wave. We shall write the plane wave in terms of a wave vector

$$\mathbf{k} + \mathbf{K}_h = 2\pi[(h_1 + p_1)\mathbf{b}_1 + (h_2 + p_2)\mathbf{b}_2 + (h_3 + p_3)\mathbf{b}_3] \quad \text{(A3-1)}$$

Here h_1, h_2, h_3 are integers, defining $\mathbf{K}_h$ as in Eq. (5-2), and p_1, p_2, p_3 give the reduced wave vector, which equals $2\pi(p_1\mathbf{b}_1 + p_2\mathbf{b}_2 + p_3\mathbf{b}_3)$. In

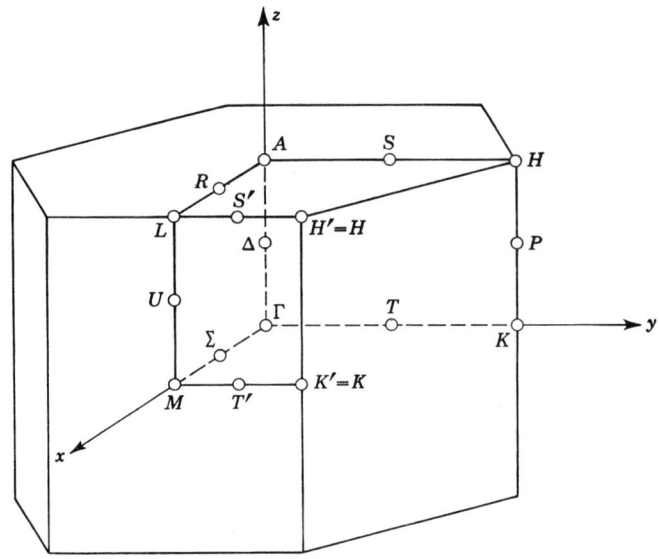

FIG. A3-1. Brillouin zone for the hexagonal Bravais lattice, with symmetry points following notation of Herring.

terms of the wave vector $\mathbf{k} + \mathbf{K}_h$ of Eq. (A3-1), our function ψ is given by $e^{i(\mathbf{k}+\mathbf{K}_h)\cdot\mathbf{r}}$, as in a single term from Eq. (5-5).

We now apply one of the operations of Eq. (2-13) to this function $e^{i(\mathbf{k}+\mathbf{K}_h)\cdot\mathbf{r}}$. As an illustration, let us find $\{X_1|\mathbf{R}_n\}\psi(\mathbf{r})$. From Eq. (2-13) we find

$$\begin{aligned}\{X_1|\mathbf{R}_n\}\psi(\mathbf{r}) &= \exp\{2\pi i[(h_1 + p_1)\mathbf{b}_1 + (h_2 + p_2)\mathbf{b}_2 + (h_3 + p_3)\mathbf{b}_3] \\ &\qquad \cdot [(\xi - \eta + n_1)\mathbf{t}_1 + (\xi + n_2)\mathbf{t}_2 + (\zeta + n_3 + \tfrac{1}{2})\mathbf{t}_3]\} \\ &= \exp\{2\pi i[(h_1 + p_1)(\xi - \eta + n_1) + (h_2 + p_2)(\xi + n_2) \\ &\qquad + (h_3 + p_3)(\zeta + n_3 + \tfrac{1}{2})]\} \\ &= \exp\{2\pi i[(h_1 + p_1 + h_2 + p_2)\xi - (h_1 + p_1)\eta \\ &\qquad + (h_3 + p_3)\zeta]\}\exp[2\pi i(n_1p_1 + n_2p_2 + n_3p_3)]\exp[\pi i(h_3 + p_3)] \quad \text{(A3-2)}\end{aligned}$$

Let us examine the three factors of Eq. (A3-2). First we have a plane wave, with a wave vector $2\pi[(h_1 + p_1 + h_2 + p_2)\mathbf{b}_1 - (h_1 + p_1)\mathbf{b}_2 + (h_3 + p_3)\mathbf{b}_3]$. This modified wave vector could be obtained from the original wave vector of Eq. (A3-1) by the application of the operator X_{-1} of the point group, since X_{-1} is the inverse operation to X_1, and we know from Sec. 5-3 that the wave vector transforms according to the inverse operation to that applying to the coordinate. This fact is not immediately obvious in this case because the space vectors are expressed in terms of the t's, and the reciprocal vectors in terms of the b's, which are not parallel to the t's; in rectangular coordinates a proof is simple.

The second factor $\exp[2\pi i(n_1 p_1 + n_2 p_2 + n_3 p_3)]$ is a constant, the same for all operations of the space group having the same primitive translation vector $\mathbf{R}_n$. It is equivalent to the factor

$$\exp[i\Sigma(p)(k + K_h)_p R_{np}] = \exp[i(\mathbf{k} + \mathbf{K}_h) \cdot \mathbf{R}_n]$$

of Eq. (5-16). In this factor, we note that since $\exp i(\mathbf{K}_h \cdot \mathbf{R}_n) = 1$, the factor can be rewritten in the form $e^{i\mathbf{k}\cdot\mathbf{R}_n}$. The third factor of Eq. (A3-2), $\exp[\pi i(h_3 + p_3)]$, arises from the nonprimitive translation, and is found only for those operations associated with nonprimitive translations. It is equivalent to the factor $\exp[i\Sigma(p)(k + K_h)_p \tau_p^i]$, or $\exp[i(\mathbf{k} + \mathbf{K}_h) \cdot \boldsymbol{\tau}^i]$, of Eq. (5-16). We thus see the equivalence of the expression of Eq. (A3-2) with the general expression of Eq. (5-16).

In a similar way we can find the effect of any one of the operations of the space group on a plane wave. Since the effect of $\mathbf{R}_n$ results in every case in the same factor $\exp[2\pi i(n_1 p_1 + n_2 p_2 + n_3 p_3)]$, we may omit this factor, setting $\mathbf{R}_n = 0$, and finding $\{R_i|0\}\psi$. In the following equations we give the effects of the unprimed operators on the plane wave $\psi = e^{i(\mathbf{k}+\mathbf{K}_h)\cdot\mathbf{r}}$.

$$\{X_0|0\}\psi = \exp 2\pi i[(h_1 + p_1)\xi + (h_2 + p_2)\eta + (h_3 + p_3)\zeta]$$

$$\{X_1|0\}\psi = \exp 2\pi i[(h_1 + p_1 + h_2 + p_2)\xi - (h_1 + p_1)\eta + (h_3 + p_3)\zeta]e^{\pi i(h_3+p_3)}$$

$$\{X_{-1}|0\}\psi = \exp 2\pi i[-(h_2 + p_2)\xi + (h_1 + p_1 + h_2 + p_2)\eta + (h_3 + p_3)\zeta]e^{\pi i(h_3+p_3)}$$

$$\{X_2|0\}\psi = \exp 2\pi i[(h_2 + p_2)\xi - (h_1 + p_1 + h_2 + p_2)\eta + (h_3 + p_3)\zeta]$$

$$\{X_{-2}|0\}\psi = \exp 2\pi i[-(h_1 + p_1 + h_2 + p_2)\xi + (h_1 + p_1)\eta + (h_3 + p_3)\zeta]$$

$$\{X_3|0\}\psi = \exp 2\pi i[-(h_1 + p_1)\xi - (h_2 + p_2)\eta + (h_3 + p_3)\zeta]e^{\pi i(h_3+p_3)}$$

$$\{Y_0|0\}\psi = \exp 2\pi i[(h_1 + p_1 + h_2 + p_2)\xi - (h_2 + p_2)\eta + (h_3 + p_3)\zeta]$$

$$\{Y_1|0\}\psi = \exp 2\pi i[(h_2 + p_2)\xi + (h_1 + p_1)\eta + (h_3 + p_3)\zeta]e^{\pi i(h_3+p_3)}$$

$$\{Y_{-1}|0\}\psi = \exp 2\pi i[(h_1+p_1)\xi - (h_1+p_1+h_2+p_2)\eta$$
$$+ (h_3+p_3)\zeta]e^{\pi i(h_3+p_3)}$$
$$\{Y_2|0\}\psi = \exp 2\pi i[-(h_1+p_1)\xi + (h_1+p_1+h_2+p_2)\eta$$
$$+ (h_3+p_3)\zeta]$$
$$\{Y_{-2}|0\}\psi = \exp 2\pi i[-(h_2+p_2)\xi - (h_1+p_1)\eta + (h_3+p_3)\zeta]$$
$$\{Y_3|0\}\psi = \exp 2\pi i[-(h_1+p_1+h_2+p_2)\xi + (h_2+p_2)\eta$$
$$+ (h_3+p_3)\zeta]e^{\pi i(h_3+p_3)} \quad \text{(A3-3)}$$

For the primed operators with even subscripts, which are associated with nonprimitive translations, the equations are as in Eq. (A3-3), except that $(h_3+p_3)\zeta$ is to be replaced by $-(h_3+p_3)\zeta$, and we have the factor $e^{\pi i(h_3+p_3)}$; for the primed operators with odd subscripts, which have no nonprimitive translations, we again replace $(h_3+p_3)\zeta$ by $-(h_3+p_3)\zeta$, but we omit the factor $e^{\pi i(h_3+p_3)}$.

We may proceed as in Eq. (A3-3), operating on our plane wave with each of the operators of the space group, which equal in number the product of the number of operations in the point group (24 in this case), and the number of unit cells in the repeating range of the crystal, since one of the translations R_n leads from the origin to the corresponding point of each of the unit cells. We see that the result will involve not more than 24 different functions, one coming from each of the operations of the point group, since the translation R_n appears only in the factor $\exp 2\pi i(n_1p_1 + n_2p_2 + n_3p_3)$ which multiplies the function. If $\mathbf{k} + \mathbf{K}_h$ is a general point of reciprocal space, with no special symmetry properties, there will be just 24 different functions, their wave vectors forming a star, in the sense of Sec. 5-4. These 24 plane waves form basis functions for a 24-dimensional irreducible representation of the space group, the matrix elements of the representation being formed as in Eq. (A3-2).

If $\mathbf{k}$ is a point with special symmetry in the Brillouin zone, certain transformed wave vectors will have reduced wave vectors identical with the reduced wave vectors of some other transformed wave vectors, so that the star will contain fewer than 24 distinct wave vectors (that is, wave vectors with different reduced wave vectors). Then, as shown in Sec. 5-4, we have the so-called group of the wave vector, the subgroup of the space group which transforms a given reduced wave vector into itself. This group will transform those of the 24 basis functions which correspond to the same reduced wave vector among themselves. We can make linear combinations of these basis functions corresponding to the same reduced wave vector, so as to obtain basis functions for irreducible representations of the group of the wave vector. These linear combinations of plane waves are the symmetrized plane waves which we desire. Their virtue is that the Hamiltonian has no nondiagonal matrix elements between symmetrized plane waves corresponding to different irreducible representations of the group of the wave vector, or to different partners in the same irreducible representation.

We shall find these linear combinations, or symmetrized plane waves, by the projection-operator method, as discussed for a simple case in Sec. 5-5. The fundamental equation of that method is Eq. (5-26), or

$$f_{jl}^{(p)} = \Sigma(R_i)\Gamma_p(R_i)^*_{jl}R_i\psi \tag{A3-4}$$

which was discussed in Sec. 5-5. Here R_i stands for one of the operations of the space group, which we have been writing in a more explicit form as $\{R_i|\mathbf{R}_n\}$. If the function ψ is a plane wave, then, as we have seen, the functions $R_i\psi$, or more explicitly $\{R_i|\mathbf{R}_n\}\psi$, will be other plane waves. Hence the functions defined by Eq. (A3-4) will be basis functions for irreducible representations of the space group. If instead of the complete space group we choose the subgroup which leaves the reduced wave vector invariant (that is, the group of the wave vector), then the linear combinations of plane waves defined by Eq. (A3-4) will be the symmetrized plane waves which we desire.

Let us note first that in using this procedure, we can dispense with the primitive vectors $\mathbf{R}_n$. For we see from Eq. (A3-2) that the matrix elements of all operators corresponding to a given $\mathbf{R}_n$ contain the common factor $\exp 2\pi i(n_1p_1 + n_2p_2 + n_3p_3)$, which is the only way in which $\mathbf{R}_n$ is contained in the result. Thus this factor will be contained in the matrix elements $\Gamma_p(\{R_i|\mathbf{R}_n\})$, contained in the equation describing the effect of operating with the operator $\{R_i|\mathbf{R}_n\}$ on one of the basis functions u_m of the pth irreducible representations of the group, which is

$$\{R_i|\mathbf{R}_n\}u_m = \Sigma(k)\Gamma_p(\{R_i|\mathbf{R}_n\})_{km}u_k \tag{A3-5}$$

by the general definition of basis functions, as given in Volume 1, Eq. (8-15). When we take the complex conjugate of this matrix element for use in Eq. (A3-4), we shall have the factor $\exp[-2\pi i(n_1p_1 + n_2p_2 + n_3p_3)]$ multiplying $R_i\psi$, or $\{R_i|\mathbf{R}_n\}\psi$. But since $\{R_i|\mathbf{R}_n\}\psi$ contains the factor $\exp[2\pi i(n_1p_1 + n_2p_2 + n_3p_3)]$, this factor will cancel, and each term of the sum of Eq. (A3-4) will be independent of $\mathbf{R}_n$. When we sum over all values of $\mathbf{R}_n$, as is included in the summation over R_i in Eq. (A3-4), we merely have as many identical terms as there are unit cells in the repeating volume of the crystal. This means that the functions $f_{jl}^{(p)}$ found from Eq. (A3-3) are merely a constant times the functions which we should have found if we used only the operations corresponding to $\mathbf{R}_n = 0$. Hence for our present purposes of using the projection operators to find symmetrized plane waves, we may assume $\mathbf{R}_n = 0$ from the outset.

We now require the matrix elements $\Gamma_p(\{R_i|0\})_{jl}$ for these irreducible representations, corresponding to the various symmetry points $\mathbf{k}$ within the Brillouin zone. If by some device we can set up basis functions for the irreducible representations, we can proceed immediately to get the matrix elements, as we did in Sec. 5-5 for the case C_{4v}. The problems under discussion are simple enough so that it is practical to set up such basis functions by inspection. Once we have done so, and have computed the matrix elements, then we can use Eq. (A3-4) to find basis functions in a systematic manner. As we shall carry out the process, these basis functions found by Eq. (A3-4) will be identical with those which we set up by inspection. We shall illustrate this method later.

The fundamental procedure which we shall use then writes the basis functions by Eq. (A3-4) as linear combinations of operators of the group operating on a plane wave ψ. To test whether this is in fact a basis function, we may in turn operate on it with one of the operators of the group, and see whether Eq. (A3-5) is satisfied. Since u_m is already of the form of a sum of terms like $\{R_j|0\}\psi$, where $\{R_j|0\}$ is one of the operators, we see that in order to find $\{R_i|0\}u_m$ we must have the multiplication table: we must be able to write $\{R_i|0\}\{R_j|0\}u_m$ as one of the operators of the group, such as $\{R_k|0\}$, operating on u_m.

By use of Eq. (A3-3), we can straightforwardly find the effect of successive application of two operators of the space group on ψ. Let us illustrate by two specific examples, which we shall take to be the products $\{X_2|0\}\{X_1|0\}\psi$ and $\{X_2'|0\}\{X_1|0\}\psi$, which will illustrate different points. For the first case, we note from Eq. (A3-3) that the operator $\{X_2|0\}$ has a coefficient for ξ which was the coefficient of η in the original function, and it has a coefficient for η which is the negative of the sum of the coefficients of ξ and η in the original function. The coefficient of ζ is unchanged. Hence the resulting plane wave is

$$\{X_2|0\}\{X_1|0\}\psi$$
$$= \exp 2\pi i[-(h_1 + p_1)\xi - (h_2 + p_2)\eta + (h_3 + p_3)\zeta]e^{\pi i(h_3+p_3)}$$
$$= \{X_3|0\}\psi \qquad (A3\text{-}6)$$

Similarly, we have

$$\{X_2'|0\}\{X_1|0\}\psi$$
$$= \exp 2\pi i[-(h_1 + p_1)\xi - (h_2 + p_2)\eta - (h_3 + p_3)\zeta]e^{2\pi i(h_3+p_3)}$$
$$= e^{2\pi i p_3}\{X_3'|0\}\psi \qquad (A3\text{-}7)$$

We note in the first place that in each case the subscript of the product function follows from the multiplication table of the point group C_{Nv}, which is

$$\begin{aligned} X_q X_p &= X_{q+p} \\ X_q Y_p &= Y_{-q+p} \\ Y_p X_q &= Y_{p+q} \\ Y_q Y_p &= X_{-q+p} \end{aligned} \qquad (A3\text{-}8)$$

as given in Volume 1, Eqs. (8-7), (8-9), (8-11), and (8-13). In these expressions, if the subscript of the product function does not lie within the prescribed range (-2 to 3 for $N = 6$), we must add or subtract integral multiples of N to bring it within this range. We note, furthermore, that we have examples of a general rule, which holds for the group D_{Nh}, namely, that the product of two unprimed operators, or two primed ones, is unprimed, while the product of an unprimed and a primed operator is primed.

As for the factor $e^{2\pi i p_3}$, we can verify that the following are the only cases in which such a factor appears:

$$\begin{array}{ll} \text{unprimed odd} \times \text{unprimed odd:} & e^{2\pi i p_3} \\ \text{unprimed odd} \times \text{primed odd:} & e^{-2\pi i p_3} \\ \text{primed even} \times \text{unprimed odd:} & e^{2\pi i p_3} \\ \text{primed even} \times \text{primed odd:} & e^{-2\pi i p_3} \end{array} \qquad \text{(A3-9)}$$

Other cases are like Eq. (A3-6), involving no additional factor. These factors arise because the product of two operators of the space group, like $\{X'_2|0\}\{X_1|0\}$, can lead to an operator of the space group involving a non-vanishing translation. In the particular case of Eq. (A3-7), we have the result

$$\{X'_2|0\}\{X_1|0\} = \{X'_3|t_3\} \qquad \text{(A3-10)}$$

which is the operator associated with the displacement $\mathbf{R}_n = t_3$, corresponding to $n_1 = n_2 = 0$, $n_3 = 1$. This displacement would result in a factor $\exp 2\pi i(n_1 p_1 + n_2 p_2 + n_3 p_3) = \exp 2\pi i p_3$ in the result of operating on the plane wave with the operator $\{X'_3|t_3\}$, just as we have found. We see from Eq. (A3-9) that in every case where a factor arises, it is of the form similar to Eq. (A3-10), corresponding to a displacement of $\mathbf{R}_n = \pm t_3$. We shall find similar results in all cases of nonsymmorphic space groups. This is demanded by the general discussion of Sec. 1-3, in which we pointed out that the translation involved in Eq. (1-14), involved in the simultaneous operation of two operators of the space group, had to be of the form $\tau^k + \mathbf{R}_l$, as given in Eq. (1-15); namely, the nonprimitive translation associated with the product operation of the point group, plus a translation $\mathbf{R}_l$ of the Bravais lattice. In our present case this translation $\mathbf{R}_l$ is $\pm t_3$.

We now have the machinery required for a discussion of the methods used in deriving and verifying the irreducible representations. In Tables A3-2 through A3-14 (at the end of this section) we present matrix elements and character tables for the various symmetry points of the space groups D_{6h}^4 and C_{6v}^4. Some of these can be determined by fairly well specified procedures, while others have been arrived at more or less by trial and error. Thus, it is well known that the group of the wave vector is one of the familiar point groups, and that the irreducible representations of these point groups can be of great help in building up the irreducible representations of the space group. For Γ, as a first instance, the group of the wave vector is the complete group D_{6h} (that is, for the space group D_{6h}^4), and the matrix elements given in Table A3-2 are those determined in a familiar way for that group, as discussed in Volume 1, Chap. 8 and Appendix 12, of this work. For Δ, given in Table A3-11, the group of the wave vector is C_{6v}. Here, however, on account of the quantity $\alpha = e^{\pi i p_3}$ appearing in the matrix elements, the situation is different from that met in the point group. Let us therefore use this as an example of the method which we can use to arrive at the matrix elements, and to verify their correctness.

Let us specifically try to arrive at the matrix elements of the one-dimensional irreducible representation Δ_1, given in Table A3-11, all of which would be unity for the point group C_{6v}. The reason why we cannot use this set of matrix elements for the space group comes from Eq. (A3-9), in which it is stated that the multiplication of two unprimed odd operators introduces a factor $e^{2\pi i p_3}$, which is not met in the point group. This does not concern us at the point Γ, where $p_3 = 0$. Here, however, it means that the operators X_1, X_{-1}, X_3, Y_1, Y_{-1}, Y_3 must be treated differently from the operators with even subscripts. It seems plausible, then, that we may have to treat differently the function $(X_0 + X_2 + X_{-2} + Y_0 + Y_2 + Y_{-2})\psi$, where $\psi = e^{i(\mathbf{k}+\mathbf{K}_h)\cdot\mathbf{r}}$, and the function $(X_1 + X_{-1} + X_3 + Y_1 + Y_{-1} + Y_3)\psi$, which would arise from Eq. (A3-4) if all matrix elements $\Gamma_p(R_i)_{jl}$ were equal to unity. If we let the first of these functions be ψ_1, the second ψ_2, we find by application of the multiplication table that the effect of any of the operators with an even subscript on ψ_1 gives ψ_1, and on ψ_2 gives ψ_2. However, the application of an operator with odd subscript on ψ_2 gives $e^{2\pi i p_3}\psi_2$, though this same operator applied to ψ_1 gives ψ_1.

We then note that if we take functions $\psi_1 \pm e^{-\pi i p_3}\psi_2$, the operation of any one of the operators on this function gives a constant times the function, so that we have a basis function for a one-dimensional irreducible representation. The only operators which might give trouble when applied to these functions are those with odd subscripts. Thus let us verify the correctness of our functions by letting X_1 operate on these functions. We have

$$X_1(\psi_1 \pm e^{-\pi i p_3}\psi_2) = X_1(X_0 + X_2 + X_{-2} + Y_0 + Y_2 + Y_{-2})e^{i(\mathbf{k}+\mathbf{K}_h)\cdot\mathbf{r}}$$
$$\pm e^{-\pi i p_3}X_1(X_1 + X_{-1} + X_3 + Y_1 + Y_{-1} + Y_3)e^{i(\mathbf{k}+\mathbf{K}_h)\cdot\mathbf{r}} \quad \text{(A3-11)}$$

We apply the multiplication rules of Eqs. (A3-8) and (A3-9), and find

$$(X_1 + X_3 + X_{-1} + Y_{-1} + Y_1 + Y_3)e^{i(\mathbf{k}+\mathbf{K}_h)\cdot\mathbf{r}}$$
$$\pm e^{-\pi i p_3}e^{2\pi i p_3}(X_2 + X_0 + X_{-2} + Y_0 + Y_{-2} + Y_2)e^{i(\mathbf{k}+\mathbf{K}_h)\cdot\mathbf{r}}$$
$$= \pm e^{\pi i p_3}(\psi_1 \pm e^{-\pi i p_3}\psi_2) \quad \text{(A3-12)}$$

Hence we have verified that we have a basis function for a one-dimensional irreducible representation, for either the + or − sign. The + sign gives the function for Δ_1 in Table A3-11, the − sign gives that for Δ_2. Furthermore, when we recall that $\alpha^* = e^{-\pi i p_z}$, we see that the coefficients involved in the linear combinations $\psi_1 \pm e^{-\pi i p_z}\psi_2$ are just the conjugates of the matrix elements of Table A3-11, as is required by Eq. (A3-4).

This is a simple example of the sort of procedure which must be used in deriving the matrix elements which we have tabulated, and the method used to verify their correctness. The reader will find in each case that if he sets up basis functions by use of Eq. (A3-4), using the matrix elements given in our tables, and then applies the operators of the group of the wave vector to these functions, he will find the matrix elements tabulated, thereby completely verifying the tables.

In some cases of symmetry points on the boundaries of the Brillouin zone, the dimensionalities of the irreducible representations, and the whole nature of these representations, are quite different from those suggested by the ordinary representations found for the group of the wave vector. Thus, as a striking case, we have the point A for the group D_{6h}^4. Here the group of the wave vector is the complete group D_{6h}, but instead of having eight one-dimensional irreducible representations, and four two-dimensional, as at Γ, we have two two-dimensional representations and a four-dimensional representation. This is already known from the character tables in the work of Herring (*loc. cit.*). If it were not known already, it would be useful at least to be able to verify that we had found the correct number of representations. For this purpose the theorem that the sum of the squares of the dimensionalities of the irreducible representations equals the number of operations in the group of the wave vector[1] is very convenient. Thus, at Γ, with 24 operations, we have

$$8(1^2) + 4(2^2) = 24$$

as the sum of the squares of the dimensionalities. At A we have

$$2(2^2) + 4^2 = 24$$

The reader can verify easily that this theorem is obeyed in each case.

In the case of the present space groups, which have already been discussed by Herring and others, there is no trouble in finding the group of the wave vector in each case. It is useful to see, however, that there is a general procedure by which this can be established. We start with the operations of the space group, in the form of Eq. (A3-3). We substitute into each of these expressions the values of p_1, p_2, p_3 describing the symmetry point in question. We can then see at once which wave functions

---

[1] For a proof of this theorem, see J. C. Slater, G. F. Koster, and J. H. Wood, *Phys. Rev.*, **126**:1307 (1962), appendix 1.

correspond to the same reduced wave vector. For example, to take a somewhat complicated case, let us take the point K, given in Table A3-8, where $p_1 = -\frac{1}{3}$, $p_2 = \frac{2}{3}$, $p_3 = 0$. From Eq. (A3-3), the function $\{X_0|0\}\psi$ is given by

$$e^{2\pi i(h_1\xi+h_2\eta+h_3\zeta)}e^{2\pi i(-\frac{1}{3}\xi+\frac{2}{3}\eta)} \qquad (A3\text{-}13)$$

where the first factor is a periodic function repeating in each unit cell, of the form $e^{i\mathbf{K}_h\cdot\mathbf{r}}$, where $\mathbf{K}_h$ is one of the vectors of the reciprocal lattice, while the second factor is that embodying the reduced wave vector. In a similar way we find that $\{X_2|0\}\psi$ is given by

$$e^{2\pi i[(h_2+1)\xi-(h_1+h_2+1)\eta+h_3\zeta]}e^{2\pi i(-\frac{1}{3}\xi+\frac{2}{3}\eta)} \qquad (A3\text{-}14)$$

with a different periodic function from Eq. (A3-13), but with the same factor involving the reduced wave vector. Thus the operator $\{X_2|0\}$ belongs to the group of the wave vector. In a similar way one can determine which operators belong to this group and which ones do not.

From the matrix elements of Tables A3-2 to A3-14, we can find the compatibility relations, which are given in Table A3-15. As was shown by Bouckaert, Smoluchowski, and Wigner,[1] the condition for compatibility of representations along a line of symmetry directions, and a point of symmetry directions lying on the line, is that the sum of the characters of the compatible representations should agree, for the group of the wave vectors along the line. For example, let us consider the compatibility of Γ and Δ. At Γ, the factor $\alpha = e^{\pi i p_3}$ found in Table A3-11 equals unity, since p_3 equals zero. Hence, for instance, the matrix elements (or characters) for Δ_1 equal 1 for all operators of the group, as we see from Table A3-11. From Table A3-2, we have this situation for the representations Γ_1^+, Γ_2^-, which differ only in the primed operations, which are not included in the group of the wave vector along Δ. Hence Δ_1 is compatible with Γ_1^+ and Γ_2^-, as is indicated in Table A3-15. Again, we see from Table A3-2 that the characters of the representations Γ_5^+ and Γ_5^- for the operations X_0, X_1, X_{-1}, X_2, X_{-2}, X_3, respectively, are $2, -1, -1, -1, -1, 2$, respectively, and for the Y's we have zero. These are the same as those found in Table A3-11 for the representation Δ_5, which is then compatible with Γ_5^+ and Γ_5^-. These examples are typical of those used in establishing Table A3-15. In case a two-dimensional representation at a symmetry point is compatible with two one-dimensional representations along the related symmetry direction, the sum of the matrix elements of the one-dimensional representations will equal the characters of the two-dimensional representation.

In Sec. 10-3, we give a discussion of the type of symmetry found in the

[1] *Phys. Rev.*, **50**:58 (1936).

neighborhood of each atom of the crystal, in the hexagonal close-packed structure. The results are given in Table 10-1. The information in this table can be derived in the following manner. We set up wave functions of the appropriate symmetry type, using the projection-operator method and the matrix elements given in the tables to follow. Specifically, this table is for the point Γ, for which we then use Table A3-2. Then we expand the wave function in power series about the position of one of the atoms, as we did in Sec. 5-5 for the two-dimensional case. The only difference between the present case and that one is that here the atoms are not at the origin, whereas in Sec. 5-5 the expansion was about the origin. When we have found the expansions of the wave function, we can tell by inspection which of the symmetry types of the point group D_{3h}, which gives the symmetry about an atom in the hexagonal close-packed structure, we are dealing with.

In Table 10-1 we also indicate which irreducible representations are bonding, which antibonding. This can be found easily, by considering the inversion symmetry about the origin, which is the midpoint between two atoms. The inversion operator, from Eq. (A3-3), is $\{X'_3|0\}$. An irreducible representation which is odd on inversion necessarily has a node midway between the atoms, and is antibonding, whereas one which is even on inversion is bonding. This situation regarding bonding is different from that found in a diatomic molecule, where a π state is bonding if it is odd on inversion. Here there is no quantization of the component of orbital angular momentum about the axis joining the two atoms. A wave function will be made up of states which would have all types of symmetry about the axis: σ, π, δ, etc. The leading term will ordinarily be of the σ type, so that the question as to whether the orbital is bonding or antibonding is determined by the inversion symmetry of this component, leading to bonding for the function even on inversion.

In further sections of this Appendix we shall not refer specifically to methods of investigating the symmetry about atomic sites, though there are a number of references to such symmetry in the discussion of energy bands in Chap. 10. In each case the symmetry is to be investigated as we have described it here, expanding a symmetrized plane wave of the appropriate symmetry type in power series about the atomic position, and identifying the resulting function in terms of the basis functions for the point group of operations which carry an atom at a special position into itself.

Table A3-1
Atomic positions in crystal structures with space groups D_{6h}^4 and C_{6v}^4. Atoms are at positions $\xi t_1 + \eta t_2 + \zeta t_3$.

Hexagonal-close-packed structure. D_{6h}^4, two atoms per unit cell, at $\xi = \frac{1}{3}$, $\eta = \frac{2}{3}$, $\zeta = \frac{1}{4}$ and $\xi = \frac{2}{3}$, $\eta = \frac{1}{3}$, $\zeta = \frac{3}{4}$.

Nickel arsenide structure. D_{6h}^4, two atoms of Ni and two of As per unit cell. Ni atoms at ξ, η, ζ given by 0, 0, 0 and 0, 0, $\frac{1}{2}$, As atoms at $\frac{1}{3}$, $\frac{2}{3}$, $\frac{1}{4}$ and $\frac{2}{3}$, $\frac{1}{3}$, $\frac{3}{4}$.

Sodium arsenide structure. D_{6h}^4, Na$_3$As. Six atoms of sodium and two of arsenic per unit cell. Na at 0, 0, $\frac{1}{4}$ and 0, 0, $\frac{3}{4}$, also at $\pm(\frac{1}{3}, \frac{2}{3}, u; \frac{2}{3}, \frac{1}{3}, u + \frac{1}{2})$, where $u = 0.583$. Arsenics at $\frac{1}{3}$, $\frac{2}{3}$, $\frac{1}{4}$ and $\frac{2}{3}$, $\frac{1}{3}$, $\frac{3}{4}$.

Graphite structure. C_{6v}^4, four atoms per unit cell, at 0, 0, 0; 0, 0, $\frac{1}{2}$; $\frac{1}{3}$, $\frac{2}{3}$, 0; $\frac{2}{3}$, $\frac{1}{3}$, $\frac{1}{2}$.

Wurtzite structure. C_{6v}^4. Two atoms of Zn and two of S per unit cell. Zn at $\frac{1}{3}$, $\frac{2}{3}$, 0 and $\frac{2}{3}$, $\frac{1}{3}$, $\frac{1}{2}$, S at $\frac{1}{3}$, $\frac{2}{3}$, u and $\frac{2}{3}$, $\frac{1}{3}$, $u + \frac{1}{2}$, where u is close to 0.375.

Table A3-2
Matrix elements and characters of the irreducible representations for the space groups D_{6h}^4 and C_{6v}^4 at point Γ, $\mathbf{k} = 0$. The notations are those of Herring for D_{6h}^4. For D_{6h}^4, the matrix elements of the primed operators are equal to those of the unprimed for the first irreducible representation of each pair tabulated, and are the negative of those of the unprimed for the second representation tabulated. For C_{6v}^4 we have only the unprimed operators. The abbreviation ω stands for $e^{2\pi i/3}$. For C_{6v}^4, the representations may be denoted $\Gamma_1, \Gamma_2, \Gamma_3, \Gamma_4, \Gamma_5, \Gamma_6$ in the order stated.

	X_0	X_1	X_{-1}	X_2	X_{-2}	X_3	Y_0	Y_1	Y_{-1}	Y_2	Y_{-2}	Y_3
Γ_1^+, Γ_2^-	1	1	1	1	1	1	1	1	1	1	1	1
Γ_2^+, Γ_1^-	1	1	1	1	1	1	-1	-1	-1	-1	-1	-1
Γ_3^-, Γ_4^+	1	-1	-1	1	1	-1	-1	1	1	-1	-1	1
Γ_4^-, Γ_3^+	1	-1	-1	1	1	-1	1	-1	-1	1	1	-1
$(\Gamma_5^+, \Gamma_5^-)_{11}$	1	ω	ω^2	ω^2	ω	1	0	0	0	0	0	0
$(\Gamma_5^+, \Gamma_5^-)_{21}$	0	0	0	0	0	0	1	ω	ω^2	ω^2	ω	1
$(\Gamma_5^+, \Gamma_5^-)_{12}$	0	0	0	0	0	0	1	ω^2	ω	ω	ω^2	1
$(\Gamma_5^+, \Gamma_5^-)_{22}$	1	ω^2	ω	ω	ω^2	1	0	0	0	0	0	0
$\chi(\Gamma_5^+, \Gamma_5^-)$	2	-1	-1	-1	-1	2	0	0	0	0	0	0
$(\Gamma_6^-, \Gamma_6^+)_{11}$	1	$-\omega^2$	$-\omega$	ω	ω^2	-1	0	0	0	0	0	0
$(\Gamma_6^-, \Gamma_6^+)_{21}$	0	0	0	0	0	0	1	$-\omega^2$	$-\omega$	ω	ω^2	-1
$(\Gamma_6^-, \Gamma_6^+)_{12}$	0	0	0	0	0	0	1	$-\omega$	$-\omega^2$	ω^2	ω	-1
$(\Gamma_6^-, \Gamma_6^+)_{22}$	1	$-\omega$	$-\omega^2$	ω^2	ω	-1	0	0	0	0	0	0
$\chi(\Gamma_6^-, \Gamma_6^+)$	2	1	1	-1	-1	-2	0	0	0	0	0	0

Table A3-3

Matrix elements and characters for the irreducible representations along Σ, propagation in the x direction, for D_{6h}^4. This is the direction $-\tfrac{1}{2} < p_1 \leq \tfrac{1}{2}$, $p_2 = p_3 = 0$.

	X_0	Y_0	X'_0	Y'_0
Σ_1	1	1	1	1
Σ_2	1	-1	-1	1
Σ_3	1	1	-1	-1
Σ_4	1	-1	1	-1

For C_{6v}^4, for which we do not have the primed operations, there are only two representations, Σ_1 and Σ_2, with matrix elements for the operations X_0, Y_0 as for Σ_1 and Σ_2 above.

Table A3-4

Matrix elements and characters for the irreducible representations at M, for D_{6h}^4. This is the point $p_1 = \tfrac{1}{2}$, $p_2 = p_3 = 0$. Upper or lower signs to be used together.

	X_0	X_3	Y_0	Y_3	X'_0	X'_3	Y'_0	Y'_3
$M_1^\pm$	1	1	± 1	± 1	± 1	± 1	1	1
$M_2^\pm$	1	-1	∓ 1	± 1	∓ 1	± 1	1	-1
$M_3^\pm$	1	-1	± 1	∓ 1	∓ 1	± 1	-1	1
$M_4^\pm$	1	1	∓ 1	∓ 1	± 1	± 1	-1	-1

For C_{6h}^4, for which we do not have the primed operations, we have the four irreducible representations $M_1^\pm$, $M_2^\pm$, with matrix elements for the unprimed operations as given above.

Table A3-5

Matrix elements and characters of the irreducible representations for D_{6h}^4 along the line R. Here $-\tfrac{1}{2} < p_1 \leq \tfrac{1}{2}$, $p_2 = 0$, $p_3 = \tfrac{1}{2}$.

	X_0	Y_0	X'_0	Y'_0
R_1	1	1	1	1
R_2	1	-1	-1	1
R_3	1	1	-1	-1
R_4	1	-1	1	-1

For C_{6v}^4, for which we do not have the primed operations, there are only two representations, R_1 and R_2, with matrix elements for the operations X_0, Y_0 as for R_1 and R_2 above.

Table A3-6

Matrix elements and characters of the irreducible representations for D_{6h}^4 at the point L. Here $p_1 = \frac{1}{2}$, $p_2 = 0$, $p_3 = \frac{1}{2}$.

	X_0	X_3	Y_0	Y_3	X_0'	X_3'	Y_0'	Y_3'
$(L_1)_{11}$	1	0	1	0	1	0	1	0
$(L_1)_{21}$	0	1	0	1	0	1	0	1
$(L_1)_{12}$	0	−1	0	−1	0	1	0	1
$(L_1)_{22}$	1	0	1	0	−1	0	−1	0
$\chi(L_1)$	2	0	2	0	0	0	0	0
$(L_2)_{11}$	1	0	−1	0	1	0	−1	0
$(L_2)_{21}$	0	1	0	−1	0	1	0	−1
$(L_2)_{12}$	0	−1	0	1	0	1	0	−1
$(L_2)_{22}$	1	0	−1	0	−1	0	1	0
$\chi(L_2)$	2	0	−2	0	0	0	0	0

For C_{6v}^4, for which we do not have the primed operations, the matrix elements are given below:

	X_0	X_3	Y_0	Y_3
L_1	1	i	1	i
L_2	1	$-i$	1	$-i$
L_3	1	$-i$	−1	i
L_4	1	i	−1	$-i$

This is simply a limiting case of U, given in Table A3-13.

Table A3-7

Matrix elements and characters for the irreducible representations along T, propagation in the y direction, and T', for D_{6h}^4. For T, $p_1 = -p_2/2$, $-\frac{2}{3} < p_2 \leq \frac{2}{3}$, $p_3 = 0$. For T', $p_1 = -p_2/2 + \frac{1}{2}$, $-\frac{1}{3} < p_2 \leq \frac{1}{3}$, $p_3 = 0$.

	X_0	Y_3	X_0'	Y_3'
T_1, T_1'	1	1	1	1
T_2, T_2'	1	−1	−1	1
T_3, T_3'	1	1	−1	−1
T_4, T_4'	1	−1	1	−1

For C_{6v}^4, for which we do not have the primed operations, there are only two representations, T_1 and T_2, or T_1' and T_2', respectively, with matrix elements for the operations X_0 and Y_3 as for T_1 and T_2, or T_1' and T_2', above.

Table A3-8

Matrix elements and characters for the irreducible representations for D_{6h}^4 at the point K. Here $p_1 = -\frac{1}{3}$, $p_2 = \frac{2}{3}$, $p_3 = 0$. The abbreviation ω stands for $e^{2\pi i/3}$.

	X_0	X_2	X_{-2}	Y_1	Y_{-1}	Y_3	X_0'	X_2'	X_{-2}'	Y_1'	Y_{-1}'	Y_3'
K_1	1	1	1	1	1	1	1	1	1	1	1	1
K_2	1	1	1	-1	-1	-1	-1	-1	-1	1	1	1
K_3	1	1	1	-1	-1	-1	1	1	1	-1	-1	-1
K_4	1	1	1	1	1	1	-1	-1	-1	-1	-1	-1
$(K_5)_{11}$	1	ω^2	ω	0	0	0	1	ω^2	ω	0	0	0
$(K_5)_{21}$	0	0	0	1	ω	ω^2	0	0	0	1	ω	ω^2
$(K_5)_{12}$	0	0	0	1	ω^2	ω	0	0	0	1	ω^2	ω
$(K_5)_{22}$	1	ω	ω^2	0	0	0	1	ω	ω^2	0	0	0
$\chi(K_5)$	2	-1	-1	0	0	0	2	-1	-1	0	0	0
$(K_6)_{11}$	1	ω^2	ω	0	0	0	-1	$-\omega^2$	$-\omega$	0	0	0
$(K_6)_{21}$	0	0	0	1	ω	ω^2	0	0	0	-1	$-\omega$	$-\omega^2$
$(K_6)_{12}$	0	0	0	1	ω^2	ω	0	0	0	-1	$-\omega^2$	$-\omega$
$(K_6)_{22}$	1	ω	ω^2	0	0	0	-1	$-\omega$	$-\omega^2$	0	0	0
$\chi(K_6)$	2	-1	-1	0	0	0	-2	1	1	0	0	0

The point K', where $p_1 = p_2 = \frac{1}{3}$, $p_3 = 0$, has the same symmetry as K, and the same representations can be used. For the case C_{6v}^4, where we do not have the primed operations, we have only the representations K_1, K_2, and K_5, whose matrix elements for the unprimed operations are the same as those given in the table.

Table A3-9

Matrix elements and characters for the irreducible representations for D_{6h}^4 along the lines S and S'. For S, $p_1 = -p_2/2$, $-\frac{2}{3} < p_2 \leq \frac{2}{3}$, $p_3 = \frac{1}{2}$. For S', $p_1 = -p_2/2 + \frac{1}{2}$, $-\frac{1}{3} < p_2 \leq \frac{1}{3}$, $p_3 = \frac{1}{2}$.

	X_0	Y_3	X_0'	Y_3'
$(S_1)_{11}$	1	0	1	0
$(S_1)_{21}$	0	1	0	1
$(S_1)_{12}$	0	-1	0	1
$(S_1)_{22}$	1	0	-1	0
$\chi(S_1)$	2	0	0	0

Table A3-9 (Continued)

For the group C_{6v}^4, for which we have only the unprimed operations, the representations are as follows:

	X_0	Y_3
S_1	1	i
S_2	1	$-i$

Table A3-10

Matrix elements and characters for the irreducible representations for D_{6h}^4 at the points H and H'. At H, $p_1 = -\frac{1}{3}$, $p_2 = \frac{2}{3}$, $p_3 = \frac{1}{2}$; at H', $p_1 = p_2 = \frac{1}{3}$, $p_3 = \frac{1}{2}$. They have the same symmetry and representations.

	X_0	X_2	X_{-2}	Y_1	Y_{-1}	Y_3	X_0'	X_2'	X_{-2}'	Y_1'	Y_{-1}'	Y_3'
$(H_1)_{11}$	1	1	1	0	0	0	1	1	1	0	0	0
$(H_1)_{21}$	0	0	0	-1	-1	-1	0	0	0	-1	-1	-1
$(H_1)_{12}$	0	0	0	1	1	1	0	0	0	-1	-1	-1
$(H_1)_{22}$	1	1	1	0	0	0	-1	-1	-1	0	0	0
$\chi(H_1)$	2	2	2	0	0	0	0	0	0	0	0	0
$(H_2)_{11}$	1	ω^2	ω	0	0	0	1	ω^2	ω	0	0	0
$(H_2)_{21}$	0	0	0	ω	ω^2	1	0	0	0	ω	ω^2	1
$(H_2)_{12}$	0	0	0	$-\omega$	$-\omega^2$	-1	0	0	0	ω	ω^2	1
$(H_2)_{22}$	1	ω	ω^2	0	0	0	-1	$-\omega$	$-\omega^2$	0	0	0
$\chi(H_2)$	2	-1	-1	0	0	0	0	$-i\sqrt{3}$	$i\sqrt{3}$	0	0	0
$(H_3)_{11}$	1	ω^2	ω	0	0	0	-1	$-\omega^2$	$-\omega$	0	0	0
$(H_3)_{21}$	0	0	0	ω	ω^2	1	0	0	0	$-\omega$	$-\omega^2$	-1
$(H_3)_{12}$	0	0	0	$-\omega^2$	$-\omega$	-1	0	0	0	$-\omega^2$	$-\omega$	-1
$(H_3)_{22}$	1	ω	ω^2	0	0	0	1	ω	ω^2	0	0	0
$\chi(H_3)$	2	-1	-1	0	0	0	0	$i\sqrt{3}$	$-i\sqrt{3}$	0	0	0

Table A3-10 (Continued)

For the group C_{6v}^4, where we do not have the primed operators, the representations are a special case of P, which is taken up in Table A3-14. The matrix elements and characters are given below.

	X_0	X_2	X_{-2}	Y_1	Y_{-1}	Y_3
H_1	1	1	1	i	i	i
H_2	1	1	1	$-i$	$-i$	$-i$
$(H_3)_{11}$	1	ω^2	ω	0	0	0
$(H_3)_{21}$	0	0	0	i	$i\omega$	$i\omega^2$
$(H_3)_{12}$	0	0	0	i	$i\omega^2$	$i\omega$
$(H_3)_{22}$	1	ω	ω^2	0	0	0
$\chi(H_3)$	2	-1	-1	0	0	0

Table A3-11

Matrix elements and characters of the irreducible representations for the space groups D_{6h}^4 and C_{6v}^4 for propagation along the z direction, Δ. There are no primed operations in this direction. We have $p_1 = p_2 = 0$, $-\tfrac{1}{2} < p_3 \le \tfrac{1}{2}$. The two groups D_{6h}^4 and C_{6v}^4 are identical as far as this direction is concerned. The abbreviation ω stands for $e^{2\pi i/3}$, and α stands for $e^{\pi i p_3}$. The case $p_3 = 0$ corresponds to Γ, and the case $p_3 = \tfrac{1}{2}$, $\alpha = i$, corresponds to A. The notations are those of Herring.

	X_0	X_1	X_{-1}	X_2	X_{-2}	X_3	Y_0	Y_1	Y_{-1}	Y_2	Y_{-2}	Y_3
Δ_1	1	α	α	1	1	α	1	α	α	1	1	α
Δ_2	1	$-\alpha$	$-\alpha$	1	1	$-\alpha$	1	$-\alpha$	$-\alpha$	1	1	$-\alpha$
Δ_3	1	α	α	1	1	α	-1	$-\alpha$	$-\alpha$	-1	-1	$-\alpha$
Δ_4	1	$-\alpha$	$-\alpha$	1	1	$-\alpha$	-1	α	α	-1	-1	α
$(\Delta_5)_{11}$	1	$\omega\alpha$	$\omega^2\alpha$	ω^2	ω	α	0	0	0	0	0	0
$(\Delta_5)_{21}$	0	0	0	0	0	0	1	$\omega\alpha$	$\omega^2\alpha$	ω^2	ω	α
$(\Delta_5)_{12}$	0	0	0	0	0	0	1	$\omega^2\alpha$	$\omega\alpha$	ω	ω^2	α
$(\Delta_5)_{22}$	1	$\omega^2\alpha$	$\omega\alpha$	ω	ω^2	α	0	0	0	0	0	0
$\chi(\Delta_5)$	2	$-\alpha$	$-\alpha$	-1	-1	2α	0	0	0	0	0	0
$(\Delta_6)_{11}$	1	$-\omega^2\alpha$	$-\omega\alpha$	ω	ω^2	$-\alpha$	0	0	0	0	0	0
$(\Delta_6)_{21}$	0	0	0	0	0	0	1	$-\omega^2\alpha$	$-\omega\alpha$	ω	ω^2	$-\alpha$
$(\Delta_6)_{12}$	0	0	0	0	0	0	1	$-\omega\alpha$	$-\omega^2\alpha$	ω^2	ω	$-\alpha$
$(\Delta_6)_{22}$	1	$-\omega\alpha$	$-\omega^2\alpha$	ω^2	ω	$-\alpha$	0	0	0	0	0	0
$\chi(\Delta_6)$	2	α	α	-1	-1	-2α	0	0	0	0	0	0

Table A3-12

Matrix elements and characters for the irreducible representations for the point A, D_{6h}^4. This is the point $p_1 = p_2 = 0$, $p_3 = \frac{1}{2}$.

	X_0	X_1	X_{-1}	X_2	X_{-2}	X_3	Y_0	Y_1	Y_{-1}	Y_2	Y_{-2}	Y_3	X'_0	X'_1	X'_{-1}	X'_2	X'_{-2}	X'_3	Y'_0	Y'_1	Y'_{-1}	Y'_2	Y'_{-2}	Y'_3
$(A_1)_{11}$	1	i	0	1	1	i	1	i	0	1	1	i	0	0	0	0	0	0	0	0	0	0	0	0
$(A_1)_{21}$	0	0	0	0	0	0	0	0	0	0	0	0	0	0	0	0	0	0	0	0	0	0	0	0
$(A_1)_{12}$	0	0	0	0	0	0	0	0	0	0	0	0	0	0	0	0	0	0	0	0	0	0	0	0
$(A_1)_{22}$	1	$-i$	0	1	1	$-i$	1	$-i$	0	1	1	$-i$	0	0	0	0	0	0	0	0	0	0	0	0
$\chi(A_1)$	2	0	0	2	2	0	2	0	0	2	2	0	0	0	0	0	0	0	0	0	0	0	0	0
$(A_2)_{11}$	-1	i	0	1	1	i	-1	$-i$	0	-1	-1	i	0	0	0	0	0	0	0	0	0	0	0	0
$(A_2)_{21}$	0	0	0	0	0	0	0	0	0	0	0	0	0	0	0	0	0	0	0	0	0	0	0	0
$(A_2)_{12}$	0	0	0	0	0	0	0	0	0	0	0	0	0	0	0	0	0	0	0	0	0	0	0	0
$(A_2)_{22}$	-1	$-i$	0	1	1	$-i$	-1	$-i$	0	-1	-1	$-i$	0	0	0	0	0	0	0	0	0	0	0	0
$\chi(A_2)$	-2	0	0	2	2	0	-2	0	0	-2	-2	0	0	0	0	0	0	0	0	0	0	0	0	0
$(A_3)_{11}$	1	$i\omega$	$i\omega^2$	ω^2	ω	i	0	$i\omega$	$i\omega^2$	ω^2	ω	0	0	0	0	0	0	$-i$	0	$-i\omega$	$-i\omega^2$	ω^2	ω	0
$(A_3)_{21}$	0	0	0	0	0	0	0	0	0	0	0	0	0	0	0	0	0	0	0	0	0	0	0	0
$(A_3)_{31}$	0	0	0	0	0	0	0	0	0	0	0	0	0	0	0	0	0	0	0	0	0	0	0	0
$(A_3)_{41}$	0	0	0	0	0	0	0	0	0	0	0	0	0	0	0	0	0	0	0	0	0	0	0	0
$(A_3)_{12}$	0	0	0	0	0	0	0	0	0	0	0	0	0	0	0	0	0	0	0	0	0	0	0	0
$(A_3)_{22}$	1	$i\omega^2$	$i\omega$	ω	ω^2	$-i$	0	$i\omega^2$	$i\omega$	ω	ω^2	0	0	0	0	0	0	$-i$	0	$-i\omega^2$	$-i\omega$	ω	ω^2	0
$(A_3)_{32}$	0	0	0	0	0	0	0	0	0	0	0	0	0	0	0	0	0	0	0	0	0	0	0	0
$(A_3)_{42}$	0	0	0	0	0	0	0	0	0	0	0	0	0	0	0	0	0	0	0	0	0	0	0	0
$(A_3)_{13}$	0	0	0	0	0	0	0	0	0	0	0	0	1	$i\omega$	$i\omega^2$	ω^2	ω	i	0	$i\omega$	$i\omega^2$	ω^2	ω	0
$(A_3)_{23}$	0	0	0	0	0	0	0	0	0	0	0	0	0	0	0	0	0	0	0	0	0	0	0	0
$(A_3)_{33}$	1	$-i\omega$	$-i\omega^2$	ω^2	ω	$-i$	0	$-i\omega$	$-i\omega^2$	ω^2	ω	0	0	0	0	0	0	0	0	0	0	0	0	0
$(A_3)_{43}$	0	0	0	0	0	0	0	0	0	0	0	0	0	0	0	0	0	0	0	0	0	0	0	0

Table A3-12 (Continued)

	X_0	X_1	X_{-1}	X_2	X_{-2}	X_3	Y_0	Y_1	Y_{-1}	Y_2	Y_{-2}	Y_3	X_0'	X_1'	X_{-1}'	X_2'	X_{-2}'	X_3'	Y_0'	Y_1'	Y_{-1}'	Y_2'	Y_{-2}'	Y_3'
$(A_3)_{14}$	0	0	0	0	0	0	0	0	0	0	0	0	0	0	0	0	0	0	1	$i\omega^2$	$i\omega$	ω	ω^2	i
$(A_3)_{24}$	0	0	0	0	0	0	0	0	0	0	0	0	1	$i\omega^2$	$i\omega$	ω	ω^2	i	0	0	0	0	0	0
$(A_3)_{34}$	1	$-i\omega^2$	$-i\omega$	ω	ω^2	$-i$	0	0	0	0	0	0	0	0	0	0	0	0	0	0	0	0	0	0
$(A_3)_{44}$	0	0	0	0	0	0	i	$-i\omega^2$	$-i\omega$	ω	ω^2	$-i$	0	0	0	0	0	0	0	0	0	0	0	0
$\chi(A_3)$	4	0	0	-2	-2	0	0	0	0	0	0	0	0	0	0	0	0	0	0	0	0	0	0	0

For the group C_{6v}^4, the point A is merely a special case of Δ, obtained by setting $\alpha = 1$. Hence we have the following matrix elements and characters:

	X_0	X_1	X_{-1}	X_2	X_{-2}	X_3	Y_0	Y_1	Y_{-1}	Y_2	Y_{-2}	Y_3
A_1	1	i	i	1	1	i	1	i	i	1	1	i
A_2	1	$-i$	$-i$	1	1	$-i$	1	$-i$	$-i$	1	1	$-i$
A_3	1	i	i	1	1	i	-1	$-i$	$-i$	-1	-1	$-i$
A_4	1	$-i$	$-i$	1	1	$-i$	-1	i	i	-1	-1	i
$(A_5)_{11}$	1	$i\omega$	$i\omega^2$	ω^2	ω	i	0	0	0	0	0	0
$(A_5)_{21}$	0	0	0	0	0	0	1	$i\omega$	$i\omega^2$	ω^2	ω	i
$(A_5)_{12}$	0	0	0	0	0	0	1	$i\omega^2$	$i\omega$	ω	ω^2	i
$(A_5)_{22}$	1	$i\omega^2$	$i\omega$	ω	ω^2	i	0	0	0	0	0	0
$\chi(A_5)$	2	$-i$	$-i$	-1	-1	$2i$	0	0	0	0	0	0
$(A_6)_{11}$	1	$-i\omega^2$	$-i\omega$	ω	ω^2	$-i$	0	0	0	0	0	0
$(A_6)_{21}$	0	0	0	0	0	0	1	$-i\omega^2$	$-i\omega$	ω	ω^2	$-i$
$(A_6)_{12}$	0	0	0	0	0	0	1	$-i\omega$	$-i\omega^2$	ω^2	ω	$-i$
$(A_6)_{22}$	1	$-i\omega$	$-i\omega^2$	ω^2	ω	$-i$	0	0	0	0	0	0
$\chi(A_6)$	2	i	i	-1	-1	$-2i$	0	0	0	0	0	0

Table A3-13

Matrix elements and characters of the irreducible representations for D_{6h}^4 and C_{6v}^4 along the line U. The two space groups are identical along this line. Here $p_1 = \frac{1}{2}$, $p_2 = 0$, $-\frac{1}{2} < p_3 \leq \frac{1}{2}$. The abbreviation α stands for $e^{\pi i p_3}$.

	X_0	X_3	Y_0	Y_3
U_1	1	α	1	α
U_2	1	$-\alpha$	1	$-\alpha$
U_3	1	$-\alpha$	-1	α
U_4	1	α	-1	$-\alpha$

Table A3-14

Matrix elements and characters for the irreducible representations for D_{6h}^4 and C_{6v}^4 for the points P. Here $p_1 = -\frac{1}{3}$, $p_2 = \frac{2}{3}$, $-\frac{1}{2} < p_3 \leq \frac{1}{2}$. The abbreviation α stands for $e^{\pi i p_3}$.

	X_0	X_2	X_{-2}	Y_1	Y_{-1}	Y_3
P_1	1	1	1	α	α	α
P_2	1	1	1	$-\alpha$	$-\alpha$	$-\alpha$
$(P_3)_{11}$	1	ω^2	ω	0	0	0
$(P_3)_{21}$	0	0	0	α	$\omega\alpha$	$\omega^2\alpha$
$(P_3)_{12}$	0	0	0	α	$\omega^2\alpha$	$\omega\alpha$
$(P_3)_{22}$	1	ω	ω^2	0	0	0
$\chi(P_3)$	2	-1	-1	0	0	0

Table A3-15
Compatibility relations for D_{6h}^4

Γ_1^+	Γ_1^-	Γ_2^+	Γ_2^-	Γ_3^+	Γ_3^-	Γ_4^+	Γ_4^-	Γ_5^+	Γ_5^-	Γ_6^+	Γ_6^-
Δ_1	Δ_3	Δ_3	Δ_1	Δ_2	Δ_4	Δ_4	Δ_2	Δ_5	Δ_5	Δ_6	Δ_6
Σ_1	Σ_2	Σ_4	Σ_3	Σ_3	Σ_4	Σ_2	Σ_1	Σ_1, Σ_4	Σ_2, Σ_3	Σ_2, Σ_3	Σ_1, Σ_4
T_1	T_2	T_4	T_3	T_2	T_1	T_3	T_4	T_1, T_4	T_2, T_3	T_2, T_3	T_1, T_4

K_1	K_2	K_3	K_4	K_5	K_6	A_1	A_2	A_3
T_1	T_2	T_4	T_3	T_1, T_4	T_2, T_3	Δ_1, Δ_2	Δ_3, Δ_4	Δ_5, Δ_6
P_1	P_2	P_2	P_1	P_3	P_3			

M_1^+	M_1^-	M_2^+	M_2^-	M_3^+	M_3^-	M_4^+	M_4^-	H_1	H_2	H_3
T_1	T_2	T_3	T_4	T_2	T_1	T_4	T_3	S_1	S_1	S_1
Σ_1	Σ_2	Σ_2	Σ_1	Σ_3	Σ_4	Σ_4	Σ_3	P_1, P_2	P_3	P_3
U_1	U_4	U_3	U_2	U_2	U_3	U_4	U_1			

L_1	L_2	S_1	S_1
U_1, U_2	U_3, U_4	R_1, R_3	R_2, R_4

Table A3-15 (Continued)
Compatibility relations for C_{6v}^4

Γ_1	Γ_2	Γ_3	Γ_4	Γ_5	Γ_6	K_1	K_2	K_3
Δ_1	Δ_3	Δ_4	Δ_2	Δ_5	Δ_6	T_1	T_2	T_1, T_2
Σ_1	Σ_2	Σ_2	Σ_1	Σ_1, Σ_2	Σ_1, Σ_2	P_1	P_2	P_3
T_1	T_2	T_1	T_2	T_1, T_2	T_1, T_2			

M_1^+	M_1^-	M_2^+	M_2^-	A_1	A_2	A_3	A_4	A_5	A_6
T_1	T_2	T_1	T_2	Δ_1	Δ_2	Δ_3	Δ_4	Δ_5	Δ_6
Σ_1	Σ_2	Σ_2	Σ_1	S_1	S_2	S_2	S_1	S_1, S_2	S_1, S_2
U_1	U_4	U_3	U_2	R_1	R_1	R_2	R_2	R_1, R_2	R_1, R_2

H_1	H_2	H_3	L_1	L_2	L_3	L_4
S_1	S_2	S_1, S_2	U_1	U_2	U_3	U_4
P_1	P_2	P_3	S_1	S_2	S_1	S_2
			R_1	R_1	R_2	R_2

A3-2. The Space Groups O_h^5 ($Fm3m$), O_h^7 ($Fd3m$), and T_d^2 ($F\bar{4}3m$). The space groups O_h^5 ($Fm3m$), found with the face-centered cubic structure (Sec. 2-2), as well as the sodium chloride and fluorite structures (Secs. 3-2 and 3-3), O_h^7 ($Fd3m$), found with the diamond and spinel structures (Secs. 2-4 and 3-4), and T_d^2 ($F\bar{4}3m$), found with the zinc-blende structure (Sec. 3-2), all have the face-centered cubic Bravais lattice, and can be conveniently treated together, using methods equivalent to those used for the hexagonal space groups D_{6h}^4 ($P6_3/mmc$) and C_{6v}^4 ($P6_3mc$) in the preceding section. Positions of atoms in the structures listed above are given in Table A3-16, at the end of this section. There is one difference between the treatment of these cubic space groups in the International Tables and the hexagonal and other groups: the International Tables use a nonprimitive cubic unit cell for all the cubic space groups, instead of using the primitive unit cell. We shall find it more convenient, however, to use the primitive unit cell.

For this primitive unit cell, the fundamental vectors are given in Eq. (1-19), and the vectors b of the reciprocal lattice are given in Eq. (5-10). The Brillouin zone, with the notations for the symmetry directions as given by Bouckaert, Smoluchowski, and Wigner (*loc. cit.*), is given in Fig. A3-2. The point group of operations for the space groups O_h^5 and O_h^7 is O_h, and that for T_d^2 is T_d. We can conveniently give these point groups together. In Table 2-2 we have given the 24 operations of the point group T_d. The additional 24 operations found in O_h, but not present in T_d, are formed from the operations of T_d plus an inversion. We indicate the 24 operations of T_d in Table 2-2 by symbols $R_1 \cdots R_{24}$. An additional operation of O_h formed from one of these operations plus the inversion is denoted by a primed symbol. The multiplication table of the point group T_d is given in Table A3-17. We form the multiplication table of the point group O_h from that of Table A3-17 as follows. We find the subscript of the product operation from Table A3-17. Then

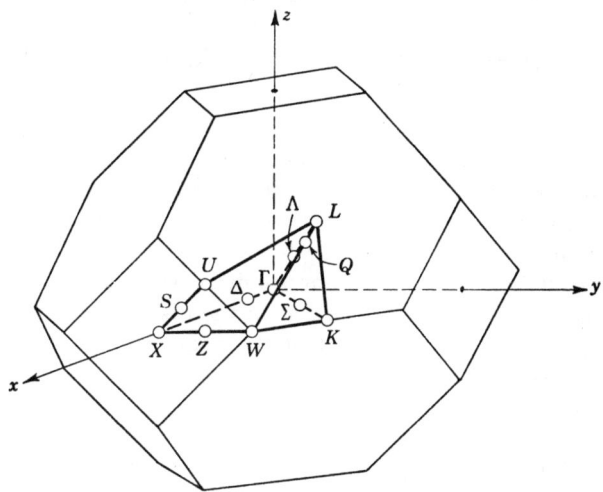

FIG. A3-2. Brillouin zone for the face-centered cubic Bravais lattice, with symmetry points following the notation of Bouckaert, Smoluchowski, and Wigner.

if both factors are unprimed, or both are primed, the product operation is unprimed; if one is primed, the other unprimed, the product is primed.

The space groups O_h^5 and T_d^2 are symmorphic. Hence the operations of the space group, corresponding to no primitive translation, are identical with the operations of the point group, and we have the same multiplication table. These two space groups, then, are much simpler than the hexagonal groups treated earlier, or than the diamond case, which have nonsymmorphic space groups. For the space group O_h^7, there are nonprimitive translations associated with some of the operations of the point group. If we use Case 1, as mentioned in Table A3-16, there is no nonprimitive translation associated with the unprimed operations, those found in the group T_d. However, associated with the primed operations we have the nonprimitive translation

$$\tau = \frac{a}{4}(\mathbf{i} + \mathbf{j} + \mathbf{k}) = \tfrac{1}{4}(\mathbf{t}_1 + \mathbf{t}_2 + \mathbf{t}_3) \tag{A3-15}$$

where the t's are given in Eq. (1-19). If we use Case 2, we have nonprimitive translations as follows:

$$\tau_1 = -\frac{a}{4}(\mathbf{j} + \mathbf{k}) = -\tfrac{1}{2}\mathbf{t}_1 \text{ associated with } R_2, R_8, R_{11}, R_{16}, R_{17}, R_{20}$$

$$\tau_2 = -\frac{a}{4}(\mathbf{i} + \mathbf{k}) = -\tfrac{1}{2}\mathbf{t}_2 \text{ associated with } R_3, R_6, R_{12}, R_{13}, R_{18}, R_{22}$$

$$\tau_3 = -\frac{a}{4}(\mathbf{i} + \mathbf{j}) = -\tfrac{1}{2}\mathbf{t}_3 \text{ associated with } R_4, R_7, R_{10}, R_{14}, R_{15}, R_{24}$$

$$\tag{A3-16}$$

No nonprimitive translations are associated with R_1, R_5, R_9, R_{19}, R_{21}, R_{23}. Each primed operation has a nonprimitive translation which is the negative of that associated with the unprimed operation with the same subscript.

To study the effect of these nonprimitive translations on the space group, we can most readily proceed as in the discussion of the hexagonal groups, by investigating the effect of the operations of the space group on a plane wave

$$\begin{aligned}\psi &= \exp 2\pi i[(h_1 + p_1)\mathbf{b}_1 + (h_2 + p_2)\mathbf{b}_2 + (h_3 + p_3)\mathbf{b}_3] \cdot \mathbf{r} \\ &= \exp[(h_1 + p_1)\xi + (h_2 + p_2)\eta + (h_3 + p_3)\zeta]\end{aligned} \tag{A3-17}$$

where as in the hexagonal case the quantities p_1, p_2, p_3 refer to the reduced wave vector, the h's are integers, and the radius vector $\mathbf{r}$ is given by

$$\mathbf{r} = \xi \mathbf{t}_1 + \eta \mathbf{t}_2 + \zeta \mathbf{t}_3 \tag{A3-18}$$

In Table A3-18 we give the effects of the operators of the group T_d on a plane wave of the type given in Eq. (A3-17). The primed operators found in the space group O_h^5 have the same effect as these operators of T_d, except that they change the sign of the exponent. The nonprimitive translations occurring in the space group O_h^7 introduce additional factors of

$$e^{i(\mathbf{k}+\mathbf{K}_h)\cdot\tau} = \exp\frac{2\pi i(h_1 + p_1 + h_2 + p_2 + h_3 + p_3)}{4} \tag{A3-19}$$

where $\mathbf{k} + \mathbf{K}_h$ is the wave vector, and τ is the nonprimitive translation given in Eq. (A3-15), for the primed operations but not for the unprimed operations, provided we treat the group according to Case 1. If we use

Case 2, we have factors of

$$e^{i(\mathbf{k}+\mathbf{K}_h)\cdot\boldsymbol{\tau}_1} = e^{-\pi i(h_1+p_1)} \text{ associated with } R_2, R_8, R_{11}, R_{16}, R_{17}, R_{20}$$
$$e^{i(\mathbf{k}+\mathbf{K}_h)\cdot\boldsymbol{\tau}_2} = e^{-\pi i(h_2+p_2)} \text{ associated with } R_3, R_6, R_{12}, R_{13}, R_{18}, R_{22} \quad (\text{A3-20})$$
$$e^{i(\mathbf{k}+\mathbf{K}_h)\cdot\boldsymbol{\tau}_3} = e^{-\pi i(h_3+p_3)} \text{ associated with } R_4, R_7, R_{10}, R_{14}, R_{15}, R_{24}$$

and the conjugates of these factors associated with the corresponding primed operations. These additional factors result in added terms in the multiplication table for the space group O_h^7. In Table A3-19 we enumerate the product operations in which these factors are found, and give the factors. This table is the same whether we use Case 1 or Case 2.

We are now in position to verify and make use of the matrix elements of the operations of the group of the wave vector, for the symmetry points in the Brillouin zone. These matrix elements are given in Tables A3-20 to A3-30 for the three space groups considered. In Table A3-31 we give the compatibility relations.

Table A3-16
Positions of atoms in structures having space groups O_h^5, O_h^7, and T_d^2

Face-centered cubic structure. O_h^5 ($Fm3m$). One atom per primitive unit cell, at origin.

Sodium chloride structure. O_h^5 ($Fm3m$). One atom of sodium and one of chlorine per primitive unit cell. Sodium atom at origin, chlorine at $(a/2)\mathbf{i} = \frac{1}{2}(-\mathbf{t}_1 + \mathbf{t}_2 + \mathbf{t}_3)$.

Fluorite structure. O_h^5 ($Fm3m$). CaF_2. One atom of calcium and two of fluorine per primitive unit cell. Calcium atom at origin, fluorines at $\pm(a/4)(\mathbf{i} + \mathbf{j} + \mathbf{k})$ or $\pm\frac{1}{4}(\mathbf{t}_1 + \mathbf{t}_2 + \mathbf{t}_3)$.

Diamond structure. O_h^7 ($Fd3m$). Two atoms per primitive unit cell. Two methods of describing the unit cell are used in International Tables, both of which we carry through. Case 1, one atom at origin, one at $(a/4)(\mathbf{i} + \mathbf{j} + \mathbf{k})$, or $\frac{1}{4}(\mathbf{t}_1 + \mathbf{t}_2 + \mathbf{t}_3)$. Case 2, atoms at $\pm(a/8)(\mathbf{i} + \mathbf{j} + \mathbf{k})$, or $\pm\frac{1}{8}(\mathbf{t}_1 + \mathbf{t}_2 + \mathbf{t}_3)$.

Spinel structure. O_h^7 ($Fd3m$). Al_2MgO_4. Four aluminums, two magnesiums, and eight oxygens per primitive unit cell. Using Case 1, aluminums at $a(\frac{5}{8}\mathbf{i} + \frac{5}{8}\mathbf{j} + \frac{5}{8}\mathbf{k})$, $a(\frac{5}{8}\mathbf{i} + \frac{7}{8}\mathbf{j} + \frac{7}{8}\mathbf{k})$, $a(\frac{7}{8}\mathbf{i} + \frac{5}{8}\mathbf{j} + \frac{7}{8}\mathbf{k})$, $a(\frac{7}{8}\mathbf{i} + \frac{7}{8}\mathbf{j} + \frac{5}{8}\mathbf{k})$, or $(\frac{5}{8}\mathbf{t}_1 + \frac{5}{8}\mathbf{t}_2 + \frac{5}{8}\mathbf{t}_3)$, $(\frac{7}{8}\mathbf{t}_1 + \frac{5}{8}\mathbf{t}_2 + \frac{5}{8}\mathbf{t}_3)$, $(\frac{5}{8}\mathbf{t}_1 + \frac{7}{8}\mathbf{t}_2 + \frac{5}{8}\mathbf{t}_3)$, $(\frac{5}{8}\mathbf{t}_1 + \frac{5}{8}\mathbf{t}_2 + \frac{7}{8}\mathbf{t}_3)$. Magnesiums at origin, and $(a/4)(\mathbf{i} + \mathbf{j} + \mathbf{k})$, or $\frac{1}{4}(\mathbf{t}_1 + \mathbf{t}_2 + \mathbf{t}_3)$. Oxygens at $ua(\mathbf{i} + \mathbf{j} + \mathbf{k})$, $ua(\mathbf{i} - \mathbf{j} - \mathbf{k})$, $ua(-\mathbf{i} + \mathbf{j} - \mathbf{k})$, $ua(-\mathbf{i} - \mathbf{j} + \mathbf{k})$, and $(a/4)(\mathbf{i} + \mathbf{j} + \mathbf{k}) - ua(\mathbf{i} + \mathbf{j} + \mathbf{k})$, $(a/4)(\mathbf{i} + \mathbf{j} + \mathbf{k}) - ua(\mathbf{i} - \mathbf{j} - \mathbf{k})$, $(a/4)(\mathbf{i} + \mathbf{j} + \mathbf{k}) - ua(-\mathbf{i} + \mathbf{j} - \mathbf{k})$, $(a/4)(\mathbf{i} + \mathbf{j} + \mathbf{k}) - ua(-\mathbf{i} - \mathbf{j} + \mathbf{k})$, or at $u(\mathbf{t}_1 + \mathbf{t}_2 + \mathbf{t}_3)$, $u(-3\mathbf{t}_1 + \mathbf{t}_2 + \mathbf{t}_3)$, $u(\mathbf{t}_1 - 3\mathbf{t}_2 + \mathbf{t}_3)$, $u(\mathbf{t}_1 + \mathbf{t}_2 - 3\mathbf{t}_3)$, and at $\frac{1}{4}(\mathbf{t}_1 + \mathbf{t}_2 + \mathbf{t}_3) - u(\mathbf{t}_1 + \mathbf{t}_2 + \mathbf{t}_3)$, $\frac{1}{4}(\mathbf{t}_1 + \mathbf{t}_2 + \mathbf{t}_3) - u(-3\mathbf{t}_1 + \mathbf{t}_2 + \mathbf{t}_3)$, $\frac{1}{4}(\mathbf{t}_1 + \mathbf{t}_2 + \mathbf{t}_3) - u(\mathbf{t}_1 - 3\mathbf{t}_2 + \mathbf{t}_3)$, $\frac{1}{4}(\mathbf{t}_1 + \mathbf{t}_2 + \mathbf{t}_3) - u(\mathbf{t}_1 + \mathbf{t}_2 - 3\mathbf{t}_3)$, where u is a parameter equal to about 0.390. Using Case 2, aluminums at $(5a/8)(\mathbf{i} + \mathbf{j} + \mathbf{k}) + (a/8)(-\mathbf{i} - \mathbf{j} - \mathbf{k})$, $(5a/8)(\mathbf{i} + \mathbf{j} + \mathbf{k}) + (a/8)(-\mathbf{i} + \mathbf{j} + \mathbf{k})$, $(5a/8)(\mathbf{i} + \mathbf{j} + \mathbf{k}) + (a/8)(\mathbf{i} - \mathbf{j} + \mathbf{k})$, $(5a/8)(\mathbf{i} + \mathbf{j} + \mathbf{k}) + (a/8)(\mathbf{i} + \mathbf{j} - \mathbf{k})$, or at $(\frac{5}{8})(\mathbf{t}_1 + \mathbf{t}_2 + \mathbf{t}_3) + \frac{1}{8}(-\mathbf{t}_1 - \mathbf{t}_2 - \mathbf{t}_3)$, $\frac{5}{8}(\mathbf{t}_1 + \mathbf{t}_2 + \mathbf{t}_3) + \frac{1}{8}(\mathbf{t}_1 - \mathbf{t}_2 - \mathbf{t}_3)$, $\frac{5}{8}(\mathbf{t}_1 + \mathbf{t}_2 + \mathbf{t}_3) + \frac{1}{8}(-\mathbf{t}_1 + \mathbf{t}_2 - \mathbf{t}_3)$, $\frac{5}{8}(\mathbf{t}_1 + \mathbf{t}_2 + \mathbf{t}_3) + \frac{1}{8}(-\mathbf{t}_1 - \mathbf{t}_2 + \mathbf{t}_3)$. Magnesiums at $\pm(a/8)(\mathbf{i} + \mathbf{j} + \mathbf{k})$, or at $\pm\frac{1}{8}(\mathbf{t}_1 + \mathbf{t}_2 + \mathbf{t}_3)$. Oxygens at $\pm[(a/8)(\mathbf{i} + \mathbf{j} + \mathbf{k}) - ua(\mathbf{i} + \mathbf{j} + \mathbf{k})]$, $\pm[(a/8)(\mathbf{i} + \mathbf{j} + \mathbf{k}) - ua(\mathbf{i} - \mathbf{j} - \mathbf{k})]$, $\pm[(a/8)(\mathbf{i} + \mathbf{j} + \mathbf{k}) - ua(-\mathbf{i} + \mathbf{j} - \mathbf{k})]$, $\pm[(a/8)(\mathbf{i} + \mathbf{j} + \mathbf{k}) - ua(-\mathbf{i} - \mathbf{j} + \mathbf{k})]$, or at $\pm[\frac{1}{8}(\mathbf{t}_1 + \mathbf{t}_2 + \mathbf{t}_3) - u(\mathbf{t}_1 + \mathbf{t}_2 + \mathbf{t}_3)]$, $\pm[\frac{1}{8}(\mathbf{t}_1 + \mathbf{t}_2 + \mathbf{t}_3) - u(-3\mathbf{t}_1 + \mathbf{t}_2 + \mathbf{t}_3)]$, $\pm[\frac{1}{8}(\mathbf{t}_1 + \mathbf{t}_2 + \mathbf{t}_3) - u(\mathbf{t}_1 - 3\mathbf{t}_2 + \mathbf{t}_3)]$, $\pm[\frac{1}{8}(\mathbf{t}_1 + \mathbf{t}_2 + \mathbf{t}_3) - u(\mathbf{t}_1 + \mathbf{t}_2 - 3\mathbf{t}_3)]$.

Zinc-blende structure. T_d^2 ($F\bar{4}3m$). ZnS. One atom of zinc and one of sulfur per primitive unit cell, the zinc at the origin, the sulfur at $(a/4)(\mathbf{i} + \mathbf{j} + \mathbf{k})$, or $\frac{1}{4}(\mathbf{t}_1 + \mathbf{t}_2 + \mathbf{t}_3)$.

Table A3-17
Multiplication table for Group T_d. Table gives k, where $R_i R_j = R_k$

$i \backslash j$	1	2	3	4	5	6	7	8	9	10	11	12	13	14	15	16	17	18	19	20	21	22	23	24
1	1	2	3	4	5	6	7	8	9	10	11	12	13	14	15	16	17	18	19	20	21	22	23	24
2	2	1	4	3	7	8	5	6	12	11	10	9	14	13	21	22	24	23	20	19	15	16	18	17
3	3	4	1	2	8	7	6	5	10	9	12	11	20	19	16	15	23	24	14	13	22	21	17	18
4	4	3	2	1	6	5	8	7	11	12	9	10	19	20	22	21	18	17	13	14	16	15	24	23
5	5	8	6	7	9	12	10	11	1	4	2	3	18	24	14	20	16	22	23	17	19	13	21	15
6	6	7	5	8	11	10	12	9	4	1	3	2	17	23	20	14	21	15	24	18	13	19	16	22
7	7	6	8	5	12	9	11	10	2	3	1	4	23	17	13	19	22	16	18	24	20	14	15	21
8	8	5	7	6	10	11	9	12	3	2	4	1	24	18	19	13	15	21	17	23	14	20	22	16
9	9	11	12	10	1	3	4	2	5	7	8	6	22	15	24	17	20	13	21	16	23	18	19	14
10	10	12	11	9	3	1	2	4	8	6	5	7	21	16	18	23	13	20	22	15	17	24	14	19
11	11	9	10	12	4	2	1	3	6	8	7	5	15	22	23	18	14	19	16	21	24	17	13	20
12	12	10	9	11	2	4	3	1	7	5	6	8	16	21	17	24	19	14	15	22	18	23	20	13
13	13	14	19	20	16	15	22	21	24	23	18	17	2	1	8	7	9	10	4	3	6	5	11	12
14	14	13	20	19	22	21	16	15	17	18	23	24	1	2	6	5	12	11	3	4	8	7	10	9
15	15	22	16	21	18	23	17	24	20	13	19	14	9	11	3	1	6	8	12	10	2	4	7	5
16	16	21	15	22	24	17	23	18	13	20	14	19	10	12	1	3	7	5	11	9	4	2	6	8
17	17	23	24	18	14	20	19	13	22	16	15	21	7	6	9	12	4	1	8	5	10	11	3	2
18	18	24	23	17	20	14	13	19	15	21	22	16	8	5	11	10	1	4	7	6	12	9	2	3
19	19	20	13	14	21	22	15	16	23	24	17	18	3	4	7	8	11	12	1	2	5	6	9	10
20	20	19	14	13	15	16	21	22	18	17	24	23	4	3	5	6	10	9	2	1	7	8	12	11
21	21	16	22	15	23	18	24	17	19	14	20	13	12	10	4	2	8	6	9	11	1	3	5	7
22	22	15	21	16	17	24	18	23	14	19	13	20	11	9	2	4	5	7	10	12	3	1	8	6
23	23	17	18	24	19	13	14	20	21	15	16	22	6	7	10	11	2	3	5	8	9	12	1	4
24	24	18	17	23	13	19	20	14	16	22	21	15	5	8	12	9	3	2	6	7	11	10	4	1

Table A3-18
Effect of operators of T_d on a plane wave

$R_1\psi = \exp 2\pi i[(h_1 + p_1)\xi + (h_2 + p_2)\eta + (h_3 + p_3)\zeta]$
$R_2\psi = \exp 2\pi i[-(h_1 + p_1)\xi + (-h_1 - p_1 + h_3 + p_3)\eta + (-h_1 - p_1 + h_2 + p_2)\zeta]$
$R_3\psi = \exp 2\pi i[(-h_2 - p_2 + h_3 + p_3)\xi + (-h_2 - p_2)\eta + (h_1 + p_1 - h_2 - p_2)\zeta]$
$R_4\psi = \exp 2\pi i[(h_2 + p_2 - h_3 - p_3)\xi + (h_1 + p_1 - h_3 - p_3)\eta + (-h_3 - p_3)\zeta]$
$R_5\psi = \exp 2\pi i[(h_3 + p_3)\xi + (h_1 + p_1)\eta + (h_2 + p_2)\zeta]$
$R_6\psi = \exp 2\pi i[(h_1 + p_1 - h_2 - p_2)\xi + (-h_2 - p_2 + h_3 + p_3)\eta + (-h_2 - p_2)\zeta]$
$R_7\psi = \exp 2\pi i[(-h_3 - p_3)\xi + (h_2 + p_2 - h_3 - p_3)\eta + (h_1 + p_1 - h_3 - p_3)\zeta]$
$R_8\psi = \exp 2\pi i[(-h_1 - p_1 + h_2 + p_2)\xi + (-h_1 - p_1)\eta + (-h_1 - p_1 + h_3 + p_3)\zeta]$
$R_9\psi = \exp 2\pi i[(h_2 + p_2)\xi + (h_3 + p_3)\eta + (h_1 + p_1)\zeta]$
$R_{10}\psi = \exp 2\pi i[(h_1 + p_1 - h_3 - p_3)\xi + (-h_3 - p_3)\eta + (h_2 + p_2 - h_3 - p_3)\zeta]$
$R_{11}\psi = \exp 2\pi i[(-h_1 - p_1 + h_3 + p_3)\xi + (-h_1 - p_1 + h_2 + p_2)\eta + (-h_1 - p_1)\zeta]$
$R_{12}\psi = \exp 2\pi i[(-h_2 - p_2)\xi + (h_1 + p_1 - h_2 - p_2)\eta + (-h_2 - p_2 + h_3 + p_3)\zeta]$
$R_{13}\psi = \exp 2\pi i[(-h_2 - p_2 + h_3 + p_3)\xi + (h_1 + p_1 - h_2 - p_2)\eta + (-h_2 - p_2)\zeta]$
$R_{14}\psi = \exp 2\pi i[(h_2 + p_2 - h_3 - p_3)\xi + (-h_3 - p_3)\eta + (h_1 + p_1 - h_3 - p_3)\zeta]$
$R_{15}\psi = \exp 2\pi i[(-h_3 - p_3)\xi + (h_1 + p_1 - h_3 - p_3)\eta + (h_2 + p_2 - h_3 - p_3)\zeta]$
$R_{16}\psi = \exp 2\pi i[(-h_1 - p_1 + h_2 + p_2)\xi + (-h_1 - p_1 + h_3 + p_3)\eta + (-h_1 - p_1)\zeta]$
$R_{17}\psi = \exp 2\pi i[(-h_1 - p_1 + h_3 + p_3)\xi + (-h_1 - p_1)\eta + (-h_1 - p_1 + h_2 + p_2)\zeta]$
$R_{18}\psi = \exp 2\pi i[(-h_2 - p_2)\xi + (-h_2 - p_2 + h_3 + p_3)\eta + (h_1 + p_1 - h_2 - p_2)\zeta]$
$R_{19}\psi = \exp 2\pi i[(h_1 + p_1)\xi + (h_3 + p_3)\eta + (h_2 + p_2)\zeta]$
$R_{20}\psi = \exp 2\pi i[(-h_1 - p_1)\xi + (-h_1 - p_1 + h_2 + p_2)\eta + (-h_1 - p_1 + h_3 + p_3)\zeta]$
$R_{21}\psi = \exp 2\pi i[(h_3 + p_3)\xi + (h_2 + p_2)\eta + (h_1 + p_1)\zeta]$
$R_{22}\psi = \exp 2\pi i[(h_1 + p_1 - h_2 - p_2)\xi + (-h_2 - p_2)\eta + (-h_2 - p_2 + h_3 + p_3)\zeta]$
$R_{23}\psi = \exp 2\pi i[(h_2 + p_2)\xi + (h_1 + p_1)\eta + (h_3 + p_3)\zeta]$
$R_{24}\psi = \exp 2\pi i[(h_1 + p_1 - h_3 - p_3)\xi + (h_2 + p_2 - h_3 - p_3)\eta + (-h_3 - p_3)\zeta]$

Table A3-19
Factors in multiplication table for group O_h^7

For the group O_h^7, in contrast to O_h^5 and T_d^2, we have nonprimitive translations associated with some of the operations. This has the following effect on the multiplication table of the space group. We are to find the product operator of the point group, according to Table A3-17 and the explanation in the text. Then there are additional factors as follows:

1. First operator primed, second operator one of R_2, R_8, R_{11}, R_{16}, R_{17}, R_{20}, factor is $e^{-2\pi i p_1}$. If the second operator is one of R_2', R_8', R_{11}', R_{16}', R_{17}', R_{20}', factor is $e^{2\pi i p_1}$.

2. First operator primed, second operator one of R_3, R_6, R_{12}, R_{13}, R_{18}, R_{22}, factor is $e^{-2\pi i p_2}$. If the second operator is one of R_3', R_6', R_{12}', R_{13}', R_{18}', R_{22}', factor is $e^{2\pi i p_2}$.

3. First operator primed, second operator one of R_4, R_7, R_{10}, R_{14}, R_{15}, R_{24}, factor is $e^{-2\pi i p_3}$. If the second operator is one of R_4', R_7', R_{10}', R_{14}', R_{15}', R_{24}', factor is $e^{2\pi i p_3}$.

4. In all other cases, no additional factor.

Table A3-20
Matrix elements and characters at point Γ, cubic groups (explanation follows table)

	R_1	R_2	R_3	R_4	R_5	R_6	R_7	R_8	R_9	R_{10}	R_{11}	R_{12}	R_{13}	R_{14}	R_{15}	R_{16}	R_{17}	R_{18}	R_{19}	R_{20}	R_{21}	R_{22}	R_{23}	R_{24}
Γ_1	1	1	1	1	1	1	1	1	1	1	1	1	1	1	1	1	1	1	1	1	1	1	1	1
Γ_2	1	1	1	1	1	1	1	1	1	1	1	1	−1	−1	−1	−1	−1	−1	−1	−1	−1	−1	−1	−1
$(\Gamma_{12})_{11}$	1	1	1	1	$-\tfrac{1}{2}$	$-\tfrac{1}{2}$	$-\tfrac{1}{2}$	$-\tfrac{1}{2}$	$-\tfrac{1}{2}$	$-\tfrac{1}{2}$	$-\tfrac{1}{2}$	$-\tfrac{1}{2}$	1	1	$-\tfrac{1}{2}$	$-\tfrac{1}{2}$	$-\tfrac{1}{2}$	$-\tfrac{1}{2}$	1	1	$-\tfrac{1}{2}$	$-\tfrac{1}{2}$	$-\tfrac{1}{2}$	$-\tfrac{1}{2}$
$(\Gamma_{12})_{21}$	0	0	0	0	$-\tfrac{\sqrt{3}}{2}$	$-\tfrac{\sqrt{3}}{2}$	$\tfrac{\sqrt{3}}{2}$	$\tfrac{\sqrt{3}}{2}$	$-\tfrac{\sqrt{3}}{2}$	$-\tfrac{\sqrt{3}}{2}$	$\tfrac{\sqrt{3}}{2}$	$\tfrac{\sqrt{3}}{2}$	0	0	$-\tfrac{\sqrt{3}}{2}$	$-\tfrac{\sqrt{3}}{2}$	$\tfrac{\sqrt{3}}{2}$	$\tfrac{\sqrt{3}}{2}$	0	0	$-\tfrac{\sqrt{3}}{2}$	$-\tfrac{\sqrt{3}}{2}$	$\tfrac{\sqrt{3}}{2}$	$\tfrac{\sqrt{3}}{2}$
$(\Gamma_{12})_{12}$	0	0	0	0	$-\tfrac{\sqrt{3}}{2}$	$-\tfrac{\sqrt{3}}{2}$	$\tfrac{\sqrt{3}}{2}$	$\tfrac{\sqrt{3}}{2}$	$\tfrac{\sqrt{3}}{2}$	$\tfrac{\sqrt{3}}{2}$	$-\tfrac{\sqrt{3}}{2}$	$-\tfrac{\sqrt{3}}{2}$	0	0	$-\tfrac{\sqrt{3}}{2}$	$-\tfrac{\sqrt{3}}{2}$	$\tfrac{\sqrt{3}}{2}$	$\tfrac{\sqrt{3}}{2}$	0	0	$-\tfrac{\sqrt{3}}{2}$	$-\tfrac{\sqrt{3}}{2}$	$\tfrac{\sqrt{3}}{2}$	$\tfrac{\sqrt{3}}{2}$
$(\Gamma_{12})_{22}$	1	1	1	1	$-\tfrac{1}{2}$	$-\tfrac{1}{2}$	$-\tfrac{1}{2}$	$-\tfrac{1}{2}$	$-\tfrac{1}{2}$	$-\tfrac{1}{2}$	$-\tfrac{1}{2}$	$-\tfrac{1}{2}$	−1	−1	$\tfrac{1}{2}$	$\tfrac{1}{2}$	$\tfrac{1}{2}$	$\tfrac{1}{2}$	−1	−1	$\tfrac{1}{2}$	$\tfrac{1}{2}$	$\tfrac{1}{2}$	$\tfrac{1}{2}$
$\chi(\Gamma_{12})$	2	2	2	2	−1	−1	−1	−1	−1	−1	−1	−1	0	0	0	0	0	0	0	0	0	0	0	0
$(\Gamma'_{25})_{11}$	1	1	−1	−1	0	0	0	0	0	0	0	0	1	1	0	0	0	0	1	1	0	0	0	0
$(\Gamma'_{25})_{21}$	0	0	0	0	1	1	−1	0	0	1	−1	−1	0	0	1	1	−1	−1	0	0	−1	−1	1	1
$(\Gamma'_{25})_{31}$	0	0	0	0	0	0	0	1	1	−1	0	0	0	0	0	0	0	0	0	0	0	0	0	0
$(\Gamma'_{25})_{12}$	0	0	0	0	1	1	1	1	−1	0	0	0	0	0	1	1	−1	0	0	0	−1	−1	1	0
$(\Gamma'_{25})_{22}$	1	1	−1	−1	0	0	0	0	0	0	0	0	0	0	0	0	0	0	0	0	0	0	0	0
$(\Gamma'_{25})_{32}$	0	0	0	0	0	0	0	0	0	0	1	−1	0	0	0	0	0	1	0	0	0	0	0	0
$(\Gamma'_{25})_{13}$	0	0	0	0	0	0	0	0	1	1	0	0	0	0	0	0	0	0	0	0	0	0	0	0
$(\Gamma'_{25})_{23}$	0	0	0	0	0	0	0	−1	0	0	1	1	0	0	0	0	0	0	0	0	0	0	0	0
$(\Gamma'_{25})_{33}$	1	1	1	1	0	0	0	0	0	0	0	0	1	−1	0	0	0	−1	−1	1	0	0	0	1
$\chi(\Gamma'_{25})$	3	3	−1	−1	0	0	0	0	0	0	0	0	1	1	−1	−1	−1	−1	1	1	−1	−1	−1	−1
$(\Gamma'_{15})_{11}$	1	1	−1	−1	0	0	0	0	0	0	0	0	−1	−1	0	0	0	0	1	1	0	0	0	0
$(\Gamma'_{15})_{21}$	0	0	0	0	1	1	−1	0	0	1	−1	−1	0	0	−1	−1	1	1	0	0	−1	−1	1	1
$(\Gamma'_{15})_{31}$	0	0	0	0	0	0	0	1	1	−1	0	0	0	0	0	0	0	0	0	0	0	0	0	0
$(\Gamma'_{15})_{12}$	0	0	0	0	1	1	1	1	−1	0	0	0	0	0	−1	−1	1	0	0	0	−1	−1	1	0
$(\Gamma'_{15})_{22}$	1	1	−1	−1	0	0	0	0	0	0	0	0	0	0	0	0	0	0	0	0	0	0	0	0
$(\Gamma'_{15})_{32}$	0	0	0	0	0	0	0	0	0	0	1	−1	0	0	0	0	0	1	0	0	0	0	0	0
$(\Gamma'_{15})_{13}$	0	0	0	0	0	0	0	0	1	1	0	0	0	0	0	0	0	0	0	0	0	0	0	0
$(\Gamma'_{15})_{23}$	0	0	0	0	0	0	0	−1	0	0	1	1	0	0	0	0	0	0	0	0	0	0	0	0
$(\Gamma'_{15})_{33}$	1	1	1	1	0	0	0	0	0	0	0	0	−1	1	0	0	0	1	−1	1	0	0	0	−1
$\chi(\Gamma'_{15})$	3	3	−1	−1	0	0	0	0	0	0	0	0	−1	−1	1	1	1	1	1	1	−1	−1	−1	−1

The matrix elements and characters are given for the unprimed operations. For the irreducible representations Γ_1, Γ_2, Γ_{12}, Γ'_{25}, and Γ'_{15} listed (in the notation of Bouckaert, Smoluchowski, and Wigner), the matrix elements for the primed operations are identical with those for the unprimed operations. For the operations Γ'_2, Γ'_1, Γ'_{12}, Γ_{15}, and Γ_{25}, the matrix elements and characters for the unprimed operations are identical with those for Γ_1, Γ_2, Γ_{12}, Γ'_{25}, and Γ'_{15}, respectively, but those for the primed operations are the negatives of those for the unprimed operations. These same notations are used by Herring [J. Franklin Inst., **233**:525 (1942)] for the group O_h^7 ($Fd3m$). For the group T_d^2 ($F\bar{4}3m$), where we have only the operators of the point group T_d, the complete table of matrix elements is given in the table, following the convention of Parmenter [Phys. Rev., **100**:573 (1955)]. In this case, the five irreducible representations of the table are denoted in succession as Γ_1, Γ_2, Γ_{12}, Γ_{15}, Γ_{24}.

Table A3-21

Matrix elements and characters at point Δ, propagation along x axis, for groups O_h^5 and O_h^7. For Δ, $p_1 = 0$, $p_2 = p_3 = p$. At Γ, $p = 0$. At X, $p = \frac{1}{2}$. The abbreviation α stands for $e^{\pi i p}$, and this value appears only for the group O_h^7, treated either according to Case 1 or Case 2. For O_h^5, α is to be replaced by unity.

	R_1	R_2	R_{19}	R_{20}	R_3'	R_4'	R_{13}'	R_{14}'
Δ_1	1	1	1	1	α	α	α	α
Δ_2'	1	1	1	1	$-\alpha$	$-\alpha$	$-\alpha$	$-\alpha$
Δ_2	1	1	-1	-1	α	α	$-\alpha$	$-\alpha$
Δ_1'	1	1	-1	-1	$-\alpha$	$-\alpha$	α	α
$(\Delta_5)_{11}$	1	-1	0	0	$-\alpha$	α	0	0
$(\Delta_5)_{21}$	0	0	1	-1	0	0	$-\alpha$	α
$(\Delta_5)_{12}$	0	0	1	-1	0	0	α	$-\alpha$
$(\Delta_5)_{22}$	1	-1	0	0	α	$-\alpha$	0	0
$\chi(\Delta_5)$	2	-2	0	0	0	0	0	0

For the group T_d^2, the matrix elements are as follows:

	R_1	R_2	R_{19}	R_{20}
Δ_1	1	1	1	1
Δ_2	1	1	-1	-1
Δ_3	1	-1	1	-1
Δ_4	1	-1	-1	1

Table A3-22

Matrix elements and characters for propagation along the direction Λ, $x = y = z$, groups O_h^5, O_h^7, and T_d^2. Along Λ, $p_1 = p_2 = p_3 = p$. At Γ, $p = 0$; at L, $p = \frac{1}{2}$.

	R_1	R_5	R_9	R_{19}	R_{21}	R_{23}
Λ_1	1	1	1	1	1	1
Λ_2	1	1	1	-1	-1	-1
$(\Lambda_3)_{11}$	1	$-\frac{1}{2}$	$-\frac{1}{2}$	1	$-\frac{1}{2}$	$-\frac{1}{2}$
$(\Lambda_3)_{21}$	0	$\frac{\sqrt{3}}{2}$	$-\frac{\sqrt{3}}{2}$	0	$-\frac{\sqrt{3}}{2}$	$\frac{\sqrt{3}}{2}$
$(\Lambda_3)_{12}$	0	$-\frac{\sqrt{3}}{2}$	$\frac{\sqrt{3}}{2}$	0	$-\frac{\sqrt{3}}{2}$	$\frac{\sqrt{3}}{2}$
$(\Lambda_3)_{22}$	1	$-\frac{1}{2}$	$-\frac{1}{2}$	-1	$\frac{1}{2}$	$\frac{1}{2}$
$\chi(\Lambda_3)$	2	-1	-1	0	0	0

Table A3-23

Matrix elements and characters for propagation along the direction Σ, $x = y$, groups O_h^5 and O_h^7. Along Σ, $p_1 = p_2 = p/2$, $p_3 = p$. At Γ, $p = 0$. At K, $p = \frac{3}{4}$. The quantity α is an abbreviation for $e^{\pi i p}$. It appears only for O_h^7, and is to be replaced by unity for O_h^5.

	R_1	R_{23}	R_4'	R_{24}'
Σ_1	1	1	α	α
Σ_2	1	-1	$-\alpha$	α
Σ_3	1	1	$-\alpha$	$-\alpha$
Σ_4	1	-1	α	$-\alpha$

For the group T_d^2, the matrix elements are as follows:

	R_1	R_{23}
Σ_1	1	1
Σ_2	1	-1

The matrix elements for the point K have the same form as for Σ, with representations K_1, K_2, K_3, K_4 for O_h^5 and O_h^7, K_1 and K_2 for T_d^2.

Table A3-24

Matrix elements and characters for point X, group O_h^5. Here $p_1 = 0$, $p_2 = p_3 = \frac{1}{2}$.

	R_1	R_2	R_3	R_4	R_{13}	R_{14}	R_{19}	R_{20}	R_1'	R_2'	R_3'	R_4'	R_{13}'	R_{14}'	R_{19}'	R_{20}'
X_1	1	1	1	1	1	1	1	1	1	1	1	1	1	1	1	1
X_2	1	1	1	1	−1	−1	−1	−1	1	1	1	1	−1	−1	−1	−1
X_3	1	1	−1	−1	−1	−1	1	1	1	1	−1	−1	−1	−1	1	1
X_4	1	1	−1	−1	1	1	−1	−1	1	1	−1	−1	1	1	−1	−1
X_1'	1	1	1	1	−1	−1	−1	−1	−1	−1	−1	−1	1	1	1	1
X_2'	1	1	1	1	1	1	1	1	−1	−1	−1	−1	−1	−1	−1	−1
X_3'	1	1	−1	−1	1	1	−1	−1	−1	−1	1	1	−1	−1	1	1
X_4'	1	1	−1	−1	−1	−1	1	1	−1	−1	1	1	1	1	−1	−1
$(X_5)_{11}$	1	−1	−1	1	0	0	0	0	1	−1	−1	1	0	0	0	0
$(X_5)_{21}$	0	0	0	0	−1	1	1	−1	0	0	0	0	−1	1	1	−1
$(X_5)_{12}$	0	0	0	0	1	−1	1	−1	0	0	0	0	1	−1	1	−1
$(X_5)_{22}$	1	−1	1	−1	0	0	0	0	1	−1	1	−1	0	0	0	0
$\chi(X_5)$	2	−2	0	0	0	0	0	0	2	−2	0	0	0	0	0	0
$(X_5')_{11}$	1	−1	1	−1	0	0	0	0	−1	1	−1	1	0	0	0	0
$(X_5')_{21}$	0	0	0	0	1	−1	1	−1	0	0	0	0	−1	1	−1	1
$(X_5')_{12}$	0	0	0	0	−1	1	1	−1	0	0	0	0	1	−1	−1	1
$(X_5')_{22}$	1	−1	−1	1	0	0	0	0	−1	1	1	−1	0	0	0	0
$\chi(X_5')$	2	−2	0	0	0	0	0	0	−2	2	0	0	0	0	0	0

For the group T_d^2, where we have only the unprimed operators, we have representations X_1, X_2, X_3, X_4, X_5, whose matrix elements are the same as those above for the unprimed operators.

Table A3-25

Matrix elements and characters for point X, group O_h^7. Here $p_1 = 0$, $p_2 = p_3 = \frac{1}{2}$.

	R_1	R_2	R_3	R_4	R_{13}	R_{14}	R_{19}	R_{20}	R_1'	R_2'	R_3'	R_4'	R_{13}'	R_{14}'	R_{19}'	R_{20}'
$(X_1)_{11}$	1	1	1	1	1	1	1	1	0	0	0	0	0	0	0	0
$(X_1)_{21}$	0	0	0	0	0	0	0	0	1	1	−1	−1	−1	−1	1	1
$(X_1)_{12}$	0	0	0	0	0	0	0	0	1	1	1	1	1	1	1	1
$(X_1)_{22}$	1	1	−1	−1	−1	−1	1	1	0	0	0	0	0	0	0	0
$\chi(X_1)$	2	2	0	0	0	0	2	2	0	0	0	0	0	0	0	0
$(X_2)_{11}$	1	1	1	1	−1	−1	−1	−1	0	0	0	0	0	0	0	0
$(X_2)_{21}$	0	0	0	0	0	0	0	0	1	1	−1	−1	1	1	−1	−1
$(X_2)_{12}$	0	0	0	0	0	0	0	0	1	1	1	1	−1	−1	−1	−1
$(X_2)_{22}$	1	1	−1	−1	1	1	−1	−1	0	0	0	0	0	0	0	0
$\chi(X_2)$	2	2	0	0	0	0	−2	−2	0	0	0	0	0	0	0	0
$(X_3)_{11}$	1	−1	0	0	0	0	1	−1	1	−1	0	0	0	0	1	−1
$(X_3)_{21}$	0	0	1	−1	−1	1	0	0	0	0	1	−1	−1	1	0	0
$(X_3)_{12}$	0	0	1	−1	1	−1	0	0	0	0	−1	1	−1	1	0	0
$(X_3)_{22}$	1	−1	0	0	0	0	−1	1	−1	1	0	0	0	0	1	−1
$\chi(X_3)$	2	−2	0	0	0	0	0	0	0	0	0	0	0	0	2	−2
$(X_4)_{11}$	1	−1	0	0	0	0	1	−1	−1	1	0	0	0	0	−1	1
$(X_4)_{21}$	0	0	1	−1	−1	1	0	0	0	0	−1	1	1	−1	0	0
$(X_4)_{12}$	0	0	1	−1	1	−1	0	0	0	0	1	−1	1	−1	0	0
$(X_4)_{22}$	1	−1	0	0	0	0	−1	1	1	−1	0	0	0	0	−1	1
$\chi(X_4)$	2	−2	0	0	0	0	0	0	0	0	0	0	0	0	−2	2

Table A3-26

Matrix elements and characters for point L, groups O_h^5 and O_h^7. Here $p_1 = p_2 = p_3 = \frac{1}{2}$. For T_d^2, representations L_1, L_2, L_3 have same matrix elements as Λ_1, Λ_2, Λ_3 in Table A3-22.

	R_1	R_5	R_9	R_{19}	R_{21}	R_{23}	R'_1	R'_5	R'_9	R'_{19}	R'_{21}	R'_{23}
L_1	1	1	1	1	1	1	1	1	1	1	1	1
L_2	1	1	1	-1	-1	-1	1	1	1	-1	-1	-1
L'_1	1	1	1	-1	-1	-1	-1	-1	-1	1	1	1
L'_2	1	1	1	1	1	1	-1	-1	-1	-1	-1	-1
$(L_3)_{11}$	1	$-\frac{1}{2}$	$-\frac{1}{2}$	1	$-\frac{1}{2}$	$-\frac{1}{2}$	1	$-\frac{1}{2}$	$-\frac{1}{2}$	1	$-\frac{1}{2}$	$-\frac{1}{2}$
$(L_3)_{21}$	0	$-\frac{\sqrt{3}}{2}$	$\frac{\sqrt{3}}{2}$	0	$-\frac{\sqrt{3}}{2}$	$\frac{\sqrt{3}}{2}$	0	$\frac{\sqrt{3}}{2}$	$-\frac{\sqrt{3}}{2}$	0	$-\frac{\sqrt{3}}{2}$	$\frac{\sqrt{3}}{2}$
$(L_3)_{12}$	0	$-\frac{\sqrt{3}}{2}$	$\frac{\sqrt{3}}{2}$	0	$-\frac{\sqrt{3}}{2}$	$\frac{\sqrt{3}}{2}$	0	$-\frac{\sqrt{3}}{2}$	$\frac{\sqrt{3}}{2}$	0	$-\frac{\sqrt{3}}{2}$	$\frac{\sqrt{3}}{2}$
$(L_3)_{22}$	1	$-\frac{1}{2}$	$-\frac{1}{2}$	-1	$\frac{1}{2}$	$\frac{1}{2}$	1	$-\frac{1}{2}$	$-\frac{1}{2}$	-1	$\frac{1}{2}$	$\frac{1}{2}$
$\chi(L_3)$	2	-1	-1	0	0	0	2	-1	-1	0	0	0
$(L_4)_{11}$	1	$-\frac{1}{2}$	$-\frac{1}{2}$	1	$-\frac{1}{2}$	$-\frac{1}{2}$	-1	$\frac{1}{2}$	$\frac{1}{2}$	-1	$\frac{1}{2}$	$\frac{1}{2}$
$(L_4)_{21}$	0	$-\frac{\sqrt{3}}{2}$	$\frac{\sqrt{3}}{2}$	0	$-\frac{\sqrt{3}}{2}$	$\frac{\sqrt{3}}{2}$	0	$-\frac{\sqrt{3}}{2}$	$\frac{\sqrt{3}}{2}$	0	$\frac{\sqrt{3}}{2}$	$-\frac{\sqrt{3}}{2}$
$(L_4)_{12}$	0	$-\frac{\sqrt{3}}{2}$	$\frac{\sqrt{3}}{2}$	0	$-\frac{\sqrt{3}}{2}$	$\frac{\sqrt{3}}{2}$	0	$\frac{\sqrt{3}}{2}$	$-\frac{\sqrt{3}}{2}$	0	$\frac{\sqrt{3}}{2}$	$-\frac{\sqrt{3}}{2}$
$(L_4)_{22}$	1	$-\frac{1}{2}$	$-\frac{1}{2}$	-1	$\frac{1}{2}$	$\frac{1}{2}$	-1	$\frac{1}{2}$	$\frac{1}{2}$	1	$-\frac{1}{2}$	$-\frac{1}{2}$
$\chi(L_4)$	2	-1	-1	0	0	0	-2	1	1	0	0	0

Table A3-27

Matrix elements for propagation along directions Z, groups O_h^5. Here $p_1 = p/2$, $p_2 = \frac{1}{2}$, $p_3 = (p+1)/2$. At X, $p = 0$. At W, $p = \frac{1}{2}$.

	R_1	R_3	R'_2	R'_4
Z_1	1	1	1	1
Z_2	1	1	-1	-1
Z_3	1	-1	-1	1
Z_4	1	-1	1	-1

For group O_h^7, the matrix elements and characters along Z are as follows, where α is an abbreviation for $e^{\pi i(p+1)/2}$:

Table A3-27 (Continued)

	R_1	R_3	R_2'	R_4'
$(Z_1)_{11}$	1	1	0	0
$(Z_1)_{21}$	0	0	$-\alpha$	α
$(Z_1)_{12}$	0	0	α	α
$(Z_1)_{22}$	1	-1	0	0
$\chi(Z_1)$	2	0	0	0

For group T_d^2, the matrix elements are as follows:

	R_1	R_3
Z_1	1	1
Z_2	1	-1

Table A3-28

Matrix elements and characters for the point W, group O_h^5. Here $p_1 = \frac{1}{4}$, $p_2 = \frac{1}{2}$, $p_3 = \frac{3}{4}$.

	R_1	R_3	R_{15}	R_{16}	R_2'	R_4'	R_{21}'	R_{22}'
W_1	1	1	1	1	1	1	1	1
W_1'	1	1	-1	-1	-1	-1	1	1
W_2	1	1	1	1	-1	-1	-1	-1
W_2'	1	1	-1	-1	1	1	-1	-1
$(W_3)_{11}$	1	-1	0	0	1	-1	0	0
$(W_3)_{21}$	0	0	-1	1	0	0	1	-1
$(W_3)_{12}$	0	0	1	-1	0	0	1	-1
$(W_3)_{22}$	1	-1	0	0	-1	1	0	0
$\chi(W_3)$	2	-2	0	0	0	0	0	0

Table A3-28 (Continued)

For the group O_h^7, the matrix elements are as follows:

	R_1	R_3	R_{15}	R_{16}	R'_2	R'_4	R'_{21}	R'_{22}
$(W_1)_{11}$	1	1	1	1	0	0	0	0
$(W_1)_{21}$	0	0	0	0	1	−1	−i	i
$(W_1)_{12}$	0	0	0	0	i	i	i	i
$(W_1)_{22}$	1	−1	−i	i	0	0	0	0
$\chi(W_1)$	2	0	$1-i$	$1+i$	0	0	0	0
$(W_2)_{11}$	1	1	−1	−1	0	0	0	0
$(W_2)_{21}$	0	0	0	0	1	−1	i	−i
$(W_2)_{12}$	0	0	0	0	i	i	−i	−i
$(W_2)_{22}$	1	−1	i	−i	0	0	0	0
$\chi(W_2)$	2	0	$-1+i$	$-1-i$	0	0	0	0

For group T_d^2, the matrix elements are as follows:

	R_1	R_3	R_{15}	R_{16}
W_1	1	1	1	1
W_2	1	1	−1	−1
W_3	1	−1	i	−i
W_4	1	−1	−i	i

Table A3-29

Matrix elements for propagation in direction S, O_h^5 and O_h^7. Along this direction, $p_1 = p = 2p_2 - 1 = 2p_3 - 1$. At X, $p = 0$. At U, $p = \frac{1}{4}$. Matrix elements at U are as at S.

	R_1	R_{19}	R'_2	R'_{20}
S_1	1	1	1	1
S_2	1	−1	−1	1
S_3	1	1	−1	−1
S_4	1	−1	1	−1

For group T_d^2, with only the operations R_1 and R_{19}, matrix elements are as follows:

	R_1	R_{19}
S_1	1	1
S_2	1	−1

Table A3-30

Matrix elements for propagation in direction Q, O_h^5 and O_h^7. Here $p_1 = p/2 + \frac{1}{4}$, $p_2 = \frac{1}{2}$, $p_3 = -p/2 + \frac{3}{4} = 1 - p_1$. At W, $p = 0$. At L, $p = \frac{1}{2}$.

	R_1	R'_{21}
Q_1	1	1
Q_2	1	-1

For group T_d^2, Q is not a symmetry direction.

Table A3-31
Compatibility relations for groups O_h^5, O_h^7, and T_d^2

	Γ_1	Γ_2	Γ'_1	Γ'_2	Γ_{12}	Γ'_{12}	Γ_{15}	Γ'_{15}	Γ_{25}	Γ'_{25}
O_h^5 and O_h^7	Δ_1	Δ_2	Δ'_1	Δ'_2	$\Delta_1\Delta_2$	$\Delta'_1\Delta'_2$	$\Delta_1\Delta_5$	$\Delta'_1\Delta_5$	$\Delta_2\Delta_5$	$\Delta'_2\Delta_5$
	Λ_1	Λ_2	Λ_2	Λ_1	Λ_3	Λ_3	$\Lambda_1\Lambda_3$	$\Lambda_2\Lambda_3$	$\Lambda_2\Lambda_3$	$\Lambda_1\Lambda_3$
	Σ_1	Σ_4	Σ_2	Σ_3	$\Sigma_1\Sigma_4$	$\Sigma_2\Sigma_3$	$\Sigma_1\Sigma_3\Sigma_4$	$\Sigma_2\Sigma_3\Sigma_4$	$\Sigma_1\Sigma_2\Sigma_4$	$\Sigma_1\Sigma_2\Sigma_3$

	Γ_1	Γ_2	Γ_{12}	Γ_{15}	Γ_{25}
T_d^2	Δ_1	Δ_2	$\Delta_1\Delta_2$	$\Delta_1\Delta_3\Delta_4$	$\Delta_2\Delta_3\Delta_4$
	Λ_1	Λ_2	Λ_3	$\Lambda_1\Lambda_3$	$\Lambda_2\Lambda_3$
	Σ_1	Σ_2	$\Sigma_1\Sigma_2$	$\Sigma_1\Sigma_2$	$\Sigma_1\Sigma_2$

	X_1	X_2	X_3	X_4	X'_1	X'_2	X'_3	X'_4	X_5	X'_5
O_h^5	Δ_1	Δ_2	Δ'_2	Δ'_1	Δ'_1	Δ'_2	Δ_2	Δ_1	Δ_5	Δ_5
	Z_1	Z_1	Z_4	Z_4	Z_2	Z_2	Z_3	Z_3	Z_2Z_3	Z_1Z_4
	S_1	S_4	S_1	S_4	S_2	S_3	S_2	S_3	S_2S_3	S_1S_4

	X_1	X_2	X_3	X_4			X_1	X_2	X_3	X_4	X_5
O_h^7	$\Delta_1\Delta'_2$	$\Delta'_1\Delta_2$	Δ_5	Δ_5	T_d^2		Δ_1	Δ_2	Δ_1	Δ_2	$\Delta_3\Delta_4$
	Z_1	Z_1	Z_1	Z_1			Z_1	Z_1	Z_2	Z_2	Z_1Z_2
	S_1S_3	S_2S_4	S_3S_4	S_1S_2			S_1	S_2	S_1	S_2	S_1S_2

Table A3-31 (Continued)

	L_1	L_2	L_1'	L_2'	L_3	L_4
O_h^6 and O_h^7	Λ_1	Λ_2	Λ_2	Λ_1	Λ_3	Λ_3
	Q_1	Q_2	Q_1	Q_2	Q_1Q_2	Q_1Q_2

	W_1	W_2	W_1'	W_2'	W_3
O_h^5	Z_1	Z_2	Z_2	Z_1	Z_3Z_4
	Q_1	Q_2	Q_1	Q_2	Q_1Q_2

	W_1	W_2
O_h^7	Z_1	Z_1
	Q_1Q_2	Q_1Q_2

	W_1	W_2	W_3	W_4
T_d^2	Z_1	Z_1	Z_2	Z_2

A3-3. The Space Groups O_h^9 ($Im3m$), O_h^{10} ($Ia3d$), and T_h^7 ($Ia3$). The space groups O_h^9 ($Im3m$), found in the body-centered cubic structure (Sec. 2-2), O_h^{10} ($Ia3d$), found in the garnets (Sec. 3-4), and T_h^7 ($Ia3$), found in the Tl_2O_3 structure (Sec. 3-4), all have the body-centered cubic Bravais lattice, and can conveniently be treated together. The group O_h^9 is symmorphic, and the familiar body-centered cubic structure has one atom per unit cell, at the origin. The group O_h^{10} on the other hand is nonsymmorphic, and there are eight different nonprimitive translations found for different operations of the group (including the case of zero nonprimitive translation). It is one of the most complicated of space groups. A typical garnet is calcium aluminum orthosilicate, $Ca_3Al_2(SiO_4)_3$. In the primitive unit cell there are four molecular constituents; that is, 12 Ca's, 8 Al's, 12 Si's, and 48 O's. In the familiar iron garnets, such for example as yttrium iron garnet, the calciums are replaced by yttrium, and both the aluminums and silicons by iron. The group T_h^7, like O_h^{10}, is nonsymmorphic. Since it has the point group T_h, rather than O_h as the other two groups have, there are only 24 operations in the point group, rather than the 48 in O_h. There are four nonprimitive translations, which are the same as those of O_h^{10} for those operations included in both groups, so that there is a good deal of similarity between the two groups. In the unit cell of the Tl_2O_3 structure, there are 16 Tl's, 24 O's. The atomic positions in the garnet and Tl_2O_3 structures are given in Table A3-32.

The primitive unit cell has vectors given in Eq. (1-18), and the vectors of the reciprocal lattice are given in Eq. (5-9). The Brillouin zone, with the notation of Bouckaert, Smoluchowski, and Wigner for the symmetry points, is given in Fig. A3-3. The operations of the point group O_h are given in Table 2-2. We have the operations $R_1 \cdots R_{24}$ included in the point group T_d, and the additional operations $R_1' \cdots R_{24}'$, each of which is one of the operations $R_1 \cdots R_{24}$ plus an inversion. The point group T_h consists of the operations $R_1 \cdots R_{12}, R_1' \cdots R_{12}'$.

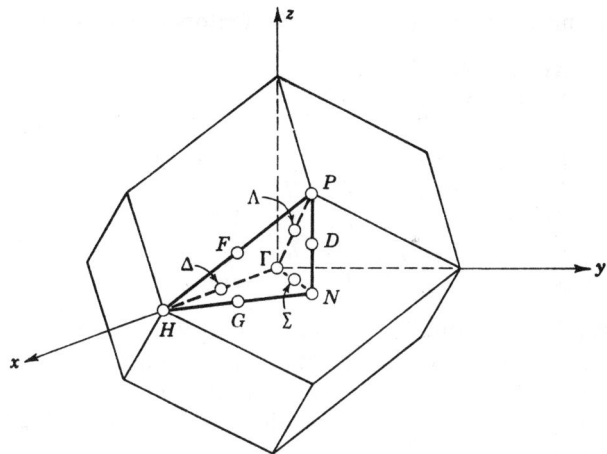

Fig. A3-3. Brillouin zone for body-centered cubic Bravais lattice, with symmetry notations of Bouckaert, Smoluchowski, and Wigner.

The nonprimitive translations in the space group O_h^{10} are as follows:

$$\begin{aligned}
& R_1, R_5, R_9, R'_1, R'_5, R'_9: \text{none} \\
& R_2, R_8, R_{11}, R'_2, R'_8, R'_{11}: \tfrac{1}{2}(\mathbf{t}_1 + \mathbf{t}_2) \\
& R_3, R_6, R_{12}, R'_3, R'_6, R'_{12}: \tfrac{1}{2}(\mathbf{t}_2 + \mathbf{t}_3) \\
& R_4, R_7, R_{10}, R'_4, R'_7, R'_{10}: \tfrac{1}{2}(\mathbf{t}_1 + \mathbf{t}_3) \\
& R_{14}, R_{15}, R_{24}, R'_{14}, R'_{15}, R'_{24}: \tfrac{1}{2}\mathbf{t}_1 \\
& R_{16}, R_{17}, R_{20}, R'_{16}, R'_{17}, R'_{20}: \tfrac{1}{2}\mathbf{t}_2 \\
& R_{13}, R_{18}, R_{22}, R'_{13}, R'_{18}, R'_{22}: \tfrac{1}{2}\mathbf{t}_3 \\
& R_{19}, R_{21}, R_{23}, R'_{19}, R'_{21}, R'_{23}: \tfrac{1}{2}(\mathbf{t}_1 + \mathbf{t}_2 + \mathbf{t}_3)
\end{aligned} \quad \text{(A3-21)}$$

As we have mentioned earlier, the nonprimitive translations for T_h^7 are the same as those of O_h^{10}, for the operators included in the group.

The effect of an operation of the point group on a plane wave is given in Table A3-33. Here, as in the earlier discussion, the radius vector $\mathbf{r}$ is given by $\xi \mathbf{t}_1 + \eta \mathbf{t}_2 + \zeta \mathbf{t}_3$, and the initial plane wave is $\exp 2\pi i[(h_1 + p_1)\xi + (h_2 + p_2)\eta + (h_3 + p_3)\zeta]$, where h_1, h_2, h_3 are integers, and p_1, p_2, p_3 refer to the reduced wave vector. Table A3-33 is different from Table A3-18, which gives the same information for the face-centered cubic case, because the vectors $\mathbf{t}_1$, $\mathbf{t}_2$, $\mathbf{t}_3$ are different in the two cases, giving a different significance to ξ, η, ζ. An operation of the space group O_h^{10} or T_h^7, associated with no primitive translations, will multiply the transformed plane wave by additional factors, on account of the nonprimitive translations. These factors are as follows:

$$
\begin{aligned}
&R_1,\ R_5,\ R_9,\ R'_1,\ R'_5,\ R'_9: \text{unity} \\
&R_2,\ R_8,\ R_{11},\ R'_2,\ R'_8,\ R'_{11}: e^{\pi i(h_1+p_1+h_2+p_2)} \\
&R_3,\ R_6,\ R_{12},\ R'_3,\ R'_6,\ R'_{12}: e^{\pi i(h_2+p_2+h_3+p_3)} \\
&R_4,\ R_7,\ R_{10},\ R'_4,\ R'_7,\ R'_{10}: e^{\pi i(h_1+p_1+h_3+p_3)} \\
&R_{14},\ R_{15},\ R_{24},\ R'_{14},\ R'_{15},\ R'_{24}: e^{\pi i(h_1+p_1)} \\
&R_{16},\ R_{17},\ R_{20},\ R'_{16},\ R'_{17},\ R'_{20}: e^{\pi i(h_2+p_2)} \\
&R_{13},\ R_{18},\ R_{22},\ R'_{13},\ R'_{18},\ R'_{22}: e^{\pi i(h_3+p_3)} \\
&R_{19},\ R_{21},\ R_{23},\ R'_{19},\ R'_{21},\ R'_{23}: e^{\pi i(h_1+p_1+h_2+p_2+h_3+p_3)}
\end{aligned}
\qquad (A3\text{-}22)
$$

The multiplication table for the point group O_h is given in Table A3-17. This table applies here as well. For the space groups O_h^{10} and T_h^7 there are extra factors which must be used in the multiplication table, on account of the nonprimitive translations just described in Eq. (A3-22). These factors are given in Table A3-34.

We can now proceed as in Sec. A3-2 to find the matrix elements of the irreducible representations along the various symmetry directions. For the point Γ, we can use Table A3-20 without change, for O_h^9 and O_h^{10}. For T_h^7, which has a different point group, the matrix elements are given in Table A3-35. For Δ, Λ, and Σ, we can use Tables A3-21, A3-22, and A3-23 for the space group O_h^9. For O_h^{10} and T_h^7 the corresponding matrix elements are given in Table A3-36. They differ from simpler cases in that we must use different factors $e^{i(\mathbf{k}+\mathbf{K}_h)\cdot\boldsymbol{\tau}}$ for the different operators. These factors are the same ones given in Eq. (A3-22) (setting the h's equal to zero), with the appropriate values of the p's. The points on the surface of the Brillouin zone are handled quite differently for the three groups, and the matrix elements are given in Tables A3-37 to A3-42, at the end of this section. Compatibility relations are given in Table A3-43.

Since the points H, at the extremity of the line Δ, for propagation along the x axis, and P, at the extremity of Λ, for propagation along the 111 direction, in the case of O_h^{10}, are as complicated as any representations of space groups which have been derived, it may be worthwhile indicating how the matrix elements can be found. As we note, at H the point group is O_h, with 48 operations, and at P it is T_d, with 24 operations. As the tables show, we find at H a six-dimensional irreducible representation, and three two-dimensional representations, and at P we have a four-dimensional and two two-dimensional irreducible representations. If one has not already carried through the calculations, however, these facts would not be known. Let us describe how the results can be obtained, without any previous knowledge.

The first step, in each case, is to take the first one-dimensional irreducible representation, along Δ or Λ, respectively, from Table A3-36, putting in the values of matrix elements for the particular wave vector in question. These give basis functions for the group of the wave vector along Δ or Λ, respectively, but the group at H or P involves more operations, and the function so found is not a basis function for a one-dimensional representation of the complete space group. Rather, when we apply all operations of the group to this one function, we find in the case of H six independent functions, and in the case of P four independent functions, which form basis functions for a six- or a four-dimensional representation, respectively. We must next ask whether these representations are reducible or not, since we do not know a priori what dimensionalities to look for.

To answer this question, we can use an easily proved theorem. We set up the matrix elements for the representation (36 for the six-dimensional, 16 for the four-dimensional representation, as given in Tables A3-37 and A3-38, respectively). If the representations are irreducible, these sets of matrix elements, regarded as vectors, must be orthogonal to each other; if they are reducible, the matrix elements will not all be orthogonal. In each case we verify at once, as the reader can from the tables, that they are orthogonal, so that the representations are irreducible.

We then have found how to set up, by projection operators, 36 linearly independent combinations of the 48 basis functions of the regular representation, for the case of H; for P, 16 linearly independent combinations of the 24 of the regular representation. The remaining basis functions for all other irreducible representations must come from $48 - 36 = 12$ remaining linear combinations in the case of H, or $24 - 16 = 8$ in the case of P. As a step toward deriving these remaining linear combinations, the following method may be used. We take the original plane-wave function ψ, and orthogonalize it to all 36 basis functions of the six-dimensional irreducible representation in the case of H, or to all 16 of the four-dimensional representation in the case of P, by use of Schmidt orthogonalization. Then we start with this function, apply all the operators of the group to it, and in this way find 12 basis functions for a reducible representation in the case of H, 8 basis functions in the case of P. The matrix elements of these representations are next computed, and as expected, they are not all orthogonal, in accordance with the theorem stated above. We recall that the sum of the squares of the dimensionalities of the remaining irreducible representations must equal the number of basis functions, or 12 for H, 8 for P, showing at once that these remaining representations must be reducible.

These remaining representations, however, involving as they do a relatively small number of independent functions, prove to be simple enough to reduce by only a moderate amount of study and manipulation.

A practical method is to consider only a few of the key operators, make linear combinations of the basis functions to secure irreducible representations as far as these operators are concerned, and then to try out the remaining operators. It will probably be found that the linear combinations of basis functions which have been arrived at are in fact the desired functions, and the matrix elements for the resulting two-dimensional irreducible representations can be found. In all cases the basis functions are found from the representations by the method of projection operators.

These two cases of H and P are the only two which are difficult enough to require special treatment. The case of T_h^7, which is somewhat simpler, can be handled by similar methods. It does not seem likely that any more difficult cases will be met in any further space groups whose symmetry properties may need to be investigated in the future, so that it seems probable that the simple and straightforward methods used here will suffice to find the symmetry properties of all 230 space groups, when it becomes worthwhile to investigate all of them.

Table A3-32
Positions of atoms in structures having space groups O_h^{10} and T_h^7

O_h^{10}, *garnet structure* (values x/a, y/a, z/a given).
  Calciums at the special positions $\pm(\tfrac{1}{8}, 0, \tfrac{1}{4}; \tfrac{1}{4}, \tfrac{1}{8}, 0; 0, \tfrac{1}{4}, \tfrac{1}{8}; \tfrac{5}{8}, 0, \tfrac{1}{4}; \tfrac{1}{4}, \tfrac{5}{8}, 0; 0, \tfrac{1}{4}, \tfrac{5}{8})$.
  Aluminums at the special positions $0, 0, 0; \tfrac{1}{2}, \tfrac{1}{2}, 0; 0, \tfrac{1}{2}, \tfrac{1}{2}; \tfrac{1}{2}, 0, \tfrac{1}{2}; \tfrac{1}{4}, \tfrac{1}{4}, \tfrac{1}{4}; \tfrac{3}{4}, \tfrac{3}{4}, \tfrac{1}{4}; \tfrac{1}{4}, \tfrac{3}{4}, \tfrac{3}{4}; \tfrac{3}{4}, \tfrac{1}{4}, \tfrac{3}{4}$.
  Silicons at the special positions $\pm(\tfrac{3}{8}, 0, \tfrac{1}{4}; \tfrac{1}{4}, \tfrac{3}{8}, 0; 0, \tfrac{1}{4}, \tfrac{3}{8}; \tfrac{7}{8}, 0, \tfrac{1}{4}; \tfrac{1}{4}, \tfrac{7}{8}, 0; 0, \tfrac{1}{4}, \tfrac{7}{8})$.
  Oxygens at general positions, derived from an initial set of values ξ, η, ζ by all 48 operations of the space group. In most garnets, the values of ξ, η, ζ are close to 0.04, 0.05, and 0.65, respectively. In yttrium iron garnet, they are 0.0274, 0.0572, and 0.6492, respectively.

T_h^7, Tl$_2$O$_3$ *structure.*
  One set of thalliums at special positions $\tfrac{1}{4}, \tfrac{1}{4}, \tfrac{1}{4}; \tfrac{1}{4}, \tfrac{3}{4}, \tfrac{3}{4}; \tfrac{3}{4}, \tfrac{1}{4}, \tfrac{3}{4}; \tfrac{3}{4}, \tfrac{3}{4}, \tfrac{1}{4}$.
  A second set of thalliums at special positions $\pm(u, 0, \tfrac{1}{4}; \tfrac{1}{4}, u, 0; 0, \tfrac{1}{4}, u; -u, \tfrac{1}{2}, \tfrac{1}{4}; \tfrac{1}{4}, -u, \tfrac{1}{2}; \tfrac{1}{2}, \tfrac{1}{4}, -u)$.
  Oxygens at general positions, derived from an initial set of values ξ, η, ζ by all 24 operations of the space group. For (Fe,Mn)$_2$O$_3$, we have $u = -0.030$, $\xi = 0.385$, $\eta = 0.145$, $\zeta = 0.380$.

Table A3-33
Effect of operators of T_d on a plane wave, body-centered cubic Bravais lattice

$R_1\psi = \exp 2\pi i[(h_1 + p_1)\xi + (h_2 + p_2)\eta + (h_3 + p_3)\zeta]$

$R_2\psi = \exp 2\pi i[(-h_1 - p_1 - h_2 - p_2 - h_3 - p_3)\xi + (h_3 + p_3)\eta + (h_2 + p_2)\zeta]$

$R_3\psi = \exp 2\pi i[(h_3 + p_3)\xi + (-h_1 - p_1 - h_2 - p_2 - h_3 - p_3)\eta + (h_1 + p_1)\zeta]$

$R_4\psi = \exp 2\pi i[(h_2 + p_2)\xi + (h_1 + p_1)\eta + (-h_1 - p_1 - h_2 - p_2 - h_3 - p_3)\zeta]$

$R_5\psi = \exp 2\pi i[(h_3 + p_3)\xi + (h_1 + p_1)\eta + (h_2 + p_2)\zeta]$

$R_6\psi = \exp 2\pi i[(h_1 + p_1)\xi + (h_3 + p_3)\eta + (-h_1 - p_1 - h_2 - p_2 - h_3 - p_3)\zeta]$

$R_7\psi = \exp 2\pi i[(-h_1 - p_1 - h_2 - p_2 - h_3 - p_3)\xi + (h_2 + p_2)\eta + (h_1 + p_1)\zeta]$

$R_8\psi = \exp 2\pi i[(h_2 + p_2)\xi + (-h_1 - p_1 - h_2 - p_2 - h_3 - p_3)\eta + (h_3 + p_3)\zeta]$

$R_9\psi = \exp 2\pi i[(h_2 + p_2)\xi + (h_3 + p_3)\eta + (h_1 + p_1)\zeta]$

$R_{10}\psi = \exp 2\pi i[(h_1 + p_1)\xi + (-h_1 - p_1 - h_2 - p_2 - h_3 - p_3)\eta + (h_2 + p_2)\zeta]$

$R_{11}\psi = \exp 2\pi i[(h_3 + p_3)\xi + (h_2 + p_2)\eta + (-h_1 - p_1 - h_2 - p_2 - h_3 - p_3)\zeta]$

$R_{12}\psi = \exp 2\pi i[(-h_1 - p_1 - h_2 - p_2 - h_3 - p_3)\xi + (h_1 + p_1)\eta + (h_3 + p_3)\zeta]$

$R_{13}\psi = \exp 2\pi i[(h_3 + p_3)\xi + (h_1 + p_1)\eta + (-h_1 - p_1 - h_2 - p_2 - h_3 - p_3)\zeta]$

$R_{14}\psi = \exp 2\pi i[(h_2 + p_2)\xi + (-h_1 - p_1 - h_2 - p_2 - h_3 - p_3)\eta + (h_1 + p_1)\zeta]$

$R_{15}\psi = \exp 2\pi i[(-h_1 - p_1 - h_2 - p_2 - h_3 - p_3)\xi + (h_1 + p_1)\eta + (h_2 + p_2)\zeta]$

$R_{16}\psi = \exp 2\pi i[(h_2 + p_2)\xi + (h_3 + p_3)\eta + (-h_1 - p_1 - h_2 - p_2 - h_3 - p_3)\zeta]$

$R_{17}\psi = \exp 2\pi i[(h_3 + p_3)\xi + (-h_1 - p_1 - h_2 - p_2 - h_3 - p_3)\eta + (h_2 + p_2)\zeta]$

$R_{18}\psi = \exp 2\pi i[(-h_1 - p_1 - h_2 - p_2 - h_3 - p_3)\xi + (h_3 + p_3)\eta + (h_1 + p_1)\zeta]$

$R_{19}\psi = \exp 2\pi i[(h_1 + p_1)\xi + (h_3 + p_3)\eta + (h_2 + p_2)\zeta]$

$R_{20}\psi = \exp 2\pi i[(-h_1 - p_1 - h_2 - p_2 - h_3 - p_3)\xi + (h_2 + p_2)\eta + (h_3 + p_3)\zeta]$

$R_{21}\psi = \exp 2\pi i[(h_3 + p_3)\xi + (h_2 + p_2)\eta + (h_1 + p_1)\zeta]$

$R_{22}\psi = \exp 2\pi i[(h_1 + p_1)\xi + (-h_1 - p_1 - h_2 - p_2 - h_3 - p_3)\eta + (h_3 + p_3)\zeta]$

$R_{23}\psi = \exp 2\pi i[(h_2 + p_2)\xi + (h_1 + p_1)\eta + (h_3 + p_3)\zeta]$

$R_{24}\psi = \exp 2\pi i[(h_1 + p_1)\xi + (h_2 + p_2)\eta + (-h_1 - p_1 - h_2 - p_2 - h_3 - p_3)\zeta]$

Table A3-34

Factors in multiplication table for groups O_h^{10} and T_h^7. To find the multiplication table of the space group, we first find the product operator of the point group, according to Table A3-17. Then we multiply by an additional factor, equal to $\exp 2\pi i \times$ (quantity in table below). The first factor stands at the top of the column, and the results are the same whether it is primed or not. The second factor is found down the left side of the table.

	$R_1 R_5 R_9$	$R_2 R_8 R_{11}$	$R_3 R_6 R_{12}$	$R_4 R_7 R_{10}$	$R_{14} R_{15} R_{24}$	$R_{16} R_{17} R_{20}$	$R_{13} R_{18} R_{22}$	$R_{19} R_{21} R_{23}$
R_1	0	0	0	0	0	0	0	0
R_2	0	0	p_2	$-p_3$	$-p_3$	0	p_2	0
R_3	0	$-p_1$	0	p_3	p_3	$-p_1$	0	0
R_4	0	p_2	$-p_2$	0	0	p_1	$-p_2$	0
R_5	0	0	0	0	0	p_3	0	0
R_6	0	p_3	$-p_1$	$-p_2$	$-p_2$	0	$-p_1$	0
R_7	0	0	0	p_2	p_2	p_3	p_1	0
R_8	0	$-p_3$	0	0	0	$-p_3$	0	0
R_9	0	0	0	0	0	$-p_2$	0	0
R_{10}	0	$-p_2$	0	p_1	p_1	$-p_2$	0	0
R_{11}	0	p_2	0	0	0	0	$-p_3$	0
R_{12}	0	0	p_3	$-p_1$	$-p_1$	0	p_3	0
R_{13}	0	p_3	$-p_2$	$-p_1-p_2$	p_3	0	$-p_1-p_2$	$-p_2$
R_{14}	0	$-p_3$	$-p_2-p_3$	p_1	0	$-p_2-p_3$	p_1	$-p_3$
R_{15}	0	$-p_2-p_3$	p_1	$-p_3$	$-p_2-p_3$	p_1	0	$-p_3$
R_{16}	0	p_2	$-p_1$	$-p_1-p_3$	0	$-p_1-p_3$	0	$-p_1$
R_{17}	0	$-p_1$	$-p_1-p_3$	p_2	$-p_1-p_2$	p_3	p_2	$-p_1$
R_{18}	0	$-p_1-p_2$	p_3	$-p_2$	$-p_1-p_2$	p_3	0	0
R_{19}	0	p_1+p_3	p_2+p_3	p_1+p_2	p_1	p_3	p_2	$p_1+p_2+p_3$
R_{20}	0	$-p_1-p_3$	p_2	$-p_1$	p_1	p_2	0	$-p_1$
R_{21}	0	p_2+p_3	p_1+p_2	p_1+p_3	p_3	p_2	p_1	$p_1+p_2+p_3$
R_{22}	0	$-p_2$	$-p_1-p_2$	p_3	0	p_1	p_3	$-p_2$
R_{23}	0	p_1+p_2	p_1+p_3	p_2+p_3	p_2	p_1	p_3	$p_1+p_2+p_3$
R_{24}	0	p_1	$-p_3$	$-p_2-p_3$	p_1	0	$-p_2-p_3$	$-p_3$

Table A3-34 (Continued)

	$R_1R_5R_9$	$R_2R_8R_{11}$	$R_3R_6R_{12}$	$R_4R_7R_{10}$	$R_{14}R_{15}R_{24}$	$R_{16}R_{17}R_{20}$	$R_{13}R_{18}R_{22}$	$R_{19}R_{21}R_{23}$
R'_1	0	$-p_1-p_2$	$-p_2-p_3$	$-p_1-p_3$	$-p_1$	$-p_2$	$-p_3$	$-p_1-p_2-p_3$
R'_2	0	p_1+p_2	$-p_3$	p_1	p_1+p_2	$-p_3$	0	p_1
R'_3	0	p_2	p_2+p_3	$-p_1$	0	p_2+p_3	$-p_1$	p_2
R'_4	0	$-p_2$	p_3	p_1+p_3	$-p_2$	$-p_1$	p_1+p_3	p_3
R'_5	0	$-p_1-p_3$	p_1-p_2	$-p_2-p_3$	$-p_3$	$-p_1$	$-p_2$	$-p_1-p_2-p_3$
R'_6	0	$-p_1$	p_2	p_2+p_3	$-p_1$	$-p_2$	p_2+0	p_2
R'_7	0	p_1+p_3	$-p_2$	p_3	p_1+p_3	$-p_2$	$-p_3$	p_3
R'_8	0	p_1	p_1+p_2	$-p_3$	0	p_1+p_2	$-p_1$	p_1
R'_9	0	$-p_2-p_3$	$-p_1-p_3$	$-p_1-p_2$	$-p_2$	$-p_3$	$-p_2$	$-p_1-p_2-p_3$
R'_{10}	0	p_3	p_1+p_3	p_1	$-p_3$	p_1+p_3	p_1+0	p_3
R'_{11}	0	p_2+p_3	p_1	p_1+p_2	p_2+p_3	$-p_1$	p_3	p_1
R'_{12}	0	$-p_1$	p_3	p_2	$-p_1$	p_1	0	p_2
R'_{13}	0	p_1	0	0	$-p_2$	0	$-p_2$	0
R'_{14}	0	0	$-p_2$	p_2	p_1	$-p_3$	p_2	0
R'_{15}	0	0	p_2	0	0	0	0	0
R'_{16}	0	$-p_3$	0	$-p_3$	$-p_3$	$-p_3$	$-p_2$	0
R'_{17}	0	p_2	0	0	p_1	p_2	0	0
R'_{18}	0	0	$-p_1$	p_3	0	0	$-p_1$	0
R'_{19}	0	0	0	0	p_3	0	0	0
R'_{20}	0	0	$-p_3$	0	0	0	$-p_3$	0
R'_{21}	0	0	0	$-p_1$	$-p_1$	p_3	0	0
R'_{22}	0	0	0	0	0	0	0	0
R'_{23}	0	0	0	0	0	0	0	0
R'_{24}	0	$-p_2$	0	0	0	$-p_2$	p_1	0

391

Table A3-35

Matrix elements and characters for T_h^7, at origin, point Γ. Below we list the matrix elements and characters for the irreducible representations of the point group T. For T_h, each of these representations goes into two, which we shall designate by superscripts $+$ and $-$. In the $+$ representation, the matrix elements for the primed operators equal those for the unprimed; in the $-$ representation, they are the negatives of those for the unprimed. The abbreviation ω stands for $e^{2\pi i/3}$.

	R_1	R_2	R_3	R_4	R_5	R_6	R_7	R_8	R_9	R_{10}	R_{11}	R_{12}
Γ_1	1	1	1	1	1	1	1	1	1	1	1	1
Γ_2	1	1	1	1	ω	ω	ω	ω	ω^2	ω^2	ω^2	ω^2
Γ_3	1	1	1	1	ω^2	ω^2	ω^2	ω^2	ω	ω	ω	ω
$(\Gamma_4)_{11}$	1	1	-1	-1	0	0	0	0	0	0	0	0
$(\Gamma_4)_{21}$	0	0	0	0	1	-1	-1	1	0	0	0	0
$(\Gamma_4)_{31}$	0	0	0	0	0	0	0	0	1	-1	1	-1
$(\Gamma_4)_{12}$	0	0	0	0	0	0	0	0	1	-1	-1	1
$(\Gamma_4)_{22}$	1	-1	1	-1	0	0	0	0	0	0	0	0
$(\Gamma_4)_{32}$	0	0	0	0	1	1	-1	-1	0	0	0	0
$(\Gamma_4)_{13}$	0	0	0	0	1	-1	1	-1	0	0	0	0
$(\Gamma_4)_{23}$	0	0	0	0	0	0	0	0	1	1	-1	-1
$(\Gamma_4)_{33}$	1	-1	-1	1	0	0	0	0	0	0	0	0
$\chi(\Gamma_4)$	3	-1	-1	-1	0	0	0	0	0	0	0	0

Table A3-36

Matrix elements and characters at points Δ, propagation along x axis, for group O_h^{10}. For Δ, $-p_1 = p_2 = p_3 = p$.

	R_1	R_2	R_{19}	R_{20}	R_3'	R_4'	R_{13}'	R_{14}'
Δ_1	1	1	$e^{\pi i p}$	$e^{\pi i p}$	$e^{2\pi i p}$	1	$e^{\pi i p}$	$e^{-\pi i p}$
Δ_2'	1	1	$e^{\pi i p}$	$e^{\pi i p}$	$-e^{2\pi i p}$	-1	$-e^{\pi i p}$	$-e^{-\pi i p}$
Δ_2	1	1	$-e^{\pi i p}$	$-e^{\pi i p}$	$e^{2\pi i p}$	1	$-e^{\pi i p}$	$-e^{-\pi i p}$
Δ_1'	1	1	$-e^{\pi i p}$	$-e^{\pi i p}$	$-e^{2\pi i p}$	-1	$e^{\pi i p}$	$e^{-\pi i p}$
$(\Delta_5)_{11}$	1	-1	0	0	$-e^{2\pi i p}$	1	0	0
$(\Delta_5)_{21}$	0	0	$e^{\pi i p}$	$-e^{\pi i p}$	0	0	$-e^{\pi i p}$	$e^{-\pi i p}$
$(\Delta_5)_{12}$	0	0	$e^{\pi i p}$	$-e^{\pi i p}$	0	0	$e^{\pi i p}$	$-e^{-\pi i p}$
$(\Delta_5)_{22}$	1	-1	0	0	$e^{2\pi i p}$	-1	0	0
$\chi(\Delta_5)$	2	-2	0	0	0	0	0	0

Table A3-36 (Continued)

Matrix elements and characters at points Δ for group T_h^7:

	R_1	R_2	R_3'	R_4'
Δ_1	1	1	$e^{2\pi i p}$	1
Δ_2	1	1	$-e^{2\pi i p}$	-1
Δ_3	1	-1	$-e^{2\pi i p}$	1
Δ_4	1	-1	$e^{2\pi i p}$	-1

Matrix elements and characters at points Λ, propagation along the 111 direction, for group O_h^{10}. For Λ, $p_1 = p_2 = p_3 = p$. The abbreviation ω stands for $e^{2\pi i/3}$.

	R_1	R_5	R_9	R_{19}	R_{21}	R_{23}
Λ_1	1	1	1	$e^{3\pi i p}$	$e^{3\pi i p}$	$e^{3\pi i p}$
Λ_2	1	1	1	$-e^{3\pi i p}$	$-e^{3\pi i p}$	$-e^{3\pi i p}$
$(\Lambda_3)_{11}$	1	ω	ω^2	0	0	0
$(\Lambda_3)_{21}$	0	0	0	$e^{3\pi i p}$	$\omega e^{3\pi i p}$	$\omega^2 e^{3\pi i p}$
$(\Lambda_3)_{12}$	0	0	0	$e^{3\pi i p}$	$\omega^2 e^{3\pi i p}$	$\omega e^{3\pi i p}$
$(\Lambda_3)_{22}$	1	ω^2	ω	0	0	0
$\chi(\Lambda_3)$	3	-1	-1	0	0	0

Matrix elements at points Λ, for group T_h^7.

	R_1	R_5	R_9
Λ_1	1	1	1
Λ_2	1	ω	ω^2
Λ_3	1	ω^2	ω

Matrix elements and characters at points Σ, propagation along the 111 direction, for group O_h^{10}. For Σ, $p_1 = p_2 = 0$, $p_3 = p$.

	R_1	R_{23}	R_4'	R_{24}'
Σ_1	1	$e^{\pi i p}$	$e^{\pi i p}$	1
Σ_2	1	$-e^{\pi i p}$	$-e^{\pi i p}$	1
Σ_3	1	$e^{\pi i p}$	$-e^{\pi i p}$	-1
Σ_4	1	$-e^{\pi i p}$	$e^{\pi i p}$	-1

Matrix elements at Σ, group T_h^7.

	R_1	R_4'
Σ_1	1	$e^{\pi i p}$
Σ_2	1	$-e^{\pi i p}$

Table A3-37

Matrix elements and characters at point H, boundary of zone along direction Δ, where $-p_1 = p_2 = p_3 = \tfrac{1}{2}$. For group O_h^9, the representations, elements, and characters are identical with those at Γ, the symbols being the same except that H replaces Γ. For group T_h^7, the matrix elements for the unprimed operators $R_1 \cdots R_{12}$ are given below. For the primed operators, the matrix elements are the same as for the unprimed for the representations given by the $+$ sign, the negatives of those for the representations given by the $-$ sign. The abbreviation ω stands for $e^{2\pi i/3}$.

	R_1	R_2	R_3	R_4	R_5	R_6	R_7	R_8	R_9	R_{10}	R_{11}	R_{12}
H_1^+, H_1^-	1	-1	-1	-1	1	-1	-1	-1	1	-1	-1	-1
H_2^+, H_2^-	1	-1	-1	-1	ω	$-\omega$	$-\omega$	$-\omega$	ω^2	$-\omega^2$	$-\omega^2$	$-\omega^2$
H_3^+, H_3^-	1	-1	-1	-1	ω^2	$-\omega^2$	$-\omega^2$	$-\omega^2$	ω	$-\omega$	$-\omega$	$-\omega$
$(H_4^+, H_4^-)_{11}$	1	-1	1	1	0	0	0	0	0	0	0	0
$(H_4^+, H_4^-)_{21}$	0	0	0	0	1	1	1	-1	0	0	0	0
$(H_4^+, H_4^-)_{31}$	0	0	0	0	0	0	0	0	1	1	-1	1
$(H_4^+, H_4^-)_{12}$	0	0	0	0	0	0	0	0	1	1	1	-1
$(H_4^+, H_4^-)_{22}$	1	1	-1	1	0	0	0	0	0	0	0	0
$(H_4^+, H_4^-)_{32}$	0	0	0	0	1	-1	1	1	0	0	0	0
$(H_4^+, H_4^-)_{13}$	0	0	0	0	1	1	-1	1	0	0	0	0
$(H_4^+, H_4^-)_{23}$	0	0	0	0	0	0	0	0	1	-1	1	1
$(H_4^+, H_4^-)_{33}$	1	1	1	-1	0	0	0	0	0	0	0	0
$\chi(H_4^+, H_4^-)$	3	1	1	1	0	0	0	0	0	0	0	0

Table A3-37 (Continued)

Matrix elements and characters at H, O_h^{10}.

	R_1	R_2	R_3	R_4	R_5	R_6	R_7	R_8	R_9	R_{10}	R_{11}	R_{12}	R_{13}	R_{14}	R_{15}	R_{16}	R_{17}	R_{18}	R_{19}	R_{20}	R_{21}	R_{22}	R_{23}	R_{24}
$(H_1)_{11}$	1	−1	−1	1	1	−1	−1	1	1	−1	−1	1	0	0	0	0	0	0	0	0	0	0	0	0
$(H_1)_{21}$	0	0	0	0	0	0	0	0	0	0	0	0	1	1	1	1	1	1	−1	−1	−1	−1	−1	−1
$(H_1)_{12}$	0	0	0	0	0	0	0	0	0	0	0	0	−1	−1	−1	−1	−1	−1	1	1	1	1	1	1
$(H_1)_{22}$	1	−1	−1	−1	1	−1	−1	−1	1	−1	−1	−1	0	0	0	0	0	0	0	0	0	0	0	0
$\chi(H_1)$	2	−2	−2	−2	2	−2	−2	−2	2	−2	−2	−2	0	0	0	0	0	0	0	0	0	0	0	0
$(H_2)_{11}$	1	−1	−1	1	3	−3	−3	−3	3^2	−3^2	−3^2	3^2	0	0	0	0	0	0	0	0	0	0	0	0
$(H_2)_{21}$	0	0	0	0	0	0	0	0	0	0	0	0	1	1	3	3	3	3	−1	−1	−3^2	3^2	−3^2	3^2
$(H_2)_{12}$	0	0	0	0	0	0	0	0	0	0	0	0	−1	−1	−3^2	−3^2	−3^2	−3^2	1	1	3^2	−3^2	3^2	−3^2
$(H_2)_{22}$	1	−1	−1	−1	3^2	−3^2	−3^2	−3^2	3	−3	−3	−3	0	0	0	0	0	0	0	0	0	0	0	0
$\chi(H_2)$	2	−2	−2	−2	−1	1	1	1	−1	1	1	1	0	0	0	0	0	0	0	0	0	0	0	0
$(H_3)_{11}$	1	−1	−1	1	3^2	−3^2	−3^2	3^2	3	−3	−3	3	0	0	0	0	0	0	0	0	0	0	0	0
$(H_3)_{21}$	0	0	0	0	0	0	0	0	0	0	0	0	1	1	3^2	3^2	3^2	3^2	−1	−1	−3	3	−3	3
$(H_3)_{12}$	0	0	0	0	0	0	0	0	0	0	0	0	−1	−1	−3	−3	−3	−3	1	1	3^2	−3^2	3^2	−3^2
$(H_3)_{22}$	1	−1	−1	−1	3	−3	−3	−3	3^2	−3^2	−3^2	−3^2	0	0	0	0	0	0	0	0	0	0	0	0
$\chi(H_3)$	2	−2	−2	−2	−1	1	1	1	−1	1	1	1	0	0	0	0	0	0	0	0	0	0	0	0

Table A3-37 (Continued)

The abbreviation α stands for $i\sqrt{3}$.

	R'_1	R'_2	R'_3	R'_4	R'_5	R'_6	R'_7	R'_8	R'_9	R'_{10}	R'_{11}	R'_{12}	R'_{13}	R'_{14}	R'_{15}	R'_{16}	R'_{17}	R'_{18}	R'_{19}	R'_{20}	R'_{21}	R'_{22}	R'_{23}	R'_{24}
$(H_1)_{11}$	-1	1	1	1	-1	1	1	1	-1	1	1	1	0	0	0	0	0	0	0	0	0	0	0	0
$(H_1)_{21}$	0	0	0	0	0	0	0	0	0	0	0	0	1	1	1	1	1	1	-1	1	-1	1	-1	1
$(H_1)_{12}$	0	0	0	0	0	0	0	0	0	0	0	0	1	1	1	1	1	1	-1	1	-1	1	-1	1
$(H_1)_{22}$	1	-1	-1	-1	1	-1	-1	-1	1	-1	-1	-1	0	0	0	0	0	0	0	0	0	0	0	0
$\chi(H_1)$	0	0	0	0	0	0	0	0	0	0	0	0	0	0	0	0	0	0	0	0	0	0	0	0
$(H_2)_{11}$	-1	1	1	1	-3^2	3^2	3^2	3^2	-3^2	3^2	3^2	3^2	0	0	0	0	0	0	0	0	0	0	0	0
$(H_2)_{21}$	0	0	0	0	0	0	0	0	0	0	0	0	1	1	3^2	3^2	3^2	3^2	-1	1	-3^2	3^2	-3^2	3^2
$(H_2)_{12}$	0	0	0	0	3	3	3	3	3	3	3	3	1	1	3	3	3	3	-1	1	-3	3	-3	3
$(H_2)_{22}$	1	-1	-1	-1	3	-3	-3	-3	3	-3	-3	-3	0	0	0	0	0	0	0	0	0	0	0	0
$\chi(H_2)$	0	0	0	0	$-\alpha$	α	α	α	α	$-\alpha$	$-\alpha$	$-\alpha$	0	0	0	0	0	0	0	0	0	0	0	0
$(H_3)_{11}$	-1	1	1	1	-3^2	3^2	3^2	3^2	-3^2	3^2	3^2	3^2	0	0	0	0	0	0	0	0	0	0	0	0
$(H_3)_{21}$	0	0	0	0	0	0	0	0	0	0	0	0	1	1	3^2	3^2	3^2	3^2	-1	1	-3^2	3^2	-3^2	3^2
$(H_3)_{12}$	0	0	0	0	3	3	3	3	-3	-3	-3	-3	1	1	3	3	3	3	-1	1	-3	3	-3	3
$(H_3)_{22}$	1	-1	-1	-1	3	-3	-3	-3	3^2	-3^2	-3^2	-3^2	0	0	0	0	0	0	0	0	0	0	0	0
$\chi(H_3)$	0	0	0	0	α	$-\alpha$	$-\alpha$	$-\alpha$	$-\alpha$	α	α	α	0	0	0	0	0	0	0	0	0	0	0	0

Table A3-37 (Continued)

	R_1	R_2	R_3	R_4	R_5	R_6	R_7	R_8	R_9	R_{10}	R_{11}	R_{12}	R_{13}	R_{14}	R_{15}	R_{16}	R_{17}	R_{18}	R_{19}	R_{20}	R_{21}	R_{22}	R_{23}	R_{24}
$(H_4)_{11}$	1	1	0	0	0	0	0	0	0	0	0	0	0	0	0	0	0	0	i	i	0	0	0	0
$(H_4)_{21}$	0	0	$-i$	1	0	0	0	0	0	0	0	0	i	$-i$	0	0	0	0	0	0	0	0	0	0
$(H_4)_{31}$	0	0	0	0	1	0	0	1	0	0	0	0	0	0	0	0	i	0	0	0	0	0	i	0
$(H_4)_{41}$	0	0	0	0	0	$-i$	$-i$	0	0	0	0	0	0	0	0	0	0	i	0	0	0	0	0	$-i$
$(H_4)_{51}$	0	0	0	0	0	0	0	0	1	0	1	0	0	0	0	i	0	0	0	0	$-i$	0	0	0
$(H_4)_{61}$	0	0	0	0	0	0	0	0	0	1	0	-1	0	0	$-i$	0	0	0	0	0	0	$-i$	0	0
$(H_4)_{12}$	0	0	$-i$	1	0	0	0	0	0	0	0	0	$-i$	i	0	0	0	0	0	0	0	0	0	0
$(H_4)_{22}$	1	1	0	0	0	0	0	0	0	0	0	0	0	0	0	0	0	$-i$	$-i$	0	0	0	0	i
$(H_4)_{32}$	0	0	0	0	0	$-i$	$-i$	0	0	0	0	0	0	0	0	0	0	$-i$	0	0	0	0	0	0
$(H_4)_{42}$	0	0	0	0	1	0	0	1	0	0	0	0	0	0	$-i$	0	$-i$	0	0	0	0	i	0	0
$(H_4)_{52}$	0	0	0	0	0	0	0	0	0	1	0	-1	0	0	0	$-i$	0	0	$-i$	0	i	0	i	0
$(H_4)_{62}$	0	0	0	0	0	0	0	0	1	0	1	0	0	0	0	0	0	0	0	$-i$	0	0	0	0
$(H_4)_{13}$	0	0	0	0	0	0	0	0	1	0	1	0	0	0	0	0	i	$-i$	0	0	0	i	0	0
$(H_4)_{23}$	0	0	0	0	0	0	0	0	0	1	0	1	0	0	0	0	0	0	0	0	0	$-i$	0	i
$(H_4)_{33}$	0	0	1	0	0	0	0	0	0	0	0	0	0	0	$-i$	0	0	0	0	0	$-i$	0	0	0
$(H_4)_{43}$	0	0	0	-1	0	1	0	0	0	0	0	0	0	0	0	i	0	0	0	0	0	0	0	0
$(H_4)_{53}$	0	0	0	0	0	0	0	0	0	0	0	0	0	$-i$	0	0	0	0	i	0	0	0	0	0
$(H_4)_{63}$	0	0	0	0	0	0	$-i$	1	0	0	0	0	0	0	0	0	0	0	0	i	0	0	0	0

Table A3-37 (Continued)

	R_1	R_2	R_3	R_4	R_5	R_6	R_7	R_8	R_9	R_{10}	R_{11}	R_{12}	R_{13}	R_{14}	R_{15}	R_{16}	R_{17}	R_{18}	R_{19}	R_{20}	R_{21}	R_{22}	R_{23}	R_{24}
$(H_4)_{14}$	0	0	0	0	0	0	0	0	0	-1	1	0	0	0	0	0	i	0	0	0	0	0	0	$-i$
$(H_4)_{24}$	0	0	0	0	0	0	0	0	1	0	0	1	0	0	0	0	0	$-i$	0	0	0	0	$-i$	0
$(H_4)_{34}$	1	1	1	-1	0	0	0	0	0	0	0	0	0	0	$-i$	i	0	0	0	0	$-i$	$-i$	0	0
$(H_4)_{44}$	1	0	1	0	0	0	0	1	0	0	0	0	0	$-i$	0	0	0	0	$-i$	i	0	0	0	0
$(H_4)_{54}$	0	0	0	0	1	0	-1	1	0	0	0	0	$-i$	0	0	0	0	0	0	0	0	0	0	0
$(H_4)_{64}$	0	0	0	0	1	1	0	0	0	0	0	0	0	0	0	0	0	0	0	0	0	0	0	0
$(H_4)_{15}$	0	0	0	0	1	0	1	0	0	0	0	0	$-i$	0	i	0	0	0	0	0	i	0	0	0
$(H_4)_{25}$	0	0	0	0	0	-1	0	-1	0	0	0	0	0	0	0	i	0	0	0	i	0	$-i$	0	0
$(H_4)_{35}$	0	0	0	0	0	0	0	0	1	1	0	0	0	i	0	0	0	0	i	0	0	0	0	0
$(H_4)_{45}$	0	0	0	1	0	0	0	0	0	0	-1	1	0	0	0	0	0	0	0	0	0	0	0	0
$(H_4)_{55}$	-1	0	0	0	0	0	0	0	0	0	0	0	0	0	0	0	i	$-i$	0	0	0	0	i	$-i$
$(H_4)_{65}$	0	-1	1	0	0	0	0	0	0	0	0	0	0	0	0	0	0	0	0	0	0	0	0	0
$(H_4)_{16}$	0	0	0	0	0	1	0	-1	0	0	0	0	0	0	0	$-i$	0	0	0	0	0	i	0	0
$(H_4)_{26}$	0	0	0	0	1	0	1	0	0	0	0	0	i	0	$-i$	0	0	0	0	$-i$	$-i$	0	0	0
$(H_4)_{36}$	0	0	0	0	0	0	0	0	0	0	-1	1	0	0	0	0	0	0	0	0	0	0	0	0
$(H_4)_{46}$	0	0	1	0	0	0	0	0	1	1	0	0	0	$-i$	0	0	0	0	i	0	0	0	0	0
$(H_4)_{56}$	0	-1	1	0	0	0	0	0	0	0	0	0	0	0	0	0	i	i	0	0	0	0	$-i$	$-i$
$(H_4)_{66}$	1	0	0	1	0	0	0	0	0	0	0	0	0	0	0	0	0	0	0	0	0	0	0	0
$\chi(H_4)$	6	2	2	2	0	0	0	0	0	0	0	0	0	0	0	0	0	0	0	0	0	0	0	0

Table A3-37 (Continued)

	R'_1	R'_2	R'_3	R'_4	R'_5	R'_6	R'_7	R'_8	R'_9	R'_{10}	R'_{11}	R'_{12}	R'_{13}	R'_{14}	R'_{15}	R'_{16}	R'_{17}	R'_{18}	R'_{19}	R'_{20}	R'_{21}	R'_{22}	R'_{23}	R'_{24}
$(H_4)_{11}$	0	0	-1	1	0	0	0	0	0	0	0	0	i	$-i$	0	0	0	0	0	0	0	0	0	0
$(H_4)_{21}$	1	1	-1	0	0	0	0	0	0	0	0	0	0	$-i$	0	0	0	0	$-i$	$-i$	0	0	0	0
$(H_4)_{31}$	0	0	0	0	1	-1	1	0	0	0	0	0	0	0	0	0	$-i$	$-i$	0	0	0	0	0	$-i$
$(H_4)_{41}$	0	0	0	0	1	0	0	1	0	0	0	0	0	0	$-i$	0	0	0	0	0	0	$-i$	$-i$	0
$(H_4)_{51}$	0	0	0	0	0	0	0	0	0	0	0	-1	0	0	0	0	0	0	0	0	0	0	0	0
$(H_4)_{61}$	0	0	0	0	0	0	0	0	1	0	1	0	0	0	0	$-i$	0	0	0	0	$-i$	0	0	0
$(H_4)_{12}$	1	1	0	0	0	0	0	0	0	0	0	0	$-i$	0	0	0	0	0	$-i$	$-i$	0	0	0	0
$(H_4)_{22}$	0	0	-1	1	0	0	0	0	0	0	0	0	i	$-i$	0	0	0	0	0	0	0	$-i$	0	0
$(H_4)_{32}$	0	0	0	0	1	0	0	1	0	0	0	0	0	0	0	0	0	$-i$	0	0	0	$-i$	0	$-i$
$(H_4)_{42}$	0	0	0	0	0	-1	1	0	0	0	0	0	0	0	0	$-i$	$-i$	0	0	0	0	0	0	0
$(H_4)_{52}$	0	0	0	0	0	0	0	0	1	1	1	0	0	0	0	0	0	0	0	0	0	0	0	0
$(H_4)_{62}$	0	0	0	0	0	0	0	0	0	0	0	-1	0	0	$-i$	0	0	0	0	0	0	0	0	0
$(H_4)_{13}$	0	0	0	0	0	0	0	0	0	-1	1	0	0	0	0	0	$-i$	0	0	0	0	0	0	$-i$
$(H_4)_{23}$	0	0	0	0	0	0	0	0	1	0	0	1	0	$-i$	0	$-i$	0	$-i$	0	0	0	0	$-i$	0
$(H_4)_{33}$	1	1	0	-1	0	0	0	0	0	0	0	0	0	0	$-i$	0	0	0	0	0	0	0	0	0
$(H_4)_{43}$	0	0	0	0	0	0	0	0	0	0	0	0	0	$-i$	0	0	0	0	0	$-i$	$-i$	0	0	0
$(H_4)_{53}$	0	0	0	0	0	0	-1	1	0	0	0	0	0	0	0	0	0	0	$-i$	0	0	0	0	0
$(H_4)_{63}$	0	0	0	0	1	1	0	0	0	0	0	0	$-i$	0	0	0	0	0	0	0	0	0	0	0

399

Table A3-37 (Continued)

	R'_1	R'_2	R'_3	R'_4	R'_5	R'_6	R'_7	R'_8	R'_9	R'_{10}	R'_{11}	R'_{12}	R'_{13}	R'_{14}	R'_{15}	R'_{16}	R'_{17}	R'_{18}	R'_{19}	R'_{20}	R'_{21}	R'_{22}	R'_{23}	R'_{24}
$(H_4)_{14}$	0	0	0	0	0	0	0	0	1	0	0	1	0	0	0	0	0	$-i$	0	0	0	0	$-i$	0
$(H_4)_{24}$	0	0	0	0	0	0	0	0	0	1	1	0	0	0	0	0	i	0	0	0	0	0	0	$-i$
$(H_4)_{34}$	1	0	1	0	0	0	0	0	0	0	0	0	0	0	0	0	0	0	0	0	$-i$	$-i$	0	0
$(H_4)_{44}$	0	1	0	-1	0	0	0	0	0	0	0	0	0	0	$-i$	$-i$	0	0	0	0	0	0	0	0
$(H_4)_{54}$	0	0	0	0	1	1	0	0	0	0	0	0	$-i$	$-i$	0	0	0	0	0	0	0	0	0	0
$(H_4)_{64}$	0	0	0	0	0	0	-1	1	0	0	0	0	0	0	0	0	0	0	$-i$	$-i$	0	0	0	0
$(H_4)_{15}$	0	0	0	0	0	1	0	-1	0	0	0	0	0	0	0	$-i$	0	0	0	0	0	$-i$	0	0
$(H_4)_{25}$	0	0	0	0	1	0	-1	0	0	0	0	0	0	0	$-i$	0	0	0	0	0	$-i$	0	0	0
$(H_4)_{35}$	0	0	0	0	0	0	0	0	0	1	-1	0	$-i$	0	0	0	0	0	0	$-i$	0	0	0	0
$(H_4)_{45}$	0	0	0	0	0	0	0	0	1	0	0	-1	0	$-i$	0	0	0	0	$-i$	0	0	0	0	0
$(H_4)_{55}$	0	-1	0	1	0	0	0	0	0	0	0	0	0	0	0	0	0	$-i$	0	0	0	0	0	$-i$
$(H_4)_{65}$	1	0	-1	0	0	0	0	0	0	0	0	0	0	0	0	0	$-i$	0	0	0	0	0	$-i$	0
$(H_4)_{16}$	0	0	0	0	1	0	1	0	0	0	0	0	0	0	$-i$	0	0	0	0	0	$-i$	0	0	0
$(H_4)_{26}$	0	0	0	0	0	1	0	-1	0	0	0	0	0	0	0	$-i$	0	0	0	0	0	$-i$	0	0
$(H_4)_{36}$	0	0	0	0	0	0	0	0	1	1	0	0	0	$-i$	0	0	0	0	$-i$	0	0	0	0	0
$(H_4)_{46}$	0	0	0	1	0	0	0	0	0	0	0	1	$-i$	0	0	0	0	0	0	$-i$	0	0	0	0
$(H_4)_{56}$	1	0	0	0	0	0	0	0	0	0	-1	0	0	0	0	0	0	0	0	0	0	0	$-i$	0
$(H_4)_{66}$	0	-1	1	0	0	0	0	0	0	0	0	0	0	0	0	0	0	$-i$	0	0	0	0	0	$-i$
$\chi(H_4)$	0	0	0	0	0	0	0	0	0	0	0	0	0	0	0	0	0	0	0	0	0	0	0	0

Table A3-38

Matrix elements and characters at point P, boundary of zone along direction Λ, where $p_1 = p_2 = p_3 = \frac{1}{4}$. The group of the wave vector is T_d, the operations $R_1 \cdots R_{24}$, for O_h^9. Hence the matrix elements and characters are as given in Table A3-20, omitting the primed operations. The five irreducible representations, denoted there as $\Gamma_1, \Gamma_2, \Gamma_{12}, \Gamma'_{25}, \Gamma'_{15}$, are denoted here as P_1, P_2, P_3, P_4, P_5, respectively, according to the conventions of Bouckaert, Smoluchowski, and Wigner. For the group T_h^7, with only the operations $R_1 \cdots R_{12}$, the matrix elements are given below; a and b stand for $1/\sqrt{3}$ and $\sqrt{2/3}$.

	R_1	R_2	R_3	R_4	R_5	R_6	R_7	R_8	R_9	R_{10}	R_{11}	R_{12}
$(P_1)_{11}$	1	a	a	a	1	a	a	a	1	a	a	a
$(P_1)_{21}$	0	b	$b\omega^2$	$b\omega$	0	$b\omega$	b	$b\omega^2$	0	$b\omega^2$	$b\omega$	b
$(P_1)_{12}$	0	b	$b\omega$	$b\omega^2$	0	$b\omega$	$b\omega^2$	b	0	$b\omega^2$	b	$b\omega$
$(P_1)_{22}$	1	$-a$	$-a$	$-a$	ω^2	$-a\omega^2$	$-a\omega^2$	$-a\omega^2$	ω	$-a\omega$	$-a\omega$	$-a\omega$
$\chi(P_1)$	2	0	0	0	$-\omega$	$-i\omega$	$-i\omega$	$-i\omega$	$-\omega^2$	$i\omega^2$	$i\omega^2$	$i\omega^2$
$(P_2)_{11}$	1	$-a$	$-a$	$-a$	1	$-a$	$-a$	$-a$	1	$-a$	$-a$	$-a$
$(P_2)_{21}$	0	b	$b\omega$	$b\omega^2$	0	$b\omega^2$	b	$b\omega$	0	$b\omega$	$b\omega^2$	b
$(P_2)_{12}$	0	b	$b\omega^2$	$b\omega$	0	$b\omega^2$	$b\omega$	b	0	$b\omega$	b	$b\omega^2$
$(P_2)_{22}$	1	a	a	a	ω	$a\omega$	$a\omega$	$a\omega$	ω^2	$a\omega^2$	$a\omega^2$	$a\omega^2$
$\chi(P_2)$	2	0	0	0	$-\omega^2$	$-i\omega^2$	$-i\omega^2$	$-i\omega^2$	$-\omega$	$i\omega$	$i\omega$	$i\omega$
$(P_3)_{11}$	1	a	a	a	ω^2	$a\omega^2$	$a\omega^2$	$a\omega^2$	ω	$a\omega$	$a\omega$	$a\omega$
$(P_3)_{21}$	0	b	$b\omega^2$	$b\omega$	0	b	$b\omega^2$	$b\omega$	0	b	$b\omega^2$	$b\omega$
$(P_3)_{12}$	0	b	$b\omega$	$b\omega^2$	0	b	$b\omega$	$b\omega^2$	0	b	$b\omega$	$b\omega^2$
$(P_3)_{22}$	1	$-a$	$-a$	$-a$	ω	$-a\omega$	$-a\omega$	$-a\omega$	ω^2	$-a\omega^2$	$-a\omega^2$	$-a\omega^2$
$\chi(P_3)$	2	0	0	0	-1	$-i$	$-i$	$-i$	-1	i	i	i

Table A3-38 (Continued)

Point P for O_h^{10}. Abbreviations: $p = (1+i)/2$, $q = (1-i)/2$, $r = (1+i)/\sqrt{2}$, $s = (1-i)/\sqrt{2}$.

	R_1	R_2	R_3	R_4	R_5	R_6	R_7	R_8	R_9	R_{10}	R_{11}	R_{12}	R_{13}	R_{14}	R_{15}	R_{16}	R_{17}	R_{18}	R_{19}	R_{20}	R_{21}	R_{22}	R_{23}	R_{24}
$(P_1)_{11}$	1	1	0	0	$-p$	q	$-p$	$-p$	$-q$	p	$-q$	$-q$	$-i$	-1	q	$-q$	q	$-q$	0	0	$-q$	p	q	p
$(P_1)_{21}$	0	0	1	$-i$	$-p$	$-q$	q	$-p$	p	$-q$	$-q$	$-p$	0	0	$-p$	$-p$	$-q$	$-q$	1	1	$-p$	q	$-q$	p
$(P_1)_{12}$	0	0	1	i	q	$-p$	p	$-q$	$-q$	$-p$	q	$-q$	0	0	$-p$	$-p$	q	q	$-i$	i	p	$-p$	$-q$	$-p$
$(P_1)_{22}$	1	-1	0	0	$-q$	$-p$	q	q	$-p$	$-q$	p	p	1	i	q	$-q$	q	$-q$	0	0	q	$-p$	$-q$	$-p$
$\chi(P_1)$	2	0	0	0	-1	$-i$	$-i$	$-i$	-1	i	i	i	$2q$	$-2q$	$2q$	$-2q$	$2q$	$-2q$	0	0	0	0	0	0
$(P_2)_{11}$	1	1	0	0	$-p$	q	$-p$	$-p$	$-q$	p	$-q$	$-q$	i	1	$-q$	q	$-q$	q	0	0	q	$-p$	$-q$	$-p$
$(P_2)_{21}$	0	0	1	$-i$	$-p$	$-q$	q	$-p$	p	$-q$	$-q$	$-p$	0	0	p	p	q	q	-1	-1	$-p$	q	$-q$	p
$(P_2)_{12}$	0	0	1	i	q	$-p$	p	$-q$	$-q$	$-p$	q	$-q$	0	0	p	p	q	$-q$	i	$-i$	p	$-p$	$-q$	$-p$
$(P_2)_{22}$	1	-1	0	0	$-q$	$-p$	q	q	$-p$	$-q$	p	p	-1	$-i$	$-q$	q	$-q$	q	0	0	$-q$	p	q	$-p$
$\chi(P_2)$	2	0	0	0	-1	$-i$	$-i$	$-i$	-1	i	i	i	$-2q$	$2q$	$-2q$	$2q$	$-2q$	$2q$	0	0	0	0	0	0

Table A3-38 (Continued)

	R_1	R_2	R_3	R_4	R_5	R_6	R_7	R_8	R_9	R_{10}	R_{11}	R_{12}	R_{13}	R_{14}	R_{15}	R_{16}	R_{17}	R_{18}	R_{19}	R_{20}	R_{21}	R_{22}	R_{23}	R_{24}
$(P_3)_{11}$	1	0	0	0	1	0	0	0	1	0	0	0	0	0	0	0	0	0	$-s$	0	$-s$	0	$-s$	0
$(P_3)_{21}$	0	1	0	0	0	0	1	0	0	0	0	1	0	0	s	0	0	s	0	s	0	0	0	0
$(P_3)_{31}$	0	0	1	0	0	0	0	1	0	1	0	0	0	s	0	s	s	0	0	0	0	s	0	0
$(P_3)_{41}$	0	0	0	1	0	1	0	0	0	0	1	0	s	0	0	0	0	0	0	0	0	0	0	s
$(P_3)_{12}$	0	1	0	0	0	0	0	1	0	0	1	0	0	0	0	$-s$	0	0	0	0	0	0	0	0
$(P_3)_{22}$	1	0	0	0	1	$-i$	0	0	0	$-i$	0	0	0	0	0	0	$-s$	0	$-s$	0	0	r	0	$-r$
$(P_3)_{32}$	0	0	i	0	1	0	0	0	0	0	0	i	r	0	$-r$	0	0	0	0	0	0	0	s	0
$(P_3)_{42}$	0	0	0	0	1	0	$-i$	0	1	0	0	0	0	$-r$	0	0	0	r	0	0	s	0	0	0
$(P_3)_{13}$	0	0	1	0	0	1	0	0	0	0	0	1	$-s$	0	0	0	0	$-s$	0	0	0	$-s$	0	0
$(P_3)_{23}$	0	0	0	i	0	0	0	0	1	0	0	0	0	0	0	0	0	0	0	s	0	0	s	0
$(P_3)_{33}$	1	0	0	0	0	0	0	0	0	0	0	0	0	r	r	0	0	0	s	$-r$	s	0	0	r
$(P_3)_{43}$	0	$-i$	0	0	1	0	0	i	0	i	0	$-i$	0	0	r	$-r$	$-r$	0	0	0	0	0	0	0
$(P_3)_{14}$	0	0	0	1	0	0	1	0	0	1	0	0	0	$-s$	$-s$	0	0	$-s$	0	0	0	0	0	$-s$
$(P_3)_{24}$	0	0	$-i$	0	1	$-i$	0	0	0	0	0	0	$-r$	0	0	r	r	0	0	0	s	0	0	0
$(P_3)_{34}$	0	i	0	0	0	0	0	$-i$	1	0	0	0	0	0	0	0	0	$-r$	s	r	0	$-r$	0	0
$(P_3)_{44}$	1	0	0	0	1	$-i$	$-i$	i	0	$-i$	$-i$	$-i$	0	0	0	0	0	0	0	r	0	0	s	0
$\chi(P_3)$	4	0	0	0	1	0	0	0	1	0	0	0	0	0	0	0	0	0	0	0	0	0	0	0

Table A3-39

Matrix elements and characters at point N, boundary of zone along direction Σ, where $p_1 = p_2 = 0$, $p_3 = \frac{1}{2}$.

		R_1	R_4	R_{23}	R_{24}	R_1'	R_4'	R_{23}'	R_{24}'
O_h^9	N_1	1	1	1	1	1	1	1	1
	N_2	1	−1	−1	1	1	−1	−1	1
	N_3	1	−1	1	−1	1	−1	1	−1
	N_4	1	1	−1	−1	1	1	−1	−1
	N_1'	1	−1	1	−1	−1	1	−1	1
	N_2'	1	1	−1	−1	−1	−1	1	1
	N_3'	1	1	1	1	−1	−1	−1	−1
	N_4'	1	−1	−1	1	−1	1	1	−1

		R_1	R_4	R_{23}	R_{24}	R_1'	R_4'	R_{23}'	R_{24}'
O_h^{10}	$(N_1)_{11}$	1	0	0	1	1	0	0	1
	$(N_1)_{21}$	0	1	−1	0	0	−1	1	0
	$(N_1)_{12}$	0	1	1	0	0	1	1	0
	$(N_1)_{22}$	1	0	0	−1	−1	0	0	1
	$\chi(N_1)$	2	0	0	0	0	0	0	2
	$(N_2)_{11}$	1	0	0	1	−1	0	0	−1
	$(N_2)_{21}$	0	1	−1	0	0	1	−1	0
	$(N_2)_{12}$	0	1	1	0	0	−1	−1	0
	$(N_2)_{22}$	1	0	0	−1	1	0	0	−1
	$\chi(N_2)$	2	0	0	0	0	0	0	−2

		R_1	R_4	R_1'	R_4'
T_h^7	$(N_1)_{11}$	1	0	1	0
	$(N_1)_{21}$	0	1	0	−1
	$(N_1)_{12}$	0	1	0	1
	$(N_1)_{22}$	1	0	−1	0
	$\chi(N_1)$	2	0	0	0

Table A3-40

Matrix elements and characters along directions F, joining H and P. Here $p_2 = p_3 = p$, $p_1 = 1 - 3p$. At H, $p = \frac{1}{2}$; at P, $p = \frac{1}{4}$. For O_h^9, the matrix elements are as for O_h^{10}, given below, except that the factors involving p are to be set equal to unity. ω stands for $e^{2\pi i/3}$.

		R_1	R_6	R_{10}	R_{19}	R_{22}	R_{24}
	F_1	1	$e^{2\pi i p}$	$e^{-2\pi i p}$	$e^{-\pi i p}$	$e^{\pi i p}$	$e^{-3\pi i p}$
	F_2	1	$e^{2\pi i p}$	$e^{-2\pi i p}$	$-e^{-\pi i p}$	$-e^{\pi i p}$	$-e^{-3\pi i p}$
O_h^{10}	$(F_3)_{11}$	1	$\omega e^{2\pi i p}$	$\omega^2 e^{-2\pi i p}$	0	0	0
	$(F_3)_{21}$	0	0	0	$e^{-\pi i p}$	$\omega e^{\pi i p}$	$\omega^2 e^{-3\pi i p}$
	$(F_3)_{12}$	0	0	0	$e^{-\pi i p}$	$\omega^2 e^{\pi i p}$	$\omega e^{-3\pi i p}$
	$(F_3)_{22}$	1	$\omega^2 e^{2\pi i p}$	$\omega e^{-2\pi i p}$	0	0	0
	$\chi(F_3)$	2	$-e^{2\pi i p}$	$-e^{-2\pi i p}$	0	0	0

		R_1	R_6	R_{10}
	F_1	1	$e^{2\pi i p}$	$e^{-2\pi i p}$
T_h^7	F_2	1	$\omega e^{2\pi i p}$	$\omega^2 e^{-2\pi i p}$
	F_3	1	$\omega^2 e^{2\pi i p}$	$\omega e^{-2\pi i p}$

Table A3-41

Matrix elements and characters along directions D, joining P and N. Here $p_1 = p_2 = p$, $p_3 = \frac{1}{2} - p$. At P, $p = \frac{1}{4}$; at N, $p = 0$.

		R_1	R_4	R_{23}	R_{24}
	D_1	1	1	1	1
O_h^9	D_2	1	1	-1	-1
	D_3	1	-1	-1	1
	D_4	1	-1	1	-1

		R_1	R_4	R_{23}	R_{24}
	$(D_1)_{11}$	1	1	0	0
O_h^{10}	$(D_1)_{21}$	0	0	1	1
	$(D_1)_{12}$	0	0	$-e^{2\pi i p}$	$e^{2\pi i p}$
	$(D_1)_{22}$	1	-1	0	0
	$\chi(D_1)$	2	0	0	0

Table A3-41 (Continued)

T_h^7		R_1	R_4
	D_1	1	1
	D_2	1	-1

Table A3-42

Matrix elements and characters along directions G, joining H and N. Here $p_1 = -p_2 = -p$, $p_3 = \frac{1}{2}$. At H, $p = \frac{1}{2}$. At N, $p = 0$.

O_h^9		R_1	R_{24}	R_4'	R_{23}'
	G_1	1	1	1	1
	G_2	1	-1	-1	1
	G_3	1	1	-1	-1
	G_4	1	-1	1	-1

O_h^{10}		R_1	R_{24}	R_4'	R_{23}'
	$(G_1)_{11}$	1	0	0	1
	$(G_1)_{21}$	0	$e^{\pi i p}$	$e^{\pi i p}$	0
	$(G_1)_{12}$	0	$e^{\pi i p}$	$-e^{\pi i p}$	0
	$(G_1)_{22}$	1	0	0	-1
	$\chi(G_1)$	2	0	0	0

T_h^7		R_1	R_4'
	G_1	1	$ie^{\pi i p}$
	G_2	1	$-ie^{\pi i p}$

Table A3-43
Compatibility relations, O_h^9, O_h^{10}, and T_h^7

	Γ_1	Γ_2	Γ_1'	Γ_2'	Γ_{12}	Γ_{12}'	Γ_{15}	Γ_{15}'	Γ_{25}	Γ_{25}'
O_h^9 and O_h^{10}	Δ_1	Δ_2	Δ_1'	Δ_2'	$\Delta_1\Delta_2$	$\Delta_1'\Delta_2'$	$\Delta_1\Delta_5$	$\Delta_1'\Delta_5$	$\Delta_2\Delta_5$	$\Delta_2'\Delta_5$
	Λ_1	Λ_2	Λ_2	Λ_1	Λ_3	Λ_3	$\Lambda_1\Lambda_3$	$\Lambda_2\Lambda_3$	$\Lambda_2\Lambda_3$	$\Lambda_1\Lambda_3$
	Σ_1	Σ_4	Σ_2	Σ_3	$\Sigma_1\Sigma_4$	$\Sigma_2\Sigma_3$	$\Sigma_1\Sigma_3\Sigma_4$	$\Sigma_2\Sigma_3\Sigma_4$	$\Sigma_1\Sigma_2\Sigma_4$	$\Sigma_1\Sigma_2\Sigma_3$

Table A3-43 (Continued)

	Γ_1^+	Γ_1^-	Γ_2^+	Γ_2^-	Γ_3^+	Γ_3^-	Γ_4^+	Γ_4^-
T_h^7	Δ_1	Δ_2	Δ_1	Δ_2	Δ_1	Δ_2	$\Delta_2\Delta_3\Delta_4$	$\Delta_1\Delta_3\Delta_4$
	Λ_1	Λ_1	Λ_2	Λ_2	Λ_3	Λ_3	$\Lambda_1\Lambda_2\Lambda_3$	$\Lambda_1\Lambda_2\Lambda_3$
	Σ_1	Σ_2	Σ_1	Σ_2	Σ_1	Σ_2	$\Sigma_1\Sigma_2$	$\Sigma_1\Sigma_2$

	H_1	H_2	H_1'	H_2'	H_{12}	H_{12}'	H_{15}	H_{15}'	H_{25}	H_{25}'
O_h^9	Δ_1	Δ_2	Δ_1'	Δ_2'	$\Delta_1\Delta_2$	$\Delta_1'\Delta_2'$	$\Delta_1\Delta_5$	$\Delta_1'\Delta_5$	$\Delta_2\Delta_5$	$\Delta_2'\Delta_5$
	F_1	F_2	F_2	F_1	F_3	F_3	F_1F_3	F_2F_3	F_2F_3	F_1F_3
	G_1	G_4	G_2	G_3	G_1G_4	G_2G_3	$G_1G_3G_4$	$G_2G_3G_4$	$G_1G_2G_4$	$G_1G_2G_3$

	H_1	H_2	H_3	H_4
O_h^{10}	Δ_5	Δ_5	Δ_5	$\Delta_1\Delta_1'\Delta_2\Delta_2'\Delta_5$
	F_1F_2	F_3	F_3	$F_1F_2F_3$
	G_1	G_1	G_1	G_1

	H_1^+	H_1^-	H_2^+	H_2^-	H_3^+	H_3^-	H_4^+	H_4^-
T_h^7	Δ_4	Δ_3	Δ_4	Δ_3	Δ_4	Δ_3	$\Delta_1\Delta_2\Delta_3$	$\Delta_1\Delta_2\Delta_4$
	F_1	F_1	F_2	F_2	F_3	F_3	$F_1F_2F_3$	$F_1F_2F_3$
	G_1	G_2	G_1	G_2	G_1	G_2	G_1G_2	G_1G_2

	P_1	P_2	P_3	P_4	P_5		P_1	P_2	P_3
O_h^9	Λ_1	Λ_2	Λ_3	$\Lambda_1\Lambda_3$	$\Lambda_2\Lambda_3$	O_h^{10}	Λ_3	Λ_3	$\Lambda_1\Lambda_2\Lambda_3$
	F_1	F_2	F_3	F_1F_3	F_2F_3		F_3	F_3	$F_1F_2F_3$
	D_1	D_2	D_1D_2	$D_1D_3D_4$	$D_2D_3D_4$		D_1	D_1	D_1

	P_1	P_2	P_3		N_1	N_2	N_3	N_4	N_1'	N_2'	N_3'	N_4'
T_h^7	$\Lambda_1\Lambda_2$	$\Lambda_1\Lambda_2$	$\Lambda_2\Lambda_3$	O_h^9	Σ_1	Σ_2	Σ_3	Σ_4	Σ_1	Σ_2	Σ_3	Σ_4
	F_1F_3	F_1F_2	F_2F_3		D_1	D_3	D_4	D_2	D_4	D_2	D_1	D_3
	D_1D_2	D_1D_2	D_1D_2		G_1	G_3	G_2	G_4	G_4	G_2	G_3	G_1

	N_1	N_2		N_1
O_h^{10}	$\Sigma_1\Sigma_2$	$\Sigma_3\Sigma_4$	T_h^7	$\Sigma_1\Sigma_2$
	D_1	D_1		D_1D_2
	G_1	G_1		G_1G_2

A3-4. The Space Groups O_h^1 ($Pm3m$) and T_h^6 ($Pa3$). The space groups O_h^1 and T_h^6 have the simple cubic Bravais lattice; O_h^1 is the one leading to the simple cubic structure, in which there is one atom per unit cell, at the origin, a structure which does not occur in actual crystals. It is also the space group for the cesium chloride structure (Sec. 3-2), found in CsCl and a variety of compounds, and for the perovskite structure (Sec. 3-4), found in $CaTiO_3$ and many other cases. The space group T_h^6 is that found in the pyrite structure (Sec. 3-3), as in FeS_2 and other compounds. In Table A3-44 we give atomic positions in these structures. The group O_h^1 is symmorphic, but T_h^6 is nonsymmorphic. The point group O_h, consisting of operations $R_1 \cdots R_{24}$ and $R'_1 \cdots R'_{24}$, and T_h, consisting of

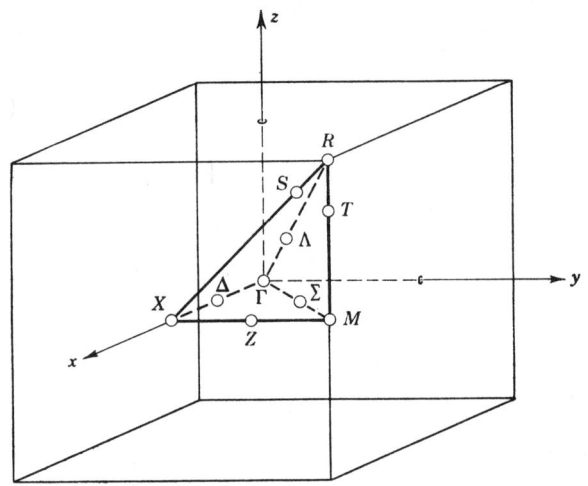

FIG. A3-4. Brillouin zone for the simple cubic Bravais lattice, with symmetry points following the notation of Bouckaert, Smoluchowski, and Wigner.

$R_1 \cdots R_{12}$ and $R'_1 \cdots R'_{12}$, have already been discussed in previous sections.

The primitive unit cell has vectors $t_1 = a\mathbf{i}$, $t_2 = a\mathbf{j}$, $t_3 = a\mathbf{k}$, and the vectors of the reciprocal lattice are $b_1 = \mathbf{i}/a$, $b_2 = \mathbf{j}/a$, $b_3 = \mathbf{k}/a$. The Brillouin zone, with the notations of Bouckaert, Smoluchowski, and Wigner for the symmetry points, is given in Fig. A3-4, identical with Fig. 5-3. The nonprimitive translations in the space group T_h^6 are the following:

$$\begin{aligned} &R_1,\ R_5,\ R_9,\ R'_1,\ R'_5,\ R'_9\text{: none} \\ &R_2,\ R_8,\ R_{11},\ R'_2,\ R'_8,\ R'_{11}\text{: } \tfrac{1}{2}(t_1 + t_2) \\ &R_3,\ R_6,\ R_{12},\ R'_3,\ R'_6,\ R'_{12}\text{: } \tfrac{1}{2}(t_2 + t_3) \\ &R_4,\ R_7,\ R_{10},\ R'_4,\ R'_7,\ R'_{10}\text{: } \tfrac{1}{2}(t_1 + t_3) \end{aligned} \quad (A3\text{-}23)$$

The effects of the operators of the point group T_d on a plane wave, for the simple cubic Bravais lattice, are given in Table A3-45. The factors involved in the nonsymmorphic space group T_h^6 are the same as those tabulated in Eq. (A3-22) for the space group T_h^7, which has the same nonprimitive translations when expressed in terms of t_1, t_2, t_3 (though not when expressed in terms of rectangular coordinates). The factors involved in the multiplication table for T_h^7 are different from those for T_h^6, on account of the different Bravais lattice, and are given in Table A3-46.

The matrix elements for the representations of O_h^1 at Γ, and in the interior of the Brillouin zone, namely, along the directions Δ, Λ, and Σ, are the same as for O_h^5 and O_h^9, given already, and the compatibility relations are the same as in those cases. For T_h^6 the matrix elements at Γ, Λ, and Σ are the same as for T_h^7, but those for Δ are different, and are given in Table A3-47. The compatibility relations between Γ, Δ, Λ, and Σ for T_h^6 are the same as for T_h^7. In Tables A3-48 to A3-53 we give matrix elements for O_h^1 and T_h^6 on the surface of the Brillouin zone, for which they are different from other cases, and in Table A3-54 we give the compatibility relations at the points X, R, and M, on the surface of the zone.

None of these tables are harder to derive than those in earlier sections of this Appendix. There is one method, however, which has not been described earlier, and which is useful in such cases as the two-dimensional representations at R, M, Z, T, and S. Let us use Z as an example, for the group T_h^6, with operations R_1, R_3, R_2', R_4'. On account of the rule giving the sum of the squares of dimensionalities of irreducible representations, we must have here either four one-dimensional representations or one two-dimensional representation. We may easily eliminate the possibility of one-dimensional representations. Let us suppose that a basis function for a one-dimensional representation was $(R_1 + a_3R_3 + a_2R_2' + a_4R_4')\psi$, where ψ is a plane wave. We then let each of the operators R_3, R_2', R_4' operate on this function, using the multiplication table to find the result. If the representation were one-dimensional, the results should be a_3^*, a_2^*, and a_4^*, respectively, times the original function, where these coefficients are arrived at by use of the projection operators. In this way we set up three equations for the a's, which are found at once to be inconsistent with each other, thereby showing that no one-dimensional irreducible representations exist.

We must then have a single two-dimensional representation. The function written above must be one of the basis functions, and we may write down the second, the two functions taking the form

$$\phi_1 = (R_1 + a_3R_3 + a_2R_2' + a_4R_4')\psi$$
$$\phi_2 = (b_3R_3 + b_2R_2' + b_4R_4')\psi \quad \text{(A3-24)}$$

Then we must find, for instance,

$$R_3\phi_1 = a_3^*\phi_1 + b_3^*\phi_2 \tag{A3-25}$$

with similar relations for R_2', R_4'. In this way we derive three equations for the a's and b's, which are now found to be consistent with each other, and which permit the solutions which we have tabulated in Table A3-53. This provides a simple and straightforward way to find the irreducible representations. An extension of this method can be used to derive most of the two-dimensional irreducible representations.

Table A3-44
Positions of atoms in structures having space groups O_h^1 and T_h^6

O_h^1, *cesium chloride structure* (values x/a, y/a, z/a given).
  Cesium at origin, *chlorine* at ½, ½, ½.

O_h^1, *perovskite structure*, $CaTiO_3$.
  Calcium at ½, ½, ½. *Titanium* at origin.
  Oxygens at ½, 0, 0; 0, ½, 0; 0, 0, ½.

T_h^6, *pyrite structure*, FeS_2.
  Iron at 000; 0, ½, ½; ½, 0, ½; ½, ½, 0.
  Sulfur at $\pm(u, u, u; u + \frac{1}{2}, -u + \frac{1}{2}, -u; -u, u + \frac{1}{2}, -u + \frac{1}{2}; -u + \frac{1}{2}, -u, u + \frac{1}{2})$ where for FeS_2 the parameter u equals 0.386.

Table A3-45
Effect of operators of T_d on a plane wave, simple cubic Bravais lattice

$R_1\psi = \exp 2\pi i[(h_1 + p_1)\xi + (h_2 + p_2)\eta + (h_3 + p_3)\zeta]$
$R_2\psi = \exp 2\pi i[(h_1 + p_1)\xi - (h_2 + p_2)\eta - (h_3 + p_3)\zeta]$
$R_3\psi = \exp 2\pi i[-(h_1 + p_1)\xi + (h_2 + p_2)\eta - (h_3 + p_3)\zeta]$
$R_4\psi = \exp 2\pi i[-(h_1 + p_1)\xi - (h_2 + p_2)\eta + (h_3 + p_3)\zeta]$
$R_5\psi = \exp 2\pi i[(h_3 + p_3)\xi + (h_1 + p_1)\eta + (h_2 + p_2)\zeta]$
$R_6\psi = \exp 2\pi i[-(h_3 + p_3)\xi - (h_1 + p_1)\eta + (h_2 + p_2)\zeta]$
$R_7\psi = \exp 2\pi i[(h_3 + p_3)\xi - (h_1 + p_1)\eta - (h_2 + p_2)\zeta]$
$R_8\psi = \exp 2\pi i[-(h_3 + p_3)\xi + (h_1 + p_1)\eta - (h_2 + p_2)\zeta]$
$R_9\psi = \exp 2\pi i[(h_2 + p_2)\xi + (h_3 + p_3)\eta + (h_1 + p_1)\zeta]$
$R_{10}\psi = \exp 2\pi i[-(h_2 + p_2)\xi + (h_3 + p_3)\eta - (h_1 + p_1)\zeta]$
$R_{11}\psi = \exp 2\pi i[-(h_2 + p_2)\xi - (h_3 + p_3)\eta + (h_1 + p_1)\zeta]$
$R_{12}\psi = \exp 2\pi i[(h_2 + p_2)\xi - (h_3 + p_3)\eta - (h_1 + p_1)\zeta]$
$R_{13}\psi = \exp 2\pi i[-(h_1 + p_1)\xi - (h_3 + p_3)\eta + (h_2 + p_2)\zeta]$
$R_{14}\psi = \exp 2\pi i[-(h_1 + p_1)\xi + (h_3 + p_3)\eta - (h_2 + p_2)\zeta]$
$R_{15}\psi = \exp 2\pi i[(h_3 + p_3)\xi - (h_2 + p_2)\eta - (h_1 + p_1)\zeta]$
$R_{16}\psi = \exp 2\pi i[-(h_3 + p_3)\xi - (h_2 + p_2)\eta + (h_1 + p_1)\zeta]$
$R_{17}\psi = \exp 2\pi i[-(h_2 + p_2)\xi + (h_1 + p_1)\eta - (h_3 + p_3)\zeta]$
$R_{18}\psi = \exp 2\pi i[(h_2 + p_2)\xi - (h_1 + p_1)\eta - (h_3 + p_3)\zeta]$
$R_{19}\psi = \exp 2\pi i[(h_1 + p_1)\xi + (h_3 + p_3)\eta + (h_2 + p_2)\zeta]$
$R_{20}\psi = \exp 2\pi i[(h_1 + p_1)\xi - (h_3 + p_3)\eta - (h_2 + p_2)\zeta]$
$R_{21}\psi = \exp 2\pi i[(h_3 + p_3)\xi + (h_2 + p_2)\eta + (h_1 + p_1)\zeta]$
$R_{22}\psi = \exp 2\pi i[-(h_3 + p_3)\xi + (h_2 + p_2)\eta - (h_1 + p_1)\zeta]$
$R_{23}\psi = \exp 2\pi i[(h_2 + p_2)\xi + (h_1 + p_1)\eta + (h_3 + p_3)\zeta]$
$R_{24}\psi = \exp 2\pi i[-(h_2 + p_2)\xi - (h_1 + p_1)\eta + (h_3 + p_3)\zeta]$

Table A3-46
Factors in multiplication table for group T_h^6
(see Table A3-34 for explanation)

	$R_1 R_5 R_9$	$R_2 R_8 R_{11}$	$R_3 R_6 R_{12}$	$R_4 R_7 R_{10}$
R_1	0	0	0	0
R_2	0	p_1	$-p_3$	$p_1 - p_3$
R_3	0	$-p_1 + p_2$	p_2	$-p_1$
R_4	0	$-p_2$	$-p_2 + p_3$	p_3
R_5	0	0	0	0
R_6	0	$-p_1$	$-p_1 + p_2$	p_2
R_7	0	p_3	$-p_2$	$-p_2 + p_3$
R_8	0	$p_1 - p_3$	p_1	$-p_3$
R_9	0	0	0	0
R_{10}	0	$-p_2 + p_3$	p_3	$-p_2$
R_{11}	0	$-p_3$	$p_1 - p_3$	p_1
R_{12}	0	p_2	$-p_1$	$-p_1 + p_2$
R_1'	0	$-p_1 - p_2$	$-p_2 - p_3$	$-p_1 - p_3$
R_2'	0	p_2	p_2	0
R_3'	0	0	p_3	p_3
R_4'	0	p_1	0	p_1
R_5'	0	$-p_1 - p_3$	$-p_1 - p_2$	$-p_2 - p_3$
R_6'	0	p_3	0	p_3
R_7'	0	p_1	p_1	0
R_8'	0	0	p_2	p_2
R_9'	0	$-p_2 - p_3$	$-p_1 - p_3$	$-p_1 - p_2$
R_{10}'	0	0	p_1	p_1
R_{11}'	0	p_2	0	p_2
R_{12}'	0	p_3	p_3	0

Table A3-47
Matrix elements and characters at points Δ, propagation along x axis, for group $T_h'^6$. Here $p_1 = p$, $p_2 = p_3 = 0$.

	R_1	R_2	R_3'	R_4'
Δ_1	1	$e^{\pi i p}$	1	$e^{\pi i p}$
Δ_2	1	$e^{\pi i p}$	-1	$-e^{\pi i p}$
Δ_3	1	$-e^{\pi i p}$	1	$-e^{\pi i p}$
Δ_4	1	$-e^{\pi i p}$	-1	$e^{\pi i p}$

Table A3-48

Matrix elements and characters at point X, boundary of Brillouin zone along x axis, for group T_h^6. Here $p_1 = \frac{1}{2}$, $p_2 = p_3 = 0$.

	R_1	R_2	R_3	R_4	R_1'	R_2'	R_3'	R_4'
$(X_1)_{11}$	1	i	0	0	0	0	1	i
$(X_1)_{21}$	0	0	1	i	1	i	0	0
$(X_1)_{12}$	0	0	1	$-i$	1	$-i$	0	0
$(X_1)_{22}$	1	$-i$	0	0	0	0	1	$-i$
$\chi(X_1)$	2	0	0	0	0	0	2	0
$(X_2)_{11}$	1	i	0	0	0	0	-1	$-i$
$(X_2)_{21}$	0	0	1	i	-1	$-i$	0	0
$(X_2)_{12}$	0	0	1	$-i$	-1	i	0	0
$(X_2)_{22}$	1	$-i$	0	0	0	0	-1	i
$\chi(X_2)$	2	0	0	0	0	0	-2	0

Table A3-49

Matrix elements and characters at point X, boundary of Brillouin zone along x axis, for group O_h^1.

	R_1	R_2	R_3	R_4	R_{13}	R_{14}	R_{19}	R_{20}	R_1'	R_2'	R_3'	R_4'	R_{13}'	R_{14}'	R_{19}'	R_{20}'
X_1	1	1	1	1	1	1	1	1	1	1	1	1	1	1	1	1
X_2	1	1	1	1	-1	-1	-1	-1	1	1	1	1	-1	-1	-1	-1
X_3	1	1	-1	-1	-1	-1	1	1	1	1	-1	-1	-1	-1	1	1
X_4	1	1	-1	-1	1	1	-1	-1	1	1	-1	-1	1	1	-1	-1
X_1'	1	1	1	1	-1	-1	-1	-1	-1	-1	-1	-1	1	1	1	1
X_2'	1	1	1	1	1	1	1	1	-1	-1	-1	-1	-1	-1	-1	-1
X_3'	1	1	-1	-1	1	1	-1	-1	-1	-1	1	1	-1	-1	1	1
X_4'	1	1	-1	-1	-1	-1	1	1	-1	-1	1	1	1	1	-1	-1
$(X_5)_{11}$	1	-1	-1	1	0	0	0	0	1	-1	-1	1	0	0	0	0
$(X_5)_{21}$	0	0	0	0	-1	1	1	-1	0	0	0	0	-1	1	1	-1
$(X_5)_{12}$	0	0	0	0	1	-1	1	-1	0	0	0	0	1	-1	1	-1
$(X_5)_{22}$	1	-1	1	-1	0	0	0	0	1	-1	1	-1	0	0	0	0
$\chi(X_5)$	2	-2	0	0	0	0	0	0	2	-2	0	0	0	0	0	0
$(X_5')_{11}$	1	-1	1	-1	0	0	0	0	-1	1	-1	1	0	0	0	0
$(X_5')_{21}$	0	0	0	0	1	-1	1	-1	0	0	0	0	-1	1	-1	1
$(X_5')_{12}$	0	0	0	0	-1	1	1	-1	0	0	0	0	1	-1	-1	1
$(X_5')_{22}$	1	-1	-1	1	0	0	0	0	-1	1	1	-1	0	0	0	0
$\chi(X_5')$	2	-2	0	0	0	0	0	0	-2	2	0	0	0	0	0	0

Table A3-50

Matrix elements and characters at point R, boundary of Brillouin zone along 111 direction, for group T_h^6. For group O_h^1, matrix elements are as at Γ, with same notation, replacing Γ by R. At R, $p_1 = p_2 = p_3 = \frac{1}{2}$. Each of the representations given below for T_h^6 really gives two representations, with superscripts $\pm$. For the $+$ case, the matrix elements for the primed operators equal those for the unprimed, while for the $-$ case they have opposite signs. The abbreviation ω stands for $e^{2\pi i/3}$, a for $1/\sqrt{3}$, b for $\sqrt{2/3}$.

	R_1	R_2	R_3	R_4	R_5	R_6	R_7	R_8	R_9	R_{10}	R_{11}	R_{12}
$(R_1)_{11}$	1	ia	ia	ia	1	ia	ia	ia	1	ia	ia	ia
$(R_1)_{21}$	0	b	$b\omega^2$	$b\omega$	0	$b\omega$	b	$b\omega^2$	0	$b\omega^2$	$b\omega$	b
$(R_1)_{12}$	0	$-b$	$-b\omega$	$-b\omega^2$	0	$-b\omega$	$-b\omega^2$	$-b$	0	$-b\omega^2$	$-b$	$-b\omega$
$(R_1)_{22}$	1	$-ia$	$-ia$	$-ia$	ω^2	$-ia\omega^2$	$-ia\omega^2$	$-ia\omega^2$	ω	$-ia\omega$	$-ia\omega$	$-ia\omega$
$\chi(R_1)$	2	0	0	0	$-\omega$	ω	ω	ω	$-\omega^2$	$-\omega^2$	$-\omega^2$	$-\omega^2$
$(R_2)_{11}$	1	$-ia$	$-ia$	$-ia$	1	$-ia$	$-ia$	$-ia$	1	$-ia$	$-ia$	$-ia$
$(R_2)_{21}$	0	b	$b\omega$	$b\omega^2$	0	$b\omega^2$	b	$b\omega$	0	$b\omega$	$b\omega^2$	b
$(R_2)_{12}$	0	$-b$	$-b\omega^2$	$-b\omega$	0	$-b\omega^2$	$-b\omega$	$-b$	0	$-b\omega$	$-b$	$-b\omega^2$
$(R_2)_{22}$	1	ia	ia	ia	ω	$ia\omega$	$ia\omega$	$ia\omega$	ω^2	$ia\omega^2$	$ia\omega^2$	$ia\omega^2$
$\chi(R_2)$	2	0	0	0	$-\omega^2$	ω^2	ω^2	ω^2	$-\omega$	$-\omega$	$-\omega$	$-\omega$
$(R_3)_{11}$	1	ia	ia	ia	ω^2	$ia\omega^2$	$ia\omega^2$	$ia\omega^2$	ω	$ia\omega$	$ia\omega$	$ia\omega$
$(R_3)_{21}$	0	b	$b\omega^2$	$b\omega$	0	b	$b\omega^2$	$b\omega$	0	b	$b\omega^2$	$b\omega$
$(R_3)_{12}$	0	$-b$	$-b\omega$	$-b\omega^2$	0	$-b$	$-b\omega$	$-b\omega^2$	0	$-b$	$-b\omega$	$-b\omega^2$
$(R_3)_{22}$	1	$-ia$	$-ia$	$-ia$	ω	$-ia\omega$	$-ia\omega$	$-ia\omega$	ω^2	$-ia\omega^2$	$-ia\omega^2$	$-ia\omega^2$
$\chi(R_3)$	2	0	0	0	-1	1	1	1	-1	-1	-1	-1

Table A3-51

Matrix elements and characters at point M, boundary of Brillouin zone along 110 direction, for group O_h^1. Here $p_1 = p_2 = \frac{1}{2}$, $p_3 = 0$.

	R_1	R_2	R_3	R_4	R_{17}	R_{18}	R_{23}	R_{24}	R_1'	R_2'	R_3'	R_4'	R_{17}'	R_{18}'	R_{23}'	R_{24}'
M_1	1	1	1	1	1	1	1	1	1	1	1	1	1	1	1	1
M_2	1	1	1	1	−1	−1	−1	−1	1	1	1	1	−1	−1	−1	−1
M_3	1	−1	−1	1	−1	−1	1	1	1	−1	−1	1	−1	−1	1	1
M_4	1	−1	−1	1	1	1	−1	−1	1	−1	−1	1	1	1	−1	−1
M_1'	1	1	1	1	−1	−1	−1	−1	−1	−1	−1	−1	1	1	1	1
M_2'	1	1	1	1	1	1	1	1	−1	−1	−1	−1	−1	−1	−1	−1
M_3'	1	−1	−1	1	1	1	−1	−1	−1	1	1	−1	−1	−1	1	1
M_4'	1	−1	−1	1	−1	−1	1	1	−1	1	1	−1	1	1	−1	−1
$(M_5)_{11}$	1	1	−1	−1	0	0	0	0	1	1	−1	−1	0	0	0	0
$(M_5)_{21}$	0	0	0	0	1	−1	1	−1	0	0	0	0	1	−1	1	−1
$(M_5)_{12}$	0	0	0	0	−1	1	1	−1	0	0	0	0	−1	1	1	−1
$(M_5)_{22}$	1	−1	1	−1	0	0	0	0	1	−1	1	−1	0	0	0	0
$\chi(M_5)$	2	0	0	−2	0	0	0	0	2	0	0	−2	0	0	0	0
$(M_5')_{11}$	1	−1	1	−1	0	0	0	0	−1	1	−1	1	0	0	0	0
$(M_5')_{21}$	0	0	0	0	−1	1	1	−1	0	0	0	0	1	−1	−1	1
$(M_5')_{12}$	0	0	0	0	1	−1	1	−1	0	0	0	0	−1	1	−1	1
$(M_5')_{22}$	1	1	−1	−1	0	0	0	0	−1	−1	1	1	0	0	0	0
$\chi(M_5')$	2	0	0	−2	0	0	0	0	−2	0	0	2	0	0	0	0

Table A3-52

Matrix elements and characters at point M, for group T_h^6.

	R_1	R_2	R_3	R_4	R_1'	R_2'	R_3'	R_4'
$(M_1)_{11}$	1	i	0	0	0	0	1	i
$(M_1)_{21}$	0	0	1	i	1	i	0	0
$(M_1)_{12}$	0	0	−1	$-i$	1	i	0	0
$(M_1)_{22}$	1	i	0	0	0	0	−1	$-i$
$\chi(M_1)$	2	$2i$	0	0	0	0	0	0
$(M_2)_{11}$	1	$-i$	0	0	0	0	1	$-i$
$(M_2)_{21}$	0	0	1	$-i$	1	$-i$	0	0
$(M_2)_{12}$	0	0	−1	i	1	$-i$	0	0
$(M_2)_{22}$	1	$-i$	0	0	0	0	−1	i
$\chi(M_2)$	2	$-2i$	0	0	0	0	0	0

Table A3-53

Matrix elements and characters for irreducible representations along directions on surface of Brillouin zone, O_h^1 and T_h^6.

Direction Z, connecting points X and M. Here $p_1 = \frac{1}{2}$, $p_2 = p$, $p_3 = 0$.

O_h^1		R_1	R_3	R_2'	R_4'
	Z_1	1	1	1	1
	Z_2	1	1	-1	-1
	Z_3	1	-1	-1	1
	Z_4	1	-1	1	-1

T_h^6		R_1	R_3	R_2'	R_4'
	$(Z_1)_{11}$	1	$-1/\sqrt{3}\,e^{\pi i p}$	$-1/\sqrt{3}\,e^{\pi i p}$	$-i/\sqrt{3}$
	$(Z_1)_{21}$	0	$-i\omega^2\sqrt{2/3}\,e^{\pi i p}$	$-i\omega\sqrt{2/3}\,e^{\pi i p}$	$\sqrt{2/3}$
	$(Z_1)_{12}$	0	$i\omega\sqrt{2/3}\,e^{\pi i p}$	$i\omega^2\sqrt{2/3}\,e^{\pi i p}$	$-\sqrt{2/3}$
	$(Z_1)_{22}$	1	$1/\sqrt{3}\,e^{\pi i p}$	$1/\sqrt{3}\,e^{\pi i p}$	$i/\sqrt{3}$
	$\chi(Z_1)$	2	0	0	0

Direction T, connecting points M and R. Here $p_1 = p_2 = \frac{1}{2}$, $p_3 = p$.

O_h^1		R_1	R_4	R_{23}	R_{24}	R_2'	R_3'	R_{17}'	R_{18}'
	T_1	1	1	1	1	1	1	1	1
	T_1'	1	1	-1	-1	-1	-1	1	1
	T_2	1	1	-1	-1	1	1	-1	-1
	T_2'	1	1	1	1	-1	-1	-1	-1
	$(T_5)_{11}$	1	-1	0	0	1	-1	0	0
	$(T_5)_{21}$	0	0	1	-1	0	0	1	-1
	$(T_5)_{12}$	0	0	1	-1	0	0	-1	1
	$(T_5)_{22}$	1	-1	0	0	-1	1	0	0
	$\chi(T_5)$	2	-2	0	0	0	0	0	0

T_h^6		R_1	R_4	R_2'	R_3'
	$(T_1)_{11}$	1	$-1/\sqrt{3}\,e^{\pi i p}$	$-i/\sqrt{3}$	$-1/\sqrt{3}\,e^{\pi i p}$
	$(T_1)_{21}$	0	$-i\omega^2\sqrt{2/3}\,e^{\pi i p}$	$\sqrt{2/3}$	$-i\omega\sqrt{2/3}\,e^{\pi i p}$
	$(T_1)_{12}$	0	$i\omega\sqrt{2/3}\,e^{\pi i p}$	$-\sqrt{2/3}$	$i\omega^2\sqrt{2/3}\,e^{\pi i p}$
	$(T_1)_{22}$	1	$1/\sqrt{3}\,e^{\pi i p}$	$i/\sqrt{3}$	$1/\sqrt{3}\,e^{\pi i p}$
	$\chi(T_1)$	2	0	0	0

Table A3-53 (Continued)

Direction S, connecting points X and R. Here $p_1 = \frac{1}{2}$, $p_2 = p_3 = p$.

O_h^1

	R_1	R_{19}	R_2'	R_{20}'
S_1	1	1	1	1
S_2	1	-1	-1	1
S_3	1	1	-1	-1
S_4	1	-1	1	-1

T_h^6

	R_1	R_2'
S_1	1	$e^{\pi i p}$
S_2	1	$-e^{\pi i p}$

Table A3-54
Compatibility relations, O_h^1 and T_h^6

O_h^1

X_1	X_2	X_3	X_4	X_1'	X_2'	X_3'	X_4'	X_5	X_5'
Δ_1	Δ_2	Δ_2'	Δ_1'	Δ_1'	Δ_2'	Δ_2	Δ_1	Δ_5	Δ_5
Z_1	Z_1	Z_4	Z_4	Z_2	Z_2	Z_3	Z_3	$Z_2 Z_3$	$Z_1 Z_4$
S_1	S_4	S_1	S_4	S_2	S_3	S_2	S_3	$S_2 S_3$	$S_1 S_4$

R_1	R_2	R_{12}	R_{15}	R_{25}	R_1'	R_2'	R_{12}'	R_{15}'	R_{25}'
Λ_1	Λ_2	Λ_3	$\Lambda_1 \Lambda_3$	$\Lambda_2 \Lambda_3$	Λ_2	Λ_1	Λ_3	$\Lambda_2 \Lambda_3$	$\Lambda_1 \Lambda_3$
S_1	S_4	$S_1 S_4$	$S_1 S_3 S_4$	$S_1 S_2 S_4$	S_2	S_3	$S_2 S_3$	$S_2 S_3 S_4$	$S_1 S_2 S_3$
T_1	T_2	$T_1 T_2$	$T_1 T_5$	$T_2 T_5$	T_1'	T_2'	$T_1' T_2'$	$T_1' T_5$	$T_2' T_5$

M_1	M_2	M_3	M_4	M_1'	M_2'	M_3'	M_4'	M_5	M_5'
Σ_1	Σ_4	Σ_1	Σ_4	Σ_2	Σ_3	Σ_2	Σ_3	$\Sigma_2 \Sigma_3$	$\Sigma_1 \Sigma_4$
Z_1	Z_1	Z_3	Z_3	Z_2	Z_2	Z_4	Z_4	$Z_2 Z_4$	$Z_1 Z_3$
T_1	T_2	T_2'	T_1'	T_1'	T_2'	T_2	T_1	T_5	T_5

T_h^6

X_1	X_2	R_1^+	R_1^-	R_2^+	R_2^-	R_3^+	R_3^-	M_1	M_2
$\Delta_1 \Delta_3$	$\Delta_2 \Delta_4$	$\Lambda_1 \Lambda_3$	$\Lambda_1 \Lambda_3$	$\Lambda_1 \Lambda_2$	$\Lambda_1 \Lambda_2$	$\Lambda_2 \Lambda_3$	$\Lambda_2 \Lambda_3$	$\Sigma_1 \Sigma_2$	$\Sigma_1 \Sigma_2$
Z_1	Z_1	$S_1 S_2$	$S_1 S_2$	$S_1 S_2$	$S_1 S_2$	$S_1 S_2$	$S_1 S_2$	Z_1	Z_1
$S_1 S_2$	$S_1 S_2$	T_1	T_1	T_1	T_1	T_1	T_1	T_1	T_1

A3-5. The Space Groups C_{3i}^2 ($R\bar{3}$), D_3^7 ($R32$), D_{3d}^5 ($R\bar{3}m$), and D_{3d}^6 ($R\bar{3}c$). The space groups considered in this section have the rhombohedral or trigonal Bravais lattice, and can be conveniently considered together. A good many important elements and compounds crystallize in one or another of these space groups. Thus the ilmenite structure (Sec. 3-5) has the space group C_{3i}^2 ($R\bar{3}$), the AlF$_3$ structure (Sec. 3-6) has D_3^7 ($R32$), the arsenic structure (Sec. 2-6), the CdCl$_2$ structure (Sec. 3-3), and the LaOF structure (Sec. 3-6) have the space group D_{3d}^5 ($R\bar{3}m$), and the calcite and corundum structures (Sec. 3-5) have the space group D_{3d}^6 ($R\bar{3}c$). Positions of atoms in the structures listed above are given in Table A3-55.

The three fundamental vectors t_1, t_2, t_3 of the primitive unit cell are of equal magnitude, and make equal angles of α with each other. In rectangular coordinates they are given in Eqs. (1-21) and (1-22). The vectors b_1, b_2, b_3 of the reciprocal lattice are given in Eqs. (5-12) and

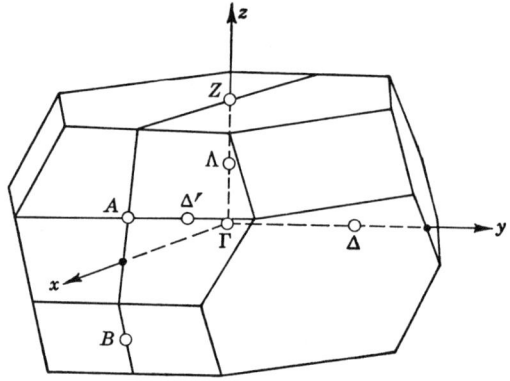

FIG. A3-5. Brillouin zone for the rhombohedral Bravais lattice, with symmetry points following the notation of Koster.

(5-13). The Brillouin zone is shown in Fig. A3-5, with convenient notations, partly following G. F. Koster, *Solid State Phys.*, **5**:173 (1957) for the symmetry directions. The inclined hexagonal faces are perpendicular bisectors of vectors $\pm b_1$, $\pm b_2$, $\pm b_3$. The rectangular faces are perpendicular bisectors of the vectors $\pm (b_1 + b_2)$, $\pm (b_2 + b_3)$, $\pm (b_1 + b_3)$, and the top and bottom faces are perpendicular bisectors of the vectors $\pm (b_1 + b_2 + b_3)$. The height of the Brillouin zone is $3r'$, and the dividing lines between the rectangular faces and the inclined hexagonal faces come at distance r' and $2r'$ from the top and bottom faces, where r' is introduced in the discussion of Eq. (5-12). The symmetry points indicated by A and B in Fig. A3-5 are the intersections of the vectors $\pm b_1$, etc., and $\pm (b_1 + b_2)$, etc., with the faces of the Brillouin zone.

The point groups C_{3i} and D_3, which are found in the first two space groups considered, may be taken as special cases of D_{3d}, met with the remaining two. Consequently, we shall carry through the case of D_{3d}, and shall specialize for the other cases. The operations of the point group D_{3d}, in cylindrical coordinates, may be taken to be defined by

$$X_0\psi(r,\phi,z) = \psi(r,\phi,z)$$
$$X_{\pm 1}\psi(r,\phi,z) = \psi\left(r, \phi \pm \frac{\pi}{3}, -z\right)$$
$$X_{\pm 2}\psi(r,\phi,z) = \psi\left(r, \phi \pm \frac{2\pi}{3}, z\right)$$
$$X_3\psi(r,\phi,z) = \psi(r, \phi + \pi, -z)$$
$$Y_0\psi(r,\phi,z) = \psi(r, -\phi, z) \quad\quad\quad (A3\text{-}26)$$
$$Y_{\pm 1}\psi(r,\phi,z) = \psi\left(r, -\phi \pm \frac{\pi}{3}, -z\right)$$
$$Y_{\pm 2}\psi(r,\phi,z) = \psi\left(r, -\phi \pm \frac{2\pi}{3}, z\right)$$
$$Y_3\psi(r,\phi,z) = \psi(r, -\phi + \pi, -z)$$

The operations of C_{3i} may be described in terms of Eq. (A3-26) as the X operations listed there; the operations of D_3 may be described as X_0, $X_{\pm 2}$, $Y_{\pm 1}$, Y_3. In Table A3-56 we describe the operations of D_{3d} in terms of their effect on a function of the vector $\mathbf{r}$ expressed in the form $\mathbf{r} = \xi \mathbf{t}_1 + \eta \mathbf{t}_2 + \zeta \mathbf{t}_3$. In Table A3-57 we show the effect of these operations on a plane wave of the form

$$\psi = \exp 2\pi i[(h_1 + p_1)\mathbf{b}_1 + (h_2 + p_2)\mathbf{b}_2 + (h_3 + p_3)\mathbf{b}_3] \cdot \mathbf{r}$$

where the reduced wave vector is $2\pi(p_1\mathbf{b}_1 + p_2\mathbf{b}_2 + p_3\mathbf{b}_3)$, and the h's are integers. These operations combine according to the multiplication table given in Eq. (A3-8).

The space groups under consideration are symmorphic, except for the case of D_{3d}^6 ($R\bar{3}c$). Consequently, the multiplication table as just stated can be used for the space groups as well as for the point groups. For the case of D_{3d}^6 ($R\bar{3}c$), a nonprimitive translation $\frac{1}{2}(\mathbf{t}_1 + \mathbf{t}_2 + \mathbf{t}_3)$ is associated with each of the Y operations; no nonprimitive translation is associated with the X's. Consequently, the effect of one of the Y operations on a plane wave is found from Table A3-57 by multiplying the value given by that table by the factor

$$e^{\pi i(h_1+p_1+h_2+p_2+h_3+p_3)} \quad\quad\quad (A3\text{-}27)$$

The effect of these nonprimitive translations on the multiplication table for the space group D_{3d}^6 ($R\bar{3}c$) is found by using Eq. (A3-8) and multiplying the result by an extra factor as follows:

Any Y operating on an X with an odd subscript, factor

$$e^{-2\pi i(p_1+p_2+p_3)}$$

Any Y operating on a Y with an even subscript, factor   (A3-28)

$$e^{2\pi i(p_1+p_2+p_3)}$$

In all other cases there is no extra factor.

We now give in Tables A3-58 to A3-64 the matrix elements for the various irreducible representations of the four space groups under consideration, and in Table A3-65 the compatibility relations. Since there is no generally accepted convention for naming the irreducible representations of these space groups, we have introduced names for them.

Table A3-55
Positions of atoms in structures having space groups C_{3i}^2, D_3^7, D_{3d}^5, and D_{3d}^6

Ilmenite structure C_{3i}^2 ($R\bar{3}$). FeTiO$_3$. Iron atoms at positions $\xi = u$, $\eta = u$, $\zeta = u$, and $\xi = -u$, $\eta = -u$, $\zeta = -u$, where $u = 0.358$; titanium atoms for $\xi = v$, $\eta = v$, $\zeta = v$, and $\xi = -v$, $\eta = -v$, $\zeta = -v$, where $v = 0.142$; and oxygens at general positions $\pm(\xi\eta\zeta,\eta\zeta\xi,\zeta\xi\eta)$, where $\xi = 0.555$, $\eta = -0.055$, and $\zeta = 0.250$. Other similar compounds have similar parameters.

AlF$_3$ structure D_3^7 ($R32$). Aluminum atoms at uuu and $\overline{uuu}$ (that is, $\xi = u$, $\eta = u$, and $\zeta = u$, and $\xi = -u$, $\eta = -u$, $\zeta = -u$), where $u = 0.237$. One type of fluorine at $Ov\bar{v}$, $\bar{v}Ov$, $v\bar{v}O$, where $v = 0.430$. Another type of fluorine at $\frac{1}{2}$, $w\bar{w}$; $\bar{w}$, $\frac{1}{2}$, w; $w\bar{w}$, $\frac{1}{2}$, where $w = 0.070$. Other similar compounds have similar parameters.

Arsenic structure D_{3d}^5 ($R\bar{3}m$). Arsenic atoms at uuu and $\overline{uuu}$, where $u = 0.226$ for arsenic.

CdCl$_2$ structure D_{3d}^5 ($R\bar{3}m$). Cadmium atom at origin, chlorine atoms at uuu and $\overline{uuu}$, where u is approximately $\frac{1}{4}$.

LaOF structure D_{3d}^5 ($R\bar{3}m$). Lanthanum atoms at uuu, $\overline{uuu}$, where $u = 0.242$. Fluorines in similar positions with $u = 0.122$; oxygens in similar positions, with $u = 0.370$.

Calcite structure D_{3d}^6 ($R\bar{3}c$), CaCO$_3$. Calcium atoms at origin, and $\frac{1}{2}$, $\frac{1}{2}$, $\frac{1}{2}$. Carbons at $\pm(\frac{1}{4},\frac{1}{4},\frac{1}{4})$. Oxygens at $\pm(\frac{1}{4} - u, \frac{1}{4} + u, \frac{1}{4}; \frac{1}{4} + u, \frac{1}{4}, \frac{1}{4} - u;$ $\frac{1}{4}, \frac{1}{4} - u, \frac{1}{4} + u)$, where $u = 0.243$.

Corundum structure D_{3d}^6 ($R\bar{3}c$). Al$_2$O$_3$. Aluminum atoms at $\pm(\frac{1}{4} + u, \frac{1}{4} + u,$ $\frac{1}{4} + u; \frac{1}{4} - u, \frac{1}{4} - u, \frac{1}{4} - u)$, with $u = 0.105$. Oxygens at $\pm(\frac{1}{4} - v, \frac{1}{4} + v,$ $\frac{1}{4}; \frac{1}{4} + v, \frac{1}{4}, \frac{1}{4} - v; \frac{1}{4}, \frac{1}{4} - v, \frac{1}{4} + v)$, with $v = 0.303$.

App. 3] SYMMETRY PROPERTIES FOR 20 GROUPS 421

Table A3-56
Operations of the point group D_{3d}

$$X_0\psi(\mathbf{r}) = \psi(\xi t_1 + \eta t_2 + \zeta t_3)$$
$$X_1\psi(\mathbf{r}) = \psi(-\eta t_1 - \zeta t_2 - \xi t_3)$$
$$X_{-1}\psi(\mathbf{r}) = \psi(-\zeta t_1 - \xi t_2 - \eta t_3)$$
$$X_2\psi(\mathbf{r}) = \psi(\zeta t_1 + \xi t_2 + \eta t_3)$$
$$X_{-2}\psi(\mathbf{r}) = \psi(\eta t_1 + \zeta t_2 + \xi t_3)$$
$$X_3\psi(\mathbf{r}) = \psi(-\xi t_1 - \eta t_2 - \zeta t_3)$$
$$Y_0\psi(\mathbf{r}) = \psi(\xi t_1 + \zeta t_2 + \eta t_3)$$
$$Y_1\psi(\mathbf{r}) = \psi(-\zeta t_1 - \eta t_2 - \xi t_3)$$
$$Y_{-1}\psi(\mathbf{r}) = \psi(-\eta t_1 - \xi t_2 - \zeta t_3)$$
$$Y_2\psi(\mathbf{r}) = \psi(\eta t_1 + \xi t_2 + \zeta t_3)$$
$$Y_{-2}\psi(\mathbf{r}) = \psi(\zeta t_1 + \eta t_2 + \xi t_3)$$
$$Y_3\psi(\mathbf{r}) = \psi(-\xi t_1 - \zeta t_2 - \eta t_3)$$

Table A3-57
Effect of operators of D_{3d} on a plane wave

$$X_0\psi = \exp 2\pi i[(h_1 + p_1)\xi + (h_2 + p_2)\eta + (h_3 + p_3)\zeta]$$
$$X_1\psi = \exp 2\pi i[-(h_3 + p_3)\xi - (h_1 + p_1)\eta - (h_2 + p_2)\zeta]$$
$$X_{-1}\psi = \exp 2\pi i[-(h_2 + p_2)\xi - (h_3 + p_3)\eta - (h_1 + p_1)\zeta]$$
$$X_2\psi = \exp 2\pi i[(h_2 + p_2)\xi + (h_3 + p_3)\eta + (h_1 + p_1)\zeta]$$
$$X_{-2}\psi = \exp 2\pi i[(h_3 + p_3)\xi + (h_1 + p_1)\eta + (h_2 + p_2)\zeta]$$
$$X_3\psi = \exp 2\pi i[-(h_1 + p_1)\xi - (h_2 + p_2)\eta - (h_3 + p_3)\zeta]$$
$$Y_0\psi = \exp 2\pi i[(h_1 + p_1)\xi + (h_3 + p_3)\eta + (h_2 + p_2)\zeta]$$
$$Y_1\psi = \exp 2\pi i[-(h_3 + p_3)\xi - (h_2 + p_2)\eta - (h_1 + p_1)\zeta]$$
$$Y_{-1}\psi = \exp 2\pi i[-(h_2 + p_2)\xi - (h_1 + p_1)\eta - (h_3 + p_3)\zeta]$$
$$Y_2\psi = \exp 2\pi i[(h_2 + p_2)\xi + (h_1 + p_1)\eta + (h_3 + p_3)\zeta]$$
$$Y_{-2}\psi = \exp 2\pi i[(h_3 + p_3)\xi + (h_2 + p_2)\eta + (h_1 + p_1)\zeta]$$
$$Y_3\psi = \exp 2\pi i[-(h_1 + p_1)\xi - (h_3 + p_3)\eta - (h_2 + p_2)\zeta]$$

Table A3-58
Matrix elements and characters at point Γ, groups C_{3i}^2, D_3^7, D_{3d}^5, and D_{3d}^6. Here $p_1 = p_2 = p_3 = 0$. The abbreviation ω stands for $e^{2\pi i/3}$.

C_{3i}^2

	X_0	X_1	X_{-1}	X_2	X_{-2}	X_3
Γ_1	1	1	1	1	1	1
Γ_2	1	ω	ω^2	ω^2	ω	1
Γ_3	1	ω^2	ω	ω	ω^2	1
Γ_4	1	$-\omega^2$	$-\omega$	ω	ω^2	-1
Γ_5	1	$-\omega$	$-\omega^2$	ω^2	ω	-1
Γ_6	1	-1	-1	1	1	-1

Table A3-58 (Continued)

D_3^7

	X_0	X_2	X_{-2}	Y_1	Y_{-1}	Y_3
Γ_1	1	1	1	1	1	1
Γ_2	1	1	1	-1	-1	-1
$(\Gamma_3)_{11}$	1	ω	ω^2	0	0	0
$(\Gamma_3)_{21}$	0	0	0	ω^2	ω	1
$(\Gamma_3)_{12}$	0	0	0	ω	ω^2	1
$(\Gamma_3)_{22}$	1	ω^2	ω	0	0	0
$\chi(\Gamma_3)$	2	-1	-1	0	0	0

D_{3d}^5 and D_{3d}^6

	X_0	X_1	X_{-1}	X_2	X_{-2}	X_3	Y_0	Y_1	Y_{-1}	Y_2	Y_{-2}	Y_3
Γ_1	1	1	1	1	1	1	1	1	1	1	1	1
Γ_2	1	1	1	1	1	1	-1	-1	-1	-1	-1	-1
Γ_3	1	-1	-1	1	1	-1	1	-1	-1	1	1	-1
Γ_4	1	-1	-1	1	1	-1	-1	1	1	-1	-1	1
$(\Gamma_5)_{11}$	1	$-\tfrac{1}{2}$	$-\tfrac{1}{2}$	$-\tfrac{1}{2}$	$-\tfrac{1}{2}$	1	1	$-\tfrac{1}{2}$	$-\tfrac{1}{2}$	$-\tfrac{1}{2}$	$-\tfrac{1}{2}$	1
$(\Gamma_5)_{21}$	0	$-\tfrac{\sqrt{3}}{2}$	$\tfrac{\sqrt{3}}{2}$	$\tfrac{\sqrt{3}}{2}$	$-\tfrac{\sqrt{3}}{2}$	0	0	$\tfrac{\sqrt{3}}{2}$	$-\tfrac{\sqrt{3}}{2}$	$-\tfrac{\sqrt{3}}{2}$	$\tfrac{\sqrt{3}}{2}$	0
$(\Gamma_5)_{12}$	0	$\tfrac{\sqrt{3}}{2}$	$-\tfrac{\sqrt{3}}{2}$	$-\tfrac{\sqrt{3}}{2}$	$\tfrac{\sqrt{3}}{2}$	0	0	$\tfrac{\sqrt{3}}{2}$	$-\tfrac{\sqrt{3}}{2}$	$-\tfrac{\sqrt{3}}{2}$	$\tfrac{\sqrt{3}}{2}$	0
$(\Gamma_5)_{22}$	1	$-\tfrac{1}{2}$	$-\tfrac{1}{2}$	$-\tfrac{1}{2}$	$-\tfrac{1}{2}$	1	-1	$\tfrac{1}{2}$	$\tfrac{1}{2}$	$\tfrac{1}{2}$	$\tfrac{1}{2}$	-1
$\chi(\Gamma_5)$	2	-1	-1	-1	-1	2	0	0	0	0	0	0
$(\Gamma_6)_{11}$	1	$\tfrac{1}{2}$	$\tfrac{1}{2}$	$-\tfrac{1}{2}$	$-\tfrac{1}{2}$	-1	1	$\tfrac{1}{2}$	$\tfrac{1}{2}$	$-\tfrac{1}{2}$	$-\tfrac{1}{2}$	-1
$(\Gamma_6)_{21}$	0	$-\tfrac{\sqrt{3}}{2}$	$\tfrac{\sqrt{3}}{2}$	$-\tfrac{\sqrt{3}}{2}$	$\tfrac{\sqrt{3}}{2}$	0	0	$\tfrac{\sqrt{3}}{2}$	$-\tfrac{\sqrt{3}}{2}$	$\tfrac{\sqrt{3}}{2}$	$-\tfrac{\sqrt{3}}{2}$	0
$(\Gamma_6)_{12}$	0	$\tfrac{\sqrt{3}}{2}$	$-\tfrac{\sqrt{3}}{2}$	$\tfrac{\sqrt{3}}{2}$	$-\tfrac{\sqrt{3}}{2}$	0	0	$\tfrac{\sqrt{3}}{2}$	$-\tfrac{\sqrt{3}}{2}$	$\tfrac{\sqrt{3}}{2}$	$-\tfrac{\sqrt{3}}{2}$	0
$(\Gamma_6)_{22}$	1	$\tfrac{1}{2}$	$\tfrac{1}{2}$	$-\tfrac{1}{2}$	$-\tfrac{1}{2}$	-1	-1	$-\tfrac{1}{2}$	$-\tfrac{1}{2}$	$\tfrac{1}{2}$	$\tfrac{1}{2}$	1
$\chi(\Gamma_6)$	2	1	1	-1	-1	-2	0	0	0	0	0	0

Table A3-59

Matrix elements and characters for propagation along Λ, z axis. Here $p_1 = p_2 = p_3$. At Γ, p's equal zero. At Z, $p_1 = p_2 = p_3 = \frac{1}{2}$.

C_{3i}^2 and D_3^7. Abbreviation ω stands for $e^{2\pi i/3}$.

	X_0	X_2	X_{-2}
Λ_1	1	1	1
Λ_2	1	ω	ω^2
Λ_3	1	ω^2	ω

D_{3d}^5 and D_{3d}^6. The abbreviation α stands for $e^{\pi i(p_1+p_2+p_3)}$. For the group D_{3d}^5, α is to be replaced by unity.

	X_0	X_2	X_{-2}	Y_0	Y_2	Y_{-2}
Λ_1	1	1	1	α	α	α
Λ_2	1	1	1	$-\alpha$	$-\alpha$	$-\alpha$
$(\Lambda_3)_{11}$	1	$-\frac{1}{2}$	$-\frac{1}{2}$	α	$-\frac{\alpha}{2}$	$-\frac{\alpha}{2}$
$(\Lambda_3)_{21}$	0	$\frac{\sqrt{3}}{2}$	$-\frac{\sqrt{3}}{2}$	0	$\frac{-\alpha\sqrt{3}}{2}$	$\frac{\alpha\sqrt{3}}{2}$
$(\Lambda_3)_{12}$	0	$\frac{-\sqrt{3}}{2}$	$\frac{\sqrt{3}}{2}$	0	$\frac{-\alpha\sqrt{3}}{2}$	$\frac{\alpha\sqrt{3}}{2}$
$(\Lambda_3)_{22}$	1	$-\frac{1}{2}$	$-\frac{1}{2}$	$-\alpha$	$\frac{\alpha}{2}$	$\frac{\alpha}{2}$
$\chi(\Lambda_3)$	2	-1	-1	0	0	0

Table A3-60

Matrix elements and characters for point Z, $p_1 = p_2 = p_3 = \frac{1}{2}$. For C_{3i}^2, D_3^7, and D_{3d}^5, the symmetry is identical with Γ, and the same table can be used, replacing the symbols Γ for the representations by Z. The abbreviation ω stands for $e^{2\pi i/3}$.

D_{3d}^6

	X_0	X_1	X_{-1}	X_2	X_{-2}	X_3	Y_0	Y_1	Y_{-1}	Y_2	Y_{-2}	Y_3
$(Z_1)_{11}$	1	0	0	1	1	0	1	0	0	1	1	0
$(Z_1)_{21}$	0	1	1	0	0	1	0	1	1	0	0	1
$(Z_1)_{12}$	0	1	1	0	0	1	0	-1	-1	0	0	-1
$(Z_1)_{22}$	1	0	0	1	1	0	-1	0	0	-1	-1	0
$\chi(Z_1)$	2	0	0	2	2	0	0	0	0	0	0	0
$(Z_2)_{11}$	1	ω	ω^2	ω^2	ω	1	0	0	0	0	0	0
$(Z_2)_{21}$	0	0	0	0	0	0	1	$-\omega$	$-\omega^2$	ω^2	ω	-1
$(Z_2)_{12}$	0	0	0	0	0	0	-1	$-\omega^2$	$-\omega$	$-\omega$	$-\omega^2$	-1
$(Z_2)_{22}$	1	$-\omega^2$	$-\omega$	ω	ω^2	-1	0	0	0	0	0	0
$\chi(Z_2)$	2	$i\sqrt{3}$	$-i\sqrt{3}$	-1	-1	0	0	0	0	0	0	0
$(Z_3)_{11}$	1	ω^2	ω	ω	ω^2	1	0	0	0	0	0	0
$(Z_3)_{21}$	0	0	0	0	0	0	1	$-\omega^2$	$-\omega$	ω	ω^2	-1
$(Z_3)_{12}$	0	0	0	0	0	0	-1	$-\omega$	$-\omega^2$	$-\omega^2$	$-\omega$	-1
$(Z_3)_{22}$	1	$-\omega$	$-\omega^2$	ω^2	ω	-1	0	0	0	0	0	0
$\chi(Z_3)$	2	$-i\sqrt{3}$	$i\sqrt{3}$	-1	-1	0	0	0	0	0	0	0

Table A3-61

Matrix elements and characters for propagation along Δ, y axis, and Δ', direction parallel to y axis passing through A. For the group C_{3i}^2, these are not symmetry directions. For Δ, $p_1 = 0$, $p_2 = -p_3$. For Δ', $p_1 = \frac{1}{2}$, $p_2 = -p_3$. Point A, $p_1 = \frac{1}{2}$, $p_2 = p_3 = 0$.

D_3^7, D_{3d}^5, and D_{3d}^6

	X_0	Y_2
Δ_1	1	1
Δ_2	1	-1

Table A3-62

Matrix elements for propagation along a direction in the xz plane. Here $p_2 = p_3$. For Γ, $p_1 = p_2 = p_3 = 0$. For A, $p_1 = \frac{1}{2}$, $p_2 = p_3 = 0$. For B, $p_1 = 0$, $p_2 = p_3 = -\frac{1}{2}$. For C_{3i}^2 and D_3^7 the general propagation in the xz plane is not a symmetry direction.

D_{3d}^5 and D_{3d}^6		X_0	Y_0
	Even	1	1
	Odd	1	−1

Table A3-63

Matrix elements and characters for the point A. Here $p_1 = \frac{1}{2}$, $p_2 = p_3 = 0$.

C_{3i}^2		X_0	X_3		D_3^7		X_0	Y_3
	A_1	1	1			A_1	1	1
	A_2	1	−1			A_2	1	−1

D_{3d}^5		X_0	X_3	Y_0	Y_3		D_{3d}^6		X_0	X_3	Y_0	Y_3
	A_1	1	1	1	1			$(A_1)_{11}$	1	0	0	1
	A_2	1	1	−1	−1			$(A_1)_{21}$	0	1	1	0
	A_3	1	−1	−1	1			$(A_1)_{12}$	0	1	−1	0
	A_4	1	−1	1	−1			$(A_1)_{22}$	1	0	0	−1
								$\chi(A_1)$	2	0	0	0

Table A3-64

Matrix elements and characters for the point B. Here $p_1 = 0$, $p_2 = p_3 = -\frac{1}{2}$. For C_{3i}^2 and D_3^7 the symmetry situation is the same as at the point A. For D_{3d}^5 and D_{3d}^6, we have the same situation as for D_{3d}^5 at point A. The matrix elements are then as follows for these two cases:

	X_0	X_3	Y_0	Y_3
B_1	1	1	1	1
B_2	1	1	−1	−1
B_3	1	−1	−1	1
B_4	1	−1	1	−1

Table A3-65
Compatibility relations for C_{3i}^2, D_3^7, D_{3d}^5, and D_{3d}^6

	Γ_1, Z_1	Γ_2, Z_2	Γ_3, Z_3	Γ_4, Z_4	Γ_5, Z_5	Γ_6, Z_6				
C_{3i}^2	Λ_1	Λ_3	Λ_2	Λ_2	Λ_3	Λ_1				

	Γ_1, Z_1	Γ_2, Z_2	Γ_3, Z_3	A_1	A_2					
D_3^7	Λ_1	Λ_2	Λ_3	Δ_1'	Δ_2'					

	Γ_1, Z_1	Γ_2, Z_2	Γ_3, Z_3	Γ_4, Z_4	Γ_5, Z_5	Γ_6, Z_6	A_1	A_2	A_3	A_4
D_{3d}^5	Λ_1	Λ_2	Λ_1	Λ_2	Λ_3	Λ_3	Δ_1'	Δ_2'	Δ_1'	Δ_2'
	Δ_1	Δ_2	Δ_2	Δ_1	$\Delta_1\Delta_2$	$\Delta_1\Delta_2$				

	Γ_1	Γ_2	Γ_3	Γ_4	Γ_5	Γ_6	Z_1	Z_2	Z_3	A_1
D_{3d}^6	Λ_1	Λ_2	Λ_1	Λ_2	Λ_3	Λ_3	$\Lambda_1\Lambda_2$	Λ_3	Λ_3	$\Delta_1'\Delta_2'$
	Δ_1	Δ_2	Δ_2	Δ_1	$\Delta_1\Delta_2$	$\Delta_1\Delta_2$				

A3-6. The Space Groups D_{3d}^3 ($P\bar{3}m1$) and D_3^4 ($P3_121$). These two space groups, though they belong to the trigonal system, have the hexagonal Bravais lattice, so that their discussion is similar to that of the space groups taken up in Sec. A3-1. We find the group D_{3d}^3 ($P\bar{3}m1$) in the CdI_2 structure (Sec. 3-3) and in the La_2O_3 and the AlB_2 structures (Sec. 3-6). The group D_3^4 ($P3_121$) is found in the selenium structure (Sec. 2-7). The fundamental vectors and the Brillouin zone are as given in Sec. A3-1. The positions of the atoms in these various structures are given in Table A3-66.

The space group D_{3d}^3 is particularly simple, since it is symmorphic. The operators of the point group D_{3d}, in the notation used in Sec. A3-1, are X_0, $X_{\pm 1}'$, $X_{\pm 2}$, X_3', Y_0, $Y_{\pm 1}'$, $Y_{\pm 2}$, Y_3'. The irreducible representations of the space group at the point Γ can be found from Table A3-2, and removing the superscript $\pm$ on the symbols giving the representations, and adding primes to the operators with odd subscripts. The direction Σ is given by Table A3-3, retaining the appropriate operators. For M we use Table A3-4, and for R we use Table A3-5, in each case retaining the appropriate operators. The case of L is different from Table A3-6, and is given in Table A3-67. T is as in Table A3-7, K as in Table A3-8, again retaining the appropriate operators, and S is given in Table A3-68.

For Δ we use Table A3-11, setting $\alpha = 1$, and retaining the appropriate operators. For U we have Table A3-13, setting $\alpha = 1$, and for P and H we have Table A3-14, again setting $\alpha = 1$ in each case and retaining appropriate operators. This space group, in other words, is so simple that we do not have to present a special discussion.

The other case, D_3^4, is somewhat more complicated, since it is nonsymmorphic, and it does not follow directly from cases already discussed. From Eq. (2-19) we can find the effect of the operators of the space group on a plane wave. We have

$$\{X_0|0\}\psi = \exp 2\pi i[(h_1 + p_1)\xi + (h_2 + p_2)\eta + (h_3 + p_3)\zeta]$$
$$\{X_1|0\}\psi = \exp 2\pi i[(h_2 + p_2)\xi - (h_1 + p_1 + h_2 + p_2)\eta$$
$$+ (h_3 + p_3)\zeta]e^{(2\pi i/3)(h_3+p_3)}$$
$$\{X_{-1}|0\}\psi = \exp 2\pi i[-(h_1 + p_1 + h_2 + p_2)\xi + (h_1 + p_1)\eta$$
$$+ (h_3 + p_3)\zeta]e^{-(2\pi i/3)(h_3+p_3)}$$
$$\{Y_0'|0\}\psi = \exp 2\pi i[(h_1 + p_1 + h_2 + p_2)\xi - (h_2 + p_2)\eta \qquad \text{(A3-29)}$$
$$- (h_3 + p_3)\zeta]$$
$$\{Y_1'|0\}\psi = \exp 2\pi i[-(h_1 + p_1)\xi + (h_1 + p_1 + h_2 + p_2)\eta$$
$$- (h_3 + p_3)\zeta]e^{(2\pi i/3)(h_3+p_3)}$$
$$\{Y_{-1}'|0\}\psi = \exp 2\pi i[-(h_2 + p_2)\xi - (h_1 + p_1)\eta - (h_3 + p_3)\zeta]$$
$$e^{-(2\pi i/3)(h_3+p_3)}$$

For the multiplication table of the space group, we have additional factors arising from the nonprimitive translations, as follows:

$$X_1X_1, \ X_{-1}Y_1', \ Y_1'X_1, \ Y_{-1}'Y_1': e^{2\pi i p_3}$$
$$X_{-1}X_{-1}, \ X_1Y_{-1}', \ Y_{-1}'X_{-1}, \ Y_1'Y_{-1}': e^{-2\pi i p_3} \qquad \text{(A3-30)}$$
$$\text{Other cases: } 1$$

We now give in Tables A3-69 to A3-72 the matrix elements and characters for the various symmetry directions, and in Table A3-73 the compatibility relations.

Table A3-66
Positions of atoms in structures having space groups D_{3d}^3 and D_3^4

CdI$_2$ *structure*. D_{3d}^3. Cd atom at origin, I at special positions $\frac{1}{3}$, $\frac{2}{3}$, u, and $\frac{2}{3}$, $\frac{1}{3}$, $-u$, with u approximately 0.25.

La$_2$O$_3$ *structure*. D_{3d}^3. La atoms at $\frac{1}{3}$, $\frac{2}{3}$, u and $\frac{2}{3}$, $\frac{1}{3}$, $-u$, with u about 0.23. Oxygen atoms at origin, and at $\frac{1}{3}$, $\frac{2}{3}$, v, and $\frac{2}{3}$, $\frac{1}{3}$, $-v$, where v is about 0.63.

AlB$_2$ *structure*. D_{3d}^3. Aluminum atom at origin, boron at $\frac{1}{3}$, $\frac{2}{3}$, u, and $\frac{2}{3}$, $\frac{1}{3}$, $-u$, with $u = \frac{1}{2}$.

Selenium *structure*. D_3^4. Atoms at u, 0, $\frac{1}{3}$; 0, u, $\frac{2}{3}$; $-u$, $-u$, 0, where u is about 0.217.

Table A3-67

Matrix elements for the irreducible representations at the point L, for D_{3d}^3. Here $p_1 = \frac{1}{2}$, $p_2 = 0$, $p_3 = \frac{1}{2}$.

	X_0	X_3'	Y_0	Y_3'
L_1	1	1	1	1
L_2	1	-1	1	-1
L_3	1	-1	-1	1
L_4	1	1	-1	-1

Table A3-68

Matrix elements for the irreducible representations along the line S, for D_{3d}^3. For S, $p_1 = -p_2/2$, $-\frac{2}{3} < p_2 \leq \frac{2}{3}$, $p_3 = \frac{1}{3}$.

	X_0	Y_3'
S_1	1	1
S_2	1	-1

Table A3-69

Matrix elements and characters of the irreducible representations of the space group D_3^4 at point Γ, $\mathbf{k} = 0$. The abbreviation ω stands for $e^{2\pi i/3}$.

	X_0	X_1	X_{-1}	Y_0'	Y_1'	Y_{-1}'
Γ_1	1	1	1	1	1	1
Γ_2	1	1	1	-1	-1	-1
$(\Gamma_3)_{11}$	1	ω	ω^2	0	0	0
$(\Gamma_3)_{21}$	0	0	0	1	ω	ω^2
$(\Gamma_3)_{12}$	0	0	0	1	ω^2	ω
$(\Gamma_3)_{22}$	1	ω^2	ω	0	0	0
$\chi(\Gamma_3)$	2	-1	-1	0	0	0

Table A3-70

Matrix elements and characters of the irreducible representations of the space group D_3^4 at the points Σ, M, R, L. At Σ, $-\frac{1}{2} < p_1 \leq \frac{1}{2}$, $p_2 = p_3 = 0$. At M, $p_1 = \frac{1}{2}$, $p_2 = p_3 = 0$. At R, $-\frac{1}{2} < p_1 \leq \frac{1}{2}$, $p_2 = 0$, $p_3 = \frac{1}{2}$. At L, $p_1 = \frac{1}{2}$, $p_2 = 0$, $p_3 = \frac{1}{2}$.

	X_0	Y_0'
Σ_1, M_1, R_1, L_1	1	1
Σ_2, M_2, R_2, L_2	1	-1

Table A3-71
Matrix elements and characters for the irreducible representations for D_3^4 along Δ and P. For Δ, $p_1 = p_2 = 0$, $-\frac{1}{2} < p_3 \leq \frac{1}{2}$. For P, $p_1 = -\frac{1}{3}$, $p_2 = \frac{2}{3}$, $-\frac{1}{2} < p_3 \leq \frac{1}{2}$. The points K and H are special cases of P, corresponding to $p_3 = 0$ and $\frac{1}{2}$, respectively. The abbreviation ω stands for $e^{2\pi i/3}$.

	X_0	X_1	X_{-1}
Δ_1, P_1	1	$e^{(2\pi i/3)p_3}$	$e^{-(2\pi i/3)p_3}$
Δ_2, P_2	1	$\omega e^{(2\pi i/3)p_3}$	$\omega^2 e^{-(2\pi i/3)p_3}$
Δ_3, P_3	1	$\omega^2 e^{(2\pi i/3)p_3}$	$\omega e^{-(2\pi i/3)p_3}$

Table A3-72
Matrix elements and characters for the irreducible representations for D_3^4 at point A. This is the point $p_1 = p_2 = 0$, $p_3 = \frac{1}{2}$. The abbreviation ω stands for $e^{2\pi i/3}$.

	X_0	X_1	X_{-1}	Y_0'	Y_1'	Y_{-1}'
A_1	1	-1	-1	1	-1	-1
A_2	1	-1	-1	-1	1	1
$(A_3)_{11}$	1	$-\omega^2$	$-\omega$	0	0	0
$(A_3)_{21}$	0	0	0	1	$-\omega^2$	$-\omega$
$(A_3)_{12}$	0	0	0	1	$-\omega$	$-\omega^2$
$(A_3)_{22}$	1	$-\omega$	$-\omega^2$	0	0	0
$\chi(A_3)$	2	1	1	0	0	0

Table A3-73
Compatibility relations for D_3^4

Γ_1	Γ_2	Γ_3	A_1	A_2	A_3
Δ_1	Δ_1	$\Delta_2 \Delta_3$	Δ_2	Δ_2	$\Delta_1 \Delta_3$
Σ_1	Σ_2	$\Sigma_1 \Sigma_2$			

A3-7. The Space Group D_{4h}^{14} ($P4_2/mnm$). The space group D_{4h}^{14} ($P4_2/mnm$), having the point group D_{4h}, and the primitive tetragonal Bravais lattice, is found in the rutile structure (Sec. 3-3), occurring in TiO_2 and similar compounds. The three fundamental vectors t_1, t_2, t_3 of the primitive unit cell are given by $t_1 = a\mathbf{i}$, $t_2 = a\mathbf{j}$, $t_3 = c\mathbf{k}$, and the vectors $\mathbf{b}_1, \mathbf{b}_2, \mathbf{b}_3$ of the reciprocal lattice are $\mathbf{b}_1 = \mathbf{i}/a$, $\mathbf{b}_2 = \mathbf{j}/a$, $\mathbf{b}_3 = \mathbf{k}/c$.

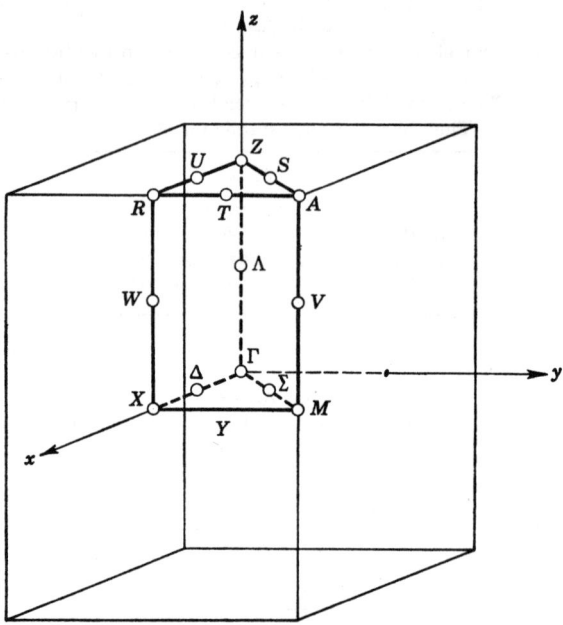

FIG. A3-6. Brillouin zone for primitive tetragonal Bravais lattice, with symmetry points following the notation of Koster.

The Brillouin zone, and the symmetry points in it, are given in Fig. A3-6. The positions of the atoms in the rutile structure are as follows:

Titanium at the origin of the cell, and at the point for which x/a, y/a, z/c are $\frac{1}{2}$, $\frac{1}{2}$, $\frac{1}{2}$

Oxygen at $\pm(uu0; u + \frac{1}{2}, \frac{1}{2} - u, \frac{1}{2})$, where for TiO_2 the parameter u is 0.31

The 16 operations of the point group D_{4h} are X_0, $X_{\pm 1}$, X_2, Y_0, $Y_{\pm 1}$, Y_2, X'_0, $X'_{\pm 1}$, X'_2, Y'_0, $Y'_{\pm 1}$, Y'_2, defined as in Volume 1, Eq. (A12-63). That is, we have

$$X_q \psi(r,\phi,z) = \psi\left(r, \phi + \frac{2\pi q}{N}, z\right)$$
$$Y_q \psi(r,\phi,z) = \psi\left(r, -\phi + \frac{2\pi q}{N}, z\right)$$
$$X'_q \psi(r,\phi,z) = \psi\left(r, \phi + \frac{2\pi q}{N}, -z\right)$$
$$Y'_q \psi(r,\phi,z) = \psi\left(r, -\phi + \frac{2\pi q}{N}, -z\right)$$

(A3-31)

where $N = 4$. For the present purpose it is more convenient to express

App. 3]   SYMMETRY PROPERTIES FOR 20 GROUPS   431

the operations in rectangular coordinates. We then have

$$\begin{aligned}
X_0\psi(x,y,z) &= \psi(x,y,z) & X'_0\psi(x,y,z) &= \psi(x,y,-z) \\
X_1\psi(x,y,z) &= \psi(-y,x,z) & X'_1\psi(x,y,z) &= \psi(-y,x,-z) \\
X_{-1}\psi(x,y,z) &= \psi(y,-x,z) & X'_{-1}\psi(x,y,z) &= \psi(y,-x,-z) \\
X_2\psi(x,y,z) &= \psi(-x,-y,z) & X'_2\psi(x,y,z) &= \psi(-x,-y,-z) \\
Y_0\psi(x,y,z) &= \psi(x,-y,z) & Y'_0\psi(x,y,z) &= \psi(x,-y,-z) \\
Y_1\psi(x,y,z) &= \psi(y,x,z) & Y'_1\psi(x,y,z) &= \psi(y,x,-z) \\
Y_{-1}\psi(x,y,z) &= \psi(-y,-x,z) & Y'_{-1}\psi(x,y,z) &= \psi(-y,-x,-z) \\
Y_2\psi(x,y,z) &= \psi(-x,y,z) & Y'_2\psi(x,y,z) &= \psi(-x,y,-z)
\end{aligned} \quad \text{(A3-32)}$$

The multiplication table for the point group is as given in Eq. (A3-8), namely,

$$X_q X_p = X_{q+p} \qquad X_q Y_p = Y_{-q+p} \qquad Y_q X_p = Y_{q+p} \qquad Y_q Y_p = X_{-q+p}$$

In this expression, if the subscript of the product function does not have one of the allowed values $0, \pm 1, 2$, we are to add or subtract integral multiples of 4 to secure a value within this range. In the space group D_{4h}^{14}, we have no nonprimitive translation associated with operators X with even subscripts, or Y with odd subscripts, but the nonprimitive translation $\frac{1}{2}(t_1 + t_2 + t_3)$ associated with all other operators.

The effect of the operators of Eq. (A3-32) on a plane wave $\exp 2\pi i[(h_1 + p_1)\xi + (h_2 + p_2)\eta + (h_3 + p_3)\zeta]$, where the vector position $\mathbf{r}$ is given by $\xi t_1 + \eta t_2 + \zeta t_3$, is given in Eq. (A3-33) below:

$$\begin{aligned}
X_0\psi &= \exp 2\pi i[(h_1 + p_1)\xi + (h_2 + p_2)\eta + (h_3 + p_3)\zeta] \\
X_1\psi &= \exp 2\pi i[(h_2 + p_2)\xi - (h_1 + p_1)\eta + (h_3 + p_3)\zeta] \\
X_{-1}\psi &= \exp 2\pi i[-(h_2 + p_2)\xi + (h_1 + p_1)\eta + (h_3 + p_3)\zeta] \\
X_2\psi &= \exp 2\pi i[-(h_1 + p_1)\xi - (h_2 + p_2)\eta + (h_3 + p_3)\zeta] \\
Y_0\psi &= \exp 2\pi i[(h_1 + p_1)\xi - (h_2 + p_2)\eta + (h_3 + p_3)\zeta] \\
Y_1\psi &= \exp [(h_2 + p_2)\xi + (h_1 + p_1)\eta + (h_3 + p_3)\zeta] \\
Y_{-1}\psi &= \exp [-(h_2 + p_2)\xi - (h_1 + p_1)\eta + (h_3 + p_3)\zeta] \\
Y_2\psi &= \exp 2\pi i[-(h_1 + p_1)\xi + (h_2 + p_2)\eta + (h_3 + p_3)\zeta]
\end{aligned} \quad \text{(A3-33)}$$

For the primed operations, the sign of $h_3 + p_3$ is to be changed. An additional factor $\exp \pi i(h_1 + p_1 + h_2 + p_2 + h_3 + p_3)$ occurs for operations X with odd or Y with even subscripts in the space group, on account of the nonprimitive translations. In the multiplication table of the space group, there is an additional factor, arising from the nonprimitive translation, if the first operation in the product is an odd-subscripted X or an even-subscripted Y. The factor is independent of which odd X or even Y we have, but depends on the second operation in the following way:

$$X_0 \text{ or } Y_1: 1$$
$$X_1 \text{ or } Y_2: \exp[2\pi i(p_2 + p_3)]$$
$$X_{-1} \text{ or } Y_0: \exp[2\pi i(p_1 + p_3)]$$
$$X_2 \text{ or } Y_{-1}: \exp[-2\pi i(p_1 + p_2)] \quad (A3\text{-}34)$$
$$X'_0 \text{ or } Y'_1: \exp(-2\pi i p_3)$$
$$X'_1 \text{ or } Y'_2: \exp(2\pi i p_2)$$
$$X'_{-1} \text{ or } Y'_0: \exp(2\pi i p_1)$$
$$X'_2 \text{ or } Y'_{-1}: \exp[-2\pi i(p_1 + p_2 + p_3)]$$

We now give in Tables A3-74 to A3-87 the matrix elements and characters for the various symmetry directions, and in Table A3-88 the compatibility relations.

Table A3-74

Matrix elements and characters at point Γ, $p_1 = p_2 = p_3 = 0$, for group D_{4h}^{14}. Only the matrix elements for the unprimed operators are tabulated. For the primed operators, the elements are the same as for the unprimed for the representations with superscript $+$, and the negatives of these values for the representations with superscript $-$.

	X_0	X_1	X_{-1}	X_2	Y_0	Y_1	Y_{-1}	Y_2
$\Gamma_1^\pm$	1	1	1	1	1	1	1	1
$\Gamma_2^\pm$	1	1	1	1	-1	-1	-1	-1
$\Gamma_3^\pm$	1	-1	-1	1	1	-1	-1	1
$\Gamma_4^\pm$	1	-1	-1	1	-1	1	1	-1
$(\Gamma_5^\pm)_{11}$	1	0	0	-1	1	0	0	-1
$(\Gamma_5^\pm)_{21}$	0	-1	1	0	0	1	-1	0
$(\Gamma_5^\pm)_{12}$	0	1	-1	0	0	1	-1	0
$(\Gamma_5^\pm)_{22}$	1	0	0	-1	-1	0	0	1
$\chi(\Gamma_5^\pm)$	2	0	0	-2	0	0	0	0

Table A3-75

Matrix elements at points Δ, $p_1 = p, p_2 = p_3 = 0, 0 < p < \frac{1}{2}, \omega = \exp(-i\pi p)$.

	X_0	Y_0	X_0'	Y_0'
Δ_1	1	ω	1	ω
Δ_2	1	$-\omega$	1	$-\omega$
Δ_3	1	$-\omega$	-1	ω
Δ_4	1	ω	-1	$-\omega$

Table A3-76

Matrix elements and characters at point X, boundary of zone along direction Δ. Here $p_1 = \frac{1}{2}, p_2 = p_3 = 0$.

	X_0	X_2	Y_0	Y_2	X_0'	X_2'	Y_0'	Y_2'
$(X_1)_{11}$	1	1	0	0	1	1	0	0
$(X_1)_{21}$	0	0	1	1	0	0	1	1
$(X_1)_{12}$	0	0	-1	1	0	0	-1	1
$(X_1)_{22}$	1	-1	0	0	1	-1	0	0
$\chi(X_1)$	2	0	0	0	2	0	0	0
$(X_2)_{11}$	1	-1	0	0	-1	1	0	0
$(X_2)_{21}$	0	0	1	-1	0	0	-1	1
$(X_2)_{12}$	0	0	-1	-1	0	0	1	1
$(X_2)_{22}$	1	1	0	0	-1	-1	0	0
$\chi(X_2)$	2	0	0	0	-2	0	0	0

Table A3-77

Matrix elements and characters for points Λ, $p_1 = p_2 = 0, p_3 = p$, for group D_{4h}^{14}. $\omega = \exp(-i\pi p)$.

	X_0	X_1	X_{-1}	X_2	Y_0	Y_1	Y_{-1}	Y_2
Λ_1	1	ω	ω	1	ω	1	1	ω
Λ_2	1	ω	ω	1	$-\omega$	-1	-1	$-\omega$
Λ_3	1	$-\omega$	$-\omega$	1	ω	-1	-1	ω
Λ_4	1	$-\omega$	$-\omega$	1	$-\omega$	1	1	$-\omega$
$(\Lambda_5)_{11}$	1	0	0	-1	ω	0	0	$-\omega$
$(\Lambda_5)_{21}$	0	$-\omega$	ω	0	0	1	-1	0
$(\Lambda_5)_{12}$	0	ω	$-\omega$	0	0	1	-1	0
$(\Lambda_5)_{22}$	1	0	0	-1	$-\omega$	0	0	ω
$\chi(\Lambda_5)$	2	0	0	-2	0	0	0	0

Table A3-78
Matrix elements and characters for point Z, boundary of zone along direction Λ, for group D_{4h}^{14}. Here $p_1 = p_2 = 0$, $p_3 = \frac{1}{2}$.

	X_0	X_1	X_{-1}	X_2	Y_0	Y_1	Y_{-1}	Y_2	X_0'	X_1'	X_{-1}'	X_2'	Y_0'	Y_1'	Y_{-1}'	Y_2'
$(Z_1)_{11}$	1	0	0	1	0	1	1	0	1	0	0	1	0	1	1	0
$(Z_1)_{21}$	0	1	1	0	1	0	0	1	0	1	1	0	1	0	0	1
$(Z_1)_{12}$	0	−1	−1	0	−1	0	0	−1	0	1	1	0	1	0	0	1
$(Z_1)_{22}$	1	0	0	1	0	1	1	0	−1	0	0	−1	0	−1	−1	0
$\chi(Z_1)$	2	0	0	2	0	2	2	0	0	0	0	0	0	0	0	0
$(Z_2)_{11}$	1	0	0	−1	0	−1	1	0	−1	0	0	1	0	1	−1	0
$(Z_2)_{21}$	0	−1	1	0	1	0	0	−1	0	1	−1	0	−1	0	0	1
$(Z_2)_{12}$	0	−1	1	0	−1	0	0	1	0	−1	1	0	−1	0	0	1
$(Z_2)_{22}$	1	0	0	−1	0	1	−1	0	1	0	0	−1	0	1	−1	0
$\chi(Z_2)$	2	0	0	−2	0	0	0	0	0	0	0	0	0	2	−2	0
$(Z_3)_{11}$	1	0	0	1	0	−1	−1	0	1	0	0	1	0	−1	−1	0
$(Z_3)_{21}$	0	−1	−1	0	1	0	0	1	0	−1	−1	0	1	0	0	1
$(Z_3)_{12}$	0	1	1	0	−1	0	0	−1	0	−1	−1	0	1	0	0	1
$(Z_3)_{22}$	1	0	0	1	0	−1	−1	0	−1	0	0	−1	0	1	1	0
$\chi(Z_3)$	2	0	0	2	0	−2	−2	0	0	0	0	0	0	0	0	0
$(Z_4)_{11}$	1	0	0	−1	0	1	−1	0	−1	0	0	1	0	−1	1	0
$(Z_4)_{21}$	0	1	−1	0	1	0	0	−1	0	−1	1	0	−1	0	0	1
$(Z_4)_{12}$	0	1	−1	0	−1	0	0	1	0	1	−1	0	−1	0	0	1
$(Z_4)_{22}$	1	0	0	−1	0	−1	1	0	1	0	0	−1	0	−1	1	0
$\chi(Z_4)$	2	0	0	−2	0	0	0	0	0	0	0	0	0	−2	2	0

Table A3-79
Matrix elements for points Σ, $p_1 = p_2 = p$, $p_3 = 0$, for group D_{4h}^{14}, also for S, $p_1 = p_2 = p$, $p_3 = \frac{1}{2}$.

	X_0	Y_1	X_0'	Y_1'
Σ_1, S_1	1	1	1	1
Σ_2, S_2	1	1	−1	−1
Σ_3, S_3	1	−1	−1	1
Σ_4, S_4	1	−1	1	−1

Table A3-80

Matrix elements and characters for the point M, $p_1 = p_2 = \frac{1}{2}$, $p_3 = 0$. In this table, the matrix elements and characters for the M_5 representation are as listed for the $+$ superscript, while the values of the primed elements change sign for the $-$ superscript.

	X_0	X_1	X_{-1}	X_2	Y_0	Y_1	Y_{-1}	Y_2	X'_0	X'_1	X'_{-1}	X'_2	Y'_0	Y'_1	Y'_{-1}	Y'_2
M_1^+	1	i	i	1	i	1	1	i	1	i	i	1	i	1	1	i
M_1^-	1	$-i$	$-i$	1	$-i$	1	1	$-i$	1	$-i$	$-i$	1	$-i$	1	1	$-i$
M_2^+	1	$-i$	$-i$	1	i	-1	-1	i	1	$-i$	$-i$	1	i	-1	-1	i
M_2^-	1	i	i	1	$-i$	-1	-1	$-i$	1	i	i	1	$-i$	-1	-1	$-i$
M_3^+	1	i	i	1	i	1	1	i	-1	$-i$	$-i$	-1	$-i$	-1	-1	$-i$
M_3^-	1	$-i$	$-i$	1	$-i$	1	1	$-i$	-1	i	i	-1	i	-1	-1	i
M_4^+	1	$-i$	$-i$	1	i	-1	-1	i	-1	i	i	-1	$-i$	1	1	$-i$
M_4^-	1	i	i	1	$-i$	-1	-1	$-i$	-1	$-i$	$-i$	-1	i	1	1	i
$(M_5^\pm)_{11}$	1	0	0	-1	0	-1	1	0	1	0	0	-1	0	-1	1	0
$(M_5^\pm)_{21}$	0	-1	1	0	1	0	0	-1	0	-1	1	0	-1	0	0	1
$(M_5^\pm)_{12}$	0	-1	1	0	-1	0	0	1	0	-1	1	0	1	0	0	-1
$(M_5^\pm)_{22}$	1	0	0	-1	0	1	-1	0	1	0	0	-1	0	1	-1	0
$\chi(M_5^\pm)$	2	0	0	-2	0	0	0	0	2	0	0	-2	0	0	0	0

Table A3-81

Matrix elements for directions W, $p_1 = \frac{1}{2}$, $p_2 = 0$, $p_3 = p$, for group D_{4h}^{14}. For point X, $p = 0$. For point R, $p = \frac{1}{2}$. $0 < p < \frac{1}{2}$, $\omega = \exp(-i\pi p)$.

	X_0	X_2	Y_0	Y_2
$(W)_{11}$	1	1	0	0
$(W)_{21}$	0	0	ω	ω
$(W)_{12}$	0	0	$-\omega^*$	ω^*
$(W)_{22}$	1	-1	0	0
$\chi(W)$	2	0	0	0

Table A3-82

Matrix elements and characters for point R, $p_1 = p_3 = \frac{1}{2}$, $p_2 = 0$, for group D_{4h}^{14}.

	X_0	X_2	Y_0	Y_2	X_0'	X_2'	Y_0'	Y_2'
$(R_1)_{11}$	1	1	0	0	1	1	0	0
$(R_1)_{21}$	0	0	1	1	0	0	1	1
$(R_1)_{12}$	0	0	1	−1	0	0	−1	1
$(R_1)_{22}$	1	−1	0	0	−1	1	0	0
$\chi(R_1)$	2	0	0	0	0	2	0	0
$(R_2)_{11}$	1	1	0	0	−1	−1	0	0
$(R_2)_{21}$	0	0	1	1	0	0	−1	−1
$(R_2)_{12}$	0	0	1	−1	0	0	1	−1
$(R_2)_{22}$	1	−1	0	0	1	−1	0	0
$\chi(R_2)$	2	0	0	0	0	−2	0	0

Table A3-83

Matrix elements and characters at points V, joining M and A, group D_{4h}^{14}. Here $p_1 = p_2 = \frac{1}{2}$, $p_3 = p$. At M, $p = 0$; at A, $p = \frac{1}{2}$. $0 < p < \frac{1}{2}$, $\omega = \exp(-i\pi p)$, $\lambda = i\omega$.

	X_0	X_1	X_{-1}	X_2	Y_0	Y_1	Y_{-1}	Y_2
V_1^+	1	λ	λ	1	λ	1	1	λ
V_1^-	1	$-\lambda$	$-\lambda$	1	$-\lambda$	1	1	$-\lambda$
V_2^+	1	$-\lambda$	$-\lambda$	1	λ	−1	−1	λ
V_2^-	1	λ	λ	1	$-\lambda$	−1	−1	$-\lambda$
$(V_3)_{11}$	1	0	0	−1	0	1	−1	0
$(V_3)_{21}$	0	ω	$-\omega$	0	ω	0	0	$-\omega$
$(V_3)_{12}$	0	ω^*	$-\omega^*$	0	$-\omega^*$	0	0	ω^*
$(V_3)_{22}$	1	0	0	−1	0	−1	1	0
$\chi(V_3)$	2	0	0	−2	0	0	0	0

Table A3-84
Matrix elements and characters for point A, $p_1 = p_2 = p_3 = \frac{1}{2}$, for group D_{4h}^{14}.

	X_0	X_1	X_{-1}	X_2	Y_0	Y_1	Y_{-1}	Y_2	X'_0	X'_1	X'_{-1}	X'_2	Y'_0	Y'_1	Y'_{-1}	Y'_2
$(A_1)_{11}$	1	0	0	1	0	1	1	0	1	0	0	1	0	1	1	0
$(A_1)_{21}$	0	1	1	0	1	0	0	1	0	1	1	0	1	0	0	1
$(A_1)_{12}$	0	1	1	0	1	0	0	1	0	−1	−1	0	−1	0	0	−1
$(A_1)_{22}$	1	0	0	1	0	1	1	0	−1	0	0	−1	0	−1	−1	0
$\chi(A_1)$	2	0	0	2	0	2	2	0	0	0	0	0	0	0	0	0
$(A_2)_{11}$	1	0	0	−1	0	−1	1	0	−1	0	0	1	0	1	−1	0
$(A_2)_{21}$	0	−1	1	0	1	0	0	−1	0	1	−1	0	−1	0	0	1
$(A_2)_{12}$	0	1	−1	0	1	0	0	−1	0	1	−1	0	1	0	0	−1
$(A_2)_{22}$	1	0	0	−1	0	1	−1	0	1	0	0	−1	0	1	−1	0
$\chi(A_2)$	2	0	0	−2	0	0	0	0	0	0	0	0	0	2	−2	0
$(A_3)_{11}$	1	0	0	1	0	−1	−1	0	1	0	0	1	0	−1	−1	0
$(A_3)_{21}$	0	−1	−1	0	1	0	0	1	0	−1	−1	0	1	0	0	1
$(A_3)_{12}$	0	−1	−1	0	1	0	0	1	0	1	1	0	−1	0	0	−1
$(A_3)_{22}$	1	0	0	1	0	−1	−1	0	−1	0	0	−1	0	1	1	0
$\chi(A_3)$	2	0	0	2	0	−2	−2	0	0	0	0	0	0	0	0	0
$(A_4)_{11}$	1	0	0	−1	0	1	−1	0	−1	0	0	1	0	−1	1	0
$(A_4)_{21}$	0	1	−1	0	1	0	0	−1	0	−1	1	0	−1	0	0	1
$(A_4)_{12}$	0	−1	1	0	1	0	0	−1	0	−1	1	0	1	0	0	−1
$(A_4)_{22}$	1	0	0	−1	0	−1	1	0	1	0	0	−1	0	−1	1	0
$\chi(A_4)$	2	0	0	−2	0	0	0	0	0	0	0	0	0	−2	2	0

Table A3-85
Matrix elements and characters at points U, connecting Z and R, group D_{4h}^{14}. Here $p_1 = p$, $p_2 = 0$, $p_3 = \frac{1}{2}$. At Z, $p = 0$; at R, $p = \frac{1}{2}$. $0 < p < \frac{1}{2}$, $\omega = \exp(-i\pi p)$.

	X_0	Y_0	X'_0	Y'_0
U_{11}	1	0	1	0
U_{21}	0	ω	0	ω
U_{12}	0	$-\omega^*$	0	ω^*
U_{22}	1	0	−1	0
$\chi(U)$	2	0	0	0

Table A3-86

Matrix elements at points T, connecting R and A, group D_{4h}^{14}. Here $p_1 = p_3 = \frac{1}{2}$, $p_2 = p$. At R, $p = 0$; at A, $p = \frac{1}{2}$.

	X_0	Y_2	X_0'	Y_2'
T_{11}	1	0	1	0
T_{21}	0	1	0	1
T_{12}	0	−1	0	1
T_{22}	1	0	−1	0
$\chi(T)$	2	0	0	0

Table A3-87

Matrix elements at points Y, connecting X and M, group D_{4h}^{14}. Here $p_1 = \frac{1}{2}$, $p_2 = p$, $p_3 = 0$. At X, $p = 0$; at M, $p = \frac{1}{2}$. $0 < p < \frac{1}{2}$. $\omega = \exp(i\pi p)$.

	X_0	Y_2	X_0'	Y_2'
Y_1	1	ω	1	ω
Y_2	1	ω	−1	$-\omega$
Y_3	1	$-\omega$	1	$-\omega$
Y_4	1	$-\omega$	−1	ω

Table A3-88

Compatibility relations for group D_{4h}^{14}.

Γ_1^+	Γ_1^-	Γ_2^+	Γ_2^-	Γ_3^+	Γ_3^-	Γ_4^+	Γ_4^-	Γ_5^+	Γ_5^-
Δ_1	Δ_4	Δ_2	Δ_3	Δ_1	Δ_4	Δ_2	Δ_3	$\Delta_1\Delta_2$	$\Delta_3\Delta_4$
Λ_1	Λ_1	Λ_2	Λ_2	Λ_3	Λ_3	Λ_4	Λ_4	Λ_5	Λ_5
Σ_1	Σ_2	Σ_4	Σ_3	Σ_4	Σ_3	Σ_1	Σ_2	$\Sigma_1\Sigma_4$	$\Sigma_2\Sigma_3$

X_1	X_2	Z_1	Z_2	Z_3	Z_4
$\Delta_1\Delta_2$	$\Delta_3\Delta_4$	$\Lambda_1\Lambda_4$	Λ_5	$\Lambda_2\Lambda_3$	Λ_5
W	W	U	U	U	U
Y_1Y_3	Y_2Y_4	S_1S_2	S_1S_3	S_3S_4	S_2S_4

M_1^+	M_1^-	M_2^+	M_2^-	M_3^+	M_3^-	M_4^+	M_4^-	M_5^+	M_5^-
Σ_1	Σ_1	Σ_4	Σ_4	Σ_2	Σ_2	Σ_3	Σ_3	$\Sigma_1\Sigma_4$	$\Sigma_2\Sigma_3$
V_1^+	V_1^-	V_2^+	V_2^-	V_1^+	V_1^-	V_2^+	V_2^-	V_3	V_3
Y_1	Y_3	Y_1	Y_3	Y_2	Y_4	Y_2	Y_4	Y_1Y_3	Y_2Y_4

SYMMETRY PROPERTIES FOR 20 GROUPS

Table A3-88 (Continued)

R_1	R_2	A_1	A_2	A_3	A_4
U	U	$V_1^+ V_1^-$	V_3	$V_2^+ V_2^-$	V_3
T	T	T	T	T	T
W	W	$S_1 S_2$	$S_1 S_3$	$S_3 S_4$	$S_2 S_4$

A3-8. The Space Group D_{2h}^{16} (Pnma). The space group D_{2h}^{16} (Pnma) has the point group D_{2h}, a simple orthorhombic Bravais lattice, and is

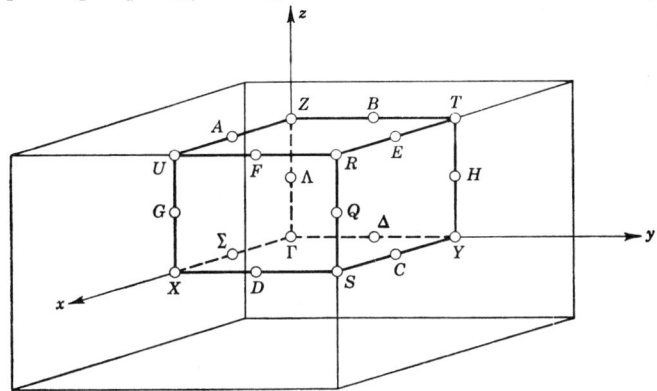

FIG. A3-7. Brillouin zone for simple orthorhombic Bravais lattice, with symmetry points following the notation of Koster.

found in the aragonite structure (Sec. 3-6). Positions of the atoms are described in Eq. (3-8) and the following discussion. The Brillouin zone is given in Fig. A3-7. The operations of the space group are given in Eq. (3-7). The effect of the operations on a plane wave is given by the equations

$$R_1\psi = \exp 2\pi i[(h_1 + p_1)\xi + (h_2 + p_2)\eta + (h_3 + p_3)\zeta]$$
$$R_2\psi = \exp 2\pi i[(h_1 + p_1)\xi - (h_2 + p_2)\eta - (h_3 + p_3)\zeta]$$
$$e^{\pi i(h_1+p_1+h_2+p_2+h_3+p_3)}$$
$$R_3\psi = \exp 2\pi i[-(h_1 + p_1)\xi + (h_2 + p_2)\eta - (h_3 + p_3)\zeta]$$
$$e^{\pi i(h_2+p_2)}$$
$$R_4\psi = \exp 2\pi i[-(h_1 + p_1)\xi - (h_2 + p_2)\eta + (h_3 + p_3)\zeta]$$
$$e^{\pi i(h_1+p_1+h_3+p_3)} \quad \text{(A3-35)}$$
$$R_5\psi = \exp 2\pi i[-(h_1 + p_1)\xi - (h_2 + p_2)\eta - (h_3 + p_3)\zeta]$$
$$R_6\psi = \exp 2\pi i[-(h_1 + p_1)\xi + (h_2 + p_2)\eta + (h_3 + p_3)\zeta]$$
$$e^{\pi i(h_1+p_1+h_2+p_2+h_3+p_3)}$$
$$R_7\psi = \exp 2\pi i[(h_1 + p_1)\xi - (h_2 + p_2)\eta + (h_3 + p_3)\zeta]e^{\pi i(h_2+p_2)}$$
$$R_8\psi = \exp 2\pi i[(h_1 + p_1)\xi + (h_2 + p_2)\eta - (h_3 + p_3)\zeta]$$
$$e^{\pi i(h_1+p_1+h_3+p_3)}$$

The multiplication table for the point group is given in Table A3-89. The additional factors in the multiplication table for the space group arising from the nonprimitive translations are complicated enough to present in the form of a table, Table A3-90. The matrix elements and characters for the various symmetry directions are given in Tables A3-91 to A3-105, and the compatibility relations in Table A3-106.

Table A3-89
Multiplication table for the point group D_{2h}. The table gives k, where $R_i R_j = R_k$.

i \ j	1	2	3	4	5	6	7	8
1	1	2	3	4	5	6	7	8
2	2	1	4	3	6	5	8	7
3	3	4	1	2	7	8	5	6
4	4	3	2	1	8	7	6	5
5	5	6	7	8	1	2	3	4
6	6	5	8	7	2	1	4	3
7	7	8	5	6	3	4	1	2
8	8	7	6	5	4	3	2	1

Table A3-90
Extra factors in multiplication table for space group D_{2h}^{16}, arising from nonprimitive translations. The columns R_1R_5, R_2R_6, etc., give the first factor in the product, while the rows labeled $R_1, R_2, \ldots$ give the second factor. The factor is exp $2\pi i \times$ (quantity tabulated).

	R_1R_5	R_2R_6	R_3R_7	R_4R_8
R_1	0	0	0	0
R_2	0	p_1	0	p_1
R_3	0	$-p_1 + p_2 - p_3$	p_2	$-p_1 - p_3$
R_4	0	$-p_2 + p_3$	$-p_2$	p_3
R_5	0	$-p_1 - p_2 - p_3$	$-p_2$	$-p_1 - p_3$
R_6	0	$p_2 + p_3$	p_2	p_3
R_7	0	0	0	0
R_8	0	p_1	0	p_1

Table A3-91
Matrix elements for irreducible representations of group D_{2h}^{16} at point Γ, $p_1 = p_2 = p_3 = 0$.

	R_1	R_2	R_3	R_4	R_5	R_6	R_7	R_8
Γ_1	1	1	1	1	1	1	1	1
Γ_2	1	1	1	1	-1	-1	-1	-1
Γ_3	1	-1	1	-1	1	-1	1	-1
Γ_4	1	-1	1	-1	-1	1	-1	1
Γ_5	1	1	-1	-1	1	1	-1	-1
Γ_6	1	1	-1	-1	-1	-1	1	1
Γ_7	1	-1	-1	1	1	-1	-1	1
Γ_8	1	-1	-1	1	-1	1	1	-1

Table A3-92
Matrix elements for irreducible representations of group D_{2h}^{16} along line Σ, $-\frac{1}{2} < p_1 \leq \frac{1}{2}$, $p_2 = p_3 = 0$; A, $-\frac{1}{2} < p_1 \leq \frac{1}{2}$, $p_2 = 0$, $p_3 = \frac{1}{2}$; C, $-\frac{1}{2} < p_1 \leq \frac{1}{2}$, $p_2 = \frac{1}{2}$, $p_3 = 0$; and E, $-\frac{1}{2} < p_1 \leq \frac{1}{2}$, $p_2 = p_3 = \frac{1}{2}$.

	R_1	R_2	R_7	R_8
Σ_1, A_1, C_1, E_1	1	$e^{\pi i p_1}$	1	$e^{\pi i p_1}$
Σ_2, A_2, C_2, E_2	1	$-e^{\pi i p_1}$	1	$-e^{\pi i p_1}$
Σ_3, A_3, C_3, E_3	1	$e^{\pi i p_1}$	-1	$-e^{\pi i p_1}$
Σ_4, A_4, C_4, E_4	1	$-e^{\pi i p_1}$	-1	$e^{\pi i p_1}$

Table A3-93
Matrix elements for irreducible representations of group D_{2h}^{16} along line Δ, $p_1 = 0$, $-\frac{1}{2} < p_2 \leq \frac{1}{2}$, $p_3 = 0$.

	R_1	R_3	R_6	R_8
Δ_1	1	$e^{\pi i p_2}$	$e^{\pi i p_2}$	1
Δ_2	1	$-e^{\pi i p_2}$	$e^{\pi i p_2}$	-1
Δ_3	1	$e^{\pi i p_2}$	$-e^{\pi i p_2}$	-1
Δ_4	1	$-e^{\pi i p_2}$	$-e^{\pi i p_2}$	1

Table A3-94
Matrix elements and characters for irreducible representation of group D_{2h}^{16} along line B, $p_1 = 0$, $-\frac{1}{2} < p_2 \leq \frac{1}{2}$, $p_3 = \frac{1}{2}$.

	R_1	R_3	R_6	R_8
$(B_1)_{11}$	1	$e^{\pi i p_2}$	0	0
$(B_1)_{21}$	0	0	$-e^{\pi i p_2}$	1
$(B_1)_{12}$	0	0	$e^{\pi i p_2}$	1
$(B_1)_{22}$	1	$-e^{\pi i p_2}$	0	0
$\chi(B_1)$	2	0	0	0

Table A3-95

Matrix elements and characters for irreducible representations of group D_{2h}^{16} along line D, $p_1 = \frac{1}{2}$, $-\frac{1}{2} < p_2 \leq \frac{1}{2}$, $p_3 = 0$.

	R_1	R_3	R_6	R_8
$(D_1)_{11}$	1	0	$e^{\pi i p_2}$	0
$(D_1)_{21}$	0	$e^{\pi i p_2}$	0	1
$(D_1)_{12}$	0	$e^{\pi i p_2}$	0	-1
$(D_1)_{22}$	1	0	$-e^{\pi i p_2}$	0
$\chi(D_1)$	2	0	0	0

Table A3-96

Matrix elements for irreducible representations of group D_{2h}^{16} along line F, for which $p_1 = \frac{1}{2}$, $-\frac{1}{2} < p_2 \leq \frac{1}{2}$, $p_3 = \frac{1}{2}$.

	R_1	R_3	R_6	R_8
F_1	1	$e^{\pi i p_2}$	$ie^{\pi i p_2}$	i
F_2	1	$e^{\pi i p_2}$	$-ie^{\pi i p_2}$	$-i$
F_3	1	$-e^{\pi i p_2}$	$ie^{\pi i p_2}$	$-i$
F_4	1	$-e^{\pi i p_2}$	$-ie^{\pi i p_2}$	i

Table A3-97

Matrix elements for irreducible representations of group D_{2h}^{16} along lines Λ, $p_1 = p_2 = 0$, $-\frac{1}{2} < p_3 \leq \frac{1}{2}$, and G, $p_1 = \frac{1}{2}$, $p_2 = 0$, $-\frac{1}{2} < p_3 \leq \frac{1}{2}$.

	R_1	R_4	R_6	R_7
Λ_1, G_1	1	$e^{\pi i p_3}$	$e^{\pi i p_3}$	1
Λ_2, G_2	1	$-e^{\pi i p_3}$	$e^{\pi i p_3}$	-1
Λ_3, G_3	1	$e^{\pi i p_3}$	$-e^{\pi i p_3}$	-1
Λ_4, G_4	1	$-e^{\pi i p_3}$	$-e^{\pi i p_3}$	1

Table A3-98

Matrix elements and characters for irreducible representations for group D_{2h}^{16} along lines H, $p_1 = 0$, $p_2 = \frac{1}{2}$, $-\frac{1}{2} < p_3 \leq \frac{1}{2}$, and Q, $p_1 = p_2 = \frac{1}{2}$, $-\frac{1}{2} < p_3 \leq \frac{1}{2}$.

	R_1	R_4	R_6	R_7
$(H_1)_{11}, (Q_1)_{11}$	1	$e^{\pi i p_3}$	0	0
$(H_1)_{21}, (Q_1)_{21}$	0	0	$-e^{\pi i p_3}$	1
$(H_1)_{12}, (Q_1)_{12}$	0	0	$e^{\pi i p_3}$	1
$(H_1)_{22}, (Q_1)_{22}$	1	$-e^{\pi i p_3}$	0	0
$\chi(H_1), \chi(Q_1)$	2	0	0	0

Table A3-99
Matrix elements and characters for irreducible representations of group D_{2h}^{16} at point X, $p_1 = \frac{1}{2}$, $p_2 = p_3 = 0$.

	R_1	R_2	R_3	R_4	R_5	R_6	R_7	R_8
$(X_1)_{11}$	1	0	0	1	0	1	1	0
$(X_1)_{21}$	0	1	1	0	1	0	0	1
$(X_1)_{12}$	0	-1	1	0	1	0	0	-1
$(X_1)_{22}$	1	0	0	-1	0	-1	1	0
$\chi(X_1)$	2	0	0	0	0	0	2	0
$(X_2)_{11}$	1	0	0	1	0	-1	-1	0
$(X_2)_{21}$	0	1	1	0	-1	0	0	-1
$(X_2)_{12}$	0	-1	1	0	-1	0	0	1
$(X_2)_{22}$	1	0	0	-1	0	1	-1	0
$\chi(X_2)$	2	0	0	0	0	0	-2	0

Table A3-100
Matrix elements and characters for irreducible representations of group D_{2h}^{16} at point Y, $p_1 = 0$, $p_2 = \frac{1}{2}$, $p_3 = 0$.

	R_1	R_2	R_3	R_4	R_5	R_6	R_7	R_8
$(Y_1)_{11}$	1	1	0	0	0	0	1	1
$(Y_1)_{21}$	0	0	1	1	1	1	0	0
$(Y_1)_{12}$	0	0	-1	1	1	-1	0	0
$(Y_1)_{22}$	1	-1	0	0	0	0	-1	1
$\chi(Y_1)$	2	0	0	0	0	0	0	2
$(Y_2)_{11}$	1	1	0	0	0	0	-1	-1
$(Y_2)_{21}$	0	0	1	1	-1	-1	0	0
$(Y_2)_{12}$	0	0	-1	1	-1	1	0	0
$(Y_2)_{22}$	1	-1	0	0	0	0	1	-1
$\chi(Y_2)$	2	0	0	0	0	0	0	-2

Table A3-101
Matrix elements and characters for irreducible representations of group D_{2h}^{16} for the point Z, $p_1 = p_2 = 0$, $p_3 = \frac{1}{2}$.

	R_1	R_2	R_3	R_4	R_5	R_6	R_7	R_8
$(Z_1)_{11}$	1	1	0	0	0	0	1	1
$(Z_1)_{21}$	0	0	1	1	1	1	0	0
$(Z_1)_{12}$	0	0	1	−1	1	−1	0	0
$(Z_1)_{22}$	1	−1	0	0	0	0	1	−1
$\chi(Z_1)$	2	0	0	0	0	0	2	0
$(Z_2)_{11}$	1	1	0	0	0	0	−1	−1
$(Z_2)_{21}$	0	0	1	1	−1	−1	0	0
$(Z_2)_{12}$	0	0	1	−1	−1	1	0	0
$(Z_2)_{22}$	1	−1	0	0	0	0	−1	1
$\chi(Z_2)$	2	0	0	0	0	0	−2	0

Table A3-102
Matrix elements and characters for irreducible representations of group D_{2h}^{16} at point S, $p_1 = p_2 = \frac{1}{2}$, $p_3 = 0$.

	R_1	R_2	R_3	R_4	R_5	R_6	R_7	R_8
$(S_1)_{11}$	1	i	0	0	0	0	1	i
$(S_1)_{21}$	0	0	1	i	1	i	0	0
$(S_1)_{12}$	0	0	−1	$-i$	1	i	0	0
$(S_1)_{22}$	1	i	0	0	0	0	−1	$-i$
$\chi(S_1)$	2	$2i$	0	0	0	0	0	0
$(S_2)_{11}$	1	$-i$	0	0	0	0	1	$-i$
$(S_2)_{21}$	0	0	1	$-i$	1	$-i$	0	0
$(S_2)_{12}$	0	0	−1	i	1	$-i$	0	0
$(S_2)_{22}$	1	$-i$	0	0	0	0	−1	i
$\chi(S_2)$	2	$-2i$	0	0	0	0	0	0

Table A3-103

Matrix elements and characters for irreducible representations of group D_{2h}^{16} at point T, $p_1 = 0$, $p_2 = p_3 = \frac{1}{2}$.

	R_1	R_2	R_3	R_4	R_5	R_6	R_7	R_8
$(T_1)_{11}$	1	1	0	0	1	1	0	0
$(T_1)_{21}$	0	0	1	1	0	0	-1	-1
$(T_1)_{12}$	0	0	-1	-1	0	0	-1	-1
$(T_1)_{22}$	1	1	0	0	-1	-1	0	0
$\chi(T_1)$	2	2	0	0	0	0	0	0
$(T_2)_{11}$	1	-1	0	0	1	-1	0	0
$(T_2)_{21}$	0	0	1	-1	0	0	-1	1
$(T_2)_{12}$	0	0	-1	1	0	0	-1	1
$(T_2)_{22}$	1	-1	0	0	-1	1	0	0
$\chi(T_2)$	2	-2	0	0	0	0	0	0

Table A3-104

Matrix elements for irreducible representations of group D_{2h}^{16} at point U, $p_1 = \frac{1}{2}$, $p_2 = 0$, $p_3 = \frac{1}{2}$.

	R_1	R_2	R_3	R_4	R_5	R_6	R_7	R_8
U_1	1	i	1	i	1	i	1	i
U_2	1	$-i$	1	$-i$	1	$-i$	1	$-i$
U_3	1	i	1	i	-1	$-i$	-1	$-i$
U_4	1	$-i$	1	$-i$	-1	i	-1	i
U_5	1	i	-1	$-i$	1	i	-1	$-i$
U_6	1	$-i$	-1	i	1	$-i$	-1	i
U_7	1	i	-1	$-i$	-1	$-i$	1	i
U_8	1	$-i$	-1	i	-1	i	1	$-i$

Table A3-105
Matrix elements and characters of irreducible representations of group D_{2h}^{16} at point R, $p_1 = p_2 = p_3 = \frac{1}{2}$.

	R_1	R_2	R_3	R_4	R_5	R_6	R_7	R_8
$(R_1)_{11}$	1	i	0	0	0	0	1	i
$(R_1)_{21}$	0	0	1	i	1	i	0	0
$(R_1)_{12}$	0	0	-1	i	1	$-i$	0	0
$(R_1)_{22}$	1	$-i$	0	0	0	0	-1	i
$\chi(R_1)$	2	0	0	0	0	0	0	$2i$
$(R_2)_{11}$	1	$-i$	0	0	0	0	1	$-i$
$(R_2)_{21}$	0	0	1	$-i$	1	$-i$	0	0
$(R_2)_{12}$	0	0	-1	$-i$	1	i	0	0
$(R_2)_{22}$	1	i	0	0	0	0	-1	$-i$
$\chi(R_2)$	2	0	0	0	0	0	0	$-2i$

Table A3-106
Compatibility relations for group D_{2h}^{16}

Γ_1	Γ_2	Γ_3	Γ_4	Γ_5	Γ_6	Γ_7	Γ_8	X_1	X_2	Y_1	Y_2	Z_1	Z_2
Σ_1	Σ_3	Σ_2	Σ_4	Σ_3	Σ_1	Σ_4	Σ_2	$\Sigma_1\Sigma_2$	$\Sigma_3\Sigma_4$	$\Delta_1\Delta_4$	$\Delta_2\Delta_3$	$\Delta_1\Delta_4$	$\Delta_2\Delta_3$
Δ_1	Δ_3	Δ_3	Δ_1	Δ_2	Δ_4	Δ_4	Δ_2	D_1	D_1	C_1C_2	C_3C_4	A_1A_2	A_3A_4
Λ_1	Λ_3	Λ_4	Λ_2	Λ_2	Λ_4	Λ_3	Λ_1	G_1G_4	G_2G_3	H_1	H_1	B_1	B_1

S_1	S_2	T_1	T_2	U_1	U_2	U_3	U_4	U_5	U_6	U_7	U_8
C_1C_3	C_2C_4	B_1	B_1	A_1	A_2	A_3	A_4	A_3	A_4	A_1	A_2
D_1	D_1	E_1E_3	E_2E_4	F_1	F_2	F_2	F_1	F_3	F_4	F_4	F_3
Q_1	Q_1	H_1	H_1	G_1	G_4	G_3	G_2	G_2	G_3	G_4	G_1

R_1	R_2
E_1E_4	E_2E_3
F_1F_4	F_2F_3
Q_1	Q_1

A3-9. The Space Group D_{2h}^{18} (Cmca). The space group D_{2h}^{18} (Cmca) has the point group D_{2h}, like the case D_{2h}^{16} (Pnma) taken up in the preceding section, but it has the one-face-centered orthorhombic Bravais lattice, described in Sec. 2-8 in connection with the iodine structure, which has this space group. The primitive vectors t_1, t_2, t_3 are given in Eq. (2-20), and the atomic positions in the iodine structure are given in Sec. 2-8. The reciprocal vectors b_1, b_2, b_3 are given in Eq. (5-14), and the Brillouin zone is shown in Fig. A3-8. The effect of the operations of the

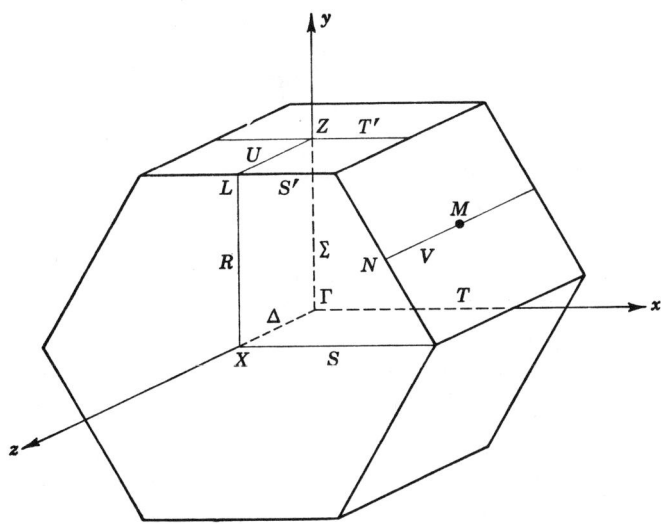

FIG. A3-8. Brillouin zone for the one-face-centered orthorhombic Bravais lattice, with symmetry points following the notation of Slater, Koster, and Wood.

space group on an arbitrary function is given in Eq. (2-22), and the effect on a plane wave is given in Eq. (A3-36) below:

$$R_1\psi = \exp 2\pi i[(h_1 + p_1)\xi + (h_2 + p_2)\eta + (h_3 + p_3)\zeta]$$
$$R_2\psi = \exp 2\pi i[-(h_1 + p_1)\xi - (h_2 + p_2)\eta + (h_3 + p_3)\zeta]$$
$$e^{\pi i(h_1+p_1+h_2+p_2+h_3+p_3)}$$
$$R_3\psi = \exp 2\pi i[-(h_2 + p_2)\xi - (h_1 + p_1)\eta - (h_3 + p_3)\zeta]$$
$$R_4\psi = \exp 2\pi i[(h_2 + p_2)\xi + (h_1 + p_1)\eta - (h_3 + p_3)\zeta]$$
$$e^{\pi i(h_1+p_1+h_2+p_2+h_3+p_3)} \quad \text{(A3-36)}$$
$$R_5\psi = \exp 2\pi i[-(h_1 + p_1)\xi - (h_2 + p_2)\eta - (h_3 + p_3)\zeta]$$
$$R_6\psi = \exp 2\pi i[(h_1 + p_1)\xi + (h_2 + p_2)\eta - (h_3 + p_3)\zeta]$$
$$e^{\pi i(h_1+p_1+h_2+p_2+h_3+p_3)}$$
$$R_7\psi = \exp 2\pi i[(h_2 + p_2)\xi + (h_1 + p_1)\eta + (h_3 + p_3)\zeta]$$
$$R_8\psi = \exp 2\pi i[-(h_2 + p_2)\xi - (h_1 + p_1)\eta + (h_3 + p_3)\zeta]$$
$$e^{\pi i(h_1+p_1+h_2+p_2+h_3+p_3)}$$

The multiplication table for the point group D_{2h} has already been given in Table A3-89. The additional factors in the multiplication table of the space group arising from the nonprimitive translations are given as follows:

First factor odd: unity
First factor even
2d is R_1 or R_7: unity
2d is R_2 or R_8: $e^{2\pi i p_3}$
2d is R_3 or R_5: $e^{-2\pi i(p_1+p_2+p_3)}$
2d is R_4 or R_6: $e^{2\pi i(p_1+p_2)}$

(A3-37)

The matrix elements of characters of the irreducible representations for the various symmetry directions are given in Tables A3-107 to A3-113, and the compatibility relations in Table A3-114.

Table A3-107

Matrix elements for the irreducible representations of the group D_{2h}^{18}, at the point Γ, $p_1 = p_2 = p_3 = 0$.

	R_1	R_2	R_3	R_4	R_5	R_6	R_7	R_8
Γ_1^+	1	1	1	1	1	1	1	1
Γ_2^+	1	−1	1	−1	1	−1	1	−1
Γ_3^+	1	1	−1	−1	1	1	−1	−1
Γ_4^+	1	−1	−1	1	1	−1	−1	1
Γ_1^-	1	1	1	1	−1	−1	−1	−1
Γ_2^-	1	−1	1	−1	−1	1	−1	1
Γ_3^-	1	1	−1	−1	−1	−1	1	1
Γ_4^-	1	−1	−1	1	−1	1	1	−1

Table A3-108

Matrix elements and characters of irreducible representations of the group D_{2h}^{18} at the points T, T', S, S'. Along T, $p_1 + p_2 = 0$, $p_3 = 0$. Along T', $p_1 + p_2 = 1$, $p_3 = 0$. Along S, $p_1 + p_2 = 0$, $p_3 = \frac{1}{2}$.

	R_1	R_3	R_6	R_8
T_1, T_1'	1	1	1	1
T_2, T_2'	1	1	−1	−1
T_3, T_3'	1	−1	1	−1
T_4, T_4'	1	−1	−1	1
$(S_1)_{11}$	1	0	1	0
$(S_1)_{21}$	0	1	0	1
$(S_1)_{12}$	0	1	0	−1
$(S_1)_{22}$	1	0	−1	0
$\chi(S_1)$	2	0	0	0

Table A3-109

Matrix elements of irreducible representations of the group D_{2h}^{18} along the directions Σ and R. Along Σ, $p_1 = p_2$, $p_3 = 0$. Along R, $p_1 = p_2$, $p_3 = \tfrac{1}{2}$.

	R_1	R_4	R_6	R_7
Σ_1	1	$e^{\pi i(p_1+p_2)}$	$e^{\pi i(p_1+p_2)}$	1
Σ_2	1	$-e^{\pi i(p_1+p_2)}$	$-e^{\pi i(p_1+p_2)}$	1
Σ_3	1	$-e^{\pi i(p_1+p_2)}$	$e^{\pi i(p_1+p_2)}$	-1
Σ_4	1	$e^{\pi i(p_1+p_2)}$	$-e^{\pi i(p_1+p_2)}$	-1
R_1	1	$-e^{\pi i(p_1+p_2)}$	$-e^{\pi i(p_1+p_2)}$	1
R_2	1	$e^{\pi i(p_1+p_2)}$	$e^{\pi i(p_1+p_2)}$	1
R_3	1	$e^{\pi i(p_1+p_2)}$	$-e^{\pi i(p_1+p_2)}$	-1
R_4	1	$-e^{\pi i(p_1+p_2)}$	$e^{\pi i(p_1+p_2)}$	-1

Table A3-110

Matrix elements of irreducible representations of group D_{2h}^{18} along directions Δ and U. Along Δ, $p_1 = p_2 = 0$, $-\tfrac{1}{2} < p_3 \leq \tfrac{1}{2}$. Along U, $p_1 = p_2 = \tfrac{1}{2}$, $-\tfrac{1}{2} < p_3 \leq \tfrac{1}{2}$.

	R_1	R_2	R_7	R_8
Δ_1	1	$e^{\pi i p_3}$	1	$e^{\pi i p_3}$
Δ_2	1	$-e^{\pi i p_3}$	1	$-e^{\pi i p_3}$
Δ_3	1	$e^{\pi i p_3}$	-1	$-e^{\pi i p_3}$
Δ_4	1	$-e^{\pi i p_3}$	-1	$e^{\pi i p_3}$
U_1	1	$-e^{\pi i p_3}$	1	$-e^{\pi i p_3}$
U_2	1	$e^{\pi i p_3}$	1	$e^{\pi i p_3}$
U_3	1	$-e^{\pi i p_3}$	-1	$e^{\pi i p_3}$
U_4	1	$e^{\pi i p_3}$	-1	$-e^{\pi i p_3}$

Table A3-111
Matrix elements and characters of irreducible representations at points N and M. group D_{2h}^{18}. At N, $p_1 = \frac{1}{2}$, $p_2 = 0$, $p_3 = \frac{1}{2}$; at M, $p_1 = \frac{1}{2}$, $p_2 = 0$, $p_3 = 0$.

	R_1	R_2	R_5	R_6
N_1	1	i	1	i
N_2	1	$-i$	1	$-i$
N_3	1	i	-1	$-i$
N_4	1	$-i$	-1	i
$(M_1)_{11}$	1	1	0	0
$(M_1)_{21}$	0	0	1	1
$(M_1)_{12}$	0	0	1	-1
$(M_1)_{22}$	1	-1	0	0
$\chi(M_1)$	2	0	0	0

Table A3-112
Matrix elements of irreducible representations of group D_{2h}^{18} at points X, L. At X, $p_1 = p_2 = 0$, $p_3 = \frac{1}{2}$. At L, $p_1 = p_2 = p_3 = \frac{1}{2}$.

	R_1	R_2	R_3	R_4	R_5	R_6	R_7	R_8
$(X_1,L_1)_{11}$	1	0	0	1	0	1	1	0
$(X_1,L_1)_{21}$	0	1	1	0	1	0	0	1
$(X_1,L_1)_{12}$	0	-1	1	0	1	0	0	-1
$(X_1,L_1)_{22}$	1	0	0	-1	0	-1	1	0
$\chi(X_1,L_1)$	2	0	0	0	0	0	2	0
$(X_2,L_2)_{11}$	1	0	0	-1	0	1	-1	0
$(X_2,L_2)_{21}$	0	1	-1	0	1	0	0	-1
$(X_2,L_2)_{12}$	0	-1	-1	0	1	0	0	1
$(X_2,L_2)_{22}$	1	0	0	1	0	-1	-1	0
$\chi(X_2,L_2)$	2	0	0	0	0	0	-2	0

Table A3-113
Matrix elements of irreducible representations of group D_{2h}^{18} at points H, V. H is general point of hexagonal face, $p_3 = \frac{1}{2}$. At V, $p_1 = \frac{1}{2}$, $p_2 = 0$.

	R_1	R_6
H_1	1	$-e^{\pi i(p_1+p_2)}$
H_2	1	$e^{\pi i(p_1+p_2)}$

	R_1	R_2
V_1	1	$e^{\pi i p_3}$
V_2	1	$-e^{\pi i p_3}$

Table A3-114
Compatibility relations, group D_{2h}^{18}

Γ_1^+	Γ_2^+	Γ_3^+	Γ_4^+	Γ_1^-	Γ_2^-	Γ_3^-	Γ_4^-	Z_1^+	Z_2^+	Z_3^+	Z_4^+	Z_1^-	Z_2^-	Z_3^-	Z_4^-
Δ_1	Δ_2	Δ_3	Δ_4	Δ_3	Δ_4	Δ_1	Δ_2	U_1	U_2	U_3	U_4	U_3	U_4	U_1	U_2
T_1	T_2	T_3	T_4	T_2	T_1	T_4	T_3	T_2'	T_1'	T_4'	T_3'	T_1'	T_2'	T_3'	T_4'
Σ_1	Σ_2	Σ_3	Σ_4	Σ_4	Σ_3	Σ_2	Σ_1	Σ_1	Σ_2	Σ_3	Σ_4	Σ_4	Σ_3	Σ_2	Σ_1

T_1	T_2	T_3	T_4	N_1	N_2	N_3	N_4	R_1	R_2	R_3	R_4
T_1'	T_2'	T_3'	T_4'	V_1	V_2	V_1	V_2	H_1	H_2	H_1	H_2

X_1	X_2	L_1	L_2	M_1
Δ_1	Δ_3	U_1	U_3	V_1
Δ_2	Δ_4	U_2	U_4	V_2
R_1	R_3	R_1	R_3	—
R_2	R_4	R_2	R_4	
S_1	S_1	S_1'	S_1'	

Appendix 4
The Momentum Eigenfunction and Fourier Expansion of the Potential

In Eq. (8-4) we give the expression for the momentum eigenfunction found in a Bloch sum. If we have an atomic eigenfunction $u(\mathbf{r})$ centered on the origin, its momentum eigenfunction is given by

$$v(\mathbf{k}) = \Omega^{-1} \int u(\mathbf{r}) e^{-i\mathbf{k}\cdot\mathbf{r}} \, dv \qquad (A4\text{-}1)$$

in which Ω is the volume of the unit cell. We shall show how to evaluate the integral in Eq. (A4-1). The atomic eigenfunction can be written in the form

$$u(\mathbf{r}) = \frac{(-1)^{(m+|m|)/2}}{(4\pi)^{1/2}} \sqrt{\frac{(2l+1)(l-|m|)!}{(l+|m|)!}} R_{nl}(r) P_l^{|m|}(\cos\theta) e^{im\phi} \qquad (A4\text{-}2)$$

as in "Quantum Theory of Atomic Structure," vol. 1, eq. (7-31). To carry out the evaluation of Eq. (A4-1), we express the plane wave as an expansion in spherical harmonics, according to the equation

$$e^{-i\mathbf{k}\cdot\mathbf{r}} = \sum_{l=0}^{\infty} \sum_{m=-l}^{l} (2l+1)(-i)^l \frac{(l-|m|)!}{(l+|m|)!} j_l(kr)$$
$$P_l^{|m|}(\cos\theta) P_l^{|m|}(\cos\theta_0) e^{-im(\phi-\phi_0)} \qquad (A4\text{-}3)$$

where θ_0, ϕ_0 are the angles defining the direction of the wave vector $\mathbf{k}$, and k is used as its magnitude.

We now insert the expansion of Eq. (A4-3) in the integral of Eq. (A4-1). In carrying out the integration, we use the orthogonality properties of the spherical harmonics, which tell us that the integral of the product of two spherical harmonics over all angles is zero, unless they both correspond to the same l and m values. Hence in the summation of Eq. (A4-3) we may omit all terms except that for which l and m are

App. 4] MOMENTUM EIGENFUNCTION AND FOURIER EXPANSION 453

equal to those of the atomic eigenfunction given in Eq. (A4-2). We then have

$$v(\mathbf{k}) = \Omega^{-1} \frac{(-1)^{(m+|m|)/2}}{(4\pi)^{1/2}} \sqrt{\frac{(2l+1)(l-|m|)!}{(l+|m|)!}} \, (2l+1)(-i)^l \frac{(l-|m|)!}{(l+|m|)!}$$
$$P_l^{|m|}(\cos\theta_0) e^{im\phi_0} \int R_{nl}(r) j_l(kr) [P_l^{|m|}(\cos\theta)]^2 \, dv \quad \text{(A4-4)}$$

We now use the result that

$$\int_0^\pi [P_l^{|m|}(\cos\theta)]^2 \sin\theta \, d\theta = \frac{2}{2l+1} \frac{(l+|m|)!}{(l-|m|)!} \quad \text{(A4-5)}$$

When we introduce the element of volume $2\pi r^2 \sin\theta \, dr \, d\theta \, d\phi$ into Eq. (A4-4), integrate over ϕ from zero to 2π, and integrate with respect to θ from 0 to π, and with respect to r from zero to infinity, we have

$$v(\mathbf{k}) = \Omega^{-1} \frac{(-1)^{(m+|m|)/2}}{(4\pi)^{1/2}} \sqrt{\frac{(2l+1)(l-|m|)!}{(l+|m|)!}} \, 4\pi(-i)^l$$
$$\left[\int_0^\infty r^2 R_{nl}(r) j_l(kr) \, dr \right] P_l^{|m|}(\cos\theta_0) e^{im\phi_0} \quad \text{(A4-6)}$$

This result verifies the statement made in the discussion of Eq. (8-5), to the effect that the momentum eigenfunction depends in the same way on angles in the $\mathbf{k}$ space that the coordinate eigenfunction does on angles in ordinary space, and that the radial part of the momentum eigenfunction is given by the radial integral in Eq. (A4-6), which is the same as that given in Eq. (8-5).

It may be that the radial atomic eigenfunction $R_{nl}(r)$ is expanded in analytic form as a linear combination of terms of the form $e^{-a_r r} r^{l+p-1}$. In this case the evaluation of the integral in Eq. (A4-6) can be easily carried out by means of the following theorem, which can be derived from an equation in G. N. Watson, "Theory of Bessel Functions," 2d ed., Cambridge University Press, New York, 1958, p. 386:

$$\int_0^\infty e^{-ar} j_l(kr) r^{l+p+1} \, dr = (-1)^p l! (2k)^l \frac{d^p}{da^p} \left[\frac{1}{(a^2+k^2)^{l+1}} \right] \quad \text{(A4-7)}$$

In addition to the momentum eigenfunction, we require the Fourier components of the spherical potential acting on an electron in an atom, given by Eq. (6-6), where $V(r)$ is the spherically symmetrical potential. This is obviously carried out in the same way just described, expanding the plane wave by Eq. (A4-3), but now on account of the spherical symmetry the quantity $W(\mathbf{k})$ is spherically symmetrical in $\mathbf{k}$ space. We evidently have

$$\begin{aligned} W(\mathbf{k}) &= \Omega^{-1} \int V(r) e^{-i\mathbf{k}\cdot\mathbf{r}} \, dv \\ &= \Omega^{-1} \, 4\pi \int_0^\infty r^2 V(r) j_0(kr) \, dr \\ &= \Omega^{-1} \, 4\pi \int_0^\infty r V(r) \frac{\sin kr}{k} \, dr \end{aligned} \qquad \text{(A4-8)}$$

where we have used the relation $j_0(x) = (\sin x)/x$.

Appendix 5
Power-series Expansions of Energy Bands near Symmetry Points

In Chap. 8 and Appendix 3 we have examined the nature of the symmetry behavior of energy bands at symmetry points, and the qualitative nature of the departure from this behavior as we go away from these points. There is, however, a general method, based on perturbation theory, which allows us to get quantitative results concerning the behavior near symmetry points, as a power-series expansion in $\mathbf{k} - \mathbf{k}_0$, if $\mathbf{k}_0$ represents the $\mathbf{k}$ vector of the symmetry point. This method was used by Seitz,[1] in studying the energy bands of lithium, and is generally known as the $\mathbf{k} \cdot \mathbf{p}$ method, for reasons which will become obvious. Though it can be applied to any symmetry point, the commonest application is to $\mathbf{k}_0 = 0$, and we shall examine that case.

We know by general principles that the wave function for a given value of $\mathbf{k}$ can be written in the form of $\exp(i\mathbf{k} \cdot \mathbf{r})$ times a periodic function of position. Furthermore, we know that the true wave functions of the periodic potential problem, for $\mathbf{k} = 0$, form a complete set of periodic functions, in terms of which any periodic function can be expanded. Let these functions be denoted as u_{n0}, where the n numbers the functions, and 0 indicates that they are for $\mathbf{k} = 0$. We then propose to use an expansion

$$u_{pk} = \Sigma(n) C_{npk} u_{n0} e^{i\mathbf{k} \cdot \mathbf{r}} \qquad (A5\text{-}1)$$

for the wave functions associated with a given value of k. To set up a secular equation for the coefficients, C_{npk}, we need the matrix elements of the Hamiltonian between the basis functions $u_{n0} e^{i\mathbf{k} \cdot \mathbf{r}}$.

The Hamiltonian operator is $-\nabla^2 + V$, where the first term is the kinetic energy (in Rydbergs), the second the potential energy. The

---

[1] F. Seitz, *Phys. Rev.*, **47**:400 (1935). See also for the general quantum-mechanical method, L. I. Schiff, "Quantum Mechanics," 2d ed., McGraw-Hill Book Company, 1955, p. 156; for the application to band theory, E. O. Kane, *J. Phys. Chem. Solids*, **8**:38 (1959).

functions u_{n0} are assumed to satisfy this equation, so that

$$(-\nabla^2 + V)u_{n0}(r) = E_{n0}u_{n0}(r) \qquad (A5\text{-}2)$$

Then we can show easily that

$$(-\nabla^2 + V)u_{n0}(r)e^{i\mathbf{k}\cdot\mathbf{r}} = [(E_{n0} + k^2 - 2i\mathbf{k}\cdot\nabla)u_{n0}]e^{i\mathbf{k}\cdot\mathbf{r}} \qquad (A5\text{-}3)$$

where the operator ∇ on the right side of Eq. (A5-3) operates only on u_{n0}. We multiply Eq. (A5-3) on the left by $u_{m0}^{*}(r)e^{-i\mathbf{k}\cdot\mathbf{r}}$, and integrate over the volume, to get the matrix element of the Hamiltonian between two of our functions. We find that this element, which can be written as H_{mn}^0, is

$$H_{mn}^0 = (E_{m0} + k^2)\delta_{mn} - 2i(\mathbf{k}\cdot\nabla)_{mn} \qquad (A5\text{-}4)$$

where the matrix element $(\mathbf{k}\cdot\nabla)_{mn}$ is to be computed with respect to the functions u_{n0}. Since the momentum operator is $-i\hbar\nabla$, in ordinary coordinates, we see that the last term in Eq. (A5-4) is proportional to the matrix element of $(\mathbf{k}\cdot\mathbf{p})$, where $\mathbf{p}$ is the momentum operator. This explains the name of the method.

If we are investigating the behavior of a nondegenerate state in the neighborhood of $\mathbf{k} = 0$, we can use second-order perturbation theory to solve our secular equation. We have

$$E_{mk} = E_{m0} + k^2 + 4\sum(j \neq m)\frac{|(\mathbf{k}\cdot\nabla)_{mj}|^2}{E_{m0} - E_{j0}} \qquad (A5\text{-}5)$$

$$u_{mk} = \left[u_{m0} + \sum(j \neq m)\frac{-2i(\mathbf{k}\cdot\nabla)_{jm}}{E_{m0} - E_{j0}}u_{j0}\right]e^{i\mathbf{k}\cdot\mathbf{r}} \qquad (A5\text{-}6)$$

We see that the second-order term in the energy will be quadratic in k, just as the term k^2 is; in a cubic crystal, on account of symmetry, the second-order term is a constant times k^2. Hence in a cubic crystal the energy of a nondegenerate state for small k will be proportional to k^2, but with a different proportionality constant from what we have with a free electron. This can be regarded as the energy of a free electron with an effective mass different from the free-electron mass. As for the wave function, in Eq. (A5-6), we see that we are adding to the function u_{m0} other functions u_{j0} which have other symmetry types, with coefficients proportional to k.

As an example, we have seen in Table A3-31 that if we start from a Γ_1 symmetry at the center of the zone, in a cubic crystal, this will join to a Δ_1 symmetry along the x axis. But not only an s-like function, but also a p_x-like function, will have this type of symmetry. In this particular case, we find that the wave function will have a contribution of p_x-like type, with an amplitude proportional to k_x. We can verify this fact, if we take the wave functions u_{m0}, and find the matrix elements of the operator ∇ with respect to them. If we take the two-dimensional case for simplicity, a basis function for the representation Γ_1 can be found as in Eq. (5-30), giving $\cos(2\pi h_1 x/a) \cos(2\pi h_2 y/a) + \cos(2\pi h_2 x/a) \cos(2\pi h_1 y/a)$. If we apply the operator $\partial/\partial x$ to this, we find $-2\pi h_1/a \sin(2\pi h_1 x/a) \cos(2\pi h_2 y/a) - 2\pi h_2/a \sin(2\pi h_2 x/a) \cos(2\pi h_1 y/a)$, which is of the correct form for a basis function for the x component of the two-dimensional p-like representation Γ_5 of Table 5-3. Hence we see that the operator $\partial/\partial x$ has nonvanishing components between the function of symmetry Γ_1 and this p-like function, and hence in Eq. (A5-6) the wave function will contain a contribution of this p_x-type function, proportional to k_x.

If we are near a degenerate state, the situation is more complicated. To have a specific case, let us assume that we are dealing with the two-dimensional square lattice, and that we wish to investigate the behavior of the doubly degenerate Γ_5 state as **k** departs from zero. We have two degenerate states at **k** = 0, with wave functions of the type given in Eq. (5-37), namely, $\sin(2\pi h_1 x/a) \cos(2\pi h_2 y/a)$ and $\cos(2\pi h_2 x/a) \sin(2\pi h_1 y/a)$. If we apply the operator $\partial/\partial x$ to the x-like function

$$\sin(2\pi h_1 x/a) \cos(2\pi h_2 y/a)$$

we get something proportional to $\cos(2\pi h_1 x/a) \cos(2\pi h_2 y/a)$, which, as we see from Eqs. (5-30) and (5-33), is a linear combination of the symmetry types Γ_1 and Γ_3. Hence this operator has nondiagonal matrix elements between the Γ_{5x} function and these two types. Similarly, the operator $\partial/\partial y$ operating on the Γ_{5x} function gives a linear combination of Γ_2 and Γ_4, so that there are nondiagonal matrix elements of $\partial/\partial y$ between Γ_{5x} and Γ_2 and Γ_4. There are no nondiagonal matrix elements of the **k · p** operator between the two degenerate p-like functions Γ_{5x} and Γ_{5y}.

We can handle this degenerate problem by a method described by Schiff (*loc. cit.*) which can be explained in terms of the method of bordered determinants, a useful method for deriving the perturbation method.[1] Let us assume that we have two unperturbed states of a problem, degenerate with each other, so that there is no nondiagonal matrix element of the exact Hamiltonian between them. They have, however, nonvanishing matrix elements to other excited states. If the degenerate states are labeled 0 and 1, the equations for the coefficients C, in such an expansion as is given in Eq. (A5-1), will be

$$\begin{aligned}
(H_{00} - E)C_0 \qquad\qquad\quad + H_{02}C_2 \quad + H_{03}C_3 + \cdots &= 0 \\
(H_{00} - E)C_1 + H_{12}C_2 \quad + H_{13}C_3 + \cdots &= 0 \\
H_{20}C_0 + H_{21}C_1 \quad + (H_{22} - E)C_2 + H_{23}C_3 + \cdots &= 0 \\
\cdots\cdots\cdots\cdots\cdots\cdots\cdots\cdots\cdots\cdots\cdots\cdots\cdots\cdots\cdots\cdots\cdots\cdots\cdots&
\end{aligned}$$
(A5-7)

---

[1] See J. C. Slater, "Quantum Theory of Atomic Structure," vol. 1, McGraw-Hill Book Company, New York, 1960, sec. 5-4.

If a complete treatment is carried through, it is found that the non-diagonal terms involving H_{pq}, where p, q are both greater than 1, do not enter into the leading terms. Let us take advantage of this fact, and disregard them from the beginning. Then any equation from the third on can be written in the form

$$H_{p0}C_0 + H_{p1}C_1 + (H_{pp} - E)C_p = 0$$
$$C_p = \frac{H_{p0}C_0 + H_{p1}C_1}{E - H_{pp}} \tag{A5-8}$$

We substitute this expression into the first two equations of Eq. (A5-7), for C_3 . . . . We see that these become two equations for C_0 and C_1, which take the form

$$(H'_{00} - E)C_0 + H'_{01}C_1 = 0$$
$$H'_{10}C_0 + (H'_{11} - E)C_1 = 0 \tag{A5-9}$$

where
$$H'_{00} = H_{00} + \sum (p \neq 0, 1) \frac{H_{0p}H_{p0}}{H_{00} - H_{pp}}$$
$$H'_{11} = H_{00} + \sum (p \neq 0, 1) \frac{H_{1p}H_{p1}}{H_{00} - H_{pp}}$$
$$H'_{01} = \sum (p \neq 0, 1) \frac{H_{0p}H_{p1}}{H_{00} - H_{pp}}$$
$$H'_{10} = \sum (p \neq 0, 1) \frac{H_{1p}H_{p0}}{H_{00} - H_{pp}}$$
(A5-10)

in which we have replaced E in the denominator by its unperturbed value H_{00}. In other words, we have reduced the problem to a secular problem with two rows and columns, which is no longer degenerate, on account of the fact that H'_{00} and H'_{11} are not equal, and which now has a non-diagonal matrix element between the two functions. We solve these equations in the usual way for a secular problem with two rows and columns.

If we apply this method to our case, in which the matrix elements of the Hamiltonian are given by Eq. (A5-4), we are to replace H_{00}, the diagonal matrix element of the Hamiltonian, by $E_0 + k^2$, and are to replace the matrix elements H_{0p}, etc., by $-2i(\mathbf{k} \cdot \nabla)_{0p}$, etc. Then we find

$$H'_{00} = E_0 + k^2 + \sum (p \neq 0, 1) \frac{-4(\mathbf{k} \cdot \nabla)_{0p}(\mathbf{k} \cdot \nabla)_{p0}}{E_0 - E_p}$$
$$H'_{11} = E_0 + k^2 + \sum (p \neq 0, 1) \frac{-4(\mathbf{k} \cdot \nabla)_{1p}(\mathbf{k} \cdot \nabla)_{p1}}{E_0 - E_p}$$
$$H'_{01} = \sum (p \neq 0, 1) \frac{-4(\mathbf{k} \cdot \nabla)_{0p}(\mathbf{k} \cdot \nabla)_{p1}}{E_0 - E_p}$$
$$H'_{10} = \sum (p \neq 0, 1) \frac{-4(\mathbf{k} \cdot \nabla)_{1p}(\mathbf{k} \cdot \nabla)_{p0}}{E_0 - E_p} \tag{A5-11}$$

We assume that the function labeled by the index 0 is the Γ_5 function of x symmetry, and that labeled by 1 is the Γ_5 of y symmetry.

Let us now take the information which we have about the basis functions and the effect of operating on them with the operator ∇, and see what properties these matrix elements must have. First we consider H'_{00}, and the quantity $(\mathbf{k} \cdot \nabla)_{0p}$. This is a sum of two terms, $(k_x \partial/\partial x)_{0p} + (k_y \partial/\partial y)_{0p}$. This will equal the complex conjugate of $(k_x \partial/\partial x)_{p0} + (k_y \partial/\partial y)_{p0}$. For the latter form of the expression, we must first operate on the Γ_{5x} function with $\partial/\partial x$ and $\partial/\partial y$. We have seen that when we operate with $\partial/\partial x$ we get a combination of Γ_1 and Γ_3 functions. Hence this first term has nondiagonal matrix elements to all functions p which are of the Γ_1 and Γ_3 symmetry types. Similarly, the second term has nondiagonal matrix elements to all functions of the Γ_2 and Γ_4 types. In the sum over p, those particular excited functions of the Γ_1 or Γ_3 types in the summation will contribute terms in k_x^2, and those of the Γ_2 and Γ_4 types will contribute terms in k_y^2. Since these are quite different excited functions, the two quantities will have quite different coefficients. Hence we shall have $H'_{00} = E_0 + A k_x^2 + B k_y^2$. On account of symmetry, we find that $H'_{11} = E_0 + B k_x^2 + A k_y^2$.

Next we consider H'_{01} and H'_{10} in the same manner. If we take H'_{01}, we require the product $(\mathbf{k} \cdot \nabla)_{0p}(\mathbf{k} \cdot \nabla)_{p1}$. Now we have found that the operator $\partial/\partial x$ operating on the Γ_{5x} function produces Γ_1 and Γ_3 functions, and we get the same functions when the operator $\partial/\partial y$ operates on the Γ_{5y}. Hence from such excited states we get terms proportional to $k_x k_y$. Again, $\partial/\partial y$ operating on Γ_{5x} produces Γ_2 and Γ_4 functions, as does $\partial/\partial x$ operating on Γ_{5y}. These states again lead to terms in $k_x k_y$. Hence we conclude that H'_{01} is proportional to $k_x k_y$, say, equal to $C k_x k_y$. By symmetry, H'_{10} will have the same value. Hence we have a secular equation

$$\begin{vmatrix} E_0 + A k_x^2 + B k_y^2 - E & C k_x k_y \\ C k_x k_y & E_0 + B k_x^2 + A k_y^2 - E \end{vmatrix} = 0 \quad \text{(A5-12)}$$

A simple extension to the three-dimensional case, where we have the threefold degeneracy of type Γ_{15} (in the BSW notation), shows that the secular equation between the three unperturbed functions of p_x, p_y, p_z types is

$$\begin{vmatrix} E_0 + A k_x^2 + B(k_y^2 + k_z^2) - E & C k_x k_y & C k_x k_z \\ C k_x k_y & E_0 + A k_y^2 + B(k_x^2 + k_z^2) - E & C k_y k_z \\ C k_x k_z & C k_y k_z & E_0 + A k_z^2 + B(k_x^2 + k_y^2) - E \end{vmatrix} = 0 \quad \text{(A5-13)}$$

This equation was set up and discussed by Shockley.[1]

In the two-dimensional case of Eq. (A5-12), we can solve the quadratic secular equation, and find the energy as a function of k_x and k_y. It is

---

[1] W. Shockley, *Phys. Rev.*, **78**:173 (1950).

more informative to give the energy as a function of k, the magnitude of the $\mathbf{k}$ vector, and θ, the angle between the vector and the x axis, so that $k_x = k \cos \theta$, $k_y = k \sin \theta$. Then the solution of the quadratic gives

$$E - E_0 = k^2 \left\{ \left(\frac{A+B}{2}\right) \pm \left[\left(\frac{A-B}{2}\right)^2 \cos^2 2\theta + \frac{C^2}{4} \sin^2 2\theta \right]^{1/2} \right\} \quad (A5\text{-}14)$$

From Eq. (A5-14) we see that the energy is proportional to k^2 for small $\mathbf{k}$. However, on account of the square root, it cannot be expanded in power series in k_x and k_y. We can understand the nature of the energy as

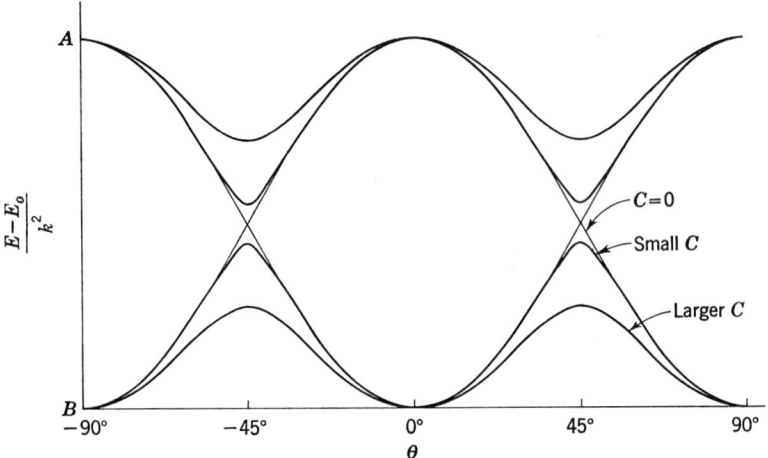

Fig. A5-1. Energy as a function of angle, near a lattice point with Γ_{15} type of degeneracy, for several values of the constant C.

a function of k and θ by plotting the function of angle given in Eq. (A5-14). This is done in Fig. A5-1, for several values of C. We see that for $C = 0$, the two bands cross at 45°, but for $C \neq 0$, as we should actually have, one band has an energy above the other, for all angles. The two bands come together, as Eq. (A5-14) shows, when k goes to zero.

This simple example shows how the $\mathbf{k} \cdot \mathbf{p}$ method can be applied to show the general form that the energy bands must have near a symmetry point, in terms of several parameters, such as A, B, and C in this case. Of course, if the wave functions have been determined in detail, these parameters can be computed from them. Otherwise they can be determined to give agreement with energy-band calculations carried out by other methods.

Appendix 6
Matrix Elements for the Augmented-plane-wave Method

We wish to derive the equation for the matrix elements $(H - E)_{ij}$ of the Hamiltonian operator minus the energy, for the augmented-plane-wave method, as given in Eq. (9-3). The derivation is given by J. C. Slater, *Phys. Rev.*, **51**:846 (1937). We shall take the special case where there is only one atom in a unit cell, at the origin, though the extension to the case of an arbitrary number of atoms in the unit cell is given in the reference above, and follows simply. It is assumed in the method that the potential equals zero outside nonoverlapping spheres of radius R surrounding the atoms, while it is a spherically symmetric function $V(r)$ in the sphere surrounding the origin, with periodically repeating values in all unit cells.

As basis functions for the solution of Schrödinger's equation we use augmented plane waves, equal to $e^{i\mathbf{k}_i \cdot \mathbf{r}}$ in the region between the spheres, where $\mathbf{k}_i = \mathbf{k} + \mathbf{K}_i$, $\mathbf{k}$ being the reduced wave vector, $\mathbf{K}_i$ a vector of the reciprocal lattice, and joining continuously at the surface of each sphere to a solution of the spherical wave equation within the sphere. We use atomic units, with Rydbergs as the unit of energy. The Schrödinger equation inside the sphere is then

$$-\nabla^2 u + V(r)u = Eu \qquad (A6\text{-}1)$$

For an arbitrary energy E we shall have solutions of this equation in the form of a spherical harmonic of angles, times $u_l(r)$, where l is the azimuthal quantum number. This solution is assumed to be regular at the origin, and we do not care about its behavior outside the sphere of radius R.

To get continuity between the function inside the sphere and the plane wave outside we proceed as follows. We expand the plane wave in the familiar expression

$$e^{i\mathbf{k}_i \cdot \mathbf{r}} = \sum_{l=0}^{\infty} \sum_{m=-l}^{l} (2l+1) i^l \frac{(l-|m|)!}{(l+|m|)!} j_l(k_i r) P_l^{|m|}(\cos\theta)$$

$$P_l^{|m|}(\cos\theta_i) e^{im(\phi - \phi_i)} \qquad (A6\text{-}2)$$

Here θ_i, ϕ_i represent the direction of the wave normal, and k_i is the magnitude of $\mathbf{k}_i$. The functions j_l are spherical Bessel functions. If we set $r = R$ in this expression, we have the expansion of the plane wave in spherical harmonics on the surface of the sphere. To get continuity of the function inside the sphere with the plane wave, we need then merely make a linear combination of spherical solutions of Eq. (A6-1) inside, with all values of l and m, choosing the coefficient of each term so that it will equal the coefficient of the corresponding term in Eq. (A6-2) when $r = R$. That is, the function inside the sphere is

$$\psi_i = \sum_{l=0}^{\infty} \sum_{m=-l}^{l} (2l+1) i^l \frac{(l-|m|)!}{(l+|m|)!} \frac{j_l(k_i R)}{u_l(R)} u_l(r) P_l^{|m|}(\cos\theta)$$
$$P_l^{|m|}(\cos\theta_i) e^{im(\phi-\phi_i)} \quad (A6\text{-}3)$$

A single augmented plane wave ψ_i is now defined as being equal to the plane wave $e^{i\mathbf{k}_i \cdot \mathbf{r}}$ of Eq. (A6-2) outside the sphere, and equal to the function of Eq. (A6-3) inside the sphere. We now assume that the solution of the problem can be expressed as a sum $\Sigma(i) v_i \psi_i$, where the v_i's are coefficients to be determined. As usual, to find the v's, we must satisfy the equations

$$\Sigma(j)(H-E)_{ij} v_j = 0 \quad (A6\text{-}4)$$

where we shall use the definition

$$(H-E)_{ij} = \frac{1}{\Omega} \int \psi_i^* (H-E) \psi_j \, dv \quad (A6\text{-}5)$$

in which H represents the Hamiltonian, equal to $-\nabla^2 + V(r)$ inside the sphere, $-\nabla^2$ outside the sphere. The integration is over a unit cell, of volume Ω. We note that the augmented plane waves are neither normalized nor orthogonal, so that the quantity E_{ij} is not equal to δ_{ij} times an energy.

We must notice one point at the outset. Though our function ψ_i is everywhere continuous, its first derivative is in general not continuous at the surfaces of the various spheres, where the functions join. It is not always realized that the kinetic-energy operator demands special treatment for a function with discontinuous slope. Two forms of integral are often seen for computing the kinetic energy. The more common one is $\int \psi_i^* (-\nabla^2) \psi_j \, dv$, but the other and more fundamental one is $\int \operatorname{grad} \psi_i^* \cdot \operatorname{grad} \psi_j \, dv$, found in Schrödinger's first papers. Ordinarily, one can show by integration by parts that one equals the other, but if the function has anywhere a discontinuous slope, they are no longer equal, but differ by a

App. 6] MATRIX ELEMENTS FOR THE AUGMENTED-PLANE-WAVE METHOD 463

surface integral over the surface of discontinuity. In this case the second, more fundamental form is the correct one, for it is the one which enters directly into the variation principle from which Schrödinger's equation is derived. If there is any doubt about this question, it can be easily shown that if we use the first formula, we must add a surface integral, for a discontinuous first derivative amounts to an infinite second derivative on the surface, and integrates to a finite contribution over the surface. This contribution can be found by a limiting process in which the change of slope occurs in smaller and smaller ranges of variable. When this surface integral is added to the expression $\int \psi_i^* (-\nabla^2) \psi_j \, dv$, the result becomes equal to $\int \text{grad } \psi_i^* \cdot \text{grad } \psi_j \, dv$. Let us carry through this process.

We assume that we have continuous functions ψ_i and ψ_j, with continuous derivatives through all space except on a surface S. Let this surface be closed, as our sphere is closed, and let the volume on one side be called V_1, that on the other side V_2. Let n be the normal pointing from V_1 to V_2. Then a small area of surface da will make a finite contribution to the integral $\int \psi_i^* \nabla^2 \psi_j \, dv$. If we take a small cylinder of base da, bounded by surfaces parallel to S inside the two volumes, this will make a contribution to the integral equal to

$$da \, \psi_i^* \left[\left(\frac{\partial \psi_j}{\partial n} \right)_2 - \left(\frac{\partial \psi_j}{\partial n} \right)_1 \right]$$

In other words, to the integral $\int \psi_i^* \nabla^2 \psi_j \, dv$ through the volume V_1 should be added the term $-\int \psi_i^* (\partial \psi_j / \partial n) \, da$ integrated over the bounding surface S, where $\partial/\partial n$ is the derivative with respect to the outer normal to the volume, with a similar term added to the integral over V_2. But by a standard form of Green's theorem, we know that

$$\int \psi_i^* \nabla^2 \psi_j \, dv - \int \psi_i^* \frac{\partial \psi_j}{\partial n} \, da = - \int \text{grad } \psi_i^* \cdot \text{grad } \psi_j \, dv \quad \text{(A6-6)}$$

In other words, we verify the statement that if we add the correction to the ordinary form of the kinetic energy integral arising from the discontinuous derivative, we come to the more fundamental expression $\int \text{grad } \psi_i^* \cdot \text{grad } \psi_j \, dv$ for the kinetic energy.

With this understanding of the situation, we can now proceed to the calculation of the matrix elements of Eq. (A6-5). First we consider the plane waves, and the integral over the region outside the sphere. In this region, $\text{grad } \psi_i^* \cdot \text{grad } \psi_j = (\mathbf{k}_i \cdot \mathbf{k}_j) e^{i(\mathbf{k}_j - \mathbf{k}_i) \cdot \mathbf{r}}$, and $\psi_i^* E \psi_j = E e^{i(\mathbf{k}_j - \mathbf{k}_i) \cdot \mathbf{r}}$, so that the integrand of Eq. (A6-5) is $(\mathbf{k}_i \cdot \mathbf{k}_j - E) e^{i(\mathbf{k}_j - \mathbf{k}_i) \cdot \mathbf{r}}$. If we integrate this throughout the whole cell, we get zero unless $i = j$, for we know that the integral of a plane wave like $e^{i(\mathbf{k}_j - \mathbf{k}_i) \cdot \mathbf{r}}$, where $\mathbf{k}_j - \mathbf{k}_i = \mathbf{K}_j - \mathbf{K}_i$, is zero over the cell unless $\mathbf{K}_j = \mathbf{K}_i$. Hence we have

$$\int_{\text{outside}} \psi_i^*(H - E)\psi_j \, dv = (\mathbf{k}_i \cdot \mathbf{k}_j - E)\delta_{ij}\Omega$$
$$- \int_{\text{sphere}} (\mathbf{k}_i \cdot \mathbf{k}_j - E)e^{i(\mathbf{k}_j - \mathbf{k}_i) \cdot \mathbf{r}} \, dv \quad \text{(A6-7)}$$

in which $\int$ (outside) indicates the integral over the part of the cell outside the sphere. We must then find the integral over the sphere of the exponential $e^{i(\mathbf{k}_j - \mathbf{k}_i) \cdot \mathbf{r}}$. This is easily done from the expansion of Eq. (A6-2). The only term in this expansion which does not vanish when we integrate over angles is that for $l = 0$, $m = 0$, which is $j_0(|\mathbf{k}_j - \mathbf{k}_i|r)$. In computing the integral over the sphere in Eq. (A6-7), we then require

$$\int j_0(|\mathbf{k}_j - \mathbf{k}_i|r) \, dv = 4\pi \int r^2 j_0(|\mathbf{k}_j - \mathbf{k}_i|r) \, dr$$

To evaluate this we use the relation $\int z^2 j_0(z) \, dz = z^2 j_1(z)$. Using these relationships, we find

$$\int_{\text{outside}} \psi_i^*(H - E)\psi_j \, dv = (\mathbf{k}_i \cdot \mathbf{k}_j - E)\left[\delta_{ij}\Omega - 4\pi R^2 \frac{j_1(|\mathbf{k}_j - \mathbf{k}_i|R)}{|\mathbf{k}_j - \mathbf{k}_i|}\right]$$
$$\text{(A6-8)}$$

Next we must carry out the integration over the inside of the sphere. Here the wave function is given by Eq. (A6-3), and the Hamiltonian is $-\nabla^2 + V(r)$. In this region it is more convenient to use the integration by parts given in Eq. (A6-6), according to which the integral is

$$\int_{\text{sphere}} \psi_i^*(H - E)\psi_j \, dv = \int \psi_i^*[-\nabla^2 + V(r) - E]\psi_j \, dv + \int \psi_i^* \frac{\partial \psi_j}{\partial n} \, da$$
$$\text{(A6-9)}$$

The reason why this is convenient is that within the sphere our function ψ_j is a solution of Eq. (A6-1), so that $[-\nabla^2 + V(r) - E]\psi_j = 0$. Hence we have only the surface integral. We expresss ψ_i and ψ_j in this surface integral as summations such as is given in Eq. (A6-3). When we integrate over the angles, we can show by orthogonality properties of the spherical harmonics that the result will be zero unless we have the same values of the indices l and m in both summations. Hence we may write the expression in terms of a single summation,

$$\int_{\text{sphere}} \psi_i^*(H - E)\psi_j \, dv = \int \sum_{l=0}^{\infty} \sum_{m=-l}^{l} (2l+1)^2 \left[\frac{(l - |m|)!}{(l + |m|)!}\right]^2 [P_l^{|m|}(\cos\theta)]^2$$
$$P_l^{|m|}(\cos\theta_i) P_l^{|m|}(\cos\theta_j) e^{im(\phi_j - \phi_i)} \frac{j_l(k_i R)}{u_l(R)} \frac{j_l(k_j R)}{u_l(R)} u_l'(R) u_l(R) \, da \quad \text{(A6-10)}$$

App. 6] MATRIX ELEMENTS FOR THE AUGMENTED-PLANE-WAVE METHOD 465

In Eq. (A6-10) we use the result that

$$P_l(\cos\theta_{ij}) = \sum_{m=-l}^{l} \frac{(l-|m|)!}{(l+|m|)!} P_l^{|m|}(\cos\theta_i) P_l^{|m|}(\cos\theta_j) e^{im(\phi_j-\phi_i)} \quad \text{(A6-11)}$$

where θ_{ij} is the angle between the two directions defined by the angles θ_i, ϕ_i and θ_j, ϕ_j. Furthermore, we use the result that for an integral over the surface of the sphere we have

$$\int [P_l^{|m|}(\cos\theta)]^2 \, da = 2\pi R^2 \int_0^\pi [P_l^{|m|}(\cos\theta)]^2 \sin\theta \, d\theta$$

$$= \frac{4\pi R^2}{2l+1} \frac{(l+|m|)!}{(l-|m|)!} \quad \text{(A6-12)}$$

By use of these two relations, Eq. (A6-10) is transformed into

$$\int_{\text{sphere}} \psi_i^*(H-E)\psi_j \, dv = 4\pi R^2 \sum_{l=0}^{\infty} (2l+1) P_l(\cos\theta_{ij})$$

$$\frac{j_l(k_i R) j_l(k_j R)}{u_l(R)} u_l'(R) \quad \text{(A6-13)}$$

If we combine Eqs. (A6-8) and (A6-13), we have Eqs. (9-2) and (9-3), for the case where there is only one atom in the unit cell; as we have stated, the general case follows easily.

Now we shall look for the alternative form of the matrix element given in Eq. (9-7). To obtain this, we carry out the integration outside the sphere, where we have the plane wave, by use of Eq. (A6-9), instead of by using the expression $\int \text{grad } \psi_i^* \cdot \text{grad } \psi_j \, dv$ for the kinetic energy. We have

$$(-\nabla^2 - E)e^{i\mathbf{k}_j \cdot \mathbf{r}} = (k_j^2 - E)e^{i\mathbf{k}_j \cdot \mathbf{r}}$$

The volume integral is then handled as in the derivation of Eq. (A6-8), leading to

$$\int_{\text{outside}} \psi_i^*(-\nabla^2 - E)\psi_j \, dv = (k_j^2 - E)\left[\delta_{i,j}\Omega - 4\pi R^2 \frac{j_1(|\mathbf{k}_j - \mathbf{k}_i|R)}{|\mathbf{k}_j - \mathbf{k}_i|}\right]$$

(A6-14)

For the surface integral, we proceed as in the derivation of Eq. (A6-10) and (A6-13), with two exceptions. First, outside the sphere it is the j_l's rather than the u_l's which are the radial wave functions; consequently, we must replace $u_l(R)$ and $u_l'(R)$ in Eq. (A6-13) by $j_l(k_j R)$ and $j_l'(k_j R)$, respectively. Second, the outer normal to the region outside the sphere is in the opposite direction to the outer normal to the region inside. Hence there is a difference in sign. When we take account of these two modifications in Eq. (A6-13), and combine the modified function with Eq. (A6-14), we arrive at Eq. (9-7), which we wished to prove.

Appendix 7
The Free-electron Approximation

In Sec. 10-2 we have pointed out that plane waves form a first approximation to the wave functions relating to the periodic-potential problem, and that it is useful to give the energy along various symmetry directions, and the symmetries of the corresponding symmetrized plane waves, in the free-electron approximation. We shall use the hexagonal close-packed structure as an example to illustrate the procedure, as in Sec. A3-1. In Fig. A7-1 we show the energy bands for the space group D_{6h}^4, determined in this way; it is a repetition of Fig. 10-10. We shall now show how the symmetry types indicated in this figure are to be found.

We have shown how to find the symmetrized plane waves in Sec. A3-1; they are the basis functions which we have built up, in the process of determining the irreducible representations. The energy of such a symmetrized plane wave, in the free-electron case, is proportional to the square of the wave vector. This wave vector is given in Eq. (A3-1), where the b's are given in Eq. (5-11). If we substitute Eq. (5-11) into Eq. (A3-1), we find

$$\mathbf{k} = \frac{2\pi}{a} \left[\left(p_x + \frac{2h_1 + h_2}{\sqrt{3}} \right) \mathbf{i} + (p_y + h_2)\mathbf{j} + \left(p_z + \frac{a}{c} h_3 \right) \mathbf{k} \right] \quad \text{(A7-1)}$$

where

$$p_x = \frac{2p_1 + p_2}{\sqrt{3}} \qquad p_y = p_2 \qquad p_z = \frac{a}{c} p_3 \quad \text{(A7-2)}$$

The quantities p_x, p_y, p_z are $a/2\pi$ times the rectangular components of the reduced wave vector. The energy is then proportional to $k^2 a^2/4\pi^2$. In Fig. A7-1 we have plotted this energy as a function of the magnitude of the quantity p, in the appropriate direction.

For given values of p_x, p_y, p_z, we may have any combination of the integers h_1, h_2, h_3. If we are dealing with a symmetry direction, there will ordinarily be more than one set of h's which will lead to the same energy; that is, we shall have more than one plane wave concerned in the wave function. A suitable linear combination of the plane waves corresponding to these different h's will form one of the symmetrized plane waves. In the general case, we can make as many different combinations as we find in a regular representation of the group of the wave vector. That is, we shall find each irreducible representation occurring as many times as its dimensionality. However, for small h's, such as we

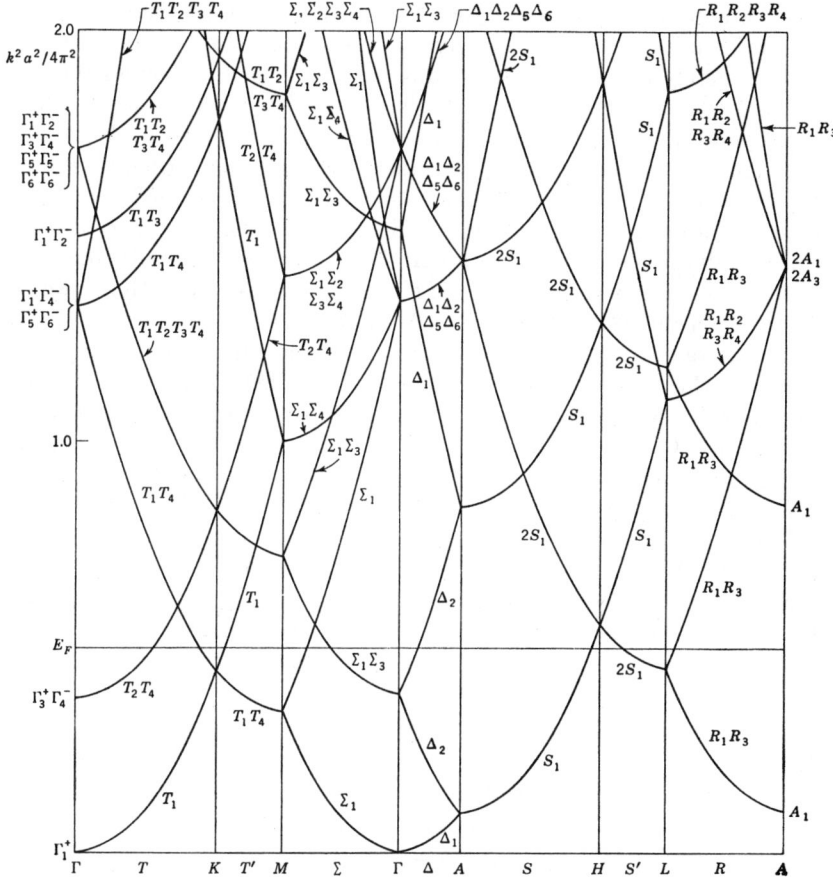

FIG. A7-1. Free-electron energy bands, hexagonal close-packed structure.

are dealing with for small energies, many of these basis functions for different irreducible representations will vanish, or pairs of them will be equal to each other, so that not all irreducible representations will occur. It is these limited sets of representations which are indicated in Fig. A7-1. Let us now illustrate the way in which we find which representations occur in any particular case.

As a first illustration, we take the lowest energy band along the direction Δ, which according to Fig. A7-1 is Δ_1, whereas in the most general case we should find one each of the representations Δ_1, Δ_2, Δ_3, Δ_4, and two each of the two-dimensional representations Δ_5 and Δ_6. This lowest band corresponds to $h_1 = h_2 = h_3 = 0$. Along Δ, we have $p_x = p_y = 0$, $p_z \neq 0$. Since the symmetry is the same for all values of p_z, we may take a value of p_z immediately adjacent to Γ, or $p_z = 0$, so that α in Table A3-11 is unity. The plane wave corresponding to $h_1 = h_2 = h_3 = 0$ is then merely unity. As we operate on it with each of the operators of Table A3-11, to set up the symmetrized plane wave, we get unity whichever operator acts on the function. We are now directed by Eq. (A3-4) to make a linear combination of these functions, with coefficients which are the conjugates of the elements in one or another row of Table A3-11. Since we are multiplying these coefficients by functions each of which equals unity, we see that only for Δ_1 do we get something different from zero: in each other case the sum of all coefficients is zero. Hence we can form only a function of symmetry Δ_1 in this case.

Let us take another somewhat more complicated illustration, to be sure that the reader understands the process. Let us take the lowest of the bands labeled $\Delta_1\Delta_2\Delta_5\Delta_6$ in Fig. A7-1. This band arises from the values $h_1 = 0$, $h_2 = -1$, $h_3 = 0$, and the other values derived from this one by the operators of the group. We use Eq. (A3-3) to arrive at the resulting plane waves. The original plane wave ψ in that equation will be $\exp 2\pi i(-\eta)$, where we have set $p_1 = p_2 = p_3 = 0$, $h_1 = h_3 = 0$, $h_2 = -1$. We now operate on this function in succession with each of the operations of the group of the wave vector, as given in Table A3-11. For the case of Δ, this includes all unprimed operations. From Eq. (A3-3), we then have

$$\begin{aligned}
\{X_0|0\}\psi &= \exp 2\pi i(-\eta) \\
\{X_1|0\}\psi &= \exp 2\pi i(-\xi) \\
\{X_{-1}|0\}\psi &= \exp 2\pi i(\xi - \eta) \\
\{X_2|0\}\psi &= \exp 2\pi i(-\xi + \eta) \\
\{X_{-2}|0\}\psi &= \exp 2\pi i(\xi) \\
\{X_3|0\}\psi &= \exp 2\pi i(\eta) \\
\{Y_0|0\}\psi &= \exp 2\pi i(-\xi + \eta) \\
\{Y_1|0\}\psi &= \exp 2\pi i(-\xi) \\
\{Y_{-1}|0\}\psi &= \exp 2\pi i(\eta) \\
\{Y_2|0\}\psi &= \exp 2\pi i(-\eta) \\
\{Y_{-2}|0\}\psi &= \exp 2\pi i(\xi) \\
\{Y_3|0\}\psi &= \exp 2\pi i(\xi - \eta)
\end{aligned} \quad (A7\text{-}3)$$

We now note from Eq. (A7-3) that we find only six independent functions from these operations, rather than twelve as in the most general case. Hence we can get only six independent symmetrized plane waves, rather than twelve as in the regular representation of the group of the wave vector. Specifically, we note that we get identical functions from the operations X_0 and Y_2, from X_1 and Y_1, from X_{-1} and Y_3, from X_2 and Y_0, from X_{-2} and Y_{-2}, and from X_3 and Y_{-1}. Hence, when we multiply these functions by the complex conjugates of one of the rows in Table A3-11, and add, we shall get zero if X_0 and Y_2 have coefficients which are equal and opposite, and similarly with the other pairs of functions enumerated above. This is the case with the two representations Δ_3 and Δ_4, which therefore cannot be obtained in this case. Furthermore, when we set up the symmetrized plane waves for the two-dimensional representations Δ_5 or Δ_6, we find that we get only one independent set of functions in each case; on account of the relations stated above, the second set of functions is identical with the first. Hence we see that, as indicated in Fig. A7-1, we find in this case only Δ_1, Δ_2, Δ_5, Δ_6. Other cases are treated in essentially the same manner, both in this and other space groups.

Appendix 8
Binding Energy of Diamond, by Method of Schmid

The method of Schmid, which we have mentioned in Sec. 10-4, is closely related to that of Hurley, Lennard-Jones, and Pople, extensively discussed in Volume 1 of this work. We recall that that method handles each bond by means of two orbitals, A and B. It sets up a singlet function for this bond, with a wave function like $[A(1)B(2) + B(1)A(2)][\alpha(1)\beta(2) - \beta(1)\alpha(2)]$, by analogy with the bonding Heitler-London function for hydrogen. It proceeds more generally than the Heitler-London method, however, in that A and B are allowed to vary arbitrarily, in order to lower the energy of the system, subject only to the condition that the A and B of one bond will be orthogonal to those of each other bond. This allows the bond function to take up a partially ionic character, as pointed out by Coulson and Fischer, again in a treatment extensively taken up in Volume 1. The method is adaptable to any problem in which individual electron-pair bonds are essentially independent of each other.

The case of diamond or similar covalent crystals is ideally suited to such a treatment. Schmid has handled the problem by an LCAO approach, building up his functions A and B out of atomic orbitals of simple analytic form. We can describe such an approach by starting with atomic $2s$ and $2p$ orbitals on the carbons. From the $2s$ and $2p$ on a given carbon, we make four tetrahedral hybrids, pointing toward the four nearest neighbors. These four tetrahedral bond orbitals on a single atom will be orthogonal to each other. Each one will overlap strongly with the similar orbital on the carbon atom at the other end of its bond, but will overlap only slightly with hybrids located on other atoms. Consequently, if we form two linear combinations of the two orbitals so set up for a single bond, one centered on each of the two atoms involved, and pointing toward each other, these combinations may be taken to be the A and B concerned in the method of Hurley, Lennard-Jones, and Pople. They will not precisely satisfy the orthogonality condition with the similar functions on other bonds, but the lack of orthogonality will not be serious. The treatment of this bond by the method of Hurley, Lennard-Jones, and Pople is equivalent to a configuration interaction between a Heitler-London treatment of the bond, and an ionic state, formed from a positive and a negative ion; it was in such language that Schmid described the result. He found that the lowest energy came for an appreciable mixture of the ionic state with a preponderantly covalent Heitler-London type of function.

It is not hard to see how to make this calculation more rigorous, by using the result of energy-band calculations. Out of the four occupied valence bands, we can construct four Wannier functions in each unit cell. These functions, as we know from the discussion of the symmetry of Wannier functions in Chap. 7, will not have the properties of s-like or p-like functions, or of tetrahedrally bonded functions. However, by making linear combinations of these Wannier functions, we can construct one bonding orbital for each of the bonds of the crystal, having the properties that it is concentrated as far as possible in the bond, that it is symmetrical about the midpoint of the bond, and that it is orthogonal to all the other similar bonding orbitals in the crystal. These are Lennard-Jones' equivalent orbitals for this case. Similarly, out of the four unoccupied conduction bands of lowest energy, we can construct four Wannier functions, from which one antibonding orbital can be set up for each of the bonds, antisymmetric about the midpoint of the bond, orthogonal to all other antibonding orbitals, and also orthogonal to all bonding orbitals.

We then take the bonding and antibonding equivalent orbitals for a given bond, and make linear combinations of them for the two functions A and B of this bond. One of the functions, say, A, will be more concentrated on one of the atoms of the bond, while the other, B, will be correspondingly concentrated on the other atom. These orbitals will have precisely the required properties of orthogonality. If we use a Hurley, Lennard-Jones, and Pople approach, with these orbitals A and B, we get a quite accurate treatment of the bonding properties of the diamond crystal, showing clearly the way in which the electron pairs in the tetrahedral bonds hold the crystal together. The calculation of Schmid, and that of Coulson, Redei, and Stocker, referred to in Sec. 10-4, represent attempts of varying degrees of accuracy to carry through essentially this calculation.

Appendix 9
Spin-Orbit and Relativistic Effects in Energy Bands

A9-1. Spin-Orbit Effects in Atoms. In the main part of this volume we have been disregarding spin-orbit and relativistic effects in our discussions of energy bands. This is justified to a first approximation, just as we can find a valid first approximation to atomic energy levels without taking these effects into account. However, as we go to atoms of higher and higher atomic number, the effects become rapidly more important, and for the heavier atoms they modify the energy bands in ways that cannot be disregarded. Consequently, in this Appendix we shall take up these effects. Fortunately, even for the heavier atoms they are small enough so that a quite good treatment can be given by perturbation methods, starting with the calculation as described in the remainder of the volume as the unperturbed case. It is only in the very heaviest atoms, those showing radioactivity, that this perturbation treatment is not fairly adequate.

The guide to the treatment of these relativistic effects in crystals is the corresponding treatment of the isolated atoms, and we shall remind the reader in this section of some of the main features of that treatment. For a more complete discussion, on which the present Appendix is based, the reader is referred to J. C. Slater, "Quantum Theory of Atomic Structure," vol. 2, chaps. 23 and 24. We shall refer to this work in the present Appendix by the abbreviation QTAS.

We can get at the theory of these effects by considering the motion of a single electron in an external electric field, such as is derived from the self-consistent-field method, and by treating this motion according to Dirac's theory of the relativistic, spinning electron. The only change required to adapt the theory of the atom to the case of a crystal is to replace the central field by a periodic potential. This is not completely accurate, but it has the same range of validity that the energy-band theory has, both being based on the approximation of a self-consistent field. Consequently, it is only this one-electron theory which will be taken up in the present Appendix.

App. 9] SPIN-ORBIT AND RELATIVISTIC EFFECTS IN ENERGY BANDS

Dirac showed (QTAS, chap. 23) that to handle Schrödinger's equation relativistically, one must replace the ordinary wave function ψ by four functions ψ_1, ψ_2, ψ_3, ψ_4, and one must replace the ordinary Schrödinger equation, a second-order partial differential equation, by first-order differential equations for ψ_1, ψ_2, ψ_3, and ψ_4, each of these equations coupling the four functions together. One finds that two of the functions, ψ_1 and ψ_2, are very small, while the other two, ψ_3 and ψ_4, are large. (If one is handling the theory of a positron instead of an electron, the role of the functions is interchanged, ψ_1 and ψ_2 being large, ψ_3 and ψ_4 small.) It is then possible, using Dirac's equations, to eliminate the two small components ψ_1 and ψ_2, incorporating their effect into second-order differential equations for ψ_3 and ψ_4, which are identical with the ordinary Schrödinger equations, except for certain small correction terms which can be interpreted as spin-orbit and relativistic terms [QTAS, eq. (23-39)]. Ordinarily, one retains only the first terms in an expansion of these correction terms in powers of $1/c^2$, where c is the velocity of light [eq. (23-40)]. The resulting Schrödinger equation is

$$\left\{ \frac{(-i\hbar\nabla + e\mathbf{A})^2}{2m_0} - e\phi - \frac{(-i\hbar\nabla)^4}{8m_0^3 c^2} + \frac{e\hbar}{2m_0}(\mathbf{B} \cdot \mathbf{\sigma}) + \frac{e\hbar}{4m_0^2 c^2} \mathbf{\sigma} \cdot [\mathbf{E} \times (-i\hbar\nabla)] \right.$$

$$\left. - \frac{e\hbar^2}{4m_0^2 c^2} \mathbf{E} \cdot \nabla \right\} \psi = E\psi \quad (A9\text{-}1)$$

In this equation, which holds for both ψ_3 and ψ_4, in a way which we shall describe later, m_0 is the electronic rest mass, e is the magnitude of its charge (a positive quantity), $\mathbf{A}$ is the vector potential of an external magnetic field, ϕ is the scalar potential of an external electric field (which in the case of the self-consistent field is taken to be the field arising from the nuclei and averaged electronic charges, corrected for exchange). The magnetic field, curl $\mathbf{A}$, is $\mathbf{B}$, and the electric field, $-\operatorname{grad} \phi$, is $\mathbf{E}$, not to be confused with the energy parameter E appearing on the right side of the equation. The quantity $\mathbf{\sigma}$ is related to the electron spin, as we shall shortly describe. We are assuming for the present purposes that the external electric and magnetic fields are independent of time.

Let us comment on the various terms in Eq. (A9-1). First we have the kinetic energy, including the vector potential term which takes account of interaction of the orbital electronic motion with an external magnetic field. The next term, $-e\phi$, is the ordinary potential energy ($-e$ being the charge). Next comes a term in ∇^4, which is the primary relativistic term, arising from the change of mass with velocity. Then comes a term in $\mathbf{B} \cdot \mathbf{\sigma}$, the scalar product of the magnetic field $\mathbf{B}$, and a quantity which proves to be the electron spin. Then we have the term in $\mathbf{\sigma} \cdot [\mathbf{E} \times (-i\hbar\nabla)]$, which is the spin-orbit interaction, as we shall describe later. Finally, the term in $\mathbf{E} \cdot \nabla$ is a nonclassical term which has no simple interpretation, but which does not depend on spin.

Now we shall come to the interpretation of the two functions ψ_3 and ψ_4. One can show that ψ_3 is the wave function for electrons with $+$, or α, spin, while ψ_4 is that for electrons with $-$, or β, spin. That is, $\psi_3^*\psi_3$ measures the probability that an electron of $+$ spin will be found in unit volume at a given point of space, while $\psi_4^*\psi_4$ gives the same probability for an electron of $-$ spin. As a result of this interpretation, we see that the normalization of the wave function must be

$$\int (\psi_3^*\psi_3 + \psi_4^*\psi_4)\, dv = 1 \qquad (A9\text{-}2)$$

If we wish to mix Dirac's and Pauli's notations, we may write the complete wave function in the form

$$\psi = \psi_3 \alpha + \psi_4 \beta \qquad (A9\text{-}3)$$

where α and β are Pauli's spin functions, and where the complete quantity ψ of Eq. (A9-3) is of the form of a spin-orbital, with contributions relating to both spin orientations. For the familiar case where we are dealing with electrons which definitely have the spin either along z or along $-z$, we should have the case $\psi_4 = 0$ or $\psi_3 = 0$, respectively, but in this Appendix we meet the general situation indicated by Eq. (A9-3).

The operator σ can now be described. It is a dimensionless vector operator proportional to the angular-momentum operator of the spin angular momentum of the electron, though its matrix elements are twice as large, equal to unity rather than $\frac{1}{2}$ unit as the spin angular momentum is. Each component operates on the functions α and β. To find the nature of these operations, we recall the general nature of an angular-momentum operator in wave mechanics. If we have an angular-momentum operator with components $(L_x)_{\text{op}}$, etc., we find [see QTAS, eq. (11-7)]

$$
\begin{aligned}
(L_x)_{\text{op}} u_{nlm} &= \tfrac{1}{2} [\sqrt{(l-m)(l+m+1)}\, u_{n,l,m+1} \\
&\quad + \sqrt{(l-m+1)(l+m)}\, u_{n,l,m-1}] \\
(L_y)_{\text{op}} u_{nlm} &= \tfrac{i}{2} [-\sqrt{(l-m)(l+m+1)}\, u_{n,l,m+1} \\
&\quad + \sqrt{(l-m+1)(l+m)}\, u_{n,l,m-1}] \\
(L_z)_{\text{op}} u_{nlm} &= m u_{nlm}
\end{aligned}
\qquad (A9\text{-}4)
$$

Here we are operating on a wave function u_{nlm} with principal quantum number n, azimuthal quantum number l, and magnetic quantum number m. We produce a linear combination of wave functions of the same

n and l values, but with m's which are one unit greater, one unit less, or equal to the original m.

If we are dealing with an electron spin, we must set $l = \frac{1}{2}$, so that m can take on only two values, $\frac{1}{2}$ and $-\frac{1}{2}$. The spin wave function corresponding to $m = \frac{1}{2}$ is α, and that corresponding to $-\frac{1}{2}$ is β. If we denote the spin operator by S, we have

$$(S_x)_{\text{op}}\alpha = \frac{1}{2}\beta \qquad (S_x)_{\text{op}}\beta = \frac{1}{2}\alpha$$
$$(S_y)_{\text{op}}\alpha = \frac{i}{2}\beta \qquad (S_y)_{\text{op}}\beta = -\frac{i}{2}\alpha \qquad \text{(A9-5)}$$
$$(S_z)_{\text{op}}\alpha = \frac{1}{2}\alpha \qquad (S_z)_{\text{op}}\beta = -\frac{1}{2}\beta$$

The angular momentum in Eqs. (A9-4) and (A9-5) is measured in units of $\hbar$. The operator $\mathbf{\sigma}$ is now twice the $\mathbf{S}$ of Eq. (A9-5), so that it does not have the factor $\frac{1}{2}$ met in each component of that equation. We should emphasize that while we have described $\mathbf{\sigma}$ by reference to the angular-momentum operators of Eq. (A9-4), and to the Pauli spin functions α and β, it is not introduced into Dirac's equations as a result of analogies with ordinary angular momentum. Rather, it appears in an entirely a priori manner, as a quantity which allows one to write the equations in a convenient manner.

We can now describe the precise interpretation of the operator $\mathbf{\sigma}$ found in Eq. (A9-1). We can most simply describe it in the following way, though this is not the method used by Dirac. We write ψ in the form of Eq. (A9-3), and assume that $\mathbf{\sigma}$ operates on the terms α and β contained in ψ according to Eq. (A9-5). We shall then have, on each side of Eq. (A9-1), an expression containing one function of coordinates times α, and another function of coordinates times β. We are to regard this as being two differential equations, one for the coefficients of α, the other for the coefficients of β. We can get these two equations from the general equation by multiplying both sides by α^* and summing over spin (for the first equation) or by β^* and summing over spin (for the second), using the orthonormal properties of α and β.

From this interpretation of Eq. (A9-1) we can find at once the method of getting matrix elements of the operator $\mathbf{\sigma}$ between two spin-orbitals $\psi = \psi_3\alpha + \psi_4\beta$ and $\psi' = \psi'_3\alpha + \psi'_4\beta$. We have

$$\int \psi^*(\sigma_x)\psi' \, dv = \Sigma(\text{spin})\int(\psi_3^*\alpha + \psi_4^*\beta)(\sigma_x)_{\text{op}}(\psi'_3\alpha + \psi'_4\beta) \, dv$$
$$= \Sigma(\text{spin})\int(\psi_3^*\alpha + \psi_4^*\beta)(\psi'_3\beta + \psi'_4\alpha) \, dv$$
$$= \int(\psi_3^*\psi'_4 + \psi_4^*\psi'_3) \, dv \qquad \text{(A9-6)}$$
$$\int \psi^*(\sigma_y)\psi' \, dv = \int(-i\psi_3^*\psi'_4 + i\psi_4^*\psi'_3) \, dv$$
$$\int \psi^*(\sigma_z)\psi' \, dv = \int(\psi_3^*\psi'_3 - \psi_4^*\psi'_4) \, dv$$

If we are interested in the average value of the operator, so that $\psi' = \psi$, we have

$$(\sigma_x)_{av} = \int(\psi_3^*\psi_4 + \psi_4^*\psi_3)\, dv$$
$$(\sigma_y)_{av} = \int(-i\psi_3^*\psi_4 + i\psi_4^*\psi_3)\, dv \qquad (A9\text{-}7)$$
$$(\sigma_z)_{av} = \int(\psi_3^*\psi_3 - \psi_4^*\psi_4)\, dv$$

Now that we understand the general method of manipulating the spin-orbitals, and the relation of Dirac's equation, Eq. (A9-1), to them, we can consider further the spin-orbit interaction, the term in $\mathbf{\sigma} \cdot [\mathbf{E} \times (-i\hbar\nabla)]$, in Eq. (A9-1). In an atom, the electric field $\mathbf{E}$ is in the radial direction, on account of the spherical symmetry of the problem. We may then write it as

$$\mathbf{E} = \frac{|\mathbf{E}|\mathbf{r}}{r} \qquad (A9\text{-}8)$$

where $|\mathbf{E}|$ is the absolute magnitude of the field, $\mathbf{r}$ is a vector along the radius, and r is its magnitude. Then we may write the spin-orbit term in Eq. (A9-1) in the form

$$\frac{e\hbar}{4m_0^2 c^2} \frac{|\mathbf{E}|}{r} \mathbf{\sigma} \cdot (\mathbf{r} \times \mathbf{p})_{op} \qquad (A9\text{-}9)$$

where $(\mathbf{p})_{op}$ is the momentum operator $-i\hbar\nabla$. But $\mathbf{r} \times \mathbf{p}$ is the orbital angular momentum $\mathbf{L}\hbar$, and $\mathbf{\sigma}\hbar/2$ is the spin-angular momentum $\mathbf{S}\hbar$, where $\mathbf{L}$ and $\mathbf{S}$ are dimensionless angular-momentum vector operators. Hence we may rewrite Eq. (A9-9) in the form

$$\frac{e\hbar^2}{2m_0^2 c^2} \frac{|\mathbf{E}|}{r} (\mathbf{S} \cdot \mathbf{L})_{op} \qquad (A9\text{-}10)$$

This has the form $\Gamma(\mathbf{S} \cdot \mathbf{L})$ postulated by Landé (QTAS, sec. 10-4). We may put the constant Γ in more convenient form by going to atomic units. We rewrite the electric field $|\mathbf{E}|$ in the form

$$|\mathbf{E}| = \frac{eZ_f(r)}{4\pi\epsilon_0 r^2} \qquad (A9\text{-}11)$$

where Z_f is a dimensionless function of r, which would be a constant, equal to Z, for the electric field of a point charge of magnitude Ze. We then convert the units of energy to Rydbergs, and the units of length to atomic units, and use the fine-structure constant

$$\alpha^2 = \frac{2}{m_0 c^2} \text{ Rydbergs} = \left(\frac{e^2}{4\pi\epsilon_0 \hbar c}\right)^2 = \frac{1}{(137)^2} \text{ approximately} \qquad (A9\text{-}12)$$

When we do so, the spin-orbit interaction of Eq. (A9-10) takes the form

$$\text{Spin-orbit interaction} = \frac{\alpha^2 Z_f(r)}{r^3} (\mathbf{S} \cdot \mathbf{L}) \qquad (A9\text{-}13)$$

In this formula, we are of course to guard against confusing the fine-structure constant α of Eq. (A9-12) with the Pauli spin function α.

For an atomic problem, we are considering the matrix elements of the spin-orbit interaction for a fixed set of quantum numbers n and l, and hence a fixed radial part of the function ψ. Consequently, we shall encounter the same radial integral of the quantity $Z_f(r)/r^3$ for every matrix element of the quantity $(\mathbf{S} \cdot \mathbf{L})_{op}$. It is the quantity

$$\left[\frac{\alpha^2 Z_f(r)}{r^3} \right]_{av} = \Gamma \text{ Rydbergs} \qquad (A9\text{-}14)$$

which determines the scale of the spin-orbit interaction. Its magnitude increases very rapidly as we go to heavier atoms, both because they have larger Z_f's (essentially proportional to atomic number) and because their wave functions and charge distributions are much more concentrated at small values of r, so that the $1/r^3$ term in Eq. (A9-14) emphasizes these large concentrations. Values of Γ are well known experimentally, from observed spin-orbit interactions, and can also be calculated with quite good accuracy from atomic self-consistent fields.

Now that we have established the form of the operator for spin-orbit interaction, we are well acquainted with the way this shows itself in the atomic spectra (QTAS, sec. 10-4 and chap. 24). The operator can be shown to have no nondiagonal matrix elements, except between two wave functions of the same component M of total angular momentum along the axis. In fact, if we use Eqs. (A9-4) and (A9-5) to find the effect of the operator $(\mathbf{S} \cdot \mathbf{L})_{op}$ on a spin-orbital of the form $u_{nlm_l}\alpha$ or $u_{n,l,m_l+1}\beta$, we find

$$(\mathbf{S} \cdot \mathbf{L})_{op} u_{nlm_l}\alpha = \tfrac{1}{2} m_l (u_{nlm_l}\alpha) + \tfrac{1}{2} \sqrt{(l-m_l)(l+m_l+1)} \, (u_{n,l,m_l+1}\beta)$$

$$(\mathbf{S} \cdot \mathbf{L})_{op} u_{n,l,m_l+1}\beta = -\tfrac{1}{2}(m_l+1)(u_{n,l,m_l+1}\beta) + \tfrac{1}{2} \sqrt{(l-m_l)(l+m_l+1)} \, (u_{nlm_l}\alpha) \qquad (A9\text{-}15)$$

which shows that the diagonal matrix element is the product of the diagonal matrix elements of spin and orbital angular momentum, $\tfrac{1}{2} m_l$ or $-\tfrac{1}{2}(m_l+1)$ for these two cases, and the nondiagonal matrix element is found only between the functions $u_{nlm_l}\alpha$ and $u_{n,l,m_l+1}\beta$, which have the same values of $m_s + m_l = M$.

We can use these simple facts to find the energy levels in the various doublet states of one-electron spectra, which we meet with the present case of a single spin. For a given l value, we know by the vector-coupling rules that we shall find J values of $l \pm \frac{1}{2}$. For each value of M, we have two basis functions, $u_{nlm_l}\alpha$ and $u_{n,l,m_l+1}\beta$. The only exception is the case where $m_l = l$, in which case the second function cannot exist. For $m_l = l$, the function $u_{nlm_l}\alpha$ has the energy $\Gamma l/2$, which then must be the energy for $J = l + \frac{1}{2}$. For any other value of M, one of the two functions has a diagonal energy $\Gamma m_l/2$, the other $-\Gamma(m_l + 1)/2$, and the nondiagonal energy is $(\Gamma/2)\sqrt{(l-m_l)(l+m_l+1)}$. If we solve the quadratic secular equation between these two values, we find one of the roots equal to $\Gamma l/2$, as before, corresponding to $J = l + \frac{1}{2}$, and the other one is $-\Gamma(l+1)/2$. This latter value is then the energy of the state $J = l - \frac{1}{2}$. These two values are those given for this case by Landé's formula

$$(\mathbf{S} \cdot \mathbf{L})_{\text{av}} = \frac{\Gamma}{2}[J(J+1) - L(L+1) - S(S+1)]$$

We can find these energies, without solving a quadratic, by using the sum rule for the case $m_l = l - 1$. The energy separation between the two states of the doublet arising from an orbital function of a given l value is $\Gamma(l + \frac{1}{2})$.

A9-2. Spin-Orbit Interaction in a Crystal Using the APW Method. We now understand the general features of spin-orbit interaction in an atom, and can go on to the application of the same methods to a crystal. This is particularly simple if we use the APW method, as has been pointed out by L. E. Johnson, J. B. Conklin, and G. W. Pratt, Jr., *Phys. Rev. Letters*, **11**:538 (1963). The reason is that, as we have mentioned earlier, we assume a potential which is spherically symmetrical inside each of the atomic spheres, and which is constant outside the spheres. The corresponding electric field is then radial within each sphere, as in an atom, and is zero outside the spheres. Consequently, there will be contributions to the spin-orbit interaction only from the regions within the spheres, and here we can still make the transformation of Eq. (A9-8), so that the spin-orbit operator still takes the form of Eq. (A9-13), where now the function Z_f is assumed to vanish outside the sphere. Furthermore, inside the spheres, the wave function is expanded as a linear combination of functions each with a given l and m_l value. Consequently, we can use Eq. (A9-15) to find the effect of the spin-orbit-interaction operator on one of the components of the wave function, having a particular l and m_l. To find the complete matrix element of the spin-orbit interaction, Eq. (A9-13), we must then carry out a sum over the various components of the wave function.

We shall write the wave function within the sphere as

$$\psi = \Sigma(l,m_l)(A_{l,m_l}\alpha + B_{l,m_l}\beta)u_l(r)Y_{m_l}(\theta,\phi) \tag{A9-16}$$

App. 9] **SPIN-ORBIT AND RELATIVISTIC EFFECTS IN ENERGY BANDS** 479

Here the $u_l(r)$'s are the radial solutions of the spherical Schrödinger equation within the sphere, corresponding to the energy of the wave function, and the $Y_{m_l l}(\theta,\phi)$'s are normalized spherical harmonics,

$$Y_{m_l l}(\theta,\phi) = \frac{(-1)^{(m_l+|m_l|)/2}}{\sqrt{4\pi}} \sqrt{\frac{(2l+1)(l-|m_l|)!}{(l+|m_l|)!}} \, P_l^{|m_l|}(\cos\theta)e^{im_l\phi} \quad (A9\text{-}17)$$

The u_l's are assumed to be the solution of the problem lacking relativistic and spin-orbit terms. Then we can write the matrix elements of the operator of Eq. (A19-13) between two such functions in the form

$$\int \psi^* \frac{\alpha^2 Z_f}{r^3} (\mathbf{S}\cdot\mathbf{L})_{\text{op}} \psi' \, dv = \int \sum (l,m_l)(A^*_{l,m_l}\alpha + B^*_{l,m_l}\beta)u_l^*(r)Y^*_{m_l l}(\theta,\phi)$$

$$\frac{\alpha^2 Z_f}{r^3} (\mathbf{S}\cdot\mathbf{L})_{\text{op}} \sum (l',m_l')(A'_{l',m_l'}\alpha + B'_{l',m_l'}\beta)u_l'(r)Y_{m_l' l'}(\theta,\phi) \, dv \quad (A9\text{-}18)$$

We use Eq. (A9-15) for the effect of the spin-orbit-interaction operator on one of the wave functions, integrate over the angles and take advantage of the orthonormal properties of the spherical harmonics, and integrate over the radius. Then the expression above can be converted into

$$\sum (l) \int \frac{\alpha^2 Z_f}{r^3} u_l^*(r)u_l'(r)r^2 \, dr \sum (m_l)[\tfrac{1}{2}m_l(A^*_{l,m_l}A'_{l,m_l} - B^*_{l,m_l}B'_{l,m_l})$$

$$+ \tfrac{1}{2}\sqrt{(l-m_l)(l+m_l+1)} \, (A^*_{l,m_l}B'_{l,m_l+1} + B^*_{l,m_l+1}A'_{l,m_l})] \quad (A9\text{-}19)$$

The expression of Eq. (A9-19) is a simple and straightforward one for the matrix elements, involving only the A's and B's, which are known if we have solved the APW problem in the absence of spin-orbit interactions, and the integrals $\int (\alpha^2 Z_f/r)u_l^* u_l' \, dr$, which have the same form as the Γ's in the atomic case, defined in Eq. (A9-14).

For a general point in the Brillouin zone, the wave function will have no symmetry properties, and all A's and B's may be expected to be nonvanishing. We shall then have terms of all the sorts indicated in Eq. (A9-19), both in the diagonal and the nondiagonal matrix elements of the spin-orbit interaction. These will involve small corrections to the energy levels. The interesting cases, however, are at symmetry points. Here, as we know, we can often find degeneracies in the absence of spin-orbit interaction. These degeneracies can be removed by the spin-orbit interaction, the resulting splittings of energy levels being closely similar to those in the atomic case. We can understand the situation best by taking up a specific example, which we shall do in the next section: the case of the splitting of a threefold degenerate state arising from an atomic p level, both at the point Γ in a cubic crystal, and as we go away from this point along the Δ direction. We shall first discuss this example in an elementary manner. Then we shall examine in later sections the relation of the resulting splitting and the remaining degeneracies to group theory. We shall be led to the theory of double groups, the types of groups encountered when we are treating systems with half-integral quantum numbers.

A9-3. Splitting of P-like State by Spin-Orbit Interaction. At the center of the Brillouin zone in a cubic crystal, we remember that an energy band arising from an atomic p state is threefold degenerate, with a symmetry described in the notation of Bouckaert, Smoluchowski, and Wigner as Γ_{15}. We may well suppose that if we consider spin-orbit interaction, we shall have the same sort of splitting which we have described in Sec. A9-1 for an atomic p state in empty space. That is, we should have the analogue of the twofold degenerate state $^2P_{1/2}$, and the fourfold degenerate $^2P_{3/2}$. We shall find that this hypothesis is correct. The twofold degenerate state for the spherical case has $J = \frac{1}{2}$, with two possible orientations in an external field, and the fourfold degenerate has $J = \frac{3}{2}$, with components $M = \frac{3}{2}, \frac{1}{2}, -\frac{1}{2}, -\frac{3}{2}$ along the axis.

Let us now proceed in an intuitive way to ask what probably happens along the direction Δ, propagation in the x (or z) direction. Let us take Δ as the z axis, and let us assume that this is the axis for quantization of the angular momentum. When the reduced wave vector is different from zero, we no longer have the analogue of spherical symmetry, but only of cylindrical symmetry, the z axis being a preferred axis. We still must expect something equivalent to quantization of the angular momentum along this axis, but the equivalent of the total angular momentum will no longer be a good quantum number. The situation is like that in a diatomic molecule, where the electronic component of angular momentum along the axis is quantized, but not the total angular momentum. As in the diatomic molecule, we should then expect that the pair of levels arising from $^2P_{3/2}$ having $M = \pm \frac{3}{2}$ will be degenerate with each other, but will no longer have the same energy as the pair with $M = \pm \frac{1}{2}$. We shall expect a splitting of the fourfold degenerate state into two doubly degenerate states. The remaining degeneracy of course arises because, in the absence of an external magnetic field, the energy does not depend on whether the angular momentum is pointing along the reduced wave vector or opposite to it.

It seems plausible that the doubly degenerate state with $M = \pm \frac{1}{2}$ arising from the $^2P_{3/2}$ might have the same type of symmetry as the doubly degenerate state with $M = \pm \frac{1}{2}$ arising from the $^2P_{1/2}$. Thus it is to be expected that there will be a nondiagonal matrix element of the Hamiltonian between these states, as well as a difference between their diagonal energies, so that it will be necessary to solve a quadratic secular equation to find the energies of the states. For the two states with $M = \pm 3/2$, however, it is clear that we have a different symmetry from those with $M = \pm 1/2$, so that the states with $M = \pm 3/2$ will not require a secular equation. This is, of course, on the assumption that other energy bands are far enough from those arising from the atomic p state so that the interactions between those bands and the ones we are considering can be neglected.

Now let us ask how we can verify these assumptions in a more analytic manner. We shall build up wave functions involving orbital and spin parts, as in Eq. (A9-3), for the propagation along the direction Δ, and shall investigate the diagonal and nondiagonal matrix elements of the spin-orbit interaction between these functions, using Eq. (A9-19). We can treat the case Γ by passing to the limit as the reduced wave vector **k** approaches zero.

We recall that the group of the wave vector along Δ is C_{4v}, which we have discussed in Chap. 5. The two irreducible representations of that group which are compatible with the irreducible representation Γ_{15} are Δ_1, a one-dimensional irreducible representation, and Δ_5, a two-dimensional representation. We shall now ask what properties the coefficients A_{l,m_l} and B_{l,m_l} in the expansion of Eq. (A9-16) must have for these symmetry types, and hence what special relations there must be in the matrix elements of Eq. (A9-19). We assume that the axis of quantization of α and β, and the axis for the spherical harmonics in Eq. (A9-16), is the reduced wave vector, or the z axis. The operations of the group C_{4v} are then rotations X_1, X_{-1}, X_2, in which ϕ is increased or decreased, and the operations Y_0, $Y_{\pm 1}$, Y_2, reflections in which ϕ changes sign, plus rotations. We ask for the effect of these operations on the spherical harmonics $Y_{l,m_l}(\theta,\phi)$, or rather on the exponential factors $e^{im\phi}$, which are the only part of the wave function affected by the operations. (For the moment we shall omit the subscript l on m_l.) We may start with two basis functions, $e^{im\phi}$ and $e^{-im\phi}$, which form basis functions for a two-dimensional representation (which may be either irreducible or reducible). We can find which ones of these will lead to the symmetry types Δ_1 or Δ_5 by finding the characters of the various operations, and comparing these with the characters of the irreducible representations of C_{4v}, as given in Table 5-3.

We see at once that $X_q e^{im\phi} = e^{\pi i m q/2} e^{im\phi}$. Hence X_q multiplies $e^{im\phi}$ by $e^{\pi i m q/2}$, and similarly it multiplies $e^{-im\phi}$ by $e^{-\pi i m q/2}$. The character of this operation is then the sum of these two diagonal matrix elements, or

$$\chi(X_q) = 2 \cos \frac{\pi m q}{2} \qquad (A9\text{-}20)$$

The operation Y_q transforms $e^{im\phi}$ into a multiple of $e^{-im\phi}$, so that it has no diagonal matrix element, and the characters of all operations Y_q are zero, for these two-dimensional representations. In Table A9-1 we give the resulting characters. We can identify the resulting irreducible representations by comparison with Table 5-3. Thus the case $m = -4$ has characters of 2 for X_0, $X_{\pm 1}$, and X_2. These are the sums of the characters of the two one-dimensional irreducible representations Γ_1 and Γ_2 (in the notation of Table 5-3), or Δ_1 and Δ_1' in the notation of Bouckaert, Smoluchowski, and Wigner, which is used in Table A9-1. Hence these two symmetry types will have functions $e^{\pm im\phi}$ with $m = 4$ in their expansions. Similarly, for $m = -3$ we have characters of 2, 0, −2, respectively, given for the two-dimensional irreducible representation Γ_5 in

Table A9-1
Characters of operations X_0, $X_{\pm 1}$, X_2 for two-dimensional representations with basis functions $e^{\pm im\phi}$, group C_{4v}, together with resulting irreducible representations
(BSW notation)

m	X_0	$X_{\pm 1}$	X_2	Irreducible representations
...	...	...	...	...
−4	2	2	2	Δ_1, Δ_1'
−3	2	0	−2	Δ_5
−2	2	−2	2	Δ_2, Δ_2'
−1	2	0	−2	Δ_5
0	2	2	2	Δ_1, Δ_1'
1	2	0	−2	Δ_5
2	2	−2	2	Δ_2, Δ_2'
3	2	0	−2	Δ_5
4	2	2	2	Δ_1, Δ_1'
...	...	...	...	...

Table 5-3, or Δ_5 in the BSW notation of Table A9-1, and $m = -2$ has 2, −2, 2, the sums of characters for the irreducible representations Γ_3 and Γ_4 in Table 5-3, or Δ_2 and Δ_2' in the BSW notation. We see in this way how we find the irreducible representations in Table A9-1.

As a result of this study, we see that the wave function in the one-dimensional irreducible representation Δ_1 of the group of the wave vector Δ can be written in the form

$$\psi = \Sigma(l,m_l) C_{l,m_l} u_l(r) Y_{m_l l}(\theta,\phi) \qquad \text{(A9-21)}$$

where m_l takes on the values −4, 0, 4, . . . , or more generally an integer times 4. Similarly, in the two-dimensional irreducible representation Δ_5 we find all odd values of m_l. If as usual we associate the first partner

App. 9] SPIN-ORBIT AND RELATIVISTIC EFFECTS IN ENERGY BANDS 483

in the basis functions of Δ_5 with $e^{i\phi}$, the second with $e^{-i\phi}$, we see more specifically that for the wave function ψ representing the first partner for Δ_5 we have $m = 1 +$ an integer times 4, while for the second partner $m = -1 +$ an integer times 4.

Let us now use these orbital functions to construct spin-orbitals. To do this, we multiply the functions ψ by spin functions α or β. We must now find the matrix elements of spin-orbit interaction between these spin-orbitals, so as to investigate the symmetry types of spin-orbitals, having the property that the spin-orbit interaction has no nondiagonal matrix elements between spin-orbitals of different symmetry types. As a first step in this process we note that in finding the diagonal matrix elements, which we find by eliminating the primes from A' and B' in Eq. (A9-19), the terms involving $A^*_{l,m_l} B_{l,m_l+1} + B^*_{l,m_l+1} A_{l,m_l}$ will all vanish, since we do not have coefficients whose m_l's differ by unity, in the expansion either of Δ_1 or of Δ_5.

Next we consider nondiagonal matrix elements. As we see from Eq. (A9-19), we can have a nonvanishing nondiagonal matrix under two circumstances. First, we may have two wave functions of the same type, that is, both derived from the same irreducible representation Δ_1 or Δ_5, each from the same partner in the latter case, and each associated with the same spin α or β. Then we shall have contributions from the terms $A^*_{l,m_l} A'_{l,m_l} - B^*_{l,m_l} B'_{l,m_l}$ of Eq. (A9-19). In our particular case of the removal of the degeneracy of the P electronic state we do not have this situation, for we have only one wave function of each type. The second case is that in which we have nonvanishing terms $A^*_{l,m_l} B'_{l,m_l+1} + B^*_{l,m_l+1} A'_{l,m_l}$. This demands that a term in the unprimed function, and a term in the primed one, have the same values of $M = m_l + m_s$, so that the coefficient A_{l,m_l}, associated with spin α and m_l, and B'_{l,m_l+1}, associated with spin β and $m_l + 1$, will both be nonvanishing; each has $M = m_l + \frac{1}{2}$. We can then find the cases in which there are such nonvanishing nondiagonal matrix elements by making a table of the exponentials times α or β, and identifying the various functions according to their M values. Such a table is given in Table A9-2.

In Table A9-2 we have arranged the functions $e^{im_l\phi}\alpha$ in one column, with $e^{im_l\phi}\beta$ in another, classified according to the value of M. For each case, we have indicated the irreducible representation Δ_1 or Δ_5 from which this function arises, according to the discussion we have given above. With Δ_5, we have specified by subscript 1 or 2 which partner in the two-dimensional representation we are dealing with. We shall then have a nondiagonal matrix element of the spin-orbit interaction between two different functions of the same M value, indicated by arrows in the table. We see that we have this situation for $M = -\frac{7}{2}, -\frac{1}{2}, \frac{1}{2}, \frac{7}{2}, \frac{9}{2}, \ldots$.

We note that the table is periodic with period of 4 units in M; we have covered two complete periods in the table, but it should be extended indefinitely both toward large positive and large negative M's. Really, then, there are only two independent cases where we have nondiagonal matrix elements, which we may take to be $\pm\frac{1}{2}$. The other cases arise from these by adding or subtracting integral multiples of 4 to M, so that they merely refer to different components in the expansion of the functions in spherical harmonics. The nondiagonal matrix elements which we have are between the function α arising from Δ_1, and $e^{i\phi}\beta$ arising from

Table A9-2

Spin-orbital components of form $e^{im\phi}$ times α or β, formed from orbital functions Δ_1 or Δ_5. The two partners of Δ_5 are indicated by $(\Delta_5)_1$ and $(\Delta_5)_2$, respectively. At left we have M. Next we give the type of orbital function from which the spin-orbital is formed, then the spin-orbital formed from α, next that formed from β, with the orbital function, and the type from which it arises. Finally, we have the symbol representing the irreducible representation of the double group to which this symmetry type belongs (described in Sec. A9-7). Arrows indicate nonvanishing nondiagonal matrix elements.

M	Spin-orbitals and symmetry types				Irreducible representation
$-\frac{7}{2}$	(Δ_1)	$e^{-4i\phi}\alpha \leftrightarrow e^{-3i\phi}\beta$		$(\Delta_5)_1$	$(\Delta_6)_1$
$-\frac{5}{2}$	$(\Delta_5)_1$	$e^{-3i\phi}\alpha$			$(\Delta_7)_1$
$-\frac{3}{2}$			$e^{-i\phi}\beta$	$(\Delta_5)_2$	$(\Delta_7)_2$
$-\frac{1}{2}$	$(\Delta_5)_2$	$e^{-i\phi}\alpha \leftrightarrow \beta$		(Δ_1)	$(\Delta_6)_2$
$\frac{1}{2}$	(Δ_1)	$\alpha \leftrightarrow e^{i\phi}\beta$		$(\Delta_5)_1$	$(\Delta_6)_1$
$\frac{3}{2}$	$(\Delta_5)_1$	$e^{i\phi}\alpha$			$(\Delta_7)_1$
$\frac{5}{2}$			$e^{3i\phi}\beta$	$(\Delta_5)_2$	$(\Delta_7)_2$
$\frac{7}{2}$	$(\Delta_5)_2$	$e^{3i\phi}\alpha \leftrightarrow e^{4i\phi}\beta$		(Δ_1)	$(\Delta_6)_2$
$\frac{9}{2}$	(Δ_1)	$e^{4i\phi}\alpha \leftrightarrow e^{5i\phi}\beta$		$(\Delta_5)_1$	$(\Delta_6)_1$
...	...	...		...	...

the first partner of Δ_5; and between the function β arising from Δ_1, and the function $e^{-i\phi}\alpha$ arising from the second partner of Δ_5.

We may now consider the implications of these facts, in the matter of the secular equation which must be solved between the various functions. In the first place, the function formed from the first partner of Δ_5 and the spin function α, with $M = \frac{3}{2}$, has no nondiagonal matrix elements, so that it is a suitable function as it stands. Going with this is the function with $M = -\frac{3}{2}$, formed from the second partner of Δ_5 and the spin function β. Since one of these functions arises from the other by reversal of all angular momenta, we expect that the two will be degenerate with each other, and will form basis functions for a two-dimensional irreducible representation, in some sense which we have not yet explored.

These two states are labeled $(\Delta_7)_1$ and $(\Delta_7)_2$ in the right-hand column of Table A9-2; we shall later show in Table A9-7 that they are basis functions for an irreducible representation Δ_7 of the double group. The other cases labeled $(\Delta_7)_1$ and $(\Delta_7)_2$ in the table merely are other components of these same two functions, with M differing by 4 from those already discussed.

Next we consider the other cases, with $M = \pm \frac{1}{2}$, and the other values related by addition or subtraction of 4. Here we have two functions of each type, to be determined as solutions of a quadratic secular equation. Here again we have a function with $M = \frac{1}{2}$, and one with $M = -\frac{1}{2}$, which will be expected to be degenerate. Consequently, we have labeled these two types of functions as $(\Delta_6)_1$ and $(\Delta_6)_2$, and we shall show later that they form two basis functions for another two-dimensional irreducible representation. All this fits in with the qualitative discussion given earlier in this section of the effect of spin-orbit interactions on the energy levels.

As the reduced wave vector approaches zero, and we approach the point Γ in the Brillouin zone, we shall expect that the same sort of situation will be found which we know exists for the atom. Namely, the energies of the states $(\Delta_7)_{1,2}$, corresponding to $M = \pm \frac{3}{2}$, and of one of the states $(\Delta_6)_{1,2}$, will approach each other, leading at the point Γ to a fourfold degenerate state, like the $^2P_{3/2}$ in the atomic case, with $M = \frac{3}{2}$, $\frac{1}{2}$, $-\frac{1}{2}$, $-\frac{3}{2}$. The other state $(\Delta_6)_{1,2}$ will remain separated from the fourfold degenerate state, and is analogous to the $^2P_{1/2}$ in the atomic case. These will lead to a fourfold degenerate or four-dimensional irreducible representation of the double group, and a doubly degenerate or two-dimensional irreducible representation of the double group. We shall see later, in Table A9-11, that these are the representations known as Γ_8^- and Γ_6^-, respectively, of the double group of O_h.

We have been using terms in this discussion which we have not defined: double groups, with irreducible representations of a type which we have not discussed. In the next section we go on to explain what they are. It is clear that we are meeting new types of symmetry, new irreducible representations, and we must start from the beginning to understand them.

A9-4. Double Groups and Their Applications. The point groups operate on functions of the coordinates, and consist of rotations and reflections which leave one point in space invariant. They are fundamental to a study of the space groups. As is well known, the continuous group of all rotations has one irreducible representation for each odd dimensionality $2l + 1$, where l is an integer going from zero to infinity. The basis functions can be taken to be the $2l + 1$ spherical harmonics corresponding to the values of m_l from $-l$ to l. In atomic problems with integral quantum numbers, we can associate the various irreducible representations with different values of the quantum number equal to the total angular momentum of the system.

This formulation of the problem has no place for the half-integral quantum numbers. In quantum mechanics we are familiar with half-integral angular momenta l, such that m, going from $-l$ to l by integral steps, has an even rather than an odd number of values, so that we meet irreducible representations of even dimensionality. We cannot use ordinary functions of space as basis functions for such irreducible representations. We must rather use spin-orbitals, composed of a function of space, and a spin function, such as we have set up in Eq. (A9-3). The operations which we must consider are those in which we are rotating both the space and spin functions. The reason is that the spin-orbit interaction depends on the angle between spin and orbital angular momentum, and it will commute with an operation only if both spin and orbital functions are rotated or reflected in the same way, so as not to change the angle between them. We must, then, investigate the rotation of the spin functions, and must consider the possibility that the spin is quantized along other directions than the z axis.

There is an immediate complication which arises as soon as we consider the rotation of axes as applied to the functions α and β. Let us first consider only rotation about the z axis, so that the axis of quantization remains fixed. In many ways the functions α and β resemble ordinary spherical harmonics set up, in some sort of way, for half-integral l's, in particular for $l = \frac{1}{2}$, leading to $m = \pm \frac{1}{2}$. Since the dependence of a spherical harmonic on ϕ is $e^{im\phi}$, we may suppose that α acts like the function $e^{i\phi/2}$, and β like $e^{-i\phi/2}$. The complication then arises if we make a rotation through an angle 2π. If ϕ increases by 2π, the function $e^{i\phi/2}$ is multiplied by $e^{\pi i} = -1$, so that α (and similarly β) changes sign on a rotation through 2π. We must rotate through 4π to come back to the original value of the function. This complication is the origin of the idea of the double group. For every rotation lying within a range 2π (for example, for every rotation in the range from 0 to 2π, or from $-\pi$ to π), we must have another rotation greater or less than this by 2π, whose effect on any function of coordinates will be identical with that of the original rotation, but whose effect on a spin function will give the negative of the value given by the original rotation.

In particular, suppose we are considering a point group of the ordinary type, consisting of a finite set of rotations, or rotations plus inversions, leaving a point invariant, and transforming a Hamiltonian function into itself. If we let the operations of this group operate on a function of spin, corresponding to a half-integral angular momentum, we shall find as in the preceding paragraph that for each operation involving a rotation, there will be another one corresponding to a rotation greater by

2π, which will give a function of opposite sign when applied to a basis function involving a spin. We shall have, then, twice as many operations to consider as in the ordinary group, which is the reason for describing it as a double group. We must proceed over again from the beginning in discussing this double group. We must find its multiplication table, prove that it in fact is a group, investigate its classes, and find its irreducible representations and basis functions for them. These quantities do not follow trivially from those of the single group. We can see, however, that each of the basis functions and irreducible representations found for the single group, and involving integral quantum numbers, and functions of the coordinates only, will also supply basis functions and irreducible representations for the double group. The reason is trivial: the new operations of the double group have the same effect on a function of coordinates alone that the original operations had. Aside from these irreducible representations, however, there will be other irreducible representations, called the extra representations, whose basis functions must be written in a form involving the spin functions. It is some of these extra irreducible representations which we have mentioned in the preceding section. These extra representations have been discussed by various writers.[1]

The study of the double groups is quite extensive, and we take it up in detail in Secs. A9-5 to A9-10. There we not only investigate the way of finding the operations of the point and space groups on spin-orbitals, and the properties of groups of such operations, but also the irreducible representations and basis functions for many cases met in actual crystals, including the specific cases we have taken up in the preceding sections. There is one simple principle which our general knowledge of group theory will tell us at once. We recall that the sum of the squares of the dimensionalities of the irreducible representations of a group equals the number of operations in the group. Now a double group has twice as many operations as the corresponding single group. Hence the sum of the squares of the dimensionalities of the extra representations of the double group must equal the number of operations in the single group, and hence must equal the sum of the squares of the dimensionalities of the ordinary representations of the single group. In general, however, the extra individual irreducible representations of the double group will be quite different in properties from the ordinary representations. For instance, in the group C_{4v}, we have eight operations in the single group. The ordinary representations, as we see from Table 5-3, include four one-dimensional irreducible representations, and one two-dimensional, the sum of squares being $1^2 + 1^2 + 1^2 + 1^2 + 2^2 = 8$. The extra represen-

[1] See for instance W. Opechowski, *Physica*, **7**:552 (1940); R. J. Elliott, *Phys. Rev.*, **96**:280 (1954); G. F. Koster, *Solid State Phys.*, **5**:173 (1957).

tations, however, consist of two two-dimensional irreducible representations, Δ_6 and Δ_7, as mentioned in the preceding section. The sum of squares, $2^2 + 2^2$, again is 8.

In a similar way, the single group of O_h has 48 operations, and 10 irreducible representations, 4 one-dimensional, 2 two-dimensional, and 4 three-dimensional, leading to $4(1^2) + 2(2^2) + 4(3^2) = 48$. The extra representations, however, comprise 4 two-dimensional and 2 four-dimensional, leading to $4(2^2) + 2(4^2) = 48$. The two-dimensional, and four-dimensional irreducible representations Γ_6^- and Γ_8^- which we have mentioned in Sec. A9-3 are two of these six irreducible representations.

It is clear how the knowledge of the double groups is to be used. We are to build up basis functions for the irreducible representations of the double groups, constructed out of functions of the coordinates alone, which are basis functions for the irreducible representations of the single groups, multiplied by spin functions α and β. We have seen examples of this in Table A9-2. Some cases are not as simple as that described there, and we may have to use projection operators to find the desired basis functions for representations of the double groups. These projection operators can be constructed in the usual way from the matrix elements of the irreducible representations given in Secs. A9-7 to A9-9. When we have set up these symmetry functions, then the Hamiltonian function, including the spin-orbit interaction, will have nondiagonal matrix elements only between functions which are basis functions for the same irreducible representation of the double group, and further are basis functions for the same partner in the same irreducible representation. Thus we achieve the same sort of simplification in the problem that we get in the case where there is no spin, by the use of the irreducible representations of the single groups.

We shall not go far in describing the actual cases where spin-orbit interaction is important in crystals, but shall mention a few of the best known. Perhaps the most famous case is the top of the valence band in silicon and germanium. Here we have the case which we have been using as an example in this Appendix. In the absence of spin-orbit interaction, the highest energy level in the valence band of these crystals comes at the point Γ, and has the Γ_{25}' symmetry, arising from atomic p electrons. This splits, along the direction Δ, into the nondegenerate Δ_2', and the doubly degenerate Δ_5 symmetries. Spin-orbit interaction splits the threefold degenerate level at Γ, into the higher Γ_6^- level, twofold degenerate like the atomic $^2P_{1/2}$, and the lower Γ_8^-, fourfold degenerate like the atomic $^2P_{3/2}$. The total number of levels, six, is of course twice what it is in the spinless case, on account of the doubling of levels arising from the two orientations of the spin. Along the direction Δ, the twofold degenerate level joins to the twofold degenerate Δ_6, while the fourfold degenerate Γ_8^- splits into two twofold degenerate levels, one Δ_6 and one

Δ_7. These separate from each other more and more as we go to larger **k** values. In Appendix 5 we have shown how these levels can be expanded in power series in **k**, by the **k · p** method.

We have mentioned that the twofold degenerate state lies above the fourfold degenerate, whereas in the case of the single p electron which we took up in Sec. A9-1, the $^2P_{3/2}$ had an energy lying above the $^2P_{1/2}$. The reason for the discrepancy is that in the valence band of silicon or germanium, the energy levels we are speaking of are those arising from a missing electron or hole in an otherwise filled band. It is known from the atomic problem that in this case the multiplet is inverted, the lower J value lying above the higher J value, instead of lower as in the normal case. The reason, in an elementary way, is that we are dealing with energies subtracted from the energy of a closed shell, so that they come in with the opposite sign to that of a one-electron system.

This same situation is met in another famous case of spin-orbit interaction in crystals, namely, the splitting of the x-ray levels in a crystal. When a crystal is ionized by an impinging electron, an electron is knocked out of one of its inner shells, and an electron then falls down into this vacancy from an outer shell. The missing electron in the inner shell is analogous to the hole in the valence band in the preceding example, and again it leads to two levels, corresponding to $l \pm \frac{1}{2}$ in the atomic case. Again we find the level of lower J value lying above that of higher J value.

There is one point which we should mention, before leaving these problems. In Sec. A9-1 we have described the other relativistic terms in the Hamiltonian, aside from the spin-orbit interaction. We have not carried further the discussion of these other effects. They are in a way less interesting than the spin-orbit interaction, in that they do not introduce new symmetry problems into the basis functions and Schrödinger's equations. They are nevertheless important terms in the Hamiltonian, for heavy atoms, and must be included in the calculations. They represent operators whose matrix elements, with respect to unperturbed basis functions calculated on a nonrelativistic basis, can be easily found. They would normally be introduced into a secular equation in which the spin-orbit interaction was also included, and this whole secular equation would be solved to find the complete relativistic and spin-orbit effect on the energy bands. While there has been little calculation of this type made so far, it is bound to come very rapidly, as crystals containing heavy atoms are investigated theoretically. We have encountered one interesting case where these terms are important, in Sec. 10-7, in our discussion of the energy bands of PbTe.[1]

---

[1] For further discussion of this case, see L. E. Johnson, J. B. Conklin, and G. W. Pratt, Jr., *Phys. Rev. Letters*, **11**:538 (1963). For other cases see F. Herman, C. D. Kuglin, K. F. Cuff, and R. L. Kortum, *Phys. Rev. Letters*, **11**:541 (1963).

A9-5. The Effect of a Rotation on the Spin Functions. We shall now start our detailed discussion of double groups by considering the effect of a rotation of axes on the Pauli functions α and β. We can derive the results by analogy from the case of the single groups. It has been shown by Wigner[1] that if we rotate axes according to the Euler angles a, b, c (which we shall define in a moment), a spherical harmonic $Y_{lm}(\theta,\phi)$ will be transformed into a linear combination of spherical harmonics $Y_{lm'}(\theta,\phi)$ with all values of m' from $-l$ to l. This is the explicit statement of the fact mentioned in the preceding section, that this set of spherical harmonics forms a set of basis functions for an irreducible representation of the rotation group. This transformation can be written in the form given below:

$$Y_{lm}(\theta',\phi') = \Sigma(m') D^{(l)}(R)^*_{mm'} Y_{lm'}(\theta,\phi) \qquad (A9\text{-}22)$$

where the coefficients of the transformation are given by

$$D^{(l)}(abc)_{mm'} = \sum_{(t)} \frac{(-1)^t \sqrt{(l+m)!(l-m)!(l+m')!(l-m')!}}{(l-m-t)!(l+m'-t)!t!(t+m-m')!}$$
$$e^{i(ma+m'c)} \left(\cos\frac{b}{2}\right)^{2l+m'-m-2t} \left(\sin\frac{b}{2}\right)^{2t+m-m'} \qquad (A9\text{-}23)$$

In Eq. (A9-22) the rotation is characterized by the symbol R, which is equivalent to the three Euler angles a, b, c of Eq. (A9-23). The index t is to run over those values for which all the factorials are those of positive integers or zero. The Euler angles are indicated in Fig. A9-1. The rotation of axes is defined as follows. One first rotates the coordinates about the z axis, in a positive direction, so as to carry x toward y, through an angle c, leading to axes x', y', $z' = z$. Next one rotates about the new y' axis, called the line of nodes, through an angle b, leading to axes x'', $y'' = y'$, z''. Finally, one rotates about the new z'' axis through an angle a. The final rotated axes $x'''y'''z'''$ lead to a final set of spherical polar coordinates r, θ', ϕ', such that θ' is the colatitude measured from the z''' axis and ϕ' is the angle of rotation from the x''' axis in the $x'''y'''$ plane, rotating in a positive direction about the z''' axis.

We shall shortly convert this statement of Euler's angles into rectangular coordinates, making it easier to understand. Before doing so, however, let us proceed with Eq. (A9-23). If the spin can really be considered like an angular-momentum vector of quantum number $\frac{1}{2}$, we should be able to find the transformation functions for the spin by specializing Eqs. (A9-22) and (A9-23) for the case $l = \frac{1}{2}$, $m = \pm\frac{1}{2}$. We find from

---

[1] See E. P. Wigner, "Group Theory and Its Application to the Quantum Mechanics of Atomic Spectra," Academic Press Inc., New York, 1959, pp. 154, 167, 357–359.

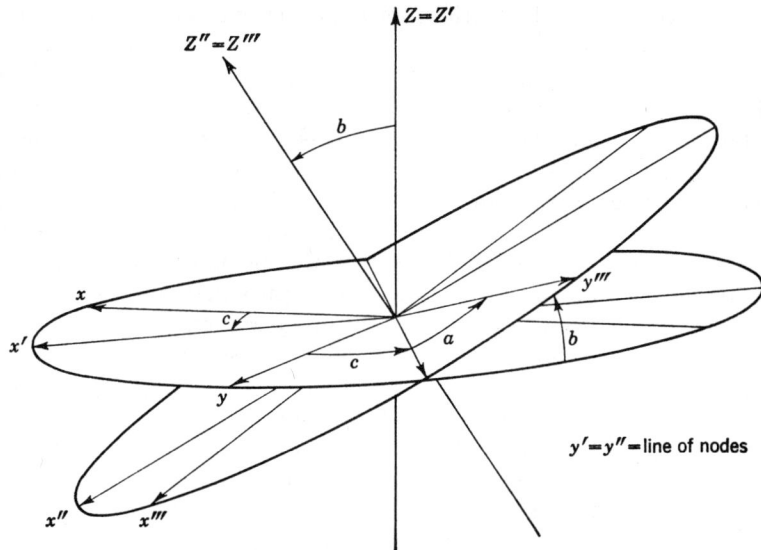

Fig. A9-1. The Euler angles.

Eq. (A9-23) that

$$D^{1/2}(abc)_{1/2,1/2} = \cos\frac{b}{2} e^{i(a+c)/2}$$

$$D^{1/2}(abc)_{1/2,-1/2} = \sin\frac{b}{2} e^{i(a-c)/2}$$

$$D^{1/2}(abc)_{-1/2,1/2} = -\sin\frac{b}{2} e^{i(-a+c)/2}$$

$$D^{1/2}(abc)_{-1/2,-1/2} = \cos\frac{b}{2} e^{i(-a-c)/2}$$

(A9-24)

If we symbolize the rotation operator as R, we expect that the value of $R\alpha$ will be the corresponding function of (θ', ϕ'), just as with a function of coordinates we should have

$$Rf(\theta,\phi) = f(\theta',\phi') \qquad (A9\text{-}25)$$

Hence, from Eq. (A9-22), we expect that we shall have

$$R\alpha = D^{1/2}(abc)^*_{1/2,1/2}\alpha + D^{1/2}(abc)^*_{1/2,-1/2}\beta$$
$$R\beta = D^{1/2}(abc)^*_{-1/2,1/2}\alpha + D^{1/2}(abc)^*_{-1/2,-1/2}\beta$$

(A9-26)

from which we have

$$R\alpha = \cos\frac{b}{2} e^{-i(a+c)/2}\alpha + \sin\frac{b}{2} e^{-i(a-c)/2}\beta$$
$$R\beta = -\sin\frac{b}{2} e^{-i(-a+c)/2}\alpha + \cos\frac{b}{2} e^{-i(-a-c)/2}\beta$$

(A9-27)

These are the standard equations for the effect of a rotation on the Pauli spin functions.

While we have merely derived these equations by analogy with the ordinary equations for the transformation of a spherical harmonic under a rotation of coordinates, there is sound reason for using the same method for the spin functions. We recall that there is a close relation between the rotation and the angular-momentum operators,[1] so that the rotation operators can be written explicitly in terms of angular-momentum operators. Furthermore, we know that the angular-momentum operators for the electron spin satisfy the same basic equations (commutation rules, etc.) as ordinary angular-momentum operators, so that the matrix elements of spin angular momentum follow the same formulas as those for ordinary orbital angular momentum. Hence one could carry through a rigorous proof that the effect of a rotation on the spin functions would have to be given by Eq. (A9-27), as given above.

We have mentioned that it is more convenient to describe our rotations in terms of rectangular coordinates than in terms of Euler's angles. Therefore let us investigate the form which Eq. (A9-27) takes when we describe our rotation in the form

$$Rf(x,y,z) = f(a_{11}x + a_{12}y + a_{13}z, a_{21}x + a_{22}y + a_{23}z, a_{31}x + a_{32}y + a_{33}z) \tag{A9-28}$$

in which the a's represent an orthogonal transformation with determinant of unity. For this purpose we shall throw the transformation of coordinates described by Euler's angles into the form given by Eq. (A9-28). First we are directed to rotate about the z axis in the positive direction, through an angle c. This transforms the coordinates x, y, z into

$$\begin{aligned} x' &= x \cos c + y \sin c \\ y' &= -x \sin c + y \cos c \\ z' &= z \end{aligned} \tag{A9-29}$$

Next we rotate about the y' axis through an angle b. This transforms x', y', z' into

$$\begin{aligned} x'' &= x' \cos b - z' \sin b \\ y'' &= y' \\ z'' &= x' \sin b + z' \cos b \end{aligned} \tag{A9-30}$$

Finally, we rotate about z'' through an angle a:

$$\begin{aligned} x''' &= x'' \cos a + y'' \sin a \\ y''' &= -x'' \sin a + y'' \cos a \\ z''' &= z'' \end{aligned} \tag{A9-31}$$

---

[1] For a discussion, see for instance J. C. Slater, "Quantum Theory of Atomic Structure," vol. 2, McGraw-Hill Book Company, New York, 1960, chap. 19.

We now combine these equations, so as to write x''', y''', z''' in terms of x, y, z. We find $x''' = a_{11}x + a_{12}y + a_{13}z$, etc., where

$$\begin{aligned}
a_{11} &= \cos a \cos b \cos c - \sin a \sin c \\
a_{12} &= \cos a \cos b \sin c + \sin a \cos c \\
a_{13} &= -\cos a \sin b \\
a_{21} &= -\sin a \cos b \cos c - \cos a \sin c \\
a_{22} &= -\sin a \cos b \sin c + \cos a \cos c \\
a_{23} &= \sin a \sin b \\
a_{31} &= \sin b \cos c \\
a_{32} &= \sin b \sin c \\
a_{33} &= \cos b
\end{aligned} \qquad (A9\text{-}32)$$

If we are given the rotation in rectangular coordinates, in terms of a_{ij}, we can use Eq. (A9-32) to solve for a, b, c, which can then be substituted into Eq. (A9-27) to find the transformation of the spin functions. If $a_{33}^2 \neq 1$, we can solve Eqs. (A9-32) to give

$$\begin{aligned}
\sin a &= \pm \frac{a_{23}}{\sqrt{1 - a_{33}^2}} & \cos a &= \mp \frac{a_{13}}{\sqrt{1 - a_{33}^2}} \\
\sin b &= \pm \sqrt{1 - a_{33}^2} & \cos b &= a_{33} \\
\sin c &= \pm \frac{a_{32}}{\sqrt{1 - a_{33}^2}} & \cos c &= \pm \frac{a_{31}}{\sqrt{1 - a_{33}^2}}
\end{aligned} \qquad (A9\text{-}33)$$

The upper signs are all to be used together, or the lower signs. If $a_{33}^2 = 1$, so that the denominators in Eq. (A9-33) become zero, we must proceed differently. We have two cases, $a_{33} = \pm 1$. We have the following situation:

Case 1, $a_{33} = 1$

$$a_{11} = a_{22} = \cos(a + c) \qquad a_{12} = -a_{21} = \sin(a + c) \qquad (A9\text{-}34)$$

Case 2, $a_{33} = -1$

$$a_{11} = -a_{22} = -\cos(a - c) \qquad a_{12} = a_{21} = \sin(a - c)$$

In these latter cases, only $a + c$ or $a - c$ can be determined, but examination of Eq. (A9-27) shows that in these cases it is only these combinations which will appear in the transformation.

We can now find the transformations from the a_{ij}'s, but when we carry the process through, we find that there is an ambiguity in the result. From Eqs. (A9-33) or (A9-34) we find the angles a, b, c, up to an additive constant of 2π. However, in Eq. (A9-27), it is half of these angles which appear. Hence, if any one of the angles is increased by 2π, the coefficients of Eq. (A9-27) change sign. This is the ambiguity which we have referred to in Sec. A9-4: corresponding to each operator of the point group we find two operators of the double group. We shall denote one of these operators as R, the other as $\bar{R}$; we take up later the question as to which of the two to denote as R, which as $\bar{R}$. We find on examination that the ambiguity of sign resulting from the $\pm$ in Eq. (A9-33) does not introduce any additional ambiguity into the final results; we can just as well use only the upper sign in Eq. (A9-33), and we shall get the same final answers.

A9-6. Examples of Rotation of Coordinates. Now we shall carry through some specific examples to illustrate the application of these methods. Let us first carry out a rotation about the z axis, which therefore does not change the axis of quantization. This is such a rotation as we meet in the operations of the group C_N, where the operator X_q is defined by the equation

$$X_q \psi(\phi) = \psi\left(\phi + \frac{2\pi q}{N}\right) \tag{A9-35}$$

If we put this into rectangular coordinates, we have $x = r \cos \phi$, $y = r \sin \phi$, $z = z$ (where r, ϕ, z are cylindrical coordinates). Thus $\psi(\phi + 2\pi q/N)$ stands for

$$\psi\left[r \cos\left(\phi + \frac{2\pi q}{N}\right), r \sin\left(\phi + \frac{2\pi q}{N}\right), z\right] \tag{A9-36}$$

When we work this out, it becomes

$$\psi\left(x \cos \frac{2\pi q}{N} - y \sin \frac{2\pi q}{N}, x \sin \frac{2\pi q}{N} + y \cos \frac{2\pi q}{N}, z\right) \tag{A9-37}$$

so that we have

$$\begin{array}{lll} a_{11} = \cos \dfrac{2\pi q}{N} & a_{12} = -\sin \dfrac{2\pi q}{N} & a_{13} = 0 \\ a_{21} = \sin \dfrac{2\pi q}{N} & a_{22} = \cos \dfrac{2\pi q}{N} & a_{23} = 0 \\ a_{31} = 0 & a_{32} = 0 & a_{33} = 1 \end{array} \tag{A9-38}$$

We then have Case 1 of Eq. (A9-34), leading to

$$\cos(a + c) = \cos \frac{2\pi q}{N} \qquad \sin(a + c) = -\sin \frac{2\pi q}{N} \tag{A9-39}$$

from which the obvious solution is $a + c = -2\pi q/N$, though we could also have a value 2π greater. Since $a_{33} = \cos b = 1$, we have $b = 0$ (or 2π). If we choose $b = 0$, $a + c = -2\pi q/N$, and substitute into Eq. (A9-27), we have

$$X_q \alpha = e^{\pi i q/N} \alpha \qquad X_q \beta = e^{-\pi i q/N} \beta \tag{A9-40}$$

These are just what we should expect from a quantum number of ½ for α, $-\tfrac{1}{2}$ for β; the general rule for the effect of X_q operating on a wave function like $e^{im\phi}$ is that $X_q e^{im\phi} = e^{2\pi i m q/N} e^{im\phi}$. We also have, however, the other case which arises when either the angle b, or $a + c$, is increased by 2π, in which case

$$\bar{X}_q \alpha = -e^{\pi i q/N}\alpha \qquad \bar{X}_q \beta = -e^{-\pi i q/N}\beta \tag{A9-41}$$

It is obvious in this case that it is natural to define the unbarred and barred operators as we have done in Eqs. (A9-40) and (A9-41), since the unbarred operators then agree with the ordinary rotation operators of the single group.

In the single group C_N, as we know, the multiplication table is very simple. If the operator X_q merely multiplies a basis function $e^{im\phi}$ by the factor $e^{2\pi i m q/N}$, and another operator X_p multiplies it by the factor $e^{2\pi i m p/N}$,

Table A9-3

Multiplication table for the double group C_3. The table gives the product of the operator along the left (first factor) and that at the top (second factor).

	X_0	X_1	X_{-1}	$\bar{X}_0$	$\bar{X}_1$	$\bar{X}_{-1}$
X_0	X_0	X_1	X_{-1}	$\bar{X}_0$	$\bar{X}_1$	$\bar{X}_{-1}$
X_1	X_1	$\bar{X}_{-1}$	X_0	$\bar{X}_1$	X_{-1}	$\bar{X}_0$
X_{-1}	X_{-1}	X_0	$\bar{X}_1$	$\bar{X}_{-1}$	$\bar{X}_0$	X_1
$\bar{X}_0$	$\bar{X}_0$	$\bar{X}_1$	$\bar{X}_{-1}$	X_0	X_1	X_{-1}
$\bar{X}_1$	$\bar{X}_1$	X_{-1}	$\bar{X}_0$	X_1	$\bar{X}_{-1}$	X_0
$\bar{X}_{-1}$	$\bar{X}_{-1}$	$\bar{X}_0$	X_1	X_{-1}	X_0	$\bar{X}_1$

it is obvious that the two operators operating in succession multiply it by $e^{2\pi i m(p+q)/N}$, so that $X_q X_p = X_p X_q = X_{p+q}$. In applying this rule, we must remember that if the sum $p + q$ lies outside the restricted range $-N/2 < q \le N/2$, we are to bring it back into the range by adding or subtracting integral multiples of N. But now we see that we have a difficulty with the double group. Changing q by N, according to Eqs. (A9-40) and (A9-41), will introduce an extra factor $e^{\pi i N/N} = e^{\pi i} = -1$ into the resulting function. In other words, if we add or subtract N to bring an operator back into the allowed range, we must change from an unbarred to a barred operator, or vice versa. We have, in fact,

$$X_{q\pm N} = \bar{X}_q \tag{A9-42}$$

This will, then, modify the multiplication table of a double group, introducing bars in some cases of product operations, not in others. By using this rule, we find that the multiplication table for the double group C_3, as a simple example, is as given in Table A9-3.

We note that among the products arising from two unbarred operators there are two barred operators, namely, $X_1 X_1 = \bar{X}_{-1}$, $X_{-1} X_{-1} = \bar{X}_1$. Thus we cannot use the multiplication table for the single group without change for the double group. We note one simplifying feature, however: the product of an unbarred and a barred operator is the same as the product of the two corresponding unbarred operators, except with a change from unbarred to barred, or vice versa. In other words, changing from barred to unbarred operators is like a change of sign. This means that the multiplication table for unbarred operators, which is only a fourth as large as the complete multiplication table, really furnishes complete information. In other words, it is no more complicated to write down the multiplication table for a double group than for the corresponding single group. We shall henceforth take advantage of this fact in writing multiplication tables.

The double group C_3 is isomorphic with the single group C_6; according to Eq. (A9-42), we could label the barred operators $\bar{X}_0$, $\bar{X}_1$, $\bar{X}_{-1}$ by the symbols X_3, X_{-2}, X_2, respectively, and we should then have the operators X_0, $X_{\pm 1}$, $X_{\pm 2}$, X_3 of the single group C_6. The multiplication table of Table A9-3 follows at once from that for the single group C_6. As in that case, the group is Abelian, each operator forms a class by itself, and there are six one-dimensional irreducible representations. Since the single group C_3 has three one-dimensional irreducible representations, this means that there are also three extra one-dimensional irreducible representations for the double group. In general, we see that the double group C_N is isomorphic with the single group C_{2N}.

Next let us take up quite a different sort of rotation, one in which the axis of quantization changes. Thus, let us take a cyclic permutation of axes, in which z shifts to x, x shifts to y, y to z, so that

$$Rf(x,y,z) = f(y,z,x) \tag{A9-43}$$

As we compare with Eq. (A9-28), we see that this is the case

$$\begin{aligned} a_{11} &= a_{13} = a_{21} = a_{22} = a_{32} = a_{33} = 0 \\ a_{12} &= a_{23} = a_{31} = 1 \end{aligned} \tag{A9-44}$$

We may now use Eq. (A9-33) to derive the Euler angles, and if we choose the upper signs, we have

$$\begin{aligned} \sin a &= 1 & \cos a &= 0 \\ \sin b &= 1 & \cos b &= 0 \\ \sin c &= 0 & \cos c &= 1 \end{aligned} \tag{A9-45}$$

The solutions may be taken to be $a = \pi/2$, $b = \pi/2$, $c = 0$. Then in Eq. (A9-27) we have

$$R\alpha = \cos\frac{\pi}{4} e^{-\pi i/4}\alpha + \sin\frac{\pi}{4} e^{-\pi i/4}\beta$$
$$= \frac{1-i}{2}(\alpha + \beta)$$
$$R\beta = -\sin\frac{\pi}{4} e^{\pi i/4}\alpha + \cos\frac{\pi}{4} e^{\pi i/4}\beta \qquad \text{(A9-46)}$$
$$= \frac{1+i}{2}(-\alpha + \beta)$$

Since we have rotated the z axis into the x direction, we should suppose that the resulting spin functions denoted as $R\alpha$ and $R\beta$, respectively, in Eq. (A9-46) would represent spins pointing along the $\pm x$ directions. We shall shortly show that this is the case.

Similarly, let us rotate so that z goes into y, y into x, x into z. Here as in Eq. (A9-44) we have all the a's equal to zero except now a_{13}, a_{21}, a_{32}, which are unity. From Eq. (A9-33) we have in this case

$$\begin{array}{ll}\sin a = 0 & \cos a = -1 \\ \sin b = 1 & \cos b = 0 \\ \sin c = 1 & \cos c = 0\end{array} \qquad \text{(A9-47)}$$

from which $a = -\pi$, $b = c = \pi/2$. Then from Eq. (A9-27) we find

$$R\alpha = \frac{1+i}{2}\alpha + \frac{-1+i}{2}\beta = \frac{1+i}{2}(\alpha + i\beta)$$
$$R\beta = \frac{1+i}{2}\alpha + \frac{1-i}{2}\beta = \frac{1+i}{2}(\alpha - i\beta) \qquad \text{(A9-48)}$$

These functions should represent spin functions along y and $-y$, respectively. In choosing the value of a, from Eq. (A9-47), we might well have hesitated as to whether to choose $-\pi$, as we did, or π; one would have given R, the other $\bar{R}$. We have made the choice in such a way that this operator, and the preceding one worked out in Eq. (A9-46), which belong to the same class in the point group O_h, will have the same character of unity. We shall see later that this forms one guide as to the choice of the primed and unprimed operations.

In Sec. A9-1, we have seen that ψ_3 represents the wave function for electrons of $+$ spin, or associated with the spin function α, while the function ψ_4 is the wave function for electrons of $-$ spin. Thus as in Eq. (A9-3) we can write the complete wave function in the form $\psi_3\alpha + \psi_4\beta$. From the results of Eqs. (A9-46) and (A9-48), we should expect that if $\psi_4 = 0$, the spin is along z; if $\psi_3 = 0$, it is along $-z$; if $\psi_3 = \psi_4$, it is along x; if $\psi_3 = -\psi_4$, along $-x$; if $\psi_4 = i\psi_3$, along y; and if $\psi_4 = -i\psi_3$, along $-y$. We shall shortly show in an independent way that this is true.

If ψ_3 and ψ_4 are proportional to the same functions of x, y, z, then their ratio will remain fixed over all space, and this means that the spin orientation will be the same everywhere. If on the other hand we wish to represent a situation in which the spin orientation varies from point to point, ψ_3 and ψ_4 must be different functions of x, y, and z. Such a situation arises in case we are dealing with spin waves in which the spin orientation varies in a helical manner as we pass through a crystal, a problem which we shall have to take up in a later volume of this series. In this Appendix, however, we shall not have to deal with that situation. Let us then assume that

$$\psi_3 = \Psi_3 u(x,y,z) \qquad \psi_4 = \Psi_4 u(x,y,z) \qquad \text{(A9-49)}$$

where Ψ_3, Ψ_4 are constants, and where we have the normalization conditions

$$\int u^*(x,y,z) u(x,y,z) \, dv = 1$$
$$\Psi_3^* \Psi_3 + \Psi_4^* \Psi_4 = 1 \qquad \text{(A9-50)}$$

We now find from Eq. (A9-7) that the x, y, z components of the average of the operator $\boldsymbol{\sigma}$ are given by

$$\begin{aligned}(\sigma_x)_{av} &= \Psi_3^* \Psi_4 + \Psi_4^* \Psi_3 \\ (\sigma_y)_{av} &= -i\Psi_3^* \Psi_4 + i\Psi_4^* \Psi_3 \\ (\sigma_z)_{av} &= \Psi_3^* \Psi_3 - \Psi_4^* \Psi_4\end{aligned} \qquad \text{(A9-51)}$$

These quantities are proportional to the rectangular components of the spin angular momentum or magnetic moment. Since they are independent of position, we may discard the average value subscript in referring to them. We then verify that if $\Psi_3 = 1$, $\Psi_4 = 0$, corresponding to a zero coefficient for the Pauli function β, we have $\sigma_z = 1$, $\sigma_x = \sigma_y = 0$. Similarly, if $\Psi_3 = 0$, $\Psi_4 = 1$, we have $\sigma_z = -1$. If $\Psi_3 = \Psi_4 = 1/\sqrt{2}$, $\sigma_x = 1$, $\sigma_y = \sigma_z = 0$, while if $\Psi_3 = -\Psi_4 = 1/\sqrt{2}$, $\sigma_x = -1$. Again, if $\Psi_4 = i\Psi_3 = 1/\sqrt{2}$, $\sigma_y = 1$, $\sigma_x = \sigma_z = 0$, while if $\Psi_4 = -i\Psi_3 = 1/\sqrt{2}$, $\sigma_y = -1$. These are the cases mentioned in Eqs. (A9-46) and (A9-48). The fact that the wave functions in those cases differ from those considered here by being multiplied by a complex constant of magnitude unity makes no difference in Eq. (A9-51).

In other words, we can check the rotated wave functions of Eqs. (A9-46) and (A9-48), and show that the spins in fact point in the directions we expect. It is easy by using the same method to set up a linear combination of α and β representing a spin pointing in any arbitrary direction. Thus, let σ_x, σ_y, σ_z, which are three components of a unit vector, have arbitrary values. Since $\Psi_3^*\Psi_3 + \Psi_4^*\Psi_4 = 1$, we see from Eq. (A9-51) that

$$\sigma_z = \Psi_3^*\Psi_3 - (1 - \Psi_3^*\Psi_3) = 2\Psi_3^*\Psi_3 - 1$$
$$\Psi_3^*\Psi_3 = \frac{1 + \sigma_z}{2} \qquad \text{(A9-52)}$$

Now let $\Psi_4/\Psi_3 = a + bi$, where a and b are real. Then from Eq. (A9-51) we have

$$\sigma_x = \Psi_3^*\Psi_3(a + bi + a - bi) = 2a\Psi_3^*\Psi_3 = a(1 + \sigma_z)$$

so that

$$a = \frac{\sigma_x}{1 + \sigma_z} \qquad \text{(A9-53)}$$

Similarly, we have

$$\sigma_y = \Psi_3^*\Psi_3(-ia + b + ia + b) = 2b\Psi_3^*\Psi_3$$
$$b = \frac{\sigma_y}{1 + \sigma_z} \qquad \text{(A9-54)}$$

We thus know the absolute magnitude of Ψ_3, and the ratio Ψ_4/Ψ_3. This is as much as we can find from σ_x, σ_y, σ_z. If in particular we choose Ψ_3 to be real, we have

$$\Psi_3 = \sqrt{\frac{1 + \sigma_z}{2}} \qquad \Psi_4 = \frac{\sigma_x + i\sigma_y}{\sqrt{2(1 + \sigma_z)}} \qquad \text{(A9-55)}$$

These formulas will be useful in our future work, to allow us to set up wave functions corresponding to spins pointing in arbitrary directions.

So far we have been considering the rotations and their effect on the spin functions α and β. We also have operations including reflections, or inversions. It is most convenient to write each such operation as the product of an inversion and a rotation. Since we have found how to handle a rotation, we need only investigate in addition the effect of an inversion on the spin functions. The answer is very simple: an inversion leaves the functions invariant. The reason for this can be seen easily from the discussion we have just been giving. Suppose we are dealing with a central-field problem, as a simple example. Then both of the functions ψ_3 and ψ_4 will be either even or odd on inversion. The inversion operation then will not change the expressions of Eq. (A9-51), and hence will not affect the spin. This is not a proof, but it suggests the correctness of the result, which can be proved rigorously in general. Hence we can show that the effect of any operation involving both an inversion and a rotation on the spin functions α and β is identical with that of the rotation alone. This gives us all the information we need to discuss the double groups completely.

A9-7. The Double Groups C_{Nv}. We are now ready to proceed with some actual cases, and shall first consider the general double group C_{Nv}, and in particular the examples of C_{2v}, C_{3v}, and C_{4v}. The single group has not only the operations X_q, defined in Eq. (A9-35), which are met in the group C_N, but also the operations Y_q, defined by

$$Y_q\psi(\phi) = \psi\left(-\phi + \frac{2\pi q}{N}\right) \tag{A9-56}$$

or in rectangular coordinates

$$Y_q\psi(x,y,z) = \psi\left(x\cos\frac{2\pi q}{N} + y\sin\frac{2\pi q}{N}, x\sin\frac{2\pi q}{N} - y\cos\frac{2\pi q}{N}, z\right) \tag{A9-57}$$

This is an improper rotation, consisting of a rotation and an inversion. If we denote the corresponding rotation alone as Y'_q, we have

$$Y'_q\psi(x,y,z) = \psi\left(-x\cos\frac{2\pi q}{N} - y\sin\frac{2\pi q}{N}, -x\sin\frac{2\pi q}{N} + y\cos\frac{2\pi q}{N}, -z\right) \tag{A9-58}$$

which corresponds to

$$\begin{array}{lll} a_{11} = -\cos\dfrac{2\pi q}{N} & a_{12} = -\sin\dfrac{2\pi q}{N} & a_{13} = 0 \\ a_{21} = a_{12} & a_{22} = -a_{11} & a_{23} = 0 \\ & a_{31} = a_{32} = 0 & a_{33} = -1 \end{array} \tag{A9-59}$$

We have an example of Case 2 of Eq. (A9-34), leading to $b = \pi$,

$$\cos\frac{2\pi q}{N} = \cos(a - c)$$

$$\sin\frac{2\pi q}{N} = -\sin(a - c)$$

so that $a - c = -2\pi q/N$. Then from Eq. (A9-27) we have

$$Y'_q\alpha = e^{\pi i q/N}\beta \qquad Y'_q\beta = -e^{-\pi i q/N}\alpha \tag{A9-60}$$

In Eq. (A9-60), coupled with Eq. (A9-40), we have the statement of the effect of the operations of the group on the functions α and β. We have a relation

$$Y'_{q+N} = \bar{Y}'_q \tag{A9-61}$$

analogous to Eq. (A9-42).

We shall now use Eqs. (A9-40), (A9-42), (A9-60), and (A9-61) to give the effect of the operators X_q and Y_q on α and β, for the groups C_{2v},

C_{3v}, and C_{4v}; we remember that since the inversion does not affect the spin functions, the effect of Y_q is the same as that of Y'_q. We shall present these results in the form of matrix elements:

$$R\alpha = R_{11}\alpha + R_{21}\beta \qquad R\beta = R_{12}\alpha + R_{22}\beta \qquad \text{(A9-62)}$$

where R is one of the operators, and R_{11}, etc., are the matrix elements of

Table A9-4

Matrix elements of the two-dimensional irreducible representations of the double groups C_{2v}, C_{3v}, C_{4v}, using α and β as basis functions. The matrix elements for the barred operators are the negatives of those for the unbarred operators.

		X_0	X_1	Y_0	Y_1
	11	1	i	0	0
C_{2v}	21	0	0	1	i
	12	0	0	-1	i
	22	1	$-i$	0	0
	χ	2	0	0	0

		X_0	X_1	X_{-1}	Y_0	Y_1	Y_{-1}
	11	1	$e^{\pi i/3}$	$e^{-\pi i/3}$	0	0	0
C_{3v}	21	0	0	0	1	$e^{\pi i/3}$	$e^{-\pi i/3}$
	12	0	0	0	-1	$-e^{-\pi i/3}$	$-e^{\pi i/3}$
	22	1	$e^{-\pi i/3}$	$e^{\pi i/3}$	0	0	0
	χ	2	1	1	0	0	0

		X_0	X_1	X_{-1}	X_2	Y_0	Y_1	Y_{-1}	Y_2
	11	1	$e^{\pi i/4}$	$e^{-\pi i/4}$	i	0	0	0	0
C_{4v}	21	0	0	0	0	1	$e^{\pi i/4}$	$e^{-\pi i/4}$	i
	12	0	0	0	0	-1	$-e^{-\pi i/4}$	$-e^{\pi i/4}$	i
	22	1	$e^{-\pi i/4}$	$e^{\pi i/4}$	$-i$	0	0	0	0
	χ	2	$\sqrt{2}$	$\sqrt{2}$	0	0	0	0	0

the operator. We give the results for the three groups in question in Table A9-4.

We notice that each of the operators of the group, acting on the two basis functions α and β, produces a linear combination of these two functions. Thus they form basis functions for an irreducible representation of the group. Each double group has such a two-dimensional irreducible representation, for which the two spin functions are basis functions, just as each single group has a totally symmetrical one-dimensional irreducible representation in which the basis function is unchanged under each of the operations of the group. We note that in Table A9-4 we have given

Table A9-5

Multiplication tables for the double groups C_{2v}, C_{3v}, and C_{4v}. The table gives the product of the operator along the left (first factor) and that at the top (second factor). For the product of one unbarred and one barred operator, barred and unbarred operators are to be interchanged in the products. The products of two barred operators are identical with those of two unbarred operators.

		X_0	X_1	Y_0	Y_1
C_{2v}	X_0	X_0	X_1	Y_0	Y_1
	X_1	X_1	$\bar{X}_0$	$\bar{Y}_1$	Y_0
	Y_0	Y_0	Y_1	$\bar{X}_0$	$\bar{X}_1$
	Y_1	Y_1	$\bar{Y}_0$	X_1	$\bar{X}_0$

		X_0	X_1	X_{-1}	Y_0	Y_1	Y_{-1}
C_{3v}	X_0	X_0	X_1	X_{-1}	Y_0	Y_1	Y_{-1}
	X_1	X_1	$\bar{X}_{-1}$	X_0	Y_{-1}	Y_0	$\bar{Y}_1$
	X_{-1}	X_{-1}	X_0	$\bar{X}_1$	Y_1	$\bar{Y}_{-1}$	Y_0
	Y_0	Y_0	Y_1	Y_{-1}	$\bar{X}_0$	$\bar{X}_1$	$\bar{X}_{-1}$
	Y_1	Y_1	$\bar{Y}_{-1}$	Y_0	$\bar{X}_{-1}$	$\bar{X}_0$	X_1
	Y_{-1}	Y_{-1}	Y_0	$\bar{Y}_1$	$\bar{X}_1$	X_{-1}	$\bar{X}_0$

		X_0	X_1	X_{-1}	X_2	Y_0	Y_1	Y_{-1}	Y_2
C_{4v}	X_0	X_0	X_1	X_{-1}	X_2	Y_0	Y_1	Y_{-1}	Y_2
	X_1	X_1	X_2	X_0	$\bar{X}_{-1}$	Y_{-1}	Y_0	$\bar{Y}_2$	Y_1
	X_{-1}	X_{-1}	X_0	$\bar{X}_2$	X_1	Y_1	Y_2	Y_0	$\bar{Y}_{-1}$
	X_2	X_2	$\bar{X}_{-1}$	X_1	$\bar{X}_0$	$\bar{Y}_2$	Y_{-1}	$\bar{Y}_1$	Y_0
	Y_0	Y_0	Y_1	Y_{-1}	Y_2	$\bar{X}_0$	$\bar{X}_1$	$\bar{X}_{-1}$	$\bar{X}_2$
	Y_1	Y_1	Y_2	Y_0	$\bar{Y}_{-1}$	$\bar{X}_{-1}$	$\bar{X}_0$	X_2	$\bar{X}_1$
	Y_{-1}	Y_{-1}	Y_0	$\bar{Y}_2$	Y_1	$\bar{X}_1$	$\bar{X}_2$	$\bar{X}_0$	X_{-1}
	Y_2	Y_2	$\bar{Y}_{-1}$	Y_1	$\bar{Y}_0$	X_2	$\bar{X}_{-1}$	X_1	$\bar{X}_0$

the matrix elements only for the unbarred operators of the double groups; the matrix elements for a barred operator are the negatives of those for the corresponding unbarred operator.

It is convenient next to set up multiplication tables for the groups. This is done in Table A9-5. It is simple to set up these tables, using the information which we already have. Thus, for example, let us consider the case of C_{2v}. From Table A9-4, we know that

$$X_1\alpha = i\alpha \qquad Y_0\alpha = \beta \qquad Y_1\alpha = i\beta \qquad X_1\beta = -i\beta$$
$$Y_0\beta = -\alpha \qquad Y_1\beta = i\alpha \qquad \text{(A9-63)}$$

Then, for instance, we have $X_1 X_1 \alpha = X_1(i\alpha) = -\alpha = \bar{X}_0\, \alpha$, as indicated in the table. We can verify this by computing the effect of $X_1 X_1$ on β. We have $X_1 X_1 \beta = X_1(-i\beta) = -\beta = \bar{X}_0 \beta$. Similar methods suffice to find each entry in the tables. As we have mentioned in the discussion of the group C_N in Sec. A9-6, it is only necessary to tabulate the products of unbarred operators in presenting the multiplication table, since putting a bar on an operator changes the product operator from barred to unbarred, or vice versa.

Once we have the multiplication table, we can find the classes of operations. We recall that if R_i^{-1} is the inverse operator to R_i, all operators $R_i^{-1} R_j R_i$ fall in the same class with R_j. By proceeding in this way, we find the following classes in the cases we are using as illustrations:

C_{2v}: five classes: X_0; $\bar{X}_0$; X_1 and $\bar{X}_1$; Y_0 and $\bar{Y}_0$; Y_1 and $\bar{Y}_1$
C_{3v}: six classes: X_0; $\bar{X}_0$; X_1 and X_{-1}; $\bar{X}_1$ and $\bar{X}_{-1}$;
$\qquad\qquad Y_0,\ \bar{Y}_1,\ \text{and}\ \bar{Y}_{-1};\ \bar{Y}_0,\ Y_1,\ \text{and}\ Y_{-1}$ (A9-64)
C_{4v}: seven classes: X_0; $\bar{X}_0$; X_1 and X_{-1}; $\bar{X}_1$ and $\bar{X}_{-1}$;
$\qquad X_2$ and $\bar{X}_2$; Y_0, Y_2, $\bar{Y}_0$, and $\bar{Y}_2$; Y_1, Y_{-1}, $\bar{Y}_1$, and $\bar{Y}_{-1}$

In contrast to this, for the single groups we have the following classes:

C_{2v}: four classes: X_0, X_1, Y_0, and Y_1
C_{3v}: three classes: X_0; X_1 and X_{-1}; Y_0, Y_1, and Y_{-1} (A9-65)
C_{4v}: five classes: X_0; X_1 and X_{-1}; X_2; Y_0 and Y_2; Y_1 and Y_{-1}

We can now use this information to deduce the number of extra irreducible representations for each of the double groups. We recall that the number of irreducible representations for any group equals the number of classes, and the sum of the squares of the dimensionalities of the irreducible representations equals the number of operators in the group. Thus for the single group C_{2v} we have four one-dimensional irreducible representations; for the single group of C_{3v} we have two one-dimensional and one two-dimensional irreducible representations; for the single group of C_{4v} we have four one-dimensional and one two-dimensional irreducible representations. For the double group C_{2v} we have five irreducible representations, namely, the four found for the single group and an extra two-dimensional irreducible representation. For the double group C_{3v} we have six irreducible representations, namely, the three found for the single group and two extra one-dimensional and one extra two-dimensional irreducible representations. For the double group C_{4v} we have seven irreducible representations, namely, the five found for the single group and two extra two-dimensional irreducible representations. We shall now proceed to find basis functions and matrix elements for these extra representations.

We have pointed out that we already have basis functions for one extra two-dimensional irreducible representation in each case, namely, the functions α and β, and we have tabulated the matrix elements in Table A9-4. For C_{2v}, this is the only extra irreducible representation. For C_{3v}, we must have two extra one-dimensional irreducible representations, and for C_{4v} an extra two-dimensional representation. We must now look at the problem more physically, to see how to set up basis functions for these extra representations, and hence to find the matrix elements.

The two functions α and β of course correspond to components $M_S = \pm\frac{1}{2}$ of spin angular momentum along the axis. It seems rather obvious that if we are looking for further basis functions, we should build up functions corresponding to angular momentum $\pm\frac{3}{2}$, or if necessary even greater half-integral values. Let us explore this possibility, using first the case C_{3v} as an example. We know that the functions of coordinates $e^{i\phi}$ and $e^{-i\phi}$ correspond to $M_L = \pm 1$, respectively. If we then set up functions $e^{i\phi}\alpha$ and $e^{-i\phi}\beta$, we should expect that they would correspond to total angular momentum along the axis, or M, of $\pm\frac{3}{2}$, respectively. Let us then try these as basis functions. To find the effect of one of the operators on one of these functions, we must know its effect both on the function of coordinates and on the function of spin. We have already investigated the effect on the spin. As for the coordinates, we have

$$X_q e^{\pm i\phi} = e^{\pm 2\pi i q/N} e^{\pm i\phi}$$
$$Y_q e^{\pm i\phi} = e^{\pm 2\pi i q/N} e^{\mp i\phi} \quad \text{(A9-66)}$$

Let us call our two functions $\psi_1 = e^{i\phi}\alpha$ and $\psi_2 = e^{-i\phi}\beta$, respectively, and let us examine the effect of the operators of the double group of C_{3v} on them. We have

$$\begin{aligned}
X_1\psi_1 &= e^{2\pi i/3}e^{i\phi}e^{\pi i/3}\alpha = -e^{i\phi}\alpha = -\psi_1 \\
X_1\psi_2 &= -\psi_2 \quad X_{-1}\psi_1 = -\psi_1 \quad X_{-1}\psi_2 = -\psi_2 \\
Y_0\psi_1 &= \psi_2 \quad Y_0\psi_2 = -\psi_1 \quad Y_1\psi_1 = -\psi_2 \quad Y_1\psi_2 = \psi_1 \\
Y_{-1}\psi_1 &= -\psi_2 \quad Y_{-1}\psi_2 = \psi_1
\end{aligned} \quad \text{(A9-67)}$$

The effects of the barred operators on the functions ψ_1 and ψ_2 are the negatives of those of the unbarred operators. We see, then, that ψ_1 and ψ_2 form basis functions for a two-dimensional representation of the double group C_{3v}. This is, however, a reducible representation. If we take new basis functions $\psi_1 \pm i\psi_2$, we find that each of these combinations forms a basis function for a one-dimensional irreducible representation. These, then, are the two extra one-dimensional irreducible representations which we are seeking, for C_{3v}. The matrix elements are given in Table A9-6.

Though we have found two one-dimensional irreducible representations, nevertheless we can prove that on account of time reversal they must be degenerate with each other, so that as far as energy is concerned, the situation is no different from what it would have been with a two-dimensional representation. This is natural; we can see no reason why

Table A9-6

Matrix elements and characters for the extra irreducible representations of the double group C_{3v}. The representations are labeled as for the group of the wave vector along the 111 axis, in the representations of the double space groups for a cubic crystal. Basis functions for Λ_4 and Λ_5 are $\psi_1 \pm i\psi_2$, where $\psi_1 = e^{i\phi}\alpha$, $\psi_2 = e^{-i\phi}\beta$, ϕ being the angle of rotation about the axis. Basis functions for Λ_6 are α, β, where in each case the spin is quantized along the axis. Note according to Sec. 9-8 of the text that for use with the cubic space groups, we must interchange the definitions of the operators Y_0 and $\bar{Y}_0$, so that the matrix elements of Y_0 will take on the negatives of the values given in the table, which agree with Eq. (A9-60). Matrix elements for barred operators are the negatives of those tabulated.

	X_0	X_1	X_{-1}	Y_0	Y_1	Y_{-1}
Λ_4	1	-1	-1	$-i$	i	i
Λ_5	1	-1	-1	i	$-i$	$-i$
$(\Lambda_6)_{11}$	1	$e^{\pi i/3}$	$e^{-\pi i/3}$	0	0	0
$(\Lambda_6)_{21}$	0	0	0	1	$e^{\pi i/3}$	$e^{-\pi i/3}$
$(\Lambda_6)_{12}$	0	0	0	-1	$-e^{-\pi i/3}$	$-e^{\pi i/3}$
$(\Lambda_6)_{22}$	1	$e^{-\pi i/3}$	$e^{\pi i/3}$	0	0	0
$\chi(\Lambda_6)$	2	1	1	0	0	0

wave functions corresponding to $M = \frac{3}{2}$ or $-\frac{3}{2}$ should have different energies, in the absence of an external field. Herring[1] has shown that there is a simple test as to whether additional degeneracy is introduced by time reversal or not. One must construct the sum

$$\Sigma(R)\chi(R^2) \tag{A9-68}$$

where we are summing over the operators R of a group, R^2 is the square of the operator in question, which can be found from the multiplication table, and we are to sum the characters of these squared operators, for the representation in question. It can be shown that this sum must take one of three values: g, 0, or $-g$, where g is the number of operators in

[1] C. Herring, *Phys. Rev.*, 52:361 (1937).

the group. In either of the first two cases, where the sum is g or 0, a new degeneracy will be introduced by time reversal, if we are dealing with extra representations of the double group; in the third case it will not.

Let us apply this test in the present case. For C_{3v} we have 12 operators, X_0, $X_{\pm 1}$, Y_0, $Y_{\pm 1}$, and the corresponding barred operators. From the multiplication table, Table A9-5, we find that

$$X_0^2 = X_0 \quad X_1^2 = \bar{X}_{-1} \quad X_{-1}^2 = \bar{X}_1 \quad Y_0^2 = Y_1^2 = Y_{-1}^2 = \bar{X}_0 \quad \text{(A9-69)}$$

The squares of the barred operators equal the squares of the corresponding unbarred operators. Now from Table A9-6 we know that for the first one-dimensional irreducible extra representations of C_{3v} the character of X_0 is 1, and of $X_{\pm 1}$ it is -1. We do not require the characters of the Y's. The characters of the barred operators are the negatives of those of the unbarred operators. The sum of Eq. (A9-68) is now

$$2[\chi(X_0) + \chi(\bar{X}_{-1}) + \chi(\bar{X}_1) + 3\chi(\bar{X}_0)] = 2(1 + 1 + 1 - 3) = 0 \quad \text{(A9-70)}$$

Since this sum is zero, an extra degeneracy is introduced by time reversal, which in this case is the degeneracy between the two one-dimensional extra representations of C_{3v}. Just for reassurance, we may apply the same test to the two-dimensional irreducible representation of C_{3v}, which according to Table A9-4 has characters given by $\chi(X_0) = 2$, $\chi(X_1) = \chi(X_{-1}) = 1$, other characters being zero. Hence the sum in this case is $2(2 - 1 - 1 - 6) = -12$, or $-g$, so that no additional degeneracy is introduced by time reversal with this representation, which of course is correct.

Now let us take up the case of C_{4v}. We recall that there must be another two-dimensional irreducible representation, in addition to that given in Table A9-5. We naturally try the same two basis functions $\psi_1 = e^{i\phi}\alpha$ and $e^{-i\phi}\beta$ here which we have used for C_{3v}. By methods similar to those already used, we find the matrix elements given in Table A9-7. This representation is irreducible, as it is expected to be.

A9-8. Cubic Double Groups. We shall illustrate our methods further by taking two of the cubic double groups, namely, T_d and O_h. These can conveniently be treated together, since the operators of O_h are merely those of T_d, and in addition those of T_d plus an inversion. Since the inversion has no effect on the spin functions, we can get complete information about the effect of the operations on the spin functions from the double group T_d. The single group has 24 operations, $R_1 \cdots R_{24}$, defined in Table 2-2. The double group has of course 48 operations, the 24 of the single group plus the barred operations. Our first task is to find the effect of the 24 operations of the single group on the spin functions α and β.

To carry this out, we use the same methods which we have sketched previously. The operations $R_1 \cdots R_{12}$ are rotations, while $R_{13} \cdots R_{24}$ are rotations plus inversion, so that $R'_{13} \cdots R'_{24}$, which consist of the original operations plus an inversion, are pure rotations, which together with $R_1 \cdots R_{12}$ comprise the operations of the group 0. For each of these it is easy to find the angles a, b, c from Eqs. (A9-32), (A9-33), and (A9-34), and hence to find the transformations of α and β by Eq. (A9-27). For instance, we have $R_2 f(x,y,z) = f(x,-y,-z)$, so

Table A9-7

Matrix elements and characters for the extra irreducible representations of the double group C_{4v}. The representations are labeled as for the group of the wave vector along the 001 axis, in the representations of the double space groups for a cubic crystal. Basis functions for Δ_6, α and β, and for Δ_7, $e^{i\phi}\alpha$ and $e^{-i\phi}\beta$, where ϕ is the angle of rotation about the axis, and α and β are quantized with respect to this axis. Note as for C_{3v}, in Table A9-6, that for use with the cubic space groups we must change the signs of the matrix elements for Y_0. Matrix elements for barred operators are the negatives of these tabulated.

	X_0	X_1	X_{-1}	X_2	Y_0	Y_1	Y_{-1}	Y_2
$(\Delta_6)_{11}$	1	$e^{\pi i/4}$	$e^{-\pi i/4}$	i	0	0	0	0
$(\Delta_6)_{21}$	0	0	0	0	1	$e^{\pi i/4}$	$e^{-\pi i/4}$	i
$(\Delta_6)_{12}$	0	0	0	0	-1	$-e^{-\pi i/4}$	$-e^{\pi i/4}$	i
$(\Delta_6)_{22}$	1	$e^{-\pi i/4}$	$e^{\pi i/4}$	$-i$	0	0	0	0
$\chi(\Delta_6)$	2	$\sqrt{2}$	$\sqrt{2}$	0	0	0	0	0
$(\Delta_7)_{11}$	1	$-e^{-\pi i/4}$	$-e^{\pi i/4}$	$-i$	0	0	0	0
$(\Delta_7)_{21}$	0	0	0	0	1	$-e^{-\pi i/4}$	$-e^{\pi i/4}$	$-i$
$(\Delta_7)_{12}$	0	0	0	0	-1	$e^{\pi i/4}$	$e^{-\pi i/4}$	$-i$
$(\Delta_7)_{22}$	1	$-e^{\pi i/4}$	$-e^{-\pi i/4}$	i	0	0	0	0
$\chi(\Delta_7)$	2	$-\sqrt{2}$	$-\sqrt{2}$	0	0	0	0	0

that $a_{11} = 1$, $a_{22} = a_{33} = -1$, all other a's vanishing. This is an example of Case 2, Eq. (A9-34), leading to $\cos(a-c) = -1$, $\sin(a-c) = 0$, from which $a - c = \pm\pi$, as well as $b = \pm\pi$. It proves more convenient, for reasons which we shall mention later, to take one of these signs to be $+$, the other $-$. In either such case we have

$$R_2 \alpha = i\beta \qquad R_2 \beta = i\alpha \qquad (A9\text{-}71)$$

Again, we have $R_5 f(x,y,z) = f(y,z,x)$, and $R_9 f(x,y,z) = f(z,x,y)$. We have already discussed these two cases in Eqs. (A9-46) and (A9-48). As still another example, we have $R'_{13} f(x,y,z) = f(x,-z,y)$, from which $a_{11} = 1$, $a_{23} = -1$, $a_{32} = 1$. These lead according to Eq. (A9-33) to

$\sin a = -1$, $\cos a = 0$, $\sin b = 1$, $\cos b = 0$, $\sin c = 1$, $\cos c = 0$, from which we may assume $a = -\pi/2$, $b = c = \pi/2$. These lead to

$$R'_{13}\alpha = \frac{1}{\sqrt{2}}(\alpha + i\beta) \qquad R'_{13}\beta = \frac{1}{\sqrt{2}}(i\alpha + \beta) \qquad (\text{A9-72})$$

These examples are enough to show how we derive the matrix elements for the operations acting on the basis functions α and β, which are tabulated in Table A9-8. This table is appropriate both for the double group T_d and O_h. In the latter case we have the 24 operations $R_1 \cdots R_{24}$, the primed operations $R'_1 \cdots R'_{24}$ (each equal to the unprimed operation plus an inversion), and the 48 barred operations associated with these. Going from an unprimed to a primed operation makes no change in the matrix elements; going from an unbarred to a barred operation changes the sign of the matrix elements.

From this table of the effect of the operations on the basis functions α and β, we can proceed to find the multiplication table for the double groups T_d and O_h, using the same methods described in the preceding section. This multiplication table is given in Table A9-9. We give the table for the 24 operations $R_1 \cdots R_{24}$. If one of the factors is primed, the product operation is primed; if both are primed, the product is unprimed. If one of the factors is barred, the product changes from unbarred to barred or vice versa. If both factors are barred, the product is as given in the table.

In setting up the matrix elements of Table A9-8, which result in the multiplication table of Table A9-9, we have a choice of sign with each operation; as mentioned earlier, we can call a given operation either R or $\bar{R}$. We have chosen the signs so as to get the closest connection possible with the results for C_{Nv} given previously. In the first place, we can carry out a cyclic permutation of variables, changing x into y, y into z, z into x. We must expect that we shall get the same form for the matrix elements under these circumstances if at the same time we change the basis functions properly, going from α and β, which point along z, to the corresponding functions pointing along x, and then to those pointing along y. From Sec. A9-6 we know how to set up those basis functions. Our choices are made so that the matrix elements are unchanged by this type of transformation.

Next, we know that the group O_h has several subgroups of the form of C_{Nv}. We have tried as far as possible to arrange the definitions of the operations to agree with our earlier treatment of those groups. A complete agreement, however, is impossible, as we shall now show. First we have a twofold rotation about the axis $x = y$, or alternatively about the z axis. In these two cases we have the following parallelism between

Table A9-8

Matrix elements for the two-dimensional irreducible representation of the double group T_d, with basis functions α and β. This is the representation denoted as Γ_6^+ or Γ_7^- in Eq. (A9-76). The matrix elements of the barred operators are the negatives of those tabulated. For the double group O_h, the primed operations have the same matrix elements as the unprimed for Γ_6^+, the negatives of these values for Γ_7^-, so that this table suffices for that case as well. The operations are defined in Table 2-2. Table lists in order the 11, 21, 12, and 22 matrix elements, and the characters.

	R_1	R_2	R_3	R_4	R_5	R_6	R_7	R_8	R_9	R_{10}	R_{11}	R_{12}	R_{13}	R_{14}	R_{15}	R_{16}	R_{17}	R_{18}	R_{19}	R_{20}	R_{21}	R_{22}	R_{23}	R_{24}
	1	0	0	i	$\dfrac{1-i}{2}$	$\dfrac{1+i}{2}$	$\dfrac{1+i}{2}$	$\dfrac{1-i}{2}$	$\dfrac{1+i}{2}$	$\dfrac{1-i}{2}$	$\dfrac{1-i}{2}$	$\dfrac{1+i}{2}$	$\dfrac{1}{\sqrt{2}}$	$\dfrac{1}{\sqrt{2}}$	$\dfrac{1}{\sqrt{2}}$	$\dfrac{1}{\sqrt{2}}$	$\dfrac{1+i}{\sqrt{2}}$	$\dfrac{1-i}{\sqrt{2}}$	$\dfrac{-i}{\sqrt{2}}$	$\dfrac{i}{\sqrt{2}}$	$\dfrac{i}{\sqrt{2}}$	$\dfrac{-i}{\sqrt{2}}$	0	0
	0	-1	0	0	$\dfrac{1-i}{2}$	$\dfrac{-1-i}{2}$	$\dfrac{1+i}{2}$	$\dfrac{-1+i}{2}$	$\dfrac{1+i}{2}$	$\dfrac{1+i}{2}$	$\dfrac{-1-i}{2}$	$\dfrac{1-i}{2}$	$\dfrac{i}{\sqrt{2}}$	$\dfrac{-i}{\sqrt{2}}$	$\dfrac{-1}{\sqrt{2}}$	$\dfrac{-1}{\sqrt{2}}$	0	0	$\dfrac{-1}{\sqrt{2}}$	$\dfrac{-1}{\sqrt{2}}$	$\dfrac{-1}{\sqrt{2}}$	$\dfrac{-i}{\sqrt{2}}$	$\dfrac{1+i}{\sqrt{2}}$	$\dfrac{1-i}{\sqrt{2}}$
	0	1	0	0	$\dfrac{-1-i}{2}$	$\dfrac{1-i}{2}$	$\dfrac{-1+i}{2}$	$\dfrac{1+i}{2}$	$\dfrac{1+i}{2}$	$\dfrac{-1+i}{2}$	$\dfrac{1-i}{2}$	$\dfrac{-1-i}{2}$	$\dfrac{i}{\sqrt{2}}$	$\dfrac{-i}{\sqrt{2}}$	$\dfrac{1}{\sqrt{2}}$	$\dfrac{-1}{\sqrt{2}}$	0	0	$\dfrac{1}{\sqrt{2}}$	$\dfrac{1}{\sqrt{2}}$	$\dfrac{-1}{\sqrt{2}}$	$\dfrac{-i}{\sqrt{2}}$	$\dfrac{-1-i}{\sqrt{2}}$	$\dfrac{-1-i}{\sqrt{2}}$
	1	0	0	$-i$	$\dfrac{1+i}{2}$	$\dfrac{1-i}{2}$	$\dfrac{1-i}{2}$	$\dfrac{1+i}{2}$	$\dfrac{1-i}{2}$	$\dfrac{1+i}{2}$	$\dfrac{1+i}{2}$	$\dfrac{1-i}{2}$	$\dfrac{1}{\sqrt{2}}$	$\dfrac{1}{\sqrt{2}}$	$\dfrac{-1}{\sqrt{2}}$	$\dfrac{1}{\sqrt{2}}$	$\dfrac{1-i}{\sqrt{2}}$	$\dfrac{1+i}{\sqrt{2}}$	$\dfrac{i}{\sqrt{2}}$	$\dfrac{-i}{\sqrt{2}}$	$\dfrac{-i}{\sqrt{2}}$	$\dfrac{i}{\sqrt{2}}$	0	0
	2	0	0	0	1	1	1	1	1	1	1	1	$\sqrt{2}$	$\sqrt{2}$	0	$\sqrt{2}$	$\sqrt{2}$	$\sqrt{2}$	0	0	0	0	0	0

Table A9-9

Multiplication table for the double group T_d. Table gives product operation R_k or $\bar{R}_k$ of R_i (on left) and R_j (at top), $R_iR_j = R_k$. For the product of one unbarred and one barred operator, barred and unbarred operators are to be interchanged in the products. The products of two barred operators are identical with those of two unbarred operators.

	1	2	3	4	5	6	7	8	9	10	11	12	13	14	15	16	17	18	19	20	21	22	23	24
1	1	2	3	4	5	6	7	8	9	10	11	12	13	14	15	16	17	18	19	20	21	22	23	24
2	2	$\bar{1}$	$\bar{4}$	3	7	8	$\bar{5}$	$\bar{6}$	$\overline{12}$	$\overline{11}$	10	9	$\overline{14}$	13	$\overline{21}$	$\overline{22}$	$\overline{24}$	23	20	$\overline{19}$	15	16	$\overline{18}$	17
3	3	4	$\bar{1}$	$\bar{2}$	8	$\bar{7}$	6	$\bar{5}$	$\overline{10}$	9	$\overline{12}$	11	$\overline{20}$	19	$\overline{16}$	15	$\overline{23}$	$\overline{24}$	$\overline{14}$	13	22	$\overline{21}$	17	18
4	4	$\bar{3}$	2	$\bar{1}$	6	$\bar{5}$	$\bar{8}$	7	$\overline{11}$	12	9	$\overline{10}$	$\overline{19}$	$\overline{20}$	$\overline{22}$	21	$\overline{18}$	17	13	14	$\overline{16}$	15	24	$\overline{23}$
5	5	8	6	7	$\bar{9}$	12	10	11	1	$\bar{4}$	$\bar{2}$	$\bar{3}$	18	24	14	20	16	22	$\overline{23}$	$\overline{17}$	$\overline{19}$	$\overline{13}$	$\overline{21}$	$\overline{15}$
6	6	7	$\bar{5}$	$\bar{8}$	11	$\overline{10}$	12	9	4	1	3	$\bar{2}$	17	$\overline{23}$	$\overline{20}$	14	21	15	$\overline{24}$	18	$\overline{13}$	19	16	22
7	7	$\bar{6}$	8	$\bar{5}$	12	9	$\overline{11}$	10	2	$\bar{3}$	1	4	23	17	13	$\overline{19}$	$\overline{22}$	16	18	24	$\overline{20}$	14	$\overline{15}$	21
8	8	$\bar{5}$	$\bar{7}$	6	10	11	9	$\overline{12}$	3	2	$\bar{4}$	1	$\overline{24}$	18	19	13	15	$\overline{21}$	$\overline{17}$	23	14	20	$\overline{22}$	16
9	9	$\overline{11}$	$\overline{12}$	10	1	3	4	2	$\bar{5}$	7	8	6	$\overline{22}$	15	$\overline{24}$	17	$\overline{20}$	13	$\overline{21}$	16	$\overline{23}$	18	$\overline{19}$	14
10	10	$\overline{12}$	11	9	$\bar{3}$	1	2	$\bar{4}$	8	$\bar{6}$	$\bar{5}$	7	$\overline{21}$	16	18	23	13	20	22	$\overline{15}$	17	24	$\overline{14}$	$\overline{19}$
11	11	9	$\overline{10}$	$\overline{12}$	$\bar{4}$	$\bar{2}$	1	3	6	8	$\bar{7}$	$\bar{5}$	15	22	$\overline{23}$	18	14	19	$\overline{16}$	$\overline{21}$	24	$\overline{17}$	13	20
12	12	10	9	$\overline{11}$	$\bar{2}$	4	$\bar{3}$	1	7	5	6	$\bar{8}$	16	21	17	24	$\overline{19}$	14	15	22	$\overline{18}$	$\overline{23}$	20	$\overline{13}$
13	13	$\overline{14}$	19	$\overline{20}$	16	15	$\overline{22}$	$\overline{21}$	$\overline{24}$	23	18	17	2	1	8	7	9	10	$\bar{4}$	$\bar{3}$	6	5	$\overline{11}$	12
14	14	13	$\overline{20}$	$\overline{19}$	22	21	16	15	17	18	$\overline{23}$	24	1	$\bar{2}$	6	5	12	11	3	$\bar{4}$	$\bar{8}$	$\bar{7}$	10	$\bar{9}$
15	15	$\overline{22}$	$\overline{16}$	21	18	$\overline{23}$	17	$\overline{24}$	$\overline{20}$	13	19	14	9	4	3	1	6	8	$\overline{12}$	10	2	$\bar{4}$	7	5
16	16	$\overline{21}$	15	$\overline{22}$	24	17	23	18	13	20	14	$\overline{19}$	10	12	1	$\bar{3}$	7	5	11	$\bar{9}$	4	$\bar{2}$	$\bar{6}$	$\bar{8}$
17	17	23	$\overline{24}$	$\overline{18}$	14	$\overline{20}$	$\overline{19}$	13	$\overline{22}$	16	15	21	7	6	9	12	4	1	8	5	$\overline{10}$	11	$\bar{3}$	$\bar{2}$
18	18	$\overline{24}$	$\overline{23}$	17	20	14	13	19	15	$\overline{21}$	22	16	8	5	11	10	1	$\bar{4}$	$\bar{7}$	$\bar{6}$	12	$\bar{9}$	2	$\bar{3}$
19	19	$\overline{20}$	$\overline{13}$	14	$\overline{21}$	22	15	$\overline{16}$	$\overline{23}$	$\overline{24}$	$\overline{17}$	18	3	$\bar{4}$	$\bar{7}$	8	11	$\overline{12}$	$\bar{1}$	2	5	$\bar{6}$	9	10
20	20	19	14	13	$\overline{15}$	16	$\overline{21}$	22	18	$\overline{17}$	24	23	$\bar{4}$	$\bar{3}$	5	$\bar{6}$	10	$\bar{9}$	2	$\bar{1}$	7	$\bar{8}$	$\overline{12}$	$\overline{11}$
21	21	16	$\overline{22}$	$\overline{15}$	$\overline{23}$	18	24	17	$\overline{19}$	14	$\overline{20}$	$\overline{13}$	12	$\overline{10}$	4	$\bar{2}$	8	$\bar{6}$	9	11	$\bar{1}$	3	5	$\bar{7}$
22	22	15	21	16	$\overline{17}$	24	18	$\overline{23}$	14	19	$\overline{13}$	20	11	$\bar{9}$	$\bar{2}$	4	5	7	$\overline{10}$	$\overline{12}$	3	$\bar{1}$	8	$\bar{6}$
23	23	$\overline{17}$	18	$\overline{24}$	$\overline{19}$	13	$\overline{14}$	20	$\overline{21}$	$\overline{15}$	16	$\overline{22}$	$\bar{6}$	7	10	$\overline{11}$	2	$\bar{3}$	5	$\bar{8}$	9	12	$\bar{1}$	4
24	24	18	17	23	$\overline{13}$	$\overline{19}$	20	14	16	22	21	$\overline{15}$	5	$\bar{8}$	12	$\bar{9}$	$\bar{3}$	$\bar{2}$	6	7	$\overline{11}$	$\overline{10}$	$\bar{4}$	$\bar{1}$

operations of O_h, as we have described them, and the operations of the double group C_{2v}:

$$\text{Axis } x = y: \quad R_1 \to X_0 \quad R'_{24} \to X_1 \quad R'_4 \to Y_0 \quad R_{23} \to Y_1$$
$$z \text{ axis}: \quad R_1 \to X_0 \quad R_4 \to X_1 \quad R'_3 \to Y_0 \quad R'_2 \to Y_1 \quad \text{(A9-73)}$$

We may then take out of Table A9-9 the part relating to these operations, rename the operators according to Eq. (A9-73), and set up a multiplication table just for these operators. This table is given in Table A9-10. It is identical for either of the cases given in Eq. (A9-73). We should expect it to be identical with the multiplication table for C_{2v}. When we compare with that multiplication table, in Table A9-5, we see that there are a number of cases where there is disagreement between barred and unbarred operators. This disagreement would all be removed if in Table A9-5 we had renamed the operator Y_0, so that the operator which was called Y_0 in that table was called $\bar{Y}_0$ instead. Alternatively, we could have renamed the corresponding operator R'_4 or R'_3 in Table A9-9 calling them $\bar{R}'_4$ or $\bar{R}'_3$, respectively. The latter procedure is impossible, however, for there is an inherent relation between the operations R_2, R_3, R_4, demanded by the cyclic permutation symmetry of the problem, and if we were to change the definition of R'_3 and R'_4, we should have to change R'_2 as well. Since they are all concerned in the second case of Eq. (A9-73), we are not allowed to tamper with them. Hence the only way to keep consistency between the group O_h and C_{2v} is to change Y_0 in C_{2v}. As far as R_2, R_3, R_4 are concerned, they are fixed. This was

Table A9-10
Multiplication table for operators X_0, X_1, Y_0, Y_1 of double group C_{2v}, assuming identification of Eq. (A9-73) with operators of group T_d, and using multiplication table of Table A9-9 for T_d.

	X_0	X_1	Y_0	Y_1
X_0	X_0	X_1	Y_0	Y_1
X_1	X_1	$\bar{X}_0$	Y_1	$\bar{Y}_0$
Y_0	Y_0	$\bar{Y}_1$	$\bar{X}_0$	X_1
Y_1	Y_1	Y_0	$\bar{X}_1$	$\bar{X}_0$

the reason for choosing the definition of R_2 as was described in Eq. (A9-71).

In a similar way we get parallelism between certain subgroups of O_h and C_{3v} and C_{4v}. Thus we have the relations given in Eq. (A9-74):

$$\begin{aligned}
&\text{For } C_{3v} & R_1 &\to X_0 & R_9 &\to X_1 & R_5 &\to X_{-1} \\
& & R_{23} &\to Y_0 & R_{21} &\to Y_1 & R_{19} &\to Y_{-1} \\
&\text{For } C_{4v}: & R_1 &\to X_0 & R'_{17} &\to X_1 & R'_{18} &\to X_{-1} & R_4 &\to X_2 \\
& & R'_3 &\to Y_0 & R_{23} &\to Y_1 & R_{24} &\to Y_{-1} & R'_2 &\to Y_2
\end{aligned} \quad (A9\text{-}74)$$

Here again we can set up multiplication tables for the groups C_{3v} and C_{4v}, respectively, out of the results of Table A9-9. Again there are certain discrepancies between Tables A9-9 and A9-5, and again they can be removed if we define Y_0 for the groups C_{3v} and C_{4v} oppositely to what we have done. That is, in place of $Y_0\alpha = \beta$, $Y_0\beta = -\alpha$, as given in Eq. (A9-60), we must choose $Y_0\alpha = -\beta$, $Y_0\beta = \alpha$. On the other hand, we must continue to use Eq. (A9-60) for all other values of q, aside from

$q = 0$. The author is unable to say whether there is a deeper meaning in the fact that this change in the definition of Y_0 brings agreement between the various cases, or whether it is an accident. As we well know, there is no fundamental consideration determining whether a given operation is to be defined as barred or unbarred.

We have now determined the multiplication table for the groups T_d and O_h, and can next find the classes of operators in each case. We proceed as before, and find the following results:

Group T_d

Single group has five classes: R_1; R_2, R_3, R_4; $R_5 \cdots R_{12}$; $R_{13} \cdots R_{18}$; $R_{19} \cdots R_{24}$. Consequently, it has five irreducible representations, two one-dimensional, one two-dimensional, two three-dimensional.

Double group has eight classes: R_1; $\bar{R}_1$; $R_2, R_3, R_4, \bar{R}_2, \bar{R}_3, \bar{R}_4$; $R_5 \cdots R_{12}$; $\bar{R}_5 \cdots \bar{R}_{12}$; $R_{13} \cdots R_{18}$; $\bar{R}_{13} \cdots \bar{R}_{18}$; $R_{19} \cdots R_{24}, \bar{R}_{19} \cdots \bar{R}_{24}$. Consequently, it has eight irreducible representations, consisting of those for the single group, two extra two-dimensional representations, and one four-dimensional.

Group O_h

Both for the single and double groups we have twice as many classes, and twice as many irreducible representations of each dimensionality, as for T_d. The additional representations found in O_h and not in T_d differ from those in that for each representation of T_d, we have two for O_h, one having the same matrix elements for each operator R' as for the corresponding operator R, the other having a matrix element for R' which is the negative of that for R.

Our problem is then to find basis functions, and matrix elements, for the two extra two-dimensional representations of T_d, and the extra four-dimensional representation. For O_h we need only produce the additional basis functions describing the property of being even or odd on inversion. We shall carry this out in the next section.

A9-9. Basis Functions and Irreducible Representations for the Double Groups T_d and O_h. The first of the extra two-dimensional irreducible representations of T_d is already given in Table A9-8. It uses the basis functions α and β. We note that though we are dealing with cubic symmetry, we are using the z axis as a preferred axis, the axis of quantization of the functions α and β. We could, however, quantize along x or y equally well. If we use the basis functions of Eq. (A9-46), we have quantization along x, and if we use those of Eq. (A9-48), we have quantization along y. The matrix elements have the same form as in Table A9-8 except that the names of the operators are interchanged in a way which can be predicted from the cyclic permutation. In studying space groups and wave vectors in crystals, it is often convenient in such a case to take the axis of quantization along the reduced wave vector of the wave. Thus, if we are considering propagation along the x axis, rather than z, we may choose the basis functions of Eq. (A9-46).

In addition to this two-dimensional irreducible representation, we must find another two-dimensional and a four-dimensional irreducible representation. We can get at these representations by setting up basis functions according to a physical argument. The cubic symmetry comes by a perturbation of spherical symmetry. The two-dimensional representation using α and β as basis functions arises from an S function in a central-field problem. When we combine a function of coordinates which is independent of angles, with the spin function, we are proceeding as if to build up the $^2S_{\frac{1}{2}}$ wave functions for $M = \pm \frac{1}{2}$, in a central-field problem. We naturally look for further basis functions by considering atomic functions of p symmetry, which will not be split in a cubic field. We know that an atomic p function, without spin, has three components, corresponding to $M_L = 1, 0, -1$. When we couple this with a spin, with $M_S = \pm \frac{1}{2}$, we can get $M = \frac{3}{2}$ and $\frac{1}{2}$ by coupling the spin to $M_L = 1$, $M = \frac{1}{2}$ and $-\frac{1}{2}$ from $M_L = 0$, and $M = -\frac{1}{2}$ and $-\frac{3}{2}$ from $M_L = -1$. We know that if spin-orbit interaction is introduced into the atomic problem, we shall have two energy levels, $^2P_{\frac{3}{2}}$, with $M = \frac{3}{2}, \frac{1}{2}, -\frac{1}{2}, -\frac{3}{2}$, and $^2P_{\frac{1}{2}}$, with $M = \frac{1}{2}, -\frac{1}{2}$. It seems reasonable to suppose that the cubic field will not remove this degeneracy, and that the four-dimensional irreducible representation of T_d will correspond to the $^2P_{\frac{3}{2}}$ atomic state, and the additional two-dimensional representation to $^2P_{\frac{1}{2}}$. Let us then use this hint to try to set up basis functions and hence the matrix elements of the irreducible representations.

Let us start with the $^2P_{\frac{3}{2}}$. We can set up the basis functions corresponding to $M = \frac{3}{2}, -\frac{3}{2}$, as in the earlier cases of C_{3v} and C_{4v}, by combining the spin α with an orbital function corresponding to an orbital angular momentum of 1 unit along the z axis, and combining β with an orbital function with angular momentum -1. Since we are taking the z axis as the axis of quantization, an orbital p function with angular momentum 1 along the axis has a wave function $x + iy$ times a function of r, and that with an angular momentum -1 has a wave function $x - iy$ times the function of r. Hence for the basis functions corresponding to $M = \frac{3}{2}$ and $-\frac{3}{2}$ we may reasonably expect to use $(x + iy)\alpha$ and $(x - iy)\beta$. These are entirely analogous to the functions ψ_1 and ψ_2 of Eq. (A9-67). We next need functions corresponding to $M = \frac{1}{2}$ and $-\frac{1}{2}$. We can find these by using atomic vector-coupling methods, coupling the orbital angular momentum of unity with the spin angular momentum of $\frac{1}{2}$ unit. When we do so, we not only find the functions corresponding to $^2P_{\frac{3}{2}}$ for $M = \pm \frac{1}{2}$, but also the functions for $^2P_{\frac{1}{2}}$. We find the following basis functions:

Basis Functions for $^2P_{3/2}$

$$\psi_1, M = 3/2: \quad (x + iy)\alpha$$
$$\psi_2, M = 1/2: \quad z\alpha - \frac{x + iy}{2}\beta$$
$$\psi_3, M = -1/2: \quad z\beta + \frac{x - iy}{2}\alpha \quad \text{(A9-75)}$$
$$\psi_4, M = -3/2: \quad (x - iy)\beta$$

Basis Functions for $^2P_{1/2}$

$$\psi_1, M = 1/2: \quad z\alpha + (x + iy)\beta$$
$$\psi_2, M = -1/2: \quad -z\beta + (x - iy)\alpha$$

We can now use these functions and investigate the operations of the double group T_d on them. We find that they form in fact basis functions for a four-dimensional and an additional two-dimensional irreducible representation, and we find matrix elements as given in Table A9-11. The basis functions of Eq. (A9-75) are odd on inversion. Hence if we use them to form basis functions for irreducible representations of O_h, the matrix elements and characters for the primed operations are the negatives of those for the unprimed operations. We must find as well basis functions for the other irreducible representations of O_h which will be even on inversion. We must also find basis functions for the two-dimensional irreducible representation of O_h which has the same matrix elements as that of Table A9-8 but which is odd on inversion. When we do so, we have the following complete set of basis functions for the extra irreducible representations of the double group of O_h, together with the notations used for Koster by these representations:

Γ_6^+: α, β
Γ_7^+: $xy\alpha + (yz + ixz)\beta, -xy\beta + (yz - ixz)\alpha$
Γ_8^+: $(yz + ixz)\alpha, xy\alpha - \dfrac{yz + ixz}{2}\beta, xy\beta + \dfrac{yz - ixz}{2}\alpha, (yz - ixz)\beta$
Γ_6^-: $z\alpha + (x + iy)\beta, -z\beta + (x - iy)\alpha$
Γ_7^-: $xyz\alpha, xyz\beta$
Γ_8^-: $(x + iy)\alpha, z\alpha - \dfrac{x + iy}{2}\beta, z\beta + \dfrac{x - iy}{2}\alpha, (x - iy)\beta$

(A9-76)

We see that the basis functions of Eq. (A9-75) are those for Γ_8^- and Γ_6^-, respectively.

A9-10. Double Point Groups and Double Space Groups. It is not the intention of this Appendix to go thoroughly into the treatment of double space groups. However, we have gone far enough with the double point

Table A9-11

Matrix elements for irreducible representations Γ_7^+, Γ_8^+, Γ_6^-, and Γ_8^- of the double group O_h. Basis functions are given in Eq. (A9-76). Matrix elements for the barred operations are negatives of those for the unbarred. Matrix elements for the primed operations equal those of the unprimed for Γ_7^+ and Γ_8^+, while for Γ_6^- and Γ_8^- they are negatives of those for unprimed.

	R_1	R_2	R_3	R_4	R_5	R_6	R_7	R_8	R_9	R_{10}
$(\Gamma_7^+,\Gamma_6^-)_{11}$	1	0	0	i	$\frac{1-i}{2}$	$\frac{1+i}{2}$	$\frac{1+i}{2}$	$\frac{1-i}{2}$	$\frac{1+i}{2}$	$\frac{1-i}{2}$
$(\Gamma_7^+,\Gamma_6^-)_{21}$	0	i	-1	0	$\frac{1-i}{2}$	$\frac{-1-i}{2}$	$\frac{1+i}{2}$	$\frac{-1+i}{2}$	$\frac{-1+i}{2}$	$\frac{1+i}{2}$
$(\Gamma_7^+,\Gamma_6^-)_{12}$	0	i	1	0	$\frac{-1-i}{2}$	$\frac{1-i}{2}$	$\frac{-1+i}{2}$	$\frac{1+i}{2}$	$\frac{1+i}{2}$	$\frac{-1+i}{2}$
$(\Gamma_7^+,\Gamma_6^-)_{22}$	1	0	0	$-i$	$\frac{1+i}{2}$	$\frac{1-i}{2}$	$\frac{1-i}{2}$	$\frac{1+i}{2}$	$\frac{1-i}{2}$	$\frac{1+i}{2}$
$\chi(\Gamma_7^+,\Gamma_6^-)$	2	0	0	0	1	1	1	1	1	1
$(\Gamma_8^+,\Gamma_8^-)_{11}$	1	0	0	$-i$	$\frac{-1-i}{4}$	$\frac{-1+i}{4}$	$\frac{-1+i}{4}$	$\frac{-1-i}{4}$	$\frac{-1+i}{4}$	$\frac{-1-i}{4}$
$(\Gamma_8^+,\Gamma_8^-)_{21}$	0	0	0	0	$\frac{1+i}{2}$	$\frac{-1+i}{2}$	$\frac{1-i}{2}$	$\frac{-1-i}{2}$	$\frac{1+i}{2}$	$\frac{-1+i}{2}$
$(\Gamma_8^+,\Gamma_8^-)_{31}$	0	0	0	0	$\frac{1+i}{2}$	$\frac{1-i}{2}$	$\frac{1-i}{2}$	$\frac{1+i}{2}$	$\frac{-1+i}{2}$	$\frac{-1-i}{2}$
$(\Gamma_8^+,\Gamma_8^-)_{41}$	0	i	i	0	$\frac{1+i}{4}$	$\frac{-1+i}{4}$	$\frac{1-i}{4}$	$\frac{-1-i}{4}$	$\frac{-1-i}{4}$	$\frac{1-i}{4}$
$(\Gamma_8^+,\Gamma_8^-)_{12}$	0	0	0	0	$\tfrac{3}{8}(1-i)$	$\tfrac{3}{8}(-1-i)$	$\tfrac{3}{8}(1+i)$	$\tfrac{3}{8}(-1+i)$	$\tfrac{3}{8}(1-i)$	$\tfrac{3}{8}(-1-i)$
$(\Gamma_8^+,\Gamma_8^-)_{22}$	1	0	0	i	$\frac{-1+i}{4}$	$\frac{-1-i}{4}$	$\frac{-1-i}{4}$	$\frac{-1+i}{4}$	$\frac{-1-i}{4}$	$\frac{-1+i}{4}$
$(\Gamma_8^+,\Gamma_8^-)_{32}$	0	$-i$	1	0	$\frac{1-i}{4}$	$\frac{-1-i}{4}$	$\frac{1+i}{4}$	$\frac{-1+i}{4}$	$\frac{-1+i}{4}$	$\frac{1+i}{4}$
$(\Gamma_8^+,\Gamma_8^-)_{42}$	0	0	0	0	$\tfrac{3}{8}(1-i)$	$\tfrac{3}{8}(1+i)$	$\tfrac{3}{8}(1+i)$	$\tfrac{3}{8}(1-i)$	$\tfrac{3}{8}(-1-i)$	$\tfrac{3}{8}(-1+i)$
$(\Gamma_8^+,\Gamma_8^-)_{13}$	0	0	0	0	$\tfrac{3}{8}(-1-i)$	$\tfrac{3}{8}(-1+i)$	$\tfrac{3}{8}(-1+i)$	$\tfrac{3}{8}(-1-i)$	$\tfrac{3}{8}(1-i)$	$\tfrac{3}{8}(1+i)$
$(\Gamma_8^+,\Gamma_8^-)_{23}$	0	$-i$	-1	0	$\frac{-1-i}{4}$	$\frac{1-i}{4}$	$\frac{-1+i}{4}$	$\frac{1+i}{4}$	$\frac{1+i}{4}$	$\frac{-1+i}{4}$
$(\Gamma_8^+,\Gamma_8^-)_{33}$	1	0	0	$-i$	$\frac{-1-i}{4}$	$\frac{-1+i}{4}$	$\frac{-1+i}{4}$	$\frac{-1-i}{4}$	$\frac{-1+i}{4}$	$\frac{-1-i}{4}$
$(\Gamma_8^+,\Gamma_8^-)_{43}$	0	0	0	0	$\tfrac{3}{8}(1+i)$	$\tfrac{3}{8}(-1+i)$	$\tfrac{3}{8}(1-i)$	$\tfrac{3}{8}(-1-i)$	$\tfrac{3}{8}(1+i)$	$\tfrac{3}{8}(-1+i)$
$(\Gamma_8^+,\Gamma_8^-)_{14}$	0	i	-1	0	$\frac{-1+i}{4}$	$\frac{1+i}{4}$	$\frac{-1-i}{4}$	$\frac{1-i}{4}$	$\frac{1-i}{4}$	$\frac{-1-i}{4}$
$(\Gamma_8^+,\Gamma_8^-)_{24}$	0	0	0	0	$\frac{-1+i}{2}$	$\frac{-1-i}{2}$	$\frac{-1-i}{2}$	$\frac{-1+i}{2}$	$\frac{1+i}{2}$	$\frac{1-i}{2}$
$(\Gamma_8^+,\Gamma_8^-)_{34}$	0	0	0	0	$\frac{1-i}{2}$	$\frac{-1-i}{2}$	$\frac{1+i}{2}$	$\frac{-1+i}{2}$	$\frac{1-i}{2}$	$\frac{-1-i}{2}$
$(\Gamma_8^+,\Gamma_8^-)_{44}$	1	0	0	i	$\frac{-1+i}{4}$	$\frac{-1-i}{4}$	$\frac{-1-i}{4}$	$\frac{-1+i}{4}$	$\frac{-1-i}{4}$	$\frac{-1+i}{4}$
$\chi(\Gamma_8^+,\Gamma_8^-)$	4	0	0	0	-1	-1	-1	-1	-1	-1

Table A9-11 (Continued)

R_{11}	R_{12}	R_{13}	R_{14}	R_{15}	R_{16}	R_{17}	R_{18}	R_{19}	R_{20}	R_{21}	R_{22}	R_{23}	R_{24}
$\frac{1-i}{2}$	$\frac{1+i}{2}$	$\frac{-1}{\sqrt{2}}$	$\frac{-1}{\sqrt{2}}$	$\frac{-1}{\sqrt{2}}$	$\frac{-1}{\sqrt{2}}$	$\frac{-1-i}{\sqrt{2}}$	$\frac{-1+i}{\sqrt{2}}$	$\frac{i}{\sqrt{2}}$	$\frac{i}{\sqrt{2}}$	$\frac{-i}{\sqrt{2}}$	$\frac{i}{\sqrt{2}}$	0	0
$\frac{-1-i}{2}$	$\frac{1-i}{2}$	$\frac{-i}{\sqrt{2}}$	$\frac{i}{\sqrt{2}}$	$\frac{1}{\sqrt{2}}$	$\frac{-1}{\sqrt{2}}$	0	0	$\frac{1}{\sqrt{2}}$	$\frac{-1}{\sqrt{2}}$	$\frac{i}{\sqrt{2}}$	$\frac{i}{\sqrt{2}}$	$\frac{-1-i}{\sqrt{2}}$	$\frac{-1+i}{\sqrt{2}}$
$\frac{1-i}{2}$	$\frac{-1-i}{2}$	$\frac{-i}{\sqrt{2}}$	$\frac{i}{\sqrt{2}}$	$\frac{-1}{\sqrt{2}}$	$\frac{1}{\sqrt{2}}$	0	0	$\frac{-1}{\sqrt{2}}$	$\frac{1}{\sqrt{2}}$	$\frac{i}{\sqrt{2}}$	$\frac{i}{\sqrt{2}}$	$\frac{1-i}{\sqrt{2}}$	$\frac{1+i}{\sqrt{2}}$
$\frac{1-i}{2}$	$\frac{1-i}{2}$	$\frac{-1}{\sqrt{2}}$	$\frac{-1}{\sqrt{2}}$	$\frac{-1}{\sqrt{2}}$	$\frac{-1}{\sqrt{2}}$	$\frac{-1+i}{\sqrt{2}}$	$\frac{-1-i}{\sqrt{2}}$	$\frac{-i}{\sqrt{2}}$	$\frac{-i}{\sqrt{2}}$	$\frac{i}{\sqrt{2}}$	$\frac{-i}{\sqrt{2}}$	0	0
1	1	$-\sqrt{2}$	$-\sqrt{2}$	$-\sqrt{2}$	$-\sqrt{2}$	$-\sqrt{2}$	$-\sqrt{2}$	0	0	0	0	0	0
$\frac{-1-i}{4}$	$\frac{-1+i}{4}$	$\frac{-\frac{1}{2}}{\sqrt{2}}$	$\frac{-\frac{1}{2}}{\sqrt{2}}$	$\frac{-\frac{1}{2}}{\sqrt{2}}$	$\frac{-\frac{1}{2}}{\sqrt{2}}$	$\frac{1-i}{\sqrt{2}}$	$\frac{1+i}{\sqrt{2}}$	$\frac{-i}{2\sqrt{2}}$	$\frac{-i}{2\sqrt{2}}$	$\frac{i}{2\sqrt{2}}$	$\frac{-i}{2\sqrt{2}}$	0	0
$\frac{1-i}{2}$	$\frac{-1-i}{2}$	$\frac{i}{\sqrt{2}}$	$\frac{-i}{\sqrt{2}}$	$\frac{-1}{\sqrt{2}}$	$\frac{1}{\sqrt{2}}$	0	0	$\frac{1}{\sqrt{2}}$	$\frac{-1}{\sqrt{2}}$	$\frac{i}{\sqrt{2}}$	$\frac{i}{\sqrt{2}}$	0	0
$\frac{-1-i}{2}$	$\frac{-1+i}{2}$	$\frac{-1}{\sqrt{2}}$	$\frac{-1}{\sqrt{2}}$	$\frac{1}{\sqrt{2}}$	$\frac{1}{\sqrt{2}}$	0	0	$\frac{-i}{\sqrt{2}}$	$\frac{-i}{\sqrt{2}}$	$\frac{-i}{\sqrt{2}}$	$\frac{i}{\sqrt{2}}$	0	0
$\frac{-1+i}{4}$	$\frac{1+i}{4}$	$\frac{-i}{2\sqrt{2}}$	$\frac{i}{2\sqrt{2}}$	$\frac{-\frac{1}{2}}{\sqrt{2}}$	$\frac{\frac{1}{2}}{\sqrt{2}}$	0	0	$\frac{-i}{2\sqrt{2}}$	$\frac{1}{2\sqrt{2}}$	$\frac{i}{2\sqrt{2}}$	$\frac{i}{2\sqrt{2}}$	$\frac{1-i}{\sqrt{2}}$	$\frac{1+i}{\sqrt{2}}$
$\frac{3}{8}(1+i)$	$\frac{3}{8}(-1+i)$	$\frac{3i}{4\sqrt{2}}$	$\frac{-3i}{4\sqrt{2}}$	$\frac{\frac{3}{4}}{\sqrt{2}}$	$\frac{-\frac{3}{4}}{\sqrt{2}}$	0	0	$\frac{-\frac{3}{4}}{\sqrt{2}}$	$\frac{\frac{3}{4}}{\sqrt{2}}$	$\frac{3i}{4\sqrt{2}}$	$\frac{3i}{4\sqrt{2}}$	0	0
$\frac{-1+i}{4}$	$\frac{-1-i}{4}$	$\frac{\frac{1}{2}}{\sqrt{2}}$	$\frac{\frac{1}{2}}{\sqrt{2}}$	$\frac{\frac{1}{2}}{\sqrt{2}}$	$\frac{\frac{1}{2}}{\sqrt{2}}$	$\frac{-1-i}{\sqrt{2}}$	$\frac{-1+i}{\sqrt{2}}$	$\frac{-i}{2\sqrt{2}}$	$\frac{-i}{2\sqrt{2}}$	$\frac{i}{2\sqrt{2}}$	$\frac{-i}{2\sqrt{2}}$	0	0
$\frac{-1-i}{4}$	$\frac{1-i}{4}$	$\frac{-i}{2\sqrt{2}}$	$\frac{i}{2\sqrt{2}}$	$\frac{\frac{1}{2}}{\sqrt{2}}$	$\frac{-\frac{1}{2}}{\sqrt{2}}$	0	0	$\frac{\frac{1}{2}}{\sqrt{2}}$	$\frac{-\frac{1}{2}}{\sqrt{2}}$	$\frac{i}{2\sqrt{2}}$	$\frac{i}{2\sqrt{2}}$	$\frac{-1+i}{\sqrt{2}}$	$\frac{1-i}{\sqrt{2}}$
$\frac{3}{8}(-1+i)$	$\frac{3}{8}(-1-i)$	$\frac{\frac{3}{4}}{\sqrt{2}}$	$\frac{\frac{3}{4}}{\sqrt{2}}$	$\frac{-\frac{3}{4}}{\sqrt{2}}$	$\frac{-\frac{3}{4}}{\sqrt{2}}$	0	0	$\frac{-3i}{4\sqrt{2}}$	$\frac{-3i}{4\sqrt{2}}$	$\frac{-3i}{4\sqrt{2}}$	$\frac{3i}{4\sqrt{2}}$	0	0
$\frac{3}{8}(1+i)$	$\frac{3}{8}(1-i)$	$\frac{-\frac{3}{4}}{\sqrt{2}}$	$\frac{-\frac{3}{4}}{\sqrt{2}}$	$\frac{\frac{3}{4}}{\sqrt{2}}$	$\frac{\frac{3}{4}}{\sqrt{2}}$	0	0	$\frac{-3i}{4\sqrt{2}}$	$\frac{-3i}{4\sqrt{2}}$	$\frac{-3i}{4\sqrt{2}}$	$\frac{3i}{4\sqrt{2}}$	0	0
$\frac{1-i}{4}$	$\frac{-1-i}{4}$	$\frac{-i}{2\sqrt{2}}$	$\frac{i}{2\sqrt{2}}$	$\frac{-\frac{1}{2}}{\sqrt{2}}$	$\frac{\frac{1}{2}}{\sqrt{2}}$	0	0	$\frac{-\frac{1}{2}}{\sqrt{2}}$	$\frac{\frac{1}{2}}{\sqrt{2}}$	$\frac{i}{2\sqrt{2}}$	$\frac{i}{2\sqrt{2}}$	$\frac{-1+i}{\sqrt{2}}$	$\frac{-1-i}{\sqrt{2}}$
$\frac{1-i}{4}$	$\frac{-1+i}{4}$	$\frac{\frac{1}{2}}{\sqrt{2}}$	$\frac{\frac{1}{2}}{\sqrt{2}}$	$\frac{\frac{1}{2}}{\sqrt{2}}$	$\frac{\frac{1}{2}}{\sqrt{2}}$	$\frac{-1+i}{\sqrt{2}}$	$\frac{-1-i}{\sqrt{2}}$	$\frac{i}{2\sqrt{2}}$	$\frac{i}{2\sqrt{2}}$	$\frac{-i}{2\sqrt{2}}$	$\frac{i}{2\sqrt{2}}$	0	0
$\frac{3}{8}(1-i)$	$\frac{3}{8}(-1-i)$	$\frac{-3i}{4\sqrt{2}}$	$\frac{3i}{4\sqrt{2}}$	$\frac{\frac{3}{4}}{\sqrt{2}}$	$\frac{-\frac{3}{4}}{\sqrt{2}}$	0	0	$\frac{-\frac{3}{4}}{\sqrt{2}}$	$\frac{\frac{3}{4}}{\sqrt{2}}$	$\frac{-3i}{4\sqrt{2}}$	$\frac{-3i}{4\sqrt{2}}$	0	0
$\frac{1+i}{4}$	$\frac{-1+i}{4}$	$\frac{-i}{2\sqrt{2}}$	$\frac{i}{2\sqrt{2}}$	$\frac{\frac{1}{2}}{\sqrt{2}}$	$\frac{-\frac{1}{2}}{\sqrt{2}}$	0	0	$\frac{\frac{1}{2}}{\sqrt{2}}$	$\frac{-\frac{1}{2}}{\sqrt{2}}$	$\frac{i}{2\sqrt{2}}$	$\frac{i}{2\sqrt{2}}$	$\frac{1+i}{\sqrt{2}}$	$\frac{1-i}{\sqrt{2}}$
$\frac{1-i}{2}$	$\frac{1+i}{2}$	$\frac{1}{\sqrt{2}}$	$\frac{1}{\sqrt{2}}$	$\frac{-1}{\sqrt{2}}$	$\frac{-1}{\sqrt{2}}$	0	0	$\frac{-i}{\sqrt{2}}$	$\frac{-i}{\sqrt{2}}$	$\frac{-i}{\sqrt{2}}$	$\frac{i}{\sqrt{2}}$	0	0
$\frac{1+i}{2}$	$\frac{-1+i}{2}$	$\frac{-i}{\sqrt{2}}$	$\frac{i}{\sqrt{2}}$	$\frac{-1}{\sqrt{2}}$	$\frac{1}{\sqrt{2}}$	0	0	$\frac{1}{\sqrt{2}}$	$\frac{-1}{\sqrt{2}}$	$\frac{-i}{\sqrt{2}}$	$\frac{-i}{\sqrt{2}}$	0	0
$\frac{-1+i}{4}$	$\frac{-1+i}{4}$	$\frac{-\frac{1}{2}}{\sqrt{2}}$	$\frac{-\frac{1}{2}}{\sqrt{2}}$	$\frac{-\frac{1}{2}}{\sqrt{2}}$	$\frac{-\frac{1}{2}}{\sqrt{2}}$	$\frac{1+i}{\sqrt{2}}$	$\frac{1-i}{\sqrt{2}}$	$\frac{i}{2\sqrt{2}}$	$\frac{i}{2\sqrt{2}}$	$\frac{-i}{2\sqrt{2}}$	$\frac{i}{2\sqrt{2}}$	0	0
-1	-1	0	0	0	0	0	0	0	0	0	0	0	0

groups so that the reader should have no trouble reading the literature on the double space groups, or in extending the methods described here to that problem. We shall indicate in this section some of the simpler aspects of the problem. We shall illustrate by the familiar example of the splitting of the threefold degenerate p-like state at the point Γ in the body- or face-centered cubic structures, by spin-orbit interaction, which we have already mentioned several times, but now we shall describe it in the light of our treatment of the double groups.

It is obvious in the first place how to set up the operations of the double space groups. We combine the translations with the double point groups of operations. Both an operator R and $\bar{R}$ will have identical effects on a function of coordinates, but their effects on the spin functions α and β are identical with what we have found in this Appendix. Hence if we have a function of coordinates and spin, made up as we have indicated here by multiplying a function of coordinates and a spin function, we can find the effect of any operator of the space group on this function. We can set up the multiplication table of the double space group, as we have done for the double point group in this Appendix, and for the single space groups in Appendix 3. We can then deduce the irreducible representations, using simple basis functions suggested by the single groups, and can find their matrix elements.

Once we have found the irreducible representations and their matrix elements, we can set up basis functions by the method of projection operators. In particular, if we are dealing with a crystal problem, we naturally start with a plane wave multiplied by a spin function α or β, and apply the projection-operator technique to this function. The result will be a symmetrized plane wave, combined with a suitably symmetrized spin function. If the matrix elements of the irreducible representations are set up like those using the basis functions of C_{Nv}, when that group forms the group of the wave vector, the spin function arising from the projection operator will ordinarily be quantized along the reduced wave vector. We can illustrate this by a simple example. We have seen in Eq. (A9-74) that there is a parallelism between the operators X_0, X_1, X_{-1}, Y_0, Y_1, Y_{-1} of the group C_{3v} and the operators R_1, R_9, R_5, R_{23}, R_{21}, R_{19} of the group O_h or T_d. These operators form the group of the wave vector along the direction Λ, the 111 direction in the cubic crystal. If we use the matrix elements for C_{3v} as given in Table A9-4 (changing the sign of the matrix elements of Y_0 according to Sec. A9-8), and form projection operators from these matrix elements, using the two-dimensional irreducible representation of C_{3v}, and allow these projection operators to operate on α, we find that the resulting two basis functions produced by the projection operators point along the 111 and −111 directions, as determined by the methods of Sec. A9-7. Similar results are found in other cases.

From the character tables of the irreducible representations of the groups of the wave vectors in the various directions, we can deduce the compatibility relations in the usual way. Let us now consider in detail the situation at Γ in a cubic crystal, and in the directions Δ, Λ, and Σ along the 100, 111, and 110 directions, respectively. We have seen in Eq. (A9-76) a set of basis functions for the point Γ in a cubic crystal. The representation Γ_6^+, as we see from this, is the one resembling the $^2S_{1/2}$ state in an atom; its basis functions are the spin functions α and β, the orbital part of the function being totally symmetric. We have noted that we have a similar two-dimensional irreducible representation for every double group, and for the directions Δ and Λ these representations are called Δ_6 and Λ_6, respectively, which are obviously compatible with Γ_6^+.

In addition to this two-dimensional irreducible representation of the groups C_{4v} and C_{3v}, met in Δ and Λ, respectively, we recall that for C_{4v} we have another two-dimensional irreducible representation, with basis functions (provided we are taking the wave vector along z) of the form $(x + iy)\alpha$, $(x - iy)\beta$. This representation is called Δ_7. It then proves to be the case, as we can prove from the character tables, that Γ_6^+ and Γ_6^- are compatible with Δ_6, Γ_7^+ and Γ_7^- with Δ_7, and Γ_8^+ and Γ_8^- with both Δ_6 and Δ_7. Similarly, we recall that for C_{3v} there are two one-dimensional irreducible representations, called Λ_4 and Λ_5, in addition to the two-dimensional representation Λ_6. It then proves to be the case that Γ_6^+, Γ_6^-, Γ_7^+, and Γ_7^- are all compatible with Λ_6, while Γ_8^+ and Γ_8^- are compatible with Λ_4, Λ_5, and Λ_6. Since there is only one irreducible representation of C_{2v}, the group of the wave vector along Σ, namely, the two-dimensional representation with basis functions α and β, this is compatible with all the irreducible representations at Γ.

Let us now look more closely at the physical meanings of these representations and compatibilities. An atomic S state will go in the crystalline field into the irreducible representation Γ_6^+, as we have seen, with the basis functions α and β. This is compatible along each of the three symmetry directions with two-dimensional representations in which the basis functions again are α and β, or two linear combinations of these. An atomic P state, on the other hand, will be split into the two-dimensional irreducible representation Γ_6^- and the four-dimensional Γ_8^-, as we can see at once from the basis functions of Eq. (A9-76). In our earlier discussion we have seen that the Γ_6^- is analogous to the atomic state $^2P_{1/2}$, and the Γ_8^- to $^2P_{3/2}$. The two states will of course be split apart by spin-orbit interaction. Along the direction Δ, Γ_6^- will be compatible with Δ_6, while Γ_8^- will be compatible with Δ_6 and Δ_7. Since these two two-dimensional irreducible representations will have different energies, we see that the four-dimensional $^2P_{3/2}$-like state at Γ will be split into two two-dimensional states along the direction Δ. In the direction Λ it will likewise be split into two two-dimensional states: we have compatibility with Λ_4, Λ_5, and Λ_6, but we remember that the two one-dimensional irreducible representations Λ_4 and Λ_5 are degenerate on account of time reversal.

In each of these directions Δ and Λ, one of the two-dimensional representations with which Γ_8^- is compatible has the basis functions α and β, and corresponds to a component of angular momentum along the reduced wave vector of $\pm \frac{1}{2}$. The other, however, has basis functions like $(x + iy)\alpha$, $(x - iy)\beta$, with corresponding functions for the direction Λ, corresponding to components of angular momentum along the reduced wave vector of $\pm \frac{3}{2}$. The orbital functions set up by projection operators, in the form of symmetrized plane waves, will have the same symmetry properties as $x \pm iy$, etc., about the lattice points. For instance, for a simple cubic lattice of spacing a, basis functions for Δ_7 are

$$e^{ikz} \left[\sin \frac{2\pi h_1 x}{a} \cos \frac{2\pi h_2 y}{a} \pm i \cos \frac{2\pi h_2 x}{a} \sin \frac{2\pi h_2 y}{a} \right] \binom{\alpha}{\beta}$$

where h_1, h_2 are integers.

The reason, of course, for using the basis functions of the irreducible representations of the double group is that the Hamiltonian, including spin-orbit interaction, commutes with the operators of this double group, and consequently there will be no nondiagonal matrix elements of the Hamiltonian, including spin-orbit interaction, between basis functions for different irreducible representations of the double group. This fact then will be used in solving the problem of spin-orbit interaction in solids. Suppose, for example, that we had solved the problem of energy bands in a crystal, with neglect of spin-orbit interaction, using the irreducible representations of the single group. Suppose, then, that we wish to introduce the spin-orbit interaction as a perturbation. We may take the basis functions of the irreducible representations of the single group, multiply these by spin functions α or β, or by the appropriate spin functions pointing along the reduced wave vector, and construct from them basis functions for the irreducible representations of the double group.

The commonest case is the one which we have been using as an illustration, the splitting of a p-like irreducible representation of the single group by spin-orbit interaction. From Eq. (A9-76) it is clear how to set up basis functions at Γ. The spin-orbit interaction will be diagonal with respect to these states. When we are not far from the point Γ, and have small splitting between the components of this p-like state, we may disregard nondiagonal matrix elements of the spin-orbit interaction between these components and any other energy levels. Along the direction Δ, then, we have seen that we shall have two states Δ_6, one arising from each of the energy levels at Γ, and one Δ_7. Obviously, we shall have to consider the interaction between the two Δ_6's. There will be a nondiagonal matrix element of the spin-orbit interaction between them, proportional to the reduced wave vector, leading to a quadratic secular equation. Similarly, in the direction Λ, we shall have Λ_4, Λ_5, and two Λ_6's, between which we must solve a quadratic secular equation.

It is hoped that this description of some of the simpler procedures for dealing with the double groups will give the reader enough familiarity with their manipulation so that he can proceed to the more complicated cases often met in practice.

Bibliography

This Bibliography, like that in Volume 1, includes not merely the papers and books specifically referred to in the text, but a good many others besides, so that it includes a fairly complete account of the literature relating to the theory of energy bands in solids. We have not included experimental papers dealing with energy bands and the Fermi surface, even in cases where they have a bearing on the theory and its verification; it is hoped to give a bibliography of these topics in a later volume of this series. We have also not included references to the many other topics which will be taken up in later volumes, such as lattice vibrations, conductivity, optical properties, dielectric and magnetic behavior, cohesive energy, and the many-body aspects of the theory of solids. Bibliographies dealing with these topics will appear in later volumes. In general we have not duplicated references already given in the bibliographies in "Quantum Theory of Atomic Structure" and in Volume 1 of the present work.

As in the earlier bibliographies of this series, titles in French and German are ordinarily given in the original language, others in English. Cross references are included to joint authors who are not the first-listed authors, provided they are also authors of other works listed alphabetically in the Bibliography. Section numbers in this book are given for locations of papers and books referred to in the text, so that the Bibliography can serve as an index of references and names.

Abrikosov, A. A.: Green's Function for Electrons in a Metal and Analysis of the Electron Spectrum, *Zh. Eksperim. i Teor. Fiz.*, **43**:1083 (1962).

───── and L. A. Falkovskii: Theory of the Electron Energy Spectrum of Metals with a Bismuth-type Lattice, *Zh. Eksperim. i Teor. Fiz.*, **43**:1089 (1962).

Adams, E. N.: Motion of an Electron in a Perturbed Periodic Potential, *Phys. Rev.*, **85**:41 (1952); **86**:427 (1952).

─────: Crystal Momentum as a Quantum-mechanical Operator, *J. Chem. Phys.*, **21**:2013 (1953).

Adler, S. L.: Theory of Valence Band Splittings at $k = 0$ in Zincblende and Wurtzite Structures, *Phys. Rev.*, **126**:118 (1962).

Aerts, E.: Surface States of One-dimensional Crystals, I, II, III, *Physica*, **26**:1047, 1057, 1063 (1960).

Allen, G.: Band Structures of One-dimensional Crystals with Square-well Potentials, *Phys. Rev.*, **91**:531 (1953).

─────: See also P. Schwed.

Allis, W. P., and P. M. Morse: Theorie der Streuung langsamer Elektronen an Atomen, *Z. Physik*, **70**:567 (1931). (Sec. 9-3.)

Altmann, S. L.: Spherical Harmonics with the Symmetry of the Close-packed Hexagonal Lattice, *Proc. Phys. Soc. (London)*, **A69**:184 (1956).

———, C. A. Coulson, and W. Hume-Rothery: On the Relation between Bond Hybrids and the Metallic Structures, Proc. Roy. Soc. (London), **A240**:145 (1957).

——— and N. V. Cohan: Cellular Eigenvalues for Titanium Metal, Proc. Phys. Soc. (London), **71**:383 (1958). (Sec. 10-8.)

———: Equivalent Functions: Hybrids and Wannier Functions, Proc. Cambridge Phil. Soc., **54**:197 (1958).

———: The Cellular Method for a Close-packed Hexagonal Lattice, I, Theory; II, The Computations: A Program for a Digital Computer and an Application to Zirconium Metal, Proc. Roy. Soc. (London), **A244**:141, 153 (1958). (Secs. 9-1, 10-8.)

——— and C. J. Bradley: The Band Structure of Hexagonal Close-packed Metals: Zirconium, Phys. Letters, **1**:336 (1962). (Sec. 10-8.)

———: "Group Theory, Quantum Theory," Academic Press Inc., New York, 1962.

——— and C. J. Bradley: A Note on the Calculation of the Matrix Elements of the Rotation Group, Phil. Trans. Roy. Soc. (London), **A255**:193 (1963).

——— and ———: On the Symmetries of Spherical Harmonics, Phil. Trans. Roy. Soc. (London), **A255**:199 (1963).

———: The Crystallographic Point Groups as Semi-direct Products, Phil. Trans. Roy. Soc. (London), **A255**:216 (1963).

———: Semidirect Products and Point Groups, Rev. Mod. Phys., **35**:641 (1963).

Anno, T., and C. A. Coulson: The Structure of Graphite, Proc. Roy. Soc. (London), **A264**:165 (1961). (Sec. 10-8.)

Antoncik, E., and M. Trlifaj: A Note on the Group Analysis of the Wave Function of Valency Electrons in a Crystal, Czech. J. Phys., **1**:97 (1952).

———: The Electron Theory of Metallic Aluminum, Czech. J. Phys., **2**:18 (1953). (Sec. 10-8.)

———: A Contribution to the Theory of Multivalent Metals, Czech. J. Phys., **2**:31 (1953).

———: The Repulsion Potential of Unoccupied States, Czech. J. Phys., **7**:118 (1957).

———: The Use of the Repulsive Potential in the Quantum Theory of Solids, Czech. J. Phys., **9**:291 (1959).

———: Approximate Formulation of the Orthogonalized Plane-wave Method, J. Phys. Chem. Solids, **10**:314 (1959).

———: On the Approximate Formulation of the Orthogonalized Plane-wave Method, Czech. J. Phys., **10**:22 (1960).

———: A Remark on the Use of the Repulsive Potential, Czech. J. Phys., **10**:766 (1960).

———: On the Theory of Surface States, J. Phys. Chem. Solids, **21**:137 (1961).

———: On the Theory of Surface States of Semiconductors with Diamond Structure, Semiconductor Phys. Conf., Prague, 1960, Academic Press Inc., New York, 1961, p. 491.

——— and P. T. Landsberg: Overlap Integrals for Bloch Electrons, Proc. Phys. Soc. (London), **82**:337 (1963).

Appel, J.: Effect of Spin-Orbit Coupling and Other Relativistic Corrections on Donor States in Ge and Si, Phys. Rev., **133**:A280 (1964).

Ariyama, K.: Zur Elektronentheorie der Metalle, Sci. Papers Inst. Phys. Chem. Res. Tokyo, **32**:103 (1937).

———: Über die Zustände der Elektronen der zweiwertigen Metalle, Sci. Papers Inst. Phys. Chem, Res. Tokyo, **34**:344 (1938).

——— and S. Mase: Electronic Structure of Graphite and Boron Nitride, Progr. Theoret. Phys. (Kyoto), **12**:244 (1954). (Sec. 10-8.)

Asdente, M., and J. Friedel: 3d Band Structure of Cr, *Phys. Rev.*, **124**:384 (1961); **126**:2262 (1962). (Sec. 10-8.)
———: Fermi Surface for 3d Band of Cr, *Phys. Rev.*, **127**:1949 (1962). (Sec. 10-8.)
Asendorf, R. H.: Space Group of Tellurium and Selenium, *J. Chem. Phys.*, **27**:11 (1957).
Austin, B. J., V. Heine, and L. J. Sham: General Theory of Pseudopotentials, *Phys. Rev.*, **127**:276 (1962).
Bader, F., K. Ganzhorn, and U. Dehlinger: Ferromagnetismus und Bandstruktur der Übergangsmetalle, *Z. Physik*, **137**:190 (1954).
Baldock, G. R.: Electronic Bound States at the Surfaces of a Metal, *Proc. Cambridge Phil. Soc.*, **48**:457 (1952).
Balkanski, M., and J. des Cloizeaux: Structure de bandes des cristaux de type wurtzite: transitions optiques intrinsèques dans le CdS, *J. Phys. Radium*, **21**:825 (1960). (Sec. 10-8.)
——— and ———: Band Structure of CdS, *Abhandl. Deut. Akad. Wiss. Berlin, Kl. Math. Phys. Tech.*, **76** (1960). (Sec. 10-8.)
Bargmann, V.: On the Representations of the Rotation Group, *Rev. Mod. Phys.*, **34**:829 (1962).
Barrett, C. S.: "Structure of Metals," 2d ed., McGraw-Hill Book Company, New York, 1952.
Barriol, J.: La structure électronique du graphite, *J. Chim. Phys.*, **57**:837 (1960). (Sec. 10-8.)
——— and J. Metzger: Application de la méthode des orbitales moléculaires au réseau du graphite, *J. Chim. Phys.*, **57**:848 (1960). (Sec. 10-8.)
Barron, T. H. K., and C. Domb: On the Cubic and Hexagonal Close-packed Lattices, *Proc. Roy. Soc. (London)*, **A227**:447 (1955).
——— and G. Fischer: Brillouin Zones and Crystal Structure Factors, *Phil. Mag.*, **4**:826 (1959).
Bassani, F.: Energy Band Structure in Silicon Crystals by the Orthogonalized Plane-wave Method, *Phys. Rev.*, **108**:263 (1957). (Sec. 10-8.)
———: Energy Band Structure of Sodium Atoms in the Diamond Lattice, *J. Phys. Chem. Solids*, **8**:375, 379 (1959). (Sec. 10-8.)
——— and V. Celli: Energy-band Structure of Lithium Atoms in the Diamond Lattice, *Nuovo Cimento*, **11**:805 (1959). (Sec. 10-8.)
———: Energy Bands in Silicon Crystals, *Nuovo Cimento*, **13**:244 (1959). (Sec. 10-8.)
——— and V. Celli: Energy-band Structure of Solids from a Perturbation on the "Empty Lattice," *J. Phys. Chem. Solids*, **20**:64 (1961).
——— and M. Yoshimine: Electronic Band Structure of Group IV Elements and of III-V Compounds, *Phys. Rev.*, **130**:20 (1963). (Secs. 10-4, 10-8.)
——— and D. Brust: Effect of Alloying and Pressure on the Band Structure of Germanium and Silicon, *Phys. Rev.*, **131**:1524 (1963).
——— and L. Liu: Electronic Band Structure of Gray Tin, *Phys. Rev.*, **132**:2047 (1963). (Sec. 10-8.)
———: See also R. S. Knox, J. E. Robinson.
Bates, C. A., and K. W. H. Stevens: The Band Structure of Body-centered Cubic Transition Metals, *Proc. Phys. Soc. (London)*, **78**:1321 (1961).
Bayliss, N. S.: Brillouin Zones and the Mathieu Equation, *Australian J. Sci.*, **12**:12 (1949).
Behrens, E.: Über die Energiebänder der Schichtengitter vom Wismut-Typ, *Z. Physik*, **161**:279 (1961). (Sec. 10-8.)

———: Über die Energiebänder der Kettengitter vom Selen-Typ, *Z. Physik*, **163**:140 (1961). (Sec. 10-8.)
Behringer, R. E.: Note on "The Band Structure of Aluminium: a Self-consistent Calculation," *J. Phys. Chem. Solids*, **5**:145 (1958).
———: Metallic Transition in Lithium Hydride, *Phys. Rev.*, **113**:787 (1959).
Belding, E. F.: 3d Band Structure of Some Transition Elements, *Phil. Mag.*, **4**:1145 (1959). (Sec. 10-8.)
Bell, D. G., D. M. Hum, L. Pincherle, D. W. Sciama, and P. M. Woodward: The Electronic Band Structure of PbS, *Proc. Roy. Soc. (London)*, **A217**:71 (1953). (Secs. 10-7, 10-8.)
———: Group Theory and Crystal Lattices, *Rev. Mod. Phys.*, **26**:311 (1954).
———, R. Hensman, D. P. Jenkins, and L. Pincherle: A Note on the Band Structure of Silicon, *Proc. Phys. Soc. (London)*, **A67**:562 (1954). (Sec. 10-8.)
Bergson, G.: Some Aspects of Elemental Sulfur: A Study by Means of the Semiempirical MO-LCAO Method, *Arkiv. Kemi*, **16**:315 (1960). (Sec. 10-8.)
Berry, R. L., M. B. Waldron, and G. V. Raynor: Outer Brillouin Zones for the Face-centered Cubic, Body-centered Cubic, and Close-packed Hexagonal Structures, *Research*, **3**:195 (1950).
Bertaut, E. F.: Configurations de spin et théorie des groupes, *J. Phys. Radium*, **22**:321 (1961).
———: Lattice Theory of Spin Configurations, *J. Appl. Phys.*, **33** (suppl.):1138 (1962).
Bethe, H.: Theorie der Beugung von Elektronen an Kristallen, *Ann. Physik*, **87**:55 (1928). (Sec. 7-6.)
———: Termaufspaltung in Kristallen, *Ann. Physik*, **3**:133 (1929).
———: Quantitative Berechnung der Eigenfunktionen von Metallelektronen, *Helv. Phys. Acta*, **7** (suppl. 2):18 (1934).
——— and F. C. von der Lage: A Method of Determining the Energies and Wave Functions in Solids, *Phys. Rev.*, **65**:255 (1944).
———: See also A. Sommerfeld, F. von der Lage.
Betts, D. D.: Solid Harmonics as Basis Functions for Cubic Crystals, *Can. J. Phys.*, **37**:350 (1959).
Bienenstock, A., and P. P. Ewald: Structure Theories in Physical and in Fourier Space, *Kristallografiya*, **6**:820 (1961).
——— and ———: Symmetry of Fourier Space, *Acta Cryst.*, **15**:1253 (1962).
Birman, J. L.: Electronic Energy Bands in ZnS: Potential in Zincblende and Wurtzite, *Phys. Rev.*, **109**:810 (1958). (Sec. 10-8.)
———: Simplified LCAO Method for Zincblende, Wurtzite, and Mixed Crystal Structures, *Phys. Rev.*, **115**:1493 (1959). (Sec. 10-8.)
———: Calculation of Electronic Energy Bands in ZnS, *J. Phys. Chem. Solids*, **8**:35 (1959). (Sec. 10-8.)
Bloch, F.: Über die Quantenmechanik der Elektronen in Kristallgittern, *Z. Physik*, **52**:555 (1928). (Secs. 1-2, 8-1.)
Blokhin, M. A., and V. P. Sachenko: Concerning the Shape of Energy Bands in Solids, *Izv. Akad. Nauk SSSR, Ser. Fiz.*, **24**:397 (1960).
Boerner, H.: "Representations of Groups, with Special Consideration for the Needs of Modern Physics," North Holland Publishing Company, Amsterdam, 1963.
Bonch-Bruevich, V. L.: On the Theory of Impurity Bands, *Fiz. Tverd. Tela*, **1**:1213 (1959).
———: A Theory of Impurity States in Semiconductors, *Fiz. Tverd. Tela*, Suppl. II, p. 177 (1959).

———: On the Theory of Degenerate Semiconductors, *Semiconductor Conf.*, Exeter, 1962, Institute of Physics and the Physical Society, London, 1962, p. 216.
———: Theory of Heavily Doped Semiconductors, *Fiz. Tverd. Tela*, **5**:1852 (1963).
Borland, R. E.: One-dimensional Chains with Random Spacing between Atoms, *Proc. Phys. Soc. (London)*, **77**:705 (1961).
———: Existence of Energy Gaps in One-dimensional Liquids, *Proc. Phys. Soc. (London)*, **78**:926 (1961).
———: The Nature of the Electronic States in Disordered One-dimensional Systems, *Proc. Roy. Soc. (London)*, **A274**:529 (1963).
Bouckaert, L. P., R. Smoluchowski, and E. Wigner: Theory of Brillouin Zones and Symmetry Properties of Wave Functions in Crystals, *Phys. Rev.*, **50**:58 (1936). (Sec. 5-4, Chap. 5, Sec. A3-1.)
Boyle, W. S., and P. Nozieres: Band Structure and Infrared Absorption of Graphite, *Phys. Rev.*, **111**:782 (1958).
Bragg, W. H.: Die Reflexion von Röntgenstrahlen an Kristallen, *Physik. Z.*, **14**:472 (1913). (Sec. 1-1.)
———: The Intensity of Reflexion of X-rays by Crystals, *Phil. Mag.*, **27**:881 (1914). (Sec. 1-1.)
———: X-rays and Crystal Structure, *Phil. Trans. Roy. Soc. (London)*, **A215**:253 (1915).
——— and W. L. Bragg: "X-rays and Crystal Structure," 5th ed., G. Bell & Sons, Ltd., London, 1925. (Sec. 1-1.)
Bragg, W. L.: The Diffraction of Short Electromagnetic Waves by a Crystal, *Proc. Cambridge Phil. Soc.*, **17**:43 (1913). (Sec. 1-1.)
———: Atomic Arrangements in Crystals, *Phil. Mag.*, **40**:169 (1920). (Secs. 3-1, 4-1.)
———, R. W. James, and C. H. Bosanquet: The Intensity of Reflexion of X-rays by Rock-salt, I, *Phil. Mag.*, **41**:309 (1921), II, **42** :1 (1921).
———: "The Crystalline State," The Macmillan Company, New York, 1934 (reprinted 1955). (Sec. 1-1.)
———: See also W. H. Bragg.
Branda, L. I.: Hole Bands in the LiCl Crystal, *Ukr. Fiz. Zh.*, **5**:368 (1960).
Braunstein, R., and E. O. Kane: The Valence Band Structure of the III-V Compounds, *J. Phys. Chem. Solids*, **23**:1423 (1962). (Sec. 10-8.)
———: Valence Band Structure of Germanium-Silicon Alloys, *Phys. Rev.*, **130**:869 (1963). (Sec. 10-8.)
Brillouin, L.: Les électrons libres dans les métaux et la rôle des reflexions de Bragg, *Compt. Rend.*, **191**:198 (1930); *J. Phys. Radium*, **1**:377 (1930). (Secs. 5-2, 7-3.)
———: Les électrons dans les métaux et le classement des ondes de de Broglie correspondantes, *Compt. Rend.*, **191**:292 (1930).
———: "Quantenstatistik," Springer-Verlag OHG, Berlin, 1931.
———: Les électrons dans les métaux de point de vue ondulatoire, *Actualités Scientifiques et Industrielles*, no. 88, Hermann et Cie, Paris, 1934.
Brindley, G. W.: The Relation of Atomic Sizes to Interatomic Distances in Homopolar Crystals, *Z. Krist.*, **84**:169 (1932).
Brode, R. B.: Quantitative Study of the Collisions of Electrons with Atoms, *Rev. Mod. Phys.*, **5**:257 (1933). (Sec. 9-3.)
Brooks, H., and F. S. Ham: Energy Bands in Solids: The Quantum Defect Method, *Phys. Rev.*, **112**:344 (1958). (Sec. 9-5.)
Brown, E., and J. A. Krumhansl: Energy Band Structure of Lithium by a Modified Plane Wave Method, *Phys. Rev.*, **109**:30 (1958). (Sec. 10-8.)

———: Role of Orthogonalization in Determination of Valence States in Crystals, *Phys. Rev.*, **126**:421 (1962).
Brown, G. E., J. S. Langer, and G. W. Schaefer: Lamb Shift of a Tightly Bound Electron, I, Method, *Proc. Roy. Soc. (London)*, **A251**:92 (1959).
——— and D. F. Mayers: Lamb Shift of a Tightly Bound Electron, II, Calculation for the K-electron in Mercury, *Proc. Roy. Soc. (London)*, **A251**:105 (1959).
Buerger, M. J.: "Elementary Crystallography," John Wiley & Sons, Inc., New York, 1942.
Bulyanitsa, D. S., and Yu. E. Svetlov: Some Properties of Bloch-Wannier Functions, *Fiz. Tverd. Tela*, **4**:1339 (1962).
Burdick, G. A.: Topology of the Fermi Surface of Copper, *Phys. Rev. Letters*, **7**:156 (1961). (Secs. 9-5, 10-8.)
———: Energy Band Structure of Cu, *Phys. Rev.*, **129**:138 (1963). (Secs. 9-5, 10-5, 10-8.)
Busch, G. A.: Semiconducting Compounds, *Nuovo Cimento Suppl.*, **7**:696 (1958).
Butusov, Yu. M., and M. V. Kopytina: The Basis of Band Theory, *Fiz. Tverd. Tela*, **3**: 395 (1961).
Callaway, J.: Orthogonalized Plane Wave Method, *Phys. Rev.*, **97**:933 (1955).
———: Electronic Energy Bands in Fe, *Phys. Rev.*, **99**:500 (1955). (Sec. 10-8.)
———: Electronic Energy Bands in Potassium, *Phys. Rev.*, **103**:1219 (1956). (Sec. 10-8.)
——— and E. L. Haase: Electron Energy Bands in Cesium, *Phys. Rev.*, **108**:217 (1957). (Sec. 10-8.)
———: Energy Bands in Gallium Arsenide, *J. Electronics*, **2**:330 (1957). (Sec. 10-8.)
——— and M. L. Glasser: Fourier Coefficients of Crystal Potentials, *Phys. Rev.*, **112**:73 (1958).
———: Electron Energy Bands in Sodium, *Phys. Rev.*, **112**:322 (1958). (Secs. 10-1, 10-8.)
——— and D. F. Morgan, Jr.: Cohesive Energy and Wave Functions for Rubidium, *Phys. Rev.*, **112**:334 (1958). (Sec. 10-8.)
———: Electron Wave Functions in Metallic Cesium, *Phys. Rev.*, **112**:1061 (1958). (Sec. 10-8.)
———: Electron Energy Bands in Solids, *Solid State Physics*, **7**:99 (1958). Reprinted, Academic Press Inc., New York, 1964.
———: d Bands in the Body-centered Cubic Lattice, *Phys. Rev.*, **115**:346 (1959).
——— and D. M. Edwards: Cubic Field Splitting of D Levels in Metals, *Phys. Rev.*, **118**:923 (1960).
———: Electron Wave Functions in Metallic K, *Phys. Rev.*, **119**:1012 (1960). (Sec. 10-8.)
———: d Bands in Cubic Lattices, II, III, *Phys. Rev.*, **120**:731 (1960); **121**:1351 (1961).
———: Electron Wave Functions in Metallic Na, *Phys. Rev.*, **123**:1255 (1961). (Sec. 10-8.)
———: Energy Bands in Li, *Phys. Rev.*, **124**:1824 (1961). (Sec. 10-8.)
——— and W. Kohn: Electron Wave Functions in Metallic Li, *Phys. Rev.*, **127**:1913 (1962). (Sec. 10-8.)
——— and A. J. Hughes: Moment Singularity Expansion for Density of States, *Phys. Rev.*, **128**:134 (1962).
———: See also M. L. Glasser, F. Herman.
Carter, J. L., and J. A. Krumhansl: Band Structure of Graphite, *J. Chem. Phys.*, **21**:2238 (1953). (Sec. 10-8.)

Casella, R. C.: Halogen Band in Sodium Chloride, *Phys. Rev.*, **104**:1260 (1956). (Sec. 10-8.)
———: Energy-band Structure of a Hypothetical Carbon Metal, *Phys. Rev.*, **109**:54 (1958).
———: Symmetry of Wurtzite, *Phys. Rev.*, **114**:1514 (1959).
———: Toroidal Energy Surfaces in Crystals with Wurtzite Symmetry, *Phys. Rev. Letters*, **5**:371 (1960).
Cheglokov, E. I.: Structure of the Valence Band of Bismuth-type Crystals, *Izv. Vysshikh Uchebn. Zavedenii, Fiz.*, vol. 13 (1960).
Chen, Shih-kang, Chang Yi-shang, and Liou De-sen: Influence of the Chemical Bond on the Energy Bands of the Zinc Blende Structure, *Acta Phys. Sinica*, **18**:491 (1962).
Chodorow, M. I., and M. F. Manning: Energy Bands in the Body-centered Lattice, *Phys. Rev.*, **52**:731 (1937).
———: The Band Structure of Metallic Copper, *Phys. Rev.*, **55**:675 (1939). (Sec. 10-8.)
———: See also M. F. Manning.
Clogston, A. M.: Impurity States in Metals, *Phys. Rev.*, **125**:439 (1962).
Cohan, N. V., D. Pugh, and R. H. Tredgold: Band Structure of Diamond, *Proc. Phys. Soc. (London)*, **82**:65 (1963). (Sec. 10-8.)
———: See also S. L. Altmann.
Cohen, M. H., and V. Heine: Electronic Band Structures of the Alkali Metals and of the Noble Metals and Their α-phase Alloys, *Advan. Phys.*, **7**:395 (1958). (Sec. 10-8.)
——— and L. M. Falicov: Effect of Spin-Orbit Splitting on the Fermi Surfaces of the Hexagonal-close-packed Metals, *Phys. Rev. Letters*, **5**:544 (1960).
———: Energy Bands in Bi Structure, I, Nonellipsoidal Model for Electrons in Bi, *Phys. Rev.*, **121**:387 (1961).
——— and V. Heine: Cancellation of Kinetic and Potential Energy in Atoms, Molecules, and Solids, *Phys. Rev.*, **122**:1821 (1961). (Sec. 9-4.)
——— and J. C. Phillips: Dielectric Screening and Self-consistent Crystal Fields, *Phys. Rev.*, **124**:1818 (1961).
———: See also L. M. Falicov.
Corbato, F. J.: A Calculation of the Energy Bands of the Graphite Crystal by Means of the Tight-binding Method, *Proc. 1957 Carbon Conf.*, Pergamon Press, New York, 1957, p. 173. (Secs. 10-6, 10-8.)
Corciovei, A., and D. Grecu: Energy Bands in Partially Disordered Binary Alloys, *Rev. Phys., Acad. Rep. Populaire Roumaine*, **5**:157 (1960).
——— and ———: The Effect of Long-range Order on Energy Bands in Binary Alloys, *Acta Phys. Polon.*, **20**:197 (1961).
——— and ———: Bandes énergetiques électroniques des alliages partiellement désordonnés, *Compt. Rend.*, **252**:1582 (1961).
Cornwell, J. F., and E. P. Wohlfarth: An Energy Band Interpolation Scheme, with Application to Body-centered Cubic Lithium, *Nature*, **186**:379 (1960). (Secs. 10-1, 10-8.)
———: The Electronic Energy Bands of the Alkali Metals and Metallic Beryllium, *Proc. Roy. Soc. (London)*, **A261**:551 (1961). (Secs. 10-1, 10-8.)
———: The Fermi Surfaces of the Noble Metals, *Phil. Mag.*, **6**:727 (1961). (Secs. 10-1, 10-8.)
——— and E. P. Wohlfarth: The Energy Band Structure of Body Centered Iron, *J. Phys. Soc. Japan*, **17** (suppl. B-1):32 (1962). (Sec. 10-8.)

Coulson, C. A.: Energy Bands in Graphite, *Nature*, **159**:265 (1947). (Sec. 10-8.)
——— and R. Taylor: Studies in Graphite and Related Compounds, I, Electronic Band Structure in Graphite, *Proc. Phys. Soc. (London)*, **A65**:815 (1952). (Sec. 10-8.)
———: A New Approach to the Theory of Electrons in Solids, *Proc. Intl. Conf. Theoret. Physics Japan*, Tokyo, 1953, p. 629.
———: Note on the Applicability of the Free-electron Network Model to Metals, *Proc. Phys. Soc. (London)*, **A67**:608 (1954).
———: Electronic Band Structure of Solids, *Nature*, **174**:949 (1954).
———: The Free-electron Network Model for Metals, II, *Proc. Phys. Soc. (London)*, **A68**:1129 (1955).
———, L. J. Schaad, and L. Burnelle: Benzene to Graphite: The Change in Electronic Energy Levels, *Proc. Third Bien. Carbon. Conf.*, Pergamon Press, New York, 1958.
———, L. B. Redei, and D. Stocker: The Electronic Properties of Tetrahedral Intermetallic Compounds, I, Charge Distribution, *Proc. Roy. Soc. (London)*, **A270**:357 (1962). (Secs. 10-4, 10-8.)
——— and A. C. Hurley: Comments on "Hellman-Feynman Wave Functions," *J. Chem. Phys.*, **37**:448 (1962).
———: See also S. L. Altmann, T. Anno.
Coxeter, H. S. M.: The Problem of Packing a Number of Equal Nonoverlapping Circles on a Sphere, *Trans. N.Y. Acad. Sci.*, ser. II, **24**:320 (1962). (Sec. 1-5.)
Daniel, E., Effet des impuretés sur la densité électronique dans les métaux, I, II, *J. Phys. Radium*, **20**:769, 849 (1959).
———: Structure électronique des alliages dilués, *J. Phys. Radium*, **23**:602 (1962).
Dank, M., and H. B. Callen: Calculation of Energy Bands in Solids by the Integral Iteration Method, *Phys. Rev.*, **86**:622 (1952).
de Boer, J. H., and E. J. W. Verwey: Semiconductors with Partially and with Completely Filled 3d-lattice Bands, *Proc. Phys. Soc. (London)*, **49** (extra part):59 (1937).
de Carvalho, A. P.: Sur la structure des bandes de tellure, *Compt. Rend.*, **248**:778 (1959). (Sec. 10-8.)
Dehlinger, U.: Zur Elektronentheorie metallischer und halbleitender Strukturen, *J. Phys. Chem. Solids*, **1**:279 (1957).
Dekker, A. J.: "Solid State Physics," Prentice-Hall, Inc., Englewood Cliffs, N.J., 1957.
Della Riccia, G.: Band Structure of II-IV Compounds, *Semiconductor Conf.*, Exeter, 1962, Institute of Physics and Physical Society, London, 1962, p. 510.
des Cloizeaux, J.: Spectre de fréquence d'une chaine linéaire désordonnée, *J. Phys. Radium*, **18**:131 (1957).
——— and P. André: Méthode de calcul des niveaux énergetiques associés aux pièges profonds d'un cristal semiconducteur, *J. Phys. Radium*, **18**:441 (1957).
———: Étude des transitions entre états métalliques et isolants pour un gas d'électrons: application aux bandes d'impuretés et aux antiferromagnétiques, I, II, *J. Phys. Radium*, **20**:606, 751 (1959).
———: Orthogonal Orbitals and Generalized Wannier Functions, *Phys. Rev.*, **129**:554 (1963).
Dimmock, J. O., and R. G. Wheeler: Symmetry Properties of Wave Functions in Magnetic Crystals, *Phys. Rev.*, **127**:391 (1962).
———: Use of Symmetry in Determination of Magnetic Structures, *Phys. Rev.*, **130**:1337 (1963).

BIBLIOGRAPHY

———: See also G. F. Koster.
Donovan, B.: A New Calculation of Some Properties of Metallic Be, *Phil. Mag.*, **43**:868 (1952). (Sec. 10-8.)
——— and N. H. March: Momentum Distribution of Electrons in Solids, II, Some General Results with an Application to Metallic Lithium, *Proc. Phys. Soc. (London)*, **B69**:1249 (1956).
Döring, W., and V. Zehler: Gruppentheoretische Untersuchung der Elektronenbänder im Diamantgitter, *Ann. Physik*, **13**:214 (1953).
Drabble, J. R., and C. H. L. Goodman: Chemical Bonding in Bismuth Telluride, *J. Phys. Chem. Solids*, **5**:142 (1958).
Edwards, S. F.: The Electronic Structure of Disordered Systems, *Phil. Mag.*, **6**:617 (1961).
———: The Electronic Structure of Liquid Metals, *Proc. Roy. Soc. (London)*, **A267**:518 (1962).
———: The Concept of a Wave Function of a Disordered System, *J. Phys. Radium*, **23**:627 (1962).
Eilenberger, G.: Phasenfestlegung in den Wannier-Funktionen und einfache Herleitung des Ersatzoperators für Gitterelektronen im Magnetfeld, *Z. Physik*, **175**:445 (1963).
Eisenschitz, R., and P. Dean: Electronic Levels of a Model of Liquid Potassium, *Proc. Phys. Soc. (London)*, **70**:713 (1957).
Elliott, R. J.: Theory of the Effect of Spin-Orbit Coupling on Magnetic Resonance in Some Semiconductors, *Phys. Rev.*, **96**:266 (1954).
———: Spin-Orbit Coupling in Band Theory: Character Tables for Some "Double" Space Groups, *Phys. Rev.*, **96**:280 (1954). (Sec. A9-4.)
Englert, F.: Application of Group Theory to Calculation of Spin-Orbit Coupling in Crystals, *Bull. Acad. Roy. Belg. Cl. Sci.*, **43**:273 (1957).
Erdmann, J.: Zur Symmetrie der Wellenfunktionen in Kristallen mit hexagonale dichtesten Kugelpackung, *Z. Naturforsch.*, **15a**:524 (1960).
Evseev, Z. Ya., and K. B. Tolpygo: Wave-function and Energy of a NaCl Crystal with an Excess Electron, *Fiz. Tverd. Tela*, **4**:3644 (1962); **5**:2345 (1963). (Sec. 10-8.)
Ewald, P. P.: Zur Theorie der Interferenzen der Röntgenstrahlen in Kristallen, *Physik. Z.*, **14**:465 (1913).
———: Bemerkung zu der Arbeit von M. Laue: Die dreizähligzymmetrischen Röntgenstrahlaufnahmen an regulären Kristallen, *Physik. Z.*, **14**:1038 (1913).
———: Zur Begründung der Kristalloptik, I, Theorie der Dispersion; II, Theorie der Reflexion und Brechung; III, Die Kristalloptik der Röntgenstrahlen, *Ann. Physik*, **49**:1, 117 (1916); **54**:519 (1917). (Sec. 9-3.)
———: Das "reziproke Gitter" in der Strukturtheorie, *Z. Krist.*, **56**:129 (1921). (Sec. 5-1.)
———: Historisches und Systematisches zum Gebrauch des "reziproken Gitters" in der Kristallstrukturlehre, *Z. Krist.*, **93**:396 (1936). (Sec. 5-1.)
———: William Henry Bragg and the New Crystallography, *Nature*, **195**:320 (1962).
———: See also A. Bienenstock.
Ewing, D. H., and F. Seitz: On the Electronic Constitution of Crystals: LiF and LiH, *Phys. Rev.*, **50**:760 (1936). (Secs. 10-7, 10-8.)
Eyges, L.: Solution of Schrödinger Equation for Periodic Lattice, *Phys. Rev.*, I, **123**:1673 (1961); II, **126**:93 (1962).
Fajans, K., and H. Grimm: Über die Molekularvolumina der Alkalihalogenide, *Z. Physik*, **2**:299 (1920). (Sec. 4-1.)

――― and K. F. Herzfeld: Die Ionengrösse und die Gitterenergie der Alkalihalogenide, *Z. Physik*, **2**:309 (1920). (Sec. 4-1.)
――― and G. Joos: Molrefraktionen von Ionen und Molekülen im Lichte der Atomstruktur, *Z. Physik*, **23**:1 (1924).
Falicov, L. M.: The Band Structure and Fermi Surface of Magnesium, *Phil. Trans. Roy. Soc. (London)*, **A255**:55 (1962). (Secs. 10-1, 10-8.)
――― and M. H. Cohen: Spin-Orbit Coupling in Band Structure of Mg and Other hcp Metals, *Phys. Rev.*, **130**:92 (1963).
―――: See also M. H. Cohen.
Faulkner, J. S., and J. Korringa: Electron Energy Bands of One-dimensional Random Alloys, *Phys. Rev.*, **122**:390 (1961).
Feinberg, E. L.: Some Relationships between Atomic Lattices, *Phys. Z. Sowjetunion*, **8**:407 (1935).
Feuer, P.: Electronic States in Crystals under Large Over-all Perturbations, *Phys. Rev.*, **88**:92 (1952).
Firsov, Yu. A.: On the Structure of Electron Spectra of Tellurium-type Lattices, *Zh. Eksperim. i Teor. Fiz.*, **32**:1350 (1957).
Fischer, G.: Speculation on the Band Structure of the Layer Compounds GaS and GaSe, *Helv. Phys. Acta*, **36**:317 (1963).
―――: See also T. H. K. Barron.
Fletcher, G. C., and E. P. Wohlfarth: Calculation of the Density of States Curve for the 3d Electrons in Nickel, *Phil. Mag.*, **42**:106 (1951). (Sec. 10-8.)
―――: Density of States Curve for the 3d Electrons in Ni, *Proc. Phys. Soc. (London)*, **A65**:192 (1952). (Sec. 10-8.)
―――: Spin-Orbit Coupling Effects in Ferromagnetic Metals, *Acta Met.*, **1**:467 (1953).
Flodmark, S.: Electron Distribution and Energy Bands in Crystals of Metal Borides of the Type MB_6, *Arkiv Fys.*, **14**:513 (1959). (Sec. 10-8.)
―――: A Solid State spd-MOLCAO Treatment of Cubic-octahedral CaB_6 with Variation of the d Orbital Exponent, *Arkiv Fys.*, **18**:49 (1960). (Sec. 10-8.)
―――: Symmetry Reduction of Secular Matrices for Crystals, *Arkiv Fys.*, **21**:89 (1961).
Flower, M., N. H. March, and A. M. Murray: Metallic Transitions in Ionic Crystals: Some Group Theoretical Results, *Phys. Rev.*, **119**:1885 (1960).
――― and ―――: Transitions to Metallic States in Ionic Crystals, with Particular Reference to CsI, *Phys. Rev.*, **125**:1144 (1962).
Fok, M. V.: Forbidden Band Width and Effective Charge in Crystal Lattice of ZnS, *Czech. J. Phys.*, **B2**:99 (1963).
Folberth, O. G., and H. Welker: Binding and Semiconductor Properties of $A^{III}B^V$ Compounds, *J. Phys. Chem. Solids*, **8**:14, 20 (1959).
―――: Die Bindung in Kristallen mit "Normal-Valenz," unter besonderer Berücksichtigung der ZnS- und Wurtzit-Phasen, *Z. Naturforsch.*, **15a**:425 (1960).
―――: Zur Frage der chemischen Bindung in Kristallen mit DO_3-(BiF_3)-Struktur, *Z. Naturforsch.*, **15a**:739 (1960).
Fowler, W. B.: Electronic Band Structure and Wannier Exciton States in Solid Krypton, *Phys. Rev.*, **132**:1594 (1963). (Sec. 10-8.)
Freise, E. J.: Structure of Graphite, *Nature*, **193**:671 (1962).
Friedel, J.: The Distribution of Electrons around Impurities in Monovalent Metals, *Phil. Mag.*, **43**:153 (1952).
―――: Electronic Structure of Primary Solutions in Metals, *Phil. Mag.*, suppl., **3**:446 (1954).

———: Sur la structure électronique des métaux et alliages de transition et des métaux lourds, *J. Phys. Radium*, **19**:573 (1958).

———: Metallic Alloys, *Nuovo Cimento*, **7** (suppl.):387 (1958).

———: Sur la structure électronique et les propriétés magnétiques des métaux et alliages de transition, *J. Phys. Radium*, **23**:501 (1962).

———: Concept du niveau lié vertuel. *J. Phys. Radium*, **23**:692 (1962).

——— and A. Guinier (eds.): "Metallic Solid Solutions," W. A. Benjamin, Inc., New York, 1963.

———: See also M. Asdente, G. Leman, and Y. A. Rocher.

Friedrich, W., P. Knipping, and M. von Laue: Interferenzerscheinungen bei Röntgenstrahlen, *Ann. Physik*, **41**:971 (1913). (Sec. 1-1.)

Frisch, H. L., and S. P. Lloyd: Electron Levels in One-dimensional Random Lattice, *Phys. Rev.*, **120**:1175 (1960).

Fröhlich, H.: The Number of Free Electrons in a Metal, *Proc. Cambridge Phil. Soc.*, **31**:277 (1935).

———: "Elektronentheorie der Metalle," Springer-Verlag OHG, Berlin, 1936.

Fröman, A.: Relativistic Corrections in Many-electron Systems, *Rev. Mod. Phys.*, **32**:317 (1960).

Fues, E., and H. Statz: Ersatzpotentiale mit verwandtem Eigenwertspektren in Schrödinger-Gleichungen, *Z. Naturforsch.*, **7a**:2 (1952). (Secs. 9-4, 10-1.)

Fukuchi, M.: The Energy Band Structure of the Metallic Copper: The Orthogonalized Plane Wave Method, *Progr. Theoret. Phys.* (*Kyoto*), **16**:222 (1956). (Sec. 10-8.)

———: See also J. Yamashita.

Ganesan, S., and R. Srinivasan: On Houston's Method of Evaluating Integrals over the Brillouin Zone: Normalization Factor in a Simple Cubic Lattice, *Can. J. Phys.*, **40**:1153 (1962).

Ganzhorn, K.: Homöopolare und metallische Bindung beim Diamanten, *Naturwissenschaften*, **3**:62 (1952).

———: Quantenmechanik der kubisch raumzentrierten Strukturen der Übergangsmetalle, *Z. Naturforsch.*, **7a**:291 (1952).

———: Gruppentheorie und Quantenmechanik der Übergangsmetall-Strukturen, *Z. Naturforsch.*, **8a**:330 (1953).

———: See also F. Bader.

Gashimzade, F. M.: Symmetry of Energy Bands in Crystals of the SnSe and Sb_2S_3 Type, *Fiz. Tverd. Tela*, **2**:2070 (1960).

———: Symmetry of the Energy Bands in TlSe-type Crystals, *Fiz. Tverd. Tela*, **2**:3040 (1960); **4**:2282 (1962).

——— and V. E. Khartsiev: Energy Structure of Complex Semiconductors: Calculation of the Band Structure of Si, Ge, and GaAs by a Simplified Orthogonalized Plane Wave Method, *Fiz. Tverd. Tela*, **3**:1453 (1961). (Sec. 10-8.)

——— and ———: Energy Structure of Complex Semiconductors: The Valence Band Spectrum of Anisotropic Compounds of SnS Type, *Fiz. Tverd. Tela*, **4**:434 (1962). (Sec. 10-8.)

———: See also A. I. Gubanov.

Gaspar, R.: Electronic Structure of Semiconducting Selenium and Tellurium, *Acta Phys. Acad. Sci. Hung.*, **7**:289 (1957). (Sec. 10-8.)

———: The Energy Spectrum of Localized Electron Levels, *Abhandl. Deut. Akad. Wiss. Berlin, Kl. Math. Phys. Tech.*, vol. 110 (1960).

Gautier, F.: Influence de la forme de la surface de Fermi sur la distribution électronique autour d'une impureté dissoute dans le cuivre, *J. Phys. Radium*, **23**:105 (1962).

Geller, S., and M. A. Gilleo: The Crystal Structure and Ferrimagnetism of Yttrium-Iron Garnet, $Y_3Fe_2(FeO_4)_3$, *J. Phys. Chem. Solids*, **3**:30 (1957). (Sec. 3-4.)

Gibbs, J. W., and E. B. Wilson: "Vector Analysis," Yale University Press, New Haven, Conn., 1902. (Sec. 5-1.)

Glasser, M. L., and J. Callaway: Electronic Energy Bands in Lithium, *Phys. Rev.*, **109**:1541 (1958). (Sec. 10-8.)

———: Symmetry Properties of the Wurtzite Structure, *J. Phys. Chem. Solids*, **10**:229 (1959).

———: Electron Structure of Metallic Silver, *Rev. Mexicana Fiz.*, **11**:31 (1962). (Sec. 10-8.)

———: See also J. Callaway.

Glatzel, E., and H. Schlechtweg: Eine einfache Näherung zur Elektronentheorie der Übergangsmetalle, *Z. Naturforsch.*, **10a**:777 (1955).

Goldman, J. E.: Perturbed Energy Bands in Transition Alloys, *Phys. Rev.*, **82**:339 (1951).

Goldschmidt, V. M.: *Skrifter Norske Videnskaps-Akad. Oslo*, I, *Mat.-Naturv. Kl.*, 1926. (Sec. 4-1.)

———: Konstruktion von Kristallen, *Z. Tech. Phys.*, **8**(7):251 (1927). (Sec. 4-1.)

———: Crystal Structure and Chemical Constitution, *Trans. Faraday Soc.*, **25**:253 (1929). (Sec. 4-1.)

Gombas, P.: Über eine Methode zur Berechnung der Lage und Breite des Energiebandes der Valenzelektronen in Alkalimetallen, *Z. Physik*, **111**:195 (1938).

———: Bestimmung der Lage und Breite des Energiebandes der Valenzelektronen der Metalle Na, K, Rb, und Cs, *Z. Physik*, **113**:150 (1939). (Secs. 10-1, 10-8.)

Goodenough, J. B.: Theory of the Role of Covalence in the Perovskite-type Manganites La, $M(II)MnO_3$, *Phys. Rev.*, **100**:564 (1955).

———: Suggestion Concerning the Role of Wave-function Symmetry in Transition Metals and Alloys, *J. Appl. Phys.*, **29**:513 (1958).

———: Band Structure of Transition Metals and Their Alloys, *Phys. Rev.*, **120**:67 (1960).

———: Direct Cation-cation Interactions in Primarily Ionic Solids, *J. Appl. Phys.*, **31** (suppl.):539S (1960).

Goodman, C. H. L.: The Prediction of Semiconducting Properties in Inorganic Compounds, *J. Phys. Chem. Solids*, **6**:305 (1958).

———: Bonding in Cadmium Telluride, *Proc. Phys. Soc.*, **74**:489 (1959).

———: Bonding and Semiconductivity Relationships in Bi_2Te_3 and CdI_2 Type Structures, *J. Electrochem. Soc.*, **107**:564 (1960).

———: Ionic-covalent Bonding in Crystals, *Nature*, **187**:590 (1960).

———, E. Mooser, and W. B. Pearson: Ionic-covalent Bonding in Crystals, *Nature*, **192**:355 (1961).

———: See also J. R. Drabble.

Goodwin, E. T.: Electronic States at the Surfaces of Crystals, I, The Approximation of Nearly Free Electrons; II, The Approximation of Tight Binding: Finite Linear Chain of Atoms; III, The Approximation of Tight Binding: Further Extensions, *Proc. Cambridge Phil. Soc.*, **35**:205, 221, 232 (1939).

Gorin, E.: The Theoretical Constitution of Metallic Potassium, *Physik. Z. Sowjetunion*, **9**:328 (1936).

Gorzkowski, W.: Space Group of Pyrite T_h^6 ($Pa3$), *Phys. Stat. Sol.*, **3**:599 (1963).

———: Representation of Space Group O_h^3, *Phys. Stat. Sol.*, **3**:910 (1963).

Gourary, B. S., and A. E. Fein: Theory of Localized Electronic States at Point Imperfections, *J. Appl. Phys.*, **33** (suppl. 1):340 (1962).

Gousseland, G., and G. Leman: Sur la répartition spatiale des électrons dans les approximations des liaisons fortes et des électrons presque libres, *J. Phys. Radium*, **22**:65 (1961).
Greene, J. B., and M. F. Manning: Electronic Energy Bands in Face-centered Iron, *Phys. Rev.*, **63**:203 (1943). (Sec. 10-8.)
Grimley, T. B.: The Electronic Structure of Crystals Having the Sodium Chloride Type of Lattice, *Proc. Phys. Soc. (London)*, **71**:749 (1958).
——— and B. W. Holland: The Surface States of a Simple Crystal Model, *Proc. Phys. Soc. (London)*, **78**:217 (1961).
Gubanov, A. I.: Theory of Semiconductors of the Type $A^{III}B^{V}$, *Zh. Tekh. Fiz.*, **26**:2170 (1956).
———: Dependence of the Zone Structure of an Alloy of the AB_3 Type on the Degree of Order, *Zh. Tekh. Fiz.*, **28**:2109 (1958).
——— and A. A. Nranyan: Application of the Equivalent Orbital Method to the Study of Band Structure in Compounds of the Type A^{III}-B^{V}, *Fiz. Tverd. Tela*, **1**:1044 (1959). (Sec. 10-8.)
——— and F. M. Gashimzade: Investigation of the Symmetry of the Energy Bands of Electrons in Crystals of the Type $CdIn_2Se_4$, *Fiz. Tverd. Tela*, **1**:1411 (1959).
——— and ———: Structures of the Energy Bands in Semiconductors of the $CdIn_2Se_4$ Type, *Fiz. Tverd. Tela*, **2**:255 (1960).
——— and A. D. Chevychelov: Calculation of the Energy Spectrum of Strongly Anisotropic Crystals, *Fiz. Tverd. Tela*, **2**:1379 (1960).
——— and O. E. Pushkarev: The Wave Functions of the Valence Bond in Certain Crystals, *Fiz. Tverd. Tela*, **2**:1776 (1960).
———: Band Theory for Partially Ordered Systems, *Fiz. Tverd. Tela*, **3**:2154 (1961).
———: Energy Bands and Impurity Levels in Amorphous Semiconductors, *Semiconductor Conf.*, Exeter, 1962, Institute of Physics and The Physical Society, London, 1962, p. 599.
Haering, R. R.: Band Structure of Rhombohedral Graphite, *Can. J. Phys.*, **36**:352 (1958). (Sec. 10-8.)
——— and S. Mrozowski: The Band Structure and Electronic Properties of Graphite Crystals, *Progr. Semiconductors*, **5**:273 (1961).
Hall, G. G.: The Electronic Structure of Diamond, *Phil. Mag.*, **43**:338 (1952). (Sec. 10-8.)
———: The Electronic Structure of Diamond, *Phys. Rev.*, **90**:317 (1953). (Sec. 10-8.)
———: The Electronic Structure of Some Body-centered Cubic Metals, *Proc. Phys. Soc. (London)*, **A66**:1162 (1953).
———: The Form of the Effective Electronic Potential in a Crystal, *Proc. Phys. Soc. (London)*, **B69**:1124 (1956).
———: The Electronic Structure of Diamond, Silicon, and Germanium, *Phil. Mag.*, **3**:429 (1958). (Sec. 10-8.)
Ham, F. S.: The Quantum Defect Method, *Solid State Phys.*, **1**:127 (1955).
———: Band Calculations of the Shape of the Fermi Surface in the Alkali Metals, in Harrison and Webb (eds.), "The Fermi Surface," John Wiley & Sons, Inc., New York, 1960, p. 9.
——— and B. Segall: Energy Bands in Periodic Lattices: Green's Function Method, *Phys. Rev.*, **124**:1786 (1961). (Sec. 9-5.)
———: Energy Bands of Alkali Metals, I, Calculated Bands; II, Fermi Surface, *Phys. Rev.*, **128**:82, 2524 (1962). (Secs. 9-1, 9-5, 10-3, 10-8.)
———: See also H. Brooks.

Hamermesh, M.: "Group Theory," Addison-Wesley Publishing Company, Inc., Reading, Mass., 1962.

Harrison, W. A.: Cellular Method for Wave Functions in Imperfect Metal Lattices, *Phys. Rev.*, **110**:14 (1958).

———: Fermi Surface in Aluminum, *Phys. Rev.*, **116**:555 (1959). (Secs. 10-1, 10-8.)

———: Band Structure of Aluminum, *Phys. Rev.*, **118**:1182 (1960). (Secs. 10-1, 10-8.)

———: Electronic Structure of Polyvalent Metals, *Phys. Rev.*, **118**:1190 (1960). (Sec. 10-1.)

———: Bismuth Fermi Surface, *J. Phys. Chem. Solids*, **17**:171 (1960). (Secs. 10-1, 10-8.)

——— and M. B. Webb (eds.): "The Fermi Surface," John Wiley & Sons, Inc., New York, 1960. (Sec. 10-1.)

———: Electronic Structures from the one-OPW or Nearly Free-electron Point of View, in Harrison and Webb (eds.), "The Fermi Surface," John Wiley & Sons, Inc., New York, 1960, p. 28. (Sec. 10-1.)

———: The Fermi Surface, *Science*, **134**:915 (1961).

———: Band Structure and Fermi Surface of Zn, *Phys. Rev.*, **126**:497 (1962). (Secs. 10-1, 10-8.)

———: Electronic Structure and Properties of Metals, I, Formulation; II, Application to Zn, *Phys. Rev.*, **129**:2503, 2512 (1963). (Secs. 10-1, 10-8.)

———: Electronic Structure of a Series of Metals, *Phys. Rev.*, **131**:2433 (1963).

Heine, V.: The Band Structure of Aluminium, I, Determination from Experimental Data; II, The Convergence of the Orthogonalized Plane Wave Method; III, A Self-consistent Calculation, *Proc. Roy. Soc. (London)*, **A240**:340, 354, 361 (1957). (Secs. 10-3, 10-8.)

———: The Band Theory of Metals, *Nature*, **181**:525 (1958).

———: Electronic Band Structure in Alloys and Liquid Metals, in Harrison and Webb (eds.), "The Fermi Surface," John Wiley & Sons, Inc., New York, 1960, p. 279.

———: "Group Theory in Quantum Mechanics," Pergamon Press, New York, 1960.

———: On the General Theory of Surface States and Scattering of Electrons in Solids, *Proc. Phys. Soc. (London)*, **81**:300 (1963).

———: See also B. J. Austin, M. H. Cohen.

Hellmann, H.: Ein kombiniertes Näherungsverfahren zur Energieberechnung im Vielelektronenproblem, *Acta Physicochimica URSS*, I, **1**:913 (1935); II, **4**:225 (1936). (Secs. 8-4, 10-1.)

Herman, F.: Electronic Structure of the Diamond Crystal, *Phys. Rev.*, **88**:1210 (1952). (Sec. 10-8.)

——— and J. Callaway: Electronic Structure of the Germanium Crystal, *Phys. Rev.*, **89**:518 (1953). (Sec. 10-8.)

———: Calculation of the Energy Band Structures of the Diamond and Germanium Crystals by the Method of Orthogonalized Plane Waves, *Phys. Rev.*, **93**:1214 (1954). (Secs. 10-4, 10-8.)

———: Some Recent Developments in the Calculation of Crystal Energy Bands: New Results for the Germanium Crystal, *Physica*, **20**:801 (1954). (Secs. 10-4, 10-8.)

———: Speculations on the Energy Band Structure of Ge-Si Alloys, *Phys. Rev.*, **95**:847 (1954).

———, J. Callaway, and F. S. Acton: Comparison of Various Approximate Exchange Potentials, *Phys. Rev.*, **95**:371 (1954).

——: Speculations on the Energy Band Structure of Zinc-blende-type Crystals, *J. Electronics*, **1**:103 (1955). (Secs. 10-4, 10-8.)
——: The Electronic Energy Band Structure of Silicon and Germanium, *Proc. Inst. Radio Engrs.*, **43**:1703 (1955). (Sec. 10-8.)
——, M. Glicksman, and R. H. Parmenter: Semiconductor Alloys, *Progr. Semiconductors*, vol. 2, Heywood and Company, Ltd., London, 1957, p. 1.
——: Theoretical Investigation of the Electronic Energy Band Structure of Solids, *Rev. Mod. Phys.*, **30**:102 (1958). (Secs. 10-1, 10-6.)
—— and S. Skillman: Theoretical Investigation of the Energy Band Structure of Semiconductors, *Semiconductor Phys. Conf.*, Prague, 1960, Academic Press Inc., New York, 1961, p. 20.
—— and ——: "Atomic Structure Calculations," Prentice-Hall, Inc., Englewood Cliffs, N.J., 1963. (Sec. 4-2.)
——, C. D. Kuglin, K. F. Cuff, and R. L. Kortum: Relativistic Corrections to the Band Structure of Tetrahedrally Bonded Semiconductors, *Phys. Rev. Letters*, **11**:541 (1963). (Sec. A9-4.)
Herring, C.: Effect of Time-reversal Symmetry on Energy Bands of Crystals, *Phys. Rev.*, **52**:361 (1937). (Sec. A9-7.)
——: Accidental Degeneracy in the Energy Bands of Crystals, *Phys. Rev.*, **52**:365 (1937).
——: New Method for Calculating Wave Functions in Crystals, *Phys. Rev.*, **57**:1169 (1940). (Sec. 8-1.)
—— and A. G. Hill: The Theoretical Constitution of Metallic Beryllium, *Phys. Rev.*, **58**:132 (1940). (Secs. 8-1, 10-3, 10-8.)
——: Character Tables for Two Space Groups, *J. Franklin Inst.*, **233**:525 (1942). (Secs. 2-3, A3-1, A3-2.)
——: The State of d Electrons in Transition Metals, *J. Appl. Phys.*, **31** (suppl.):3S (1960).
Heywang, W., and B. Seraphin: Wasserstoffähnliches Modell der Valenz in halbleitenden Verbindungen vom Typus III-V, *Z. Naturforsch.*, **11a**:425 (1956).
Hilsum, C., and A. C. Rose-Innes: "Semiconducting III-V Compounds," Pergamon Press, New York, 1961.
Hoerni, J. A.: Application of the Free-electron Theory to Three-dimensional Networks, *J. Chem. Phys.*, **34**:508 (1961).
Hoffmann, T. A., and A. Konya: Linear Atomic Chain and the Metallic State, *J. Chem. Phys.*, **16**:172 (1948).
——: Atomic Space-lattice and the Metallic State, *J. Chem. Phys.*, **18**:989 (1950).
—— and A. Konya: Some Investigations in the Field of the Theory of Solids, I, Linear Chain of Similar Atoms, *Acta Phys. Acad. Sci. Hung.*, **1**:5 (1950).
——: Some Investigations in the Field of the Theory of Solids; II, Linear Chain of Different Atoms: Binary Systems, *Acta Phys. Acad. Sci. Hung.*, **1**:175 (1951); III, Plane and Space Lattice of Similar Atoms, **2**:97 (1952); IV, A-B-type Ordered Binary Systems in the Plane and the Space, **2**:107 (1952); V, Adsorption: Surface States, **2**:195 (1952).
Holland, B. W.: Energy Extremes for an Electron in a Periodic Field, *Proc. Phys. Soc. (London)*, **80**:557 (1962).
——: The One-electron States of Imperfect Crystals, *Phil. Mag.*, **8**:87 (1963).
——: See also T. B. Grimley.
Holmes, D. K.: An Application of the Cellular Method to Silicon, *Phys. Rev.*, **87**:782 (1952). (Sec. 10-8.)
Howarth, D. J., and H. Jones: The Cellular Method of Determining Electronic Wave

Functions and Eigenvalues in Crystals, with Applications to Sodium, *Proc. Phys. Soc. (London)*, **A65**:355 (1952). (Secs. 9-1, 10-8.)

———: Electronic Eigenvalues of Copper, *Proc. Roy. Soc. (London)*, **A220**:513 (1953). (Sec. 10-8.)

———: Application of the Augmented Plane Wave Method to Copper, *Phys. Rev.*, **99**:469 (1955). (Sec. 10-8.)

Howland, L. P.: Band Structure and Cohesive Energy of Potassium Chloride, *Phys. Rev.*, **109**:1927 (1958). (Secs. 10-7, 10-8.)

Huggins, M. L.: Atomic Radii, *Phys. Rev.*, **28**:1086 (1926). (Sec. 4-1.)

———: Solid Matter: What is it, and Why? *Sci. Monthly*, **32**:140 (1931).

———: Principles Determining the Arrangement of Atoms and Ions in Crystals, *J. Phys. Chem.*, **35**:1270 (1931).

——— and J. E. Mayer: Interatomic Distances in Crystals of the Alkali Halides, *J. Chem. Phys.*, **1**:643 (1933).

———: See also L. Pauling.

Hund, F.: Zur Theorie der schwerflüchtigen nichtleitenden Atomgitter, *Z. Physik*, **74**:1 (1932).

———: Theorie der Bewegung der Elektronen in nichtmetallischen Kristallgittern, *Z. Tech. Phys.*, **16**:331 (1935); *Physik. Z.*, **36**:725 (1935).

——— and B. Mrowka: Über die Zustände der Elektronen in einem Kristallgitter, insbesondere beim Diamant, *Sächsische Akad. Wiss. Leipzig*, **87**:185 (1935); II, Energiebänder in einfachen Gittern, **87**:325 (1935). (Secs. 10-4, 10-8.)

———: Zustände der Elektronen in Kristallgittern, *Physik. Z.*, **36**:888 (1935); *Z. Tech. Phys.*, **16**:494 (1935).

———: Über den Zusammenhang zwischen der Symmetrie eines Kristallgitters und den Zuständen seiner Elektronen, *Z. Physik*, **99**:119 (1936).

Huntington, H. B., and F. Seitz: A Self-consistent Solution to the Vacancy Problem in Metals, *Phys. Rev.*, **58**:209 (1940); **61**:325 (1942).

International Tables for X-Ray Crystallography, published for the International Union of Crystallography by the Kynoch Press, Birmingham, England, 1952. (Secs. 1-4, 2-2, 2-3, etc.)

Ioffe, A. F.: "Physics of Semiconductors," Academic Press Inc., New York, 1960.

Jacques, R.: Determination of the Electronic Wave Function and Density of Beryllium in Its Various Physical and Chemical States: Application to the Variation of the Capture Half-life of Be^7, *Cahiers Phys.*, nos. 70-72 (1956).

James, H. M.: Energy Bands and Wave Functions in Periodic Potentials, *Phys. Rev.*, **76**:1602 (1949).

———: Electronic States in Perturbed Periodic Systems, *Phys. Rev.*, **76**:1611 (1949).

——— and A. S. Ginzbarg: Band Structure in Disordered Alloys and Impurity Semiconductors, *J. Phys. Chem.*, **57**:840 (1953).

James, R. W., I. Waller, and D. R. Hartree: Existence of Zero-point Energy in the Rock-salt Lattice by an X-ray Diffraction Method, *Proc. Roy. Soc. (London)*, **A118**:334 (1928). (Sec. 4-4.)

——— and G. W. Brindley: Quantitative Study of the Reflection of X-rays by Sylvine, *Proc. Roy. Soc. (London)*, **A121**:155 (1928).

———, ———, and R. G. Wood: A Quantitative Study of the Reflection of X-rays from Crystals of Aluminium, *Proc. Roy. Soc. (London)*, **A125**:401 (1929).

———: See also W. L. Bragg.

Jenkins, D. P.: A Variation Principle for Electronic Wave Functions in Crystals, *Phil. Mag.*, **45**:93 (1954).

———: The Electronic Band Structure of Silicon, *Physica*, **20**:967 (1954). (Sec. 10-8.)

―――: Calculations on the Band Structure of Silicon *Proc. Phys. Soc. (London),* **A69**:548 (1956). (Sec. 10-8.)
―――: See also D. G. Bell.
Johnson, L. E., J. B. Conklin, and G. W. Pratt, Jr.: Relativistic Effects in the Band Structure of PbTe, *Phys. Rev. Letters,* **11**:538 (1963). (Secs. 10-7, 10-8, A9-2.)
Johnston, D. F.: The Structure of the π-band of Graphite, *Proc. Roy. Soc. (London),* **A227**:349 (1955). (Sec. 10-8.)
―――: The Effect of the Mixing of π- and σ-orbitals on the Fermi Surface in the LCAO Model for Graphite, *Proc. Roy. Soc. (London),* **A237**:48 (1956). (Sec. 10-8.)
―――: Space-group Operations and Time-reversal for a Dirac Electron in a Crystal Field, *Proc. Roy. Soc. (London),* **A243**:546 (1958).
―――: Group Theory in Solid-state Physics, *Rept. Progr. Phys.,* **23**:67 (1960).
Jones, H.: The Theory of Alloys in the γ-phase, *Proc. Roy. Soc. (London),* **A144**:225 (1934). (Secs. 7-3, 10-1, 10-8.)
―――: Applications of the Bloch Theory to the Study of Alloys and of the Properties of Bismuth, *Proc. Roy. Soc. (London),* **A147**:396 (1934). (Secs. 7-3, 10-1, 10-6, 10-8.)
―――, N. F. Mott, and H. W. B. Skinner: A Theory of the Form of the X-ray Emission Bands of Metals, *Phys. Rev.,* **45**:379 (1934).
―――: Application de la théorie électronique des métaux à l'étude des alliages, *Helv. Phys. Acta,* **7** (suppl. 2):84 (1934).
―――: Structural and Elastic Properties of Metals, *Physica,* **15**:13 (1949). (Sec. 10-1.)
―――: The Effect of Electron Concentration on the Lattice Spacings in Magnesium Solid Solutions, *Phil. Mag.,* **41**:663 (1950). (Secs. 10-1, 10-8.)
―――: "The Theory of Brillouin Zones and Electronic States in Crystals," Interscience Publishers, Inc., New York, and North Holland Publishing Company, Amsterdam, 1960. (Sec. 10-1.)
―――: See also D. J. Howarth, N. F. Mott.
Jørgensen, C. K.: Relevant and Irrelevant Symmetry Components: Are the Bloch Energy Bands the Best One-electron Functions? *Phys. Stat. Sol.,* **2**:1146 (1962).
―――: "Orbitals in Atoms and Molecules," Academic Press Inc., New York, 1962.
Junod, P.: Zones de Brillouin, liaisons chimiques et mode de conduction de Ag_2S et Ag_2Se, *Helv. Phys. Acta,* **32**:601 (1959).
Kane, E. O.: Energy Band Structure in p-type Germanium and Silicon, *J. Phys. Chem. Solids,* **1**:82 (1957). (Sec. 10-8.)
―――: Band Structure of Indium Antimonide, *J. Phys. Chem. Solids,* **1**:249 (1957). (Sec. 10-8.)
―――: The Semi-empirical Approach to Band Structure, *J. Phys. Chem. Solids,* **8**:38 (1959). (Sec. A5.)
―――: Energy Bands in Impure Semiconductors, *Semiconductor Conf.,* Exeter, 1962, Institute of Physics and The Physical Society, London, 1962, p. 252.
―――: See also R. Braunstein.
Kaplunova, E. I.: Many-electron Theory of the Hole Band in Diamond-type Crystals, *Fiz. Tverd. Tela,* **1**:177 (1959).
Kapuy, E.: Configuration Interaction for Wave Function Constructed from Orthogonal Many-electron Group Orbitals, *Acta Phys. Acad. Sci. Hung.,* **13**:345 (1961).
―――: Derivation of "Almost" Orthogonal Two-electron Orbitals, *Acta Phys. Acad. Sci. Hung.,* **13**:461 (1961).
―――: Derivation of Orthogonal Many-electron Group Orbitals and the Effect of

Small External Perturbation on a System Consisting of Loosely Coupled Electron Groups, *Acta Phys. Acad. Sci. Hung.*, **15**:177 (1962).

―――: On the Correlation Problem in the Theory of Atoms and Molecules, *Acta Phys. Acad. Sci. Hung.*, **15**:341 (1963).

Katsura, S., T. Hatta, and A. Morita: On the Conception of the Energy Band in the Perturbed Periodic Field, *Sci. Rept. Tohoku Univ.*, **34**:19 (1950).

Katz, E.: Splitting of Bands in Solids, *Phys. Rev.*, **85**:495 (1952).

Kerner, E. H.: The Band Structure of Mixed Linear Lattices, *Proc. Phys. Soc. (London)*, **A69**:234 (1956).

Kimball, G. E.: Electronic Structure of Diamond, *J. Chem. Phys.*, **3**:560 (1935). (Secs. 10-4, 10-8.)

Kittel, C., and A. H. Mitchell: Theory of Donor and Acceptor States in Silicon and Germanium, *Phys. Rev.*, **96**:1488 (1954).

―――: "Introduction to Solid-state Physics," 2d ed., John Wiley & Sons, Inc., New York, 1956.

―――― and W. Marshall: On the Number of 3d Electrons in Iron, *J. Phys. Chem. Solids*, **6**:100 (1958).

―――: "Quantum Theory of Solids," John Wiley & Sons, Inc., New York, 1963.

Klein, O.: Quelques remarques sur le traitement approximatif du problème des électrons dans un réseau cristallin par la mécanique quantique, *J. Phys. Radium*, **9**:1 (1938).

Kleinman, L., and J. C. Phillips: Crystal Potential and Energy Bands of Semiconductors, I, Self-consistent Calculations for Diamond; II, Self-consistent Calculations for Cubic Boron Nitride; III, Self-consistent Calculations for Silicon, *Phys. Rev.*, **116**:880 (1959); **117**:460 (1960); **118**:1153 (1960). (Secs. 8-4, 9-4, 10-1, 10-8.)

―――― and ―――: Covalent Binding and Charge Density in Diamond, *Phys. Rev.*, **125**:819 (1962). (Sec. 8-4.)

―――: See also J. C. Phillips.

Klemens, P. G.: Band Structure of Monovalent Metals and Their Alloys, *Australian J. Phys.*, **13**:238 (1960).

Knight, B. W., and G. A. Peterson: Solvable Three-dimensional Lattice Models, *Phys. Rev.*, **132**:1085 (1963).

Knox, R. S., and F. Bassani: Band Structure of Solid Ar, *Phys. Rev.*, **124**:652 (1961). (Sec. 10-8.)

Kobayasi, S.: Energy Band Structures of the Carborundum SiC Crystal, *J. Phys. Soc. Japan*, **11**:175 (1956). (Sec. 10-8.)

―――: Calculation of the Energy Band Structure of the β-SiC Crystal by the Orthogonalized Plane Wave Method, *J. Phys. Soc. Japan*, **13**:261 (1958). (Sec. 10-8.)

Koenig, H. D.: Calculation of Characteristic Values for Periodic Potentials, *Phys. Rev.*, **44**:657 (1933).

Kohler, M.: Elektronentheorie in Metallen beliebiger Kristallform, *Ann. Physik*, **27**:201 (1936).

Kohn, W.: Two Applications of the Variational Method to Quantum Mechanics, *Phys. Rev.*, **71**:635 (1947).

―――: A Variational Iteration Method for Solving Secular Equations, *J. Chem. Phys.*, **17**:670 (1949).

―――: Variational Methods for Periodic Lattices, *Phys. Rev.*, **87**:472 (1952).

―――― and J. Rostoker: Solution of Schrödinger's Equation in Periodic Lattices with Application to Metallic Lithium, *Phys. Rev.*, **94**:1111 (1954). (Secs. 9-5, 10-8.)

―――― and J. Luttinger: Theory of Donor States in Silicon, *Phys. Rev.*, **98**:915 (1955).
―――― and D. Schechter: Theory of Acceptor Levels in Germanium, *Phys. Rev.*, **99**:1903 (1955).
―――― and S. Michaelson: Properties of Wannier Functions, *Proc. Phys. Soc. (London)*, **72**:301 (1958).
――――: Analytic Properties of Bloch Waves and Wannier Functions, *Phys. Rev.*, **115**:809 (1959). (Sec. 7-5.)
――――: Band Structure of Semiconductors, *Semiconductor Phys. Conf.*, Prague, 1960, Academic Press Inc., New York, 1961, p. 15.
――――: Theory of the Insulating State, *Phys. Rev.*, **133**:A171 (1964).
――――: See also J. Callaway.
Korringa, J.: On the Calculation of the Energy of a Bloch Wave in a Metal, *Physica*, **13**:392 (1947). (Sec. 9-3.)
Koster, G. F.: Localized Functions in Molecules and Crystals, *Phys. Rev.*, **89**:67 (1953). (Sec. 7-5.)
―――― and J. C. Slater: Wave Functions for Impurity Levels, *Phys. Rev.*, **94**:1392 (1954); **95**:1167 (1954).
――――: Theory of Scattering in Solids, *Phys. Rev.*, **95**:1436 (1954).
―――― and J. C. Slater: Simplified Impurity Calculation, *Phys. Rev.*, **96**:1208 (1954).
――――: Extension of Hund's Rule, *Phys. Rev.*, **98**:514 (1955).
――――: Density of States Curve for Nickel, *Phys. Rev.*, **98**:901 (1955). (Sec. 10-8.)
――――: Space Groups and Their Representations, *Solid State Phys.*, **5**:173 (1957). Reprinted, Academic Press Inc., New York, 1964. (Secs. 1-3, 1-5, 5-1, A3-5, A9-4.)
――――: Matrix Elements of Symmetric Operators, *Phys. Rev.*, **109**:227 (1958).
――――: Symmetry Properties of Ga Energy Bands: Effect of Spin-Orbit Interaction, *Phys. Rev.*, **127**:2044 (1962).
――――, J. O. Dimmock, R. G. Wheeler, and H. Statz: "Properties of the Thirty-two Point Groups," Massachusetts Institute of Technology Press, Cambridge, Mass., 1963.
――――: See also J. C. Slater.
Koutecky, J.: Contribution to the Theory of the Surface Electronic States in the One-electron Approximation, *Phys. Rev.*, **108**:13 (1957).
――――: Application of the Self-consistent Field Method to the Theory of Surface States of Electrons in a Crystal, *Czech. J. Phys.*, **8**:148 (1958).
――――: On the Theory of Surface States, *J. Phys. Chem. Solids*, **14**:233 (1960).
―――― and M. Tomasek: Contribution to the Theory of the Shockley Surface States, I, General Formulation and the Case of Zero Deformation of the Potential on the Goodwin-Artmann Model, *J. Phys. Chem. Solids*, **14**:241 (1960).
―――― and ――――: Study of the Surface States of Diamond and Graphite by a Simple MO-LCAO Method, *Phys. Rev.*, **120**:1212 (1960).
――――: An Interpretation of the Conditions for the Existence of Shockley Surface States, *Czech. J. Phys.*, **11**:565 (1961).
――――: Die Theorie der Oberflächenzustände und die Quantenchemie der Kristall-oberfläche, *Phys. Stat. Sol.*, **1**:554 (1961).
―――― and M. Tomasek: Shockley Surface States of Diamond within the Framework of MO Method, *Semiconductor Phys. Conf.*, Prague, 1960, Academic Press Inc., New York, 1961, p. 495.
―――― and ――――: Shockley Surface States for a Graphite and Diamond Model with a Delta Function Potential, *Czech. J. Phys.*, **12**:48 (1962).
――――: The Shockley Surface States of Electrons in the Kronig-Penney Model of a

Linear Chain of Atoms Joined by Alternately Strong Bonds, *Czech. J. Phys.*, **12**:177 (1962).
———: Surface States of a Semi-infinite Diamond Crystal Limited by the (100) Plane, *Czech. J. Phys.*, **12**:184 (1962).
———: See also M. Tomasek.
Kovalev, O. V., and G. Ya. Lyubarskii: The Contact of Energy Bands in Crystals, *Zh. Tekh. Fiz.*, **28**:1151 (1958).
———: Degeneracy of Energy Levels in Crystals, *Fiz. Tverd. Tela*, **2**:2557 (1960).
———: Characters of Single-valued Irreducible Representations of Space Groups of a Hexagonal System, I, *Ukr. Fiz. Zh.*, **6**:353 (1961).
———: Characters of Two-valued Irreducible Representations of Space Groups of a Hexagonal System, *Ukr. Fiz. Zh.*, **6**:366 (1961).
Kozima, H.: Electron Wave Functions in Metallic Sodium, *Tech. Rept. Inst. Solid State Physics, Univ. Tokyo*, ser. B, no. 2 (1961).
Kramers, H. A.: Das Eigenwertproblem in eindimensionalen periodischen Kraftfelder, *Physica*, **2**:483 (1935).
Kronig, R. de L., and W. G. Penney: Quantum Mechanics of Electrons in Crystal Lattices, *Proc. Roy. Soc. (London)*, **A130**:499 (1931).
Krutter, H. M.: Energy Bands in Copper, *Phys. Rev.*, **48**:664 (1935). (Sec. 10-8.)
———: See also M. F. Manning.
Kucher, T. I.: Hole Bands in Crystals with NaCl-type Lattice, *Zh. Eksperim. i Teor. Fiz.*, **34**:394 (1958). (Sec. 10-8.)
———: Hole Bands in NaCl, *Zh. Eksperim. i Teor. Fiz.*, **35**:1049 (1958). (Sec. 10-8.)
——— and K. B. Tolpygo: Structure of Hole Bands in Alkali Chlorides, *Fiz. Tverd. Tela*, **2**:2301 (1960). (Sec. 10-8.)
Kudinov, E. K.: Energy Spectrum of Holes in Bi_2Te_3, *Fiz. Tverd. Tela*, **1**:1851 (1959). (Sec. 10-8.)
———: Investigation of the Hole Spectrum of Bi_2Te_3, *Fiz. Tverd. Tela*, **3**:317 (1961). (Sec. 10-8.)
Kuhn, T. S., and J. H. Van Vleck: Simplified Method of Computing the Cohesive Energies of Monovalent Metals, *Phys. Rev.*, **79**:382 (1950). (Sec. 9-5.)
Landauer, R., and J. C. Helland: Electronic Structure of Disordered One-dimensional Chains, *J. Chem. Phys.*, **22**:1655 (1954).
Landé, A.: Über die Grösse der Atome, *Z. Physik*, **1**:191 (1920). (Sec. 4-1.)
Laves, F.: Crystal Structure and Atomic Size, in "Theory of Alloy Phases," American Society for Metals, 1956, p. 124.
Lax, M., and J. C. Phillips: One-dimensional Impurity Bands, *Phys. Rev.*, **110**:41 (1958).
LeBlanc, O. H., Jr.: Band Structure and Transport of Holes and Electrons in Anthracene, *J. Chem. Phys.*, **35**:1275 (1961).
Le Corre, Y.: Les groupes cristallographiques magnétiques et leurs propriétés, *J. Phys. Radium*, **19**:750 (1958).
Lee, P. M., and N. H. March: Electrostatic Fields around Localized Defects in Metals, *Phil. Mag.*, **2**:1226 (1957).
——— and L. Pincherle: The Electronic Band Structure of Bismuth Telluride, *Proc. Phys. Soc. (London)*, **81**:461 (1963). (Sec. 10-8.)
Lehman, G. W.: Effect of Spin-Orbit Coupling on the Energy Levels in the 6d Band for Actinide Metals, *Phys. Rev.*, **116**:846 (1959).
———: Statistical Potential for Actinide Metal Energy Band Calculations, *Phys. Rev.*, **117**:1493 (1960).

Leigh, R. S.: The Augmented Plane Wave and Related Methods for Crystal Eigenvalue Problems, *Proc. Phys. Soc. (London)*, **A69**:388 (1956).

Leman, G.: Influence des conditions aux limites sur la densité électronique dans un réseau périodique, *J. Phys. Chem. Solids*, **13**:221 (1960).

———: Sur la structure électronique des impuretés métalliques dans l'approximation des liaisons fortes, *J. Phys. Chem. Solids*, **20**:50 (1961).

——— and J. Friedel: On the Description of Covalent Bonds in Diamond Lattice Structures by a Simplified Tight-binding Approximation, *J. Appl. Phys.*, **33** (suppl. 1):281 (1962).

———: Étude des structures covalentes du type cubique diamant par une méthode de liaisons fortes simplifiée, *Ann. Phys.*, **7**:505 (1962).

———: See also G. Gousseland.

Lennard-Jones, J. E., and H. J. Woods: Distribution of Electrons in a Metal, *Proc. Roy. Soc. (London)*, **A120**:727 (1928).

Libby, W. F.: Theory of Metallic Diamond, *Phys. Rev.*, **130**:548 (1963).

Lifshitz, I. M., and M. I. Kaganov: Some Problems of the Electron Theory of Metals, I, Classical and Quantum Theory of Electrons in Metals, *Usp. Fiz. Nauk*, **69**:419 (1959).

———: Structure of the Energy Spectrum of Impurity Bands in Disordered Solid Solutions, *Zh. Eksperim. i Teor. Fiz.*, **44**:1723 (1963).

Liu, L.: Effects of Spin-Orbit Coupling in Si and Ge, *Phys. Rev.*, **126**:1317 (1962).

———: See also F. Bassani, J. C. Phillips.

Loeb, A. L.: A Modular Algebra for the Description of Crystal Structures, *Acta Cryst.*, **15**:219 (1962).

Lomer, W. M.: The Valence Bands in Two-dimensional Graphite, *Proc. Roy. Soc. (London)*, **A227**:330 (1955). (Sec. 10-8.)

——— and W. Marshall: The Electronic Structure of the Metals of the First Transition Period, *Phil. Mag.*, **3**:185 (1958).

———: Electronic Structure of Chromium Group Metals, *Proc. Phys. Soc. (London)*, **80**:489 (1962). (Sec. 10-8.)

———: Transition Metal Alloys, *J. Phys. Radium*, **23**:716 (1962).

Lomont, J. S.: "Applications of Finite Groups," Academic Press Inc., New York, 1959.

Long, D.: Energy Bands in Semiconductors, *J. Appl. Phys.*, **33**:1682 (1962).

Loucks, T. L., and P. H. Cutler: Band Structure and Fermi Surface of Beryllium, *Phys. Rev.*, **133**:A819 (1964). (Secs. 10-3, 10-8.)

Löwdin, P.-O.: A Quantum Mechanical Calculation of the Cohesive Energy: The Interatomic Distance and the Elastic Constants of Some Ionic Crystals, I, II, *Arkiv Mat. Astron. Fis.*, vol. A35, no. 9 (1947); no. 30 (1948). (Sec. 4-2.)

———: Quantum Theory of Cohesive Properties of Solids, *Phil. Mag.*, suppl., **5**:1(1956); *Advan. Phys.*, **5**:1 (1956). (Sec. 4-2.)

———: Band Theory, Valence Bond, and Tight-binding Calculations, *J. Appl. Phys.*, **33** (suppl. 1):251 (1962).

Lukesh, J. S.: On the Symmetry of Graphite, *Phys. Rev.*, **80**:226 (1950).

Lundquist, N.: On the Electronic Structure of MnB, *Arkiv Fis.*, **23**:65 (1962).

McClure, J. W.: Band Structure of Graphite and de Haas–van Alphen Effect, *Phys. Rev.*, **108**:612 (1957). (Sec. 10-8.)

———: See also D. E. Soule.

MacColl, L. A.: On the Reflection of Electrons by Metallic Crystals, *Bell System Tech. J.*, **30**:888 (1951).

McIntosh, H. V.: Towards a Theory of the Crystallographic Point Groups, *J. Mol. Spectry.*, **5**:269 (1960).

———: Symmetry-adapted Functions Belonging to the Symmetric Groups, *J. Math. Phys.*, **1**:453 (1960).
———: On Matrices Which Anticommute with a Hamiltonian, *J. Mol. Spectry.*, **8**:169 (1962).
———: Symmetry-adapted Functions Belonging to the Crystallographic Groups, *J. Mol. Spectry.*, **10**:51 (1963).
Makinson, R. E. B., and A. P. Roberts: Localized Electron Eigenstates in One-dimensional Liquids, *Proc. Phys. Soc. (London)*, **79**:222 (1962).
———: See also A. P. Roberts.
Manning, M. F., and H. M. Krutter: Electronic Energy Bands in Metallic Calcium, *Phys. Rev.*, **51**:761 (1937). (Secs. 10-3, 10-8.)
——— and M. I. Chodorow: Electronic Energy Bands in Metallic Tungsten, *Phys. Rev.*, **56**:787 (1939). (Sec. 10-8.)
———: Electronic Energy Bands in Body-centered Iron, *Phys. Rev.*, **63**:190 (1943). (Sec. 10-8.)
———: See also M. I. Chodorow, J. B. Greene.
March, N. H., Momentum Distribution of Electrons in Solids: Results for Some Metals Using the Thomas-Fermi Method, *Proc. Phys. Soc. (London)*, **A67**:9 (1954).
———: On Metallic Hydrogen, *Physica*, **22**:311 (1956).
——— and A. M. Murray: Electronic Wave Functions round a Vacancy in a Metal, *Proc. Roy. Soc. (London)*, **A256**:400 (1960).
——— and ———: Self-consistent Perturbation Treatment of Impurities and Imperfections in Metals, *Proc. Roy. Soc. (London)*, **A261**:119 (1961).
——— and ———: Self-consistent Perturbation Treatment of Impurities and Imperfections in Metals, II, Second-order Perturbation Corrections, *Proc. Roy. Soc. (London)*, **A266**:559 (1962).
———: See also B. Donovan, M. Flower, P. M. Lee.
Mariot, L.: Groupes finis de symétrie et recherche de solutions de l'équation de Schrödinger, *J. Phys. Radium*, **18**:345 (1957).
———: "Group Theory and Solid State Physics," transl. by A. Nussbaum, Prentice-Hall, Inc., Englewood Cliffs, N.J., 1962.
Marshall, W., and R. J. Weiss: Electronic Structure of Transition Metals, *J. Appl. Phys.*, **30** (suppl.):220S (1959).
Mase, S.: Electronic Structure of Bismuth Type Crystals, *J. Phys. Soc. Japan*, **13**:434 (1958). (Sec. 10-8.)
———: Electronic Structure and Diamagnetism of Graphite, *J. Phys. Soc. Japan*, **13**:563 (1958). (Sec. 10-8.)
———: Electronic Structure of Bismuth Type Crystals, II, *J. Phys. Soc. Japan*, **14**:584 (1959). (Sec. 10-8.)
———: See also K. Ariyama.
Mattheiss, L. F.: Crystal Potentials in the Free-electron-exchange Approximation, *Bull. Am. Phys. Soc.*, ser. II, **8**:222 (1963). (Secs. 10-5, 10-8.)
———: Energy Bands for Solid Argon, *Phys. Rev.*, **133**:A1399 (1964). (Secs. 10-5, 10-8.)
Mattuck, R. D.: Effect of Local Order on Energy Bands in Binary Alloys, *Phys. Rev.*, **127**:738 (1962).
Matyas, Z.: The Energies of Electrons in Aluminum, *Phil. Mag.*, **39**:429 (1948). (Sec. 10-8.)
———: A New Method for Calculating the Energy Levels in Solids, *Czech. J. Phys.*, **1**:3 (1952).

Maue, A. W.: Die Oberflächenwelle in der Elektronentheorie der Metalle, *Helv. Phys. Acta*, **7** (suppl. 2):68 (1934).
―――: Die Oberflächenwellen in der Elektronentheorie der Metalle, *Z. Physik*, **94**:717 (1935).
Meijer, P. H. E., and E. Bauer: "Group Theory: The Application to Quantum Mechanics," North Holland Publishing Company, Amsterdam, 1962.
Men, A. N., and A. V. Sokolov: Applications of the Theory of Representations to Orderable Binary Systems, *Fiz. Tverd. Tela*, **5**:78 (1963).
Miasek, M.: Determination of the Valence Band in Metallic Sodium by the Parzen Variational Method, *Bull. Acad. Polon. Sci.*, Cl. III, **4**:453 (1956). (Sec. 10-8.)
―――: Tight-binding Method for Hexagonal Close-packed Structure, *Phys. Rev.*, **107**:92 (1957).
―――: The Application of the Tight Binding Method to the Investigation of Energy Bands in Hexagonal Close-packed Structure, I, II, III, *Acta Phys. Polon.*, **16**:343, 447 (1957); **17**:371 (1958).
―――: Tight Binding Method for White Tin, *Bull. Acad. Polon. Sci. Ser. Sci. Math. Astron. Phys.*, **8**:9 (1960). (Sec. 10-8.)
――― and M. Suffczynski: Space Group of White Tin, I, Symmetry Points; II, Symmetry Lines and Planes; III, Double Group, *Bull. Acad. Polon. Sci. Ser. Sci. Math. Astron. Phys.*, **9**:477, 483, 489 (1961).
―――: Empty-lattice Analysis for White Tin, *Bull. Acad. Polon. Sci. Ser. Sci. Math. Astron. Phys.*, **10**:39 (1962).
―――: Band Structure of White Tin, *Phys. Rev.*, **130**:11 (1963). (Sec. 10-8.)
Millman, J.: Electronic Energy Bands in Metallic Lithium, *Phys. Rev.*, **47**:286 (1935). (Sec. 10-8.)
Mooser, E., and W. B. Pearson: New Semiconducting Compounds, *Phys. Rev.*, **101**:492 (1956).
――― and ―――: Chemical Bond in Semiconductors, *Phys. Rev.*, **101**:1608 (1956).
――― and ―――: The Chemical Bond in Semiconductors: The Group VB to VIIB Elements and Compounds Formed between Them, *Can. J. Phys.*, **34**:1369 (1956).
――― and ―――: The Chemical Bond in Semiconductors, *J. Electronics*, **1**:629 (1956); **2**:406 (1957).
――― and ―――: Recognition and Classification of Semiconducting Compounds with Tetrahedral sp³ Bonds, *J. Chem. Phys.*, **26**:893 (1957).
――― and ―――: The Crystal Structure and Properties of the Group VB to VIIB Elements and Compounds Formed between Them, *J. Phys. Chem. Solids*, **7**:65 (1958).
――― and ―――: On the Crystal Chemistry of Normal Valence Compounds, *Acta Cryst.*, **12**:1015 (1959).
――― and ―――: The Chemical Bond in Semiconductors, *Progr. Semiconductors*, **5**:103 (1960).
――― and ―――: The Ionic Character of Chemical Bonds, *Nature*, **190**:406 (1961).
―――: See also C. H. L. Goodman.
Morita, A.: Electronic Structure of Diamond-lattice Type Crystals, *Sci. Rept. Tohoku Univ.*, **33**:92 (1949). (Sec. 10-8.)
―――: The Electronic Structure of Bismuth Crystal, *Sci. Rept. Tohoku Univ.*, **33**:144 (1949). (Sec. 10-8.)
――― and C. Horie: Electronic Structure of BaO Crystal, *Sci. Rept. Tohoku Univ.*, **36**:259 (1952). (Sec. 10-8.)
――― and K. Takahashi: Theory of Cohesive Energy of LiH Crystal, *Progr. Theoret. Phys.* (Kyoto), **19**:257 (1958).

———: Theory of Cohesive Energies and Energy-band Structures of Diamond-type Valence Crystals: The Method of SLCO, II, *Progr. Theoret. Phys.* (Kyoto), **19**:534 (1958). (Sec. 10-8.)
———: Cohesive Energies and Band Structures of Covalent and Ionic Crystals, *J. Phys. Chem. Solids*, **8**:363 (1959).
———, M. Azuma, and H. Nara: Theory of Impurity Levels, *J. Phys. Soc. Japan*, **17**:1570 (1962).
———: See also S. Katsura.
Morita, T.: Existence of Energy Gaps in Disordered Systems, *Rept. Liberal Arts Sci. Fac. Shizouka Univ.*, **3**:149 (1963).
Morrison, J. A.: On the Number of Electron Levels in a One-dimensional Random Lattice, *J. Math. Phys.*, **3**:1023 (1962).
Morse, P. M.: Quantum Mechanics of Electrons in Crystals, *Phys. Rev.*, **35**:1310 (1930).
———: Quantum Mechanics of Collision Processes, II, *Rev. Mod. Phys.*, **4**:577 (1932). (Sec. 9-3.)
———: Waves in a Lattice of Spherical Scatterers, *Proc. Natl. Acad. Sci.*, **42**:276 (1956). (Sec. 9-3.)
———: See also W. P. Allis.
Mott, N. F.: A Discussion of the Transition Metals on the Basis of Quantum Mechanics, *Proc. Phys. Soc.*, **47**:571 (1935).
——— and H. Jones: "The Theory of the Properties of Metals and Alloys," Oxford University Press, Fairlawn, N.J., 1936. (Sec. 10-1.)
———: Energy Levels in Real and Ideal Crystals, *Trans. Faraday Soc.*, **34**:822 (1938).
———: Recents progrés et difficultés de la théorie électronique des métaux, *Le Magnetisme*, **2**:1 (1940).
———: The Basis of the Electron Theory of Metals, with Special Reference to the Transition Metals, *Proc. Phys. Soc.*, **62**:416 (1949).
———: Recent Advances in the Electron Theory of Metals, *Progr. Metal Phys.*, **3**:76 (1952).
———: Note on the Electronic Structure of the Transition Metals, *Phil. Mag.*, **44**:187 (1953).
———: On the Transition to Metallic Conduction in Semiconductors, *Can. J. Phys.*, **34**:1356 (1956).
———: Presidential Address, The Physics and Chemistry of Metals, *Yearbook Phys. Soc.*, 1957, p. 1.
———: On the Transition to Metallic Conduction in Semiconductors, *Semiconductor Meeting Rept.* 5, 1957.
——— and K. W. H. Stevens: The Band Structure of the Transition Metals, *Phil. Mag.*, **2**:1364 (1957).
———: The Transition from the Metallic to the Non-metallic State, *Nuovo Cimento Suppl.*, **7**:312 (1958).
———: The Transition to the Metallic State, *Phil. Mag.*, **6**:287 (1961).
———: See also H. Jones.
Mullaney, J. F.: Optical Properties and Electronic Structure of Solid Silicon, *Phys. Rev.*, **66**:326 (1944). (Sec. 10-8.)
Müller, K.: Zur rationellen Bestimmung der irreduziblen Bestandteile einer vorgegebenen Tensordarstellung in den 32 Kristallklassen, *Abhandl. Braunschweig. Wiss. Ges.*, **9**:116 (1957).
———: Zur Konstruktion der Charaktere eigensymmetrischer Tensorverknüpfungen, *Abhandl. Braunschweig. Wiss. Ges.*, **9**:127 (1957).

Murnaghan, F. D.: "The Theory of Group Representations," The Johns Hopkins Press, Baltimore, Md., 1938.

Muto: T.: On the Electronic Structure of Alloys, *Sci. Papers Inst. Phys. Chem Res., Komagome*, **34**:377 (1938).

Niessen, K. F.: Über den Atomabstand in Kristallen tetraedrischer Struktur, *Physik. Z.*, **31**:610 (1930).

Nordheim, L.: Zur Elektronentheorie der Metalle, I, II, *Ann. Phys.*, **9**:607, 641 (1931).

Nosanow, L. H., and G. L. Shaw: Hartree Calculations for Ground State of Solid He and Other Noble Gas Crystals, *Phys. Rev.*, **128**:546 (1962).

Nosawa, R.: Electronic Eigenfunctions in Ionic Crystals, *Phys. Rev.*, **75**:1102 (1949).

Nowacki, W.: Symmetrie und physikalisch-chemische Eigenschaften kristallisierter Verbindungen, I, Die Verteilung der Kristallstrukturen über die 219 Raumgruppen; II, Die allgemeinen Bauprinzipien organischer Verbindungen, *Helv. Chim. Acta*, **23**:863 (1942); **26**:459 (1943); III, Zur Kristallchemie organischer Verbindungen, *Mitt. Naturforsch. Ges. Bern*, **2**:43 (1944).

Nranyan, A. A.: The Calculation of Integrals in the Method of Equivalent Orbits and the Estimation of the Parameters of the Valency Zones in Semiconductors of the Type $A^{III} B^{V}$, *Fiz. Tverd. Tela*, **2**:474 (1960).

———: On the Band Structure of a Crystal of the Diamond Type, *Fiz. Tverd. Tela*, **2**:1650 (1960). (Sec. 10-8.)

———: See also A. I. Gubanov.

Nussbaum, A.: Group Theory and the Energy Band Structure of Semiconductors, *Proc. Inst. Radio Engrs.*, **50**:1762 (1962).

Nutkins, M. A. E.: The Density of Electronic States in Body-centered Cubic Crystals, *Proc. Phys. Soc.*, **B69**:619 (1956).

Okada, S.: Energy Eigenvalues of an Electron in a One-dimensional Periodically Dislocated Lattice with a Defect, *Sci. Rept. Tohoku Univ.*, **39**:194 (1956).

———: An Eigenvalue Problem of an Electron in a One-dimensional Crystal Lattice with a Defect, *Sci. Rept. Tohoku Univ.*, **41**:1 (1957).

———: Symmetrical Property of Eigenfunctions of an Electron in a One-dimensional Periodic Potential Field, *Sci. Rept. Tohoku Univ.*, **42**:130 (1958).

Olbrychski, K.: Representations of the Space Group of Rutile, (TiO_2), *Bull. Acad. Polon. Sci. Ser. Sci. Math. Astron. Phys.*, **9**:537 (1961).

Opechowski, W.: Sur les groupes cristallographiques "doubles," *Physica*, **7**:552 (1940). (Sec. A9-4.)

O'Sullivan, W.: Study of the Wurtzite-type Binary Compounds, III, The Valence Band Structure of BeO, *J. Chem. Phys.*, **30**:379 (1959). (Sec. 10-8.)

Parmenter, R. H.: Electronic Energy Bands in Crystals, *Phys. Rev.*, **86**:552 (1952). (Secs. 8-1, 10-8.)

———: Energy Levels of a Disordered Alloy, *Phys. Rev.*, **97**:587 (1955); **104**:22 (1956).

———: Energy Levels of a Crystal Modified by Alloying or by Pressure, *Phys. Rev.*, **99**:1759 (1955).

———: Symmetry Properties of the Energy Bands in the Zinc Blende Structure, *Phys. Rev.*, **100**:573 (1955). (Sec. A3-2.)

Parzen, G.: Electronic Energy Bands in Metals, *Phys. Rev.*, **89**:237 (1953).

Patterson, A. L.: Über das Gibbs-Ewaldsche reziproke Gitter und den dazugehörigen Raum, *Z. Physik*, **44**:596 (1927).

Pauling, L.: The Sizes of Ions and the Structure of Ionic Crystals, *J. Am. Chem. Soc.*, **49**:765 (1927).

———: The Theoretical Prediction of the Physical Properties of Many-electron Atoms and Ions: Mole Refraction, Diamagnetic Susceptibility, and Extension in Space, *Proc. Roy. Soc. (London)*, **A114**:181 (1927). (Sec. 4-1.)

———: The Sizes of Ions and Their Influence on the Properties of Salt-like Compounds, *Z. Krist.*, **67**:377 (1928).

———: Influence of Relative Ionic Sizes on the Properties of Ionic Compounds, *J. Am. Chem. Soc.*, **50**:1036 (1928).

———: The Principles Determining the Structure of Complex Ionic Crystals, *J. Am. Chem. Soc.*, **51**:1010 (1929).

——— and J. Sherman: Screening Constants for Many-electron Atoms: The Calculation and Interpretation of X-ray Term Values, and the Calculation of Atomic Scattering Factors, *Z. Krist.*, **81**:1 (1932). (Sec. 4-1.)

——— and M. L. Huggins: Covalent Radii of Atoms and Interatomic Distances in Crystals Containing Electron-pair Bonds, *Z. Krist.*, **87**:205 (1934). (Sec. 4-1.)

———: Nature of the Interatomic Forces in Metals, *Phys. Rev.*, **54**:899 (1938).

———: Atomic Radii and Interatomic Distances in Metals, *J. Am. Chem. Soc.*, **69**:542 (1947).

——— and F. J. Ewing: The Ratio of Valence Electrons to Atoms in Metals and Intermetallic Compounds, *Rev. Mod. Phys.*, **20**:112 (1948).

———: The Metallic State, *Nature*, **161**:1019 (1948).

———: The Resonating Valence-bond Theory of Metals, *Physica*, **15**:23 (1949).

———: La valence des métaux et la structure des composés intermétalliques, *J. Chim. Phys.*, **46**:276 (1949).

———: A Resonating-valence-bond Theory of Metals and Intermetallic Compounds, *Proc. Roy. Soc. (London)*, **A196**:343 (1949).

———: Electron Transfer in Intermetallic Compounds, *Proc. Natl. Acad. Sci.*, **36**:533 (1950).

———: The Electronic Structure of Metals and Alloys, in "Theory of Alloy Phases," American Society for Metals, 1956, p. 220.

———: The Use of Atomic Radii in the Discussion of Interatomic Distances and Lattice Constants of Crystals, *Acta Cryst.*, **10**:685 (1957).

———: "The Nature of the Chemical Bond," 3d ed., Cornell University Press, Ithaca, N.Y., 1960. (Sec. 4-1.)

———: Nature of the Metallic Orbital, *Nature*, **189**:656 (1961).

———: See also B. Podolsky, J. Waser.

Peacock, T. E., and R. McWeeny: A Self-consistent Calculation of the Graphite π Band, *Proc. Phys. Soc., (London)*, **74**:385 (1959). (Sec. 10-8.)

———: The π Electron Properties of Graphite, *J. Chim. Phys.*, **57**:844 (1960).

———: The Effective Number of Electrons in the π Band of Graphite, *Proc. Phys. Soc. (London)*, **77**:1214 (1961).

———: Further Considerations of the π Electron Properties of Graphite, *Proc. Fifth Conf. on Carbon*, Pergamon Press, New York, 1962, p. 8.

Peierls, R.: Elektronentheorie der Metalle, *Ergeb. Exakt. Naturw.*, **11**:264 (1932).

———: "Quantum Theory of Solids," Oxford University Press, Fair Lawn, N.J., 1955.

Pekar, S.: Local Quantum States of an Electron in an Ideal Ionic Crystal, *J. Phys. USSR*, **10**:341 (1946).

———: "Untersuchungen über die Elektronentheorie der Kristalle," Akademie-Verlag GmbH, Berlin, 1954.

Phariseau, P.: Surface States in a One-dimensional Perfect Semi-infinite Crystal, *Physica*, **26**:737 (1960).

———: The Energy Spectrum of an Amorphous Substance, *Physica*, **26**:1185 (1960).
———: Subsurface States in One-dimensional Crystals, *Physica*, **26**:1192 (1960).
———: The Band Structure of Amorphous Solids, *Physica*, **27**:351 (1961).
———: Surface States in a Finite Crystal, *Physica*, **27**:590 (1961).
Phillips, F. C.: "An Introduction to Crystallography," Longmans Green & Co., Inc., New York, 1946.
Phillips, J. C.: Energy-band Interpolation Scheme Based on a Pseudopotential, *Phys. Rev.*, **112**:685 (1958). (Sec. 9-4.)
——— and H. B. Rosenstock: Topological Methods of Locating Critical Points, *J. Phys. Chem. Solids*, **5**:288 (1958).
——— and L. Kleinman: New Method for Calculating Wave Functions in Crystals and Molecules, *Phys. Rev.*, **116**:287 (1959). (Secs. 8-4, 9-4, 10-1.)
———: Energy Bands of Silicon and Germanium, *J. Phys. Chem. Solids*, **8**:369, 379 (1959). (Secs. 10-1, 10-8.)
———: Band Structure of Silicon, Germanium, and Related Semiconductors, *Phys. Rev.*, **125**:1931 (1962). (Sec. 10-8.)
——— and L. Kleinman: Crystal Potential and Energy Bands of Semiconductors, IV, Exchange and Correlation, *Phys. Rev.*, **128**:2098 (1962). (Sec. 9-4.)
——— and L. Liu: Bonding and Antibonding Spin-Orbit Splittings, *Phys. Rev. Letters*, **8**:94 (1962).
——— and M. H. Cohen: A New Approach to Ionic Character in Solids, *J. Appl. Phys.*, **33** (suppl. 1):293 (1962).
——— and L. F. Mattheiss: Many-electron Effects and Exchange Splittings in Nickel, *Phys. Rev. Letters*, **11**:556 (1963).
———: Fermi Surface of Ferromagnetic Nickel, *Phys. Rev.*, **133**:A1020 (1964).
———: See also M. H. Cohen, L. Kleinman, M. Lax.
Pincherle, L., and J. M. Radcliffe: Semiconducting Intermetallic Compounds, *Advan. Phys.*, **5**:271 (1956).
———: A Note on the Cellular Method for Energy Bands in Solids, *Proc. Phys. Soc. (London)*, **72**:281 (1958).
———: Band Structure Calculations in Solids, *Rept. Progr. Phys.*, **23**:355 (1960).
——— and P. M. Lee: The Electronic Band Structure of a Model of a Cubic Crystal, *Proc. Phys. Soc.*, **78**:1195 (1961).
———: The Effect of the Potential on the Band Structure of a Model of a Simple Cubic Lattice, *Semiconductor Phys. Conf.*, Prague, 1960, Academic Press Inc., 1961, p. 37.
——— and P. M. Lee: On the Momentum Eigenfunctions for the Periodic Problem, *Proc. Phys. Soc.*, **80**:305 (1962).
———: Recent Progress in Band Theory of Semiconductors, *Semiconductor Conf.*, Exeter, 1962, Institute of Physics and The Physical Society, London, 1962, p. 541.
———: See also D. G. Bell, P. M. Lee.
Podolsky, B., and L. Pauling: Momentum Distribution in Hydrogen-like Atoms, *Phys. Rev.*, **34**:109 (1929). (Sec. 8-3.)
Prokofjew, W.: Berechnung der Zahlen der Dispersionszentren des Natriums, *Z. Physik*, **58**:255 (1929). (Sec. 9-1.)
Pryce, M. H. L.: Electronic Structure of Point Defects in Crystals, *J. Appl. Phys.*, **33** (suppl. 1), 390 (1962).
Pugh, D., C. Rhys-Roberts, and R. H. Tredgold: Unstable Solutions and Surface States, *Proc. Phys. Soc. (London)*, **80**:534 (1962).
Quelle, F. W., Jr.: Energy Bands in Semiconductors, *Semiconductor Phys. Conf.*, Prague, 1960, Academic Press Inc., 1961, p. 48.

Raghavacharyulu, I. V. V.: Representations of Space Groups, *Can. J. Phys.*, **39**:830 (1961).
—— and C. B. Shrestha: Irreducible Representations of Plane Groups, *J. Mol. Spectry.*, **7**:341 (1961).
—— and I. Bhavanacharyulu: Simplifications in Finding the Allowable Representations of Space Groups, *Can. J. Phys.*, **40**:1490 (1962).
Raimes, S.: Energy Band Shapes and Band Widths in Metals, *Phil. Mag.*, **45**:727 (1954).
——: "The Wave Mechanics of Electrons in Metals," Interscience Publishers, Inc., New York, and North Holland Publishing Company, Amsterdam, 1961.
——: The Rigid-band Model, *J. Phys. Radium*, **23**:639 (1962).
Ramsauer, C.: Über den Wirkungsquerschnitt der Gasmoleküle gegenüber langsame Elektronen, *Ann. Physik*, **64**:513 (1921); **66**:545 (1921); **72**:345 (1923). (Sec. 9-3.)
Rashba, E. I.: The Energy-band Symmetry in Wurtzite-type Crystals, I, Band Symmetry without Taking into Account the Spin-Orbit Interaction, *Fiz. Tverd. Tela*, **1**:407 (1959).
—— and V. I. Sheka: The Energy Band Symmetry in Wurtzite-type Crystals, II, Band Symmetry with Allowance for Spin Interactions, *Fiz. Tverd. Tela*, suppl. II, p. 162 (1959).
Raychaudhuri, A.: Electronic Energy Bands in Model Three-Dimensional Lattices, *Z. Physik*, **148**:435 (1957).
——: The Isoelectronic Series of Semiconducting Compounds with the Zinc-blende Structure, *Proc. Natl. Inst. Sci. India*, **25**:201 (1959).
Raynor, G. V.: "Introduction to Electron Theory of Metals," The Institute of Metals, London, 1943.
——: The Band Structure of Metals, *Rept. Progr. Phys.*, **15**:173 (1952).
——: See also R. L. Berry.
Redei, L. B.: The Electronic Properties of Tetrahedral Intermetallic Compounds, II, Conduction Band in Diamond, *Proc. Roy. Soc. (London)*, **A270**:373 (1962); III, A Many-electron Model for Calculating the Electronic Excited States in IV-IV Compounds and III-V Compounds with Zinc-blende Structure, **A270**:383 (1962). (Secs. 10-4, 10-8.)
——: See also C. A. Coulson.
Reitz, J. R.: Methods of the One-electron Theory of Solids, *Solid State Phys.*, **1**:1 (1955).
——: Electronic Band Structure of Selenium and Tellurium, *Phys. Rev.*, **105**:1233 (1957). (Sec. 10-8.)
——: See also V. E. Wood.
Ridley, E. C.: A Qualitative Investigation of the Wave Functions of Metallic Uranium, *Proc. Roy. Soc. (London)*, **A247**:199 (1958). (Sec. 10-8.)
Roaf, D. J.: The Fermi Surfaces of Copper, Silver, and Gold, II, Calculation of the Fermi Surfaces, *Phil. Trans. Roy. Soc. (London)*, **A255**:135 (1962).
Roberts, A. P., and R. E. B. Makinson: Electron Eigenstates in a One-dimensional Liquid, *Proc. Phys. Soc. (London)*, **79**:630 (1962).
——: See also R. E. B. Makinson.
Robinson, J. E., F. Bassani, R. S. Knox, and J. R. Schrieffer: Screening Correction to the Slater Exchange Potential, *Phys. Rev. Letters*, **9**:215 (1962).
Rocher, Y. A., and J. Friedel: Sur la structure électronique des métaux de transition liquides et de certaines phases complexes, *J. Phys. Chem. Solids*, **21**:287 (1961).
——: L'étude de la structure électronique des métaux de terres rares, *Advan. Phys.*, **11**:233 (1962).

———: Emploi des niveaux liés virtuels dans les métaux de terres rares, *J. Phys. Chem. Solids*, **23**:1621 (1962).
Sachs, M.: "Solid State Theory," McGraw-Hill Book Company, New York, 1963.
Saffren, M. M., and J. C. Slater: An Augmented Plane Wave Method for the Periodic Potential Problem, II, *Phys. Rev.*, **92**:1126 (1953). (Sec. 9-2.)
Sah, P., and R. Eisenschitz: Electrons in Weak Interaction with a Chain of Atoms, *Proc. Phys. Soc. (London)*, **75**:700 (1960).
———: Application of the Augmented Plane Wave Method to a Liquid Model, *Proc. Phys. Soc. (London)*, **79**:1082 (1962).
Saint-James, D.: Calcul des bandes d'énergie d'un solide avec interaction spin-orbite, *Compt. Rend.*, **246**:1533 (1958).
Sasvari, K.: On the Construction of the Lattice of Ionic Crystals from the Viewpoint of Close Packing, *Acta Phys. Acad. Sci. Hung.*, **8**:245 (1957).
———: On the Determination of Ionic Radii, *Acta Phys. Acad. Sci. Hung.*, **11**:333 (1960).
———: On the Construction of the Lattice of Ionic Crystals from the Viewpoint of Close Packing, II, *Acta Phys. Acad. Sci. Hung.*, **11**:353 (1960).
Sato, H.: On the Diamagnetism of Graphite, I, Energy Levels of π-electrons, *J. Phys. Soc. Japan*, **14**:609 (1959).
——— and R. S. Toth: Effect of Additional Elements on Period of CuAu II and Origin of Long-range Superlattice, *Phys. Rev.*, **124**:1833 (1961).
——— and ———: Long-period Superlattices in Alloys, II, *Phys. Rev.*, **127**:469 (1962).
——— and ———: Fermi Surface of Alloys, *Phys. Rev. Letters*, **8**:239 (1962).
——— and ———: On the Long Period Superlattice in Alloys, *J. Phys. Soc. Japan*, **17** (suppl. B-II):262 (1962).
Sato, M.: On the Energy States of Valency Electrons in Some Metals, 1 . . . 11, *Sci. Rept. Tohoku Imperial Univ., Honda Anniv. Vol.* 136 (1936); **25**:197, 771, 829, 871 (1937); **26**:206, 341, 377 (1937); **27**:137, 278 (1938); **28**:143, 398 (1940); **29**:87 (1940).
Saxon, D. S., and R. A. Hutner: Some Electronic Properties of a One-dimensional Crystal Model, *Philips Res. Rept.*, **4**:81 (1949).
Scarf, F. L.: New Soluble Energy Band Problem, *Phys. Rev.*, **112**:1137 (1958).
Schiff, B.: A Calculation of the Eigenvalues of Electronic States in Metallic Lithium by the Cellular Method, *Proc. Phys. Soc. (London)*, **A67**:2 (1954). (Sec. 10-8.)
———: The Cellular Method for a Close-packed Hexagonal Lattice, with Application to Titanium, *Proc. Phys. Soc. (London)*, **A68**:686 (1955). (Sec. 10-8.)
———: The Cellular Method for Close-packed Hexagonal Titanium, *Proc. Phys. Soc. (London)*, **A69**:185 (1956). (Sec. 10-8.)
Schlosser, H. C.: Symmetrized Combinations of Plane Waves and Matrix Elements of the Hamiltonian for Cubic Lattices, *J. Phys. Chem. Solids*, **23**:963 (1962).
——— and P. M. Marcus: Composite Wave Variational Method for Solution of the Energy-band Problem in Solids, *Phys. Rev.*, **131**:2529 (1963). (Sec. 7-7.)
Schmid, L. A.: Calculating of the Cohesive Energy of Diamond, *Phys. Rev.*, **92**:1373 (1953). (Secs. 10-4, A8.)
———: Valence Bond Calculations, *Am. J. Phys.*, **22**:255 (1954).
Schmidt, H.: Disordered One-dimensional Crystals, *Phys. Rev.*, **105**:425 (1957).
Schön, M.: Modèle des bandes dans les sulfures phosphorescents cristallins, *J. Phys. Radium*, **17**:689 (1956).
Schönhofer, A.: Zum Wannierschen Theorem, *Z. Physik*, **168**:560 (1962).
Schubin, S., and S. Wonsowsky: On the Electron Theory of Metals, *Proc. Roy. Soc. (London)*, **A145**:159 (1934).

———: Zur Elektronentheorie der Metalle, I, II, *Physik. Z. Sowjetunion*, **7**:292 (1935); **10**:348 (1936).
Schwed, P., and G. Allen: On A. Raychaudhuri's Paper "Electronic Energy Bands in Model Three-dimensional Lattices," *Z. Physik*, **158**:623 (1960).
Sclar, N.: Energy Gaps of the II-V and (rare earth)-V Semiconductors, *J. Appl. Phys.*, **33**:2999 (1962).
Seeger, A.: The Electronic Structure of Point Defects, *J. Phys. radium*, **23**:616 (1962).
Segall, B.: Calculation of the Band Structure of "Complex" Crystals, *Phys. Rev.*, **105**:108 (1957).
———: Band Structure Calculations of Semiconductors by the Kohn-Rostocker-Korringa Method: Application to Germanium, *J. Phys. Chem. Solids*, **8**:371, 379 (1959). (Sec. 10-4, 10-8.)
———: Energy Bands of Al, *Phys. Rev.*, **124**:1797 (1961). (Secs. 10-3, 10-8.)
———: Calculated Shape of the Fermi Surface of Copper, *Phys. Rev. Letters*, **7**:154 (1961). (Secs. 9-5, 10-8.)
———: Fermi Surface and Energy Bands of Cu, *Phys. Rev.*, **125**:109 (1962). (Secs. 9-5, 10-5, 10-8.)
———: Fermi Surface of Aluminum, *Phys. Rev.*, **131**:121 (1963).
———: See also F. S. Ham.
Seitz, F.: A Matrix-algebraic Development of the Crystallographic Groups, I, II, III, IV, *Z. Krist.*, **88**:433 (1934); **90**:289 (1935); **91**:336 (1935); **94**:100 (1936). (Secs. 1-3, 5-1.)
———: Theoretical Constitution of Metallic Lithium, *Phys. Rev.*, **47**:400 (1935). (Secs. 10-8, A5.)
———: On the Reduction of Space Groups, *Ann. Math.*, **37**:17 (1936).
——— and R. P. Johnson: Modern Theory of Solids, I, II, III, *J. Appl. Phys.*, **8**:84, 186, 246 (1937).
———: "The Modern Theory of Solids," McGraw-Hill Book Company, New York, 1940.
———: "The Physics of Metals," McGraw-Hill Book Company, New York, 1943.
———: The Basic Principles of Semiconductors, *J. Appl. Phys.*, **16**:553 (1945).
———: The Contribution of Modern Physics to Metallurgy, *J. Appl. Phys.*, **19**:973 (1948).
———: See also D. H. Ewing, H. B. Huntington, E. Wigner.
Sergiescu, V.: A Statistical Justification of the Cyclic Condition in the Zone Theory of Crystals, *Studii Cercetari Fiz.*, **9**:459 (1958).
Sexl, R.: Über die Aufspaltung der Energieeigenwerte in einem Kristall und die Berechnung der Bindungsenergie, *Acta Phys. Austriaca*, **13**:476 (1960).
Shakin, C., and J. Birman: Electronic Energy Bands in ZnS: Preliminary Results, *Phys. Rev.*, **109**:818 (1958). (Sec. 10-8.)
Sheka, V. I.: Symmetry of the Energy Bands of an Electron with Spin, *Fiz. Tverd. Tela*, **2**:1211 (1960).
Shestopalov, L. M.: Relationships between Properties of Atoms and Crystals, *Fiz. Tverd. Tela*, suppl. 1, p. 228 (1959).
Shinohara, S.: The Electronic Structures around Impurities in Monovalent Metals, *J. Fac. Sci. Hokkaido Univ.*, II, **4**:377 (1955).
———: The Shielding of a Fixed Charge in a Metal, *Progr. Theoret. Phys. (Kyoto)*, **26**:420 (1961).
———: The Shielding of a Fixed Charge in a Metal, *J. Phys. Soc. Japan*, **16**:1963 (1961).

―――: Donor States and Deformation around Impurity Atoms in Semiconductors, *Nuovo Cimento,* **22**:18 (1961).
Shirokovskii, V. P.: Electron States in Crystals, *Fiz. Metallov i Metallovedenie,* **6**:3 (1958).
―――: See also A. V. Sokolov.
Shockley, W.: Electronic Energy Bands in Sodium Chloride, *Phys. Rev.,* **50**:754 (1936). (Sec. 10-8.)
―――: Energy Bands for the Face-centered Lattice, *Phys. Rev.,* **51**:129 (1937).
―――: The Empty Lattice Test of the Cellular Method in Solids, *Phys. Rev.,* **52**:866 (1937). (Sec. 9-1.)
―――: On the Surface States Associated with a Periodic Potential, *Phys. Rev.,* **56**:317 (1939).
―――: Quantum Theory of Solids, *Bell Telephone System Tech. Publ.,* **18**:645 (1939).
――― and J. Bardeen: Energy Bands and Mobilities in Monatomic Semiconductors, *Phys. Rev.,* **77**:407 (1950).
―――: Energy Band Structure in Semiconductors, *Phys. Rev.,* **78**:173 (1950). (Sec. A5.)
―――: "Electrons and Holes in Semiconductors," D. Van Nostrand Company, Inc., Princeton, N. J., 1950.
―――: See also J. C. Slater.
Shoemaker, D. P.: A Method for Calculating the Energy of a Bloch Wave in a Metal, *Physica,* **14**:34 (1949).
Shtivelman, K. Y.: The Energy Spectrum of Holes in Crystals of the Diamond Type, *Fiz. Tverd. Tela,* **2**:499 (1960).
Shubnikov, A. V.: The Inherent Symmetry of Atoms and Molecules in a Crystal, *Kristallografiya,* **3**:521 (1958).
Shulman, R. G.: Tight-bonding Calculation of Acceptor Energies in Germanium and Silicon, *J. Phys. Chem. Solids,* **2**:115 (1957).
Shuttleworth, R.: The Surface Energies of Inert-gas and Ionic Crystals, *Proc. Phys. Soc. (London),* **A62**:167 (1949).
Silverman, R. A.: Fermi Energy of Metallic Lithium, *Phys. Rev.,* **85**:227 (1952). (Sec. 10-8.)
Siota, Y.: Periodic Potential in Metals, *Sci. Rept. Tohoku Univ.,* **62**:173 (1958).
―――: Periodic Potential in Ionic Crystals, *Sci. Rept. Tohoku Univ., First Ser.,* **43**:137 (1959).
―――: Periodic Potentials of Diamond Lattices and Hexagonal Close-packed Lattices, *Sci. Rept. Tohoku Univ., First Ser.,* **45**:243 (1961).
Slater, J. C.: Note on the Structure of the Groups XO_3, *Phys. Rev.,* **38**:325 (1931).
―――: Electronic Energy Bands in Metals, *Phys. Rev.,* **45**:794 (1934). (Secs. 9-1, 10-8.)
―――: Electronic Structure of Metals, *Rev. Mod. Phys.,* **6**:209(1934). (Secs. 8-2, 8-3, 9-1.)
――― and W. Shockley: Optical Absorption by the Alkali Halides, *Phys. Rev.,* **50**:705 (1936). (Secs. 10-7, 10-8.)
―――: Damped Electron Waves in Crystals, *Phys. Rev.,* **51**:840 (1937).
―――: Wave Functions in a Periodic Potential, *Phys. Rev.,* **51**:846 (1937). (Secs. 9-2, A6.)
―――: A Simplification of the Hartree-Fock Method, *Phys. Rev.,* **81**:385 (1951). (Sec. 4-5.)
―――: Note on Superlattices and Brillouin Zones, *Phys. Rev.,* **84**:179 (1951).

———: "Quantum Theory of Matter," McGraw-Hill Book Company, New York, 1951.
———: A Soluble Problem in Energy Bands, *Phys. Rev.*, **87**:807 (1952). (Sec. 6-6.)
———: An Augmented Plane Wave Method for the Periodic Potential Problem, *Phys. Rev.*, **92**:603 (1953). (Sec. 9-2.)
——— and G. F. Koster: Simplified LCAO Method for the Periodic Potential Problem, *Phys. Rev.*, **94**:1498 (1954). (Sec. 8-4.)
———: The Electronic Structure of Solids, in Handbuch der Physik, 3d ed., vol. 19, Springer-Verlag OHG, Berlin, 1956.
———: Band Theory of Bonding in Metals, in "Theory of Alloy Phases," American Society for Metals, 1956, p. 1.
———: Band Theory, *J. Phys. Chem. Solids*, **8**:21 (1959).
———: "Quantum Theory of Atomic Structure," vols. 1 and 2, McGraw-Hill Book Company, New York, 1960. (Secs. 4-5, A5, A9.)
———, G. F. Koster, and J. H. Wood: Symmetry and Free Electron Properties of Ga Energy Bands, *Phys. Rev.*, **126**:1307 (1962). (Secs. 2-8, A3-1.)
———: "Quantum Theory of Molecules and Solids," vol. 1, Electronic Structure of Molecules, McGraw-Hill Book Company, New York, 1963. (Many references.)
———: See also G. F. Koster, M. M. Saffren.
Slonczewski, J. C., and P. R. Weiss: Band Structure of Graphite, *Phys. Rev.*, **109**:272 (1958). (Sec. 10-8.)
Smith, R. A.: "Semiconductors," Cambridge University Press, New York, 1959.
———: Physics of Semiconductors, *Nature*, **188**:632 (1960).
———: "Wave Mechanics of Crystalline Solids," John Wiley & Sons, Inc., New York, 1961.
Sokolov, A. V., and V. P. Shirokovskii: Group Theory Method in the Quantum Physics of a Solid Body (Point Symmetry), *Usp. Fiz. Nauk*, **60**:617 (1956).
——— and ———: The Description of Electronic States in a Cubic Crystal, *Fiz. Metallov i Metallovedenie*, **4**:3 (1957).
——— and ———: Group Theoretical Methods in the Quantum Physics of Solids (Spatial Symmetry), *Usp. Fiz. Nauk*, **71**:485 (1960).
———: See also A. N. Men.
Sommerfeld, A., and H. Bethe: Elektronentheorie der Metalle, Handbuch der Physik, 2d ed., vol. 24, pt. 2, Springer-Verlag OHG, Berlin, 1933.
Soule, D. E., and J. W. McClure: Band Structure and Transport Properties of Single-crystal Graphite, *J. Phys. Chem. Solids*, **8**:29 (1959).
Speiser, A.: "Die Theorie der Gruppen von endlicher Ordnung," Springer-Verlag OHG, Berlin, 1937; Dover Publications, Inc., New York, 1945.
Statz, H.: Zur Theorie der Oberflächenzuständen, *Z. Naturforsch.*, **5a**:534 (1950).
———: See also E. Fues, G. F. Koster.
Steinman, O.: Equivalent Periodic Potentials, *Helv. Phys. Acta*, **30**:515 (1957).
Stern, F.: Calculation of the Cohesive Energy of Metallic Iron, *Phys. Rev.*, **116**:1399 (1959). (Sec. 10-8.)
Sternheimer, R.: On the Compressibility of Metallic Cesium, *Phys. Rev.*, **78**:235 (1950). (Sec. 10-8.)
Stocker, D.: The Electronic Properties of Tetrahedral Intermetallic Compounds, IV, The Determination of Valence Bands by a Method of Linear Combination of Bond Orbitals, *Proc. Roy. Soc. (London)*, **A270**:397 (1962). (Secs. 10-4, 10-8.)
———: See also C. A. Coulson.
Stoner, E. C.: The Energy Distribution of States in a Single Brillouin Zone, *Proc. Leeds Phil. Soc.*, **3**:120 (1936).

Streitwolf, R. W.: Methoden zur Berechnung des Energiespektrums in Idealkristallen, *Phys. Stat. Sol.*, **2**:1595 (1962).
Strutt, M. J. O.: Wirbelströme im elliptischen Zylinder (Mathieu functions), *Ann. Physik.*, **84**:485 (1927).
———: Eigenschwingungen einer Saite mit sinusförmiger Massenverteilung (Mathieu functions), *Ann. Physik.*, **85**:129 (1928).
Suchet, J. P.: Sur la nature des liaisons dans les composés I-V, *Acta Cryst.*, **14**:651 (1961).
———: "Physique des Semiconducteurs," Dunod, Paris, 1961.
———: Règles de prévision de la semiconductibilité dans les composés d'éléments de transition, *Phys. Stat. Sol.*, **2**:167 (1962).
Suffczynski, M.: Two-center Integrals for Iron Using Wave Functions with Exchange, *Nuovo Cimento*, **2**:1320 (1955).
———: Energy Integrals for 3d Electrons in Body-centered Iron, *Acta Phys. Polon.*, **14**:493 (1955).
———: Two-center Integrals for Body-centered Iron Using Atomic Functions with Exchange, *Acta Phys. Polon.*, **15**:111 (1956).
———: Two-center Integrals in Solids, *Acta Phys. Polon.*, **15**:287 (1956).
———: Two-center Integrals over the Atomic Sphere, *Bull. Acad. Polon. Sci.*, *Cl.* 3, **4**:273 (1956).
———: On the Width of the 3d Electron Band in Iron, *Acta Phys. Polon.*, **16**:161 (1957).
———: Three-center Integrals in Iron, *Bull. Acad. Polon. Sci. Ser. Sci. Math. Astron. Phys.*, **6**:195 (1958).
———: On Some Applications of the OPW Calculation, *Bull. Acad. Polon. Sci. Ser. Sci. Math. Astron. Phys.*, **6**:481 (1958).
———: The Numerical Values of the Two-center Integrals for 3d Electrons in Transition Metals, *Bull. Acad. Polon. Sci. Ser. Sci. Math. Astron. Phys.*, **7**:285 (1959).
———: On the 3d Electron Band in Body Centered Structure, *Acta Phys. Polon.*, **16**:157 (1959).
———: The Group Theoretical Analysis of α-uranium, *J. Phys. Chem. Solids*, **16**:174 (1960).
———: Numerical Values of Two-center Integrals for 3d Electrons, *Acta Phys. Polon.*, **20**:945 (1961).
———: See also M. Miasek.
Tables of Interatomic Distances and Configurations in Molecules and Ions, The Chemical Society, London, 1958. (Appendix 1.)
Takaschke, O.: Über die mathematischen Grundlagen der quantenmechanischen Elektronentheorie in unendlichen Kristallgittern, *Ann. Physik*, **8**:76 (1961).
Takeno, S.: Energy Spectrum of Lattices with Defects, III, *Progr. theoret. Phys.*, (*Kyoto*), **28**:631 (1962).
Takimoto, N.: On the Screening of Impurity Potentials by Conduction Electrons, *J. Phys. Soc. Japan*, **14**:1142 (1959).
Takizawa, E. I.: On the Solution of Schrödinger Equation with Periodic Potential in the Three Dimensional Finite Space, *Mem. Fac. Eng. Nagoya Univ.*, **12**:59 (1960).
Tamm, I.: Über eine mögliche Art der Elektronenbindung an Kristalloberflächen, *Phys. Z. Sowjetunion*, **1**:733 (1932).
Tibbs, S. R.: Electronic Energy Bands in Metallic Copper and Silver, *Proc. Cambridge Phil. Soc.*, **34**:89 (1938). (Sec. 10-8.)
———: Electron Energy Levels in NaCl, *Trans. Faraday Soc.*, **35**:1471 (1939). (Sec. 10-8.)

Tinkham, M.: "Group Theory and Quantum Mechanics," McGraw-Hill Book Company, New York, 1964.

Tobin, M. C.: Irreducible Representations, Symmetry Coordinates, and the Secular Equation for Line Groups, *J. Mol. Spectry.*, **4**:349 (1960).

Tolpygo, K. B., and O. F. Tomasevich: Wave-functions and Energy of a Conduction Band Electron in a NaCl Crystal, *Ukr. Fiz. Zh.*, **3**:145 (1958). (Sec. 10-8.)

——— and ———: Wave-functions and Energy of a Conduction Band Electron in a NaCl Crystal, II, *Fiz. Tverd. Tela*, **2**:3110 (1960). (Sec. 10-8.)

———: The Problem of Chemical Binding and the Part It Plays in the Theory of Band and Localized States of Electrons in Semiconductors, *Semiconductor Phys. Conf.*, Prague, 1960, Academic Press Inc., 1961, p. 901.

———: See also T. I. Kucher, Z. Ya. Evseev.

Tomasek, M., and J. Koutecky: Theory of Tamm Surface States in Approximation Higher than Tight-binding Approximation, *Czech. J. Phys.*, **10**:268 (1960).

———: See also J. Koutecky.

Tovstyuk, K. D., and D. M. Gemus: Energy Spectrum Structure of CdSb-type Crystals, *Fiz. Tverd. Tela*, **5**:142 (1963).

——— and M. V. Tarnavskaya: Investigations of the Energy Spectrum in Crystals with $O_h^1 \cdots O_h^{10}$ Structures, *Fiz. Tverd. Tela*, **5**:819 (1963).

Trefftz, E.: Bestimmung der Eigenwerte einer elliptischen Differentialgleichung mit der Randbedingung der Periodizität (Schrödinger-gleichung in Metallen), *Z. angew. Math. Mech.*, **34**:1 (1954).

Trlifaj, M.: The Electron Theory of Metallic Magnesium, *Czech. J. Phys.*, **1**:110 (1952). (Sec. 10-8.)

———: See also E. Antoncik.

van Arkel, A. E.: "Molecules and Crystals in Inorganic Chemistry," Interscience Publishers, Inc., New York, 1949.

van der Waerden, B. L.: "Die gruppentheoretische Methode in der Quantenmechanik," Springer-Verlag OHG, Berlin, 1932.

Verwey, E. J. W.: "Oxidic Semi-conductors, from Semiconducting Materials," Butterworth Scientific Publications, London, 1951, p. 151.

———: See also J. H. de Boer.

von der Lage, F., and H. Bethe: Method for Obtaining Electronic Eigenfunctions and Eigenvalues in Solids with an Application to Sodium, *Phys. Rev.*, **71**:612 (1947). (Secs. 9-1, 10-8.)

———: See also H. A. Bethe.

von Laue, M.: Eine quantitative Prüfung der Theorie für die Interferenzerscheinungen an Röntgenstrahlen, *Ann. Physik*, **41**:989 (1913). (Sec. 1-1.)

———: See also W. Friedrich.

Vonsovskii, S. V., and Yu. A. Izyumov: Statistical Properties of the Electron System of Ferromagnetic Transition Metals, *Fiz. Metallov i Metallovedenie*, **10**:321 (1960).

——— and ———: Electron Theory of Transition Metals, I, *Usp. Fiz. Nauk*, **77**:377 (1962).

Voss, W.: Bedingungen für das Auftreten des Ramsauereffektes, *Z. Physik*, **83**:581 (1933).

Wainwright, T., and G. Parzen: Electronic Energy Bands in Crystals, *Phys. Rev.*, **92**:1129 (1953).

Wallace, P. R.: Band Theory of Graphite, *Phys. Rev.*, **71**:622 (1947); Erratum, **72**:258 (1947). (Sec. 10-8.)

Wannier, G. H.: Structure of Electronic Excitation Levels in Insulating Crystals, *Phys. Rev.*, **52**:191 (1937). (Sec. 6-5.)

———: Improved Wigner-Seitz Method for the Calculation of Electronic Energy Bands, *Phys. Rev.*, **53**:671 (1938).
———: On the Energy Band Structure of Insulators, *Phys. Rev.*, **76**:438 (1949).
———: "Elements of Solid State Theory," Cambridge University Press, New York, 1959.
Wasastjerne, J. A.: *Soc. Sci. Fennica, Commentationes Phys. Math.*, **38**:1 (1923). (Sec. 4-1.)
———: Electron Distribution in Atoms and Ions, *Soc. Sci. Fennica*, **6**:19 (1932–1933).
———: Wave Mechanical Significance of the Apparent Radii of Atoms and Ions, *Soc. Sci. Fennica*, **6**:21 (1932–1933).
Waser, J., and L. Pauling: Compressibilities, Force Constants, and Interatomic Distances of the Elements in the Solid State, *J. Chem. Phys.*, **18**:747 (1950).
Watanabe, H.: On the Widths of Forbidden Bands of Electronic States for Onedimensional Periodic Fields, *Sci. Rept. Tohoku Univ., First Ser.*, **43**:13 (1959).
Watts, B. R.: The Fermi Surface of Beryllium, *Phys. Letters*, **3**:284 (1963).
Welker, H., and H. Weiss: Group III–Group VC Compounds, *Solid State Physics*, **3**:1 (1956).
———: See also O. G. Folberth.
Wells, A. F.: The Geometrical Basis of Crystal Chemistry, VI, *Acta Cryst.*, **9**:23 (1956).
———: The Structures of Crystals, *Solid State Phys.*, **7**:425 (1958).
———: "Structural Inorganic Chemistry," 3d ed., Oxford University Press, Fairlawn, N.J., 1962. (Sec. 1-1.)
Weyl, H.: "Gruppentheorie und Quantenmechanik," S. Hirzel Verlag, Stuttgart, 1928; "Theory of Groups and Quantum Mechanics," Dover Publications, Inc., New York, 1950.
Wigner, E.: "Gruppentheorie und ihre Anwendung auf die Quantenmechanik der Atomspektren," Friedr. Vieweg & Sohn, Brunswick, Germany, 1931.
——— and F. Seitz: On the Constitution of Metallic Sodium, I, II, *Phys. Rev.*, **43**:804 (1933); **46**:509 (1934). (Secs. 1-5, 9-1, 10-8.)
———: On the Interaction of Electrons in Metals, *Phys. Rev.*, **46**:1002 (1934).
——— and H. B. Huntington: On the Possibility of a Metallic Modification of Hydrogen, *J. Chem. Phys.*, **3**:764 (1935).
———: On the Structure of Solid Bodies, *Sci. Monthly*, **42**:40 (1936).
———: "Group Theory and Its Applications to the Quantum Mechanics of Atomic Spectra," Academic Press Inc., New York, 1959. (Sec. A9-5.)
Williams, F. E.: Theory of the Energy Levels of Donor-Acceptor Pairs, *J. Phys. Chem. Solids*, **12**:265 (1960).
Wilson, A. H.: The Theory of Electronic Semiconductors, *Proc. Roy. Soc. (London)*, I, **A133**:458 (1931); II, **A134**:277 (1931).
———: The Theory of Metals, I, *Proc. Roy. Soc. (London)*, **A138**:594 (1932).
———: "Semiconductors and Metals: An Introduction to the Electron Theory of Metals," Cambridge University Press, New York, 1939.
———: "Theory of Metals," Cambridge University Press, New York, 1936; 2d ed., 1953.
Winston, H.: The Electronic Energy Levels of Molecular Crystals, *J. Chem. Phys.*, **19**:156 (1951).
Witte, H., and E. Wölfel: Röntgenographische Bestimmung der Elektronenverteilung in Kristallen, II, Die Elektronenverteilung im Steinsalz, *Z. Physik. Chem. (Frankfort)*, **3**:296 (1955). (Sec. 4-4.)

Wohlfarth, E. P.: Electronic Properties and Band Structure of Palladium and Platinum, *Proc. Leeds Phil. Soc.*, **5**:89 (1948).
———: The Energy Band Structure of a Linear Metal, *Proc. Phys. Soc. (London)*, **A66**:889 (1953).
———: See also J. F. Cornwell, G. C. Fletcher.
Wolff, P. A.: Theory of the Band Structure of Very Degenerate Semiconductors, *Semiconductor Conf.*, Exeter, 1962, Institute of Physics and The Physical Society, London, 1962, p. 220.
Wonsowski, S. W.: Fragen der gegenwärtigen Quantentheorie elektronischer Leiter, *Fortschr. Phys.*, **1**:239 (1954).
———: See also S. Schubin.
Wood, J. H.: Wave Functions for the Iron d Band, *Phys. Rev.*, **117**:714 (1960). (Sec. 10-8.)
———: Energy Bands in Fe via Augmented Plane Wave Method, *Phys. Rev.*, **126**:517 (1962). (Secs. 9-2, 10-5, 10-8.)
———: Energy Band Calculations for Ga, *Bull. Am. Phys. Soc.*, ser. II, **8**:222 (1963). (Secs. 10-6, 10-8.)
———: See also J. C. Slater.
Wood, V. E., and J. R. Reitz: Electronic Band Structure of Cesium Gold, *J. Phys. Chem. Solids*, **23**:229 (1962). (Sec. 10-8.)
Woodruff, T. O.: Solution of the Hartree-Fock-Slater Equations for Silicon Crystal by the Method of Orthogonalized Plane Waves, *Phys. Rev.*, **98**:1741 (1955). (Sec. 10-8.)
———: Application of the Orthogonalized Plane-wave Method to Silicon Crystal, *Phys. Rev.*, **103**:1159 (1956).
———: The Orthogonalized Plane-wave Method, *Solid State Phys.*, **4**:367 (1957). (Sec. 8-1.)
Wyckoff, R. W. G.: "The Structure of Crystals," Chemical Catalog Company, Inc., New York, 1924. (Sec. 4-1.)
———: "Crystal Structures," Interscience Publishers, Inc., New York, 1948. (Secs. 1-1, 3-1, Appendixes 1, 2.)
Yamaguchi, T.: Electronic States of Single Vacancies in Diamond, *J. Phys. Soc. Japan*, **17**:1359 (1962).
———: Electronic States of Single Interstitial Atoms in Diamond, *J. Phys. Soc. Japan*, **18**:368 (1963).
Yamaka, E., and T. Sugita: Energy Band Structure in Silicon Crystal, *Phys. Rev.*, **90**:992 (1953). (Sec. 10-8.)
Yamashita, J., and M. Kojima: On the Electronic States of the Doubly-charged Negative Ions of Oxygen in Oxide Crystals, *J. Phys. Soc. Japan*, **7**:261 (1952).
———, M. Fukuchi, and S. Wakoh: Energy Band Structure of Nickel, *J. Phys. Sci. Japan*, **18**:999 (1963). (Sec. 10-8.)
———: Electronic Structure of TiO and NiO, *J. Phys. Soc. Japan*, **18**:1010 (1963). (Sec. 10-8.)
Yamazaki, M.: Electronic Band Structure in Graphite, *J. Chem. Phys.*, **26**:930 (1957). (Sec. 10-8.)
———: Electronic Band Structure of Boron Carbide, *J. Chem. Phys.*, **27**:746 (1957). (Sec. 10-8.)
———: Group-theoretical Treatment of the Energy Bands in Metal Borides MeB_6, *J. Phys. Soc. Japan*, **12**:1 (1957).
Yosida, K., and A. Watabe: Fermi Surfaces and Spin Structures in Heavy Rare-earth Metals, *Progr. Theoret. Phys. (Kyoto)*, **28**:361 (1962).

Zachariasen, W. H.: A Set of Empirical Crystal Radii for Ions and Inert Gas Configurations, *Z. Krist.*, **80**:137 (1931). (Sec. 4-1.)
Zak, J.: Method to Obtain the Character Tables of Nonsymmorphic Space Groups, *J. Math. Phys.*, **1**:165 (1960).
———: Selection Rules for Integrals of Bloch Functions, *J. Math. Phys.*, **3**:1278 (1962).
Zehler, V.: Die Berechnung der Energiebänder im Diamantkristall, *Ann. Physik*, **13**:229 (1953).
———: See also W. Döring.
Zhilich, A. G., and V. P. Makarov: Band Structure of Cuprous Oxide, *Fiz. Tverd. Tela*, **3**:585 (1961). (Sec. 10-8.)
Ziman, J. M.: Electrons in Metals: A Short Guide to the Fermi Surface, I, The Electron Gas; II, Bands and Zones, *Contemporary Phys.*, **3**:241, 321 (1962).
Zorina, E. L.: A Note on the Mooser-Pearson Law, *Fiz. Tverd. Tela*, **2**:1936 (1960).

Index

Since the bibliography immediately preceding this index contains references to sections in which each paper is mentioned, some names referred to in the text are omitted here. The reader's attention is also called to Appendix 1, in which information is given on many compounds which are not mentioned in the Index.

Alkali halides, 57–58, 291–297
Altmann, S. L., 233
Aluminum, 269–272, 283
Aluminum boride structure, 89, 337
Aluminum fluoride structure, 85, 336
Aluminum oxide, 77, 81–84
Ammonia, 26
Antiferromagnetism, 62
Antifluorite structure, 62, 344–345
Antimony, 26, 62
APW method, 233–236, 242–246, 461–465
Aragonite structure, 91–94, 334
Argon, 24, 280–282
Arsenates, 77
Arsenic, 26, 45–49, 62
Arsenic structure, 45–49, 337
Asbestos, 76
Atomic radii, 54–55, 61, 95–115, 307–333
Atomic wave functions, radii, 101–105
Augmented-plane-wave method, 233–236, 242–246, 461–465

Barium titanate, 68, 85
Basis functions, 138–143
Bassani, F., and M. Yoshimine, 275, 279
Bell, D. G., et al., 298
Benzene, 27, 44
Beryllium, 260–269
Bethe, H., 198
  and F. von der Lage, 233
Bismuth, 26, 62, 291
Bloch, F., 6, 203–206
Bloch sum, 154–157

Body-centered cubic lattice, 15–17, 73–76, 122–123, 384–407
Body-centered cubic structure, 27–32, 58, 255–256, 346
Bonds, tetrahedral, 41, 59, 76–77, 470–471
Borates, 81
Bouckaert, L. P., R. Smoluchowski, and E. Wigner, 131, 140, 356, 368–369, 384–385, 408, 480
Bragg, W. H., and W. L. Bragg, 2, 54, 95–96, 101
Bragg, W. L., and W. H. Bragg, 2, 54, 95–96, 101
Bragg's law, 195–199
Bravais lattices, 10–23, 121–126
Brillouin zones, 121–126, 147–148, 150, 169–199, 256, 264, 348, 369, 385, 408, 418, 430, 438, 447
Bromine, 3, 24, 50
Brooks, H., 248
Burdick, G. A., 246, 283–284

Cadmium chloride structure, 65–67, 338
Cadmium iodide structure, 65–66, 336
Calcite structure, 77–81, 338
Calcium aluminum orthosilicate, 74–77
Calcium carbonate, 77–81, 91–94
Calcium titanate, 68
Carbon, 24, 26, 272–279, 470–471
Carbonates, 81
Cellular method, 228–233
Cesium, 259–261
Cesium chloride structure, 57, 342
Cesium fluoride, 110–111

559

Charge transfer, 297
Chlorine, 24
Chromium, 281
Chromium oxide, 83
Cobalt, 281
Cohen, M. H., and V. Heine, 245
Compatibility relations, 140–142, 177, 367–368, 383–384, 406–407, 417, 426, 429, 437–438, 446, 451
Conductor, 15
Conklin, J. B., L. E. Johnson, and G. W. Pratt, Jr., 298–299, 478, 489
Coordination number, 99
Copper, 241–243, 281, 283–285
Copper bromide, 58, 100, 107–108, 151
Corbato, F., 288–289
Corundum structure, 77, 81–84, 339
Coulson, C. A., 279
Cristobalite, 77
Cromer, D. T., D. Liberman, and J. M. Waber, 102
Crystal symmetry, 3–23, 118–143, 174–176, 347–451
Crystal systems, 10–14, 121–126
Cutler, P. H., and T. L. Loucks, 262

Diamond structure, 27, 40–43, 59, 62, 272–280, 345
Diffraction, x-ray, 2
Dirac, P. A. M., 473
Double groups, 485–520
   C_{2v}, C_{3v}, C_{4v}, 495, 501–506
   T_d and O_h, 506–516

Elements, crystal structures, 25–26
Elliott, R. J., 487
Energy bands, 150–151, 154–166, 208–215, 257–305
Energy contours, 171–174
Euler angles, 491
Ewald, P. P., 119, 237
Extended-zone scheme, 180–182

Face-centered cubic lattice, 17–18, 40, 43, 62, 122–123, 368–384
Face-centered cubic structure, 27–32, 59, 254, 256, 269–272, 343
Fajans, K., 97

Fermi level, 171–173
Ferrites, 69–71
Ferroelectricity, 68–69, 85
Fine-structure constant, 476
Fluorine, 24
Fluorite structure, 62–63, 73, 344–345
Fourier expansion, 144–145, 207, 452–454
Free-electron approximation, 120, 144–202, 241–246, 250–257, 263, 273, 466–469
Fues, E., and H. Statz, 245

Gallium, 24, 50–53, 125, 282, 291–292
Gallium arsenide, 58, 100, 107–108, 115, 279–290
Garnet structure, 68, 74–77, 346
General position, 29
Germanates, 77
Germanium, 24, 26, 107–108, 115, 272–280, 282, 488–489
Gibbs, J. W., 119
Goldschmidt, V. M., 98–101
Graphite structure, 27, 44–45, 48, 288–291, 339, 347–368
Green's function, 246–249, 283
Group of the wave vector, 134

Ham, F. S., 233, 246, 248, 259–260
Hamiltonian for periodic potential, 149
Hartree-Fock method, 5
Heine, V., 269
   and M. H. Cohen, 245
Helium, 24
Hellmann, H., 223
Hematite, 83
Herman, F., 102, 272–274, 279, 290, 489
Herring, C., 34, 205, 262–263
Hexagonal close-packed structure, 32–40, 44, 49, 60, 89–91, 263–268, 340, 347–368
Hexagonal lattice, 18–20, 32–40, 45, 49–50, 65, 86, 123, 347–368
Howarth, D. J., and H. Jones, 233
Howland, L. P., 294–295
Huggins, M. L., 100
Hund, F., and B. Mrowka, 277
Hurley, A. C., J. E. Lennard-Jones, and J. A. Pople, 43, 278, 470–471
Hydrogen, 24

INDEX

Ilmenite structure, 77, 84–85, 335
Insulator, 151
Intermetallic compounds, 58
International Tables of Crystallography, 10, 36, 40, 43, 49, 307–333
Inverse spinels, 71
Iodine, 3, 24, 50–53, 125
Iodine structure, 50–53, 334
Ionic radii, 97–115
Iron, 281, 285–287
Iron oxide, 83
Iron titanate, 85
Irreducible representations, 120, 133–143, 174–176, 347–451

Jacobian, 145
Johnson, L. E., J. B. Conklin, and G. W. Pratt, Jr., 298–299, 478, 489
Jones, H., and D. J. Howarth, 233

Kenney, J. F., 258–261
Kimball, G. E., 277
Kleinman, L., and J. C. Phillips, 223, 245
Kohn, W., 194, 246
Korringa, J., 237
Koster, G. F., 9, 22, 118, 193, 418, 430, 438, 447, 487
k · p method, 455–460, 489
Krutter, H. M., and M. F. Manning, 260, 269
Krypton, 24
Kuhn, T. S., and J. H. Van Vleck, 248

Landé, A., 97, 476
Lanthanum compounds, 86–89, 337–338
LCAO method, 203–206, 218, 227
Lead sulfide, 298
Lead telluride, 298–300, 489
Lennard-Jones, J. E., A. C. Hurley, and J. A. Pople, 43, 278, 470–471
Liberman, D., J. M. Waber, and D. T. Cromer, 102
Lithium, 259–261
Lithium fluoride, 296–297
Lithium hydride, 107
Lithium iodide, 110
Lithium niobate, 85
Lithium nitrate, 81

Lithium tantalate, 85
Loucks, T. L., and P. H. Cutler, 262

Madelung energy, 294
Magnesium compounds, 63
Manganous sulfide and selenate, 99
Manning, M. F., and H. M. Krutter, 260, 269
Mathieu's equation, 159–166
Mattheiss, L. F., 280–282
Metals, 27
Mica, 76
Molecular orbitals, 5
Momentum eigenfunctions, 151–166, 182–195, 209–218, 452–454
Monoclinic lattice, 14, 125
Morse, P. M., 237–240
Mrowka, B., and F. Hund, 277

Negative ions, repulsion, 108–111
Neon, 24
Nickel, 281
Nickel arsenide, 58–62, 347–368
Nickel arsenide structure, 58–62, 340
Nitrogen, 24

One-dimensional case, energy bands, 158–166
Opechowski, W., 487
OPW method, 205–206, 218–227
Order-disorder transformations, 58
Orthogonalized-plane-wave method, 205–206, 218–227
Orthorhombic lattice, 22, 50–53, 125, 438–451
Oxides, 58, 62
Oxygen, 3, 24, 26, 108–110

Packing of spheres, 30–32
Parmenter, R. H., 204
Pauling, L., 96, 98–104
  and B. Podolsky, 215
Periodic boundary conditions, 146–148
Periodic potential, 145
  Schrödinger's equation for, 148–151, 218–223, 228–249
Perovskite structure, 68–69, 343

Phillips, J. C., and L. Kleinman, 223, 245
Phosphates, 77
Phosphorus, 24, 26, 50
Plane-wave expansions, 120, 144–202
Podolsky, B., and L. Pauling, 215
Point group C_N, 147–148
   C_{3v}, 264
   C_{4v}, 134–140
   C_6, 147
   C_{6v}, 45
   D_{2h}, 50–53
   D_3, 49–50, 85
   D_{3d}, 43, 61, 66, 77–78
   D_{3h}, 62, 264
   D_{4h}, 65
   D_{6h}, 32–36, 45
   O_h, 28, 43, 59, 62, 75, 140
   T_d, 43, 59
   T_h, 64, 73
Point groups, general discussion, 10, 127–128
Polarizability, ionic, 97–98
Pople, J. A., A. C. Hurley, and J. E. Lennard-Jones, 43, 278, 470–471
Potassium, 259–261
Potassium chloride, 106–107, 112–114, 294–296
Potential in crystal, 111–117
Pratt, G. W., Jr., L. E. Johnson, and J. B. Conklin, 298–299, 478, 489
Projection operators, 135–140, 347–451
Prokofjew, W., 229
Pyrite structure, 63–64, 73, 341

Quantum-defect method, 248–249
Quartz, 77

Radii, atomic, 54–55, 61, 95–115, 307–333
   and ionic, 54–55, 95–115
   ionic, 97–115
Radon, 24
Ramsauer effect, 238–239
Reciprocal lattice, 118–120
Relativistic effects, 299–300, 472–520
Rhombohedral lattice, 20–22, 45–49, 65–67, 77–89, 124–125, 417–426
Ring of hydrogen atoms, 146–147
Rotation, effect, on plane wave, 126–133
   on spin function, 490–520
   and reflections, 4–10

Rubidium, 259–261
Ruby, 83–84
Rutile structure, 64–65, 335

Saffren, M., 233, 236
Scattered-wave method, 236–246
Schlosser, H. C., 201
Schmid, L. A., 278, 470–471
Schrödinger's equation for periodic potential, 148–151, 218–223, 228–249
Segall, B., 246, 269–272, 283–284
Seitz, F., 9, 118, 455
Selenates, 77
Selenides, 58, 62
Selenium, 24, 26, 49–50, 62
Selenium structure, 49–50
Self-consistent field, 115–117
Semiconductor, 151
Shockley, W., 232, 293, 459
Silica, 77
Silicates, 76–77
Silicon, 24, 26, 272
Simple cubic lattice, 14–15, 58, 63–64, 73, 122, 408–417
Simple cubic structure, 58, 68–69, 131–140
Smoluchowski, R., L. P. Bouckaert, and E. Wigner, 131, 140, 356, 368–369, 384–385, 408, 480
Sodium, 205–218, 224–227, 241–242, 258–260
Sodium arsenide structure, 89–91, 341, 347–368
Sodium chloride, 3, 57
Sodium chloride structure, 57–58, 99, 293–294, 343–344
Sodium nitrate, 81
Space group $C_{3i}^2(R\bar{3})$, 85, 335, 417–426
   $C_{6v}^4(P6_3mc)$, 27, 44–45, 60–62, 339, 347–368
   $D_{2h}^{16}(Pnma)$, 91–94, 334, 439–446
   $D_{2h}^{18}(Cmca)$, 27, 50–53, 91, 334, 446–451
   $D_3^4(P3_121)$ and $D_3^6(P3_221)$, 27, 335, 426–429
   $D_3^7(R32)$, 85, 336, 417–426
   $D_{3d}^3(P\bar{3}m1)$, 65–66, 77, 86–89, 336–337, 426–429
   $D_{3d}^5(R\bar{3}m)$, 27, 45–49, 65–67, 78, 86–89, 337–338, 417–426
   $D_{3d}^6(R\bar{3}c)$, 77–84, 338–339, 417–426
   $D_{4h}^{14}(P4_2/mnm)$, 64–65, 335, 429–439

INDEX 563

Space group, $D_{6h}^4(P6_3/mmc)$, 27, 32–40, 45, 61–62, 89–91, 340–341, 347–368
  $O_h^1(Pm3m)$, 58, 68–69, 131–140, 342–343, 408–417
  $O_h^5(Fm3m)$, 27–32, 43, 57, 62, 343–345, 368–384
  $O_h^7(Fd3m)$, 27, 43, 59, 68–71, 345–346, 368–384
  $O_h^9(Im3m)$, 27–32, 346, 384–407
  $O_h^{10}(Ia3d)$, 68, 74–77, 346, 384–407
  $T_d^2(F\bar{4}3m)$, 59, 342, 368–384
  $T_h^6(Pa3)$, 63–64, 73, 341, 408–417
  $T_h^7(Ia3)$, 68, 72–74, 341, 384–407
Space groups, general discussion, 6–23, 27–53, 54–94, 347–451
Special position, 29, 62, 130–131
Spin-orbit interaction, 280, 299–300, 472–520
Spinel structure, 68–71, 108–110, 345–346
Square lattice, 167–195
Stacking fault, 48
Stars of wave vectors, 130–133
Statz, H., and E. Fues, 245
Sulfates, 77
Sulfides, 58, 62
Sulfur, 24, 26, 62
Symmetrized plane waves, 133–143, 174–176, 347–451
Symmetry operations of crystals, 3–23, 118–143, 174–176, 347–351
Symmorphic space groups, 7, 27, 45, 59

Tellurides, 58, 62
Tellurium, 24, 26, 62
Tetragonal lattice, 22–23, 64–65, 125–126, 429–438
Tetrahedral bonds, 41, 59, 76–77, 470–471
Tetrahedral covalent radii, 100
Thallium oxide structure, 68, 72–74, 341
Tight-binding method, 203–206, 218–227
Tin, 24, 26, 62, 272
Titanium, 281
Titanium dioxide, 63–64
Transition elements, 280–287
Translation vectors, 4–10, 128–130
Triclinic lattice, 14, 125
Trigonal lattice, 20–22, 45–49, 65–67, 77–89, 124–125, 417–426

Unit cell, 4, 16–23
Wigner-Seitz, 16–23, 45, 51, 121, 145, 228–239

Vanadium, 281
Van der Waals forces, 24, 44, 65
Van Vleck, J. H., and T. S. Kuhn, 248
von der Lage, F., and H. Bethe, 233
von Laue, M., 2

Waber, J. M., D. Liberman, and D. T. Cromer, 102
Wannier functions, 154–166, 187–195
Wasastjerne, J. A., 97–98
Wave functions, atomic, radii, 101–105
  crystalline, 209–215
Wigner, E., 490
  and L. P. Bouckaert, and
  R. Smoluchowski, 131, 140, 356, 368–369, 384–385, 408, 480
Wigner-Seitz cell, 16–23, 45, 51, 121, 145, 228–239
Witte, H., and E. Wölfel, 112
Wölfel, E., and H. Witte, 112
Wood, J. H., 168, 235–236, 285–287, 291–292
Wurtzite structure, 57–60, 99–100, 339, 347–368
Wyckoff, R. W., 3, 55, 96, 307–346

X-ray diffraction, 2
X-rays and crystal structure, 2
Xenon, 24

Yoshimine, M., and F. Bassani, 275, 279
Ytterbium iron garnet, 75
Yttrium iron garnet, 74

Zachariasen, W. H., 98–101
Zinc, 281
Zinc-blende structure, 58–59, 62, 99–100, 279–280, 342
Zinc selenide, 58, 100, 107–108, 118
Zinc sulfide, 58

A CATALOGUE OF SELECTED
DOVER SCIENCE BOOKS

A CATALOGUE OF SELECTED
DOVER SCIENCE BOOKS

Physics: The Pioneer Science, Lloyd W. Taylor. Very thorough non-mathematical survey of physics in a historical framework which shows development of ideas. Easily followed by laymen; used in dozens of schools and colleges for survey courses. Richly illustrated. Volume 1: Heat, sound, mechanics. Volume 2: Light, electricity. Total of 763 illustrations. Total of cvi + 847pp.
60565-5, 60566-3 Two volumes, Paperbound 5.50

THE RISE OF THE NEW PHYSICS, A. d'Abro. Most thorough explanation in print of central core of mathematical physics, both classical and modern, from Newton to Dirac and Heisenberg. Both history and exposition: philosophy of science, causality, explanations of higher mathematics, analytical mechanics, electromagnetism, thermodynamics, phase rule, special and general relativity, matrices. No higher mathematics needed to follow exposition, though treatment is elementary to intermediate in level. Recommended to serious student who wishes verbal understanding. 97 illustrations. Total of ix + 982pp.
20003-5, 20004-3 Two volumes, Paperbound $6.00

INTRODUCTION TO CHEMICAL PHYSICS, John C. Slater. A work intended to bridge the gap between chemistry and physics. Text divided into three parts: Thermodynamics, Statistical Mechanics, and Kinetic Theory; Gases, Liquids and Solids; and Atoms, Molecules and the Structure of Matter, which form the basis of the approach. Level is advanced undergraduate to graduate, but theoretical physics held to minimum. 40 tables, 118 figures. xiv + 522pp.
62562-1 Paperbound $4.00

BASIC THEORIES OF PHYSICS, Peter C. Bergmann. Critical examination of important topics in classical and modern physics. Exceptionally useful in examining conceptual framework and methodology used in construction of theory. Excellent supplement to any course, textbook. Relatively advanced.
Volume 1. Heat and Quanta. Kinetic hypothesis, physics and statistics, stationary ensembles, thermodynamics, early quantum theories, atomic spectra, probability waves, quantization in wave mechanics, approximation methods, abstract quantum theory. 8 figures. x + 300pp. 60968-5 Paperbound $2.50
Volume 2. Mechanics and Electrodynamics. Classical mechanics, electro- and magnetostatics, electromagnetic induction, field waves, special relativity, waves, etc. 16 figures, viii + 260pp. 60969-3 Paperbound $2.75

FOUNDATIONS OF PHYSICS, Robert Bruce Lindsay and Henry Margenau. Methods and concepts at the heart of physics (space and time, mechanics, probability, statistics, relativity, quantum theory) explained in a text that bridges gap between semi-popular and rigorous introductions. Elementary calculus assumed. "Thorough and yet not over-detailed," *Nature*. 35 figures. xviii + 537 pp.
60377-6 Paperbound $3.50

CATALOGUE OF DOVER BOOKS

FUNDAMENTAL FORMULAS OF PHYSICS, edited by Donald H. Menzel. Most useful reference and study work, ranges from simplest to most highly sophisticated operations. Individual chapters, with full texts explaining formulae, prepared by leading authorities cover basic mathematical formulas, statistics, nomograms, physical constants, classical mechanics, special theory of relativity, general theory of relativity, hydrodynamics and aerodynamics, boundary value problems in mathematical physics, heat and thermodynamics, statistical mechanics, kinetic theory of gases, viscosity, thermal conduction, electromagnetism, electronics, acoustics, geometrical optics, physical optics, electron optics, molecular spectra, atomic spectra, quantum mechanics, nuclear theory, cosmic rays and high energy phenomena, particle accelerators, solid state, magnetism, etc. Special chapters also cover physical chemistry, astrophysics, celestian mechanics, meteorology, and biophysics. Indispensable part of library of every scientist. Total of xli + 787pp.
60595-7, 60596-5 Two volumes, Paperbound $6.00

INTRODUCTION TO EXPERIMENTAL PHYSICS, William B. Fretter. Detailed coverage of techniques and equipment: measurements, vacuum tubes, pulse circuits, rectifiers, oscillators, magnet design, particle counters, nuclear emulsions, cloud chambers, accelerators, spectroscopy, magnetic resonance, x-ray diffraction, low temperature, etc. One of few books to cover laboratory hazards, design of exploratory experiments, measurements. 298 figures. xii + 349pp.
(EBE) 61890-0 Paperbound $3.00

CONCEPTS AND METHODS OF THEORETICAL PHYSICS, Robert Bruce Lindsay. Introduction to methods of theoretical physics, emphasizing development of physical concepts and analysis of methods. Part I proceeds from single particle to collections of particles to statistical method. Part II covers application of field concept to material and non-material media. Numerous exercises and examples. 76 illustrations. x + 515pp. 62354-8 Paperbound $4.00

AN ELEMENTARY TREATISE ON THEORETICAL MECHANICS, Sir James Jeans. Great scientific expositor in remarkably clear presentation of basic classical material: rest, motion, forces acting on particle, statics, motion of particle under variable force, motion of rigid bodies, coordinates, etc. Emphasizes explanation of fundamental physical principles rather than mathematics or applications. Hundreds of problems worked in text. 156 figures. x + 364pp. 61839-0 Paperbound $2.75

THEORETICAL MECHANICS: AN INTRODUCTION TO MATHEMATICAL PHYSICS, Joseph S. Ames and Francis D. Murnaghan. Mathematically rigorous introduction to vector and tensor methods, dynamics, harmonic vibrations, gyroscopic theory, principle of least constraint, Lorentz-Einstein transformation. 159 problems; many fully-worked examples. 39 figures. ix + 462pp. 60461-6 Paperbound $3.50

THE PRINCIPLE OF RELATIVITY, Albert Einstein, Hendrick A. Lorentz, Hermann Minkowski and Hermann Weyl. Eleven original papers on the special and general theory of relativity, all unabridged. Seven papers by Einstein, two by Lorentz, one each by Minkowski and Weyl. "A thrill to read again the original papers by these giants," *School Science and Mathematics.* Translated by W. Perret and G. B. Jeffery. Notes by A. Sommerfeld. 7 diagrams. viii + 216pp.
60081-5 Paperbound $2.25

CATALOGUE OF DOVER BOOKS

EINSTEIN'S THEORY OF RELATIVITY, Max Born. Relativity theory analyzed, explained for intelligent layman or student with some physical, mathematical background. Includes Lorentz, Minkowski, and others. Excellent verbal account for teachers. Generally considered the finest non-technical account. vii + 376pp.
60769-0 Paperbound $2.75

PHYSICAL PRINCIPLES OF THE QUANTUM THEORY, Werner Heisenberg. Nobel Laureate discusses quantum theory, uncertainty principle, wave mechanics, work of Dirac, Schroedinger, Compton, Wilson, Einstein, etc. Middle, non-mathematical level for physicist, chemist not specializing in quantum; mathematical appendix for specialists. Translated by C. Eckart and F. Hoyt. 19 figures. viii + 184pp.
60113-7 Paperbound $2.00

PRINCIPLES OF QUANTUM MECHANICS, William V. Houston. For student with working knowledge of elementary mathematical physics; uses Schroedinger's wave mechanics. Evidence for quantum theory, postulates of quantum mechanics, applications in spectroscopy, collision problems, electrons, similar topics. 21 figures. 288pp.
60524-8 Paperbound $3.00

ATOMIC SPECTRA AND ATOMIC STRUCTURE, Gerhard Herzberg. One of the best introductions to atomic spectra and their relationship to structure; especially suited to specialists in other fields who require a comprehensive basic knowledge. Treatment is physical rather than mathematical. 2nd edition. Translated by J. W. T. Spinks. 80 illustrations. xiv + 257pp.
60115-3 Paperbound $2.00

ATOMIC PHYSICS: AN ATOMIC DESCRIPTION OF PHYSICAL PHENOMENA, Gaylord P. Harnwell and William E. Stephens. One of the best introductions to modern quantum ideas. Emphasis on the extension of classical physics into the realms of atomic phenomena and the evolution of quantum concepts. 156 problems. 173 figures and tables. xi + 401pp.
61584-7 Paperbound $3.00

ATOMS, MOLECULES AND QUANTA, Arthur E. Ruark and Harold C. Urey. 1964 edition of work that has been a favorite of students and teachers for 30 years. Origins and major experimental data of quantum theory, development of concepts of atomic and molecular structure prior to new mechanics, laws and basic ideas of quantum mechanics, wave mechanics, matrix mechanics, general theory of quantum dynamics. Very thorough, lucid presentation for advanced students. 230 figures. Total of xxiii + 810pp.
61106-X, 61107-8 Two volumes, Paperbound $6.00

INVESTIGATIONS ON THE THEORY OF THE BROWNIAN MOVEMENT, Albert Einstein. Five papers (1905-1908) investigating the dynamics of Brownian motion and evolving an elementary theory of interest to mathematicians, chemists and physical scientists. Notes by R. Fürth, the editor, discuss the history of study of Brownian movement, elucidate the text and analyze the significance of the papers. Translated by A. D. Cowper. 3 figures. iv + 122pp.
60304-0 Paperbound $1.50

CATALOGUE OF DOVER BOOKS

MATHEMATICAL FOUNDATIONS OF STATISTICAL MECHANICS, A. I. Khinchin. Introduction to modern statistical mechanics: phase space, ergodic problems, theory of probability, central limit theorem, ideal monatomic gas, foundation of thermodynamics, dispersion and distribution of sum functions. Provides mathematically rigorous treatment and excellent analytical tools. Translated by George Gamow. viii + 179pp. 60147-1 Paperbound $2.50

INTRODUCTION TO PHYSICAL STATISTICS, Robert B. Lindsay. Elementary probability theory, laws of thermodynamics, classical Maxwell-Boltzmann statistics, classical statistical mechanics, quantum mechanics, other areas of physics that can be studied statistically. Full coverage of methods; basic background theory. ix + 306pp. 61882-X Paperbound $2.75

DIALOGUES CONCERNING TWO NEW SCIENCES, Galileo Galilei. Written near the end of Galileo's life and encompassing 30 years of experiment and thought, these dialogues deal with geometric demonstrations of fracture of solid bodies, cohesion, leverage, speed of light and sound, pendulums, falling bodies, accelerated motion, etc. Translated by Henry Crew and Alfonso de Salvio. Introduction by Antonio Favaro. xxiii + 300pp. 60099-8 Paperbound $2.25

FOUNDATIONS OF SCIENCE: THE PHILOSOPHY OF THEORY AND EXPERIMENT, Norman R. Campbell. Fundamental concepts of science examined on middle level: acceptance of propositions and axioms, presuppositions of scientific thought, scientific law, multiplication of probabilities, nature of experiment, application of mathematics, measurement, numerical laws and theories, error, etc. Stress on physics, but holds for other sciences. "Unreservedly recommended," *Nature* (England). Formerly *Physics: The Elements*. ix + 565pp. 60372-5 Paperbound $4.00

THE PHASE RULE AND ITS APPLICATIONS, Alexander Findlay, A. N. Campbell and N. O. Smith. Findlay's well-known classic, updated (1951). Full standard text and thorough reference, particularly useful for graduate students. Covers chemical phenomena of one, two, three, four and multiple component systems. "Should rank as the standard work in English on the subject," *Nature*. 236 figures. xii + 494pp. 60091-2 Paperbound $3.50

THERMODYNAMICS, Enrico Fermi. A classic of modern science. Clear, organized treatment of systems, first and second laws, entropy, thermodynamic potentials, gaseous reactions, dilute solutions, entropy constant. No math beyond calculus is needed, but readers are assumed to be familiar with fundamentals of thermometry, calorimetry. 22 illustrations. 25 problems. x + 160pp. 60361-X Paperbound $2.00

TREATISE ON THERMODYNAMICS, Max Planck. Classic, still recognized as one of the best introductions to thermodynamics. Based on Planck's original papers, it presents a concise and logical view of the entire field, building physical and chemical laws from basic empirical facts. Planck considers fundamental definitions, first and second principles of thermodynamics, and applications to special states of equilibrium. Numerous worked examples. Translated by Alexander Ogg. 5 figures. xiv + 297pp. 60219-2 Paperbound $2.50

CATALOGUE OF DOVER BOOKS

MICROSCOPY FOR CHEMISTS, Harold F. Schaeffer. Thorough text; operation of microscope, optics, photomicrographs, hot stage, polarized light, chemical procedures for organic and inorganic reactions. 32 specific experiments cover specific analyses: industrial, metals, other important subjects. 136 figures. 264pp.
61682-7 Paperbound $2.50

OPTICKS, Sir Isaac Newton. A survey of 18th-century knowledge on all aspects of light as well as a description of Newton's experiments with spectroscopy, colors, lenses, reflection, refraction, theory of waves, etc. in language the layman can follow. Foreword by Albert Einstein. Introduction by Sir Edmund Whittaker. Preface by I. Bernard Cohen. cxxvi + 406pp. 60205-2 Paperbound $4.00

LIGHT: PRINCIPLES AND EXPERIMENTS, George S. Monk. Thorough coverage, for student with background in physics and math, of physical and geometric optics. Also includes 23 experiments on optical systems, instruments, etc. "Probably the best intermediate text on optics in the English language," *Physics Forum*. 275 figures. xi + 489pp. 60341-5 Paperbound $3.50

PHYSICAL OPTICS, Robert W. Wood. A classic in the field, this is a valuable source for students of physical optics and excellent background material for a study of electromagnetic theory. Partial contents: nature and rectilinear propagation of light, reflection from plane and curved surfaces, refraction, absorption and dispersion, origin of spectra, interference, diffraction, polarization, Raman effect, optical properties of metals, resonance radiation and fluorescence of atoms, magneto-optics, electro-optics, thermal radiation. 462 diagrams, 17 plates. xvi + 846pp.
61808-0 Paperbound $4.50

MIRRORS, PRISMS AND LENSES: A TEXTBOOK OF GEOMETRICAL OPTICS, James P. C. Southall. Introductory-level account of modern optical instrument theory, covering unusually wide range: lights and shadows, reflection of light and plane mirrors, refraction, astigmatic lenses, compound systems, aperture and field of optical system, the eye, dispersion and achromatism, rays of finite slope, the microscope, much more. Strong emphasis on earlier, elementary portions of field, utilizing simplest mathematics wherever possible. Problems. 329 figures. xxiv + 806pp. 61234-1 Paperbound $5.00

THE PSYCHOLOGY OF INVENTION IN THE MATHEMATICAL FIELD, Jacques Hadamard. Important French mathematician examines psychological origin of ideas, role of the unconscious, importance of visualization, etc. Based on own experiences and reports by Dalton, Pascal, Descartes, Einstein, Poincaré, Helmholtz, etc. xiii + 145pp. 20107-4 Paperbound $1.50

INTRODUCTION TO CHEMICAL PHYSICS, John C. Slater. A work intended to bridge the gap between chemistry and physics. Text divided into three parts: Thermodynamics, Statistical Mechanics, and Kinetic Theory; Gases, Liquids and Solids; and Atoms, Molecules and the Structure of Matter, which form the basis of the approach. Level is advanced undergraduate to graduate, but theoretical physics held to minimum. 40 tables, 118 figures. xiv + 522pp.
62562-1 Paperbound $4.00

CATALOGUE OF DOVER BOOKS

CONTRIBUTIONS TO THE FOUNDING OF THE THEORY OF TRANSFINITE NUMBERS, Georg Cantor. The famous articles of 1895-1897 which founded a new branch of mathematics, translated with 82-page introduction by P. Jourdain. Not only a great classic but still one of the best introductions for the student. ix + 211pp.
60045-9 Paperbound $2.50

ESSAYS ON THE THEORY OF NUMBERS, Richard Dedekind. Two classic essays, on the theory of irrationals, giving an arithmetic and rigorous foundation; and on transfinite numbers and properties of natural numbers. Translated by W. W. Beman. iii + 115pp.
21010-3 Paperbound $1.75

GEOMETRY OF FOUR DIMENSIONS, H. P. Manning. Part verbal, part mathematical development of fourth dimensional geometry. Historical introduction. Detailed treatment is by synthetic method, approaching subject through Euclidean geometry. No knowledge of higher mathematics necessary. 76 figures. ix + 348pp.
60182-X Paperbound $3.00

AN INTRODUCTION TO THE GEOMETRY OF N DIMENSIONS, Duncan M. Y. Sommerville. The only work in English devoted to higher-dimensional geometry. Both metric and projectiv properties of n-dimensional geometry are covered. Covers fundamental ideas of incidence, parallelism, perpendicularity, angles between linear space, enumerative geometry, analytical geometry, polytopes, analysis situs, hyperspacial figures. 60 diagrams. xvii + 196pp.
60494-2 Paperbound $2.00

THE THEORY OF SOUND, J. W. S. Rayleigh. Still valuable classic by the great Nobel Laureate. Standard compendium summing up previous research and Rayleigh's original contributions. Covers harmonic vibrations, vibrating systems, vibrations of strings, membranes, plates, curved shells, tubes, solid bodies, refraction of plane waves, general equations. New historical introduction and bibliography by R. B. Lindsay, Brown University. 97 figures. lviii + 984pp.
60292-3, 60293-1 Two volumes, Paperbound $6.00

ELECTROMAGNETIC THEORY: A CRITICAL EXAMINATION OF FUNDAMENTALS, Alfred O'Rahilly. Critical analysis and restructuring of the basic theories and ideas of classical electromagnetics. Analysis is carried out through study of the primary treatises of Maxwell, Lorentz, Einstein, Weyl, etc., which established the theory. Expansive reference to and direct quotation from these treatises. Formerly *Electromagnetics.* Total of xvii + 884pp.
60126-9, 60127-7 Two volumes, Paperbound $6.00

ELEMENTARY CONCEPTS OF TOPOLOGY, Paul Alexandroff. Elegant, intuitive approach to topology, from the basic concepts of set-theoretic topology to the concept of Betti groups. Stresses concepts of complex, cycle and homology. Shows how concepts of topology are useful in math and physics. Introduction by David Hilbert. Translated by Alan E. Farley. 25 figures. iv + 57pp.
60747-X Paperbound $1.25

CATALOGUE OF DOVER BOOKS

THE ELEMENTS OF NON-EUCLIDEAN GEOMETRY, Duncan M. Y. Sommerville. Presentation of the development of non-Euclidean geometry in logical order, from a fundamental analysis of the concept of parallelism to such advanced topics as inversion, transformations, pseudosphere, geodesic representation, relation between parataxy and parallelism, etc. Knowledge of only high-school algebra and geometry is presupposed. 126 problems, 129 figures. xvi + 274pp.
60460-8 Paperbound $2.50

NON-EUCLIDEAN GEOMETRY: A CRITICAL AND HISTORICAL STUDY OF ITS DEVELOPMENT, Roberto Bonola. Standard survey, clear, penetrating, discussing many systems not usually represented in general studies. Easily followed by non-specialist. Translated by H. Carslaw. Bound in are two most important texts: Bolyai's "The Science of Absolute Space" and Lobachevski's "The Theory of Parallels," translated by G. B. Halsted. Introduction by F. Enriques. 181 diagrams. Total of 431pp.
60027-0 Paperbound $3.00

ELEMENTS OF NUMBER THEORY, Ivan M. Vinogradov. By stressing demonstrations and problems, this modern text can be understood by students without advanced math backgrounds. "A very welcome addition," *Bulletin, American Mathematical Society.* Translated by Saul Kravetz. Over 200 fully-worked problems. 100 numerical exercises. viii + 227pp.
60259-1 Paperbound $2.50

THEORY OF SETS, E. Kamke. Lucid introduction to theory of sets, surveying discoveries of Cantor, Russell, Weierstrass, Zermelo, Bernstein, Dedekind, etc. Knowledge of college algebra is sufficient background. "Exceptionally well written," *School Science and Mathematics.* Translated by Frederick Bagemihl. vii + 144pp.
60141-2 Paperbound $1.75

A TREATISE ON THE DIFFERENTIAL GEOMETRY OF CURVES AND SURFACES, Luther P. Eisenhart. Detailed, concrete introductory treatise on differential geometry, developed from author's graduate courses at Princeton University. Thorough explanation of the geometry of curves and surfaces, concentrating on problems most helpful to students. 683 problems, 30 diagrams. xiv + 474pp.
60667-8 Paperbound $3.50

AN ESSAY ON THE FOUNDATIONS OF GEOMETRY, Bertrand Russell. A mathematical and physical analysis of the place of the a priori in geometric knowledge. Includes critical review of 19th-century work in non-Euclidean geometry as well as illuminating insights of one of the great minds of our time. New foreword by Morris Kline. xx + 201pp.
60233-8 Paperbound $2.50

INTRODUCTION TO THE THEORY OF NUMBERS, Leonard E. Dickson. Thorough, comprehensive approach with adequate coverage of classical literature, yet simple enough for beginners. Divisibility, congruences, quadratic residues, binary quadratic forms, primes, least residues, Fermat's theorem, Gauss's lemma, and other important topics. 249 problems, 1 figure. viii + 183pp.
60342-3 Paperbound $2.00

CATALOGUE OF DOVER BOOKS

AN ELEMENTARY INTRODUCTION TO THE THEORY OF PROBABILITY, B. V. Gnedenko and A. Ya. Khinchin. Introduction to facts and principles of probability theory. Extremely thorough within its range. Mathematics employed held to elementary level. Excellent, highly accurate layman's introduction. Translated from the fifth Russian edition by Leo Y. Boron. xii + 130pp.
60155-2 Paperbound $2.00

SELECTED PAPERS ON NOISE AND STOCHASTIC PROCESSES, edited by Nelson Wax. Six papers which serve as an introduction to advanced noise theory and fluctuation phenomena, or as a reference tool for electrical engineers whose work involves noise characteristics, Brownian motion, statistical mechanics. Papers are by Chandrasekhar, Doob, Kac, Ming, Ornstein, Rice, and Uhlenbeck. Exact facsimile of the papers as they appeared in scientific journals. 19 figures. v + 337pp. 6⅛ x 9¼.
60262-1 Paperbound $3.50

STATISTICS MANUAL, Edwin L. Crow, Frances A. Davis and Margaret W. Maxfield. Comprehensive, practical collection of classical and modern methods of making statistical inferences, prepared by U. S. Naval Ordnance Test Station. Formulae, explanations, methods of application are given, with stress on use. Basic knowledge of statistics is assumed. 21 tables, 11 charts, 95 illustrations. xvii + 288pp.
60599-X Paperbound $2.50

MATHEMATICAL FOUNDATIONS OF INFORMATION THEORY, A. I. Khinchin. Comprehensive introduction to work of Shannon, McMillan, Feinstein and Khinchin, placing these investigations on a rigorous mathematical basis. Covers entropy concept in probability theory, uniqueness theorem, Shannon's inequality, ergodic sources, the E property, martingale concept, noise, Feinstein's fundamental lemma, Shanon's first and second theorems. Translated by R. A. Silverman and M. D. Friedman. iii + 120pp.
60434-9 Paperbound $1.75

INTRODUCTION TO SYMBOLIC LOGIC AND ITS APPLICATION, Rudolf Carnap. Clear, comprehensive, rigorous introduction. Analysis of several logical languages. Investigation of applications to physics, mathematics, similar areas. Translated by Wiliam H. Meyer and John Wilkinson. xiv + 214pp.
60453-5 Paperbound $2.50

SYMBOLIC LOGIC, Clarence I. Lewis and Cooper H. Langford. Probably the most cited book in the literature, with much material not otherwise obtainable. Paradoxes, logic of extensions and intensions, converse substitution, matrix system, strict limitations, existence of terms, truth value systems, similar material. vii + 518pp.
60170-6 Paperbound $4.50

VECTOR AND TENSOR ANALYSIS, George E. Hay. Clear introduction; starts with simple definitions, finishes with mastery of oriented Cartesian vectors, Christoffel symbols, solenoidal tensors, and applications. Many worked problems show applications. 66 figures. viii + 193pp.
60109-9 Paperbound $2.50

CATALOGUE OF DOVER BOOKS

GUIDE TO THE LITERATURE OF MATHEMATICS AND PHYSICS, INCLUDING RELATED WORKS ON ENGINEERING SCIENCE, Nathan Grier Parke III. This up-to-date guide puts a library catalog at your fingertips. Over 5000 entries in many languages under 120 subject headings, including many recently available Russian works. Citations are as full as possible, and cross-references and suggestions for further investigation are provided. Extensive listing of bibliographical aids. 2nd revised edition. Complete indices. xviii + 436pp.
60447-0 Paperbound $3.00

INTRODUCTION TO ELLIPTIC FUNCTIONS WITH APPLICATIONS, Frank Bowman. Concise, practical introduction, from familiar trigonometric function to Jacobian elliptic functions to applications in electricity and hydrodynamics. Legendre's standard forms for elliptic integrals, conformal representation, etc., fully covered. Requires knowledge of basic principles of differentiation and integration only. 157 problems and examples, 56 figures. 115pp. 60922-7 Paperbound $1.50

THEORY OF FUNCTIONS OF A COMPLEX VARIABLE, A. R. Forsyth. Standard, classic presentation of theory of functions, stressing multiple-valued functions and related topics: theory of multiform and uniform periodic functions, Weierstrass's results with additiontheorem functions. Riemann functions and surfaces, algebraic functions, Schwarz's proof of the existence-theorem, theory of conformal mapping, etc. 125 figures, 1 plate. Total of xxviii + 855pp. $6\frac{1}{8}$ x $9\frac{1}{4}$.
61378-X, 61379-8 Two volumes, Paperbound $6.00

THEORY OF THE INTEGRAL, Stanislaw Saks. Excellent introduction, covering all standard topics: set theory, theory of measure, functions with general properties, and theory of integration emphasizing the Lebesgue integral. Only a minimal background in elementary analysis needed. Translated by L. C. Young. 2nd revised edition. xv + 343pp. 61151-5 Paperbound $3.00

THE THEORY OF FUNCTIONS, Konrad Knopp. Characterized as "an excellent introduction . . . remarkably readable, concise, clear, rigorous" by the Journal of the American Statistical Association college text.

A COURSE IN MATHEMATICAL ANALYSIS, Edouard Goursat. The entire "Cours d'analyse" for students with one year of calculus, offering an exceptionally wide range of subject matter on analysis and applied mathematics. Available for the first time in English. Definitive treatment.

VOLUME I: Applications to geometry, expansion in series, definite integrals, derivatives and differentials. Translated by Earle R. Hedrick. 52 figures. viii + 548pp. 60554-X Paperbound $5.00

VOLUME II, PART I: Functions of a complex variable, conformal representations, doubly periodic functions, natural boundaries, etc. Translated by Earle R. Hedrick and Otto Dunkel. 38 figures. x + 259pp. 60555-8 Paperbound $3.00

VOLUME II, PART II: Differential equations, Cauchy-Lipschitz method, non-linear differential equations, simultaneous equations, etc. Translated by Earle R. Hedrick and Otto Dunkel. 1 figure. viii + 300pp. 60556-6 Paperbound $3.00

CATALOGUE OF DOVER BOOKS

VOLUME III, PART I: Variation of solutions, partial differential equations of the second order. Poincaré's theorem, periodic solutions, asymptotic series, wave propagation, Dirichlet's problem in space, Newtonian potential, etc. Translated by Howard G. Bergmann. 15 figures. x + 329pp. 61176-0 Paperbound $3.50

VOLUME III, PART II: Integral equations and calculus of variations: Fredholm's equation, Hilbert-Schmidt theorem, symmetric kernels, Euler's equation, transversals, extreme fields, Weierstrass's theory, etc. Translated by Howard G. Bergmann. Note on Conformal Representation by Paul Montel. 13 figures. xi + 389pp.
61177-9 Paperbound $3.00

ELEMENTARY STATISTICS: WITH APPLICATIONS IN MEDICINE AND THE BIOLOGICAL SCIENCES, Frederick E. Croxton. Presentation of all fundamental techniques and methods of elementary statistics assuming average knowledge of mathematics only. Useful to readers in all fields, but many examples drawn from characteristic data in medicine and biological sciences. vii + 376pp.
60506-X Paperbound $2.50

ELEMENTS OF THE THEORY OF FUNCTIONS. A general background text that explores complex numbers, linear functions, sets and sequences, conformal mapping. Detailed proofs. Translated by Frederick Bagemihl. 140pp.
60154-4 Paperbound $1.50

THEORY OF FUNCTIONS, PART I. Provides full demonstrations, rigorously set forth, of the general foundations of the theory: integral theorems, series, the expansion of analytic functions. Translated by Federick Bagemihl. vii + 146pp.
60156-0 Paperbound $1.50

INTRODUCTION TO THE THEORY OF FOURIER'S SERIES AND INTEGRALS, Horatio S. Carslaw. A basic introduction to the theory of infinite series and integrals, with special reference to Fourier's series and integrals. Based on the classic Riemann integral and dealing with only ordinary functions, this is an important class text. 84 examples. xiii + 368pp. 60048-3 Paperbound $3.00

AN INTRODUCTION TO FOURIER METHODS AND THE LAPLACE TRANSFORMATION, Philip Franklin. Introductory study of theory and applications of Fourier series and Laplace transforms, for engineers, physicists, applied mathematicians, physical science teachers and students. Only a previous knowledge of elementary calculus is assumed. Methods are related to physical problems in heat flow, vibrations, eletcrical transmission, electromagnetic radiation, etc. 828 problems with answers. Formerly *Fourier Methods*. x + 289pp. 60452-7 Paperbound $2.75

INFINITE SEQUENCES AND SERIES, Konrad Knopp. Careful presentation of fundamentals of the theory by one of the finest modern expositors of higher mathematics. Covers functions of real and complex variables, arbitrary and null sequences, convergence and divergence. Cauchy's limit theorem, tests for infinite series, power series, numerical and closed evaluation of series. Translated by Frederick Bagemihl. v + 186pp. 60153-6 Paperbound $2.00

CATALOGUE OF DOVER BOOKS

INTRODUCTION TO THE DIFFERENTIAL EQUATIONS OF PHYSICS, Ludwig Hopf. No math background beyond elementary calculus is needed to follow this classroom or self-study introduction to ordinary and partial differential equations. Approach is through classical physics. Translated by Walter Nef. 48 figures. v + 154pp.
60120-X Paperbound $1.75

DIFFERENTIAL EQUATIONS FOR ENGINEERS, Philip Franklin. For engineers, physicists, applied mathematicians. Theory and application: solution of ordinary differential equations and partial derivatives, analytic functions. Fourier series, Abel's theorem, Cauchy Riemann differential equations, etc. Over 400 problems deal with electricity, vibratory systems, heat, radio; solutions. Formerly *Differential Equations for Electrical Engineers*. 41 illustrations. vii + 299pp.
60601-5 Paperbound $2.50

THEORY OF FUNCTIONS, PART II. Single- and multiple-valued functions; full presentation of the most characteristic and important types. Proofs fully worked out. Translated by Frederick Bagemihl. x + 150pp.   60157-9 Paperbound $1.50

PROBLEM BOOK IN THE THEORY OF FUNCTIONS, I. More than 300 elementary problems for independent use or for use with "Theory of Functions, I." 85pp. of detailed solutions. Translated by Lipman Bers. viii + 126pp.
60158-7 Paperbound $1.50

PROBLEM BOOK IN THE THEORY OF FUNCTIONS, II. More than 230 problems in the advanced theory. Designed to be used with "Theory of Functions, II" or with any comparable text. Full solutions. Translated by Frederick Bagemihl. 138pp.
60159-5 Paperbound $1.75

INTRODUCTION TO THE THEORY OF EQUATIONS, Florian Cajori. Classic introduction by leading historian of science covers the fundamental theories as reached by Gauss, Abel, Galois and Kronecker. Basics of equation study are followed by symmetric functions of roots, elimination, homographic and Tschirnhausen transformations, resolvents of Lagrange, cyclic equations, Abelian equations, the work of Galois, the algebraic solution of general equations, and much more. Numerous exercises include answers. ix + 239pp.   62184-7 Paperbound $2.75

LAPLACE TRANSFORMS AND THEIR APPLICATIONS TO DIFFERENTIAL EQUATIONS, N. W. McLachlan. Introduction to modern operational calculus, applying it to ordinary and partial differential equations. Laplace transform, theorems of operational calculus, solution of equations with constant coefficients, evaluation of integrals, derivation of transforms, of various functions, etc. For physics, engineering students. Formerly *Modern Operational Calculus*. xiv + 218pp.
60192-7 Paperbound $2.50

PARTIAL DIFFERENTIAL EQUATIONS OF MATHEMATICAL PHYSICS, Arthur G. Webster. Introduction to basic method and theory of partial differential equations, with full treatment of their applications to virtually every field. Full, clear chapters on Fourier series, integral and elliptic equations, spherical, cylindrical and ellipsoidal harmonics, Cauchy's method, boundary problems, method of Riemann-Volterra, many other basic topics. Edited by Samuel J. Plimpton. 97 figures. vii + 446pp.   60263-X Paperbound $3.00

CATALOGUE OF DOVER BOOKS

PRINCIPLES OF STELLAR DYNAMICS, Subrahmanyan Chandrasekhar. Theory of stellar dynamics as a branch of classical dynamics; stellar encounter in terms of 2-body problem, Liouville's theorem and equations of continuity. Also two additional papers. 50 illustrations. x + 313pp. $5\frac{5}{8}$ x $8\frac{3}{8}$.

60659-7 Paperbound $3.00

CELESTIAL OBJECTS FOR COMMON TELESCOPES, T. W. Webb. The most used book in amateur astronomy: inestimable aid for locating and identifying hundreds of celestial objects. Volume 1 covers operation of telescope, telescope photography, precise information on sun, moon, planets, asteroids, meteor swarms, etc.; Volume 2, stars, constellations, double stars, clusters, variables, nebulae, etc. Nearly 4,000 objects noted. New edition edited, updated by Margaret W. Mayall. 77 illustrations. Total of xxxix + 606pp.

20917-2, 20918-0 Two volumes, Paperbound $5.50

A SHORT HISTORY OF ASTRONOMY, Arthur Berry. Earliest times through the 19th century. Individual chapters on Copernicus, Tycho Brahe, Galileo, Kepler, Newton, etc. Non-technical, but precise, thorough, and as useful to specialist as layman. 104 illustrations, 9 portraits, xxxi + 440 pp. 20210-0 Paperbound $3.00

ORDINARY DIFFERENTIAL EQUATIONS, Edward L. Ince. Explains and analyzes theory of ordinary differential equations in real and complex domains: elementary methods of integration, existence and nature of solutions, continuous transformation groups, linear differential equations, equations of first order, non-linear equations of higher order, oscillation theorems, etc. "Highly recommended," *Electronics Industries.* 18 figures. viii + 558pp. 60349-0 Paperbound $4.00

DICTIONARY OF CONFORMAL REPRESENTATIONS, H. Kober. Laplace's equation in two dimensions for many boundary conditions; scores of geometric forms and transformations for electrical engineers, Joukowski aerofoil for aerodynamists, Schwarz-Christoffel transformations, transcendental functions, etc. Twin diagrams for most transformations. 447 diagrams. xvi + 208pp. $6\frac{1}{8}$ x $9\frac{1}{4}$.

60160-9 Paperbound $2.50

ALMOST PERIODIC FUNCTIONS, A. S. Besicovitch. Thorough summary of Bohr's theory of almost periodic functions citing new shorter proofs, extending the theory, and describing contributions of Wiener, Weyl, de la Vallée, Poussin, Stepanoff, Bochner and the author. xiii + 180pp. 60018-1 Paperbound $2.50

AN INTRODUCTION TO THE STUDY OF STELLAR STRUCTURE, S. Chandrasekhar. A rigorous examination, using both classical and modern mathematical methods, of the relationship between loss of energy, the mass, and the radius of stars in a steady state. 38 figures. 509pp. 60413-6 Paperbound $3.75

INTRODUCTION TO THE THEORY OF GROUP'S OF FINITE ORDER, Robert D. Carmichael. Progresses in easy steps from sets, groups, permutations, isomorphism through the important types of groups. No higher mathematics is necessary. 783 exercises and problems. xiv + 447pp. 60300-8 Paperbound $4.00

CATALOGUE OF DOVER BOOKS

ELEMENTARY MATHEMATICS FROM AN ADVANCED STANDPOINT: VOLUME II—GEOMETRY, Feliex Klein. Using analytical formulas, Klein clarifies the precise formulation of geometric facts in chapters on manifolds, geometric and higher point transformations, foundations. "Nothing comparable," *Mathematics Teacher*. Translated by E. R. Hedrick and C. A. Noble. 141 figures. ix + 214pp.
(USO) 60151-X Paperbound $2.25

ENGINEERING MATHEMATICS, Kenneth S. Miller. Most useful mathematical techniques for graduate students in engineering, physics, covering linear differential equations, series, random functions, integrals, Fourier series, Laplace transform, network theory, etc. "Sound and teachable," Science. 89 figures. xii + 417pp. 6 x 8½.
61121-3 Paperbound $3.00

INTRODUCTION TO ASTROPHYSICS: THE STARS, Jean Dufay. Best guide to observational astrophysics in English. Bridges the gap between elementary popularizations and advanced technical monographs. Covers stellar photometry, stellar spectra and classification, Hertzsprung-Russell diagrams, Yerkes 2-dimensional classification, temperatures, diameters, masses and densities, evolution of the stars. Translated by Owen Gingerich. 51 figures, 11 tables. xii + 164pp.
60771-2 Paperbound $2.50

INTRODUCTION TO BESSEL FUNCTIONS, Frank Bowman. Full, clear introduction to properties and applications of Bessel functions. Covers Bessel functions of zero order, of any order; definite integrals; asymptotic expansions; Bessel's solution to Kepler's problem; circular membranes; etc. Math above calculus and fundamentals of differential equations developed within text. 636 problems. 28 figures. x + 135pp.
60462-4 Paperbound $1.75

DIFFERENTIAL AND INTEGRAL CALCULUS, Philip Franklin. A full and basic introduction, textbook for a two- or three-semester course, or self-study. Covers parametric functions, force components in polar coordinates, Duhamel's theorem, methods and applications of integration, infinite series, Taylor's series, vectors and surfaces in space, etc. Exercises follow each chapter with full solutions at back of the book. Index. xi + 679pp.
62520-6 Paperbound $4.00

THE EXACT SCIENCES IN ANTIQUITY, O. Neugebauer. Modern overview chiefly of mathematics and astronomy as developed by the Egyptians and Babylonians. Reveals startling advancement of Babylonian mathematics (tables for numerical computations, quadratic equations with two unknowns, implications that Pythagorean theorem was known 1000 years before Pythagoras), and sophisticated astronomy based on competent mathematics. Also covers transmission of this knowledge to Hellenistic world. 14 plates, 52 figures. xvii + 240pp.
22332-9 Paperbound $2.50

THE THIRTEEN BOOKS OF EUCLID'S ELEMENTS, translated with introduction and commentary by Sir Thomas Heath. Unabridged republication of definitive edition based on the text of Heiberg. Translator's notes discuss textual and linguistic matters, mathematical analysis, 2500 years of critical commentary on the Elements. Do not confuse with abridged school editions. Total of xvii + 1414pp.
60088-2, 60089-0, 60090-4 Three volumes, Paperbound $9.50

CATALOGUE OF DOVER BOOKS

ASTRONOMY AND COSMOGONY, Sir James Jeans. Modern classic of exposition, Jean's latest work. Descriptive astronomy, atrophysics, stellar dynamics, cosmology, presented on intermediate level. 16 illustrations. Preface by Lloyd Motz. xv + 428pp. 60923-5 Paperbound $3.50

EXPERIMENTAL SPECTROSCOPY, Ralph A. Sawyer. Discussion of techniques and principles of prism and grating spectrographs used in research. Full treatment of apparatus, construction, mounting, photographic process, spectrochemical analysis, theory. Mathematics kept to a minimum. Revised (1961) edition. 110 illustrations. x + 358pp. 61045-4 Paperbound $3.50

THEORY OF FLIGHT, Richard von Mises. Introduction to fluid dynamics, explaining fully the physical phenomena and mathematical concepts of aeronautical engineering, general theory of stability, dynamics of incompressible fluids and wing theory. Still widely recommended for clarity, though limited to situations in which air compressibility effects are unimportant. New introduction by K. H. Hohenemser. 408 figures. xvi + 629pp. 60541-8 Paperbound $5.00

AIRPLANE STRUCTURAL ANALYSIS AND DESIGN, Ernest E. Sechler and Louis G. Dunn. Valuable source work to the aircraft and missile designer: applied and design loads, stress-strain, frame analysis, plates under normal pressure, engine mounts, landing gears, etc. 47 problems. 256 figures. xi + 420pp.
61043-8 Paperbound $3.50

PHOTOELASTICITY: PRINCIPLES AND METHODS, H. T. Jessop and F. C. Harris. An introduction to general and modern developments in 2- and 3-dimensional stress analysis techniques. More advanced mathematical treatment given in appendices. 164 figures. viii + 184pp. 6⅛ x 9¼. (USO) 60720-8 Paperbound $2.50

THE MEASUREMENT OF POWER SPECTRA FROM THE POINT OF VIEW OF COMMUNICATIONS ENGINEERING, Ralph B. Blackman and John W. Tukey. Techniques for measuring the power spectrum using elementary transmission theory and theory of statistical estimation. Methods of acquiring sound data, procedures for reducing data to meaningful estimates, ways of interpreting estimates. 36 figures and tables. Index. x + 190pp. 60507-8 Paperbound $2.50

GASEOUS CONDUCTORS: THEORY AND ENGINEERING APPLICATIONS, James D. Cobine. An indispensable reference for radio engineers, physicists and lighting engineers. Physical backgrounds, theory of space charges, applications in circuit interrupters, rectifiers, oscillographs, etc. 83 problems. Over 600 figures. xx + 606pp. 60442-X Paperbound $3.75

Prices subject to change without notice.

Available at your book dealer or write for free catalogue to Dept. Sci, Dover Publications, Inc., 180 Varick St., N.Y., N.Y. 10014. Dover publishes more than 150 books each year on science, elementary and advanced mathematics, biology, music, art, literary history, social sciences and other areas.